THE COMPLETE HANDBOOK OF MAGNETIC RECORDING

3RD EDITION

In memory of
Morten Jorgensen
1958–1982

THE COMPLETE HANDBOOK OF MAGNETIC RECORDING

3RD EDITION

FINN JORGENSEN

TAB Professional and Reference Books

Division of TAB BOOKS Inc.
Blue Ridge Summit, PA

Illustrations were drawn by the author on an Apple III computer, using program "Draw-On III" by Dr. Mel Astrahan. Input device: Apple Graphics Tablet. Printing was done on an Epson FX-80 dot matrix printer.

Manuscript was typed on an Apple III computer, using program "Applewriter III." Files were transferred to MS-DOS format, and processed into WordStar files for typesetting.

THIRD EDITION
FIRST PRINTING

Library of Congress Cataloging in Publication Data

Jorgensen, Finn.
The complete handbook of magnetic recording.

Bibliography: p.
Includes index.
1. Magnetic recorders and recording. I. Title.
TK7881.6.J66 1988 621.389′3 85-22299
ISBN 0-8306-1979-8

TAB BOOKS Inc. offers software for sale. For information and a catalog, please contact TAB Software Department, Blue Ridge Summit, PA 17294-0850.

Questions regarding the content of this book should be addressed to:

Reader Inquiry Branch
TAB BOOKS Inc.
Blue Ridge Summit, PA 17294-0214

Contents

Acknowledgments

Several friends deserve thanks for their advice and help during the preparation of the original manuscript for the first edition: Eric D. Daniel (formerly with Memorex Corporation); Cmdr. Steve S. Jauregui of the U.S. Naval Postgraduate School in Monterey; Len Johnson (formerly with Bell and Howell); and John McKnight of Magnetic Reference Laboratories. My son Morten was also of invaluable help with many editorial advices.

And through the past five years, many sincere thanks to the numerous students who have attended my courses. Their comments and critique helped shape this book.

A grateful thanks also goes to the many equipment and media manufacturers who supported this work by sponsoring in-house courses by the author.

Finally thanks go to my daughter Tina for great help with the manuscript, and to my wife Bodil for her unfailing support. To Joseph W. Judge of the Aerospace Corporation for many good suggestions and a painstaking proofreading of all the technical material, and to the staff at TAB BOOKS for patience and help in transmitting the manuscript and illustrations to the printing press.

Preface to the Third Edition

This new edition is a greatly expanded, updated, and edited version of the first *Complete Handbook of Magnetic Recording*. It is, like the first edition, written to be easily understood by the technically inclined person working in the field, and for the equipment user who wants to acquire an in-depth knowledge of magnetic recording and storage. It serves the dual purpose of a textbook and reference guide. It is divided into logical sections:

	Part	Chapters
1	Introduction	1–2
2	Fundamentals	3–6
3	Magnetic Heads	7–10
4	Magnetic Tapes and Disks	11–14
5	Tape Transports and Disk Drives	15–18
6	The Write/Read Process	19–21
7	Equalization and Coding	22–25
8	Applications	26–30
		(Appendices)

The presentation of magnetism is novel in that it builds up the reader's understanding without popular explanations or gross simplifications. Mathematics and formulas are included where they assist in understanding or in summarizing certain laws. The first four chapters cover fundamental and technical magnetization, the theory of recording and playback theory. This section is comparable to a photographer's basic course in lights and optics, which gives him the foundation for understanding lenses and films; their magnetic counterparts are covered in depth in the following sections.

The next eight chapters on magnetic heads and media (tapes, disks) cover virtually all aspects of theory, materials, fabrication, and performance. A large number of references and further reading material has been included so that the reader can probe into subjects that are of particular interest. Some chapters are quite detailed, in order to make the reader aware of the many parameters that affect the final recorder performance. This should allow the designer and end-user to better evaluate and specify magnetic heads and media.

The section on drive mechanisms is new, although some of the material is in the first edition. Important as it is, this topic has always been sort of a stepchild in magnetic recording. To date, less than 200 papers have addressed mechanisms, while over 3,000 have been written on magnetic recording as a whole.

The sections on write/read processes and equalization, detection and coding dig into the details of signal processing through a recorder/reproducer system, treating it as a communication channel. The nonspecialist reader should not have much difficulty in achieving an understanding of these topics. I have striven to explain what happens in plain language and I also have fortified the explanations with models and mathematics.

The last section on applications was frustrating to write because there are so many rapid changes in the marketplace. There has not been an astounding breakthrough in magnetic recording since the introduction of AC-bias (perpendicular recording is a reappraisal of an old scheme, reviewed in light of modern technologies). But recorder/reproducer equipment is tailor-made for specific applications, and I believe it is fair to state that improvements occur at an average annual rate of one dB (= 11 percent).

The magnetic recording industry really took off in the '70s and is growing at an unprecedented pace. The acceptance of Philips' audio cassette around 1970 allowed tape recording to establish itself into a large percentage of homes and automobiles. And in the first half of the '80s we have witnessed the explosion in disk drive applications for microcomputers, and the half-inch-wide cassette tape in home video systems; recently we have added magnetic home movies on 8mm wide tape, in cassettes.

The name of the game is to achieve the largest amount of recorded information per unit area of recording media. The next step may well be to record magnetically, but read out the information by optical means.

The technical units in the book are SI (meter, kilogram, second). The reader will notice a lack of consistency because, in several places, I gave in to overwhelming common usage, such as Gauss rather than Wb/m^2 and Oersted instead of A/m. The net effect should, I hope, minimize confusion and mistakes.

As a rule, all technical magnetization is in SI-units, while material properties are in cgs-units. This should ease the transition we sooner or later must accept.

This new edition is sprinkled with numerous examples, and a couple of small computer programs are included in order to aid in design techniques. They are written in TrueBasic, and hence are portable between different computer models.

Most chapters have several references so the people contributing to our knowledge about magnetic recording get due credit; these references are generally useful for further studies of a topic. Additional papers and textbooks may be included in a bibliography.

Chapter 1

A Colorful History

The history of inventions and the developments of magnetic recording make a very human and fascinating story. It has all the facets of a true cloak-and-dagger tale. It is a story of the many hopes and dreams of men who met the challenges to not only create saleable products, but also to advance the welfare of their fellow man as well.

We are indebted to them for their contributions to scientific advancement and better living conditions. All of us are now in touch with magnetic recording including television, homevideo, computers and best of all—music. This book would be incomplete without highlighting the history of magnetism, and one of its many aspects—magnetic recording.

EARLY HISTORY

Our recording technology is founded on magnetism and on electromagnetic induction. The earliest description of magnetism is obscure, but a mineral called magnetite was known centuries before the birth of Christ. It would attract iron, and would also magnetize a piece of iron if it was rubbed against it.

The sailor's compass could be made from a properly shaped piece of magnetite, free to turn about a pivot. It would turn in the north-south direction and was named *lodestone*, which means "waystone" or "leading stone"—pointing the way. The first to use this principle were apparently the Chinese, although they only used it to maneuver about in China; it was European sailors who first used the lodestone as a *compass* (Fig. 1-1). Another legend concerning the use of a lodestone depicts its use in defense (Fig. 1-2).

The first scientific study of magnetism was made by the Englishman William Gilbert (1540-1603), who published a classic book, *De Magnete* (Fig. 1-3). All his experiments had to be carried out using iron or steel samples that were rubbed with a lodestone. He rationalized that the earth itself was a magnet, and his experimental samples were therefore shaped

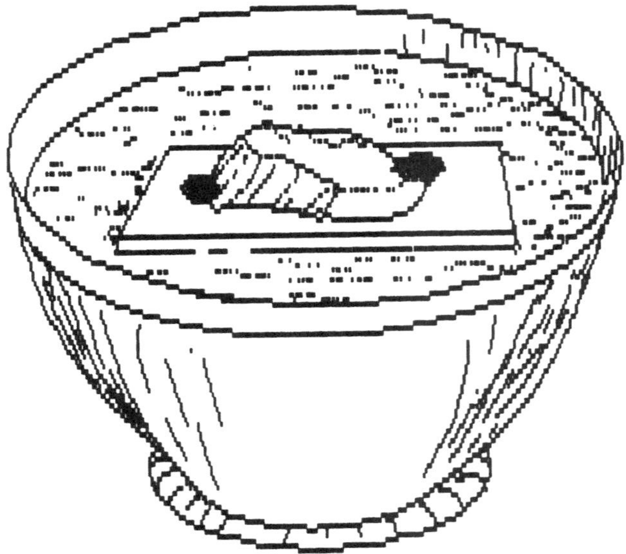

Fig. 1-1. Early sailors compass, the "lodestone."

Fig. 1-2. Legend about defense using large lodestones.

Fig. 1-3. British physician and physicist William Gilbert.

as spheres. And he did not recognize a North and a South Pole, but rather a centered source of magnetic force.

A certain amount of mysticism was associated with magnetism. In one experiment a piece of iron was carefully weighed, and placed in a box. The empty space was next filled with iron filings and their weight determined. The box was closed and stored for a few years. Then the box was opened and the weight measurements repeated: the amount of filings now weighed slightly less than before, and the iron piece more. Conclusion: lodestones eat iron powders to sustain themselves.

Gilbert believed that the lodestone possessed a soul, but discounted the old superstition that "Onyons and Garlick are at odds with the lodestone" (steersmen were forbidden to eat the vegetables when on duty!). The medical healing power of magnets was, apparently, discarded by Gilbert, as it was later on by Edison, who found "no effect from strong fields, upon himself nor upon his dog"; and yet—in recent years it has been discovered that operational scars heal much nicer when a strong magnet is placed in the bandage near the wound.

Rene Descartes (1596-1650) exorcised the soul out of the lodestone, and established a constructive philosophy. He postulated that magnetism had "threaded parts" that were channeled through the earth through pores in the poles. These parts would further seize upon the opportunity to cross any lodestone in the way.

One of Descartes students, the Swedish scientist and theologian Emanuel Swedenborg (1688-1772), envisioned an ordering within a magnetized versus a non-magnetized piece of iron (see Fig. 1-4). Today's domain theory is in remarkable agreement with this illustration (see Chapter 3).

A very important discovery occurred in 1820 when the Danish pharmacist and philosopher Hans Christian Oersted (1777-1851) found that an electric current produces a magnetic field

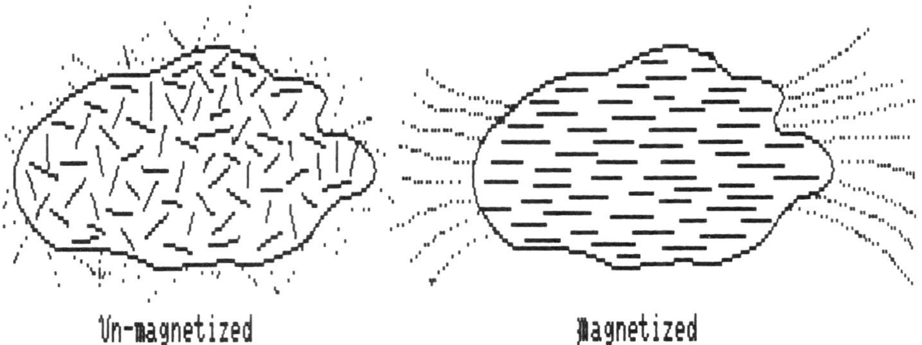

Fig. 1-4. Swedenborg's vision of a non-magnetic and a magnetic lodestone.

(Fig. 1-5). He had, like other scientists in those days, pondered over the nature of electricity and magnetism, and a possible connection between the two. A compass was not influenced by placing it near a battery's terminal, like a charged piece of amber was.

One of the many experiments gave a clue: Oersted discovered that a compass needle would jitter ever so slightly, when it was placed near a metal wire which in turn was connected to the terminals on a battery. See Fig. 1-6. More experiments revealed that a thicker wire caused a measurable deflection, and Oersted now revealed his discovery to the world's scientific com-

Fig. 1-5. Danish scientist and philosopher Hans Christian Oersted.

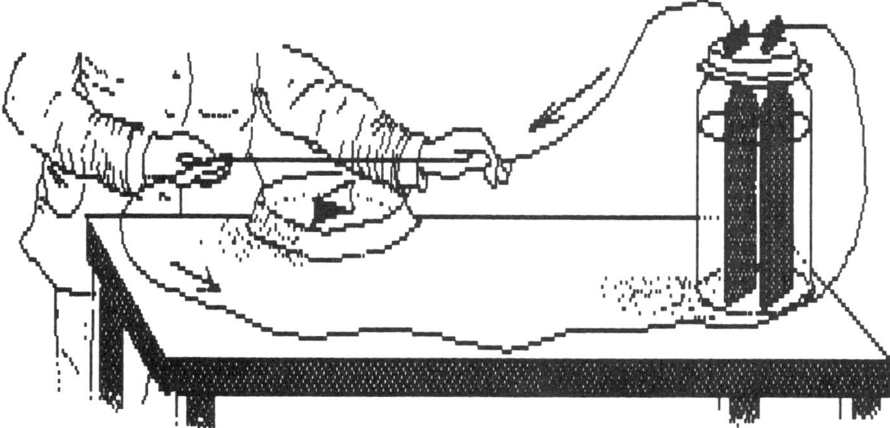

Fig. 1-6. Oersted's experimental setup to show influence of a current upon a compass needle.

munity. He also discovered that the forces acting upon a compass had circular patterns, centered at the wire.

Andre-Marie Ampere (1775-1836) found that the wire connecting the battery terminals conducted a current, and he also found that two currents have a mutual magnetic effect, and hence a force between them (Figs. 1-7, 1-8). One could now manufacture lifting electromag-

Fig. 1-7. French mathematician and physicist Andre-Marie Ampere.

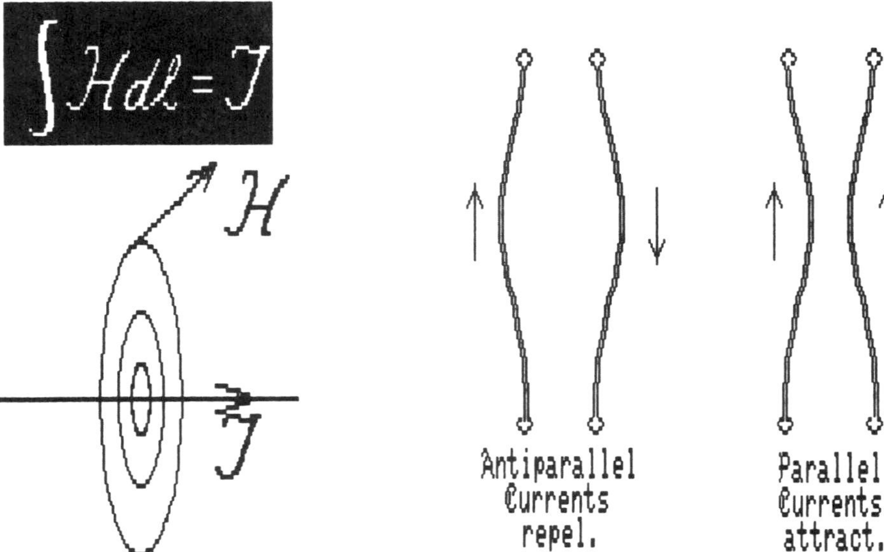

Fig. 1-8. Ampere's Law, and his discovery of currents' mutual influences.

Fig. 1-9. British experimental genius Michael Faraday.

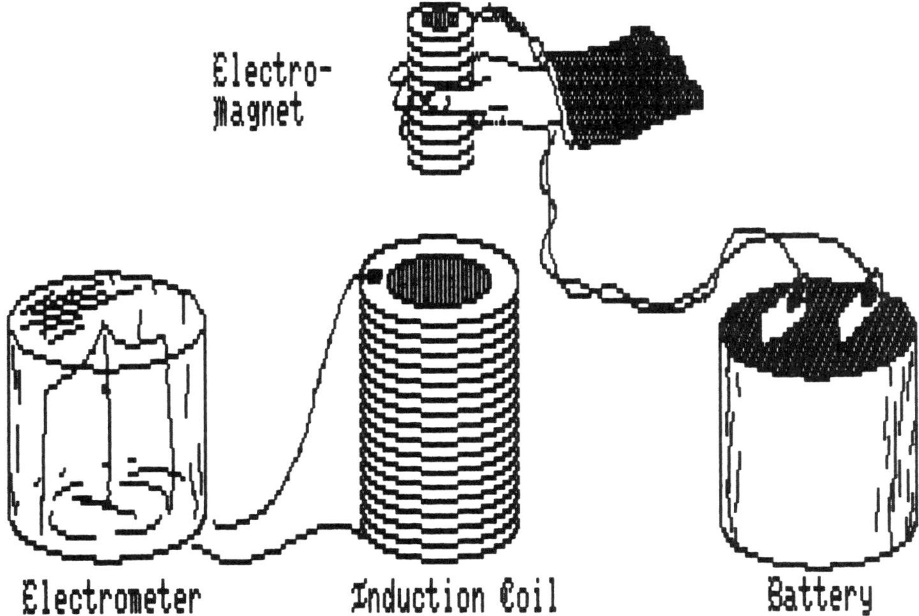

Fig. 1-10. Faraday's discovery of electromagnetic induction of voltage.

nets, and motors. Ampere also postulated that all magnetism has circulating currents as their origin; we will return to this topic in Chapter 3.

Another eleven years were to lapse before Michael Faraday (1791-1867) in 1831 discovered that a magnetic field conversely generates an electric voltage (Figs. 1-9, 1-10).

Oersted's, Ampere's and Faraday's discoveries were the necessary ingredients for the Industrial Revolution: electric generators and motors, each located where most efficient, and separated only by electric wiring.

Their discoveries also form the backbone of magnetic recording. This technology was first discussed in a paper by Oberlin Smith in 1888. He envisioned a cotton thread impregnated with steel dust, and being magnetized by passing through a current carrying coil, with the current controlled by a microphone. But he discarded the idea of using a steel wire since "the magnetic influence would probably distribute along the wire in a most totally depraved way." No experiments were conducted.

THE TELEGRAPHONE

While Oersted had enjoyed triumphal recognition, this did not quite happen for his countryman Valdemar Poulsen (1869-1942) Fig. 1-11. His invention of the Telegraphone in August 1898 has indeed today spread into every home in the Western world, but the progress has been by leaps and bounds and with numerous setbacks. During certain periods technology was lacking (wire, tape, AC-bias) and at other times manipulating business interests or national interest seemed to hold back universal progress.

Poulsen's invention was the outcome of a simple experiment where he stroked out a line along an iron plate and found that iron filings would gather along the line. The next experiment involved a strung-out piano wire and a primitive electromagnet connected to a microphone. Poulsen moved the electromagnet along the wire as he spoke into the microphone, and by

Fig. 1-11. Danish engineer and inventor Valdemar Poulsen.

later connecting the wires from the magnet to the telephone receiver he heard his voice reproduced.

A practical record/playback machine is shown in Fig. 1-12. The magnetic wire is wound from one end to the other of a brassdrum, which in turn is rotated through a belt coupling to a motor. The head assembly has two small pole pieces that protrude from its surface, grabs each side of the wire, and hence will be guided along the drum. When reaching the end it is lifted from the drum surface, and then quickly moved back to the start position. Poulsen was received with great enthusiasm at the World's Fair in Paris in 1900, and was awarded a Grand Prix for the Telegraphone.

The electromagnetic head assembly could be either single pole or double pole and Poulsen later advised how the two poles should be offset to produce a longitudinal magnetization rather than a perpendicular one. When he conceived of a disk recorder he selected a diameter of 5.25 inches, and used a needle to write and read—on both sides. So he predicted the correct dimension, opted for perpendicular recording—and double-sided. Finally, Poulsen is also credited with applying a DC-current for improvement of the recording (see Chapter 5, DC-bias). This came about by first erasing the steel wire with a permanent magnet, and then adjusting a superimposed field upon the record signal field to provide optimum quality.

In 1902 Poulsen made another invention, described in his paper "System for Producing Continuous Electric Oscillations." This was a high power, continuous oscillator operating at a frequency of many kHz (1000 Hertz, or cycles per second), up to one MHz. The oscillations were modulated by shunting the series-tuned LC-circuit with an electric arc, burning in an atmosphere of hydrogen. This system replaced many of Marconi's On-Off transmitters, which spread signals over the entire broadcast spectrum. This work soon absorbed all of Poulsen's time, and his company sold transmitters to many parts of the world, including, in 1910, Stockton, Sacramento and San Francisco. Lee de Forest's vacuum tube transmitters replaced these transmitters after 1920.

Poulsen sold his patent rights to the Telegraphone in 1905. It is, in retrospect, a shame that just a little fraction of the signal from his arc-generator did not stray over to the Telegraphone: Poulsen could then have added AC-bias to his patents.

The first decades of the twentieth century were tumultuous for the Telegraphone. Several manufacturing companies were formed, and many varieties of the recorder promoted. There were dictating machines, message repeaters, telephone answering machines, small disks were tried for tape letters, and so on. But there was no technology to improve the quality or playing time. The invention and the associated ideas were clearly too early. So the companies struggled and either lost money or changed hands.

Accusations were voiced against the American Telegraphone Company. People wondered if the president of the company was paid to suppress production which the phonograph and the telephone companies feared. Or worse, was he or others in a part with the Germans, who successfully used telegraphones onboard their submarines in World War 1. The Germans had

Fig. 1-12. Poulsen's Telegraphone.

made message recordings at normal speed and then transmitted them backwards, at higher speed.

THE MAGNETOPHONE

Whatever the cause, leadership in magnetic recording went to the Germans. In the twenties they manufactured and sold recorders with steel tape, designed by Stille. Similar machines were made in England, called the *Blattnerphone*.

In 1928 Pfleumer filed a patent for coating iron particles on a strip of paper as a recording medium, and a machine using such a tape, the German Magnetophone was exhibited in Berlin in 1935. And then—no more was heard of it until 1943.

At that time the U.S. Army Signal Corps, stationed in England, was puzzled by sometimes hearing radio broadcasts of operas and music during the middle of the night (noting no record scratches and other such deficiencies) and then hearing Hitler speaking from different parts of Germany almost within the hour.

The answer was found in 1945 in Frankfurt by one of their officers, John Mullin. He found in a radio station several A.E.G. Magnetophones, all equipped with ¼″ plastic tape, some of which later would serve for the Bing Crosby radio shows.

The sound quality was far better than any other machine in those days, and a close examination revealed that the Germans used high-frequency AC-bias. Chapter 5 will explain how it works—but it is today universally used in recorders. History will have it that W.L. Carlson and G.W. Carpenter of the U.S. Navy discovered and patented AC-bias in 1927, but it was obviously not used to any extent. The Magnetophone AC-bias patent was in 1946 granted in Germany (retroactive to 1940) to H. J. von Braunmuhl and Dr. W. Weber, and in the United States to Marvin Camras of Armour Research.

AFTER WORLD WAR II

The next three decades bring us through a rapid growth period with innovations, products, people, and companies too numerous to list in a few pages. The Minnesota Mining Manufacturing Co. (3M Co.) finished their first oxide tapes in 1947, under Dr. W. Wetzel. Ampex, founded by Alexander M. Poniatoff, started delivering audio recorders in 1948. Mincom, a division of 3M Co., pushed the state-of-the-art instrumentation recorders, and demonstrated television recording in 1951, followed by RCA in 1953.

Other early pioneers were Dr. Marvin Camras of Illinois Institute of Research (then Armour Research), Otto Kornei of Brush/Clevite, and S.J. Begun, who wrote the first book on magnetic recording. The result was an industry that flourished with a large selection of sound tape and sound film recorders.

The breakthrough in television came from Ampex where in 1955 Charles Ginburg and Ray Dolby (father of today's Dolby System) unveiled the rotating head video recorder (the readers of this book are certainly aware of the perfection this technique has achieved today). Instrumentation recording jumped ahead in 1961 when Wayne Johnson at Mincom conceived a low inertia servo controlled tape drive virtually free of timing errors (low-TDE).

Industry standards were always needed to provide interchangeability of recorded tapes. Audio tapes experienced rapid developments from full track recorded tapes to 2, 4, and 8 tracks on ¼″ tape, and now 4 tracks on 0.150″ wide tape in cassettes. Such transitions could not have happened in an orderly fashion if it were not for the cooperative work of manufacturers and standard committees. There are today numerous standard groups and of these the following played a key role in the past 40 years of developments:

- ANSI—American National Standards Institute.
- CCIR—International Radio Consultative Committee.
- IRIG—Interrange Instrumentation Group.
- NAB—National Association of Broadcasters.
- SMPTE—Society of Motion Picture and Television Engineers.

The most difficult standards to reach agreement on were those for the equalization of the recorder response. Thus, there are still several standards today for essentially the same thing (see Chapters 26-29).

INTO THE EIGHTIES

New generations of recorders are always on their way. This follows improved technologies and new methods, such as using digital encoding of data, music, or video. Home video recorders are now available that reproduce music without noise and flutter. They employ analog-to-digital converters, and the signal is recorded in its digital form. And before long digital audio recorders using small cassettes will be available for personal use.

Magnetic recording is enjoying a steady growth in the field of computer storage. The pace setting leader in this field is IBM who since the fifties has introduced techniques that have become industry standards. The most interesting device is the *diskette,* and 8″ or 5.25″ round disk for storage of digital data. It was introduced by IBM in 1970 as a time saving peripheral device with a storage capacity of a few hundred kilobytes. This number fits the requirements of a large number of small business and large home computers. The diskette, or floppy, market is therefore enjoying fast growth with about one million disk drives manufactured in 1979.

Computers are now using disk drives with tremendous storage capacities, and tape systems that are using 16 tracks across a half inch wide tape, packing about 40,000 bits per inch track length. Micro-computers are gradually switching to 3½ inch diskettes that store the same or more data as any of the 8″ diskettes did. A 2 inch diameter diskette drive is under development to replace photographic film in cameras. Home video on 8mm tapes in cassettes is replacing photographic film equipment. What next ?

An entire book could be written about the evolution of the magnetic recording industry, its people and the chain of developments. It would be most interesting reading. This chapter is well concluded with a saying an old friend of mine had about us in the industry:

"It is not really a business—it's a way of life."

REFERENCES TO CHAPTER 1

Gilbert, William, *De Magnete—On the Lodestone and Magnetic Bodies and on the Great Magnet the Earth*, England, 1600. Reprinted by Dover Publications (New York), 1957, and by Encyclopedia Britannica, *Great Books of the Western World*, Vol. 28, 1952, pp. 1-121.

Franksen, O.I., *H. C. Oersted—a man of the two cultures*, Bang Olufsen, 1981, 49 pages.

Friedlander, G.D., "Ampere: Father of Electrodynamics," *IEEE Spectrum*, Aug. 1975, pp. 75-78.

Smith, O., "Some Possible Forms of Phonograph," *The Electrical World*, Sept. 1888, 3 pages.

Poulsen, V., "The Telegraphone," *Ann. d. Phys.* Nov. 1900, Vol. 3, pp. 754-760.

Poulsen, V., "System for Producing Continuous Electric Oscillations," *Trans. el. Congress St. Louis*, 1904, Vol. II, pp. 963-971.

Holst, H., *Eleckriciteten*, Gyldendal, Copenhagen, 1906, pp. 553-561.

Fankhauser, C.K., "The Telegraphone," *Jour. Franklin Inst.*, Jan. 1909, Vol. 16, pp. 37-46.

Larsen, A., *The Telegraphone* (in Danish), Ingenioer Videnskabelige Skrifter, 1950, No. 2, 305 pages.
Shea, J.R., "The Untold Story of Tape Recording," *Tape Recording 1967* , Nos. 3, 4, 5, and 6.
Drenner, D.V.R., "The Magnetophone," *Audio Engineering*, Oct. 1947, Vol. 31, pp. 7-11.
Stille, K., "Electromagnetic Sound Recording," *Elektrotech. Zeitschrift*, Mar. 1930, Vol. 51, pp. 449-451.
Rust, N.M., "Marconi-Stille Recording and Reproducing Equipment," *Marconi Review*, Jan. 1934, Vol. 46, pp. 1-11.
Hamilton, H.E., "The Blattnerphone—Its Operation and Use," *Electronic Digest*, Dec. 1935, p. 347.
Begun, S.J., "The New Lorenz Steel Tone Tape Machine," *Electronic Communication*, July 1936, Vol. 15, pp. 62-69.
Hickman, C.N., "Sound Recording on Magnetic Tape," *Bell System Technical Journal*, July 1937, Vol. 16, pp. 165-177.
Hickman, C.N., "Magnetic Recording and Reproducing," *Bell Laboratory Record*, Sept. 1937, Vol. 16, pp. 2-7.
Volk, T., "A. E. G. Magnetophone," *A E G Mitteilung*, Sept. 1935, pp. 299-301.
Hammar, P., and Ososke, D., "The Birth of the German Magnetophone Tape Recorder 1928-1935," *dB - the Sound Engineers Magazine*, March 1982, pp. 47-52.
Camras, M., "A New Magnetic Wire Recorder," *Radio News, Radionics Section*, Nov. 1943, Vol. 30, pp. 3-5, 38, 39.
Thiele, H., "On the Origin of High-Frequency Biasing for Magnetic Audio Recording," *Jour. SMPTE*, July 1983, pp. 752-754.
Snyder, R.H., "History and Development of Stereophonic Sound Recording," *Jour. AES*, April 1953, Vol. 1, pp. 176-179.
Selsted, W.T., and Snyder, R.H., "Magnetic Recording - A Report on the State of the Art," *IRE Convention Transactions*, Sept. 1954, 8 pages.
Roizen, J., "Project Videotape," *Ampex Bulletin AI*, 1956.
Mooney, Jr., M., "The History of Magnetic Recording," *Hi-Fi Tape Recording*, 1957.
Mullin, J.T., "The Birth Of The Recording Industry," *Billboard,* Nov. 1972, 6 pages.
Dunlop, D.J., "Rocks as High-Fidelity Tape Recorders," *IEEE Trans. Magn.*, July 1977, Vol. MAG-13, No. 5, pp. 1267-1272.
Bate, G., "Bits and Genes: A Comparison of the Natural Storage of Information in DNA and Digital Magnetic Recording," *IEEE Trans. Magn.*, Sept. 1978, Vol. MAG-14, No. 5, pp. 964-965.
Hoagland, A.S., "Trends and Projections in Magnetic Recording Storage on Particulate Media," *IEEE Trans. Magn.*, Jan. 1980, Vol. 16, No. 1, pp. 26-29.
Stevens, L.D., "The Evolution of Magnetic Storage," *IBM Journal of Research and Development*, Sept. 1981, Vol. 25, No. 5, pp. 663-675.
Camras, M., "Origins of Magnetic Recording Concepts," *Jour. ASA*, Apr. 1985, Vol. 77, No. 4, pp. 1314-1319.
Mallinson, J.C., "The Next Decade in Magnetic Recording," *IEEE Trans. Magn.*, May 1985, Vol. MAG-21, No. 3, pp. 1217-1220.
Mallinson, J.C., "Recording Limitations," Chapter 5 in *Magnetic Recording*, Vol. I, Ed. by Mee, C.D. and Daniel, E.D., McGraw-Hill, 1987, pp. 337-375.

Chapter 2

Introduction to Modern Recording Equipment

Magnetic recording has come to play a major role in our modern society. Speech, music, measurement data, bookkeeping, computations, and live pictures are currently recorded, stored, and replayed on magnetic tapes or disks.

The success of magnetic recording is found in its convenience of use, low cost, and reusability of the media. It is quite paradoxical, however, that many tapes are recorded only once. When the voice of a family member or a distant friend is heard, or a favorite musical number is finally recorded, one is quite hesitant to erase it. This also applies in the recording of scientific data where the tapes and disks normally are stored in libraries. The *principles of magnetic recording* are based on the physics of magnetism, a phenomenon which relates to certain materials; magnetization of these occurs when they are placed in a magnetic field. If the material is in the group of so-called "hard" magnetic materials, it will hold its magnetization after it has been moved away from the exciting field.

Figure 2-1 is a simplified diagram of a sound recorder. An incoming sound wave is picked up by a microphone (1) and amplified (2) into a recording current, I_r (3), which flows through the winding in the record head. The ring-shaped record head has a "soft" magnetic core (so magnetization is not retained) with an air gap in front. The current, I_r, produces magnetic field lines that diverge from the air gap (4) and penetrate the tape, moving past the record head from the supply reel (5). The tape itself is a plastic ribbon coated with a "hard" magnetic material which retains its magnetization after it has passed through the field from the record head gap.

The tape passes over the playback head which also is a ring core with a front gap. The magnetic field lines (flux) from the recorded tape permeate the core and produce an induced voltage, e (6), across the winding. This voltage, after suitable amplification (7), reproduces the original sound through a speaker (8).

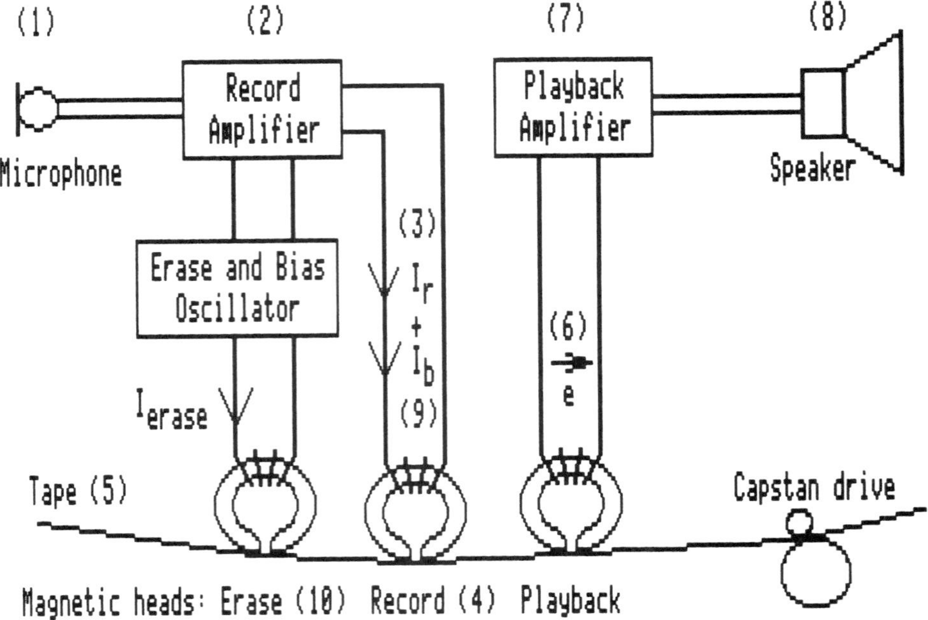

Fig. 2-1. Block diagram of a sound recorder.

This elementary record and playback process is limited by a poor fidelity in music and data recording. It is used only in *computer* applications, where the criteria for performance is the presence or absence of a signal. In high-fidelity music recordings and in instrumentation recordings, an additional *bias* current, I_b (9), is added to the record current flowing through the record head winding. The bias is a high frequency current that provides a great improvement in recording fidelity and a simultaneous reduction of background noise.

Figure 2-1 also illustrates the components that are found in all recorders:

- *Magnetic heads*—for recording (write) and playback (read); these functions may both be served by a combination record/play (write/read) head.
- A *magnetic head*—for *erasure* of any signal previously recorded on the tape. This erase head is an optional feature that is mainly used in audio and television recorders. Old information in computer disk drives is generally over-written by the recording field. Instrumentation tapes are bulk degaussed.
- A *transport mechanism*—for moving the media (tape or disk) past the magnetic heads at a uniform speed. Reeling mechanisms or motors are required for providing smooth supply and takeup of tape.
- *Record and playback amplifiers*—for processing of signals to and from the magnetic heads.
- *Control logic circuitry*—for start, stop, and fast winding of tape.
- *Power supply assemblies*—for the transport motors, solenoids, and relays, and for the amplifiers and control logic.
- The *media*—is currently not just the narrow version found in audio cassettes, or the half-inch wide VCR cartridge tapes, but comes in various grades and widths, as magnetic stripes on film and cards, and coatings on rigid disks and flexible disks (''floppies'').

Table 2-1. General Information on Magnetic Storage Devices, 1987. See Text for Details.

APPLICATION/media:	Width/dia.	Upper freq./bit rate	Speed	BPI	Trackwidth	No. tracks
AUDIO cassette	.150"	20 kHz (bias 100)	2 IPS	21,300	.021"	4 parallel
INSTRUMENTATION	.25 - 2"	4 MHz (bias 16)	120 "	66,700	.025"	16-80 ""
VIDEO, studio	1"	FM to 10 MHz	935 "	21,400	.004"	1 slanted
VHS	.5"	FM to 4.2 "	230 "	36,300	.0012"	1 """
Betamax	.5"	FM to 5.5 "	275 "	40,000	.0008"	1 """
8/mm	8/mm	FM to 5.4 "	150 "	72,000	.0008"	1 """
DIGITAL tape cassette	.15"	75 KBPS	30 "	10,000	.020"	4 parallel
tape cartridge	.25"	90 ""	90 "	10,000	.020"	8 ""
reel-to-reel	.5"	1250 ""	200 "	6,000	.050"	9 ""
tape cartridge	.5"	3000 ""	200 "	38,000	.015"	16 ""
DIGITAL floppy	5.25"	250 KBPS	360 RPM	6,000	.014"	48 TPI
floppy	5.25"	250 ""	360 "	6,000	.008"	96 "
floppy	5.25"	500 ""	600 "	10,000	.006"	192 "
Mini-disk	3.50"	250 ""	360 "	6,000	.0045"	140 "
Cartridge	5.25"	1130 ""	1500 "	18,000	.0025"	300 TPI
"Hard disk"	3.5 - 8"	2-10,000 KBPS	1000 "	10,000	.0015"	600 "
Winchester	8 - 14"	2-15,000 ""	2500 "	15,000	.0008"	1,000 "
Magneto optical	5 - 12"	?	?	15,000	.00007"	15,000 "

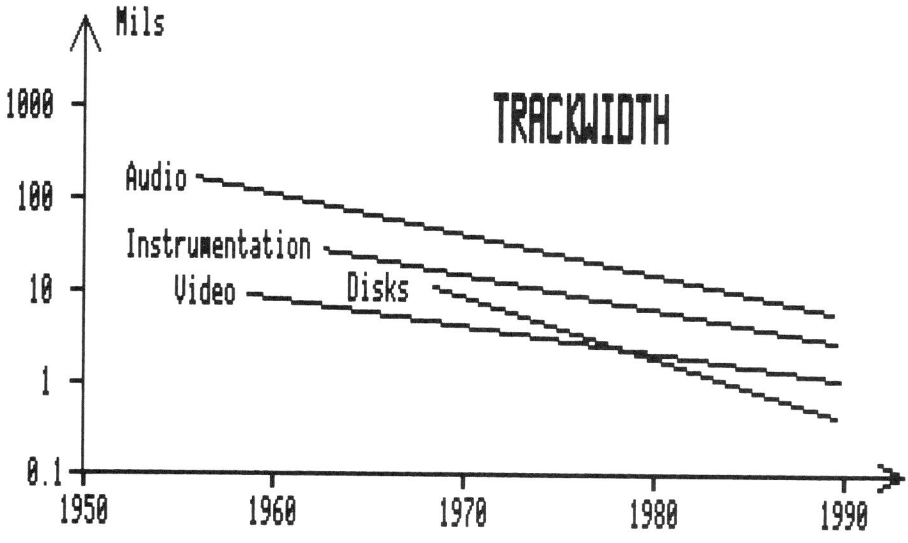

Fig. 2-2A. Projection for the future development of track width.

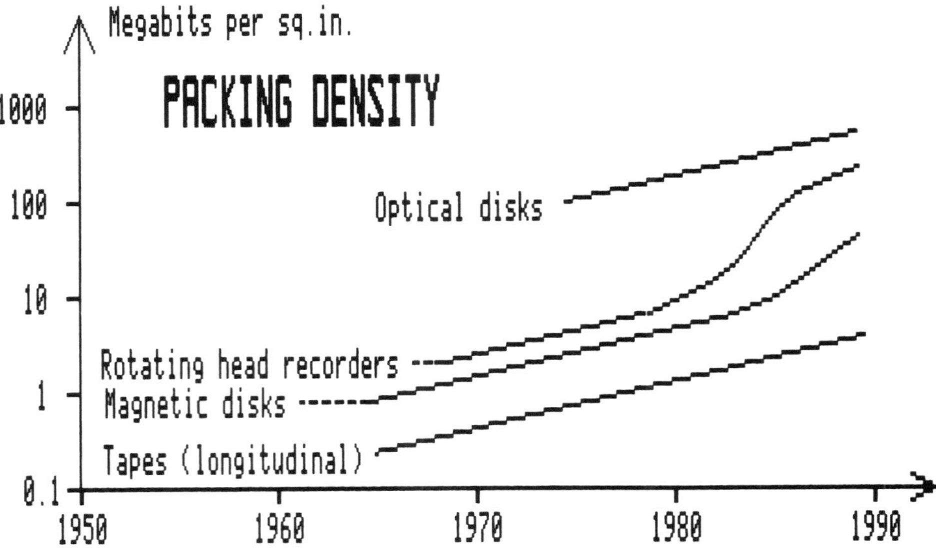

Fig. 2-2B. Projection for the future development of packing density.

Each recorder application and its media configuration dictates how that recorder is configured. Anyone familiar with recording equipment is aware of the large variety of equipment designs.

The matter of selecting, specifying, using or designing a piece of recording equipment will therefore involve tradeoff decisions. These decisions will be better, and easier to make, if the person involved has a knowledge of magnetic recording.

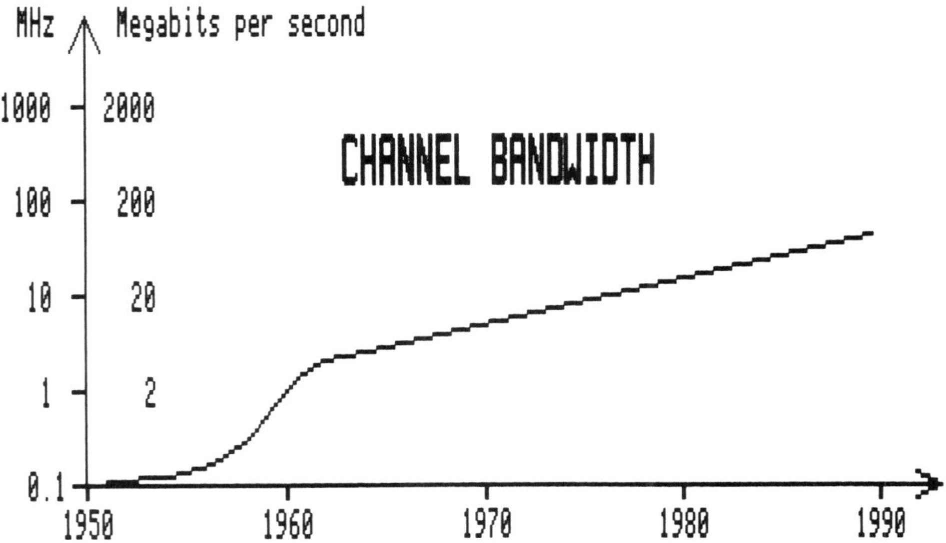

Fig. 2-2C. Projection for the future development of channel bandwidth.

It is not possible or practical to provide a survey of various recorders and data storage files within the scope of this book. So there is no detailed treatment of conventional amplifier circuits, control logics, power supplies, overall packaging, and the many possible operational features.

Table 2-1 provides the reader with a survey of the current state of the art in magnetic recording devices. The much talked about optical disks are beginning to enter the digital storage market, but may possibly find applications in 15 to 20 percent of the total market in 1990, in dollar volume. Tapes, floppies, and Winchesters will keep on growing, as will the video tape and audio cassette market, the latter two with a boost from the new digital sound recording techniques (6 hours, for example, on a VHS or BETA recorder). Add to this the new 8mm video and sound tapes, and the R-DAT.

Projections for the future developments and growth in the sound and video recording field and the data storage field are often displayed in the trade magazines. Averages of these projections are shown in Figs. 2-2A, 2-2B, 2-2C.

The main theme of this book is the nature of magnetism, recording (write), playback (read), and the performance of magnetic heads and media, in a variety of applications. Of prime concern is that the playback signal be a faithful reproduction of the input signal. The second concern is the achievement of the highest possible packing density on the media which provides the minimum storage volume (length of tape or size/number of disks).

The original (and basic) audio recorder is a good example to illustrate the signal record/reproduce process in further detail. Later chapters will deal with instrumentation, video, computer storage, coding, and modulation techniques.

FREQUENCY RESPONSE AND NOISE

When the magnetic tape leaves the record head, it has a permanent magnetic record of the sound or data signal, with flux lines extending from the surface (see Fig. 2-3). These flux

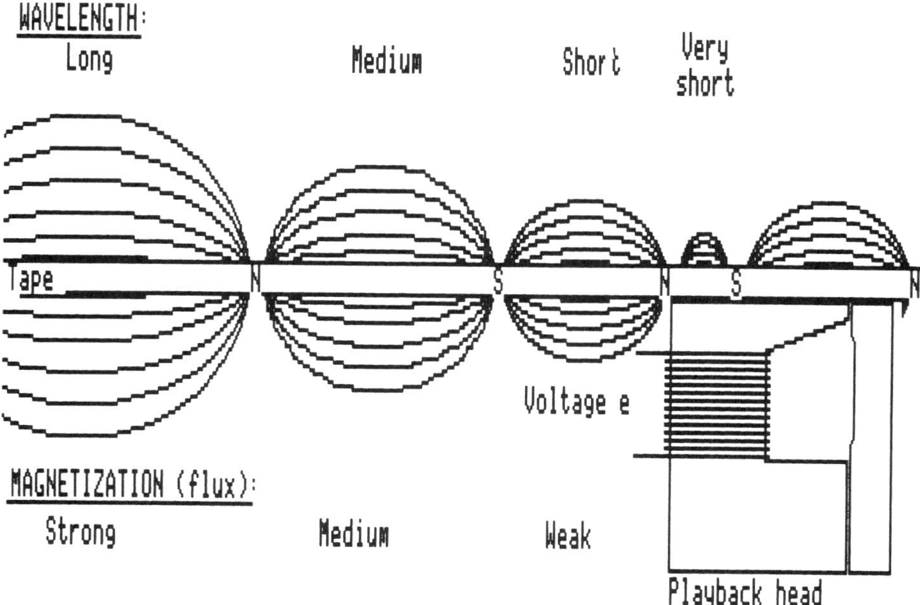

Fig. 2-3. Recorded flux pattern on a magnetic tape.

lines pass in sequence through the reproduce-head core and induce a voltage, e, which after amplification reproduces the original signal. The number of flux lines is proportional to the recorded signal strength and their duration is inversely proportional to the recorded frequency. Their duration represents a certain wavelength λ on the tape and can be expressed as:

$$\text{Wavelength } \lambda = \text{Tape speed v / Frequency f}$$

Unless stated otherwise, wavelength means "recorded" wavelength, not the electromagnetic wavelength (where v is the speed of light).

It is a common practice to run tape at slow speeds to reduce the quantity of tape needed and to design magnetic heads and tapes for good, short wavelength performance. For example, if a high-frequency response of 15 kHz is desired, as in the reproduction of high-fidelity music, the wavelength should be as short as possible to save tape. If you want to record and reproduce 15 kHz at a tape speed of 9 cm/sec (3¾ IPS), the wavelength is:

$$\begin{aligned}\text{Wavelength } \lambda &= 90\text{mm}/15{,}000 \\ &= 0.006\,\text{mm} \\ &= 6\mu\text{m } (240\mu\text{in})\end{aligned}$$

That is an extremely short length—only one quarter of the thickness of a long-play magnetic tape. And the reproduce (playback) head gap must be shorter than half of that wavelength, otherwise the high-frequency response will suffer and the fidelity of the recording will be lost.

The *induced voltage* (e) in the reproduce head winding is very low and requires amplification of 10 to 100,000 times to provide a useful output. This is particularly true in modern recorders with micro-gaps, narrow track widths (four tracks are recorded on a 3.84mm (.15 inch) wide tape in a cassette), and low tape speeds. These factors complicate the noise problem, a design consideration that is aggravated by the fact that the induced voltage is lowest at low and high frequencies, where the noise is highest. If all signal frequencies were recorded at the same level we would expect that the corresponding wavelengths on the tape would be of equal strength. But in reality, we find that the flux from the tape follows a pattern, as shown in Fig. 2-4A.

Thus the voltage (e) induced in the playback head is not only proportional to the tape flux but also to the frequency. This tends to compensate for the decreasing flux, with a voltage vs. frequency curve, as shown in Fig. 2-4B. In order to achieve a constant output voltage over the entire frequency range, more amplification must be provided for the low and high frequencies, as shown in Fig. 2-4C. This is what is technically referred to as *equalization*. From the drawing it is also seen that an otherwise flat noise level increases at the low and high ends, which is unfortunate since it emphasizes any amplifier hum and tape hiss.

It is standard practice to boost the record current in these high and low frequency regions so that less playback equalization is required. Too much boost may cause an overload condition with subsequent distortion, so the audio industry has established standards for equalization (see Chapter 28). These standards rely upon the knowledge of sound level vs. frequency in music and speech. When this knowledge does not exist, as may happen in instrumentation recording, the boost of the record current cannot be tolerated and instrumentation recorders, therefore, show an apparently worse *signal-to-noise ratio* (i.e., the separation between the maximum record level and quiescent noise level). However, the signal-to-noise spectrum should always be evaluated for the more realistic details, such as in the "weighted" signal-to-noise ratio for audio recorders and noise spectrum level for instrumentation recorders. (A weighted signal-to-noise ratio is measured with an instrument that compensates for the ear's sensitivity to low and high frequencies.)

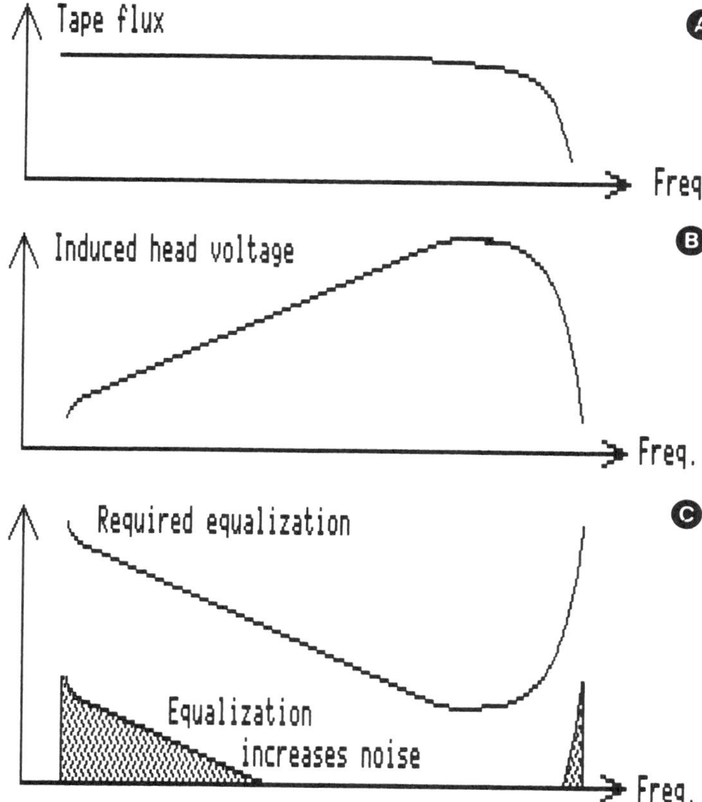

Fig. 2-4. Curves demonstrating, (A) tape flux, (B) induced voltage and (C) the required equalization.

As a guideline in evaluating audio recorder performance, the following signal-to-noise ratios are typical:

35-40 dB—Old 78 RPM, phonograph recorders.
50-65 dB—Modern LP phonograph recorders.
35-45 dB—Inexpensive home tape recorders.
45-55 dB—Good home tape recorders (two tracks).
50-65 dB—Professional tape recorders.
65-75 dB—High-quality studio tape recorders.
85-95 dB—Digital studio tape recorders.

It should be noted that signal-to-noise ratios are reduced at least 3 dB each time the track width is cut in half (for example, from 2 tracks to 4 tracks).

In high-quality audio recorders, the frequency response should cover from 20 to 16,000 Hz. This corresponds to the limits of the human ear, but consideration should be given to the fact that few musical instruments contain frequencies above 10,000 Hz, and the hearing of high frequencies generally decreases after the age of 20.

Values of signal-to-noise ratios for instrumentation recorders vary widely according to tape speed and the range of the frequency response. However, after equalization of the reproduce head voltage, the frequency response is essentially flat.

The *frequency response* of a tape recorder is never perfectly flat and may be specified as 50-16,000 Hz, plus or minus 3 dB. The term dB is an abbreviation of *decibel* (decibel is the logarithmic ratio between two signal levels: 20*LOG(LEVEL 1/LEVEL 2) and one dB corresponds fairly well to a change in sound level that can barely be noticed. As a rule of thumb, we can use the following scale:

Level change: 1 dB—Barely noticed by an expert under ideal conditions.
3 dB—noticed under normal listening.
6 dB—A definite change in sound level.

A specification of plus or minus 3 dB corresponds quite well to what the ear can tolerate. It should also be noted that frequency response applies to the record and reproduce electronics, using a good grade magnetic tape; it does not include microphones or speakers often built in or supplied with the recorder. And the requirements for the frequency response vary with the applications and program source:

Live recordings—(record companies, FM broadcast use) 20-20,000 Hz at 38 CM/S (15 IPS)).
Recordings of FM programs—30-15,000 Hz at 19 CM/S (7½ IPS).
Recordings of AM programs—50-6,000 Hz at 4.8 CM/S (1⅞ IPS).

The most frequently used *tape speeds* are indicated in the above list in CM/Sec, with the equivalent speed in IPS, inches-per-second, in parentheses. Again, very few instruments have sound signals (harmonics) above 10,000 Hz, and a speed of 4.8 CM/S (1⅞ IPS) is in many cases fully adequate for home recordings of FM programs. As the above table indicates, the frequency response does improve with increasing tape speed and it is in fact more logical to speak of how many wavelengths can be recorded per cm of tape. One thousand to four thousand wavelengths per cm of tape length is common practice for audio recorders and the present limitation is 20,000 wavelengths per cm. This requires high-quality tapes and good magnetic heads, but it is an improvement of ten times over the past twenty years. (The frequency-response requirements of instrumentation recorders and video (TV) are discussed in Chapters 29 and 31.)

NARROW TRACKS

Twenty-five years ago, all magnetic recordings were made on 6.25mm (¼″) wide tape with the full width of the tape being used for one channel (monaural). Improvements in tape quality came rapidly during the sixties, and the amateurs and high-fidelity enthusiasts took advantage of this improvement and began recording on only half the width of the tape (called half or two-track recording). After playing one side, the reel was simply turned over for the other half (Fig. 2-5). By cutting the track width in half, only half the flux is available, and the signal-to-noise ratio should suffer, but the improved tapes made up for the difference. (Otherwise, the signal-to-noise ratio would have decreased by a factor of 3 to 6 dB.)

With the advent of *stereo* (or binaural) recordings, the two tracks were used for the two channels. With further improvements in tape quality, the standards were once again changed to what now is called four (or quarter) track tapes, two tracks being used for stereo recording in one direction and the remaining two tracks for the reverse direction. Latest developments have made possible the recording and playback of four tracks on 3.8mm (0.15″) wide tape, the method used with prerecorded stereo music cassettes. (Eight-track recording is primarily used in car installations, where the noise level is somewhat high and the worse signal-to-noise ratio of the narrow track width (21 mils) is not too critical.)

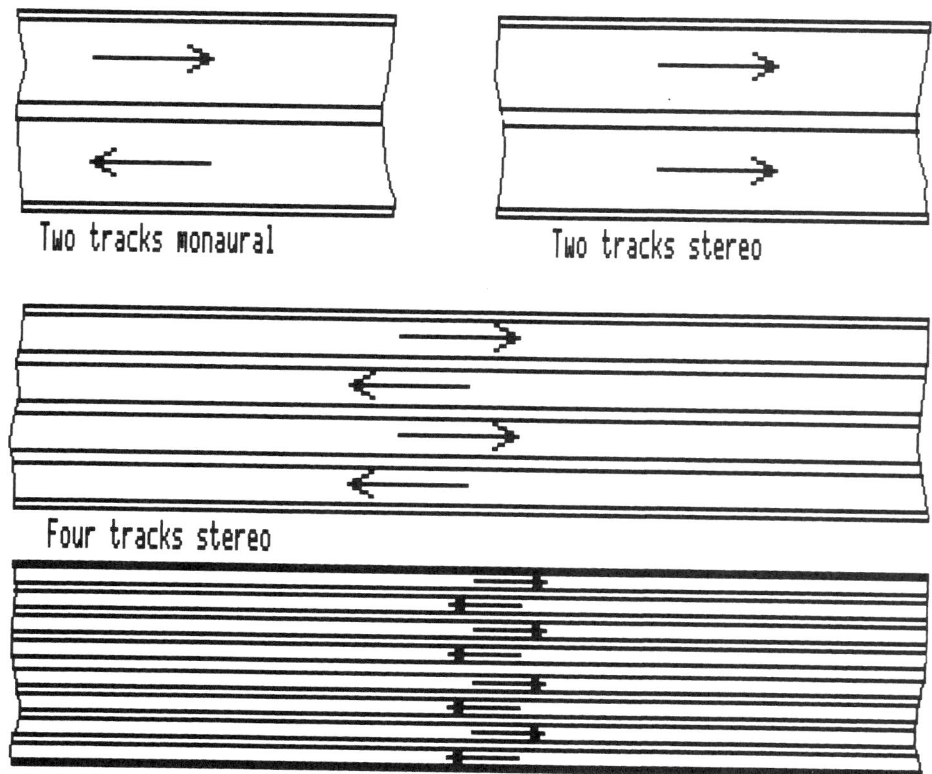

Fig. 2-5. Two, four and eight-track arrangement on a quarter-inch magnetic tape.

The improvements in tapes that made this evolution possible are dramatically illustrated in Figs. 2-6 and 2-7. They show magnetic particles from then (A) and now (B), and the refinements achieved in providing a smooth tape surface.

Today we are entering an era where magnetic recording is made in digital format on home VCR machines, 8mm video format and very small cassettes.

THE TAPE TRANSPORT

Among the variety of recorders manufactured today, there are probably no two with transports that are exactly alike. But the differences in general are reflected only in cost, quality, and style, while the basic functions are common to them all. The art of designing and mass producing recorders has come a long way over the past decade, with ever increasing performance. There have been reductions in price and many features have been added.

It is basic to all transports that they move the tape at a constant speed past the magnetic heads while recording or playing back. (Small speed variations, called flutter and wow, do exist, as will be discussed later.) All tape transports have provisions for fast winding of the tape, either for reaching a certain portion of the tape in a short time or for complete rewinding.

The tape transport mechanism in its simplest and most reliable form consists of a metal plate with three motors mounted on it, one motor for driving the tape at a constant speed and the other two for reeling the tape. Savings in the manufacturing costs of a tape transport are

sometimes sought by using only one motor with friction or clutch drives for the reeling mechanism. The layout in general is, as shown in Fig. 2-1, where the tape passes from the supply reel through the guide and over the heads; the tape speed is controlled by the capstan, against which the tape is held by a rubber pinch roller. The capstan shaft rotates at a constant speed and is a critical item in the transport mechanism. It must rotate in good bearings and should be perfectly concentric.

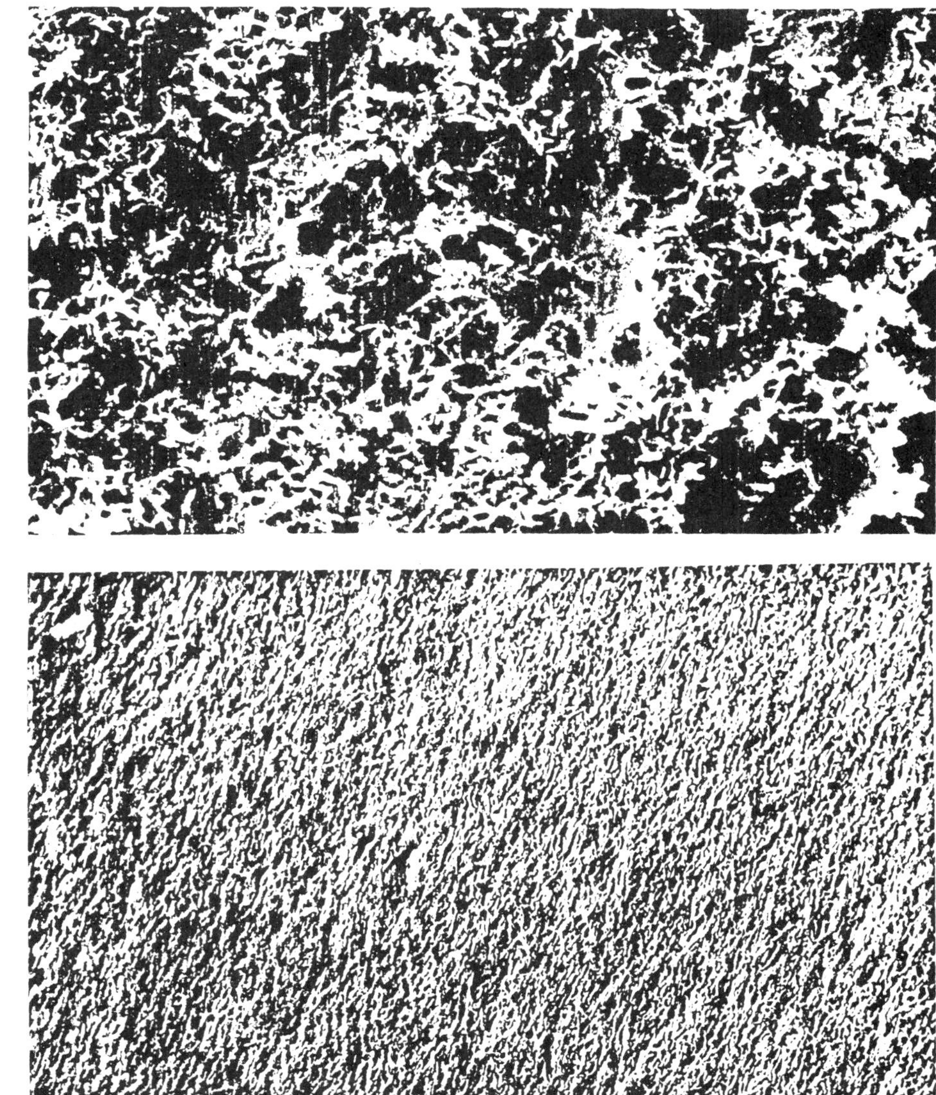

Fig. 2-7. (A) Early and (B) modern finishes of a tape surface (courtesy of Memorex Corp.).

A constant tape speed is generally best obtained by using a hysteresis-synchronous motor for the capstan drive. A hysteresis-synchronous motor follows the power supply frequency (50 or 60 Hz) which is quite accurately controlled. These motors exhibit hunting characteristics, which are usually smoothed out by a flywheel. This is obtained in the better grade recorders using the so-called inside-out hysteresis-synchronous motors. Such motors usually have two speeds, which are electrically switchable, and this, in a simple way, provides for a choice in tape economy (and/or playing time). As mentioned earlier, other tape transports utilize a single motor for the capstan drive and reel drive systems. In this arrangement the speed change is accomplished by an arrangement of pulleys or belts.

Later years have seen many recorders emerge with servo-controlled capstan speed, a technique which will be described in Chapter 27.

The selection of tape speed depends entirely upon program material bandwidth to be recorded. "Program material" bandwidth ranges from direct current to a few Hz for underwater oceanographic investigations to a few thousand Hz for taped letters and voice communications to 15,000 Hz for high-fidelity music reproduction to a few MHz for wideband recordings and up to 10 MHz for color video recording. In data recording, floppies use a data frequency of 125 kHz, while rigid disk drives approach 10 MHz.

Since most users of magnetic tape recorders have requirements for a variety of bandwidth ranges, it is generally found that recorders have from two to six speed ranges. In home entertainment recorders, this is achieved by a two or three-speed hysteresis-synchronous motor or by an appropriate pulley arrangement, and in instrumentation recorders it is achieved by the use of servo-controlled tape drive systems.

The tape transport mechanism, in addition to providing a constant speed, must ensure that good contact is maintained between the magnetic tape and the heads. This contact requires a certain amount of pressure, which can be obtained through hold-back tension on the supply reel (a felt pad against the in-going tape guide or felt pads which press the tape against the heads).

The felt pad arrangement is inexpensive but can easily cause excessive wear on the heads, which increases the cost of replacing worn-out heads. The felt pad is best used against the tape at a guide post and will, if properly adjusted, give a more constant speed throughout a reel of tape than the hold-back torque applied at the feed reel. In a single-motor tape deck, torque is provided by a slip-clutch arrangement and in a three-motor deck by reverse current to the feed-reel motor.

WOW AND FLUTTER

No capstan drive system is perfect and, as mentioned earlier, this will give rise to tape speed variations that affect the playback quality of music. (Its effect upon instrumentation data is discussed in Chapter 27.) Even minute speed variations are noticeable although it does depend on the type of program material. The human ear is particularly sensitive to speed variations during playback of pure tones, like those of bells, the flute, and the piano. Such speed changes are called *wow* and *flutter*. Speed variations up to 10 Hz are called "wow," and speed variations above 10 Hz are considered "flutter."

Wow is primarily caused by capstan shaft eccentricity or dirt buildup on the capstan, or possibly a poor reel drive system with a varying hold-back. Motor cogging and layer-to-layer adhesion in the tape pack can also cause wow. In cases of wow, tones are frequency modulated which gives rise to a singing sound. *Flutter*, on the other hand, is generated by the tape itself. Magnetic tape has elastic properties and when it moves over guides and heads, a scraping, jerky motion takes place which causes longitudinal oscillations in the tape. This destroys the otherwise pure tones and gives them a raw and harsh sound which is often mistaken for modulation noise.

Typical amounts of wow and flutter (up to 250 Hz) that can be tolerated are:

Speech	Max. 0.6 percent peak
Popular music	Max. 0.3 percent peak
Classical music	Max. 0.1 percent peak
Wow is just barely noticeable at	Max. 0.1 percent peak

where the percent peak is related to the nominal speed of the recorder.

In this area the digital recorders are superior. They do have about the same flutter in the transports, but the digitized signal stream can now be clocked into perfect synchronism with a clock, and flutter is removed.

TECHNICAL SPECIFICATIONS

Recording equipment from various manufacturers is described in terms of many specified items so a comparison between similar units can be made. A two-page listing of recorder specifications is quite common, and the media, tape or disk, may have a one-page listing. This book will, throughout its chapters, make clear to the reader what such specifications cover, and what they imply for the recorder's overall performance.

BIBLIOGRAPHY TO CHAPTER 2

Begun, S.J., *Magnetic Recording*, Rinehart & Co., New York, 1949.

Lindsay, H.W., "Precision Magnetic Tape Recorder for High-Fidelity Professional Use," *Elec. Mfg.*, Oct. 1950, Vol. 46, pp. 135-139.

Krones, F., *Die Magnetische Schallaufzeichnung*, Technischer Zeitschriftenverlagf, Wien, 1952, 216 pages.

Westmijze, W.K., "Principles of Magnetic Recording and Reproduction of Sound," *Philips Tech. Rev.*, Sept. 1953, Vol. 15, pp. 84-96.

Westmijze, W.K., "Studies in Magnetic Recording, I: Introduction," *Philips Res. Rep.*, April 1953, Vol. 8, No. 3, pp. 148-157.

Stewart, W.E., *Magnetic Recording Techniques*, McGraw-Hill, 1958, 272 pages.

Winckel, F., *Technik der Magnetspeicher*, Springer-Verlag, Berlin 1960, 614 pages.

Mee, C.D., *The Physics of Magnetic Recording*, Elsevier, New York 1964, 271 pages.

Scholz, C., *Magnetbandspeichertechnik*, VEB Verlag Technik, Berlin, 1968, 284 pages.

Schneider, C., Volz, H., *Grundlagen der Magnetischen Signalspeicherung II; Magnetbander und grundlagen der Transportwerke*, Akademie-Verlag, Berlin, 1970, 133 pages.

Mallinson, J.C., and Ragosine, V.E., "Bulk Storage Technology: Magnetic Recording," *IEEE Trans. Magn.*, Sept. 1971, Vol. MAG-7, pp. 598-600.

Altrichter, E., Boden, G., Lehmann, H., and Volz, H., *Grundlagen der Magnetischen Signalspeicherung III: Anwendung fur Fernsehen, Film, Messtechnik und Akustik sowie eine Geschichliche Entwicklung*, Akademie-Verlag, Berlin, 1972, 124 pages.

Lowman, C.E., *Magnetic Recording*, McGraw-Hill, New York, 1972.

Phillips, W.B., and McDonough, H.P., "Maximizing the Areal Density of Magnetic Recording," *IEEE COMPCON Proc.*, April 1974, pp. 101-103.

Geller, S.B., "Archival Data Storage," *Datamation*, Oct. 1974, Vol. 20, pp. 72-80.

Winckel, F., *Technik der Magnetspeicher, 2nd Edition*, Springer-Verlag, Berlin , 1977, 402 pages.

Scholz, C., *Handbuch der Magnetband-speichertechnik*, Hanser Verlag, Muenchen/Wien, 1980, 392 pages.

Patent Digest, IBM Class 360, Dynamic Magnetic Information Storage or Retrieval, The Boston Patent Co., 1984, 490 pages.

White, R.M., *Introduction to Magnetic Recording*, IEEE Press, 1985, 307 pages.

Camras, M. (Editor), *Magnetic Tape Recording*, Van Nostrand Reinhold, 1985, 443 pages.

Mee, C.D., and Daniel, E.D., *Magnetic Recording*, Vol. 1, McGraw-Hill, 1987, 514 pages, Vol. 2, 1988 408 pages, Vol. 3, 1988, 415 pages.

Mallinson, J.C., *The Foundations of Magnetic Recording*, Academic Press, 1987, 175 pages.

Part 2

Magnet Heads

Chapter 3

Fundamental Magnetism

In this chapter magnetism is introduced in a manner that will give the reader an understanding of the physics underlying magnetic recording and playback (or write and read). As the reader goes through the sections he/she will learn how electricity and magnetic materials relate to magnetic write/read heads and tapes/disks. This chapter focuses on magnetism from electric currents, and magnetism in materials, while Chapter 4 will expand on the technical aspects of magnetism when magnetic materials are applied in heads and media coatings (*media* is a common noun for tapes or disks).

OVERVIEW

We experience magnetism in two forms: as field lines from permanent magnets (tape and disk recordings), and from electric currents (in record, or write, fields). Chapters 3 and 4 will provide the reader with the necessary tools for understanding magnetism in recording and playback. The particular magnetic materials used in heads are useful because they can conduct, or guide, and concentrate magnetic fields.

We start the chapter by learning about the fields from a magnetic pole, a current carrying straight wire, a loop, and a coil. Next we form an electromagnet by inserting a piece of iron in the coil, and find that this greatly increases the field strength. To understand why, we must next learn about magnetic materials.

From physics we know that all materials may be divided into *diamagnetic, paramagnetic* and *ferromagnetic* by their magnetic properties. These properties are briefly discussed, with details postponed for Chapters 7 and 11.

The ferromagnetic materials are evaluated by tracing their hysteresis loop, which leads to an introduction of the *domain hypothesis*, and the magnetization process of ferromagnetic materials. These can roughly be divided into *soft* and *hard* materials, corresponding to their use in heads and media.

Media coatings are finally analyzed, based upon models for the individual particles they are made from. Statistical tools are well-suited for the overall properties of these assemblies of particles, and the important *switching field distribution* is introduced. This chapter concludes with a brief introduction to solid metallic coatings.

THE EARTH'S MAGNETISM

The lodestone was found to point one end toward north, and this was hence named the *north seeking pole*, or, in brief, *north pole*. The compass needle does the same, and we are, with the early definition of polarities, stuck with the fact that the earth magnet has its *South* pole located near the geographic North Pole, and its *North* pole near the South Pole (see Fig. 3-1).

The earth poles are not stationary; they have moved about quite a bit, and still do so. They are about 750 miles from the geographic poles, and move approximately one mile per year. There is also evidence that they have alternated: when molten rock containing iron cools and solidifies, it becomes magnetic. The direction of its magnetization will follow any external field that may be present, such as the earth field. It will also be of polarity opposite the external field (in accordance with Le Chatelier's Law: when a system in equilibrium is disturbed, it always reacts in such a manner as to oppose the forces which upset the balance).

During the mining of iron ore magnetizations have been found, varying in strength and direction. Later examination of these data revealed the changes in pole positions and polarities. We can correctly say that the earth was the first magnetic recorder.

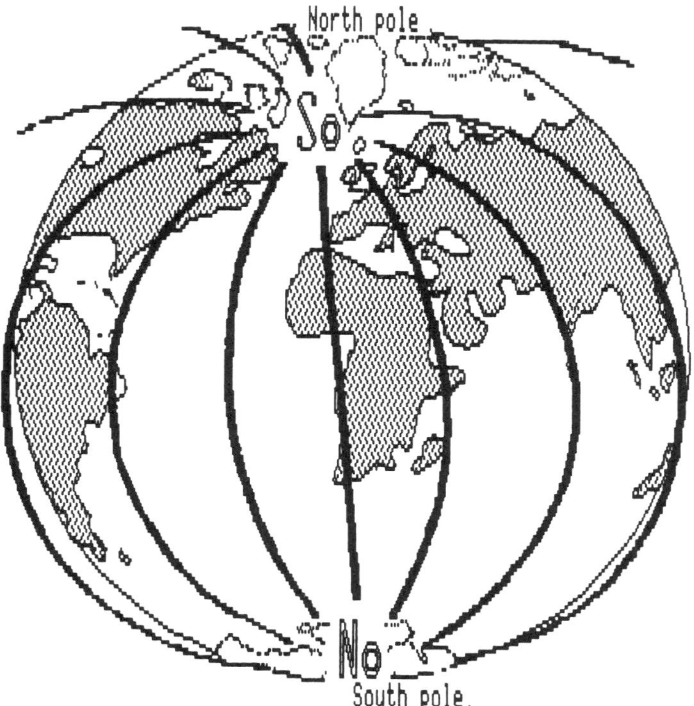

Fig. 3-1. Earth's magnetic field.

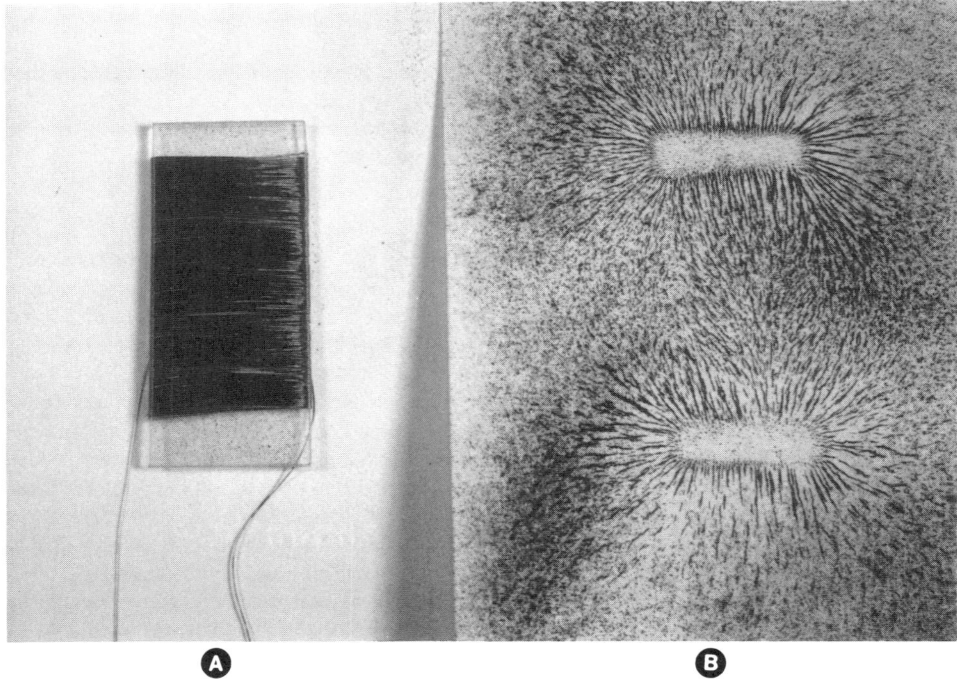

Fig. 3-2. Magnetic field lines from coil made visible using iron powder.

The origin of the earth magnetic field is obscure. All magnetization ceases to exist above a certain temperature called the *Curie temperature*, being in the order of a few hundred degrees Celsius. Most of the earth's mass is in the molten state, and hence non-magnetic. One hypothesis explains the magnetization as arising from circulating currents, that somehow are related to the earth's rotation.

A snapshot of the earth's field is provided for by the northern lights (aurora borealis). They are caused by very high-energy particles, chiefly electrons, plunging from space into the atmosphere along the outermost, closed field lines of the earth's magnetic field. When these electrons hit atoms of the high atmosphere the atoms glow in colors indicating their composition and the electron energies.

Field lines from magnets may be displayed by means of iron powder patterns. This can easily be done by the reader by first obtaining some iron powder from a hobby store. Sprinkle a small portion thereof upon a 5 × 8 inch index card, as evenly as possible. Next place this card on top of a small permanent magnet, and tap the card lightly with a finger. This will free the powder to move about, and the iron particles will behave like small compass needles: soon a field pattern develops. (Note: some magnets are so strong that they literally soak the powder to them, and destroy the patterns. Try placing a spacing between the magnet and the index card, e.g., a piece of corrugated board, or wood.) An example is shown in Fig. 3-2.

MAGNETIC FIELD STRENGTH FROM A POLE

During the latter half of the eighteenth century Charles Augustin Coulomb carried out experimental work in order to determine the field strength from magnetic poles. He had proven that the electrostatic force between two electric charges varied in strength inversely proportional to the distance squared. He proved that the same law holds for magnetic poles, and used long bar magnets in order to isolate the forces from the poles at the ends of the magnets (see Fig. 3-3). A distance of one unit between neighboring poles and ten between the end poles would introduce an error of less than two percent, assuming an "inverse squared" force law.

Magnetic field strength is defined as the force the field exerts upon a unit pole. This leads to the expression for the field strength from a pole of strength p_1:

$$H_1 = \frac{p_1}{4\pi r^2} \quad \text{A/m} \qquad \textbf{(3.1)}$$

where:

H = field strength in A/m
r = distance in m
p = pole strength in Am

Coulomb's expression for the force F upon a pole p_2 in the field H_1 then becomes $F = \mu_o H_1 p_2$, where μ_o is a constant $= 4\pi 10^{-7}$ Hy/m. We may replace the field strength H with another expression for the magnetic field, namely the equivalent number of field lines per unit area, B, by using:

$$\begin{aligned} B &= \mu_o H \quad \text{Wb/m}^2 \\ \mu_o &= 4\pi 10^{-7} \text{ Hy/m} \end{aligned} \qquad \textbf{(3.2)}$$

We can now determine the torque that a uniform field exerts upon a compass needle of length L, making an angle of α degrees with the field. The force on each pole is B*p, and the torque arm is (L/2)*sinα; the total torque is therefore:

$$T = B*p*L*\sin\alpha \qquad \textbf{(3.3)}$$

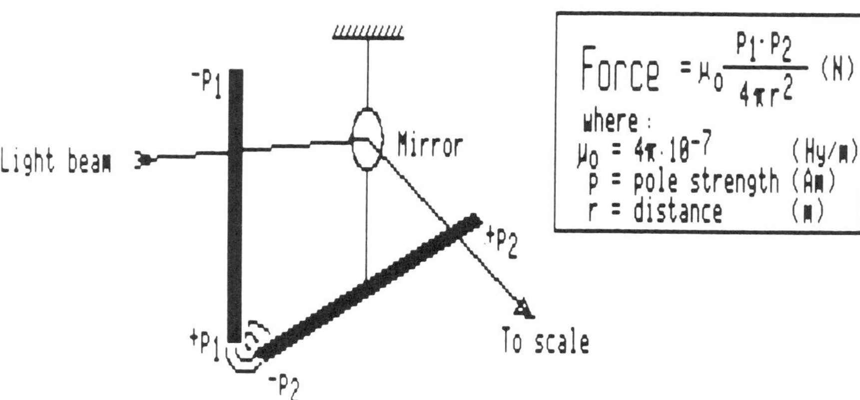

Fig. 3-3. Coulomb's experiment to find the force between two magnetic poles.

The product p∗L is defined as the *magnetic moment m*:

$$m = pL \quad Am^2 \tag{3.4}$$

We will find that this quantity corresponds to the permanent magnetization in a recorded track. We will be able to calculate this magnetization, and hence find the field associated with it.

In digital recording bits of alternating polarity represent ONEs and ZEROs. The boundary between the magnetized bits is named the *transition zone* (see Fig. 3-33), and is actually a short and wide pole, stretching from one side of the track to the other. We will return to this topic in Chapter 4.

The field strength from an infinitely long line source decreases inversely proportional to the distance (not squared, as for a point source), and the formula for the field strength that we will find useful is:

$$H = \frac{p'}{4\pi r} \quad A/m \tag{3.5}$$

where:

H = field strength in A/m
p′ = pole strength per unit length in Am/m = A
r = distance in meters

The visualization of field lines by iron powder can now be explained: These patterns show up where the field is changing. A small iron whisker placed in a field will have a pole induced in both ends. If the field strength is the same everywhere then the pulling forces in these poles are identical in strength, but in opposite directions. This will cause the dipole to turn like a compass, but without lateral motion. When the field strength varies then the pole in the strongest portion of the field will be pulled more than the pole in the weaker portion, and now the particle will move toward the stronger field.

FORCES IN MAGNETIC FIELDS

Oersted made another significant discovery: a wire in a magnetic field will seek to move when a current flows through it. Experiments showed that the force was proportional to the product of the field strength, the current, and the length of the wire, assuming the wire is placed perpendicular to the field lines. Otherwise we must multiply by the sine of the angle between wire and field directions:

$$f = B*l*i*\sin\alpha \text{ newtons} \tag{3.6}$$

where:

B = flux density in Wb/m^2
l = length of wire in m
i = current in A

The direction of force is given by a simple rule by Fleming, illustrated in Fig. 3-4.

Example 3.1: A linear motor has a circular winding with 50 turns and a diameter of 6 cm. What is the axial force if placed in a uniform field of 5000 Oe, and the coil current is 2 A?

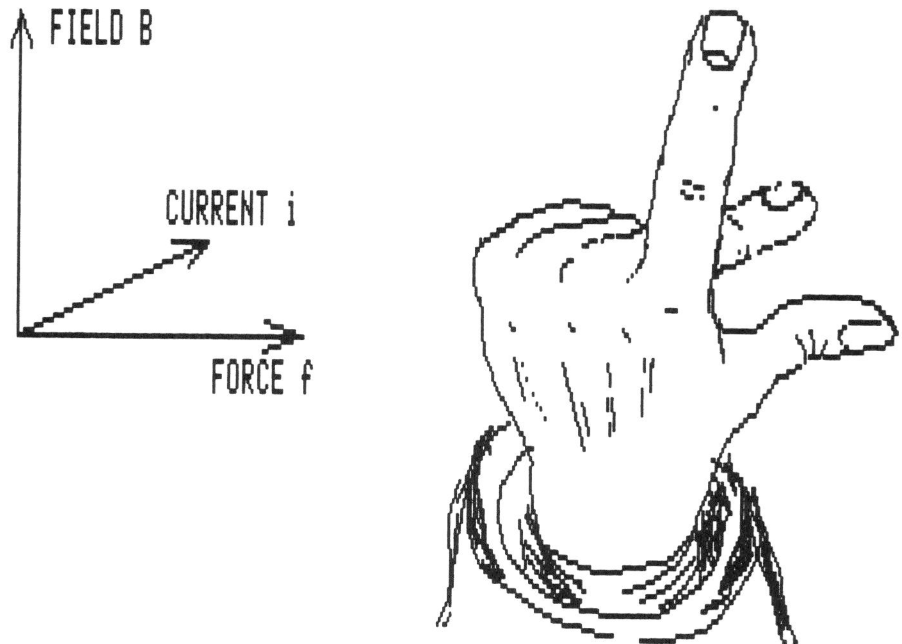

Fig. 3-4. Fleming's rule for finding the direction of the force on a current carrying wire, knowing the directions of the current i and the field B (H).

Answer: Converting the field strength gives $H = 5000*1000/4*\pi = 1.25*10^6/\pi$ A/m. Next $B = 4*\pi*10^{-7}*H = 0.5$ Wb/m^2. Finally $f = 0.5*50*\pi*0.06*2 = 9.42$ N = 9.42/9.81 kg-force = 0.96 kg-force.

MAGNETIC FIELD STRENGTH FROM A CURRENT

The field lines from the current I in a straight wire are shown in Fig. 3-5, top. The French mathematician Andre Marie Ampere showed that the field is generated by the current I, and its strength can be determined by taking the line integral along a closed path surrounding the wire:

$$\int H\,dl = I \tag{3.7}$$

The simplest path around the wire is a circle, centered at the wire, and the value of the line integral can be found to equal $H*2\pi r$, where r is the circle diameter. This results in an important mathematical expression that tells us that the field strength is inversely proportional to the distance from the wire:

$$\boxed{H = \frac{I}{2\pi r} \quad \text{A/m}} \tag{3.8}$$

where:

I = current in amperes
r = distance in meters

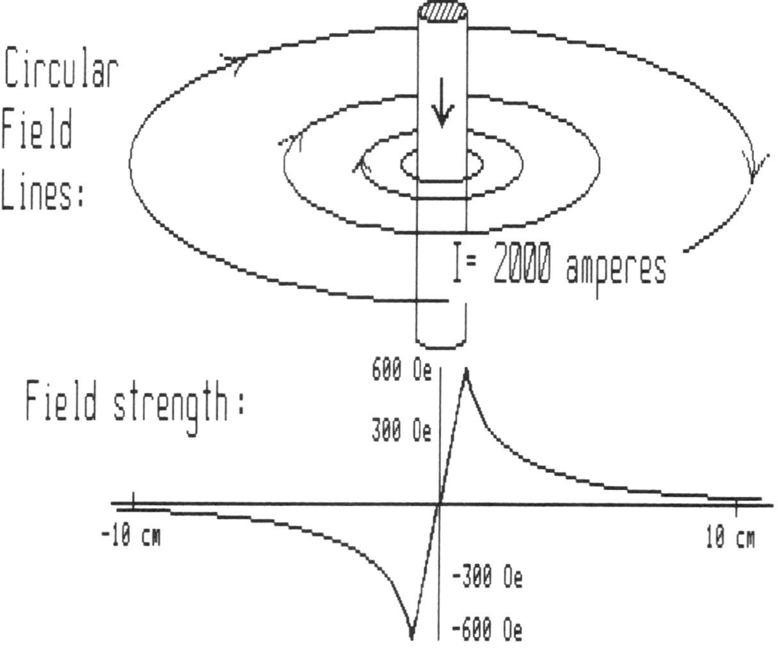

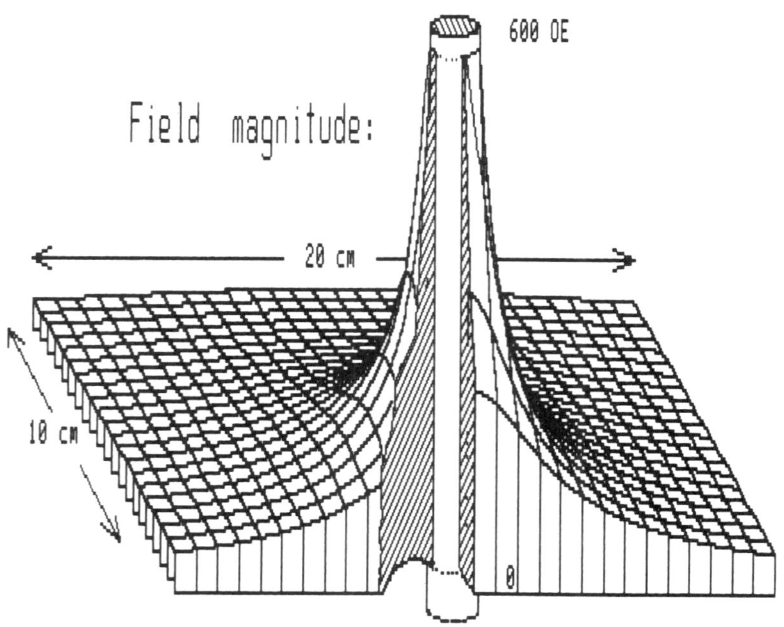

Fig. 3-5. Magnetic field from a current in a round conductor.

The field strength decreases proportionally to the distance. It appears that the field strength becomes infinite at the center of the wire (r-> 0), but bear in mind that the line integral now is taken along a path with radius -> 0, and therefore encircles a current of value -> 0. As a matter of fact, when the circular path has a radius that equals or is less than the wire radius then we find a lesser field strength:

$$H = \frac{I*r}{2\pi R^2} \tag{3.9}$$

Figure 3-5 illustrates the variation in field strength with distance from the wire. The same curve applies for the field strength from a *line pole source* (or row of poles), and the formula for that field strength is:

$$H = \frac{p/l}{4\pi r} \tag{3.10}$$

The early unit for magnetic field strength was called Oersted, which still is widely used. In this book all technical formulas and computations are in SI units, while materials will be described using cgs units; in many places the complementary unit will be shown in parentheses.

It is not difficult to translate from one unit to another. For magnetic field strength the numbers are:

$$\boxed{\begin{aligned} 1\ \text{A/m} &= 4\pi/1000\ \text{Oe} \\ 1\ \text{Oe} &= 1000/4\pi\ \text{A/m} \end{aligned}} \tag{3.11}$$

The A/m is a rather small unit, being approximately 1/80 of one Oersted. And the Oersted itself is a small unit: the earth field strength varies from 0.6 Oe near the poles to 0.3 Oe near the equator. And the field strength required to erase an ordinary audio cassette must be greater than 300 Oe ≅ 24000 A/m.

Example 3.2: A person inside an elevator carries a cassette tape in his pocket, and is leaning against the wall. Just outside are cables carrying heavy currents for the lift motor. Will the tape be erased if the distance between the tape and one cable is 10 cm, and the in-rush motor current is 2000 A?

Answer: The field strength is $H = 2000/2\pi*.1 = 10000/\pi$ A/m = $(4\pi/1000)*10000/\pi$ = 40 Oe. The tape is not erased. Figure 3-5 illustrates the circular field lines, the field strength variation and a topographic map of the field magnitude around the wire. At a distance of 1 cm the strength is 400 Oe, and that would erase most tapes of standard coercivity (around 300 Oe).

FIELD FROM A COIL

There are several instances where we wish to generate and control a magnetic field: in the gap of a recording head, or in a test instrument to measure magnetic properties of materials. These are just a few applications of what is known as *electromagnets*, i.e., where magnetism is created by currents in electric wires.

Each of the many wires in a coil generates a field, each of which adds to a final, larger field. We will find that the field lines run in patterns through the length of the coil and close themselves (Fig. 3-6, right). It is characteristic for the magnetic *field lines from currents to form closed patterns* around these currents, in contrast to electrical field lines that begin and

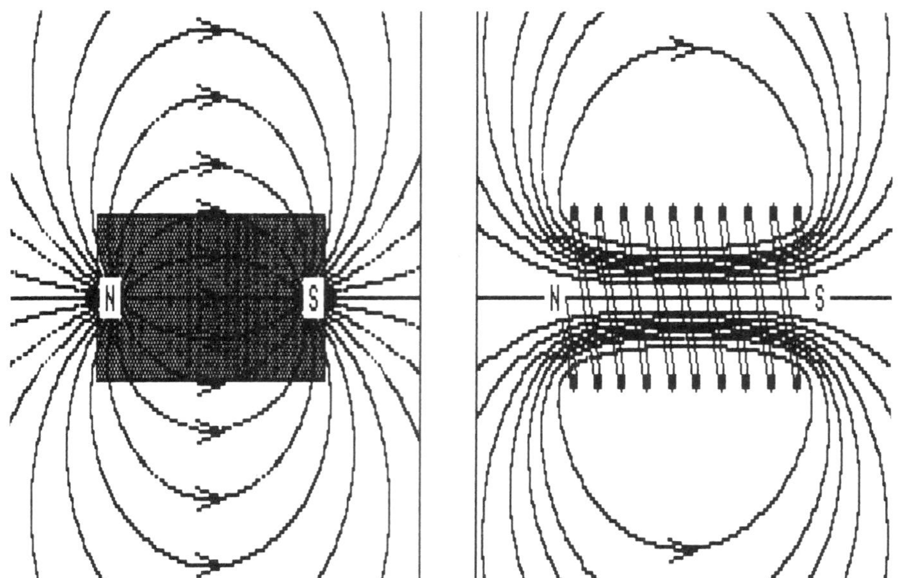

Fig. 3-6. Magnetic field lines inside and around a magnet and a solenoid.

end on electric charges. It is also in contrast to *field lines from permanent magnets, that run from pole to pole*, i.e., open paths as shown in Fig. 3-6, left.

We cannot use formula (3.8) to calculate the field strength in and about the coil since it only holds for an infinitely long wire. We must start with the field from a single, closed turn, which we arrive at by summing up the contributions from many small sections dl of this loop. Each section dl produces a field that is found by Biot-Savart's Law:

$$H = \frac{I * dl * \sin\alpha}{4\pi r^2} \quad A/m \qquad \textbf{(3.12)}$$

where:

- I = current in A
- dl = length of wire segment in m
- α = angle measured clockwise from direction of current along dl to direction of radius vector r, from dl to point P
- r = distance from dl to P, in m

For a long, straight wire we find formula (3.8), by summing (integrate) the contributions from minus to plus infinity. For a single turn loop we find for the field strength on the axis of the loop:

$$H = \frac{I * a^2}{2 * (a^2 + x^2)^{3/2}} \quad A/m \qquad \textbf{(3.13)}$$

where:

I = current in A
a = wire loop radius in m
x = distance from loop center, in m

A coil with n turns can be thought of as comprising n single loops. The field intensity well inside the coil is proportional to the current as well as the number of turns n, and is inversely proportional to the coil length l. The computed field pattern is illustrated in Fig. 3-6, right. The field strength in the center of the coil can be calculated. For a coil with many, evenly wound turns (n) and length (l) much greater than its diameter, it is:

$$H = \frac{ni}{l} \text{ A/m} \tag{3.14}$$

Figure 3-6 illustrates another interesting point: it appears that we must deal with two distinct sources of magnetic fields. From the earth, the lodestone, and other permanent magnets we have field lines that are "open," and directed from a north pole toward a south pole (left illustration). And from a current in a straight wire, or wound into a coil, we have "closed" field lines (right illustration).

When we move away from the permanent magnet and the coil, as shown in Fig. 3-7, we find that the fields look very much alike. Further away, in Fig. 3-8, they are identical, and are both named the *magnetic dipole field*. Near the dipole (magnet or current-coil) this field decreases inversely proportional to the distance cubed.

Similar to the way the magnet would turn in a magnetic field, so will the current carrying coil. And the torque is found from:

$$T = B*i*A*n*\sin\alpha \text{ Nm} \tag{3.15}$$

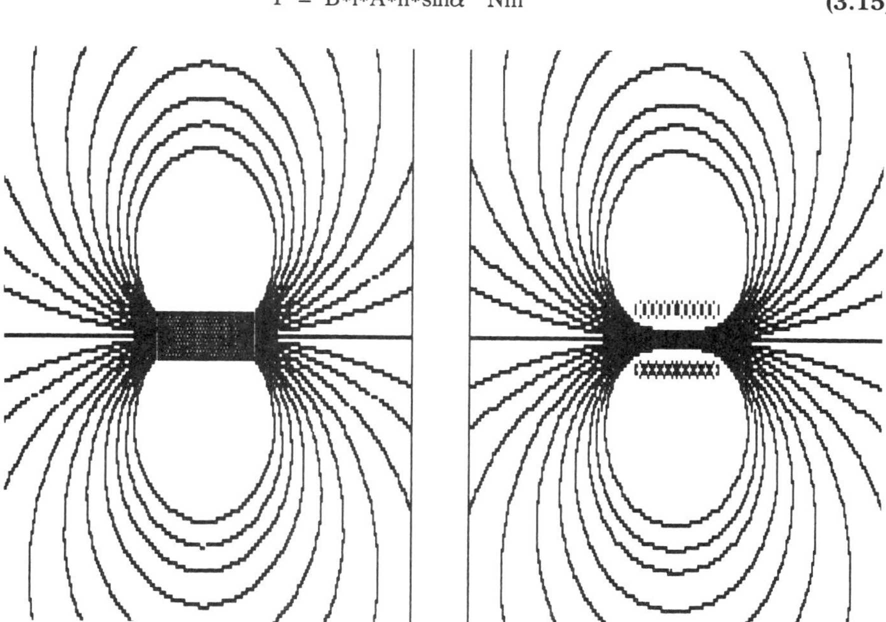

Fig. 3-7. Field lines from a permanent magnet and from a solenoid; near field.

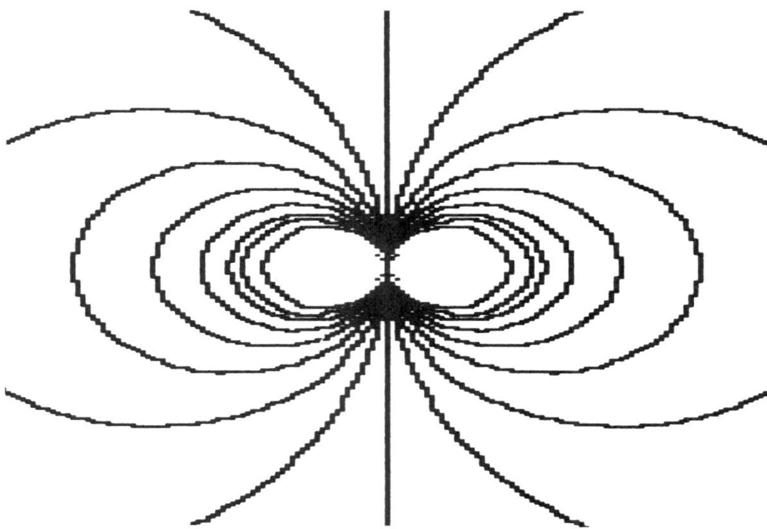

Fig. 3-8. Magnetic field lines from a magnetic dipole. The field from a permanent magnet dipole is identical to the field from a solenoid, provided they have the same magnetic moment. The field strength from a dipole decreases with the distance to the third power.

where:

B = flux density in Wb/m^2
i = current in A
A = area of the winding "window"
n = number of turns
α = angle between field direction and coil axis

This formula has the same form as (3.3). Since the behavior of the magnet and the coil are the same, and the torque equations are alike, then we say they are models of each other. The *dipole moment* of the current carrying coil is:

$$j = iAn \quad Am^2 \tag{3.16}$$

We will use the identity between the dipole moments when we seek to find the read voltage from a magnetized media coating. Finding the flux lines that link through the coil winding on the read head is exceedingly difficult if we try to follow the field lines from the magnetization in the coating. By substituting the permanent magnetization dipoles with the equivalent coil dipoles we find the problem readily solved by using the analogy to coupled coils. We shall return to this topic in Chapter 6.

Example 3.3: A uniformly wound coil has n = 100 turns, a length of 10 cm and a current I = 2 A running through the wire. Find the flux density at the coil center.

Answer: $H = ni/l = 100*2/0.1 = 2{,}000$ A/m. $B = \mu_o H = 4\pi 10^{-7}*2{,}000 = 8\pi*10^{-4}$ Wb/m^2.

NOTE: μ_0 equals one in the cgs system; remember, however: **Never use cgs units in MKS formulas:** H = ni/l = 100*2/10 = 20 Oe is wrong! The true number is (4π/1000)*2,000 = 8π = 25.13, where we have used a conversion factor (in parentheses) to convert A/m to Oe.

INDUCTION OF VOLTAGE; FARADAY'S EXPERIMENT

A current can generate a magnetic field which in turn can magnetize a piece of iron permanently. This permanent magnet will generate an electric voltage when it is dropped through a coil (bringing about a change in the number of flux lines). Figure 1-10 shows how Faraday carried out the experiment by changing the flux using a magnetized coil.

The magnitude of the generated voltage is:

$$\boxed{e = -n \frac{d\psi}{dt} \text{ volts}} \qquad \textbf{(3.17)}$$

where:

n = number of turns
ψ = number of flux lines in webers

The unit weber is equal to volts*seconds (verify: differentiate volts*seconds with respect to time, and voltage results).

We can appreciate this voltage generation by recalling Le Chatelier's law, which states that when any system in equilibrium is disturbed, it always reacts in such a manner as to oppose the forces which upset the balance. This situation arises when the field through the coil is changed, and the electrons will start moving to generate a current that in turn creates a field that opposes the change. The minus sign in the formula expresses the reaction.

Example 3.4: The flux level from a 25 m wide track is 0.01 nWb, at a data rate that corresponds to a frequency f = 1 MHz. The read head coil has 25 turns. What is the induced voltage (disregarding losses)?

Answer: Assume a sinusoidal variation in the flux, so it can be expressed as $\psi = \psi_{max}*\sin \omega t$, where $\omega = 2\pi f$. Then we find, after differentiation: $e = -n*\psi_{max}*\omega*\cos\omega t$. The peak voltage is then $E = n*\psi_{max}*\omega = 25*0.01*10^{-9}*2*\pi*1*10^6 = 1.57$ mV.

The discovery of *electromagnetic induction* was made by Michael Faraday in 1831—eleven years after Oersted's discovery of electromagnetism. The two discoveries together laid the foundation for the Industrial Revolution of the 19th century: electric generators and motors. The circumstances and methods of their discoveries may seem trivial in this day and age, but both scientists were diligently pursuing a possible connection between magnetism and electricity, as testified by Oersted's lectures and Faraday's notebook entries.

They are also the principles behind magnetic recording and playback, as shown in Fig. 3-10. The field from the write coil is concentrated inside, and in front of its gap by the use of a magnetic material, and forms an *electromagnet*. Similarly, the flux lines from the media are collected by a magnetic core and conducted through the read coil.

We have now covered the fundamentals we need to know regarding the relations between electricity and magnetics, and their role in the write/read processes. Next we turn our attention to magnetic materials, as they are used in write and read heads, and media coatings.

ELECTROMAGNETS

The flux density near a coil is increased when the turns are wound onto a magnetic core of, for instance, iron. We are actually generating poles at the end of the iron, with polarities as shown in Fig. 3-9. The coil plus core now forms what is called an *electromagnet.*

The field patterns in and around this electromagnet are a combination of the left and right side of Fig. 3-6, and we will discuss the resulting field pattern in Chapter 4. The contribution to the field from the magnetic core (poles) adds outside the coil, but subtracts inside where they run counter to the coil field lines. They actually tend to *demagnetize* the core magnetization.

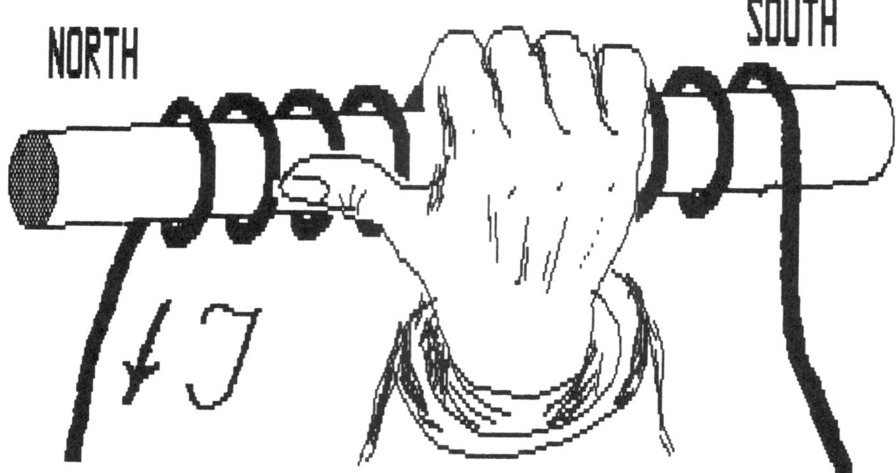

Fig. 3-9. The Right Hand Rule shows where the north pole is located: Grab the Solenoid, or electromagnet, with the right hand, and wrapping the fingers in the direction of the current. The thumb will then point toward the north pole.

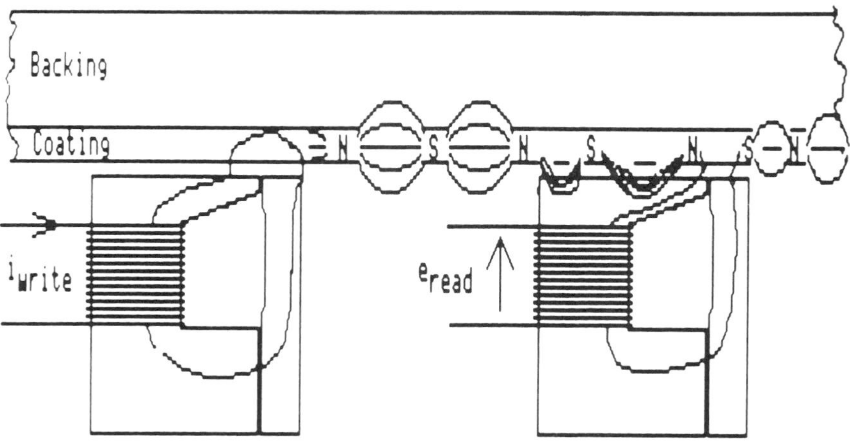

Fig. 3-10. Principles in magnetic write/read process (record/playback): Write current I_{write} produces a field that magnetizes the magnetic coating; field lines from transitions (N, S) induce a voltage e_{read} in the read head.

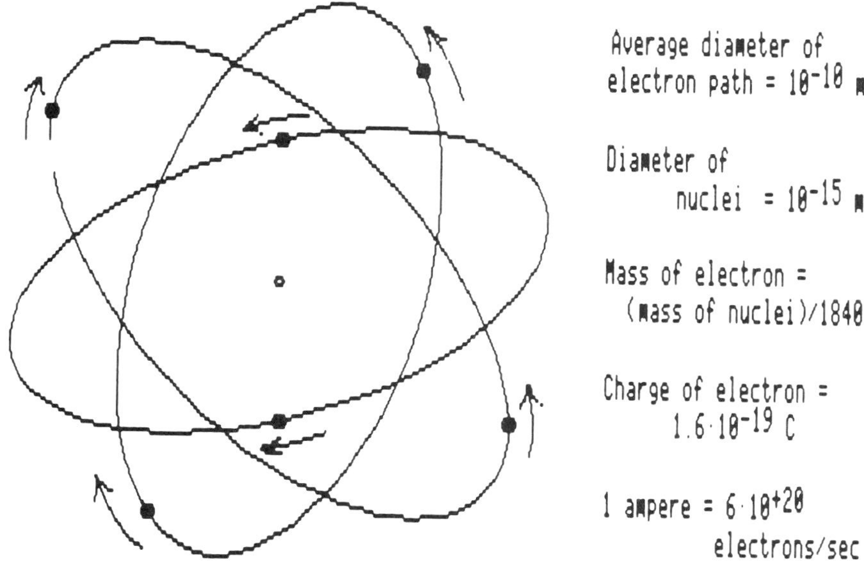

Fig. 3-11. Simplified model of an atom with 6 orbiting electrons.

ORIGIN OF MAGNETISM

The knowledge of magnetism was at the time of Valdemar Poulsen's invention very shallow and Poulsen's invention was therefore quite significant. He received a gold medal in recognition thereof at the World's Fair in Paris in 1900.

Ampere had earlier speculated that magnetism in lodestones and other magnetic rocks had its origin in "some" molecular circulating currents inside the material. Modern quantum physics explains magnetism as originating in electron-spins which produce interacting fields among themselves, giving rise to magnetization in certain materials.

Ampere was quite right in his hypothesis, when we consider a simple model of an atom, see Fig. 3-11. Electrons are moving in circular orbits around the positively charged nucleus. Such a moving charge represents a current: $i = dq/dt$, and we know that currents generate a magnetic field. The electrons are also spinning about their axes, and with their charge distributed over their surfaces these again represent currents, giving rise to an electron-spin magnetic moment.

A magnetic field will, therefore, always originate from currents. When the current runs in a loop we have a magnetic dipole (see Fig. 3-7), and it always has *two poles*: cutting the magnet in half results in two smaller magnets, each with half the original moment. Similarly, slicing a single turn loop will result in two single turn loops, each with half the original ampere-turns, and therefore also half the original moment—and each with a pair of poles.

Diamagnetism

A circular conductor with a current i running in it will generate a magnetic vector with a dipole moment $j = iA$ Am^2. Each orbiting electron will therefore generate a magnetic moment, and this would be a plausible explanation for the origin of magnetization in *all* materials. But it so happens that in diamagnetic materials two electrons travel in each orbit, in opposite directions, so no net magnetic moment is produced! The number of electrons is always even.

When a magnetic field is applied, and links through the electron orbit, then one electron will resist this field by speeding up, and the other by slowing down. Now the balance is destroyed, and the material is said to be *diamagnetic*, a property of *all* materials.

This magnetic property is very weak, and manifests itself by the fact that a diamagnetic material will be pushed away by a magnetic field. Typical diamagnetic materials are copper, lead, and water.

Paramagnetism

We now consider the other motion: the electron spin. We will find that the charge of q ($= 1.6022*10^{-19}$ coulomb) is distributed only on the surface of the electron since like charges repel themselves. The charge distribution that moves in a circular pattern around the electron's equator does again represent a current, and is associated with a magnetic moment, also called the *electron-spin*. Its magnitude equals one Bohr magneton, after the atomic physicist Niels Bohr:

$$m = 9.3*10^{-24} \text{ Am}^2 \qquad \textbf{(3.18)}$$

The electrons orbit within shells, as indicated in Fig. 3-11, are for most materials in perfect balance. Because of thermal vibrations, the axes of the spins are distributed over all possible directions, and the net magnetic moment is equal to zero.

The application of a magnetic field will unbalance the pairing of the spins, and the material will become slightly magnetic, and be attracted into the field. Air and aluminum are paramagnetic materials.

Ferro- and Ferrimagnetic Materials

It now happens that there are three materials where there are unpaired electrons: four electrons in iron, three in cobalt, and two in nickel. These materials are the main ingredients in the magnetic materials used in modern recording technology.

Owing to the presence of such uncompensated spins some of the electrons in one atom are located so closely to the nucleus of another atom that there is an exchange of electrons between the two atoms. The resulting forces are called *quantum-mechanical forces of exchange*, and are electrostatic in origin. They are associated with the exchange energy and exchange field, the latter being positive or negative. *Ferromagnetism is displayed when the exchange field is positive and has a certain definite value.*

These exchange forces cause a parallel alignment of neighboring spins, and very strong magnetization levels (10^7 Oe) are present inside these materials, named *ferromagnetic*. The spins will tend to align themselves within small volumes, called *domains*, in such a way that the domain magnetizations are pointing in different directions. The material may therefore appear non-magnetic, until influenced by an external field. We will shortly proceed by investigating what happens when such a field is applied (and discuss no further quantum-mechanics).

The spin alignment is perfect at the very lowest temperature, −273 degrees Celsius, or 0 degrees Kelvin. At higher temperatures thermal agitation causes excursions of the angle of mutual orientation, and at high enough temperatures the exchange forces lose control. The spins are now in total disarray, pointing in all directions, and the material is no longer ferromagnetic. It has become *paramagnetic*.

The temperature where this disorder happens is called the *Curie temperature* T_c. The process is reversible: when the material cools it again becomes ferromagnetic—but it has no memory of its prior magnetization.

We will encounter this phenomenon in coatings made from chromium dioxide particles (T_c = 110° Celsius), and in certain high permeability ferrites in heads (T_c near room temperature).

The remainder of this book will assume ferro- or ferri-magnetic materials, whenever magnetic materials are discussed.

MAGNETIZATION CURVE; HYSTERESIS LOOP

The field lines associated with a magnetic field H are called flux lines, given the symbol ψ. A stronger field has more flux lines running through a given area than a weak field, and it is convenient to measure the intensity in terms of *flux density* B (or incorrectly: induction B). Its unit of measure is weber/m² or tesla (gauss is used in the cgs system—abbreviated G). The conversion formula is:

$$\boxed{1\ \text{Wb/m}^2 = 10^4\ \text{G}} \tag{3.19}$$

From quantities in Wb/m² (or gauss) we can deduce that the number of flux lines for a given cross section is measured in webers (or maxwells in cgs; 1 Wb = 10^{-4} Mx).

When a magnetic material is placed in a magnetic field H then the number of flux lines inside the material is multiplied by its *relative permeability* μ_r, defined as:

$$\mu_r = \frac{\text{flux density in material}}{\text{flux density in air}} \tag{3.20}$$

If the field strength was strong enough to saturate the material, and the field turned off, then a certain amount of magnetism would remain, called the *retention*, or *remanence* B_r, in Wb/m². A field of the opposite direction is required for reduction of this remanence to zero, and its magnitude is named the *coercive force* or *coercivity* H_c, in A/m.

These characteristics are easier to remember if we point them out as we follow a complete magnetization cycle of a magnet. We will consider a ring core sample as shown in Fig. 3-12. The magnetic field from the current i through the n turns in the coil generates a field H_{core} = ni/l, and we will use this field H_{core} as the abscissa in a graph of the flux density B.

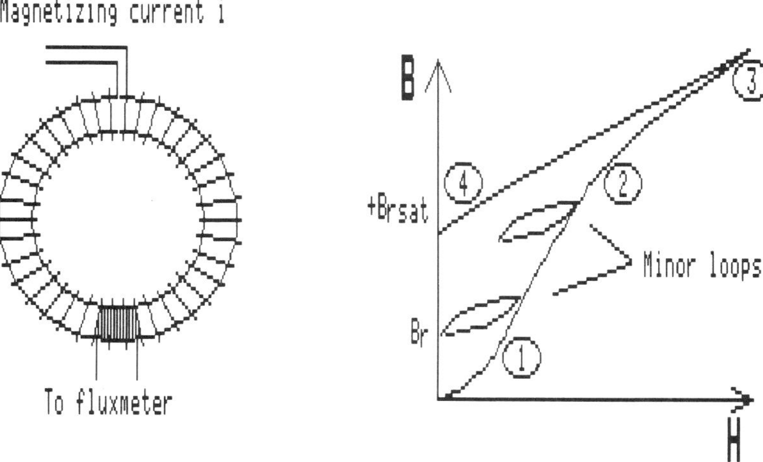

Fig. 3-12. Test ring core with magnetizing and flux sensing windings, and initial magnetization curve (1-2-3), remanence after saturation (4) and two minor hysteresis loops.

We will, in other words, plot the flux density B versus field H_{core} (B is measured with a small search coil with N turns placed on the ring core).

The ring core is initially without remanence (i.e., it is in a demagnetized state) and the flux density B will increase with the field H_{core} from the origin (0). Its increase becomes more rapid (1-2) as H_{core} increases but for large values of H_{core} it levels off (2-3); we say the material saturates (which is what happens when the record level is too high during an audio recording).

This curve is called the *initial magnetization curve*. The permeability μ (absolute value) is defined as the ratio between the flux density B inside the material and the flux density B we would have when the material is non-magnetic (e.g., wood, air).

μ is a function of both material and field strength H_{core}. The starting value at $H_{core} = 0$ is called the *initial permeability* μ_{init}, and equals the slope of the initial magnetization curve at (0,0). The permeability attains its maximum value in the range 1-2, then it decreases and eventually reaches the value of μ_0 one for very high fields H.

If we had returned the current to zero before we reached saturation then the B-value would not go back to the origin, but to a point B_r between 0 and B_{rsat}. Its value would depend upon the field level to which the magnetization was taken. These remanence values are proportional to the field values that the material was brought to, although the relationship is nonlinear. It illustrates the principle used by Valdemar Poulsen in his first recordings; the removal of the field was accomplished by moving the steel wire out of the recording field, and the field magnitude was controlled by a carbon microphone that in accordance with the sound pressure field adjusted the current from a battery through the record head winding.

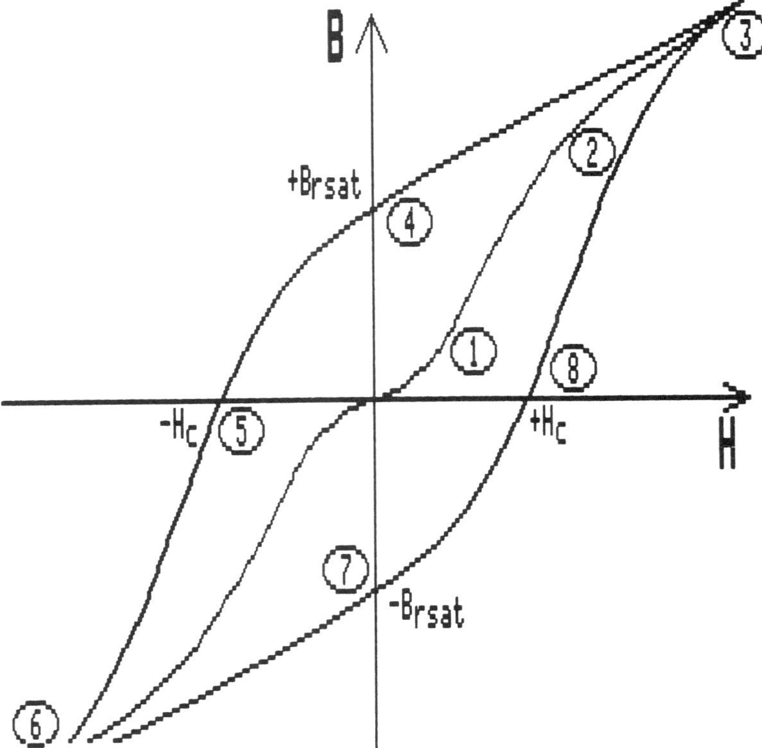

Fig. 3-13. Complete hysteresis loop for magnetic materials: 0-1-2-3 is the initial magnetization curve; 3-4-5-6-7-8-3 is the hysteresis loop.

Removal of these remanent magnetizations can be done by reversing the current, and exposing the material to a negative field. Removal of B_{rsat} would require the largest field, which is a material property named the *coercivity* H_c.

From this point the material can be saturated into the negative magnitude by further increasing the negative field. Again, removing the field brings the magnetization level to $-B_{rsat}$. A positive field of magnitude H_c is required to bring the magnetization back to zero, and further field increases will bring us back to $+B_{sat}$. The curve thus traced is called the *hysteresis loop* (see Fig. 3-13).

The full theory for the initial magnetization curve, the hysteresis loop and minor hysteresis loops is very complicated. Yet, the phenomena that take place in ferromagnetic materials can be conveniently discussed on the basis of the domain hypothesis.

WEISS' DOMAIN THEORY

In 1907 Pierre Weiss put forward the hypothesis that ferromagnetism might be the result of an unusually strong interaction between the individual atomic magnets, which in some way made them all point in the same direction. Such regions with aligned magnetic moments are called *domains*, and they may have a variety of shapes and sizes ranging from one micrometer to several centimeters, depending on the size, shape, material and the magnetic history of the sample. In most materials the preferred direction is parallel to one or the other of the major crystallographic axes.

Each domain has a very strong magnetization called *spontaneous magnetization* J_s (Wb/m^2). The direction of these domain magnetizations vary from domain to domain, and the net overall magnetization is zero in a virgin ferromagnetic material. There are various techniques for visible observation of these domains, one by applying a colloidal iron powder on the polished surface of a sample and obtain the so-called *Bitter-patterns*.

A series of *domain patterns* is shown in Fig. 3-14, and how they change under the influence of an external field. The patterns that one normally observes by the Bitter method are two-dimensional; the domains are, like crystal grains, three-dimensional.

The first column shows the simplest case where the domain orientations are parallel, the center has four preferred directions and the right column is a more typical random pattern.

A small external field H corresponding to the range 0-1 in Fig. 3-13 will cause a shift in the areas of preferred and non-preferred directions of J_s; the areas where J_s follows H will expand at the cost of those with opposing J_s. This causes a shift in the boundary between domains.

These boundaries are called *walls*, and the electron spin orientation changes gradually across the width of a wall. The wall thickness is in the order of 100-1000 Angstroms (1 Angstrom $= 10^{-10}$ meters), and the locations and pattern of walls in a material may follow crystal grain boundaries as well as other preferred patterns, dictated by the material's metallurgy, imperfections, and impurities.

The wall movements in the low field region are *reversible* and will move back if the field is removed. But, as the field is increased (range 1-2, Fig. 3-13) some wall movements are impeded by impurities and imperfections in the crystal structure; the boundary will move discontinuously to a new equilibrium position, and the magnetization process is *irreversible*. These movements often occur in jumps.

If now the field is reduced to zero the flux density does not go to zero again, but will have a certain remanence B_r (Fig. 3-12). If next the field strength is increased then the flux density goes back along a lower curve; this pattern is called a *minor hysteresis loop*. The number of minor loops within the one and only major hysteresis loop is infinite.

The irreversible wall movements were discovered early in the history of magnetism since in signal transformers and certain magnetic microphones they would generate noise spikes by

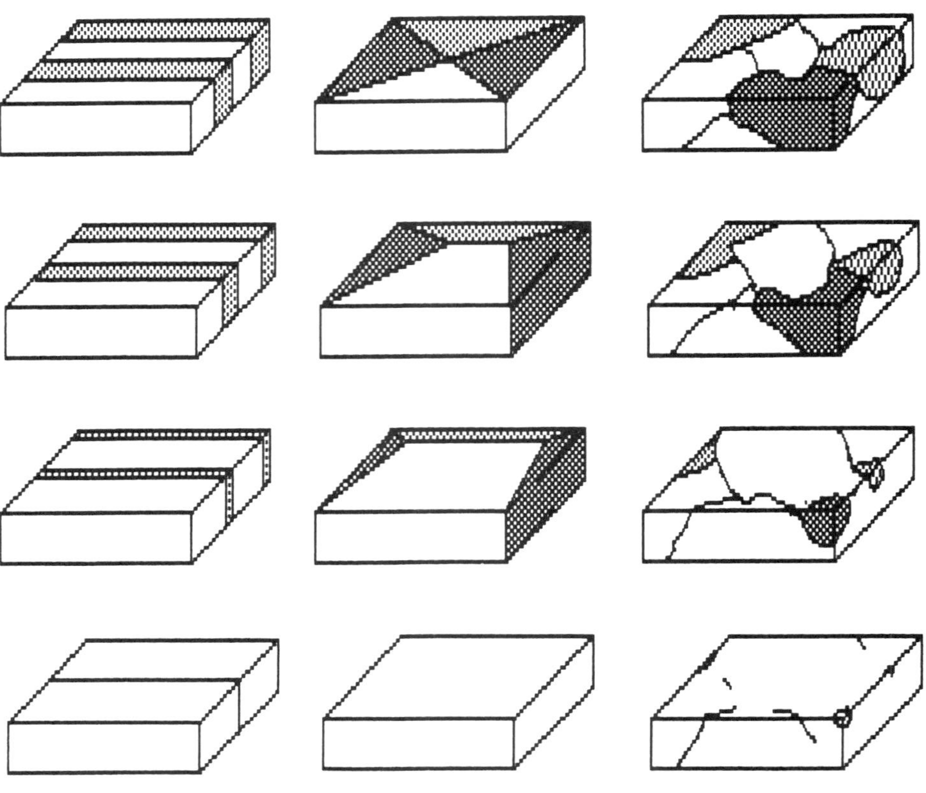

Domain patterns will remain near saturation when there is no demagnetization:

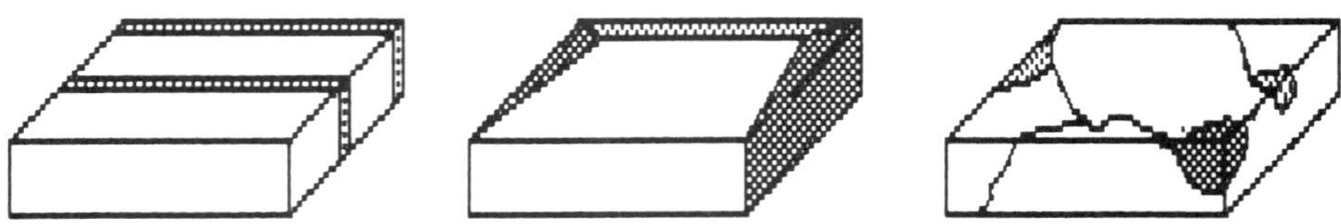

Domain patterns may return to near the virgin state for high demagnetization:

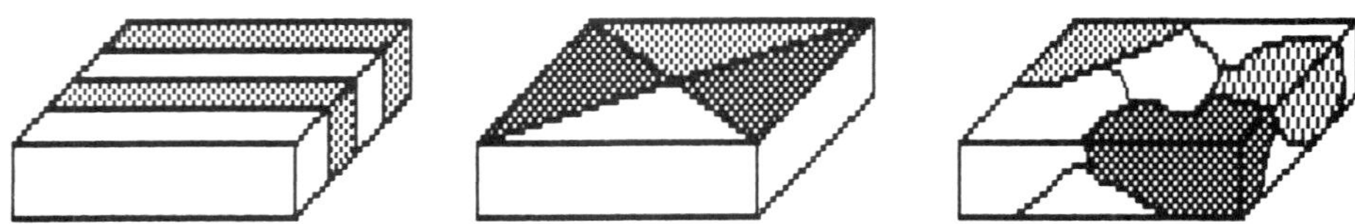

Fig. 3-14. Domain patterns in three material samples. Top four rows show changes under the influence of an increasing magnetic field; the field direction favors the clear domain volumes, which expand at the cost of the others volume. When the field is turned off the domain patterns will change little if no demagnetizing field is present (no airgaps, i.e., no free poles).

induction when they suddenly shifted to a new position. They were discovered by Barkhausen and this type of noise is named after him.

Further increase in the field strength H will eventually reduce all opposing domains to zero, and when H approaches saturation values we find that the last increase in B is caused by *rotation* of the magnetizations J_s into the field direction.

This rotation is reversible, and when the field H is removed two things can happen: the magnetization will rotate back and the walls will recur and move back toward their original positions. But the original domain pattern may not show up again due to *anistropy* (preferred orientation of magnetization in the material, caused by crystal structure, impurities, slip planes, stresses).

SOFT MAGNETIC MATERIALS

Materials used in magnetic heads are: Mu-metal, Permalloy, Alfesil, ferrites, Hot-Pressed ferrites, and deposited thin films of nickel/iron compositions. These are what we call "soft" magnetic materials, and they are easily remagnetized with low magnetic fields. Hysteresis loops for some commonly used head materials are shown in Fig. 3-15.

For small values of H_{core}, near the origin, one can use the formula:

$$\begin{aligned} B &= \mu H_{core} \\ &= \mu_r \mu_o H_{core} \end{aligned} \tag{3.21}$$

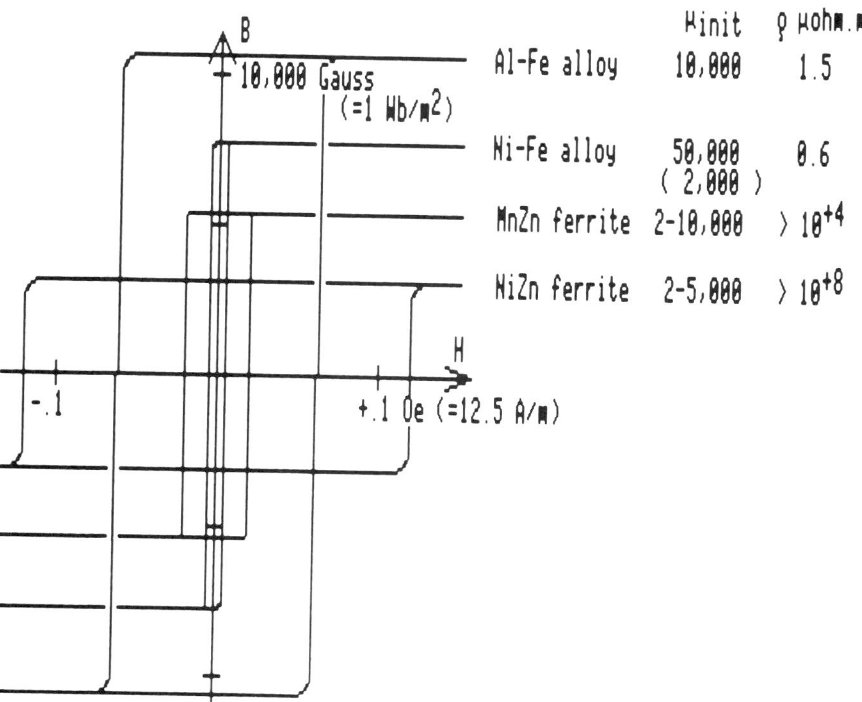

Fig. 3-15. B-H loops for common magnetically soft materials. Table of initial, relative permeability and resistivity.

where:

$\mu_o = 4\pi 10^{-7}$ Hy/m
$\mu_r = \mu_{init}$

The *relative permeability* μ_r is a material property. μ_o equals the absolute permeability of air, and $\mu_r = 1$ for air. Magnetic heads are made from materials that have values of μ_r ranging from several hundreds to several thousands. μ_r depends upon the flux level in the material, as shown in Fig. 3-16; the field levels are very low in most heads, and the formula listed above can be applied to computations of head performance (see Chapters 7 and 10). (Note: The flux density levels in short gap heads for writing on high coercivity media may approach saturation levels. Then formula (3.21) is no longer valid. See under pole tip saturation, and record head distortion.)

The cgs-system of units has one convenience to offer: $\mu_o = 1$. That simplifies the formula above to $B = \mu_r H$, or $B = H$ when in air. This means that a given number X oersteds corresponds to the same number X Gauss. The number for the gap flux density is therefore the same for the field strength.

Example 3.5: A toroidal core (see Fig. 3-12, left) has a mean radius of R = 10 cm, and a cross-sectional area of S = 2 cm^2. The core is wound with n = 200 turns, and carries a current of 1 A. The magnetic flux ψ is measured to 100 μWb. What is the relative permeability of the core material?

Answer: $H = ni/l = 200*½\pi*0.1 = 318$ A/m; $B = \psi/\text{area} = 100*10^{-6}/2*10^{-4} = 0.5$ Wb/m^2. Also $B = \mu_o\mu_r H = 4\pi 10^{-7}*\mu_r*318 = 0.5$; $\mu_r = 0.5/0.0004 = 1{,}250$.

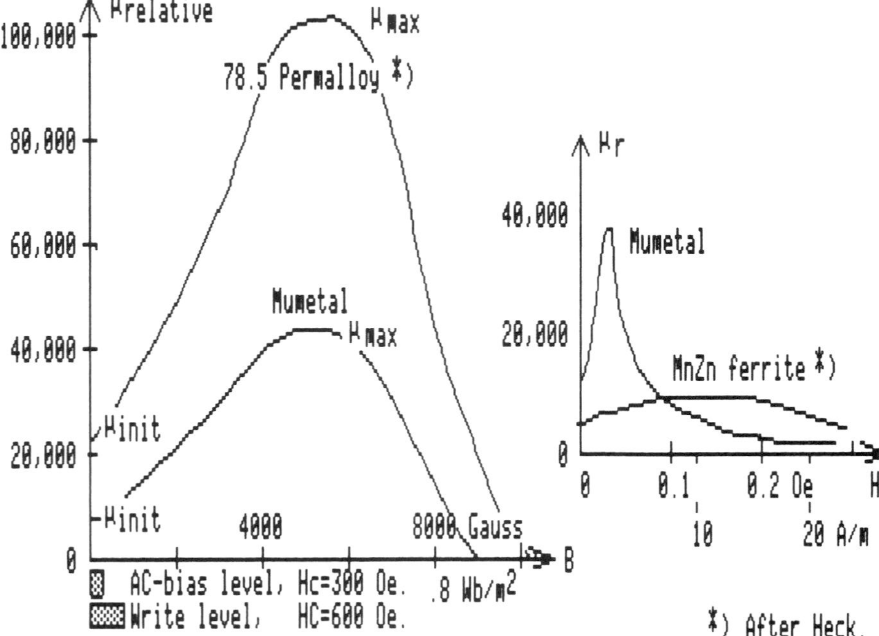

Fig. 3-16. Permeability as functions of flux density B, or field strength H.

NOTE: There are materials with a relative permeability of 1,000,000. But in usual engineering applications the material is handled in different ways, and unavoidable coldworking reduces the permeability. Ni-Fe alloys (Permalloy, Mu-metal, etc.) in finished heads seldom have relative permeabilities over 2,000 to 3,000 (see Chapter 7).

HARD MAGNETIC MATERIALS

The other group of materials, the magnetically "hard" ones, are found in the powders used for *tape* and *disk coatings*: gamma-ferric-oxide, cobalt-iron-oxide, chromium dioxide, iron and barium ferrites. Also used are deposited films composed of nickel, cobalt, phosphorus, etc.

A formula for the entire B-H loop cannot be found since the B values are many, for any single value of H_{core} (with the exception of B_{rs}, the remanence after saturation). The value of B depends upon the history of the prior magnetization process.

The B-H loop for air is a straight line with slope μ_o. The higher B-values for a magnetic material show that they have a *susceptibility* to become magnetized, and we express that as:

$$B = \chi H_{core} \quad \textbf{(3.22)}$$

where χ is named the susceptibility.

We could also write:

$$B = \mu_o H_{core} + J \quad \textbf{(3.23)}$$

$$= \mu_o (H_{core} + M) \quad \textbf{(3.24)}$$

where:

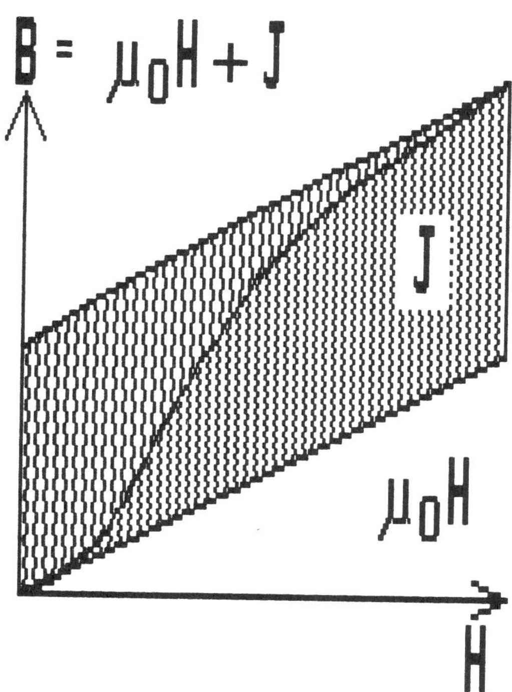

Fig. 3-17. Plotted values of B versus H equals the sum of μ_oH plus J.

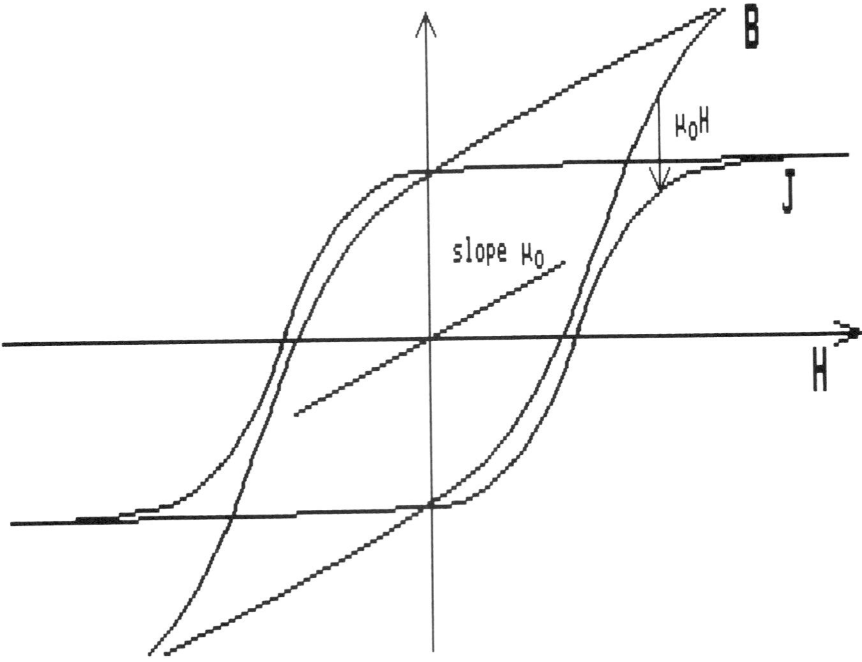

Fig. 3-18. Construction of JH-loop for BH-loop: subtract $\mu_o H$.

B = magnetic flux density in Wb/m^2
H_{core} = magnetic field in A/m
J = magnetization in Wb/m^2
M = magnetization in A/m

$\mu_o H_{core}$ is the contribution from the field H, while J (or M) is from the material's *spontaneous magnetization.* This is illustrated in Fig. 3-17.

The magnetizations are correlated by:

$$M = J/\mu_o \qquad \textbf{(3.25)}$$

A true magnetization loop for the material alone is obtained by subtracting $B = \mu_o H$ from the B-H loop. The resulting J-H loop is shown in Fig. 3-18.

Typical recording materials are shown in Fig. 3-19, showing their JH-loops. The values are typical and will be used in the rest of this book. (For exact engineering applications the designer should use the data sheet from the manufacturer of the material or his own measured values. This comment applies to head materials as well.)

DOMAINS AND WALLS

It may have puzzled the reader to learn that the spin moments line up to form strong internal magnetization, and then to also split up in groups called domains. A simple experiment will convince the reader that this formation of domains results in a lower overall energy in the magnetization system: take two bar magnets and place them next to each other. If like poles are

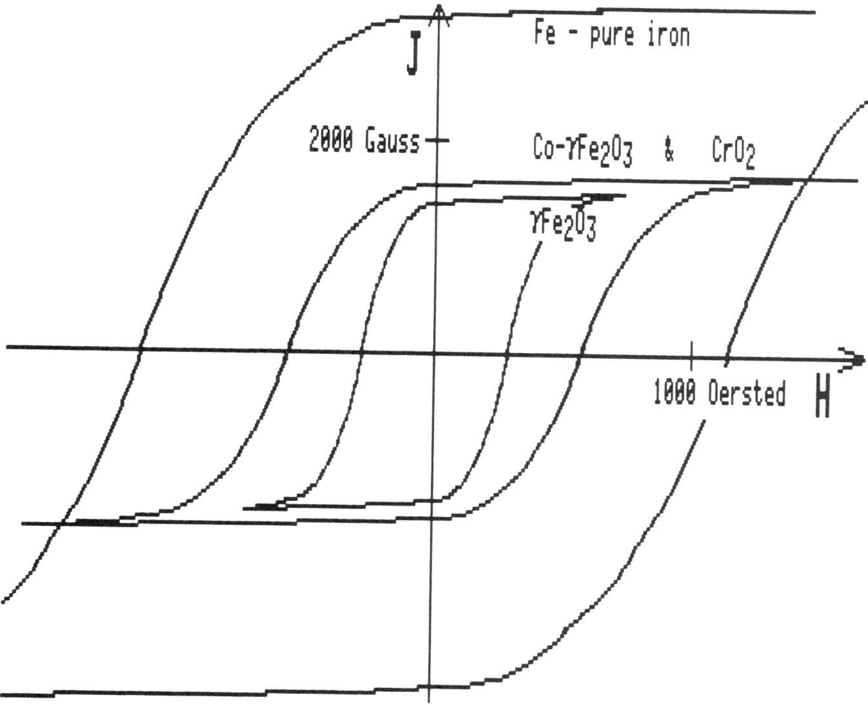

Fig. 3-19. JH-loops for particulate recording media.

next to each other then the bars will move away from each other. Turning one bar over will result in strong attraction; we even say that they serve as "keepers" for each other.

Figure 3-20 shows a similar situation: at top left, a single domain cube is shown, corresponding to the two bars parallel with poles in the same directions. Dividing itself up into two domains corresponds to the bars facing each other with opposite poles. If possible, the cube would further divide itself up into four domains, the two smaller ones called *closure domains*; that last configuration has the lowest energy of the three, the magnetostatic energy being virtually zero.

A magnet would keep on dividing itself into smaller and smaller domains, until the exchange energy starts rising again because few spins are parallel. Hence, there should be an optimum domain size for a given material, where the total energy is at a minimum.

The division into domains is not favored by the exchange energy, and this will also affect the dimension over which the magnetization direction changes, the so-called wall. Figure 3-21 illustrates how magnetization changes direction across the wall (180 degrees total). The exchange energy is inversely proportional to the wall thickness, and thus favors a thick wall. The magnetostatic energy will, however, be small for a thin wall, and an optimum wall thickness exists when the sum of two energies is minimum. This occurs for the following typical values:

Fe	300 Angstroms (120 atoms)
Ni	720 Angstroms (290 atoms)
γFe_2O_3	1000 Angstroms (approx.)

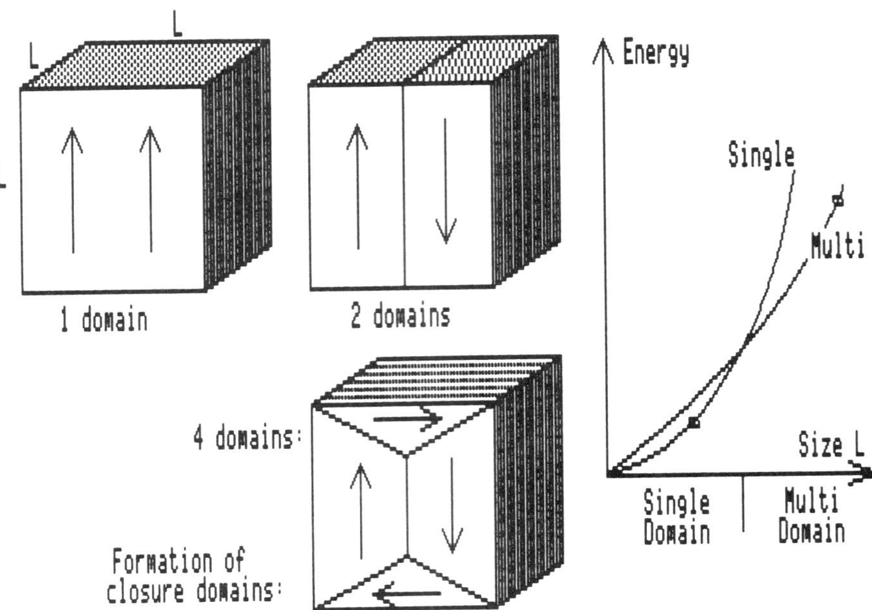

Fig. 3-20. Energy stored in single versus multi-domain structure.

An analog model to the magnetic pattern in the wall is shown in the stake fence, that has two magnets as end pieces. Twisting the fence (exchange energy) is more difficult for a short fence, while the pull between the end magnets then is less.

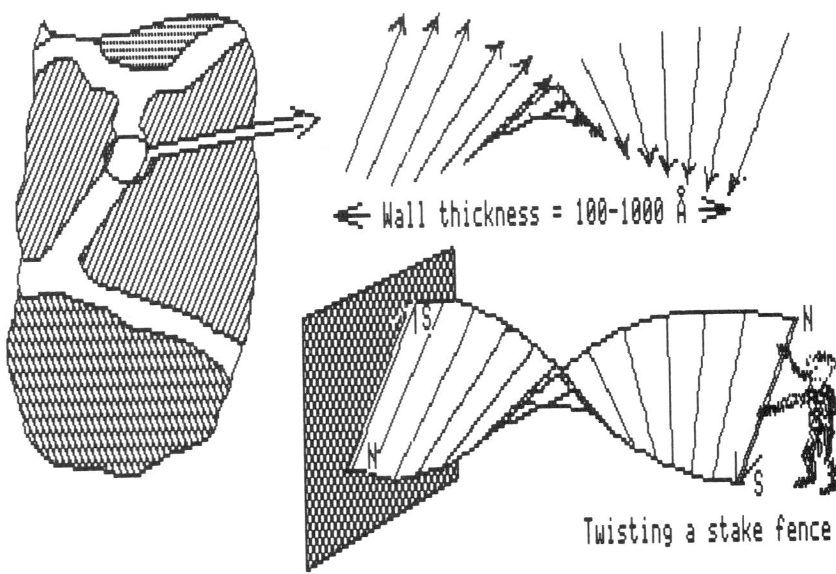

Fig. 3-21. The magnetization changes gradually across the dimension called the wall thickness. The pattern is the same as in a twisted stake fence.

MAGNETIZATION OF SMALL PARTICLES

We will now apply our knowledge about magnetism to explain what happens when a magnetic tape is recorded. The most commonly used magnetic material in the tape coating is gamma-ferric-oxide γFe_2O_3), almost equivalent to iron rust. It is manufactured in a fine powder form and dispersed in a plastic binder which is finally applied to a PET film (Polyethylene-terathplate).

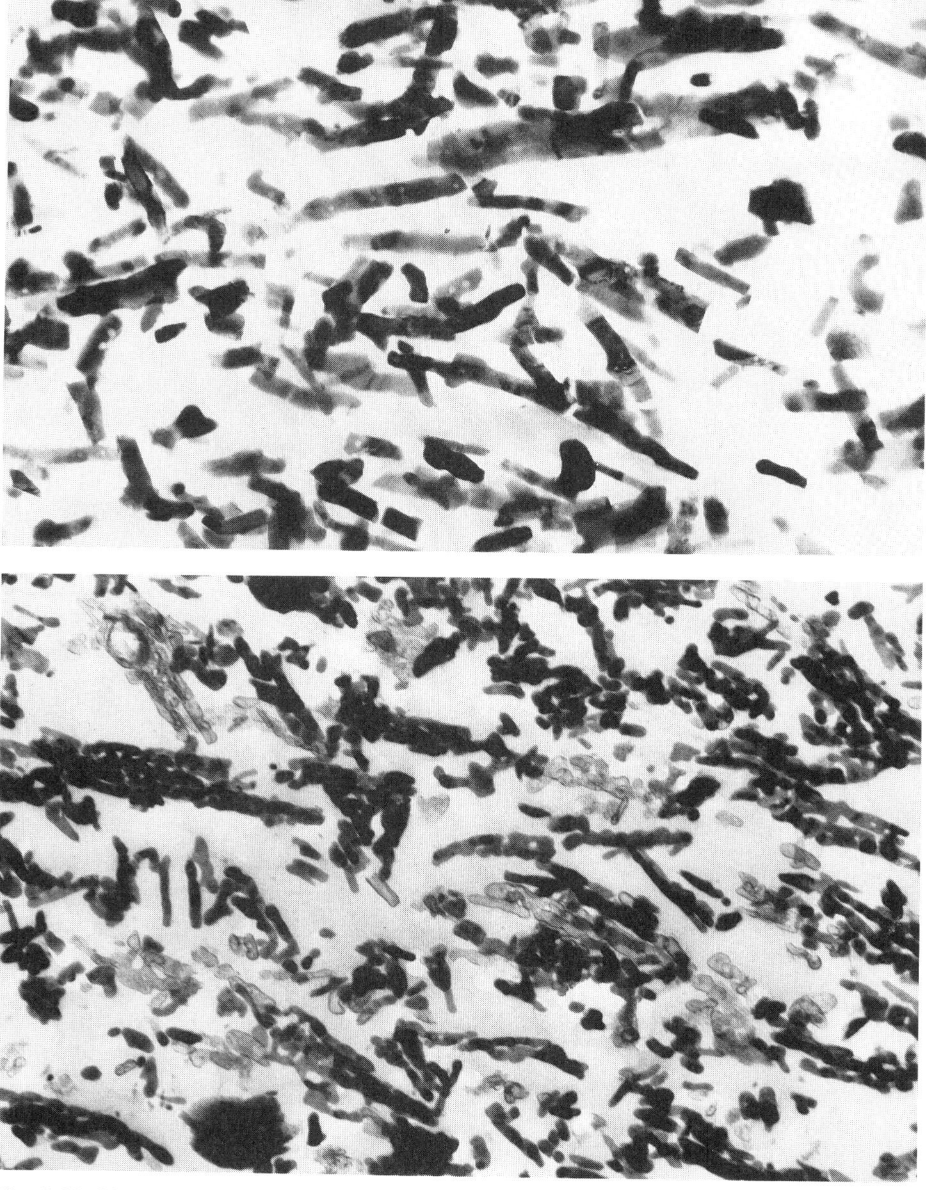

Fig. 3-22. Modern Fe_2O_3 (top), and pure Fe (bottom) particles; dimensions: |——| = 0.1 μm (courtesy Ffizer Minerals, Inc.).

The particles are cigar-shaped and are, during the coating process, aligned with a strong magnetic field in the lengthwise direction of the final tape. Another particle is pure iron (Fe); both are shown in Fig. 3-22.

Their sizes are so small that most of them have a length that is equal to, or a few times larger than the wall thicknesses. Magnetic theory provides us with a guideline to their domain structure, shown in Fig. 3-20, right: small particles have lowest energy when they are single domains. Larger particles are multi-domains.

A single domain particle is magnetized in one or the other direction and remagnetization takes place through rotation of the magnetization vector J_s. Let us assume that each particle is a single domain, with *shape anisotropy*: its preferred direction of magnetization is along the major axis; for a round particle there is no preferred direction, and its resistance to remagnetization zero ($H_c = 0$); only materials with *crystalline anisotropy*, such as cobalt treated oxides, can be round and have a high coercivity (see Chapter 11).

STONER-WOHLFARTH SINGLE-DOMAIN MODEL

A single domain particle model was proposed and discussed by Stoner and Wohlfarth in 1955. An ellipsoid shaped particle is assumed, and the shape anisotropy is the determining force in the orientation of the particle's magnetization, overriding exchange and crystalline anisotropies. The magnetization will lie along the major axis.

An external field will turn the magnetization away from the major axis. An example is shown in Fig. 3-23, where a particle is placed at a 45 degree angle to the x-axis, while a field H is applied in the negative x-axis direction. The magnetization will begin to turn into the direction of the field, but stop at a certain equilibrium position where the sum energy of the field torque and anisotropy energies is minimum (2). The magnetization will turn toward the field direction

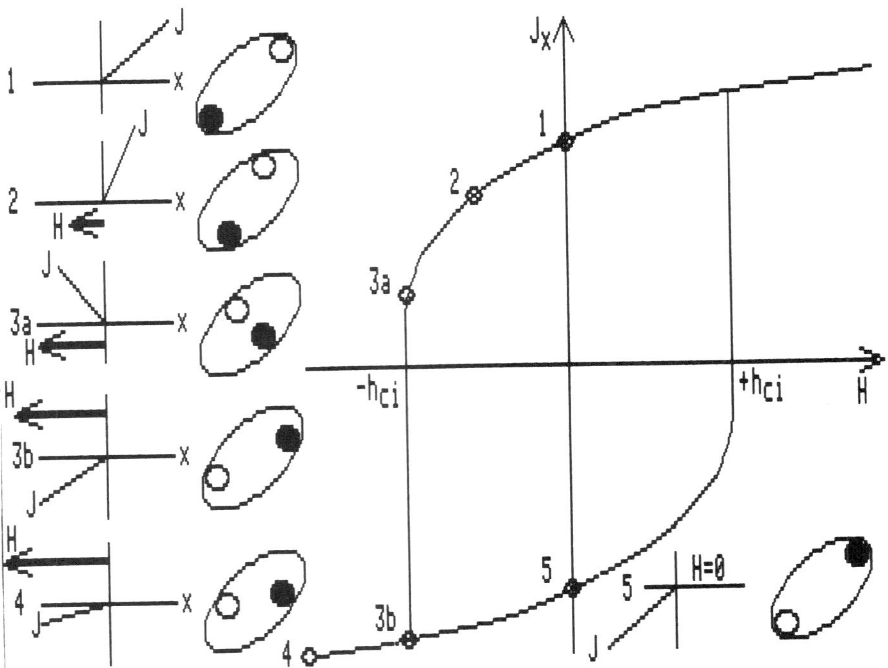

Fig. 3-23. Direction of J in particle (and mech. model), 45° to the field H.

until a critical value h_{ci} is reached where it jumps to a new direction, now close to the direction of H (3a -> 3b).

When H is made very large the magnetization J lies almost in the H direction (4). If the field is reduced to zero, then the direction changes to that of lowest anisotropy energy, which now is opposite the original direction (5). We have remagnetized the particle into the direction of the field.

Just to the right of each vector diagram are shown eggshell-like containers with two small balls inside. Imagine the two balls are kept apart by a compression spring, and at rest they will lie on a line equal to the major axis. An external force (possibly electrostatic field) is now applied so the balls will try to change positions. Again an equilibrium is reached (2), and only a certain force can overcome the spring and flip the balls' positions (3a -> 3b).

This mechanical model serves well as a tool to remember how the magnetization in a single domain particle changes under the influence of an external field.

If we make a plot of the x-component of the magnetization vector versus H we obtain the JH-loop for a single domain particle with shape anisotropy. It is shown to the right in Fig. 3-23. Similarly, if the x-components of the displacements and forces for the mechanical model were plotted, an identical loop would be traced. This justifies the analogy.

Another situation exists when the particle's major axis follows the x-direction. This situation is encouraged when the media coating is made: a magnetic field is applied to the still wet coating, and it will turn the particles like compass needles into the field direction. This process is named *orientation*.

Figure 3-24 shows how a particle with magnetization in the positive x-direction is affected by a negative H field. The magnetization vector will be "stressed" by the field, and eventually flip when the field strength equals, or exceeds, the particle coercivity h_{ci}. This is also explained by the mechanical model: when the external force has compressed the (invisible) spring be-

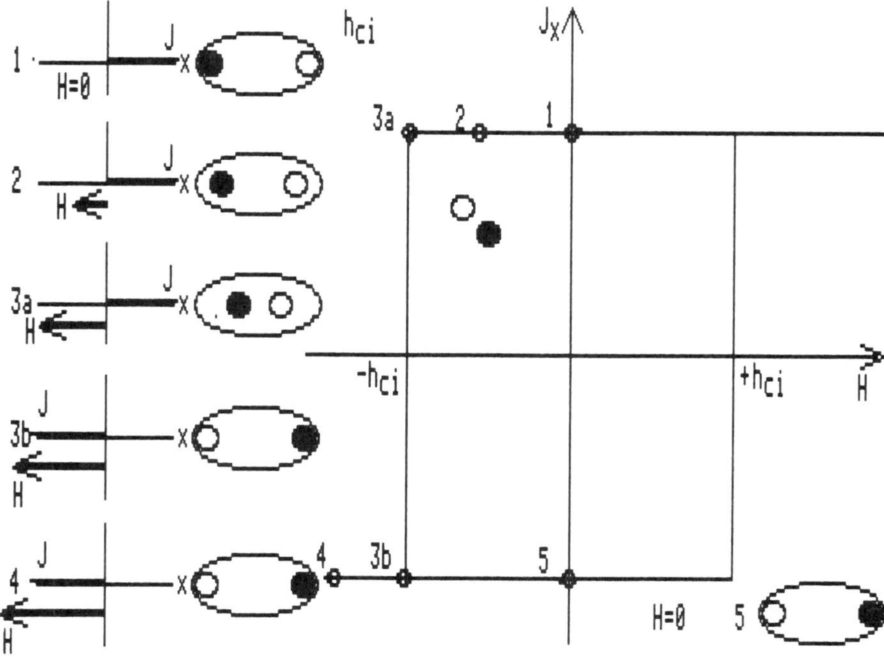

Fig. 3-24. Direction of J in particle (and mech. model), 0° to the field H.

tween the two balls to the point where their distance equals the small axis dimension in the eggshell, then they will flip positions.

The other extreme is a particle placed ninety degrees to the x-axis. It is left to the reader as an exercise to plot its J_xH-loop; its B_{rsat} value equals zero (in the X direction).

The J_xH-loop for the particle previously discussed is named a *square loop*. The JH-loop for media coatings will approach this shape at high levels of orientation.

If we prepare a coating mix, and thoroughly stir it, we can assume that the particles will be pointing in all 4π directions. The value of B_{rsat} will then equal one half of maximum magnetization J_{sat}. Orientation will increase the number of particles that are in the x-direction (normal direction of media travel), and hence increase B_{rsat}. Typical values in modern media range from 0.7 to 0.9 times J_{sat}.

Now we can explain the initial magnetization curve and the hysteresis loop for a magnetic tape (see Fig. 3-13). The initial magnetization (0-1) from the erased state (0) is reversible since the increase in B over H ($B = \mu H$) is caused by small reversible domain rotations.

At higher field levels (1-2) we have *irreversible rotations*. The rotations are completed in small ''jumps'' (each domain sort of falling into place). These ''jumps'' show up as a fuzzy line on an oscilloscope, or as crackling noise in a speaker/amplifier connected to an induction coil around a tape-wound toroid.

As we approach saturation (2-3) the additional increase in induction B takes place by *reversible rotations*, and at 3 all particle magnetizations follow the direction of H.

When H now is reduced to zero it follows that the magnetization rotates back to the state of 2, and the material is permanently magnetized. The rest of the hysteresis loop follows the path 4-5-6-7-8 as described earlier.

JACOBS-BEAN CHAIN-OF-SPHERES MODEL

The Stoner-Wohlfarth model explains very nicely the magnetization behavior of particulate media. So far in the text, however, we neglected one important fact—that of the existing fields among the particles, between themselves. These are known as *interacting fields*, and their magnitude varies from zero to h_{ci}. To this field is now added any external field, and the latter need not exceed h_{ci} to flip particle magnetizations.

It is always nice to be able to model various engineering problems, write algorithms, and let the computer calculate a solution. But in the case of interacting fields we must account for the JH-loop of each particle, and also the overall resulting interacting field from the particles poles. The resulting model is difficult, if not impossible, to program, as the computed field pattern of the interacting field shown in Fig. 3-25 illustrates. We will shortly introduce statistical concepts as a tool for solving the write process problem.

Another difficulty with the single domain particle model exists in a couple of practical conflicts. First consider Fig. 3-26 that illustrates the magnetization in a particulate medium. The bit length is 0.4 μm, which corresponds to a packing density of 25,000 bits/cm (63,500 bits/inch). This sort of recording is found in high density digital recorders (instrumentation machines), and audio is approaching this density (20,000 Hz at 15/16 IPS = 21,333 wavelength/inch = 42,667 bits/inch).

The middle of the illustration shows the outline of the particles, and their single domain magnetization pattern. Clearly, a couple of the larger particles cannot resolve the short bits.

A more appealing picture is obtained if we consider a multi-domain model introduced by Jacobs and Bean, also in the mid-fifties. They proposed that the particle magnetization is represented by small magnetization vectors, like compasses, and that each particle has a number thereof that equals their length divided by their width. The shape anisotropy is accounted for by the fact that the magnetizations will try to pattern themselves in a chain-like fashion,

Fig. 3-25. Field lines among particles in a medium coating (= interaction field). These lines were plotted at random, and line densities in this illustration do not represent field strengths; but general directions are accurate.

Recorded bit pattern (6 bits):

Magnetization in single domain medium:

Magnetization in multiple domain medium:

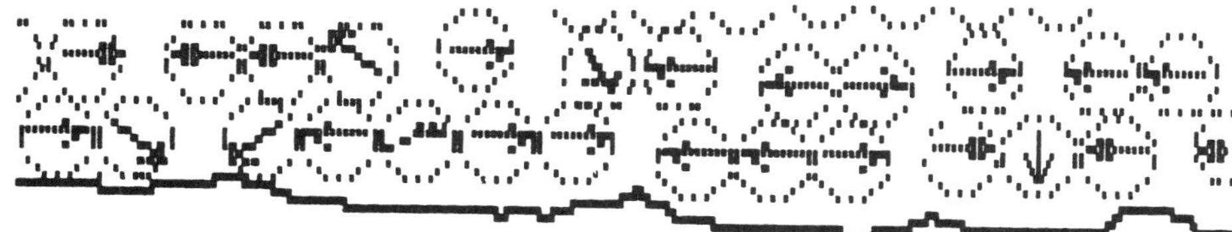

Fig. 3-26. Multiple domain media appear to have higher resolution than single domain media. Bit length = 0.4 μm (= 16 μin), λ = 0.8 μm (32 μin).

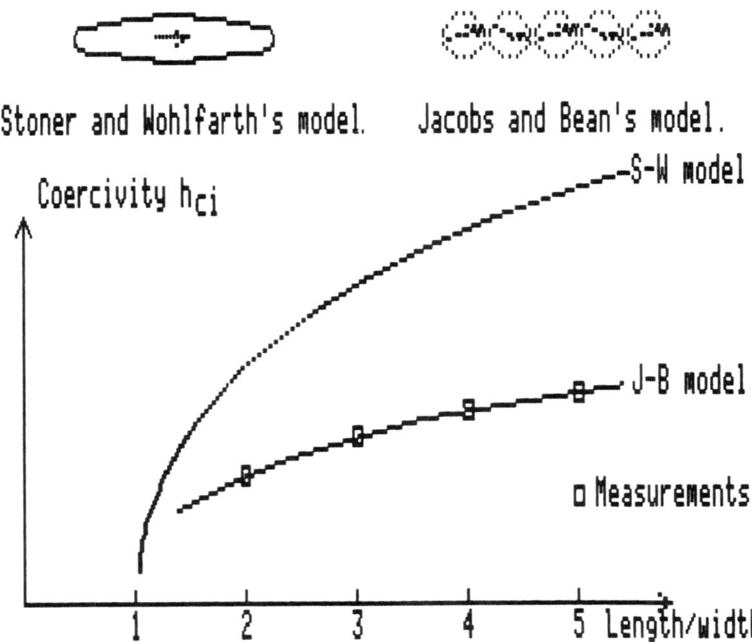

Fig. 3-27. Comparison of single versus multi-domain models.

with north poles seeking south poles. They may lie in a straight pattern, or in a zig-zag pattern, called *fanning*. And the particles'relationship with their neighbors are again through the interaction field.

This model allows for a much improved resolution, as shown in the bottom, Fig. 3-26. And coercivity measurements made on particle assemblies show close agreement with the theoretical data for the Jacobs-Bean model.

A long particle has higher coercivity than a short particle. This shape anisotropy is easily understood by the mechanical eggshell model: a long distance between the balls requires a higher force to revert them. Calculated values for an assembly of particles are shown in Fig. 3-27. Measured values fall on top the Jacobs-Bean curve.

Directional variations of the coercivity can be expected in a media coating. Stoner-Wohlfarth's model would prescribe a high value for the coercivity measured along the oriented direction, approaching zero in the direction perpendicular thereto. Jacobs-Bean's model should give almost no variation, if any, since it magnetically is an assembly of individual compasses. Measurements (ref. E.D. Daniel, pers. communication) show a 10 to 15 percent increase at angles between 30 to 60 degrees, then decreasing by 10 percent at a perpendicular field (Fig. 3-28). This again corresponds closest to the Jacobs-Bean model.

A particle configuration that closely resembles Jacobs-Bean's model has recently been found in bacteria (see references; Blakemore, et al.). During a study of a live group of bacteria under a microscope they were found to gather along one side of the droplet containing them (see Fig. 3-29). At first the light from a nearby window was thought to attract them—but they behaved the same way at night.

The investigator then placed a magnet near the droplet, and now the bacteria swam toward or away from the magnet, depending on its direction. After exposing the bacteria to a slowly decaying AC-field half swam in one direction, the other half the opposite way! A closer examination in a transmission electron microscope revealed a chain of 20 or so beads, cubic or octahedral in form, each having a diameter of approximately 500 angstrom. And the material was magnetite, very much like the materials in lodestones.

This chain of particles serves to guide the bacteria along the earth's magnetic field lines. For the location where these bacteria were found the earth field has an inclination of 60 degrees, and it would therefore guide the bacteria down into the ground. They were found about two feet below the surface, where the temperature is nice and warm during the winter above.

The same mechanism is guiding homing pigeons, and is possibly involved in other animals' navigation. This is currently under intense investigation. It also brings a thought to mind: if your neighbor cultivates homing pigeons, and they are a nuisance to you, just think what you can do with a degausser during the dark of the night?

SWITCHING FIELD DISTRIBUTION

When a group of particles is placed in a magnetic field all opposing domain magnetizations will be acted upon by the field and brought into different positions. First of all they will not

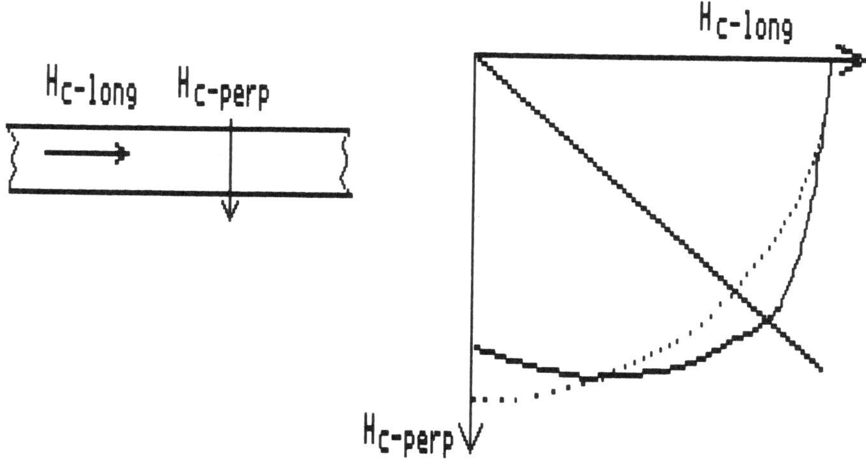

Fig. 3-28. Polar plot of variation in media coercivity.

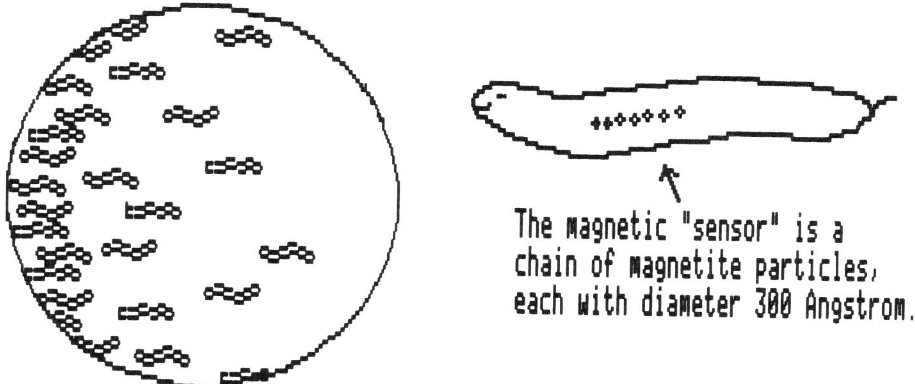

Fig. 3-29. Magnetic bacteria swim along the earth's magnetic field lines.

rotate in unison since they will exhibit various degrees of resistance to domain rotation (anisotropy). Secondly, they are also influenced by the magnetic fields from their neighbors, called the *interaction field*. If this was not the case, then they would all, more or less, rotate together, and the tape would become fully magnetized in one direction or the other. No intermediate levels would be available for dynamic range, and the tape would be good for digital pulse recording only.

It is exactly the *particle interactions* and their effect on coercivities that make these particle materials useful as a linear recording medium, using AC-bias (see Chapter 20). Their influence shows up strongly in the BH-loop for a coating.

We need to introduce yet another property of small magnetic particle which is a *size effect*: their coercivity dependence upon volume. We have seen that the coercivity is proportional to their length/width ratio. Its dependence upon volume is shown in Fig. 3-30, left. This curve is general for all particles we will encounter. There is a maximum coercivity, which occurs at a diameter near the same values as earlier listed for the wall thicknesses.

Below that critical size the coercivity falls off rapidly, and eventually becomes zero. And in this range—from maximum to zero coercivity—a disturbing behavior is found. The particles are unstable, or *superparamagnetic*. They each contain an inadequate number of atoms to support the exchange force phenomenon, which therefore at times looses its grip on the spin alignments. When this happens the particle becomes nonmagnetic!

Some time after the exchange forces gain, and the particle is again magnetic. Needless to say, this behavior is enhanced by thermal fluctuation. It is possible to calculate a time constant for the on-off behavior of the magnetism, and it is dependent upon not only the particle volume, but also its temperature. It has a range from a few seconds to a few hundred years for a 30 percent change in particle volume only.

This naturally plays a role in archival storage of tapes, and also in *print-through* signal in audio recordings (see Chapters 12 and 14).

We will here restrict the particle size to belong to the right side of the maximum, and will now also find that particles in a batch of material have different sizes. When a histogram thereof is plotted it is not surprising to find a gaussian distribution curve; occasionally with a log-normal tendency, a characteristic of some powders.

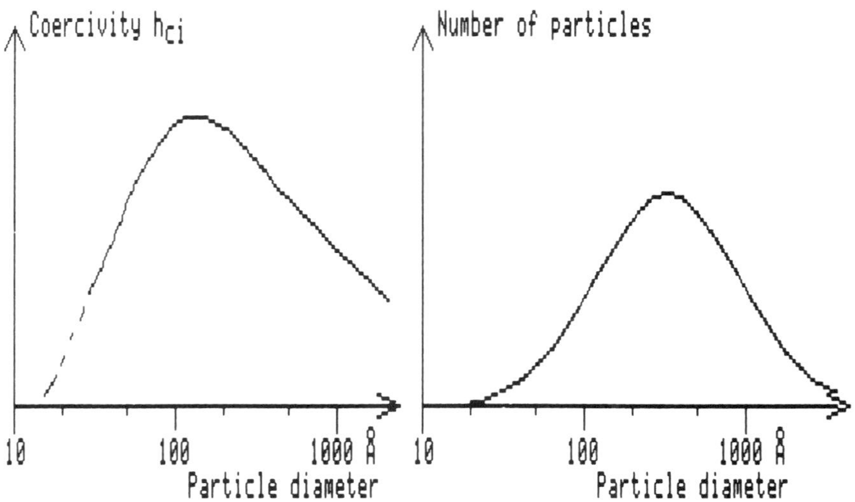

Fig. 3-30. Particle coercivity h_{ci} as function of particle size (diameter), and distribution (histogram) of particle sizes in a batch of oxide powder.

When we superimpose the coercivity versus size curve upon the histogram we obtain another gaussian distribution (see Fig. 3-31, top). Assuming an assembly that to the outside is non-magnetic then half the particles are negatively magnetized, the other half positively.

Now let us repeat the experiment shown in Fig. 3-12, with a coil form made from wound magnetic tape. Assume that it has been completely degaussed, so half the particles have positive remanence, the other half negative. As the positive going field increases in strength it will switch more and more of the negatively magnetized particles. This will lead to a rise in positive magnetization, eventually reaching J_{sat}. This level will ideally remain when the field is reduced back to zero.

Next apply an increasing negative going field, and all particles will eventually be switched, resulting in $-J_{sat}$. Reduce the field to zero, and $-J_{sat}$ remains. Another positive half cycle with a magnetic field will flip all particles to $+J_{sat}$.

We have in essence traced the ideal JH-loop for a particulate media, assuming perfect orientation of the particles and no interaction. It is a simple matter to arrive at the BH-loop by adding the values $\mu_o H$ to the J values, as shown in Fig. 3-32.

The effect of interaction and imperfect orientation is evident in the difference between the JH-loops in Figs. 3-31 and 3-18. The latter has a remanence value that is less than the saturation value.

When we reduce a measured BH-loop to the JH-loop and finally differentiate it we obtain a double peaked curve which represents a mix of the particle size effect and the interaction fields. We can add a useful number to the magnetic values such as coercivity, remanence, and saturation magnetization: the *switching field distribution*. This is the width of either peak in the differentiated JH-loop, at the half height points, as shown in Fig. 3-31, and designated Δh_r. Its usefulness will become apparent in Chapter 5, on recording.

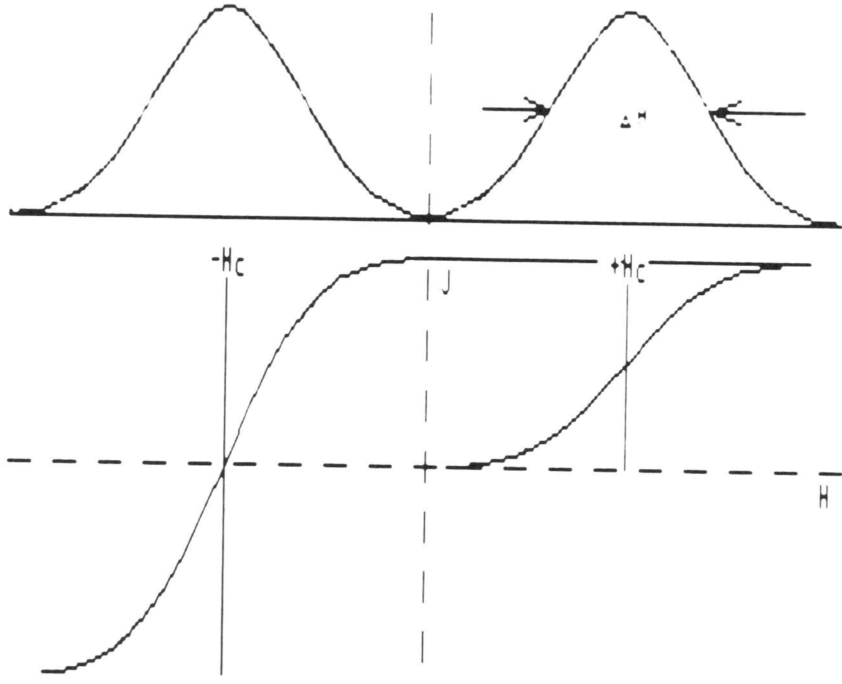

Fig. 3-31. Distributions of h_{ci}. Integration gives ideal JH-loop.

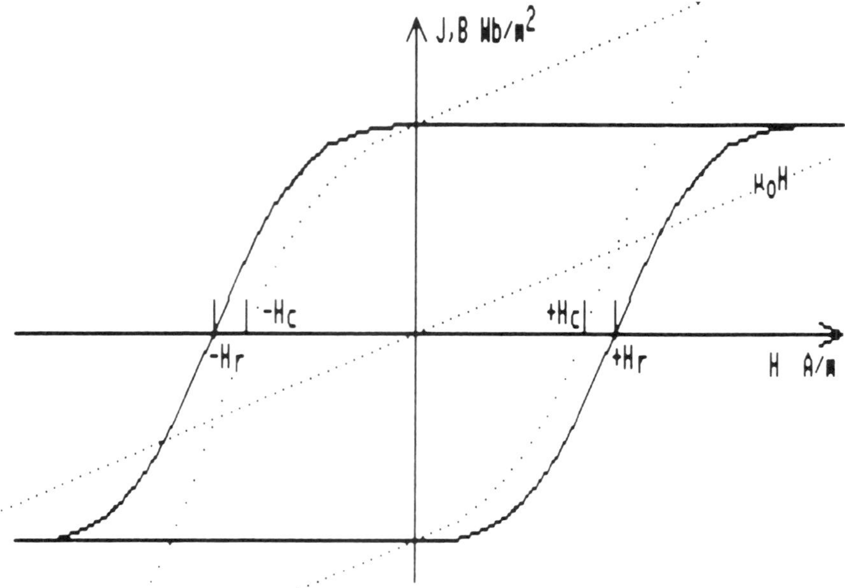

Fig. 3-32. Resulting JH-loop; add $\mu_0 H$ to obtain BH-loop (shown dotted).

MAGNETIZATION IN METALLIC FILM COATINGS

The recording surface in rigid disk drives is no longer predominately a particulate coating, it is often a deposited metal film. It is quite thin, only a few tenths of a μm (5-20 μin), as compared to a few μm (70-200 μin). It appears, from what we now know, ideal since each transition (often designated Δx) laid down by the write head represents a wall between two domains (bits).

Given a metallic coating for which the optimum wall thickness is say 400 angstroms, then the ultimate maximum packing density is $10^{+8}/400 = 250{,}000$ transitions per cm (= 635,000 flux reversals per inch, FRI) (assuming no space for the bit itself, so transitions are touching each other. That is okay since the recorded information lies in the transitions.)

In reality only 10,000 or so transitions per cm are achieved. This again has its explanation in the magnetics striving for a state of minimum energy, see Fig. 3-33, left. The magnetics bordering the transition will, given the slightest chance, form *Neel spikes*, which reduce the wall energy, similar to closure domains (Fig. 3-20). The Neel spikes are formed so north and south poles move closer to each other, thus reducing the energy stored. They may occur where there are crystalline faults or foreign particles in the metal film.

Let us consider a metal surface, that has perpendicular anisotropy. This encourages the magnetization to form perpendicularly to the surface. This transition is interesting for two reasons: the transition length Δx tends to be short since the bits' opposite magnetizations attract each other. Further, transitions with wiggles are discouraged: the energy stored in the transition (wall) is proportional to its length, which nature therefore will try to shorten. This results in a straight transition (see references in Chapter 19, Soohoo).

SWITCHING SPEEDS

We have presented various mechanisms for the change, or switching of magnetism in materials. It seems natural to conclude this chapter with an answer to the question "how fast

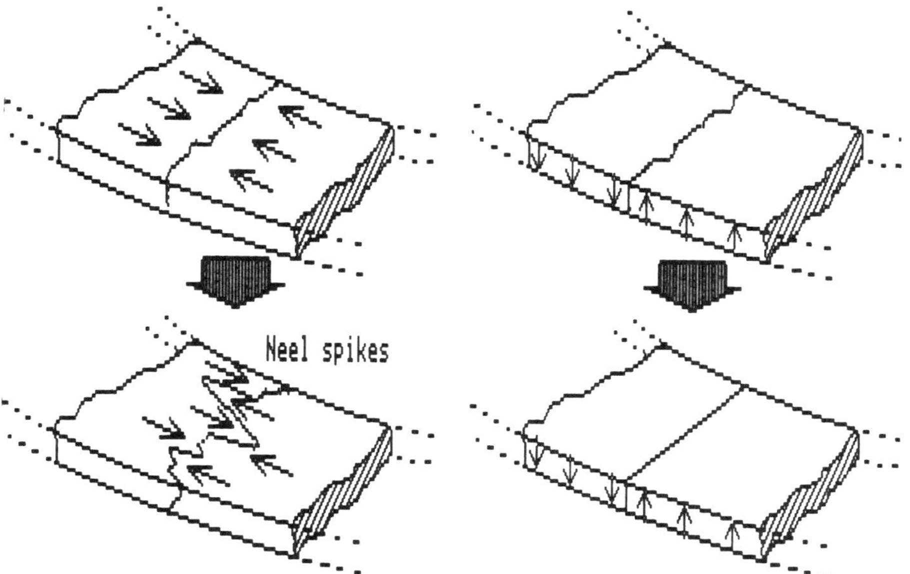

Fig. 3-33. Transition between bits in longitudinal and perpendicular magnetic media.

can we switch the magnetization in for example a coating with small particles? or in the core of a head?'' Both will eventually limit the data rate we can use in writing and reading signals.

Thornley and Williams (see reference) conducted experiments to record pulses on several particulate tape samples, using a very-high-frequency technique. They found that the recording ''efficiency'' decreased when pulses shorter than tabulated below were used:

$$t = 4.1 \text{ ns for } C_rO_2$$
$$t = 2.6 \text{ ns for } \gamma Fe_2O_3$$
$$t = 1.4 \text{ ns for Co-}\gamma Fe_2O_3$$

These numbers correspond to upper frequencies $f_u = t/2$:

$$f_u = 120 \text{ MHz for } CrO_2$$
$$f_u = 200 \text{ MHz for } \gamma Fe_2O_3$$
$$f_u = 350 \text{ MHz for Co-}\gamma Fe_2O_3$$

The change of magnetization direction in bulk materials (mumetal laminations in cores, solid ferrite cores) has a much lower cut-off frequency, which will be discussed in Chapter 7. In general it can be said that the higher the permeability the lower the cut-off frequency; values of μ_r less than 1,000 are typical in heads reaching out into the MHz range.

REFERENCES TO CHAPTER 3

Lee, E.W., *Magnetism, an Introductory Survey*, Dover 1970, 281 pages.

Weisburd, S., ''The Earth's Magnetic Hiccup,'' *Science News,* Nov. 1985, Vol. 128, pp. 218-220.

Tebble, R.S., *Magnetic Domains*, Barnes and Noble, New York, 1969, 98 pages.

Brown, W.F., ''Tutorial Paper on Dimensions and Units,'' *IEEE Trans. Magn.*, Jan. 1984, Vol. MAG-20, No. 1, pp. 112-118.

Giacoletto, L.J., "Standardized Use of SI Magnetic Units," *IEEE Trans. Magn.*, Dec. 1974, Vol. MAG-10, No. 4, pp. 1134-1136.

Stoner, E.C., and Wohlfarth, E.P., "A Mechanism of Magnetic Hysteresis in Heterogenous Alloys," *Phil. Trans. Roy. Soc.*, May 1948, Vol. 240, Ser. A, pp. 559-642.

Jacobs, I.S., and Bean, C.P., "An Approach to Elongated Fine-Particle Magnets," *The Phys. Rev.*, Nov. 1955, Vol. 100, No. 4, pp. 1060-1067.

Blakemore, R.P., and Frankel, R.B., "Magnetic Navigation in Bacteria," *Scientific American*, Dec. 1981, Vol. 245, No. 6, pp. 58-65.

Thornley, F.R.M., and Williams, J.A., "Switching Speeds in Magnetic Tapes," *IBM Jour. Res. Develop.*, Nov. 1974, pp. 576-578.

BIBLIOGRAPHY TO CHAPTER 3

Gilbert, W., *De Magnete*, Dover Publ., New York 1958.

Ewing, J.A., "New Model of Ferromagnetic Induction," *Phil. Mag.* (6), 1922, Vol. 43, pp. 493-503.

Cullity, B.D., *Introduction to Magnetic Materials*, Addison-Wesley 1972, 666 pages.

Bozorth, R.M., *Ferromagnetism*, D. van Nostrand, 1951, 968 pages.

Chikazumi, S., *Physics of Magnetism*, John Wiley & Sons, 1966, 554 pages.

Kneller, E., *Ferromagnetismus*, Springer Verlag, 1962, 792 pages.

Durand, E., *Magnetostatique*, Masson et Cie, Paris, 1968, 673 pages.

O'Reilly, W., *Rock and Mineral Magnetism*, Chapman and Hall, New York (Blackie, London), 1984, 220 pages.

Craik, D.J., and Tebble, R.S., *Ferromagnetism and Ferromagnetic Domains*, North-Holland (Wiley), New York, 1965, 319 pages.

Mapps, D.J., "Magnetic Domain Demonstrations Using the Kerr Magneto-optic Effect," *Contemp. Phys.*, March 1978, Vol. 19, No. 3, pp. 269-281.

Fuller Brown, W. Jr., *Micromagnetics*, Interscience Publ. (Wiley), New York, 1963, 143 pages.

Mattis, D.C., *The Theory of Magnetism I*, Statics and Dynamics, Springer-Verlag, New York (Berlin-Heidelberg), 1981, 300 pages.

Zijlstra, H., *Experimental Methods in Magnetism*, North-Holland Publ. (Wiley), 1967, 295 pages.

Hammond, P., *Electromagnetism for Engineers*, Pergamon Press, 1978, 290 pages.

Bolton, B., *Electromagnetism and its Applications*, Van Nostrand Reinhold, 1980, 157 pages.

Kraus, John D., *Electromagnetics*, McGraw-Hill, 1984, 775 pages.

Armstrong, R.L., and King, J.D., *The Electromagnetic Interaction*, Prentice-Hall, 1973, 493 pages.

Hayt Jr., W.H., *Engineering Electro-Magnetics*, McGraw-Hill, 1974, 496 pages.

Plonus, M.A., *Applied Electro-Magnetics*, McGraw-Hill, 1978, 615 pages.

Watson, J.K., *Applications of Magnetism*, John Wiley & Sons, 1980, 468 pages.

Lerner, E.J., "Biological Effects of Electromagnetic Fields," *IEEE Spectrum*, May 1984, pp. 57-69.

Koch, A.J., and Becker, J.J., "Permanent Magnets and Fine Particles," *Jour. Appl. Phys.*, Feb 1968, Vol. 39, No. 2, pp. 1261-1264.

Soohoo, R.F., *Magnetic Thin Films*, Harper & Row, 1965, 316 pages.

Bickford, L.R., "Magnetism During the IEEE's First One Hundred Years (1884-1984)," *IEEE Trans. Magn.*, Jan 1985, Vol. MAG-21, No. 1, pp. 2-9.

Lowther, D.A., and Silvester, P.P., *Computer-Aided Design in Magnetics*, Springer-Verlag, 1986, 324 pages.

Parton, J.E., Owen, S.J.T., and Raven, M.S., *Applied Electromagnetics*, Springer-Verlag, 1986, 288 pages.

Chapter 4

Magnetization in Media and Heads

In Chapter 3 we examined the magnetization J, which is an intrinsic material property equal to the magnetization level within each domain. The remanence B_{rs} depends upon the material composition while the coercivity H_c depends upon external factors such as shape and stresses. We will in this chapter discuss *technical magnetizations* which are the processes whereby a material's magnetization is changed by a shift in domain sizes and directions. Such changes are brought about by introducing air gaps, shaping the material, application of external fields, and interactions among small magnetic particles.

The BH-loop changes in a dramatic way when an air gap is introduced, as shown in Fig. 4-1. When we repeat our experiment from Figs. 3-12 and 3-13 on a core sample from the same material, but now with an air gap, then the BH-loop in Fig. 4-1 results. Notice the lowered remanence but unchanged coercivity. In this chapter we will examine this magnetization process, which now includes *demagnetization*.

We will examine how demagnetization reduces the recorded level in a coating. This problem has in recent years stirred up a lot of activities and arguments in the recording industry. It is argued that a perpendicular magnetization mode is preferable at short bit lengths since its demagnetization is less than in the longitudinal mode. This matter will be explained by developing the theory of demagnetization shortly, illustrated with several examples.

Magnetic heads are basically simple to analyze, as will be shown by employing an electrical circuit model for the head core. This allows for calculations of *head efficiency*, *inductance*, and the *write gap field*.

THE DEMAGNETIZING FIELD

The introduction of an air gap in the path for the magnetic flux causes the formation of *poles*, and the field H_d from these poles opposes the applied field H_a from the magnetizing cur-

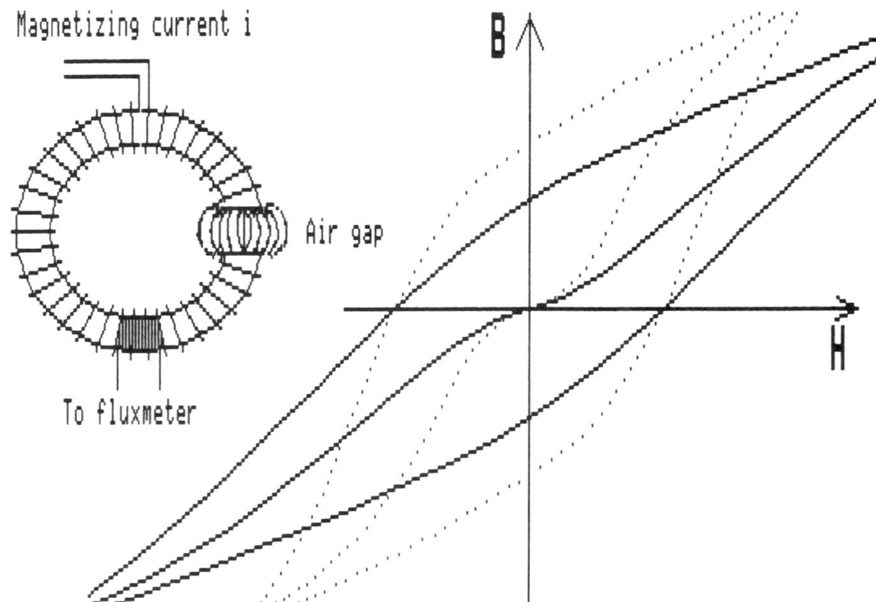

Fig. 4-1. Sheared hysteresis loop for a magnetic toroid with an air gap. The material is low permeability, high coercivity–typical for media, magnets.

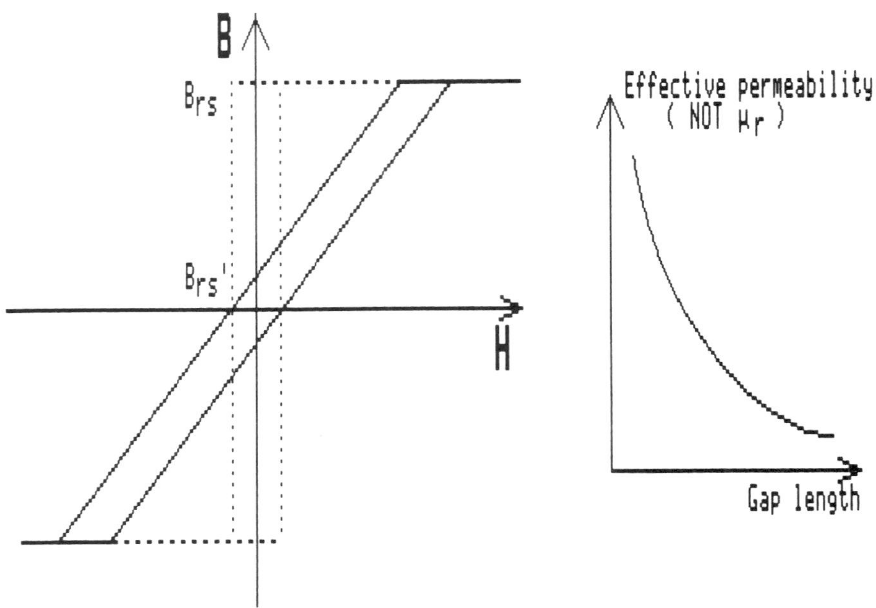

Fig. 4-2. Sheared hysteresis loop for a magnetic toroid with an air gap. The material is high permeability, low coercivity–typical for heads.

rent i. The net effect is an overall decrease in field acting upon the material by H_d to $H_{core} = H_a - H_d$, and therefore, also a decrease of the corresponding flux densities. We call this effect *demagnetization*, and the field H_d the *demagnetizing field.*

Air gaps are found in all applications of magnetics, with the one exception of some toroids and the switching cores in magnetic memories. Air gaps exist in moving coil instruments, loudspeaker magnets, motors, generators, relays, etc.

Figure 4-2 shows a BH-loop for a head material with an air gap. The severe shearing of the curve means a lower effective permeability, and at the same time a *linearizing effect* on the effective permeability. The effective permeability is still defined as the ratio between B and $\mu_0 H$, but is no longer a material property; it includes the effect of demagnetization.

Another advantage lies in the *reduction of the remanence* B_{rs} to B_{rs}', where the added superscript $'$ stands for the demagnetized value. It depends upon B_{rs}, and is a function of the air gaps and the shape of the magnet. The remanence in a soft magnetic material is by no means small (dotted curve, Fig. 4-2), and would, if not reduced, cause serious noise and distortion problems in recording. A very small air gap reduces the remanence to a tolerable level—and even then, it is good housekeeping to demagnetize a recorder's heads at regular intervals to remove any remanence. Otherwise recordings will be noisy and distorted.

MAGNETIZATION OF MAGNETS

It is instructive to examine the magnetization process of a permanent magnet, such as the bar shown in Fig. 4-3A, top. It can become permanently magnetized by an external field H_a from a coil carrying a current, or from another magnet, such as the shown horseshoe magnet. Poles will be induced, and their polarities are easy to remember since it is known that the bar magnet will be attracted to the horseshoe magnet, and that poles of opposite polarity attract each other. Hence a negative south pole is formed opposite the horseshoe's north pole, and vice versa at the other end.

These poles remain when the external field is removed (Fig. 4-3C). The field from these poles is opposite the external magnetizing field. The net field *inside* the bar magnet was therefore:

$$H_{core} = H_a - H_d$$

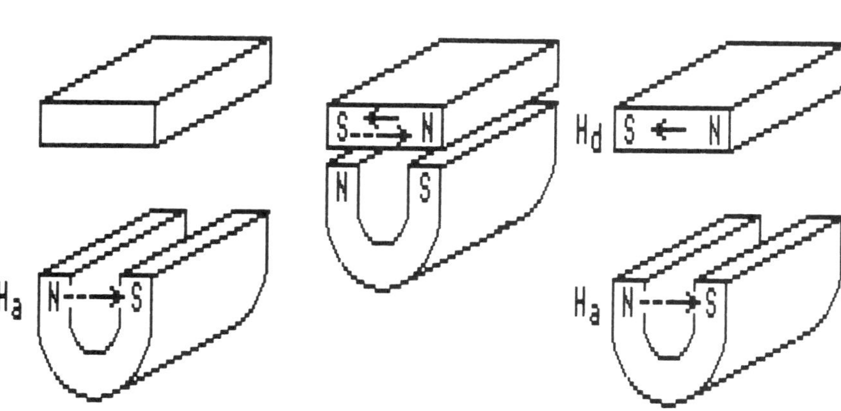

Fig. 4-3. Steps in the magnetization of a bar magnet (inducing poles).

during the magnetization process, and

$$H_{core} = -H_d$$

after removal of the external field. We want, at this point, to become aware of the fact that the field from the generated magnet is *opposite* the field that generated the magnet. We have once more encountered Le Chatelier's rule.

B- AND H- FIELDS IN A PERMANENT MAGNET

A recorded track in a magnetic recording consists of alternating magnetism along the track; this is particularly pronounced in digital recordings where the magnetization alternates between positive and negative values. (Figure 4-10 shows the bit pattern.) We are interested in finding the remanent magnetization J that exists in one such bit, or bar magnet, since this represents the flux level that is available for read out.

The relationship between B and H is established by:

$$B = \mu H \tag{4.1}$$

where:

$$\mu = \mu_o \mu_r \tag{4.2}$$

We have no problem envisioning the B and H lines from a current carrying wire, or coil—they are alike since $\mu_r = 1$ in air. Their magnitudes differ by μ_o, and here we all wish we were still using the cgs-system of units since here $\mu_o = 1$, and gauss equals oersteds in air (but NOT inside ferro- or ferri-magnetic materials).

In a toroid shaped core the B- and H- fields are also easy to envision; there are μ_r more B lines than H lines. But cutting an air gap in the flux path immediately complicates matters: we measure a decrease in the flux density B.

Let us first consider the fields inside the bar we magnetized in Fig. 4-3. There is initially no field (Fig. 4-3A). The horseshoe magnet will send a field H_a through it, directed left to right. We will designate this polarity positive, in analog to the x-axis in a standard coordinate system.

The field H_a induces the poles as shown. They will in turn set up a field H_d going from the north to the south pole. It is directed opposite to H_a, and therefore named the demagnetizing field.

The total, true field that acts upon the magnetic material is the sum of the two fields:

$$\boxed{\begin{aligned} H_{core} &= H_a + (-H_d) \\ &= H_a - H_d \end{aligned}} \tag{4.3}$$

We can now use the BH-loop measured on a closed toroid by using H_{core}, *not* H_a. The results are shown in Fig. 4-4. If we disregard the air gap then a remanence B_r results as shown in Fig. 4-4A. The introduction of an air gap hinders the B value in ever reaching the level that could result in B_r; in Fig. 4-4B a lesser value corresponding to H_{core} was reached.

What happens now is that the field $-H_d$ remains in the bar since it originates from the pole. It will move the point for the remanence B_r down in the second quadrant to B_r'.

The point where B_r' ends up is the point where the descending loop intersects a line known as the *load line*. This comes as no surprise to anyone that has worked with permanent magnet design. Only the second quadrant of the BH- or JH-loop is needed, and the portion

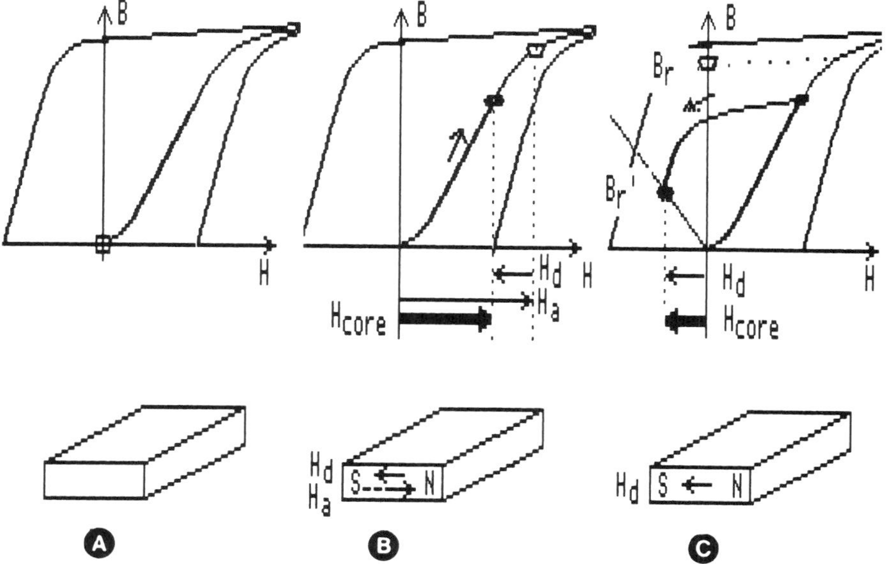

Fig. 4-4. Applied field H_a, demagnetizing field H_d and core field H_{core} during magnetization process. True remanence $B_{rs}' < B_{rs}$.

of the BH-loop that is located therein is named the *demagnetization curve*. Our task ahead is to find the load line, but let us first clear up the concepts of B- and H- fields, and associated magnetization J, in an electromagnet.

B- AND H- FIELDS IN AN ELECTROMAGNET

An air coil was shown in Fig. 3-6, right; here the B-field pattern is identical to that of the H-field. There are no poles in the air coil and the field lines are closed patterns.

When a ferromagnetic material is inserted into the coil two poles are induced. They will in turn produce the demagnetizing field that will reduce the field acting upon the material (equation (4.3)).

The field H_{core} will shift the domains into its direction; this will give rise to a magnetization J, that increases the flux density term by J:

$$B = \mu_o H_{core} + J \qquad \textbf{(4.4)}$$

The result is a net increase in the number of flux lines, inside as well as outside the material. This situation corresponds to Fig. 4-4B.

Finally, turn off the magnetizing current. H now changes to $-H_d$, the demagnetizing field from the poles. Inside the material we now have:

$$B = -\mu_o H_d + J \qquad \textbf{(4.5)}$$

Now the H lines go from pole to pole, while the B lines still are closed lines. We can summarize (and should memorize) three rules:

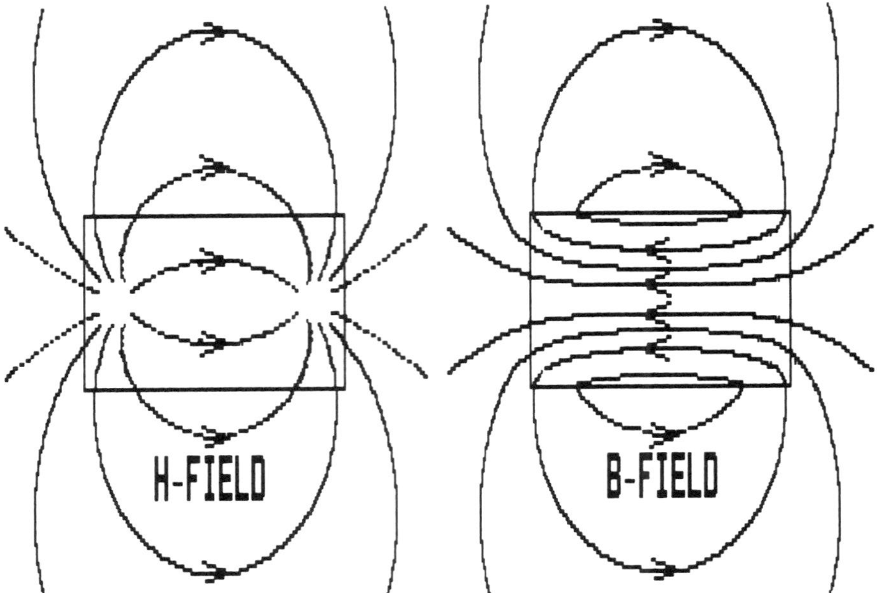

Fig. 4-5. The H- and B- fields in, and around, a permanent magnet.

H—field lines are always closed when due to currents.
H—field lines are always open when due to poles.
B—field lines are always closed.

The reader will have no difficulty in verifying the identity between B and H outside the magnet, as shown in Fig. 4-5. Inside the picture does not look right; that is because B, H, and J really are vectors, i.e., magnitude with directions. Properly written equation (4.5) is:

$$\mathbf{B} = -\mu_o \mathbf{H}_d + \mathbf{J} \qquad \textbf{(4.6)}$$

Note also that B and H_d have opposite signs. That is how we ended up in the second quadrant—and that is where most of our future discussion on magnetism will concentrate.

THE DEMAGNETIZING FACTOR N

The magnet's own field H_d needs to be better specified. We can accomplish that by first realizing that its magnitude will be proportional to the magnetization J that is being produced by the outside magnetizing field H_a. Let us introduce a multiplier N so:

$$H_d = -NJ/\mu_o \qquad \textbf{(4.7)}$$

The total field acting inside the magnet material is therefore:

$$H_{core} = H_a - NJ/\mu_o \qquad \textbf{(4.8)}$$

and the resulting expression for B becomes:

$$B = \mu_o (H_a - NJ/\mu_o) + J \quad \textbf{(4.9)}$$

When the external field is removed (which, for instance, happens when a recorded bit is moved away from the write head field), then we have:

$$H_a \rightarrow 0$$

and

$$B \rightarrow B_r' = -NJ + J$$

or

$$B_r' = (1 - N) * J \quad \textbf{(4.10)}$$

We have added the superscript ′ to tell that this is the demagnetized value. The limits for the domain of N is:

$$0 \leqq N \leqq 1$$

which corresponds to a range of B:

$$J_{rsat} \leqq B_r' \leqq 0$$

Expression (4.8) is not a formula per se, but a relationship. B_r' must be determined by graphic methods. To do this we need the graph of the BH or JH loop, plus the value of N. Figure 4-6 illustrates how the magnetization B_r' will settle down by descending along a branch of the hysteresis loop until the point where it intersects with a line with slope α, with tangent α determined by one of the expressions below.

$$B = \mu_o H + J \quad \textbf{(3.23)}$$
$$B = \mu_o (H + M) \quad \textbf{(3.24)}$$

where

$$M = J/\mu_o$$

There is disagreement regarding the use of J (magnetic polarization in Wb/m^2, or Tesla) and M (intensity of magnetization, in A/m). This disagreement is carried over into the use of the dipole moment, which then have units of Wbm, or Am^2 (Giacoletto, Chapter 3 references).

Further, some authors use M in expression (3.23) instead of J. Therefore, always check the actual units to verify which system the author works with.

	$\tan\alpha_B = (\mu_o (1/N - 1))$	for the BH-loop	**(4.11)**
or	$\tan\alpha_J = (\mu_o (1/N))$	for the JH-loop	**(4.12)**
or	$\tan\alpha_M = 1/N$	for the MH-loop	**(4.13)**

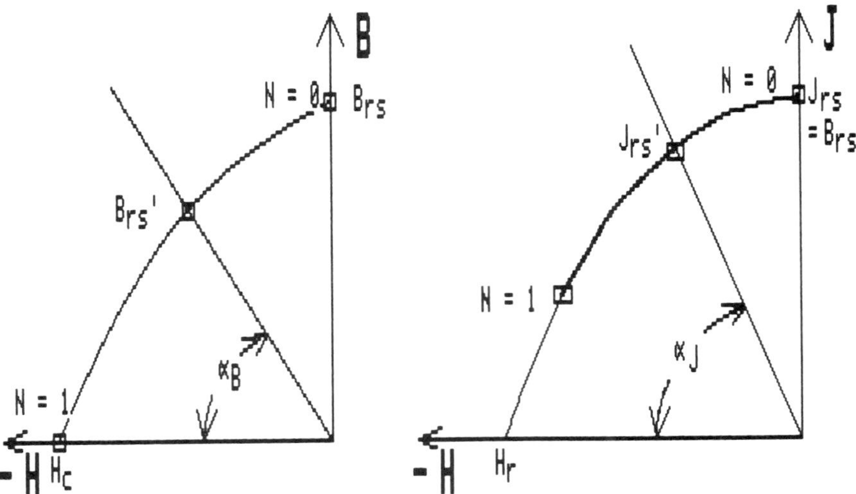

Fig. 4-6. The remanence assumes a value that is located where the descending BH- (or JH-) loop branch intersects with the demagnetization line. Note that complete demagnetization (N = 1) leaves $B_{rs}' = 0$, but $J_{rs}' > 0$.

It should be noted that these expressions are *valid (with $\mu_o = 1$) when used in conjunction with a graph with cgs-values (Oe, G)*. (The reader will find that N varies from 0 to 4π in the cgs-system. However, $\tan \alpha$ is now $(4\pi/N - 1)$ or $(4\pi/N)$, so the slopes are identical. An unusual coincidence.)

We will return to the task of plotting the load line after we have determined N for some different shapes (bits along the track, heads).

Proof: Derive formulas (4.11), (4.12), and (4.13). Assume that the JH-loop is a straight line through $(0, J_{rs})$ and $(-H_r, 0)$, the latter being the coercivity of the JH-loop (which always is larger than the coercivity H_c of the BH-loop).

Answer: Substituting J from (4.4) into (4.3): $H_{core} = H_a - H_d = H_a - NJ/\mu_o = H_a - N(B - \mu_o H_{core})/\mu_o$. Now let $H_a \rightarrow 0$: $H_{core} = -NB/\mu_o - NH_{core}$, which reduces to $B/H_{core} = -\mu_o(1/N - 1)$, i.e., a negative slope $= \mu_o(1/N - 1)$.

Formula (4.12) comes from rewriting (4.7) to $J/H = -\mu_o/N$.

Formula (4.13) is (4.12) with $M = J/\mu_o$.

When N was introduced in equation (4.6) it expressed the flux density $\mu_o H_d$ as a fraction N of the magnetization J. The latter is a material property, so if we can determine H_d then we immediately have the value of N.

It will be a number that represents the shape *anisotropy* of the magnet. Since all magnets are three-dimensional we need to determine N for the three rectangular coordinates: x, y, and z. From a textbook on magnetism we learn that these components always add up to one (or 4π in the cgs-system):

$$\boxed{N_x + N_y + N_z = 1} \quad \textbf{(4.14)}$$

By plain, logical reasoning we can now arrive at the values for N for some simple geometries: a sphere is simple—the magnetization will not know which direction to settle down to if there are no anisotropies in the material itself. The analog from Fig. 3-23 applies if we make the eggshell a perfect sphere. The result is that N = ⅓, in any direction. This will also apply to a cube.

A sheet magnetized in two different directions is shown in Fig. 4-7, and the values of N are not surprising. A sheet is nearly impossible to magnetize from side to side, but easy from one edge to the opposite.

For a closed ring core N = 0; there are no poles and therefore no demagnetizing field. But the slightest air gap will change that, and it can be shown that:

$$\boxed{N \cong L_g/L} \qquad \textbf{(4.15)}$$

where

L_g = length of the gap

and

L = overall length of the flux path ($L >> L_g$).

Example 4.1: A magnetic write head is made from NiZn with characteristics as shown in Fig. 3-15. The gap length L_g = 1 μm (40μin), and the overall flux path is 5mm (200 mils). What is the remanent magnetization if the head inadvertently was saturated?

Answer: $N = L_g/L = 1/5000$. atan $((\mu_o)(1/N - 1))$ = atan (4999) = 89.99°. A graphic determination of B_{rs}' is quite impossible. You would have to draw a graph where each divi-

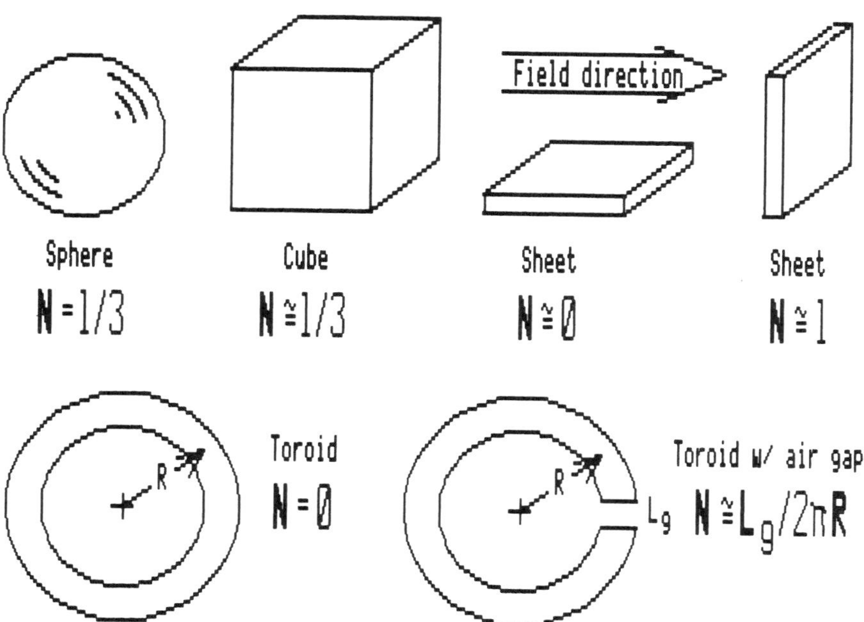

Fig. 4-7. Demagnetization factor N for various shapes.

sion of one oersted corresponds to one gauss: if we set 1 Oe = 1 cm then the resulting loop will be 2*0.12 cm wide, and 100 meters tall! Next find the intersection between the left side of the loop with the demagnetization line—?.

The general problem is shown in Fig. 4-8. Recall that the tangent to an angle in a rectangular triangle equals the ratio between the opposite side divided by the adjacent side. This immediately leads to the formula:

$$\boxed{B_{rs}' = \mu_o H_c / N} \qquad \textbf{(4.16)}$$

where:

B_{rs}' is in gauss
H_c is in oersteds
N is the MKS value (0 to 1)

which holds for materials with low coercivity and high remanence.

Returning to the problem we find that $B_{rs}' = 600$ G, clearly a damagingly high value: it will erase all information on a 300 Oe media.

Example 4.2: A permanent magnet motor with an Alnico magnet is disassembled in order to replace a bearing. When reassembled and tested its torque is far below the specified value. What happened? The length of the flux path is 8 cm and the air gap 5mm.

Answer: The BH-loop is shown in Fig. 4-9, where also the load (or demagnetization) line is shown. α_1 = atan (1/N − 1) = atan (L/L_g − 1) = atan 15 = 86°.

When the motor is taken apart the gap length increases dramatically, say to 40 cm. Now N increases to 40/8 according to the formula. That exceeds the domain for N—and remember that $N = L_g/L$ only holds for $L_g << L$. Let us set N —> 1, then α_2 —> 0°, and the load line is shown as line II.

Now reassemble the motor. The load line moves back to I, with B_{rs}' = 7000 gauss. But the remanence does not move up to its original value—there is no field present to help the

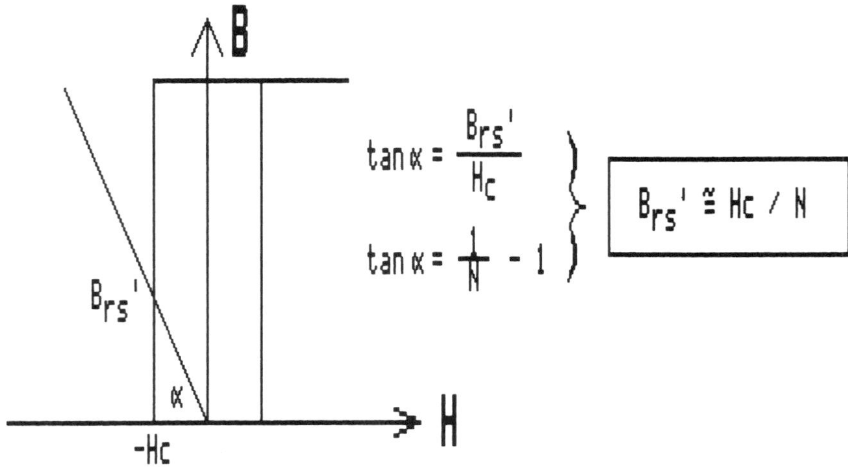

Fig. 4-8. True remanence B_{rs}' in a high remanence, low coercivity material.

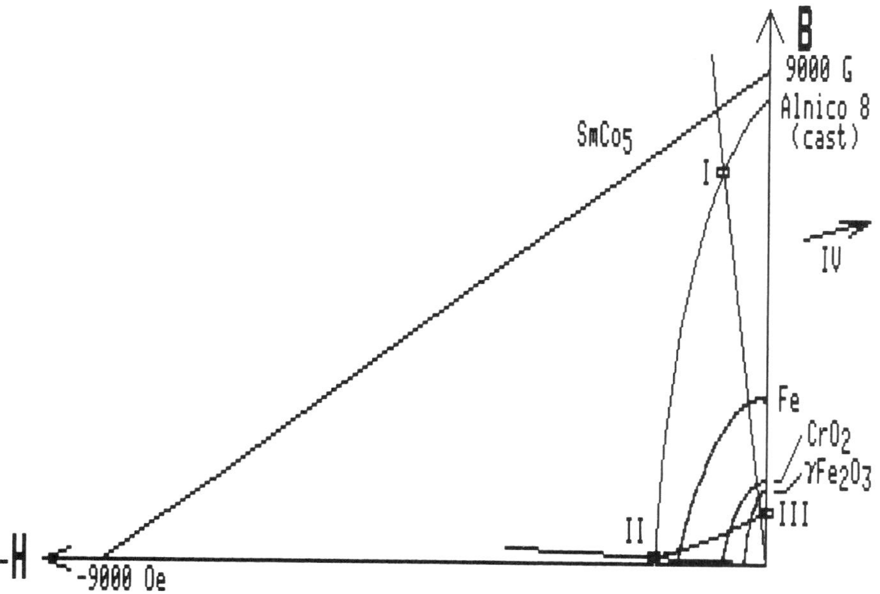

Fig. 4-9. Comparative demagnetization curves (from trade literature). Also, changes in the working point for a permanent magnet motor (I -> II -> III).

domains in shifting *). Rather, it moves along a minor loop up to point III, where B_{rs}'' only equals 900 gauss. (Note the use of superscript " for this case.) Hence the motor torque is only 900/7000 = ⅛ of its rated value.

To bring it back a magnetization process is required. A few turns of a heavy wire are wound around the pole piece(s) and a large field induced by sending many amperes through the wire, for instance by discharging a large capacitor. This brings the B value up to IV, which settles to I.

• Alnico magnets are essentially made from minute particles (with inherent high coercivity), that are sintered together. Magnetization changes therefore occur by rotations.

Figure 4-9 also provides comparisons between some magnet materials: the three particulate media materials (down in the right hand corner), and samarium cobalt. The latter material does not demagnetize, the working point merely moves up and down the BH-loop, which now is a straight line. This is advantageous, but the cost is also 100 times that of Alnico.

DEMAGNETIZATION IN A RECORDING

A realistic picture of the tape magnetization is achieved when we divide the coating up into bars, each representing a bit, or half a wavelength. This is shown in Fig. 4-10 and Fig. 4-12, where the arrows show the magnetization direction. The ideal magnetization patterns are shown in Fig. 4-11.

Our discussions will be in the *x-y plane only*; all conditions shown on this cross section (at z = 0) are the same for various other cross sections at different values of z.

Further, each bar element represents a packet of a large number of magnetic particles, normally longitudinal oriented. We will treat the magnetization of each bar as an individual magnet, leaving a discussion of interacting fields for Chapter 11. Each bar represents one bit; two bars represents one wavelength, λ.

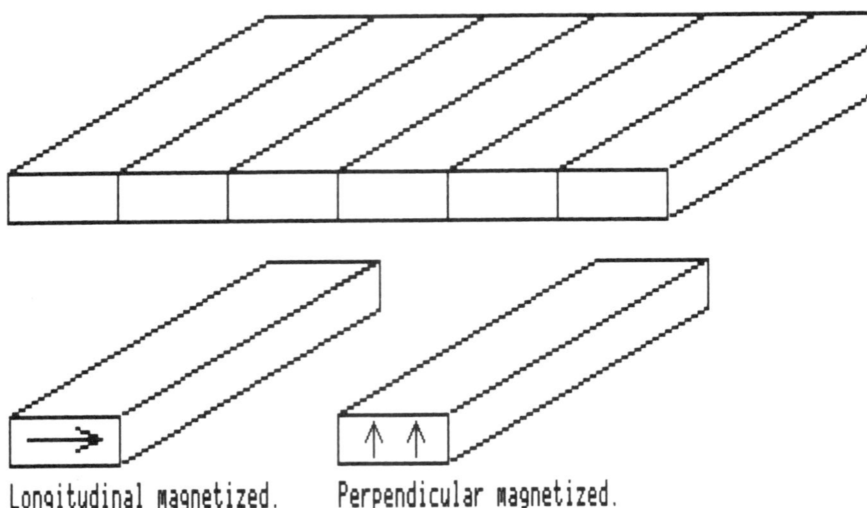

Fig. 4-10. Geometry of six bits along a track, magnetized longitudinal or perpendicular.

It is difficult to find H_d, and then N, for shapes other than the sphere and perfectly symmetrical ellipsoids. The demagnetization factors N_x and N_y can be determined as approximations to the same factors for a prolate ellipsoid. Figure 4-12 illustrates how we can stretch the ellipsoid in the z-direction and thus make N_z approach 0, and preserve the ratio between N_x and N_y, while keeping the sum of the two equal to one. Good approximations are:

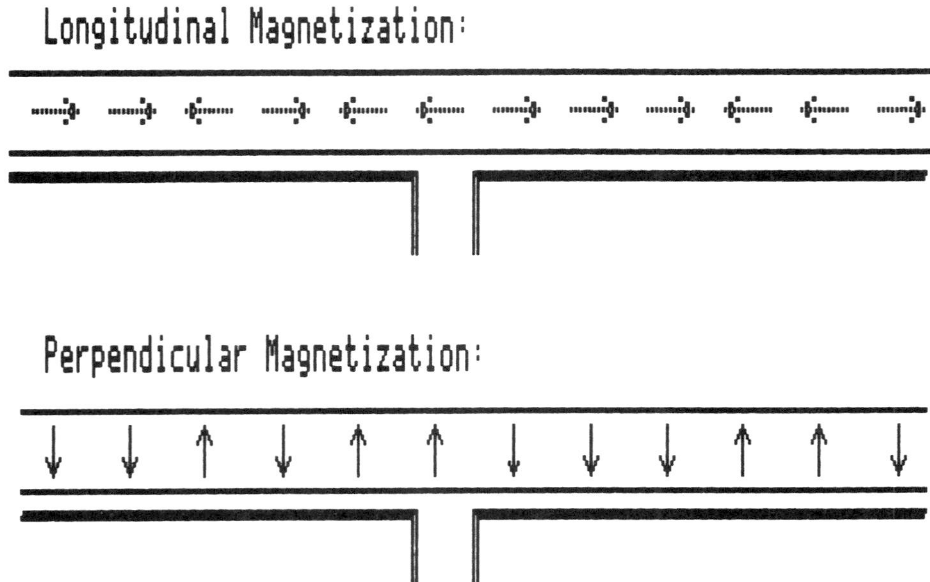

Fig. 4-11. Two magnetization modes in magnetic recording.

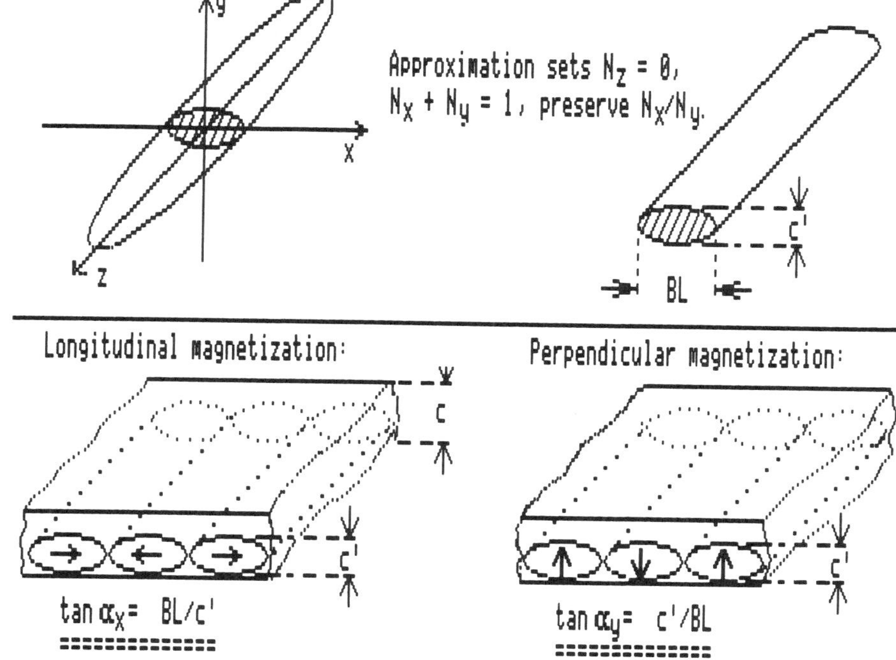

Fig. 4-12. Slope tan α for longitudinal and perpendicular magnetizations.

$$N_x = c'/(BL + c') \quad \textbf{(4.17)}$$

and

$$N_y = BL/(BL + c') \quad \textbf{(4.18)}$$

where

c' = recorded depth <= coating thickness.

Note the introduction of the *recorded depth c'*, rather than the physical thickness c of the coating. The case where $c' < c$ corresponds to the conditions *under-write* or *under-bias*, which both result in higher packing densities (better high frequency response). $c' = c$ is normal write, or normal bias. (Note: Over-write, or over-bias, does not produce a $c' > c$.)

In the remainder of the book we will primarily use the JH- or MH-loops, which tie the magnetization and the H-field together. We will further use cgs-units when describing the magnetization of materials.

When the terms from (4.17) and (4.18) are substituted in the expressions for tanα (4.11-13), we find:

BH-loop:

$$\tan \alpha_x B = \mu_o * BL/c' \quad \textbf{(4.19-X)}$$
$$\tan \alpha_y B = \mu_o * c'/BL \quad \textbf{(4.19-Y)}$$

JH-loop:

$$\tan \alpha_x J = \mu_o(1 + BL/c') \quad \textbf{(4.20-X)}$$
$$\tan \alpha_y J = \mu_o (1 + c'/BL) \quad \textbf{(4.20-Y)}$$

MH-loop:

$$\tan \alpha_x M = 1 + BL/c' \quad \textbf{(4.21-X)}$$
$$\tan \alpha_y M = 1 + c'/BL \quad \textbf{(4.21-Y)}$$

The demagnetized level (B_r', J_r' or M_r') may playback as a higher level (B_r'', J_r'' or M_r'') due to *recoil*: when the track comes in contact with the read head it sees a high permeability material, which shunts the flux through it. Or, in terms of demagnetization: the air gap from pole to pole in the bit has now been replaced with a keeper, and the demagnetization field vanishes from the inside of the coating.

When $H_d \rightarrow 0$ then the point for B_r' (or J_r') moves along a recoil line with slope $\mu_o*\mu_{rt}$ in the BH-loop ($\mu_o*(\mu_{rt} - 1)$ in the JH- or MH-loop) to the point where it intersects the vertical B- (or J-, M-axis). μ_{rt} is the relative permeability of the tape, typically between one and two in the direction of the oriented particles, but three to four when measured perpendicular thereto. This magnetization level produces the *short circuit flux* (shown as B_{rs}'' in Fig. 4-13), that is equal to ($\mu_{rt} = 1$) or greater than ($\mu_{rt} > 1$) the *open circuit flux*.

The demagnetization losses are now quite simple to determine, either by graphic methods or computations. The latter were treated earlier in the literature (Smaller, Mallinson), using either of the two models in Fig. 4-14.

Worksheets for graphical estimates of the demagnetization losses are shown in Figs. 4-15 and 4-16, using the BH-loop or the JH-loop; the methods are identical, except for line slopes.

The method is illustrated in Fig. 4-17 to determine the demagnetization losses in a longitudinally oriented tape, as function of frequency at a speed of 1⅞ IPS. The recorded depth c′ is equal to the coating thickness c.

The losses will operate on different levels of remanent maximum magnetizations in a medium, in accordance with the squareness ratios S_x and S_y. A longitudinally oriented medium will have a predominantly longitudinal remanence (except at short wavelengths). A perpendicularly oriented medium will have none or little demagnetization at short wavelengths, but very little output at long wavelengths. Only an isotropic medium will offer the best of both worlds.

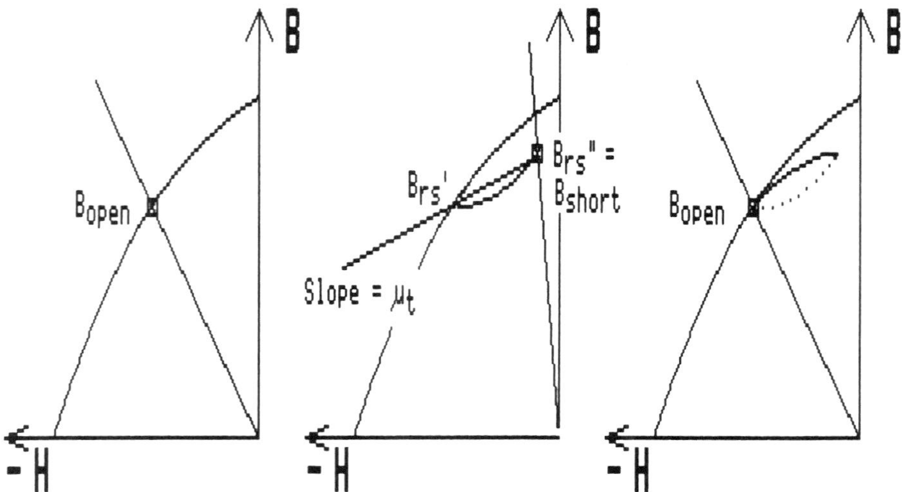

Fig. 4-13. The read head presents a smaller load (middle) for the flux, and the read flux (short circuit flux) is higher than through air (open circuit flux).

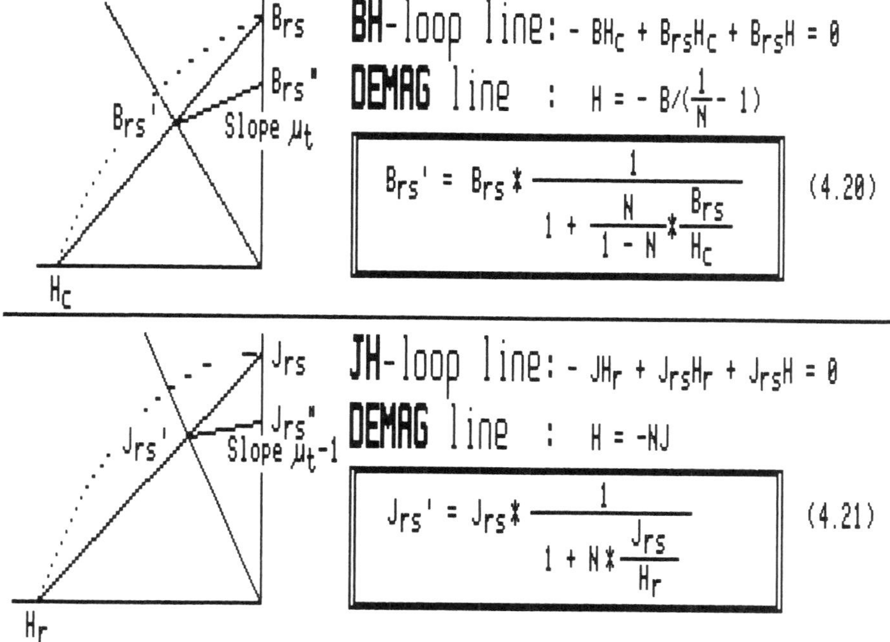

Fig. 4-14. Simple models for numerical computation of demagnetization.

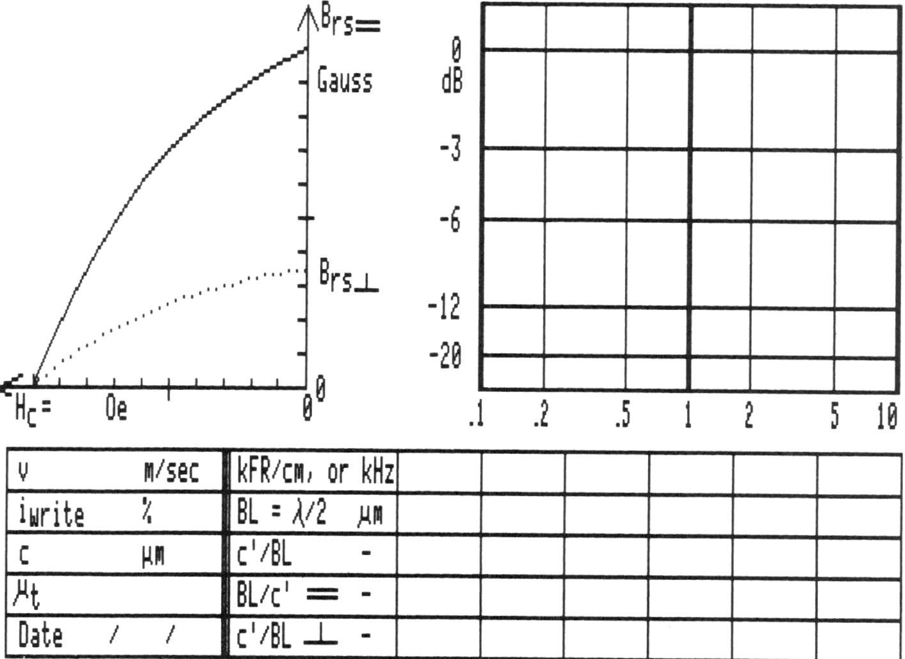

v m/sec	kFR/cm, or kHz						
iwrite %	BL = λ/2 μm						
c μm	c'/BL -						
μ_t	BL/c' = -						
Date / /	c'/BL ⊥ -						

Fig. 4-15. Demagnetization in a longitudinal anisotropic tape coating.

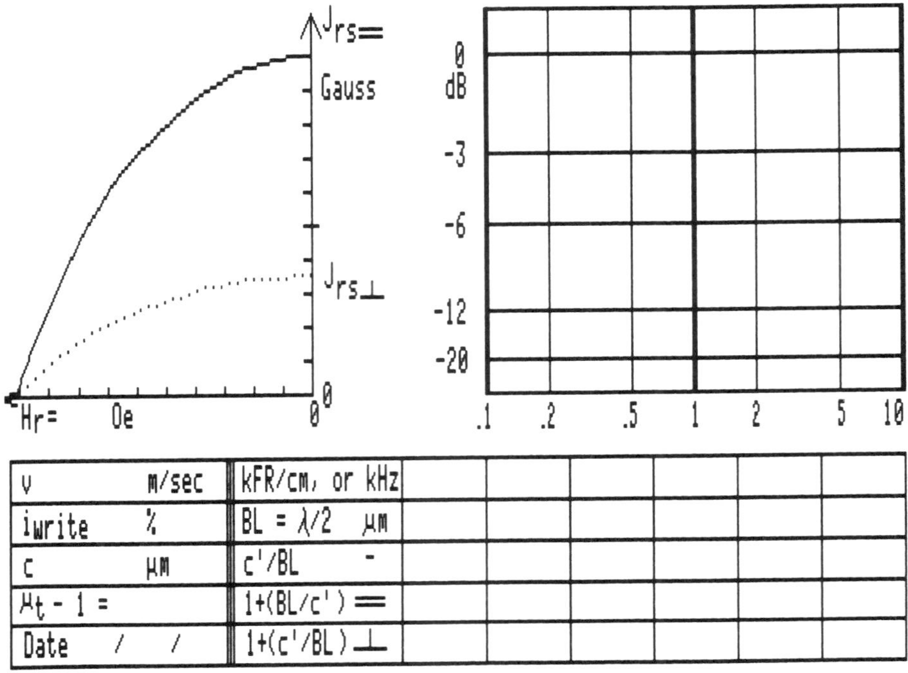

v m/sec	kFR/cm, or kHz						
i_{write} %	BL = λ/2 μm						
c μm	c'/BL -						
μ_t - 1 =	1+(BL/c') =						
Date / /	1+(c'/BL) ⊥						

Fig. 4-16. Demagnetization in a longitudinal anisotropic tape coating.

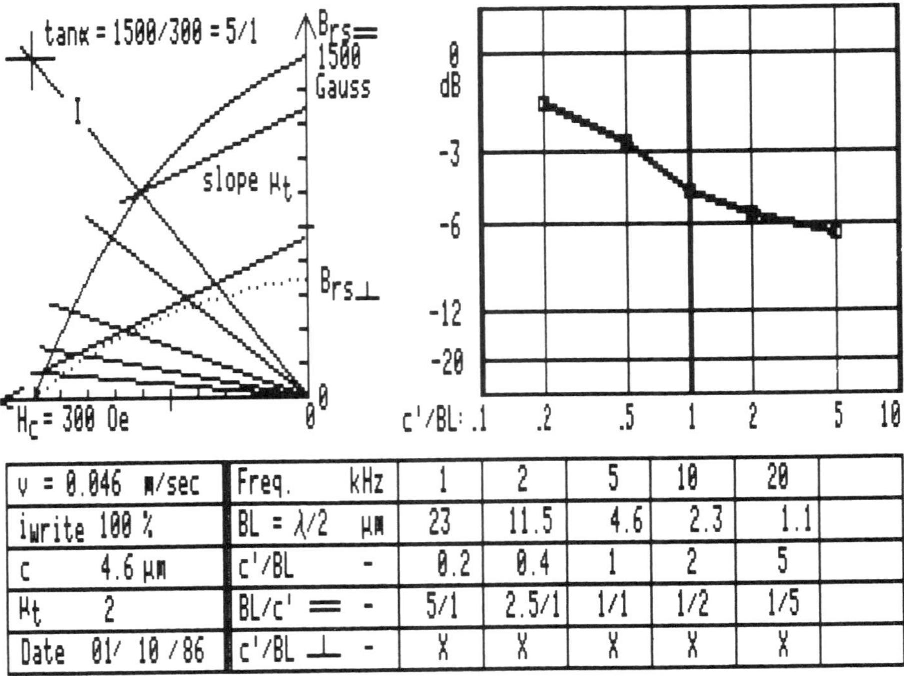

v = 0.046 m/sec	Freq. kHz	1	2	5	10	20	
i_{write} 100 %	BL = λ/2 μm	23	11.5	4.6	2.3	1.1	
c 4.6 μm	c'/BL -	0.2	0.4	1	2	5	
μ_t 2	BL/c' = -	5/1	2.5/1	1/1	1/2	1/5	
Date 01/ 10 /86	c'/BL ⊥ -	X	X	X	X	X	

Fig. 4-17. Demagnetization in a longitudinally magnetized audio recording.

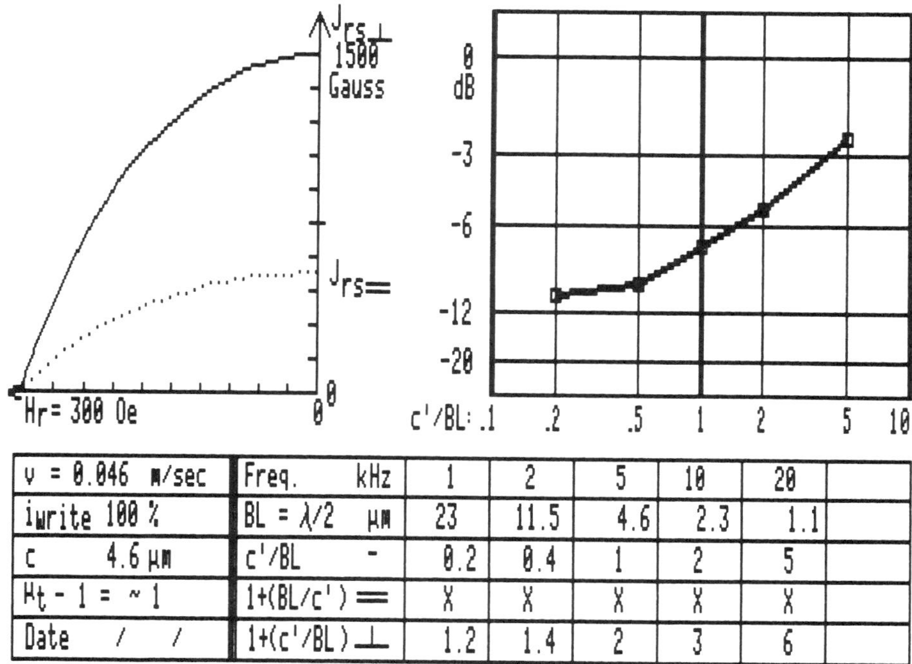

v = 0.046 m/sec	Freq. kHz	1	2	5	10	20	
iwrite 100 %	BL = λ/2 µm	23	11.5	4.6	2.3	1.1	
c 4.6 µm	c'/BL -	0.2	0.4	1	2	5	
Ht - 1 = ~1	1+(BL/c')═	X	X	X	X	X	
Date / /	1+(c'/BL)⊥	1.2	1.4	2	3	6	

Fig. 4-18. Demagnetization in a perpendicular anisotropic tape coating.

Figure 4-18 shows the demagnetization losses for a perpendicular medium, and the reader may wish to draw in the line through the origin with slopes 1.2, 1.4, 2, etc. NOTE: do not solve for the angle, it is easier to apply the definition of the tangent function, with tan α equal to the ratio between the opposite side and the adjacent side in the triangle. Then it does not matter that the B (J) and H scales have different lengths per unit of gauss or oersteds.

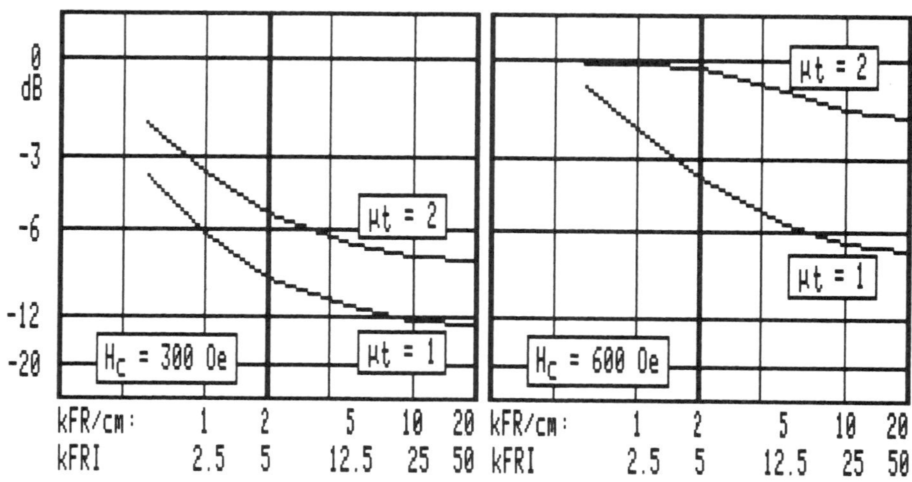

Fig. 4-19. Demagnetization losses in a 5¼ inch diskette; B_{rs} = 1500 Gauss; c = 5μm (200 μin); write level = 100%. Notice improvement with higher H_c.

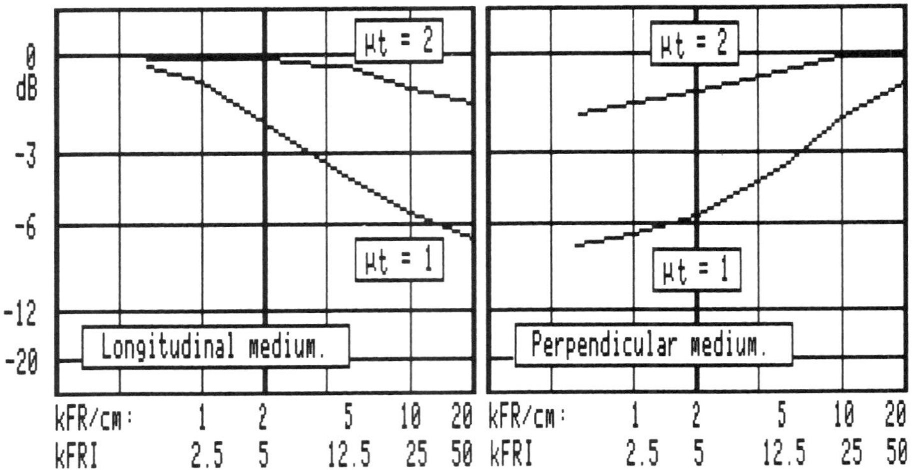

Fig. 4-20. Demagnetization losses in a 3½ inch diskette; Brs = 1500 Gauss; c = 2 μm (80 μin); write level = 100%; Hc = 600 Oe. Notice improvement at high densities with a perpendicular anisotropic medium.

Further examples are shown for a 5¼ inch floppy and a 3½ inch mini-floppy in Figs. 4-19 and 4-20. The figure captions plus data in the illustrations contain all pertinent data. We will return to a discussion of longitudinal versus perpendicular magnetization modes in Chapters 19 and 20.

CHARACTERIZATION OF MAGNETIC HEADS

The recording of a signal onto a tape or disk is accomplished by feeding a current representing the signal through the winding in the write head; this will generate the field in front of the record head gap that lays down a permanent magnetization on the coating. The flux from this magnetization will later on induce the read voltage in the windings of the playback head.

We will need to design the head so its current requirements, and input impedance, are compatible with our drive electronics. It seems logical to put a large number of turns onto the core, which will result in the lowest current requirement, and also the greatest induced voltage during read.

We will learn that the head impedance is proportional to the number of turns squared, and that the winding will exhibit a self inductance that will resonate with the combined capacitance of amplifier-, cable-, and self-capacitances. This frequency must of course lay above the highest signal frequency we plan to use.

There are finally certain losses occurring in the head core material at high frequencies. These losses will be discussed in Chapter 7, but we clearly need some method to characterize a head's impedance and current requirements, and voltage output during the read mode.

An elegant solution is found in the analogy between the current flowing in an electric circuit to the magnetic flux flowing in what we now will name a *magnetic circuit*. The key element is the definition of the *magnetic resistance* of an element (like the air gap, or a core-half) through which the flux flows, identical to the definition of the electric resistance of a wire element.

We will arrive at an equivalent electrical circuit for the head, from which we can compute efficiency, write current, read voltage, and self-inductance. The model will be extended in Chapter 7 to include high frequency losses.

MAGNETIC CIRCUITS

We can treat *magnetic circuits* as being made up of magnetic flux paths in the same way we consider electric circuits to be made up of resistors. Figure 4-21 (left) shows a battery E connected to a resistor R. The current I is easily found, using Ohm's law:

$$I = E/R \qquad \textbf{(4.22)}$$

Also shown is a ring core with relative permeability μ_r, average length L, and cross sectional area A. With a current I flowing through the winding we find the field strength:

$$H = nI/L \qquad \textbf{(4.23)}$$

where nI is the magnetomotive force in Ampere-turns. The flux ψ is:

$$\begin{aligned} \psi &= BA \\ &= \mu HA \\ &= \mu A * nI/L \\ &= nI/(L/\mu A) \end{aligned} \qquad \textbf{(4.24)}$$

This compares with an electric current I through a wire of conductivity σ, length L and cross section A:

$$I = E/(L/\sigma A) \qquad \textbf{(4.25)}$$

The similarity between (4.24) and (4.25) shows the following analogies between magnetic and electric quantities:

Electric current I = Magnetic flux ψ (Wb)
Electromotive force E = Magnetomotive force nI (Ampturns)
Electric resistance R = Magnetic resistance R_m (1/Hy)

Drawing the analog resistance network is invaluable in analyzing magnetic circuits with various substances and air gaps in the flux path, and the method will be used extensively in this book.

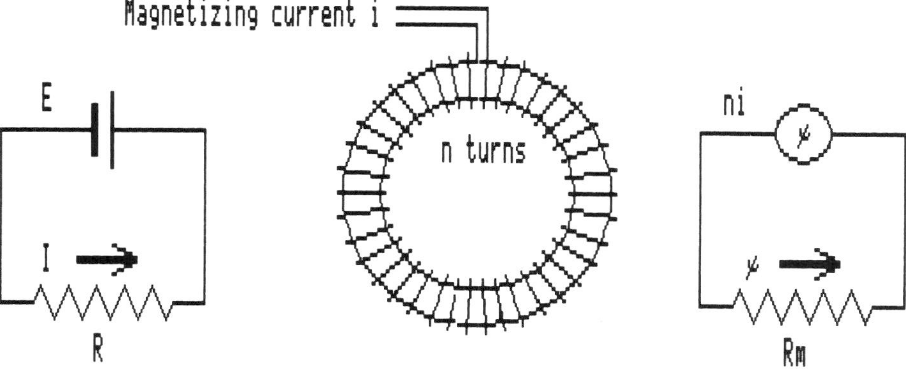

Fig. 4-21. Electric circuit, an electromagnet, and its equivalent circuit.

Two air gaps are treated like two *reluctances* (magnetic resistor) in series, while stray flux paralleling the gap is treated as a parallel reluctance (similar to parallel resistors).

Also used are *magnetic permeance* P, which is magnetic conductivity equal to the reciprocal of resistance. P values are particularly handy for parallel combinations.

EFFICIENCY OF READ HEADS

The media flux flows ideally through the read head as shown in Fig. 4-22, left. However, a portion of the flux links across the front gap, and does not contribute to the useful flux through the winding.

The total flux is:

$$\psi_{media} = \psi_{core} + \psi_{gap}$$

or

$$\psi_{core} = \psi_{media} - \psi_{gap}$$

Across the gap there exists a magnetic potential V_m:

$$\begin{aligned} V_m &= \psi * R_m \\ &= \psi_{gap} * R_{fg} \\ &= \psi_{core} * (R_{bg} + R_{c1} + R_{c2}) \\ &= \psi_{media} * (R_{fg}/ /(R_{bg} + R_{c1} + R_{c2}) \end{aligned}$$

We can now find the core flux, and determine the *efficiency* ηread:

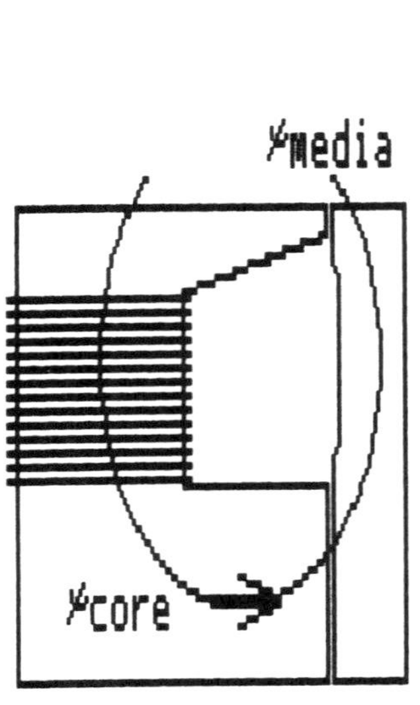

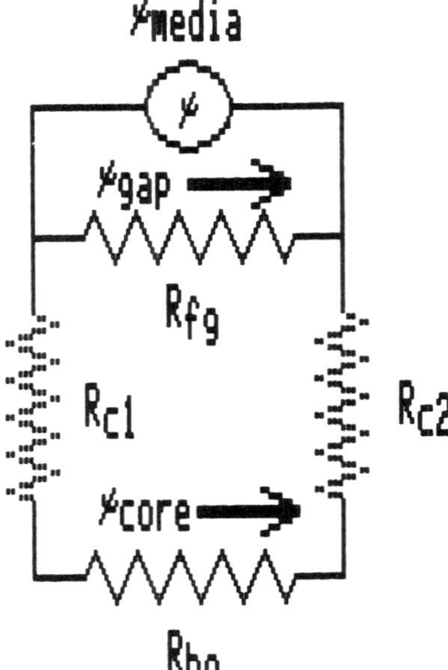

Fig. 4-22. Magnetic read head and its equivalent resistance circuit.

$$\eta_r = \psi_{core}/\psi_{media}$$

or

$$\boxed{\eta_r = R_{fg}/\Sigma R_m} \qquad \textbf{(4.26)}$$

where

$$\Sigma R_m = R_{fg} + R_{bg} + R_{c1} + R_{c2}$$
$$= \text{sum of all resistances going once around the circuit}$$

R_{fg} should include the stray fields at the gaps (mandatory in modern micro head structures). But it is generally safe to ignore the stray fields at the back gap due to its short length and large area.

The front gap stray fields are discussed in Chapter 8. The formulas provide the permeances of the various sections of the stray fields; the total magnetic resistance is equal to the reciprocal of the sum of the permeances.

A quick examination of the expression for the efficiency tells us how to make a good head:

1. *Make the front gap resistance large:*

1.1 Make the gap length L_{fg} large—this is not possible due to short wavelength losses, as illustrated in Fig. 4-23. When the gap length equals two bit lengths, then the flux through the head core is zero. A gap length equal to one bit length BL is optimum. With a packing density of 10,000 FRI (flux reversals per inch) the bit length is 100 μin. Then L_{fg} should be no longer than 100 μin.

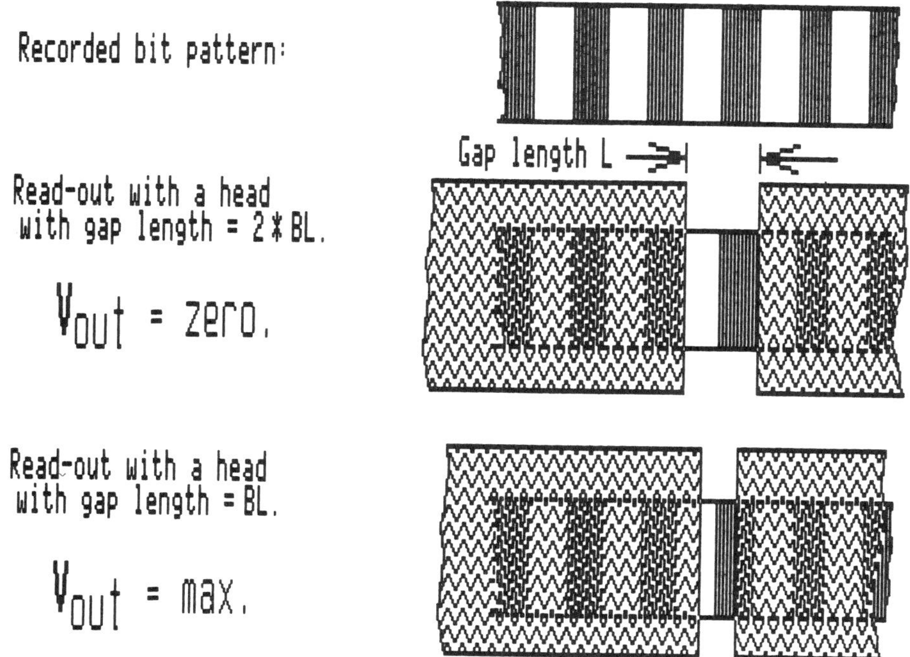

Fig. 4-23. The playback (read) head gap length must not exceed the bit length.

1.2 Make the gap depth small. This is possible for heads used in high speed disk drives, where the head assembly actually is flying over the disk surface. Then there is no wear, and the gap depth can be made as small as 5 μm (200 μin). This would be too shallow a gap depth for in-contact applications, such as tape and floppy disks. One must allow 25-75 μm (1-3 mils) for wear. (See further in several sections of Chapter 18.)

2. *Make the back gap resistance small,* i.e., short length and large area. The inclusion of a back gap is necessary in most head designs for the reason of assembly. A back gap also makes life easier in the production in terms of yield (see Chapter 9).

3. *Make the core magnetic resistance small,* i.e., use short, stubby core halves. Use high permeability material.

FIELD STRENGTH IN THE WRITE HEAD GAP

Recording of a tape or disk is accomplished when the media moves through the stray field from the gap in the write head. We need to know this field so we can determine the locations where the field strength exceeds the coercive force of the individual particles, and hence may change their magnetization (interaction fields should be included, see Chapter 9 on AC-bias recording).

The field strength components H_x and H_y have been calculated by Karlqvist (1954), and are listed in Fig. 4-24 where the field lines are also shown. The vector field strength $\mathbf{H} = \mathbf{i}\,H_x + \mathbf{j}H_y$ can now be calculated, with known values of H_g, L_{fg}, x, and y.

We need to determine the value of Hg, the field inside the gap, and will use the magnetic circuit analog.

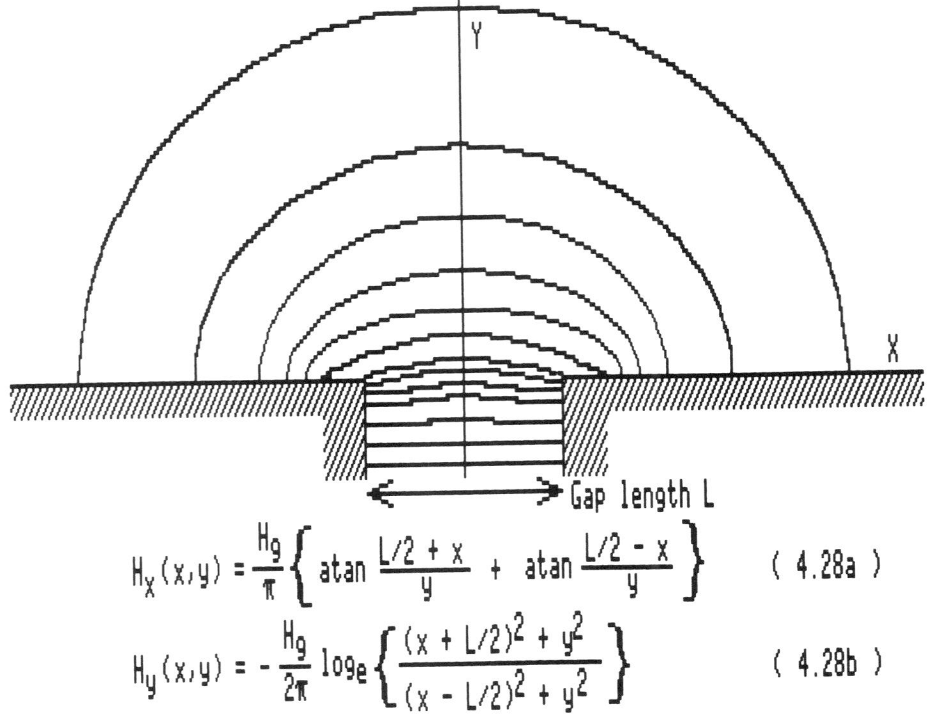

$$H_x(x,y) = \frac{H_g}{\pi}\left\{ \operatorname{atan}\frac{L/2 + x}{y} + \operatorname{atan}\frac{L/2 - x}{y} \right\} \qquad (4.28a)$$

$$H_y(x,y) = -\frac{H_g}{2\pi}\log_e\left\{ \frac{(x + L/2)^2 + y^2}{(x - L/2)^2 + y^2} \right\} \qquad (4.28b)$$

Fig. 4-24. Equations for the magnetic field in front of a gap (Ref. Karlquist).

We have:

$$H_g = B_g/\mu_o$$

where

$$B_g = \psi_g/A_g$$

Also

$$\psi_g = nI/\Sigma R_m$$

Substituting

$$\begin{aligned} H_g &= \psi_g/\mu_o * A_g \\ &= nI/\Sigma R_m \mu_o A_g \end{aligned}$$

or, after reduction:

$$H_g = (nI/L_g) * \eta_r \qquad \textbf{(4.27)}$$

The appearance of the read efficiency factor accounts for the *reciprocity* property of magnetic heads. It serves equally well in write or read process.

We need to multiply with two additional terms. First, the gap flux is only a fraction of the total front gap flux, as we interpreted it in Fig. 4-22. A portion of the flux goes through air (stray) and contributes to the formation of H_g. Consider that the total stray flux reluctance of the front gap is summed up in a term R_{stray}. Then we can easily find that the gap field needs to be multiplied with the additional term of a write efficiency:

$$\eta_w = R_{stray}/(R_{stray} + R_{fg}) \qquad \textbf{(4.28)}$$

Write heads also have lower efficiency than read heads due to flux leakage around the coil (insufficient coupling) and to shielding that may be next to the core. An efficiency η_{coupl} can be calculated for the flux from nI that reaches and goes through the front (stray plus gap), see Fig. 8-36 and formula (8.6). Also see Chapter 10, under write head efficiency.

For now we will use a term *coupling efficiency* η_{coupl}, and for the deep-gap field H_g for use in Karlqvist's formulas we find:

$$\boxed{H_g = (nI/L_g) * \eta_r * \eta_w * \eta_{coupl}} \qquad \textbf{(4.29)}$$

We may compare this with the magnitude of the field H in the center of a solenoid L meters long with n turns: H = nI/L.

This makes the formula for H_g easy to remember: The field strength H_g is equal to that of a very short solenoid with n turns, placed in the gap of length L_g, and multiplied with the product of the three efficiencies.

HEAD IMPEDANCE

The impedance of a magnetic write or read head is made up of three components: the wire resistance R, the loss resistance R_{loss} and the inductance L (here we use the same symbol as for the gap length, but that appears better than using a lower case l for length; that is often mistaken for a 1 (one)).

The inductance of an electric circuit is a measure of the magnetic flux which links the circuit when a current I flows in that circuit:

$$L = n\psi/I$$
$$= n(nI/\Sigma R_m)/I$$

or

$$\boxed{L = n^2/\Sigma R_m} \qquad \textbf{(4.30)}$$

The wire resistance and the equivalent loss resistance will be discussed in Chapter 10.

REFERENCES TO CHAPTER 4

Osbron, J.A., "Demagnetizing Factors of the General Ellipsoid," *Physical Review*, July 1945, Vol. 67, No. 11 and 12, pp. 351-357.

Westmijze, W.K., "Studies in Magnetic Recording, I: Introduction," *Philips Res. Rep.*, April 1953, Vol. 8, No. 3, pp. 148-157.

Karlqvist, O., "Calculation of the Magnetic Field in the Ferromagnetic Layer of a Magnetic Drum," *Trans. Royal Inst. Tech.*, June 1954, No. 86, pp. 3-27.

Joseph, R.I., and Schloemann, E., "Demagnetizing Field in Nonellipsoidal Bodies," *Jour. Appl. Phys.*, May 1965, Vol. 36, No. 5, pp. 1579-1593.

Mallinson, J.C., "Demagnetization Theory for Longitudinal Recording," *IEEE Trans. Magn.*, Sept. 1966, Vol. MAG-2, No. 3, pp. 233-236.

Smaller, P., "An Experimental Study of Short Wavelength Recording Phenomena," IEEE *Trans. Magn.*, Sept. 1966, Vol. MAG-2, No. 3, pp. 242-247.

Chapter 5
Recording of Signals (Write)

The magnetic recording process converts an electric current signal into an equivalent magnetization in the coating of a magnetic tape or disk. This is done with a transducer called a *record* or *write* head, that transforms the electrical signal into a magnetic field through which the coating passes. The result in the coating is a magnetic remanence proportional to the field.

We will often use the term *write* rather than *record* although the two words are identical in meaning. Write is used in the digital disk and tape industry, while record is used among audio and instrumentation engineers.

A write head is identical to a read head: it consists of a ring core with a short gap, from which the stray flux forms the recording field. This is shown in Fig. 5-1, which also sets the stage for the geometrical parameters involved: gap length L, coating thickness c, recorded depth c′ (c′ < = c) and the spacing d between coating and head. We will also need to know the values of the coating coercivity H_c, the switching field distribution Δh_r, and the deep-gap field H_g. Finally, we assume that the field and the coating magnetization do not change in the z-direction and our analysis is thus limited to the x-y plane. This assumption will need modification for very narrow track widths (see Chapter 8 on head fields).

Several observations can now be made for a coating moving through the field H, which for the time being is held constant. Considering just a small volume $\Delta vol = dx*dy*w$, where w is the track width, we find:

• Each Δvol undergoes a magnetizing field that increases to some maximum strength and then decreases to zero.

• The volumes nearest the head "see" the strongest field: the magnetization throughout the coating is not uniform.

• The direction of the field changes, or rotates, during the pass through it. The overall magnetization is not necessarily longitudinal.

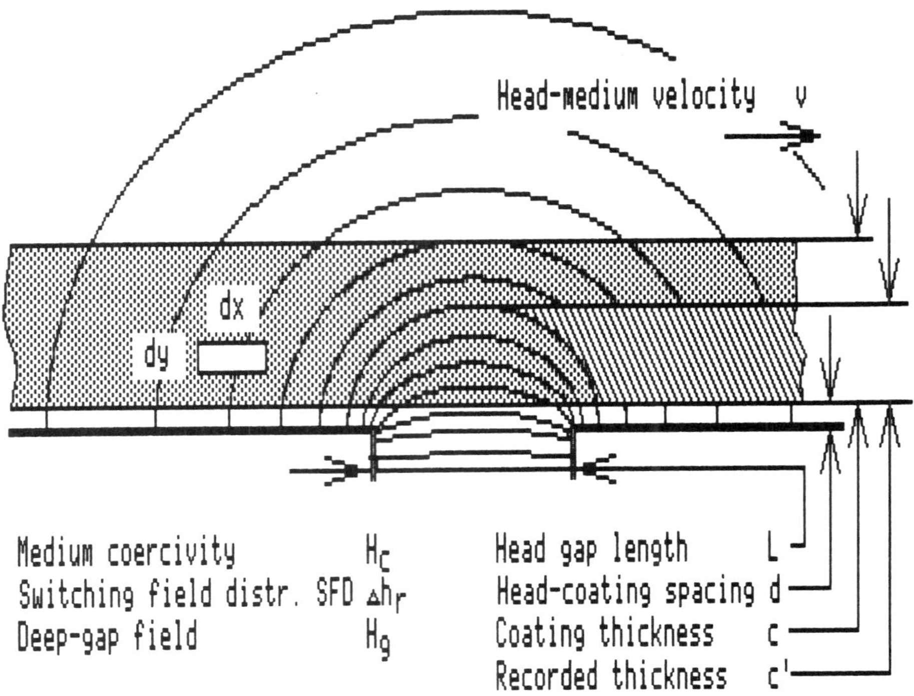

Fig. 5-1. Magnetic properties and mechanical dimensions needed to describe the magnetic recording (write) process.

• If the coating is spaced away from the head the field strength is less and a recording loss occurs.

We will start our exploration of the write process by considering a simplified model with a very thin coating so the field strength from the write head is the same in the front and the back of the coating, and that only longitudinal magnetization occurs. This will follow the earliest explanation of the write process, used until the mid-sixties. It is of educational value, but remains a weak model of the write process.

The switching field distribution is at first assumed zero, which results in a zero-length transition zone between written bits. We will next improve the model by introducing a non-zero switching field distribution, and finally full vector magnetization, i.e., longitudinal plus perpendicular magnetizations in a thick coating. We can then make estimates of the length of the transition zone Δx between two bits, and will learn that the length of the write gap has negligible influence upon the length of Δx, and hence the packing density.

The method of adding a high frequency AC-bias signal to the write signal is briefly introduced in order to familiarize the reader with the concept of a *recording zone*. This is the region where a Δvol moves away from the gap and the field strength becomes less than the switching field strength; then the particles' magnetizations will flip (following the AC-bias field frequency), and are "frozen" into a final magnetization. The length of this zone is identical to the length of the transition zone.

We complete our exploration of the write process by considering the transition zone to be a small line pole, stretching from one side of the track to the other; its field should be added to the field from the gap as a demagnetizing field (see formula (4.3)). Further details of the write/record process can be found in Chapters 19 and 20, the latter presenting a full treatment of the AC-bias model.

MEDIA MAGNETIZATION

We have experienced a remarkable increase in the amount of information bits that can be stored on a given length of tape. Computer tape recorders have gone from 800 to 1600 and 3200 BPI (bits-per-inch), and are approaching 6400 and 12,800 BPI. Instrumentation recording can now be done with an upper frequency of 4 MHz (4,000,000 Hz) at 120 IPS (inches per second tape speed). And audio has gone from 15,000 Hz at 30 IPS to 20,000 Hz at 1⅞ IPS!

This means that the smallest bit size or wave length is shorter than the coating thickness. And we are then dealing with a remanence that is perpendicular rather than longitudinal, in spite of the particle's orientation. The concept of *perpendicular remanence* was supported as early as 1963 when two engineers at Philips, Tjaden and Leyten, carried out a large scale recording experiment (see ref.).

The magnetization pattern inside a coating cannot be made visible by the Bitter method used in ordinary observations of domain patterns; the pictures that develop when iron powder is laid down on a side-cut of a magnetized coating will at best show closed flux patterns. An example is Iwasaki's observation in 1975, which led to the discovery of circular magnetization and then to his suggestion, in 1977, of using a perpendicular mode at high packing densities (see ref.).

Tjaden and Leyten fabricated a 5000:1 scale model of the coating and the head, which was then used in recording experiments. The coatings were used once only, and then cut into small blocks. Each block's level and direction of magnetization was measured and plotted. One of the results is shown in Fig. 5-2, clearly indicating variations in magnetization through the thickness of the coating.

When the bit size, or half wavelength, equals the coating thickness, the demagnetization factor is ½ for the longitudinal or perpendicular remanence. But for shorter wavelengths the

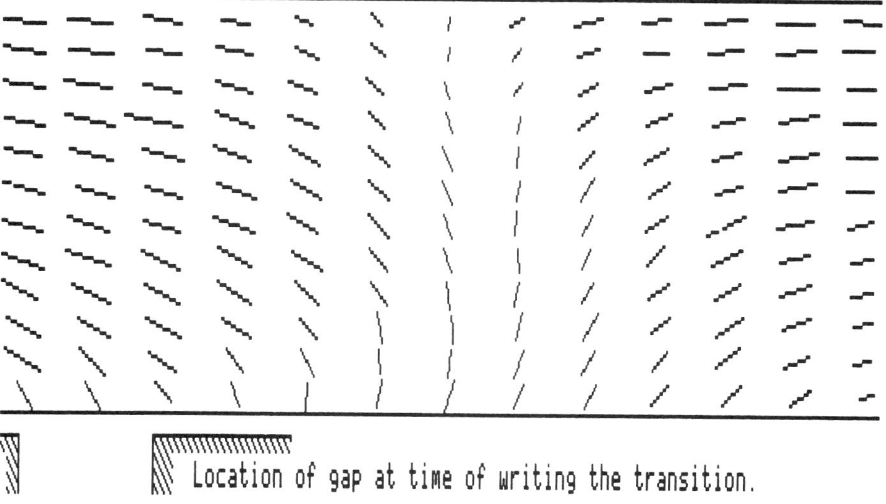

Fig. 5-2. Magnetization inside the coating near a single transition (after Tjaden and Leyten).

perpendicular demagnetization is less than the longitudinal. And a simple dipole model shows least energy for perpendicular remanence at very short wavelengths.

It would appear that oriented, acicular particles have little remanence to offer in the perpendicular direction—but each of them may be considered a chain of single domains (Jacobs-Beans model), where indeed the magnetization could be "fanned-out" toward perpendicular remanence, as shown earlier in Fig. 3-26, bottom.

The perpendicular magnetization components require some amount of phase equalization upon playback. This principle has long been used in instrumentation recorders, and recently in other units. We would expect this to be so when the recorded wavelength becomes shorter than the coating thickness. Examples of this, for a coating thickness of c = 0.2 mils, are as follows:

Instrumentation	: v = 120 IPS,	$\lambda/2 < c : f > 300{,}000$ Hz
Audio	: v = 1⅞ IPS,	$\lambda/2 < c : f > 4{,}700$ Hz
Computer disk (5¼" diskette)	: v = 25 IPS,	IFR < c : rate > 125,000 frs

(λ = wavelength)
(IFR = inches per flux reversal)
(frs = flux reversals per second)

The reader can now understand why we should not consider longitudinal recording only. It is, after all, the high packing density with associated perpendicular magnetization that is the pacing item in modern recording equipment. The end result is smaller size disks, tape reels, or recorders with longer playing time.

DIRECT RECORDING ON A VERY THIN COATING

Direct recording implies that a signal voltage is amplified and converted into a current through the record head winding. No AC-bias is applied. The tape passing over the record head is assumed to be neutral, i.e., completely erased to a zero remanence. A thin coating implies that the field strength in front of the coating is equal to the field strength in the back, as depicted in Fig. 5-3.

The initial magnetization curve (from a nonmagnetic state to saturation) is shown in Fig. 5-4. The quality of a recording depends upon the linearity between the field strength in the record gap and the magnetization (remanence B_r) left in the coating. A doubling of the field strength (below saturation) should result in a doubling of the remanence.

Figure 5-4 illustrates the relationship between the field strength H and the remanence B_r. This relationship is derived from the remanence values obtained by the increasing values of the field strength as shown in Steps 1 through 10. The remanence from an alternating field can be constructed by a graphic projection from the remanence values in Fig. 5-4 as shown in Fig. 5-5.

The pure waveform of the record current (and field) has become a highly distorted remanence signal, which is useless for any intelligent voice recording and can be used only in digital recorders, where the signal consists of ONEs and ZEROs, and in its elementary, non-coded form is written as positive or negative saturations. (Coding of the digital signal is a process that will be described in Chapters 24 and 25; its bears no significance upon the material in this chapter.)

One observes that the midrange of the remanence curve is quite linear, and if the range A'A (Fig. 5-6) could be used, then the remanence on the tape would be more linear. This was recognized by Valdemar Poulsen, and his friend and associate, Professor P.O. Pedersen, and

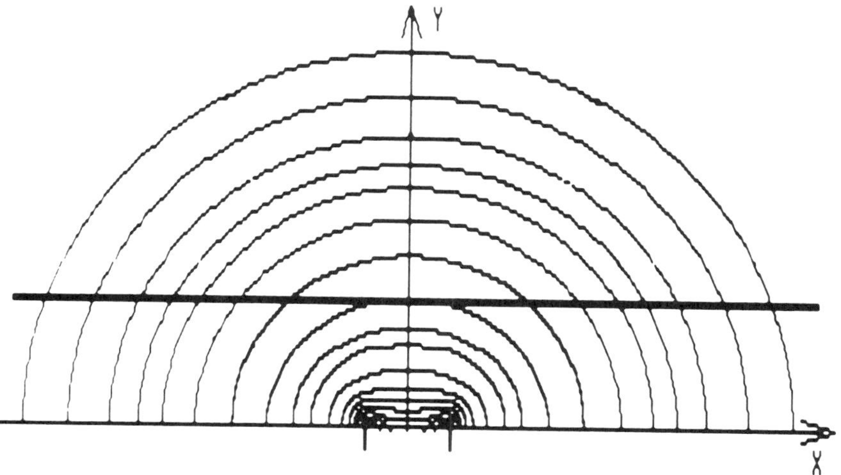

Fig. 5-3. Recording on a thin coating; the field strength in the back is equal to the field strength in front, at all values of x.

they added a DC-bias current to the signal current, as shown in Fig. 5-6. The biasing field H_{DC} moves the *working point* to midway between A′ and A, and the resultant remanence has a cleaner alternating waveform, with a superimposed DC remanence.

The presence of DC theoretically does not present a problem since the induced read voltage is proportional to the differential of the flux. But in practice the story is different. Figure 5-7 shows a coating with two imperfections: an agglomerate of particles, and a void. This causes a variation in the DC remanence, with resulting noise spikes in the read voltage. Any nonuniformity of the coating will produce noise under DC magnetized conditions. The noise level is easily 20 dB or more above that produced by a properly AC-erased coating, and this method of biasing had to be abandoned.

The method of DC-magnetizing a coating instead found its proper home in the QC department for evaluation of tapes and disks, in the examination of coating uniformity.

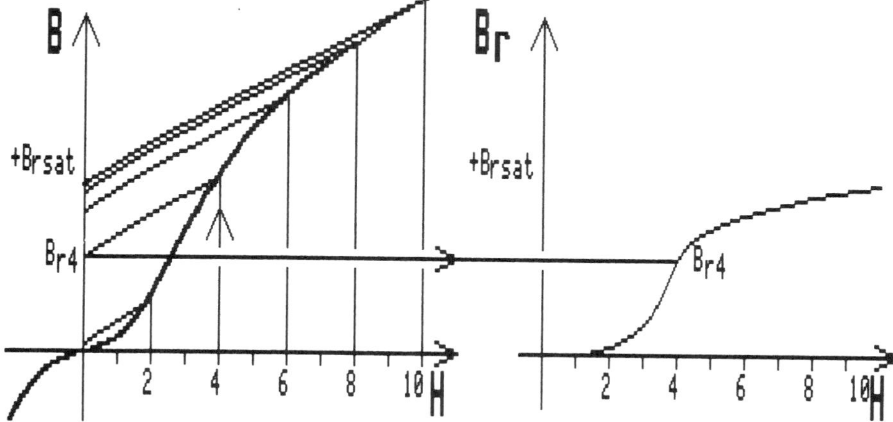

Fig. 5-4. Construction of a transfer curve for direct recording (i.e., no bias).

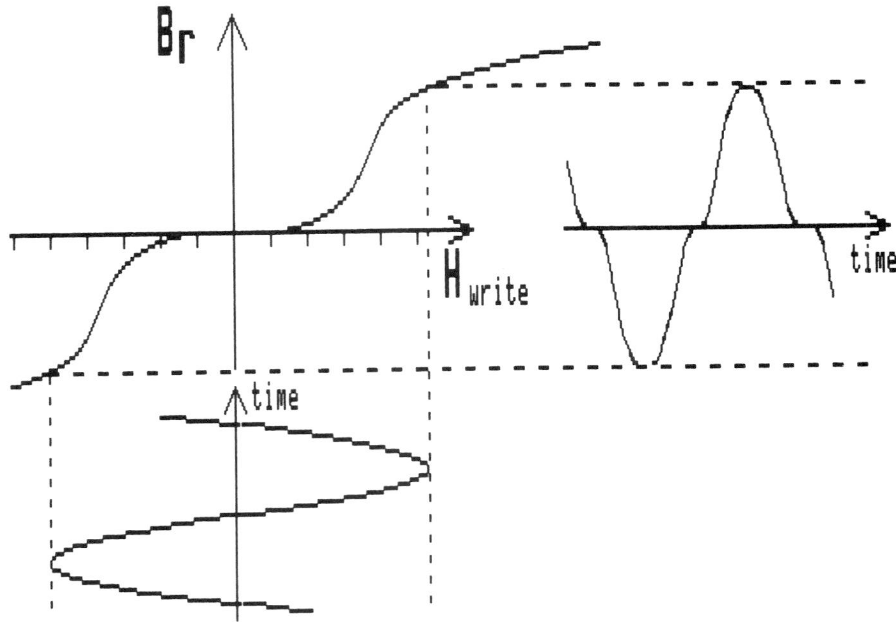

Fig. 5-5. Remanent magnetization in coating, recorded direct. Note distortion.

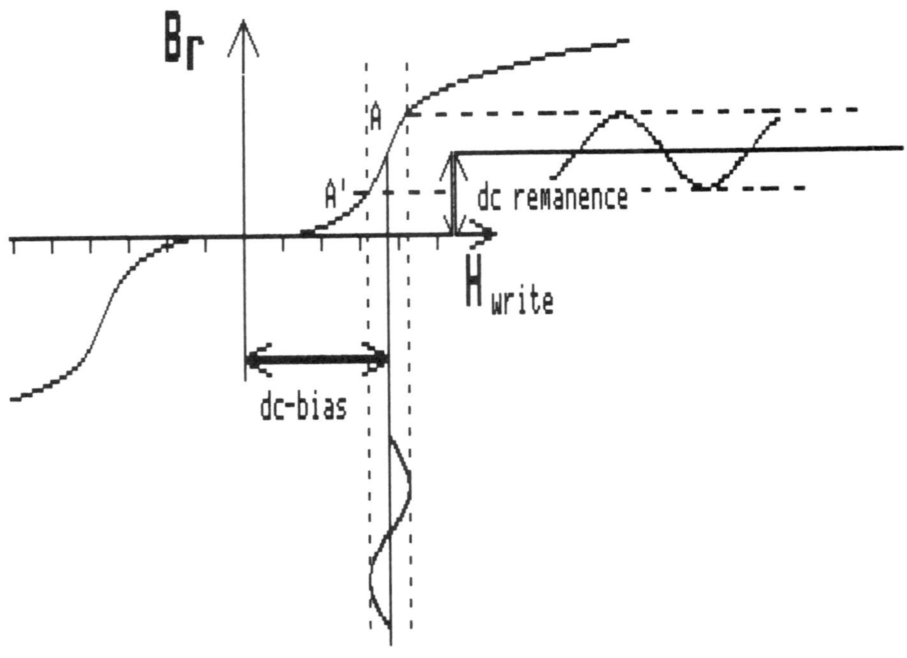

Fig. 5-6. Remanent magnetization in coating, recorded direct, with DC-bias on a demagnetized coating.

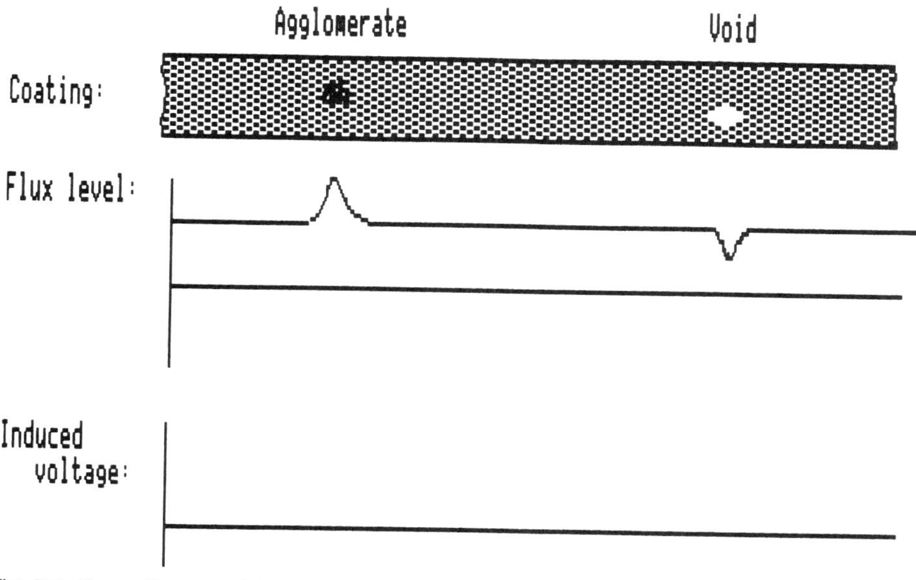

Fig. 5-7. Non-uniform particle distributions (agglomerates, voids) in a magnetic coating produce high noise voltages when coating is DC-magnetized.

The DC-bias method can be improved by writing on a pre-magnetized coating, as shown in Fig. 5-8. The coating passes over a permanent magnet that saturates the coating into the negative region, erases old information and leaves it with a remanence equal to $-B_{rs}$ before reaching the record head. A new transfer curve for the remanence B_r versus applied field H is constructed in Fig. 5-8 right, similar to the construction in Fig 5-4. This new curve has a larger linear region (B′ − B) as well as higher sensitivity (''amplification'').

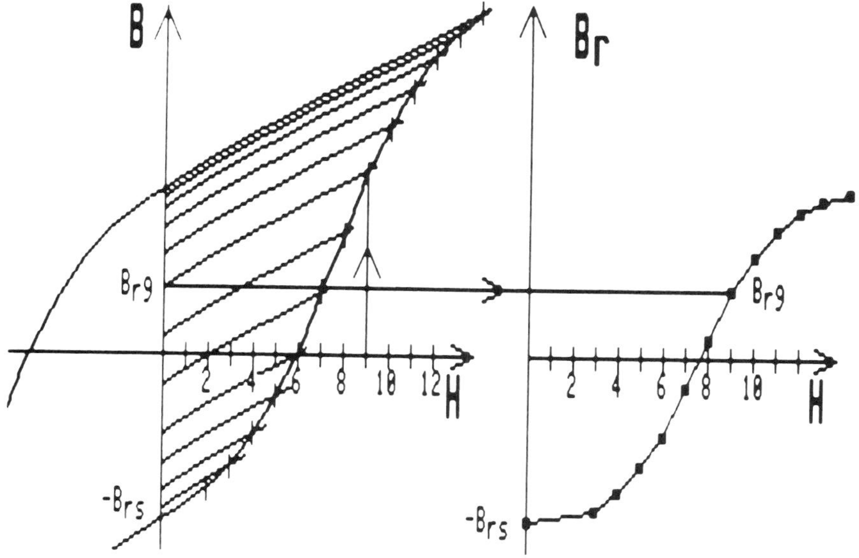

Fig. 5-8. Construction of transfer curve for direct recording on a premagnetized coating $-(B_{rs})$.

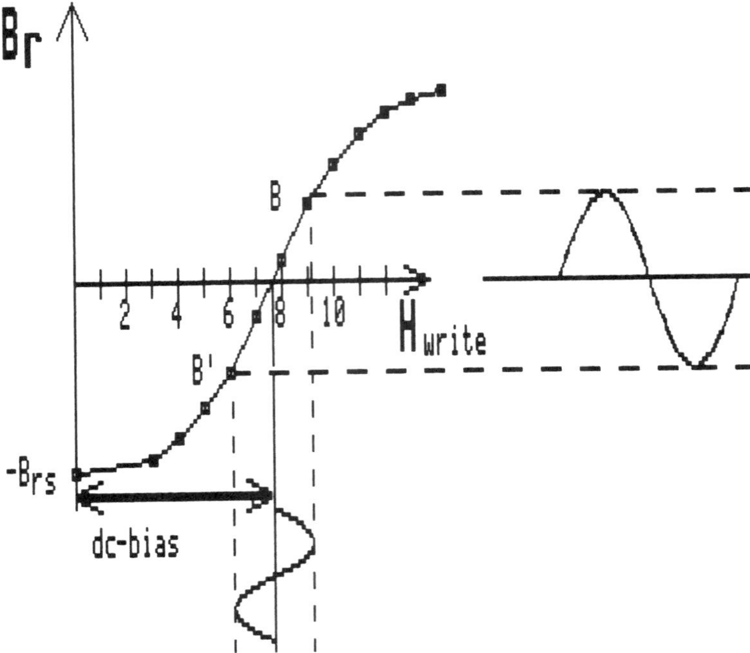

Fig. 5-9. Remanent magnetization in coating, recorded direct, with DC-bias on a premagnetized (DC-erased) coating.

A DC-bias field is again used during recording and the remanence of a sine wave signal is shown in Fig. 5-9. This method has not only higher signal remanence but the bias can also be adjusted so the DC level remanence is zero. The bias adjustment is critical, though, and a change of a few percent of the magnetic properties of the coating (also coating thickness) will throw the carefully adjusted working point off. In Chapter 12 we will find that the temperature coefficient of the coating's coercivity is 2-3 Oe/degree Celsius, and a temperature change may thus throw the carefully adjusted DC-bias point off.

All three methods (no bias, DC-bias with virgin tape, and DC-bias with pre-saturated and erased tape) are used in digital recording and some dictating recorders. But the above explanations of each write process are more qualitative than quantitative. They give us no answer about the length of the transition zone between bits, and they are quite useless if we wish to explore partial penetration recordings, or the overwrite process. (Partial penetration recording is advantageous in that it provides short transition zones, as shown later in this chapter, where overwrite of old information is discussed.)

DIRECT RECORDING, PURELY LONGITUDINAL

Identical Particles (Zero Switching Field Distribution)

The criterion for recording information onto a particulate coating is quite simple: if the field $H_x(x,y)$ at the element Δvol is greater than the switching field Δh_r then the particles within Δvol will switch magnetization in accordance with the field.

On the other hand: When Δvol has moved away from the strong field immediately above the gap, and the field $H_x(x,y)$ is less than Δh_r, then the particles will no longer follow the

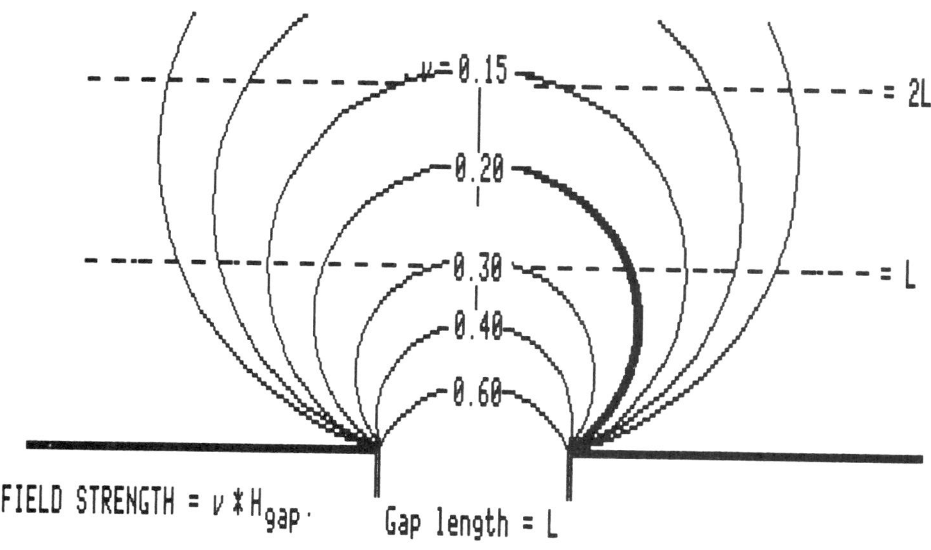

Fig. 5-10. Chart of constant longitudinal (H_x) field strength in front of gap (after Mee and Bauer).

field—their magnetizations have been *frozen* at the point where $H_x(x,y)$ became just less than Δh_r.

A "*freezing*" zone therefore exists at the trailing edge of a recording head gap; this is where transitions in digital recordings are formed. The freezing zone also corresponds to the AC-bias recording zone, shown later in this chapter.

We will now apply Fig. 5-1 in a limited sense, using only the horizontal field components to establish the location of the freezing zone, which has zero thickness since $\Delta h_r = 0$. The recording zones will therefore be sheaths with circular cross sections in the x-y plane, as shown in Fig. 5-10.

These circular patterns are contours of constant H_x field strength. They could be measured by moving a small field probe around in front of the record gap, keeping it on a trail where the horizontal field strength is equal to a constant fraction of the field strength deep in the gap, where the field is uniform (see formula (4.29)).

A recording model based upon these circular record zones was introduced in the early sixties by C.D. Mee and B.B. Bauer, and nicknamed the "bubble" or "cylinder" model (see ref.). Consider Fig. 5-10 and assume that the dark contour has a field strength that equals the intrinsic coercivity, h_{ci}, of the particles. The coating is moving left to right. When each new bit is written the polarity of the field alters, and each time all particles within the bubble alter magnetization. Just outside the bubble no particles will switch, and on the bubble sheath 50 percent will change.

This leaves an ideal transition zone of zero length, and the write model predicts an infinite number for the FRI, flux-reversals-per-inch. We will now improve upon the model by incorporating a number greater than zero for the SFD.

Normal Particle Sizes with a Non-Zero Switching Field Distribution

In a normal tape or disk coating there are particles with high and low coercivities, and the SFD (switching field distribution) is a measure of the spread in coercivities around an average value. There is currently no standard for its measurement, and it can be measured in several different ways (see ref. Koester); we will discuss two of them.

One SFD measure is obtained by finding the difference between the H-values that correspond to the 25 percent and 75 percent points on the JH- (or MH-) loop, as in Fig. 5-11A. The value of SFD is obtained by dividing this difference with the mean value H_r, and is then assigned the symbol Δh_r. (Values of SFD obtained in this fashion are very close to those obtained from another expression for SFD: 1 − S*; see ref. Koester and further Chapter 12.)

The SFD can also be expressed as Δh_c which uses the difference in H-values corresponding to the 50 percent points on the differentiated BH-loop (see Fig. 5-11B), or on the $d\psi/dt$ curve from a BH-loop measurement (see under BH-meter, Chapter 12). The value of Δ is obtained by dividing this difference with the mean value of H, which in this case is H_c. This measure for the SFD corresponds to the PW50 values for the read pulses at transitions. PW50 is the pulse width (in seconds) at the points where the amplitude is 50 percent of the peak value.

Assuming a Gaussian distribution, Δh_c encompasses 76 percent of the particles (Δh_r was 50 percent). That corresponds to 12 and 88 percentage points on the integrated $d\psi/dt$ curve. It is a curiosity that this is quite close to what a signal processing engineer would have chosen: the 10 and 90 percentage points corresponding to the rise time.

(NOTE: A statistician would have chosen points that correspond to the standard deviation σ. The SFD would in that case have encompassed 68.26 percent of the particles.)

If we assume a Gaussian distribution of the particles' h_{ci} we will find the following relations:

$$\Delta h_r = 1.18 * \sigma\text{Gaussian}$$

and $$\Delta h_r = 0.57 * \Delta h_c$$

or $$\Delta h_c = 1.75 * \Delta h_r \qquad \textbf{(5.1)}$$

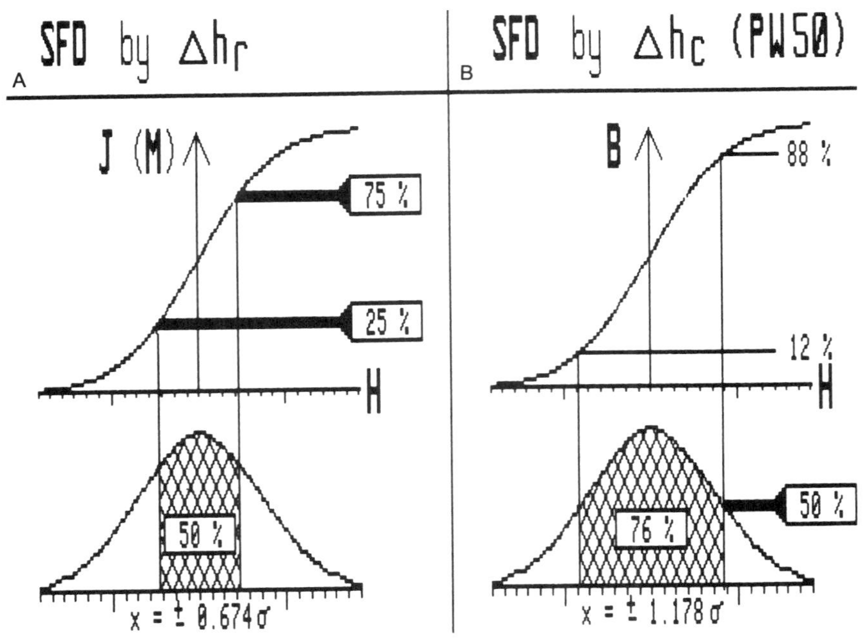

Fig. 5-11. Determination of the SFD, using Gaussian distributions.

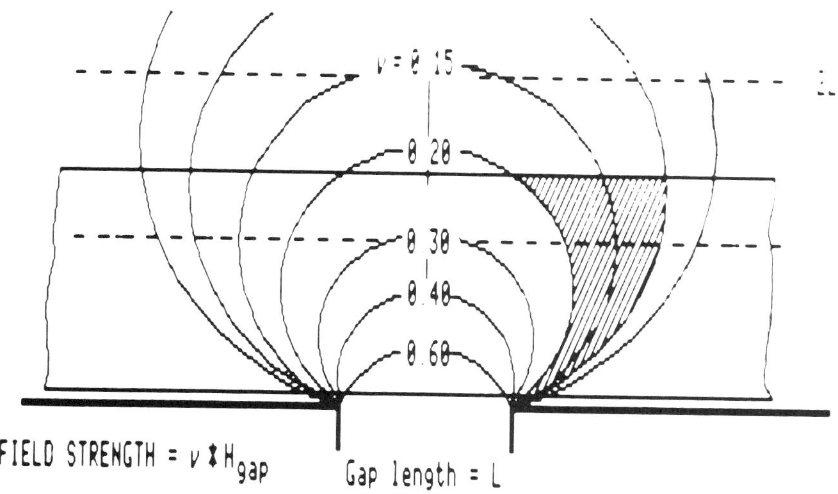

Fig. 5-12. Transition zone recorded on a longitudinal anisotropic coating.

Returning now to the writing of the transition between two bits we should find the field contours corresponding to $H_r*(1 + \text{and} - \Delta h_r/2)$. This is illustrated in Fig. 5-12.

At this point the reader may have some concern about when to use H_c or H_r, and which one of the three SFDs. H_c is the field strength required to reduce a saturated material's flux density B to zero, while H_r is the field strength required to reduce the material's magnetization J (or M) to zero, and H_r is typically a few percent larger than H_c.

The recommended value for the SFD is Δh_r which corresponds to the switching of 50 percent of the particles. Use it with the mean value H_r, or H_c increased by 5 percent.

As the coating moves to the right, out of the write field, the right transition zone remains magnetized as shown, while the zone written on the left side of the head field is erased as it enters the region of higher field strength above the gap. In general, *we do not care what happens on the ingoing side of the recording field*; the magnetization in that region is always changed when it crosses the higher field above the gap.

VECTOR MAGNETIZATION; THICK COATINGS. DETERMINATION OF THE TRANSITION ZONE LENGTH BY Δh_r

The cylinder serves well for all longitudinal only models, but it became apparent in the late sixties/early seventies that the perpendicular magnetization component should be included. One method would be to consider a longitudinal and a perpendicular write process separately, and then—somehow—combine them to a final transition. We will do that later on in the chapter; for now we will concentrate on finding the dimensions of the transition zone Δx.

It is useful to recall Fig. 3-28 that showed little directional change in a particulate coating's coercivity. In other words, its switching properties are primarily related to the field's strength, not to its direction nor to its anisotropy (which may not be true for solid, metallic coatings).

We can therefore again use the field magnitudes as criteria for finding the inside and outside surfaces of the transition zone. A chart for contours of constant field strength ($\sqrt{H_x^2 + H_y^2}$ = constant) was made by Westmijze (see ref.), and is reproduced in Figs. 5-13 and 5-14. *They are probably the most useful illustrations in this book for budgetary estimates of write conditions and resulting transition zones.*

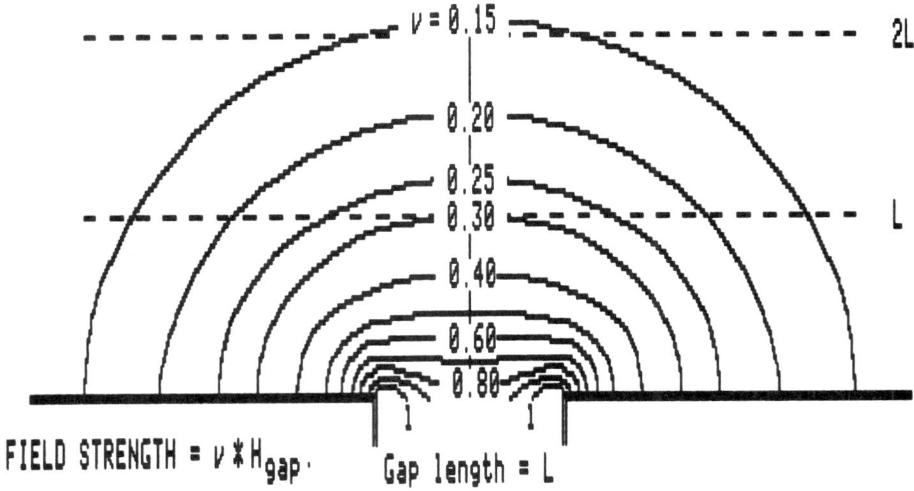

Fig. 5-13. Chart of constant field strength contours in front of gap (after Westmijze).

First of all, both charts have some horizontal lines spaced one or more *gap lengths* away from the head interface, i.e., lines at L, 2L, 3L etc. One will start with the knowledge of the gap length L (often dictated by the read resolution, i.e., L = 0.6 – 0.9 * BL, where BL = bit length). The coating thickness c and the head/coating distance d allows for drawing the front and back of coating, in proper scale to the gap length by utilizing the horizontal, evenly spaced lines. Next the write level is set, with a 100 percent level defined as that level where the contour corresponding to H_c is tangent to the coating back side.

Next observe the numbers marked ν. They represent the field strength on a given contour, relative to the deep-gap field H_g. This is the field we calculated in formula (4.29).

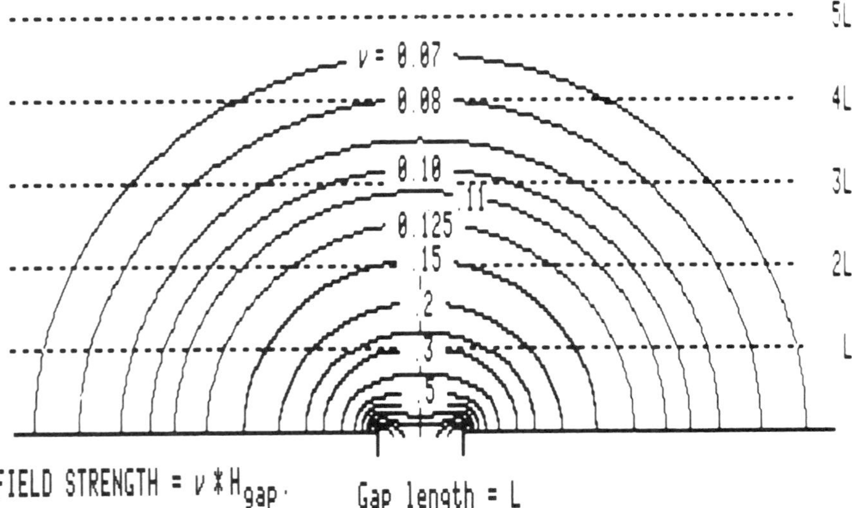

Fig. 5-14. Chart of constant field strength contours in front of gap (after Westmijze).

Finally note that most of the contours are almost circles, quite identical to the actual field lines from Fig. 4-24; only near the gap are there discrepancies. That is of little concern since most write fields have a strength that places the write zone at contour ν = 0.6, or smaller. From the foregoing we can conclude that the field directions, that are tangents to the field lines in Fig. 4-24, are also almost tangents to the magnitude contours in Figs. 5-13 and 5-14.

The transition zone Δx is now determined by drawing the contours that correspond to the fields of strengths equal to $H_r*(1 \; +\text{or}- \; \Delta h_r/2)$. *For the length of the transition zone we will choose the dimension in the very front of the coating.*

Example 5.1: Determine the length of the transition zone that results from writing bits on a coating with a SFD (Δh_r) of 0.40. The coating has a thickness of 5 μm (200 μin), the head/media spacing is d = 0.1 μm, and the gap length is L = 1.7 μm. The write level is 100 percent. Also determine the maximum FRI.

Solution: In Fig. 5-15 the coating is outlined as lines that are spaced 3L and L/17 away from the head surface. The recording zone contours are found by interpolation: The center contour that is tangent to the backside has ν = 0.105, and the lines outlining Δx are therefore located at 0.105 +or− 20 percent, or ν = 0.126 and ν = 0.084.

The actual length of Δx is determined by using the gap length dimension L as a scale = 1.7 μm. We find Δx = 1.2 * L = 2.04 μm = 82 μin. The maximum packing density, where transition follows transition, is (1,000,000 μin/in)/(82 μin/flux reversal) = 12,200 FRI.

Example 5.2: Determine the length (and shape) of the transition zone in a write process onto a 0.5 μm thick coating, spaced 0.5 μm away from the head, that has a 1 μm long gap. The write level is set at 140 percent, and the SFD is 0.50. Also, determine the required gap field strength if H_t is 600 Oe.

Solution: A 100 percent write level requires a gap field of H_g = 600/ν = 600/0.31 = 1935 Oe. At 140 percent write level the gap field must therefore be 1.40 * 1935 = 2709 Oe. The corresponding contour is found at ν = 600/2709 = 0.2214 or approximately 0.22. The transition zone is outlined by ν = 0.22*(1 +and− 0.25) = 0.22 +and− 0.055 = 0.275 and 0.165.

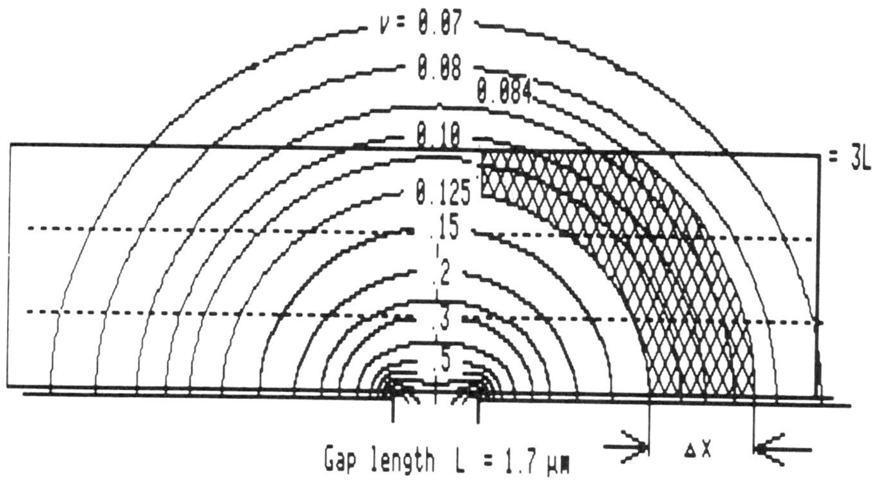

Fig. 5-15. Recorded transition zone Δx on a 5 μm, (200 μin) thick coating. SFD = 40% = +− 20%. Write level = 100% (see example 5.1).

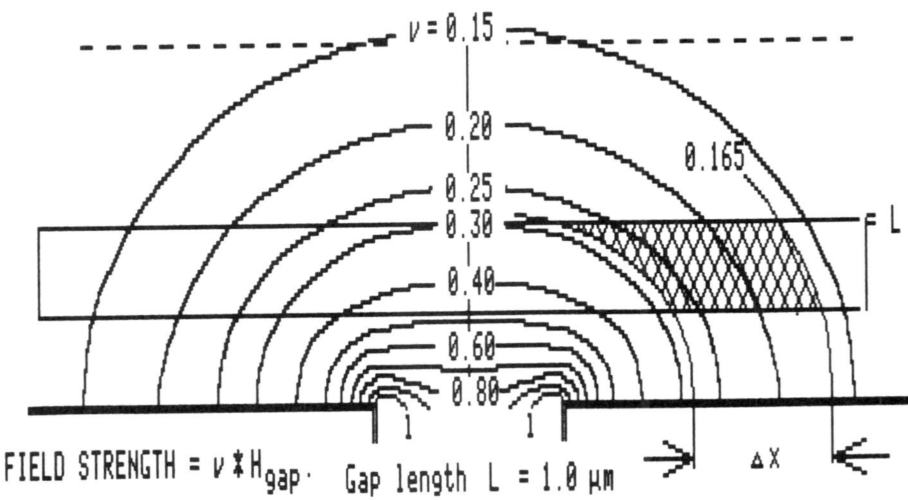

Fig. 5-16. Recorded transition zone Δx on a 0.5 μm (20 μin) thick coating. SFD = 50% = + − 25%. Write level = 140% (see example 5.2).

The transition zone is drawn in Fig. 5-16. Δx is approximately 0.9 μm, and the direction of the magnetization varies between 40 degrees in the back, and 60 degrees in the front.

The required gap field strength is $H_g = H_c/\nu = 600/0.22 = 2{,}709$ Oe = 215,575 A/m = 216 kA/m.

COMPUTATIONS OF THE TRANSITION ZONE, USING Δh_r

The steps made in examples 5.1 and 5.2 certainly lend themselves to computational methods. A short program, with graphics output, and written in TrueBasic, is included in this chapter. A program listing is found in Appendix A at the end of the book.

Let us first evaluate the effects of a variation in the SFD. That a larger SFD causes a longer Δx is obvious from the discussions above. But the pictorial presentations we made in Figs. 5-12, 5-15, and 5-16 leave the reader with an overly simplistic impression of an abrupt transition zone, even though we know that the crosshatched area represents only 50 percent of the particles involved in making up the transition between alternating magnetizations. The mathematical model for Δx has long been established as an arctangent function, and that is not revealed by the above referenced figures.

Let us see what happens when we write at a 70 percent level with SFD's that vary from 0 to 40 percent of H_c. This is illustrated in Fig. 5-17. The 0 percent SFD gives the ideal transition (although it follows the curvature of the field contour, rather than being straight and perpendicular to the coating surface). The unrecorded portion of the coating (in this case 30 percent, from the back) is shown as a checkered pattern, indicating that there are equally many particles magnetized at either polarity; therefore, that portion of the coating does not contribute to the read flux.

As the SFD is increased from 0 to 10, 20, and 40 percent, the picture becomes rather grim. At this point it should be mentioned that a modern, high quality coating has a *SFD in the vicinity of 30 percent*—which does not look too good, does it? The visual impression is supported by a program subroutine that keeps track of the magnetized levels on the front of the coating, and the results are plotted in Fig. 5-18.

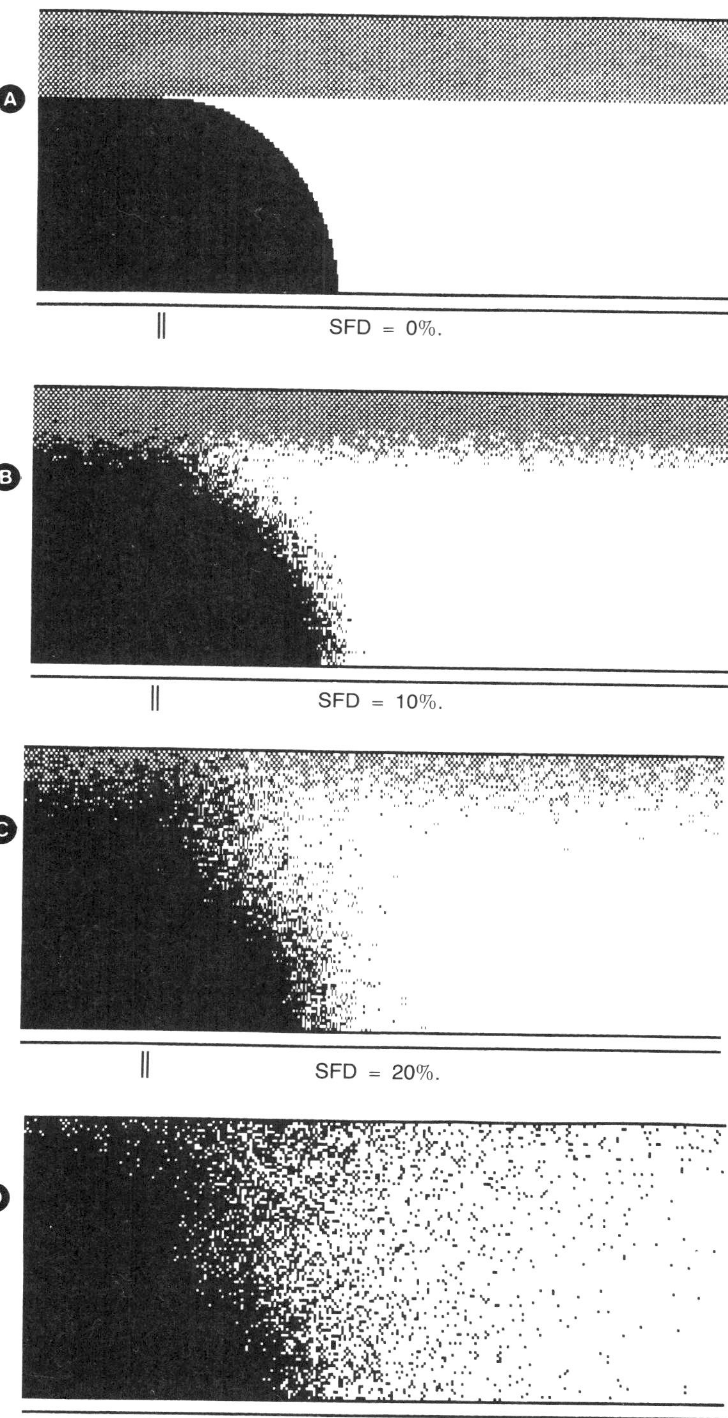

Fig. 5-17. Computed transition zones. Notice spread in magnetization with increasing switching field distribution SFD. This spread results in a gradual change in magnetization across Δx, see Fig. 5-18.

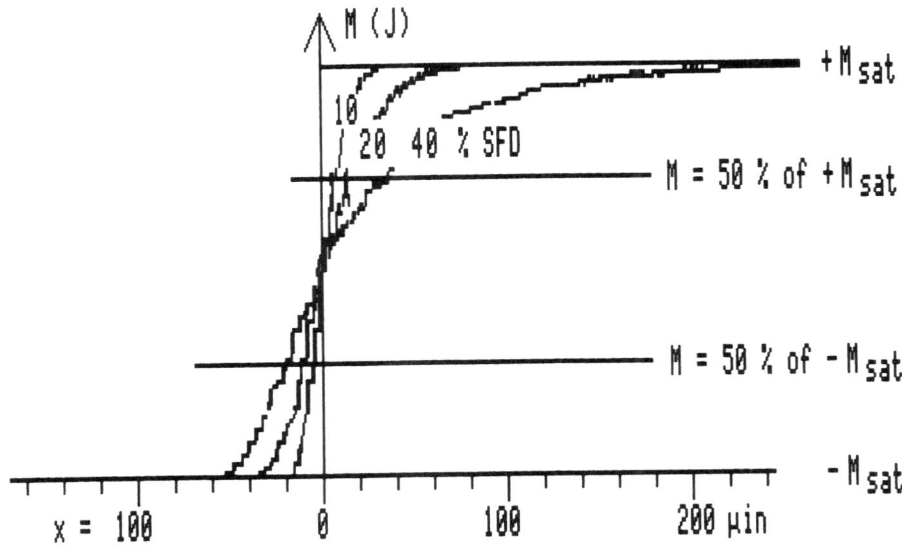

Fig. 5-18. Change in magnetization M (J) across Δx, in the front of the coating. Write level = 70%. Actual transitions are shown in Fig. 5-17.

Two horizontal lines are drawn, corresponding to SFD = Δh_r (and the 50 percent switched particles' measure). The magnetization transitions are indeed seen to follow arctangent like curves. For a 40 percent SFD we read that $\Delta x = 54\ \mu$in. Had the level been 100 percent (as in example 5.1) then $\Delta x = 77\ \mu$in; that compares with 82 μin by the graphical method.

For a given medium the transition zone length will be proportional to the write (record) level. This is illustrated in Fig. 5-19, showing the computations for a coating with SFD = 30 percent. These images graphically display the fact that increasing write levels results in longer transition zones.

The gradual change in magnetization across the transitions is shown in Fig. 5-20. At a write level of 70 percent (and SFD of 30 percent) we find $\Delta x = 38\ \mu$in.

Figure 5-21 summarizes our findings from Figs. 5-18 and 5-20. Note that the graphs are for a 200 μin thick coating—but the program in Appendix A allows the reader to explore any other record geometries; and Fig. 5-21 may be scaled for some applications.

COMPUTATIONS OF WRITE RESOLUTION

It is also necessary to appraise what happens when transitions are written close to one another. The program provides a pictorial and a quantitative answer, as shown in Fig. 5-22. The medium has thickness c, the bit length is BL = c/4, and the write level is 70 percent (i.e., underwrite).

The levels of magnetization in the front of the coating are shown to the right of each illustration of the magnetization pattern. The resolution suffers with increasing SFD, and the overall amplitude decreases. This loss is discussed at the end of this chapter, after an introduction to AC-bias recording. The interested reader will find a detailed analysis in the more advanced Chapter 19 on the write process.

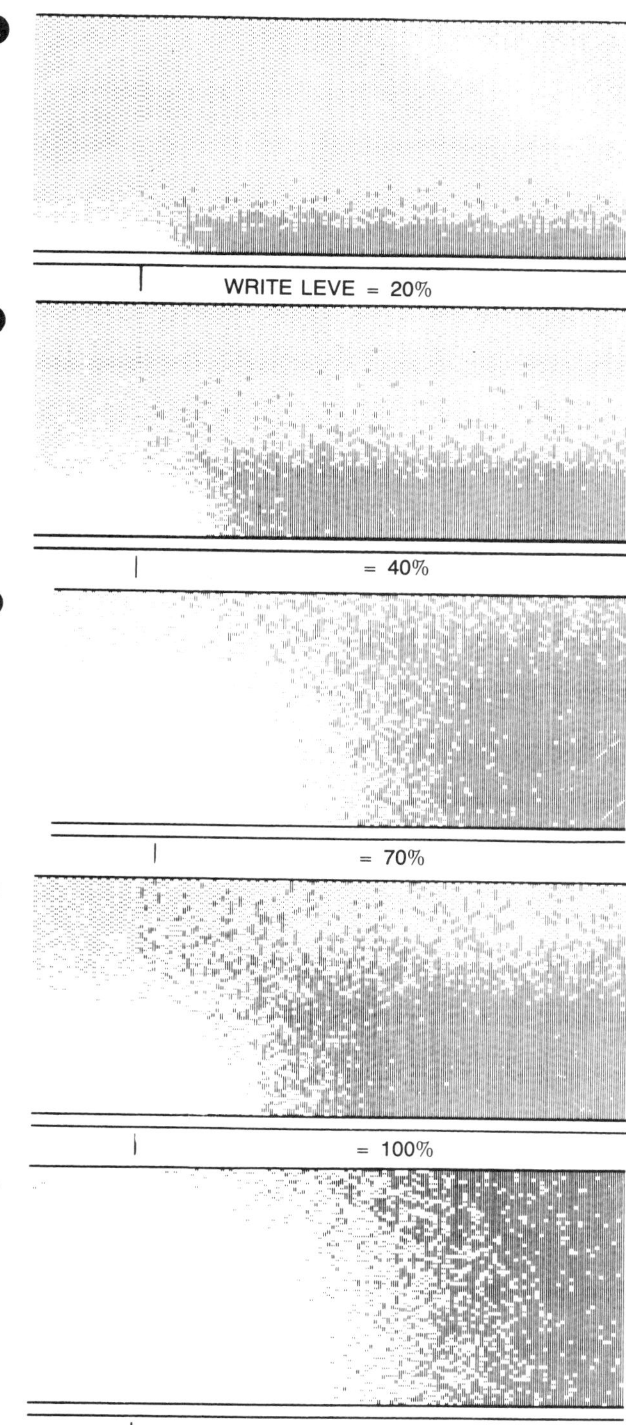

Fig. 5-19. Computed transition zones for SFD = 30 percent. Notice spread in magnetization with increasing write level. This results in a longer transition Δx, see Fig. 5-20.

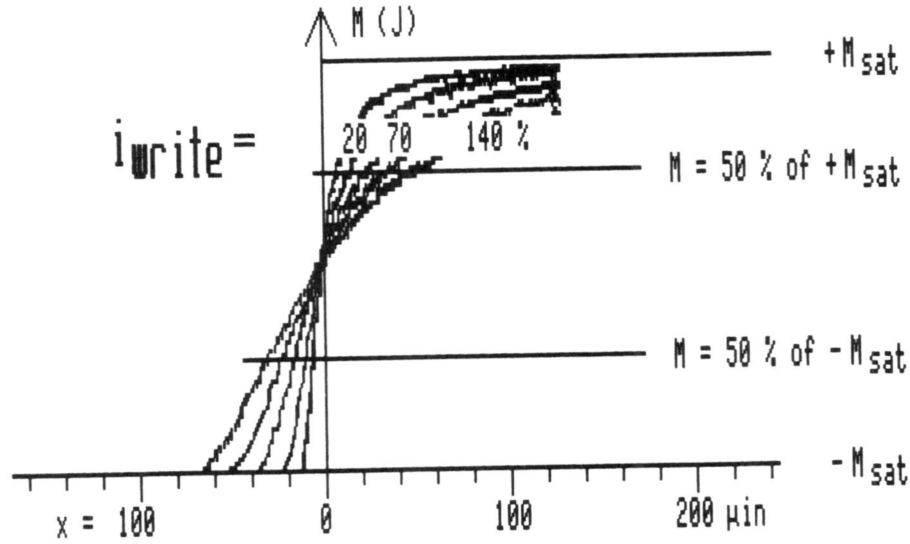

Fig. 5-20. Change in magnetization M (J) across Δx, in front of the coating. Switching field distribution SFD = 30% (see transition in Fig. 5-19).

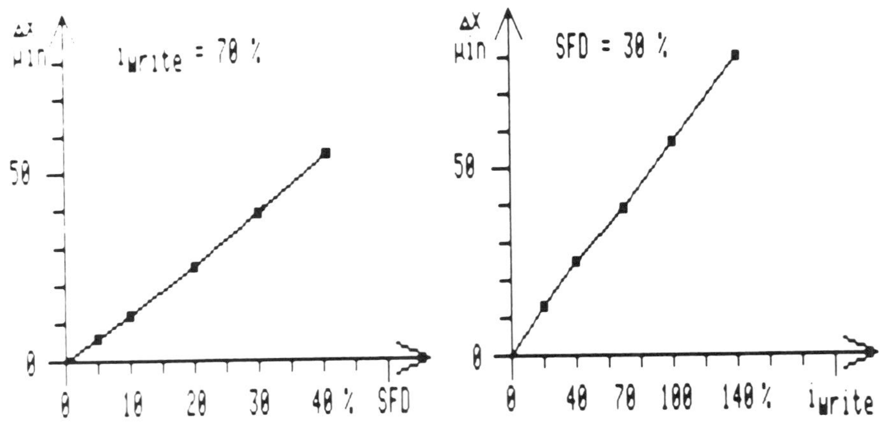

Fig. 5-21. Graphs showing increase in Δx, for increases in the SFD as well as the write level. Data is from curves in Figs. 5-18 and 5-20. c = 200 μin.

COMPUTATIONS OF ERASURE AND OVERWRITE

The program used in the prior section is readily modified to illustrate the effect of erasure. An ideally saturated coating can be displayed as alternating bits, with straight, perpendicular and thin transitions. We will expose this magnetized pattern to a head field that reaches 50 percent into the coating (i.e., with a field strength equal to the coating coercivity H_C halfway through the coating).

The result is shown in Fig. 5-23, assuming a SFD = 30%. It is seen that some high coercivity particle in the front half of the coating are not affected while, on the other hand, many low coercivity particles in the back half are erased.

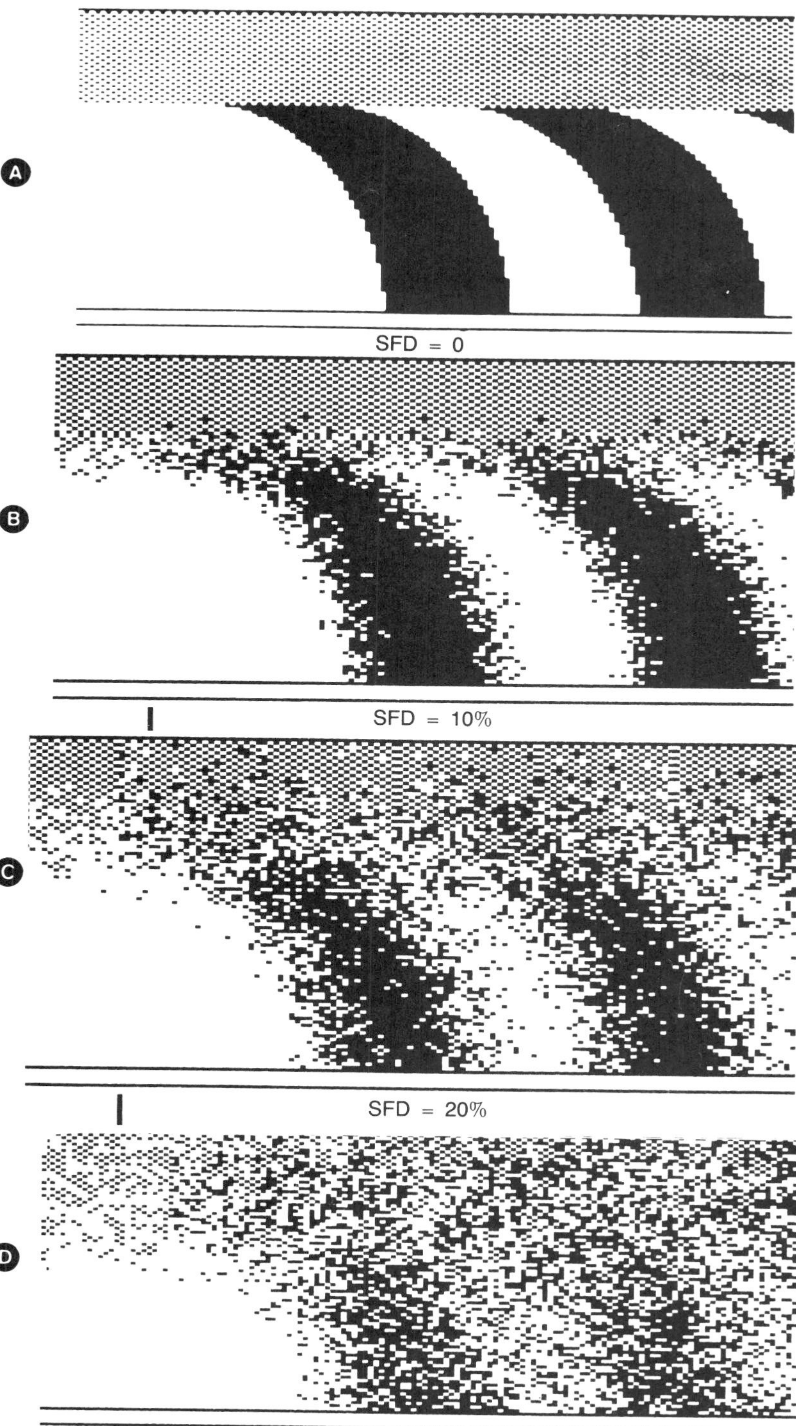

Fig. 5-22. Resolution in a bit pattern with BL = 25% of the coating thickness, and write level = 70%.

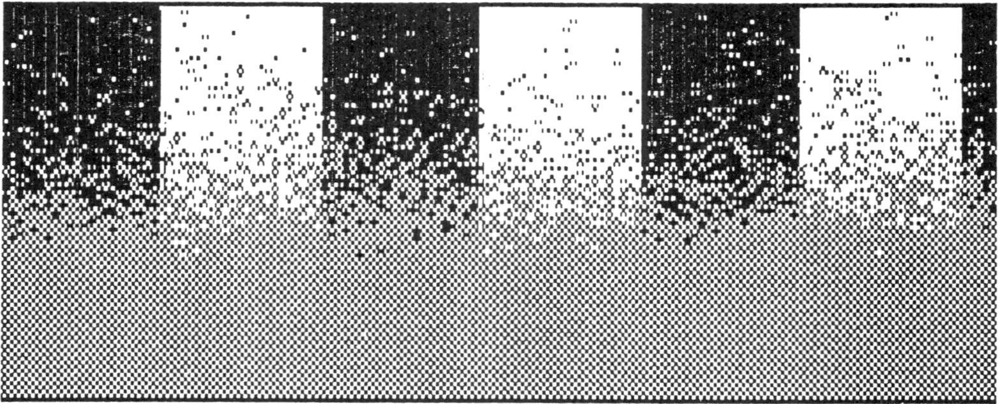

Fig. 5-23. 50% erasure of recorded coating with a SFD = 30%.

If an erase level corresponding to 100 percent was used one can extrapolate that many high coercivity particles in the coating backside remain unaffected, and the erasure is incomplete. It is generally necessary to increase the erase (or overwrite) field strength to 140 to 150 percent in order to get more than 99 percent of all the particles erased. This corresponds to an erasure of −40 dB.

The reader is referred to the section on erase heads in Chapter 8 for details on the difficulties in achieving a complete erasure during overwrite, i.e., where new data erase old.

RECORDING WITH AC-BIAS

In the late twenties, it was discovered that the addition of a high-frequency signal to a sound or data signal greatly improved the quality of magnetic recording. This technique is called AC-bias recording and is used in all the audio and instrumentation recorders of today. The action of AC-bias is not easily explained.

An elementary understanding is obtained by studying the effect upon a magnetic material by a decaying AC-field. Figure 5-24A illustrates how a magnetic material is demagnetized by an alternating field that slowly decays to zero. Any initial remanence B_r' is gradually reduced through each cycle of the field and finally reaches a value of zero. The figure uses only a couple of cycles to illustrate the principle; in reality hundreds of cycles are required for the decaying signal.

This is the principle used in *erasing* a tape, where the tape is subjected to a strong alternating field that slowly decays to zero. If a data DC-field is present during the field's decay, then the alternating cycles will end at $B_{r\text{-data}}$ (Fig. 5-24B), leaving the tape strongly magnetized.

If the DC-field, representing a digital data signal to be recorded, varies in strength, or if it represents an analog signal (varying at a slower rate than the superimposed AC-field) then the remanence will vary accordingly. This remanence will be left where the coating leaves the record head field; both of the field magnitudes decay with distance (the AC-field as well as the slower varying signal field).

For a certain value of the superimposed AC-bias field this recording process is *very linear, with a high signal-to-noise ratio*. Therefore, high-quality recordings are obtained by adding an AC-bias field to the signal field; the frequency of the AC-bias must be several times higher than the highest audio (or data) frequency to be recorded to avoid beat frequency appearing in the recording.

The AC-bias field also further reduces the tape background noise, since the alternating field causes the tape to leave the record head in a neutral condition when no signal is present. Tape coating irregularities will still cause noise, but to a much lesser degree than when recording with DC-bias, where the tape is recorded essentially as one long magnet in the absence of a signal field. Any coating irregularities show up as a noise output caused by a change in the external tape flux.

A distorted AC-bias waveform will cause excessive tape noise since the distortion very likely will contain a DC-component. The better recorders use a push-pull oscillator to develop bias, which produces an AC-field that is largely free of any DC-component and second harmonic distortion. The *amplitude* of the AC-bias field plays a major role in the quality of a recording. Each particular tape (i.e., coercivity, formulation and/or tape thickness) requires a certain bias level for optimum performance. The following general rules apply:

- A bias level that is *too low* will result in a noisy and a highly distorted recording with excessive high-frequency response; this condition is called *underbias*.

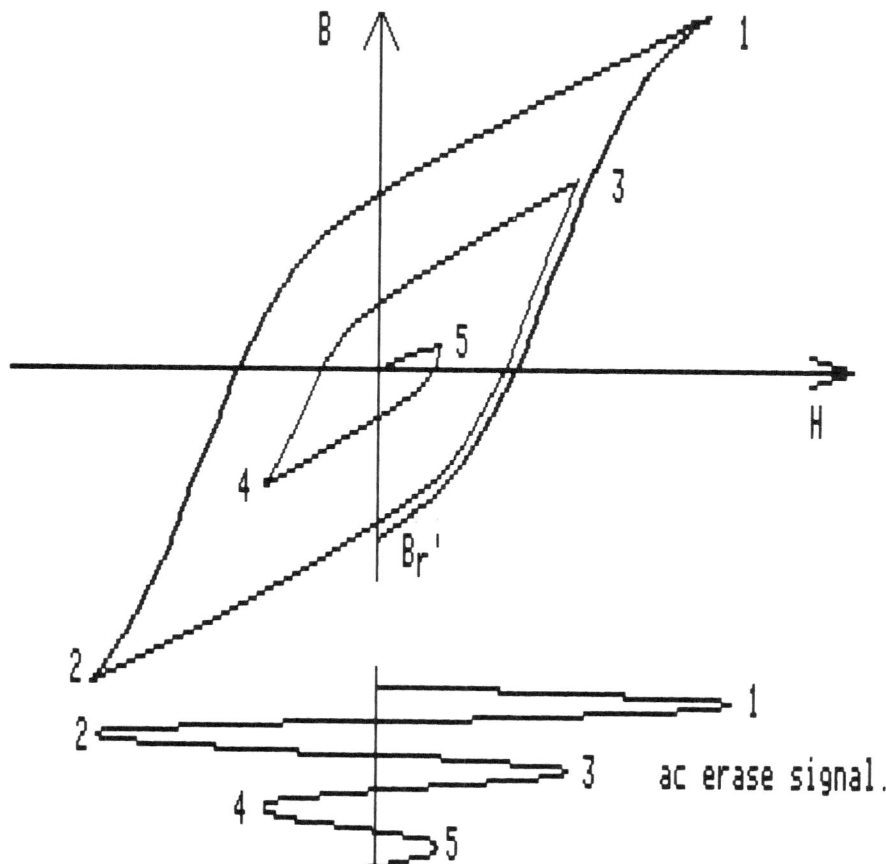

Fig. 5-24A. AC erasure of magnetic material.

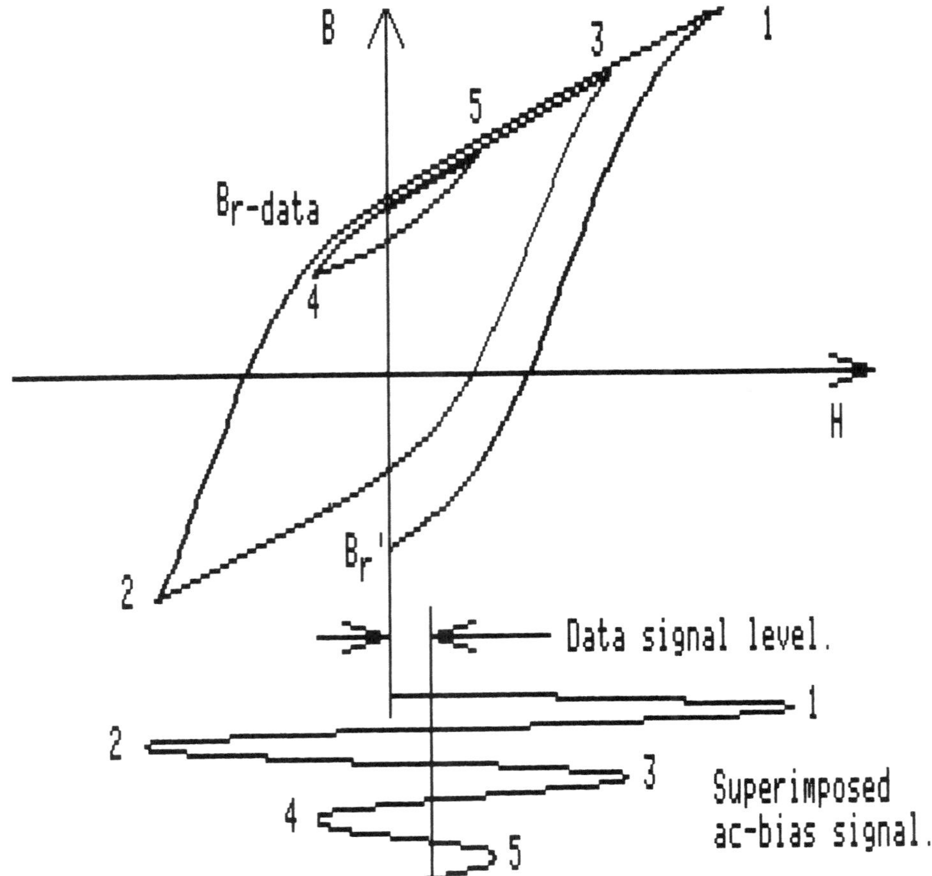

Fig. 5-24B. Recording with AC-bias, simplified.

• On the other hand, a *high bias level* will result in a quiet recording, but with a noticeable drop in the high-frequency level. This condition is called *overbias* and the bias field is now so strong that the short wavelength resolution suffers.

THE RECORDING ZONE; RECORDING LOSSES

The freezing of a particle's magnetization occurs when it crosses the head field line that has a strength equal to its coercivity h_{ci}. This leads us to the concept of a *recording zone* which, speaking in terms of field lines, has a width of SFD$*H_c$ around the line corresponding to a strength equal to H_c. This is illustrated in Fig. 5-25, where the recording zone is shown for different bias levels. The recording zone therefore has the same geometry as the transition zone discussed earlier.

The upper figure shows a recording where the recording zone reaches through the magnetic coating of the tape. When the tape approaches the gap, it enters a magnetic field of increasing strength, and as it passes over the gap, most particles ''flip'' back and forth under the influence of a saturating bias field. When the coating passes through the trailing edge of the gap, it goes through the recording zone and becomes permanently magnetized in accordance with the presence of a signal field H_{signal}.

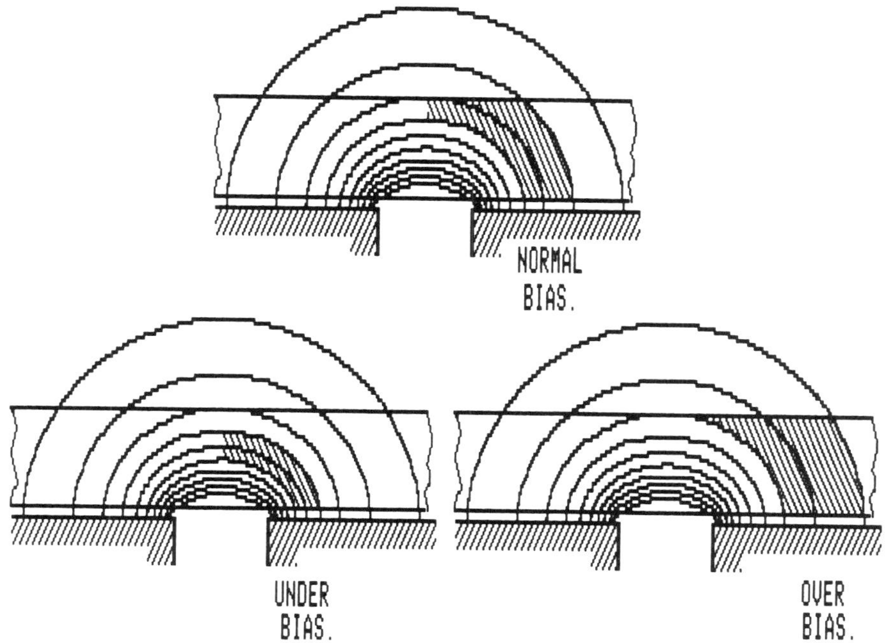

Fig. 5-25. Recording zones in AC-bias recordings (normal, under- and overbias).

The length of the recording zone determines the shortest wavelength that can be recorded on the tape for a particular tape speed. If the polarity of the data field changes 360 degrees as the tape moves through the recording zone, the net recording will be zero. This null occurs when the recorded wave length equals the length of the recording zone; this is not a very well defined zone, and will be discussed further in Chapter 19.

The length of the recording zone in a typical cassette audio recorder is normally between 75 and 200 μin for *underbias* and *overbias* settings respectively. The band edge losses (highest frequencies), due to the recording zone length, range from a few dB to 10-20 dB. They are caused by the SFD, and are not a result of bias or self-erasure (whatever that may be?). Modern tapes with a narrow range of switching fields reduce this loss.

Another loss during recording occurs when the record head is contaminated by debris and the tape is therefore spaced away from the recording head. Then only a portion of the coating is magnetized, and the loss may be analyzed using Figs. 5-15 or 5-16. A general formula for this loss as a function of the spacing is not available (as it is in playback, see next chapter).

In order to reduce the recording losses it is necessary, therefore, to have a recording zone as short as possible. This can be achieved by reducing the bias level, as shown in Fig. 5-25. But as we reduce the bias level, the tape is no longer recorded throughout its full thickness and the level at long wavelengths is reduced and the signal is thus highly distorted.

During the past years considerable effort has been devoted to the design of record heads with a narrow recording zone; some success has been reported with pole shaping (thin film heads), multiple poles (the cross field head), very-high-frequency eddy current field shaping, and perpendicular pole heads on a magnetic coating with a highly permeable undercoat. Simultaneously, work continues in developing tape and disk coatings with smaller SFD. The interested reader is referred to later chapters in this book, and their literature listings.

REFERENCES TO CHAPTER 5

Bauer, B.B., and Mee, C.D., "A New Model for Magnetic Recording," *IRE Trans. Audio*, Jan. 1961, Vol. AU-9, No. 1, pp. 139-145.

Tjaden, D.L.A., and Leyten, J., "A 5000:1 Scale Model of the Magnetic Recording Process," *Philips Technical Review*, Sept. 1963, Vol. 25, No. 11/12, pp. 319-329.

Iwasaki, S.I., and Takemura, K., "An Analysis for the Circular Mode of Magnetization in Short Wavelength Recording," *IEEE Trans. Magn.*, Vol. MAG-11, No. 5, Nov. 1975, pp. 1173-1176.

Iwasaki, S.I., and Nakamura, Y., "An Analysis for the Magnetization Mode for High Density Magnetic Recording," *IEEE Trans. Magn.*, Vol. MAG-13, No. 5, July 1977, pp. 1272-1278.

Wilson, D.M., "Effects of Switching Field Distributions and Coercivity on Magnetic Recording Properties," *IEEE Trans. Magn.*, Sept. 1975, Vol. MAG-11, No. 5, pp.1200-1202.

Koester, E., "Recommendation of a Simple and Universally Applicable Method for Measuring the Switching Field Distribution of Magnetic Recording Media," *IEEE Trans. Magn.*, Vol. MAG-20, No. 1, Jan. 1984, pp. 81-83.

Bertram, H.N., "Geometric Effects in the Magnetic Recording Process," *IEEE Trans. Magn.*, May 1984, Vol. MAG-20, No. 3, pp. 468-478.

Koester, E., Jakusch, H., and Kullman, U., "Switching Field Distribution and A.C. Bias Recording Parameters," *IEEE Trans. Magn.*, Nov. 1981, Vol. MAG-17, No. 6, pp. 2550-2552.

Dressler, D.D., and Judy, J.H, "A Study of Digitally Recorded Transitions in Thin Magnetic Films," *IEEE Trans. Magn.*, Sept. 1974, Vol. MAG-10, No. 3, pp. 674-677.

BIBLIOGRAPHY TO CHAPTER 5

Middleton, B.K., "Recording and Reproducing Processes," in *Magnetic Recording*, Vol. 1, Ed. by Mee and Daniel, McGraw-Hill, 1987, pp. 22-97.

Mee, C.D., "Applications and Limitations of the New Magnetic Recording Model," *IRE Trans. Audio*, Jan. 1962, Vol. AU-10, No. 1, pp. 161-164.

Daniel, E.D., Axon, P.E., and Frost, W.T., "A Survey of Factors Limiting the Performance of Magnetic Recording Systems," *Jour. AES*, Jan. 1957, Vol. 5, No. 1, pp. 42-52.

Westmijze, W.K., "Studies on Magnetic Recording: pt. III. The Recording Process," *Philips Res. Repts.*, July 1953, Vol. 8, pp. 245-255.

Westmijze, W.K., "The Principle of the Magnetic Recording and Reproduction of Sound," *Philips Technical Review*, Sept. 1953, Vol. 15, No. 3, pp. 84-95.

Chapter 6
Playback Theory

A recorded tape is, during playback, moved across a read (also known as reproduce or playback) head, shown in Fig. 6-1. The head's core is fabricated from a magnetically soft material with high permeability and the magnetic flux from the tape links through this core and induces a voltage in its winding.

We can easily find the signal voltage from the playback of a sinusoidal signal recorded on a very thin coating in contact with the read head. The read voltage e is, from Faraday's law:

$$\begin{aligned} e &= -n\, d\psi/dt \\ &= -n * \psi_m * \omega * \cos \omega t \\ &= -n * \psi_m * 2\pi f * \cos(2\pi x/\lambda) \end{aligned} \qquad \textbf{(6.1)}$$

where

n = number of turns on the core
ω = $2\pi f$
f = signal frequency in Hz
ψ_m = peak flux level from coating in Wb
ψ = $\psi_m \cos \omega t$
= $\psi_m \sin(2\pi x/\lambda)$
x = position along coating in m ($x = vt$)
λ = recorded signal wave length, in m
v = head-to-coating speed in m/sec

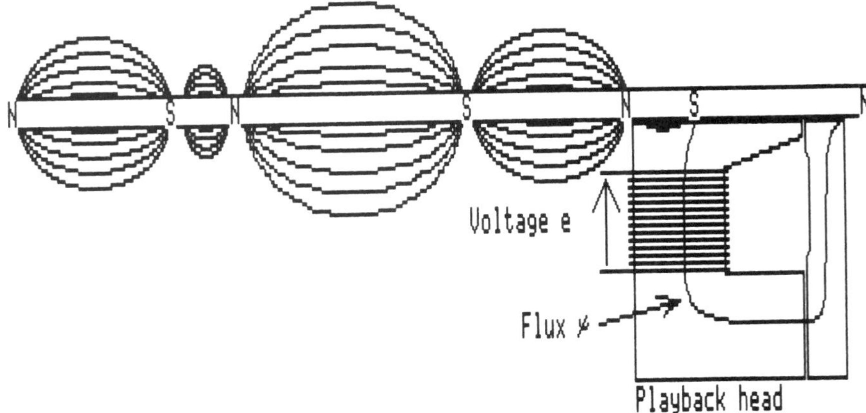

Fig. 6-1. Playback (read) of recorded data.

Hence the voltage increases proportional to ω, or the frequency f, at a rate of 6 db/octave. This is not observed in practice; it is rather like those in Fig. 6-2. The difference between a theoretical 6 dB/octave rise, and the actual voltage is made up from several losses, and in this chapter we will examine these playback losses:

- Spacing loss.
- Coating thickness loss.
- Gap length loss.

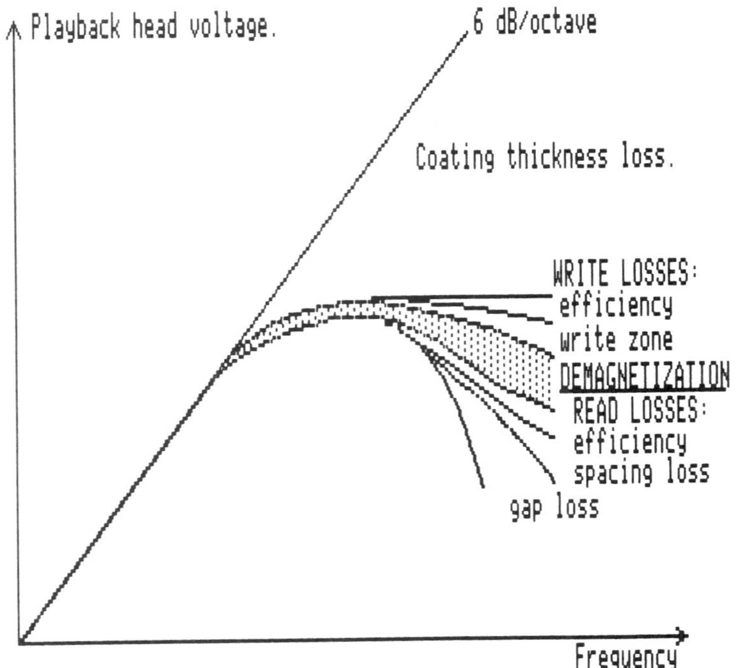

Fig. 6-2. Summary of losses in a magnetic write/read system.

- Gap alignment loss.
- Reproduce core loss.

The read, or playback, voltage from complicated magnetization patterns can only be calculated by means of a digital computer. We have shown a formula for the read voltage only for purely longitudinal magnetization recorded on a very thin coating with sine or cosine wave patterns. By writing the flux variations along the coating as $\psi = \psi_m \sin(2\pi x/\lambda)$, and substituting this flux into Faraday's formula we have the resulting ω in formula (6.1), which is the *loss free playback voltage*.

The voltage is, to no surprise, proportional to the number of turns and the flux level on the tape. It is also proportional to the recorded frequency, and would, therefore, result in a very unbalanced reproduction of the input signal. For example, the higher frequency content of an input signal would be read out at a higher voltage than would lower frequencies. Some electronic equalization will be necessary to restore the signal, a topic we will discuss in Chapter 22.

FLUX LEVEL THROUGH HEAD CORE; THICK COATING

The flux from a recorded tape is rarely purely sinusoidal, and is always comprised of a mix of longitudinal and perpendicular magnetizations. It it also apparent that equal magnetizations in front and back of a coating result in different flux levels at the read head. How does the flux actually link to and through the head? An answer to this question is no easy task when the coating thickness is comparable to the gap length.

Take for example a simple small magnetic dipole, placed in front of the read head. When alone, the field strength from a dipole is readily calculated, but the presence of the core greatly modifies the field pattern. Elaborate calculations, using magnetic field imaging and mapping, have been carried out giving excellent results in agreement with measurements, but with a loss of understanding during the mathematical derivations. Moreover the reader is reminded of Fig. 5-2 showing how the total coating magnetization can be thought of as being a large number of dipoles, each contributing to flux through the head.

If, however, we can find a method for calculating the flux into the head core from just one such dipole element, then we can sum up the fluxes from all elements in the vicinity of the head to account for the total flux.

We will therefore consider the simple system shown in Fig. 6-3. Recall that a magnetic dipole can be characterized by its *magnetic moment* $m = pl$, and that this is equivalent to the *magnetic dipole moment* $j = iAn$ from a small coil (formulas (3.4) and (3.16)). Let us therefore replace the small magnetic moment with the small coil.

Further: a magnetic head can be modelled by a small coil, of length equal to the gap length, and with n turns. Let us therefore replace the head with such a coil.

We have now reduced the problem to that of finding the flux coupling from one coil into another nearby coil. Finding an answer to this problem may be straightforward, and we must then substitute the magnetic dipole moment back into the expression we derive.

The magnetic circuit of two coupled coils is an item we all come across. Take for example the small power supply you connect to 110 Vac, which then delivers 6 volts to your transistor radio or tape recorder. Inside the power supply is a small *transformer* that changes the 110 V into 6 volts (which then is rectified to provide the 6 volts dc). The transformer is an example of two closely coupled coils.

One may also connect 6 volts ac backward through the transformer, and obtain 110 volts. This property of being able to act in both directions is called *reciprocity*. We call the device a *reciprocal transducer*; this property is found in many devices, for example piezo-electric crystals: with properly mounted electrodes you will find that an applied voltage will bend the crys-

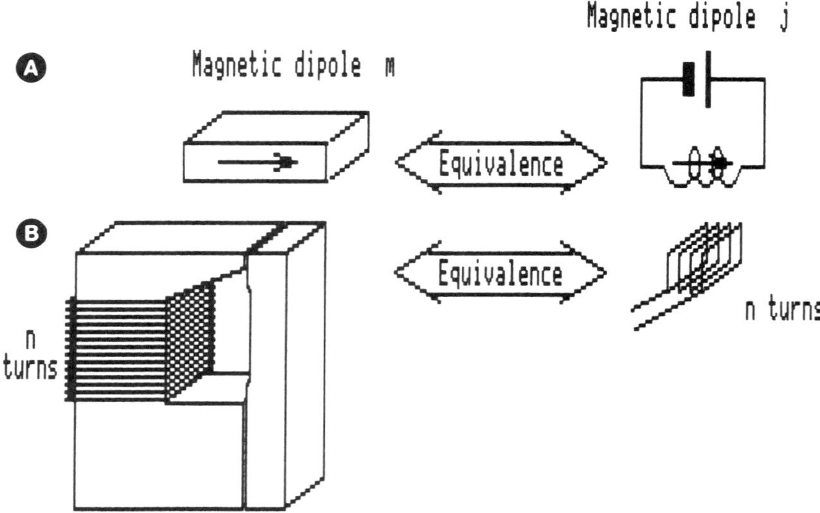

Fig. 6-3. Replacement of (A) a magnetized bit with its equivalent coil; (B) a magnetic ring core head with its equivalent coil.

tal. If you, however, bend the crystal, then a voltage is produced between the electrodes. Another example is the magnetostrictive transducers used in ultrasonic cleaning tanks, and in underwater sound producing/receiving devices. The loudspeaker in an intercom system is reciprocal; it will produce sound as well as act as a microphone.

A magnetic ring core head will record as well as read: it is a reciprocal transducer. This fact will allow us to proceed with finding the read flux.

Let us consider two coils, L_1 and L_2, with a mutual coupling M (see Fig. 6-4). A current i_1 through L_1 produces a flux:

$$\psi_2 = M * i_1 \qquad \textbf{(6.2)}$$

through L_2; and a current i_2 through L_2 produces flux:

$$\psi_1 = M * i_2 \qquad \textbf{(6.3)}$$

through L_1. When comparing Fig. 6-3 with 6-4 we see that the answer we seek is formula (6.2), flux through the head. The current i_1 is part of the dipole moment, and the unknown is therefore M.

The coupled coils we are interested in are shown in Fig. 6-5. The left and right drawings correspond to those in Fig. 6-4.

By expressing (6.3) in words we have: At the location of coil 1 a field with flux ψ_1 is produced by the current i_2, and its magnitude is $M * i_2$. This sounds familiar, and brings to mind the name Karlqvist. He found the formulas (Fig. 4-24) for the field produced at points (x,y) in front of a record head, fed with the current i (i_2).

We are seeking the flux, and can write:

$$\begin{aligned}\psi_1 &= B_1 * A_1\\ &= H_1\mu_o A_1\\ &= (H_1/i_2)\, i_2\, \mu_o A_1\\ &= H_s i_2 \mu_o A_1 \qquad \textbf{(6.4)}\end{aligned}$$

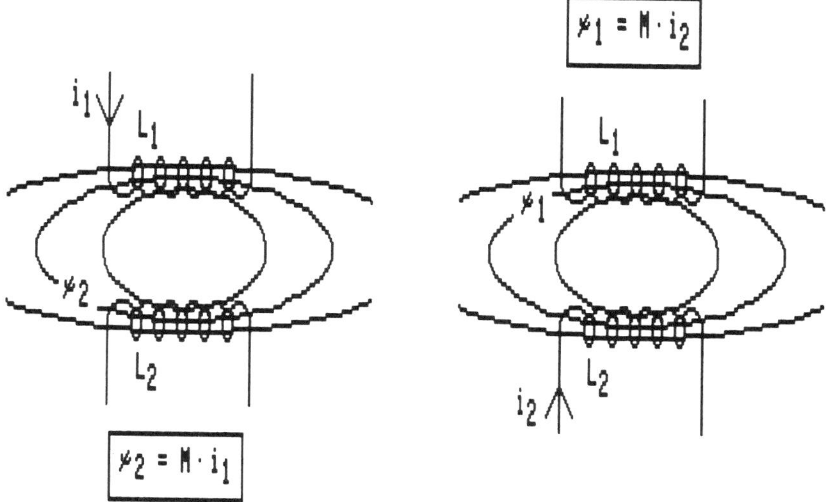

Fig. 6-4. Mutual coupling between two coils.

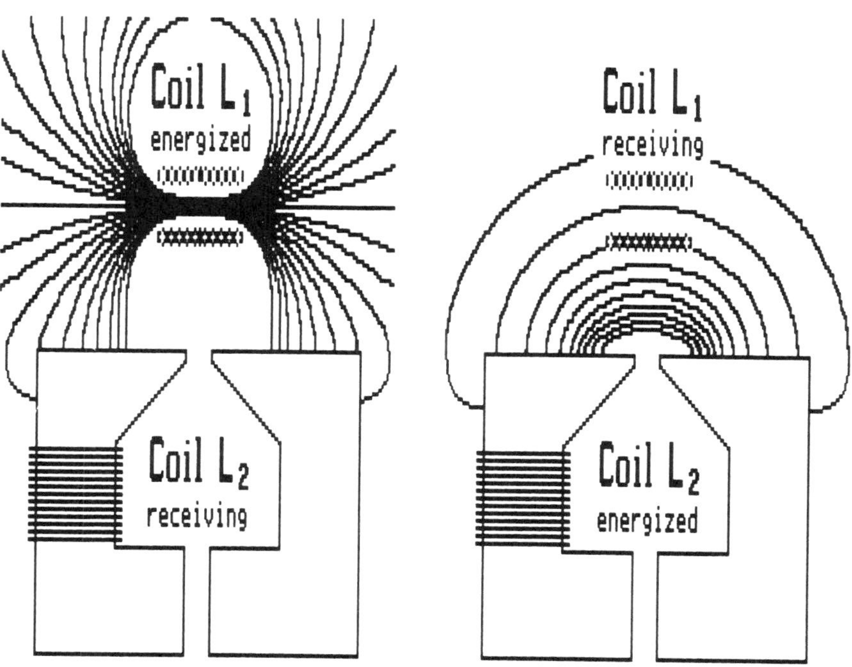

Fig. 6-5. Mutual couplings. Left: coating magnetization to head; right: head field to coating.

Comparing (6.3) and (6.4) we see then the mutual coupling coefficient equals $H_{sens} * \mu_o * A_1$, and we obtain our intermediate result by substituting into formula (6.2):

$$\psi_2 = H_s \mu_o A_1 i_1$$

The dipole moment for a single turn loop equals j = Ai, and we may now further substitute m for j. The result is a net head flux that equals the product of the *sensing field* H_s and the magnetization $j' = \mu_o m \cdot \Delta_{vol}$, where the dipole moment m is in A/m and Δ_{vol} is the volume $\Delta x \bullet y \bullet w$ of the magnetic element. Hence ψ has unit of Wb.

A more stringent mathematical development would have resulted in the vector product of the sensing field and the magnetization:

$\Delta\psi$	$= \Delta j' - H_s$	**(6.5a)**
j	$= \Delta j'_x \bullet H_{sx} + \Delta j'y \bullet H_{sy}$	**(6.5b)**
H	$= \Delta j' \; H_s \bullet \cos\beta$	**(6.5c)**

The playback flux $\Delta\psi$ is the vector dot product of the tape magnetization $\Delta j'$ and the *sensing field* H_s from the reproduce head. The second expression is convenient for computations, and the third for quick appraisals; β is the angle between the directions of the magnetization and the sensing field. Fig. 6-6 illustrates a simple example. The magnetization in a coating is typically tilted as shown. j_1 forms an angle of 90° with the sensing field direction, and its contribution to the flux is therefore zero (cos 90° = 0). j_2 is parallel to the sensing field, and its flux contribution is therefore equal to $j_2 \bullet H_s$.

The total flux is now obtained by a double summation (integration), one from minus infinity to plus infinity to account for all magnetizations along a layer in the tape, and the other from

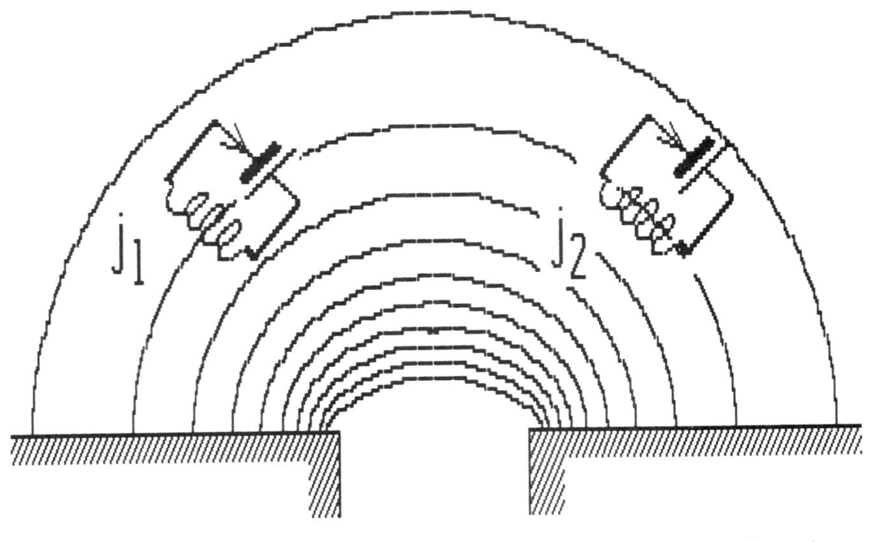

Fig. 6-6. Flux from m_1 (j_1) at zero, while flux from m_2 (j_2) is at maximum.

d to d+c′, where d is the spacing between the head and tape surfaces and c′ the recorded thickness of the coating. The last summation accounts for all the layers in the coating.

A few comments are in order. It is difficult to envision the reciprocal sensing function, but an analog example will help. Envision a VHF antenna being used by a radio amateur for transmitting (see Fig. 6-7). The antenna is made from a single dipole with added elements for directivity; this will focus the transmitted energy into a narrow beam, which should then be directed toward the geographical location where the receiver is. When the operator switches to listening he disconnects the antenna from the transmitter, and connects it to the input of the receiver. This assures a higher sensitivity toward the point where he expects to receive a reply. The sensing function does have the same pattern as the antenna when used as a transmitting device. The same holds true for the write/read (record/playback) heads.

Note also the dimensions in the formulas above. The field strength from the head gap is measured in A/m (Oe), using the expressions in Fig. 4-24; H_g equals ni/L, multiplied with the efficiency η_p (L is the gap length). The definition of the sensing function is field strength divided by a unit current of 1 Amp, which gives the sensing function the unit of m^{-1}. In using the formulas in Fig. 4-24 use $H_s = \eta_p n/L$ when figuring the sensing field.

The units for the dipole magnetization are normally given in Am^2. Here we do multiply with μ_o, and obtain Wbm; further multiplication of the magnetization with the sensing field results in units of Wb, which is correct for flux.

The playback waveforms depend upon the magnetization in the coating: longitudinal, perpendicular, or a mix of the two. A slight expansion of the situation in Fig. 6-6 provides the answer in a way that is easy to understand (see Fig. 6-8).

A single transition in a longitudinal coating is represented by two oppositely magnetized coils, both lying with their axes in the longitudinal direction. The reader may wish to trace the head coil on a small piece of paper and slide it past the two upper coils that represent the

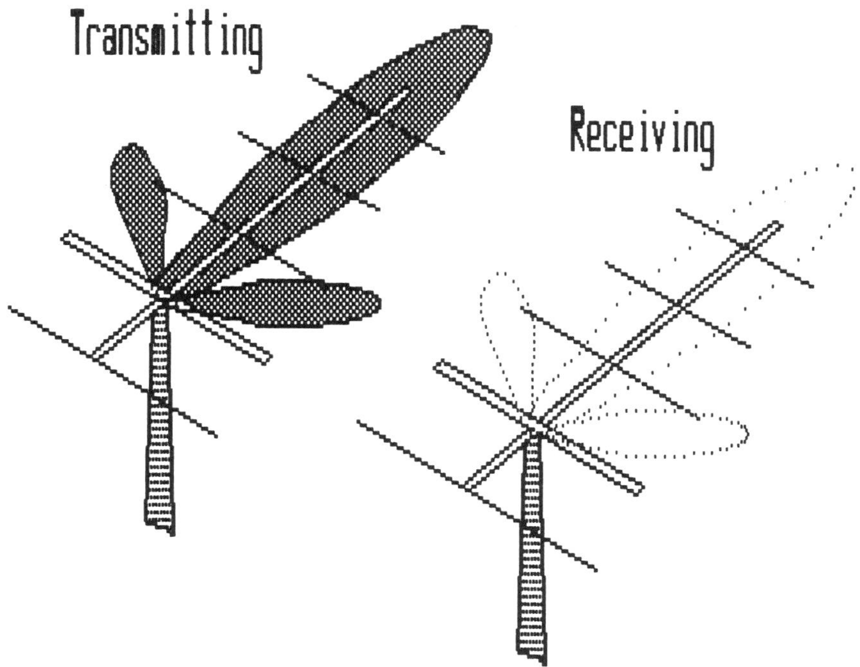

Fig. 6-7. An antenna has directivity in both operating modes.

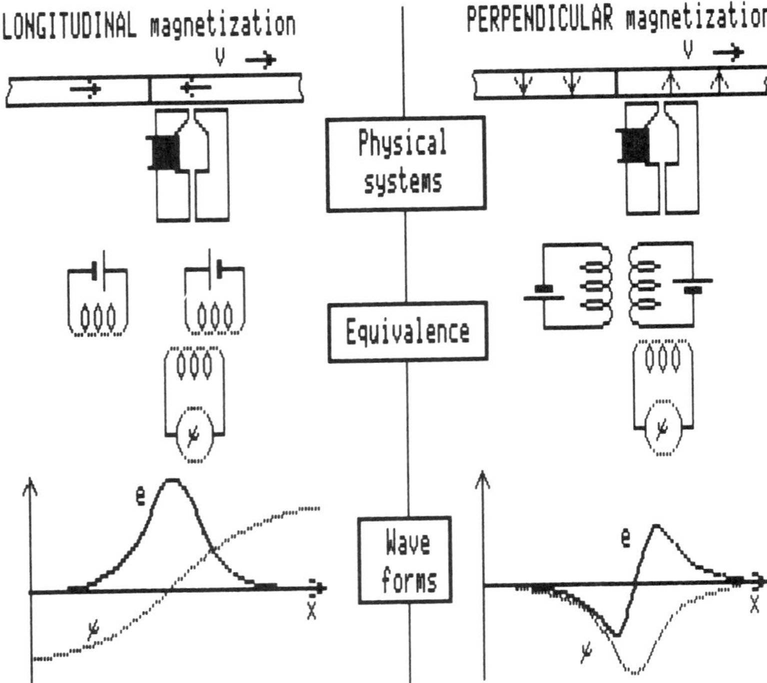

Fig. 6-8. Read waveforms.

magnetization on each side of the transition. It is then easy to observe how the flux varies, and a differentiation of this waveform provides the familiar bell-shaped voltage waveform. The same analysis can be applied for the perpendicular magnetization, where the coils now are parallel to each other with their axes perpendicular to the head coil. The voltage waveform is now S-shaped.

In a longitudinally oriented coating the longitudinal magnetization will be at least 4 times the level of the perpendicular component. Adding ¼ of the right curve to the left curve results in the real life curve, that has a flat start (possibly with undershoot), then a fast rise, followed by a slow decay. We will return to this topic in Chapter 21.

TOTAL PLAYBACK LOSSES

The read voltage can be calculated for a magnetization pattern that is sinusoidal, longitudinal, and uniform through the thickness. The result is, after a lengthy integration, a general formula for the reproduce voltage, shown in Fig. 6-9. The last four terms are playback losses, which we will discuss in separate sections.

The peak voltage is found by inserting numbers in the first portion of the formula:

$$\boxed{e_{peak} = n\psi_m 10^{-3}\eta_p\omega \ \ \mu V} \tag{6.6}$$

Example 6.1: Find the loss-free peak voltage at 1000 Hz from the head in an audio cassette player. The tape is recorded within its full thickness of 4μm, at a peak level of 160 nWb/m. The track width is 0.5mm (20 mils), the head read efficiency is 70 percent, and the number of turns equal to 280.

$$e = e_{peak} * \cos\omega t * \text{LOSSES} \qquad \underline{\underline{\mu\text{volts}}}$$

$$= n \cdot \psi_m \cdot w \cdot 10^{-3} \cdot \eta_p \cdot (2\pi v/\lambda) * \cos(2\pi vt/\lambda)$$

$$* e^{-(2\pi d/\lambda)} \cdot \frac{1-e^{-(2\pi c'/\lambda)}}{(2\pi c'/\lambda)} \cdot \frac{\sin G}{G} \cdot \frac{\sin X}{X}$$

where:

n = number of turns	λ = wavelength in meters
ψ_m = peak flux in nWb/m	d = head-to-coating distance in meters
w = track width in meters	c' = recorded thickness in meters
η_p = head efficiency (fraction)	$G = \pi L/\lambda$
v = head-to-coating speed in m/sec	$X = (\pi w \cdot \tan\beta)/\lambda$
L = gap length in meters	β = misalignment angle (see Fig. 6-15)

Fig. 6-9. General playback formula.

Answer: The flux level is $\psi_m*(c'/c)*w = 160*(4/4)*0.5*10^{-3} = 0.08$ nWb. The peak voltage is $e_p = 280*0.08*10^{-3}*0.7*2\pi*1000 = 98.5\ \mu V_{peak}$.

SPACING LOSS

This loss was first described by R.L.Wallace, Jr. (see ref.), and its magnitude is generally referred to as the "Wallace Loss":

$$\boxed{\text{Spacing Loss} = 55\ d/\lambda\ \text{dB}} \qquad \textbf{(6.7)}$$

This makes it quite simple to appraise: if the spacing d equals 20 percent of the wavelength then the loss is 11 dB; for a wave length of 125μin (15,000 Hz at 1⅞ IPS) the spacing equals 25μin (much less than the diameter of a human hair, about the size of the particles in cigarette smoke).

The spacing loss is explained by an example in Fig. 6-10: the flux from dipole j_1 produces 2.5 times less flux than the dipole j_2, even though they are of the same magnitude. The further away from the head the dipoles are, the smaller their flux contributions. The spacing loss is therefore proportional to the spacing d.

The inverse proportionality to λ is found in flux calculations from the sum of magnetizations "seen" by the head field. At short wavelengths the flux contributions along a thin layer in the coating tends to cancel each other, as shown by the flux contribution due to the top row of dipoles in Fig. 6-11.

The spacing loss in tape recorders and floppy disk drives is mainly due to coating surface roughness, provided the heads are clean and smooth without any debris formed on their surfaces. (Another cause for spacing loss may be coldworking of the core surface which causes

Spacing Loss = 55 d/λ dB

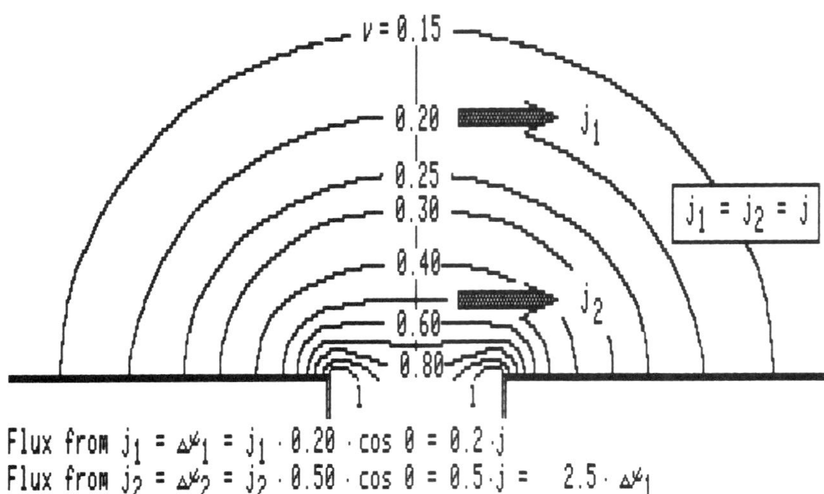

Fig. 6-10. Spacing loss explained (see text).

Effective coating thickness = .22 * λ = .44 * BL

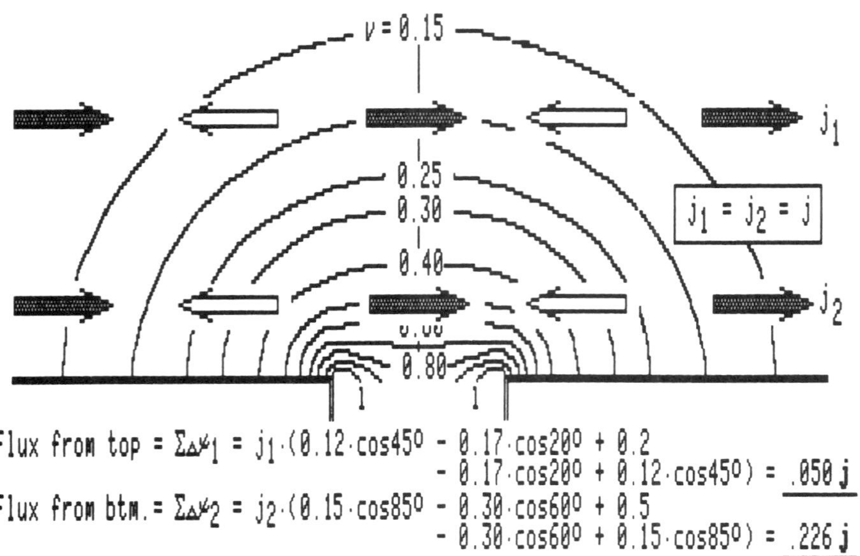

Fig. 6-11. Coating thickness loss explained (see text).

a loss of permeability, which in turn will act as a spacing loss. New heads should be broken in with a mildly abrasive disk or tape.) The quality of modern computer disks, audio, video and instrumentation tapes assures an effective coating roughness in the range of 10 – 2 μin (0.25-0.05 μm).

The spacing losses in rigid disk drives are of course controlled by the spacing between the flying head and the disk surface.

The Wallace spacing loss assumes a coating permeability of μ_r equal to one. It is in reality somewhere between 2 and 4, and will affect the sensing function in a way that increases the spacing loss (see Chapter 21 for a detailed discussion). The use of Wallace's formula in flying height measurements is therefore not valid.

The spacing loss in perpendicular read/write systems differs from the Wallace law when a soft magnetic underlayer below the coating is used. When a single pole head is used a loss in the order of 100 d/λ dB is typical, while there is not a specific rule for the use of a conventional ring core head (ref. Iwasaki, Speliotis and Yamamoto).

COATING THICKNESS LOSS

Coating thickness loss shows up as a reduction in external flux from the tape. The name is unfortunate since it is not a loss in magnetization as it implies. Having just discussed the spacing loss we can explain the thickness loss as a spacing loss for the various layers of the coating, with those furthest away from the head suffering the greatest spacing loss (see Fig. 6-11).

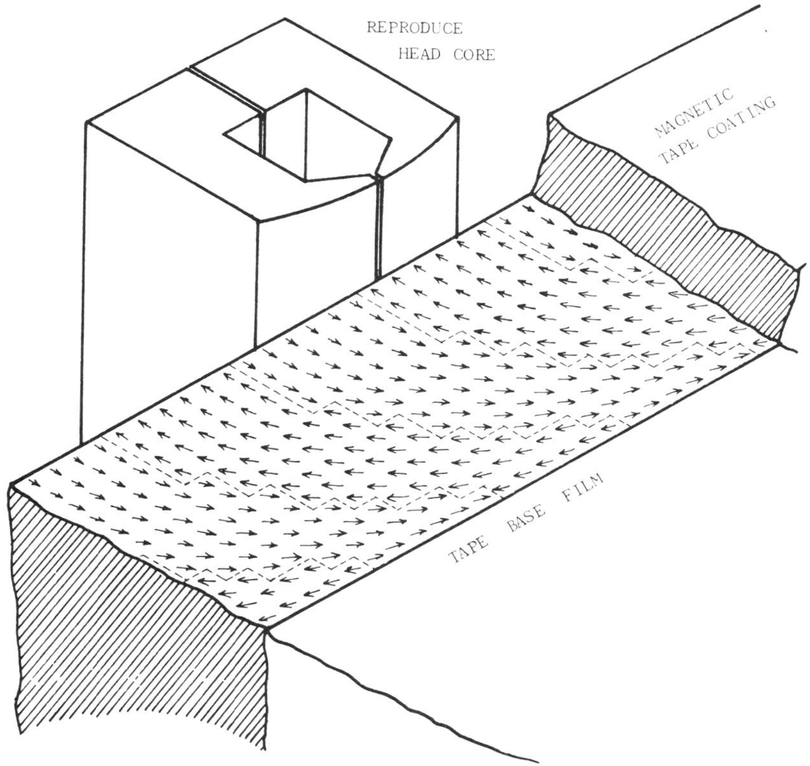

Fig. 6-12. Square-wave magnetization throughout coating.

An example will serve to explain this. Fig. 6-12 shows the magnetization pattern in a tape recorded with a square wave. There are 12 layers in the coating and the vector magnetization for each small dipole in the array is in the direction it "saw" when it left the recording zone.

The computed flux levels from each element in Fig. 6-13 clearly show how the layers away from the head contribute very little to the sum total flux. The layer nearest the head provides about 50 percent of the total flux, the next two layers the rest, while the balance more or less cancel out each other.

This picture fits a formula that states that the thickness of the layer that contributes 75 percent of the flux is (see Daniel et al., further readings to Chapter 2):

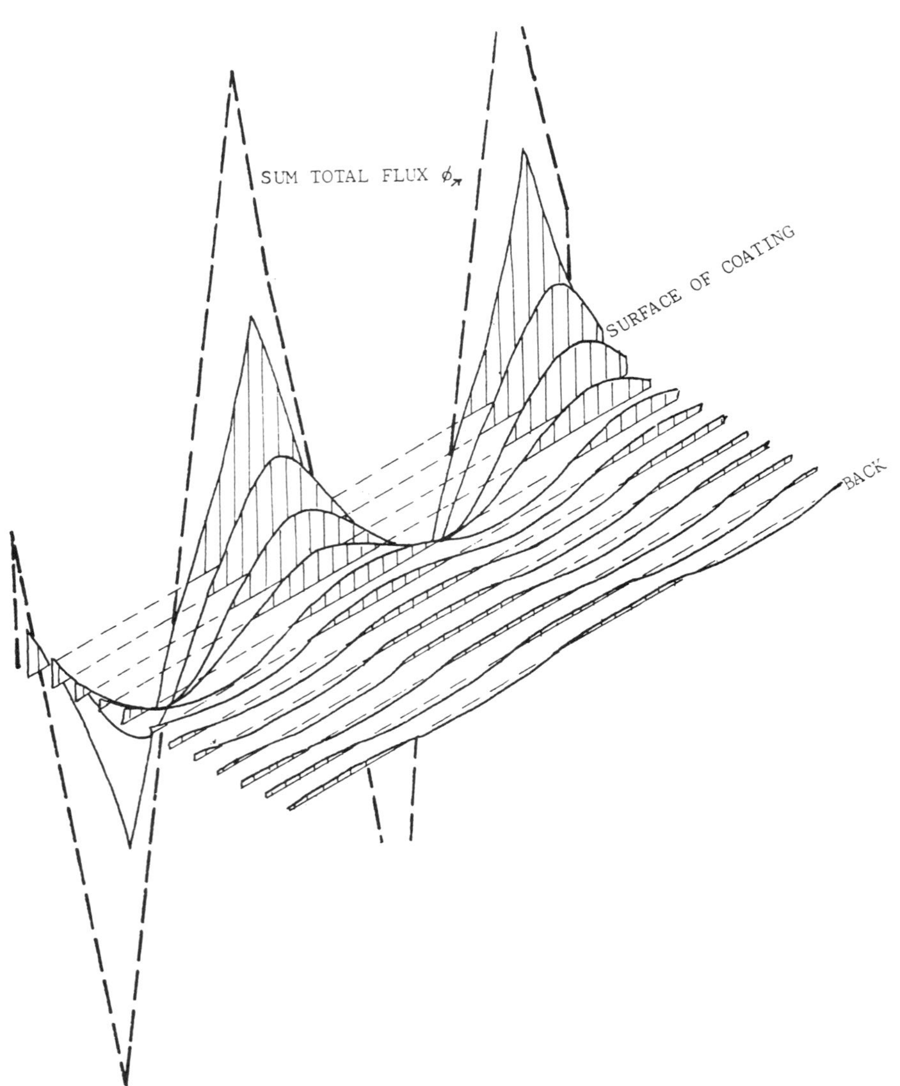

Fig. 6-13. Flux contribution for square-wave magnetization, illustrating coating thickness loss.

$$\text{Effective Coating Thickness} = 0.22 * \text{Wavelength} \quad \textbf{(6.8a)}$$
$$= 0.44 * \text{Bit Length} \quad \textbf{(6.8b)}$$

We now have three coating thicknesses to deal with: the physical coating thickness c, the recorded thickness c′, and now the effective coating thickness c_{eff}. They are related by the following inequality:

$$c \quad => \quad c' \quad => \quad c_{eff} \qquad \textbf{(6.9)}$$

A coating thickness greater than c_{eff} adds no output, but increases the demagnetization losses in longitudinal recording. (But a thick coating is preferable in the perpendicular record mode. Why?)

REPRODUCE GAP LOSS

The reproduce gap loss is caused by the finite length of the front gap in the head. A limit is reached when the recorded wavelength is equal to the gap length: the flux contributions from the two oppositely magnetized half wave lengths cancel, and the induced voltage is zero.

The length L of the gap in the sinx/x type expression for the loss is slightly longer than the measured physical gap length. An accurate loss is determined by using 1.12*L instead of L, or better: measure the magnetic gap length by recording a number of frequencies and determining the one where a null occurs. Modern thin film heads have multipliers different from 1.12 (see ref. Westmijze, and Chapter 8).

The bandwidth of a recorder is seldom extended beyond the frequency corresponding to a wavelength twice the effective gap length (f= tape speed in IPS/(2 * gap length in

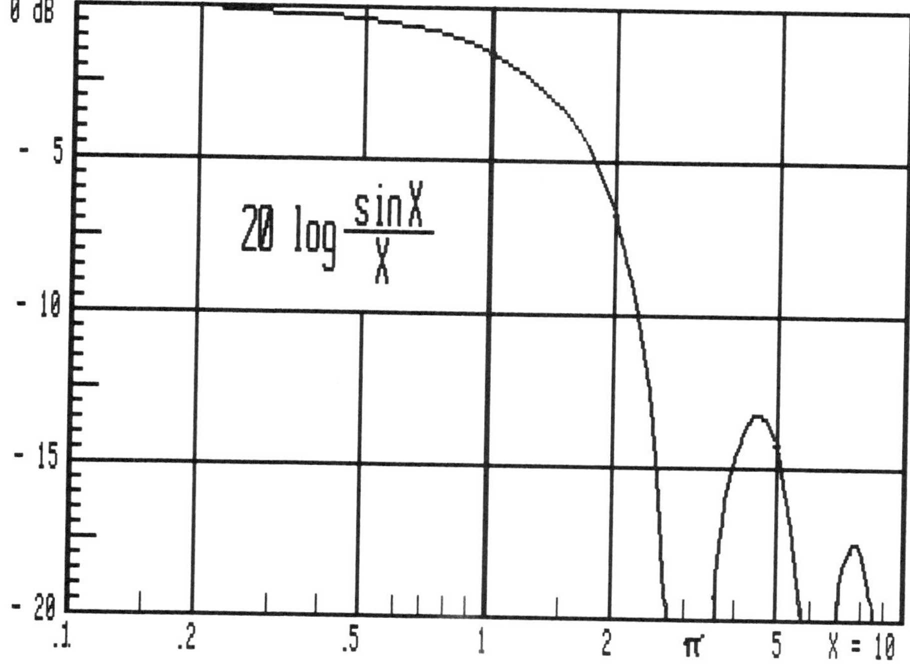

Fig. 6-14. Gap loss function sin x/x. Use $x = \pi \cdot L/\lambda$.

inches). Otherwise, high-frequency equalization will be excessive and the noise level too high. Disk drives use gap lengths ranging from 0.5 times the BL for high resolution to 1.0 times the BL for high output.

Figure 6-14 shows the loss curve as a function of λ/L.

ALIGNMENT LOSS

A loss, similar to the gap length loss, occurs if the reproduce head gap is not parallel to the recording head gap pattern. This is named alignment loss, and is a sinx/x type loss, like the gap loss.

Most recorders have their reproduce heads mounted on a plate that allows for correct adjustment of the alignment angle. It should be noted that the alignment is less critical when the track width is narrow.

Home video recorders (all formats) utilized the alignment loss as a guard against reading information from adjacent tracks. The write/read heads are mounted on a drum, 180 degrees apart, each laying down (or reading) a track that has the data for half a picture frame, two of which make up a complete interlaced picture. (See the chapter on video applications.)

The tape area is fully utilized by laying down the tracks with plus/minus 7 degrees misalignment as shown in Fig. 6-15. When playing back the videotape adjacent tracks will produce zero output due to severe misalignment loss. This principle is further expanded in the newer units with digital sound capabilities; the sound is recorded with two additional heads, also 180 degrees apart, with misalignment angles of plus/minus 15 degrees. The sound is recorded first, in depth, with a low carrier frequency. Next the video is recorded by partial-depth recording

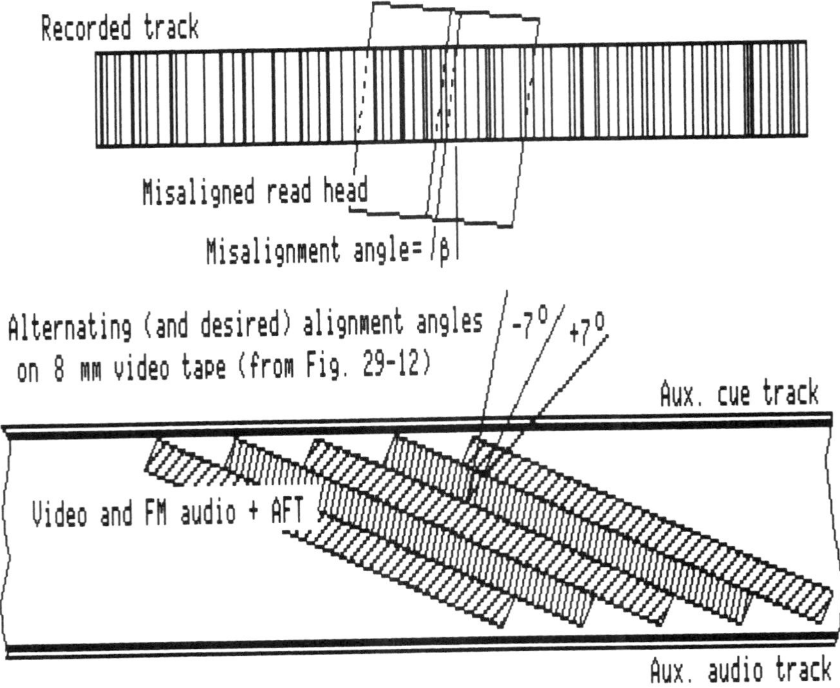

Fig. 6-15. Gap misalignment (β) and a useful application in video recording.

so it does not erase (overwrite) the digital audio. Filtering upon playback separates the signals. Note that the terms first and next apply to the sequence of heads on the head drum.

LOSSES DEPEND UPON DIMENSIONS

In the foregoing sections several losses have been discussed and explained for the playback process. Note that they all are functions of the recorded wavelength, NOT FREQUENCY! (This is true also for the losses during the record process). Playback losses are always analyzed by way of the wavelength (or bit length) in relation to the spacing, the recorded depth of the coating, the gap length, and the head alignment.

Additional losses do occur in write and read heads at higher frequencies, and will be covered in Chapter 7.

REFERENCES TO CHAPTER 6

Wallace, R.L., Jr., "The Reproduction of Magnetically Recorded Signals," *The Bell System Tech. Jour.*, Oct. 1951, Vol. 30, No. 10, pp. 1145-1173.

Westmijze, W.K., "Studies on Magnetic Recording: Pt. II. Field Configuration Around the Gap and Gap-length Formula," *Philips Res. Repts.*, 1953, Vol. 8, pp. 161-183.

Eldridge, D.F., and Daniel, E.D., "New Approaches to AC-Biased Magnetic Recording," *IRE Trans. Audio*, May-June 1962, Vol. AU-10, No. 3, pp. 72-78.

Geurst, J.A., "The Reciprocity Principle in the Theory of Magnetic Recording," *Proc. IEEE*, Nov. 1963, Vol. 51, No. 11, pp. 1573-1577.

Alstad, J., "A Novel Technique for Measuring Head-Tape Spacing," *IEEE Trans. Magn.*, Sept. 1973, Vol. MAG-9, No. 3, pp. 327-329.

Lindholm, D., "Spacing Losses in Finite Track Width Reproducing Systems," *IEEE Trans. Magn.*, Mar. 1978, Vol. MAG-14, No. 2, pp. 55-59.

Part 3

Magnetic Heads

Chapter 7

Materials for Magnetic Heads

The *magnetic heads* in a tape recorder or disk drive are the focal point of recording and playback of signals. They have one thing in common, that of having a magnetic core that either guides a concentrated field for the purpose of recording (and/or erasing), or they are used to sense the magnetic flux from a recorded tape or disk.

A traditional core structure is shown in Fig. 7-1. Thin laminations of a high permeability material are stacked and cemented together, and each half of the core is provided with a winding. The shallow front gap either generates the recording field or collects the tape flux; the deep back gap offers a minimum reluctance to the flux lines through the core. (This gap, in general, is present because of methods of fabrication of the winding and for reasons of assembly techniques. It does not affect the recording or reproduce process.)

Magnetic cores are also made from ferrites, or a combination of a ferrite core and a wear resistant pole piece, made from a magnetic iron-aluminum alloy named *Alfenol*, *Alfesil*, *Sendust*, *Spin-alloy*, or *Vacodur*. A new amorphous iron-nickel alloy named *Metglass* or *Vitrovac* has recently been introduced in some applications, in addition to other high wear resistance materials.

Cores are found in all kinds of shapes and sizes, ranging from about one inch in overall diameter to the size of the head of a pin. Figure 7-2 illustrates a few examples.

Materials for head cores are investigated in three recent books (Chen, Snelling, Boll) and in the classic by R. Bozorth. The latter is unfortunately out of print, but there are copies in libraries. Specific information can be found in many of the referenced and listed papers at the end of the chapter.

The reader may wonder: Why is there such a large variety of head designs? This and the following three chapters will examine the many performance criteria placed on a magnetic head, and where trade-offs among them are necessary. The most important ones are:

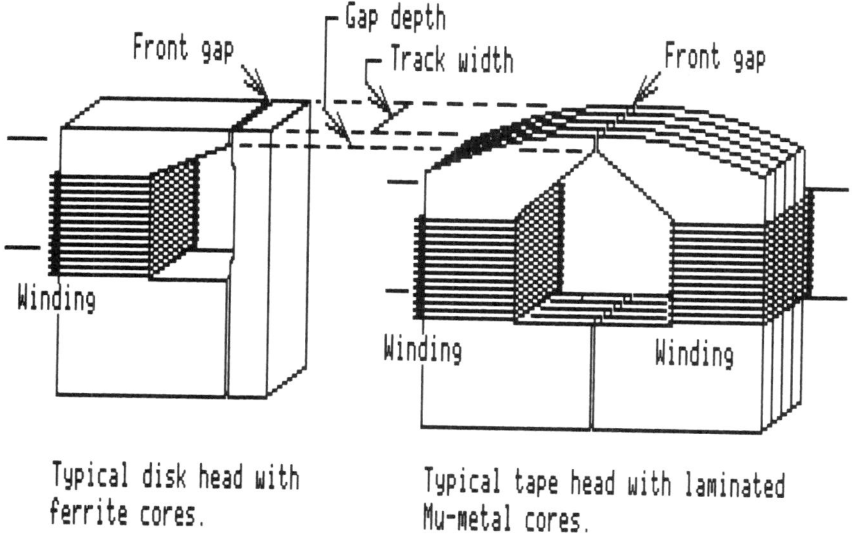

Fig. 7-1. Traditional structure of a magnetic write/read head.

- Frequency range.
- Wavelength response.
- Signal sensitivity.
- Losses (and noise).
- Useful life.
- Track configurations.
- Ease of fabrication.
- Materials and processes.
- Overall performance versus cost.

Most of these can be analyzed by a study of the magnetic circuit that represents the magnetic core with its coil. The analysis will depend on the head's use, whether for erasing, recording (writing) or playback (reading). The fabrication may be piecewise assembly of materials, or vacuum deposition of the cores, insulators, and conductive windings as integral parts.

Many head assemblies contain several tracks and consideration must be given to cross talk. This consists of unwanted signals coupled from adjacent tracks into a core; it may be stray flux from recorded tracks or from transformer coupling between core windings.

The optimized designs and performances for various applications are deferred to Chapter 10 in order that we may first learn how many modern heads are built, and why certain construction techniques are used. Examples will reflect applications in the following areas: computer data storage on disk and tape, instrumentation, audio, and video. It is the writer's hope that this arrangement of discussing head materials first, then the vital topic of fields in Chapter 8, and then construction techniques will be valuable when the book is used as a reference. This should also provide a good background for the design guides in Chapter 10.

In this chapter we will examine the magnetic properties of several materials, striving for the optimum magnetic performance of a head. That spawns a requirement for *high permeability*. An important practical concern is also how long a given head will last for a given initial cost. High sensitivity may result in head designs with a very shallow gap depth, in the range

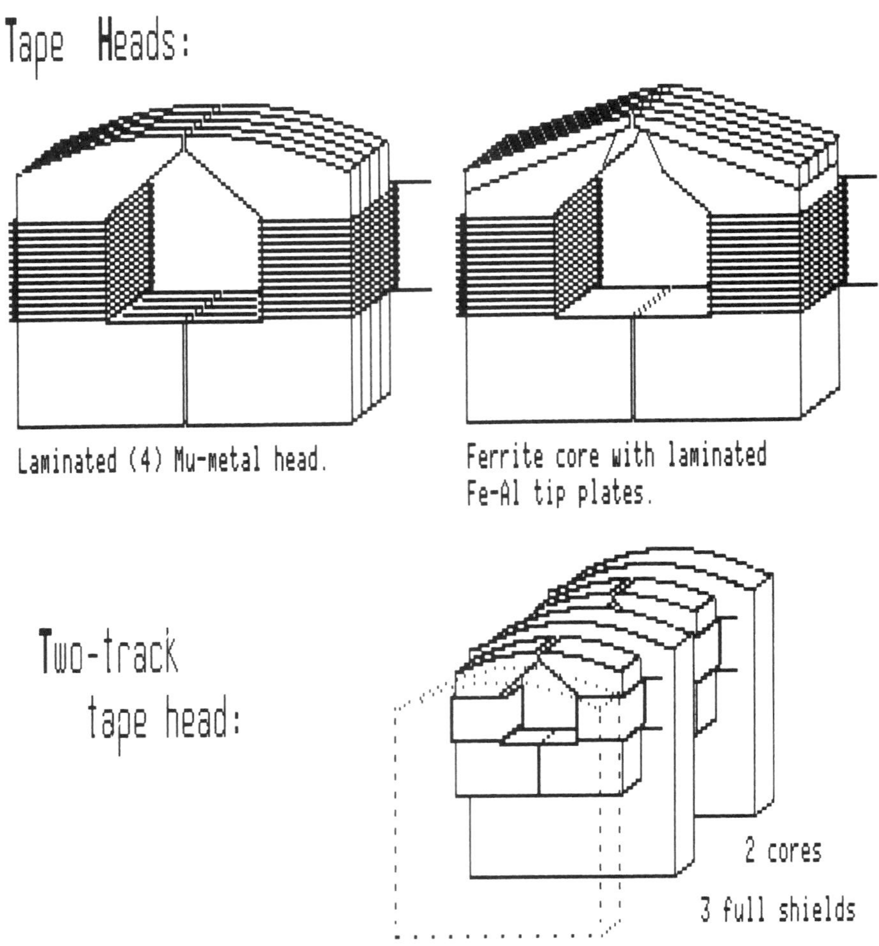

Fig. 7-2A. Examples of magnetic tape heads.

of 0.2-3 mils (5-75 μm). The abrasive action of the magnetic tape coating may wear out such heads rather quickly, and attention must be paid to the *mechanical wear properties* of the materials.

The electromagnetic performance of a head is appraised by its efficiency, impedance, and noise levels, all over a specified frequency range. We will review the efficiency and inductance (from Chapter 4), and apply the formulas to material selection. We will discover that a rather high resistive component of the head's impedance may appear at high frequencies, and by examining its cause find restrictions on the material selection. *Eddy-currents* are induced in the cores and they must be reduced by *laminating the core*, or by using a high resistivity material: a ceramic named *ferrite*.

It also becomes apparent that a small size is desirable for optimum efficiency, and that borrowing techniques from the semiconductor industry has in recent years given us a new head, the so-called *thin film head*.

Upon examining materials we also find some whose properties are field dependent, such as changing resistance value when a field crosses them. Such materials are used in flux sensing

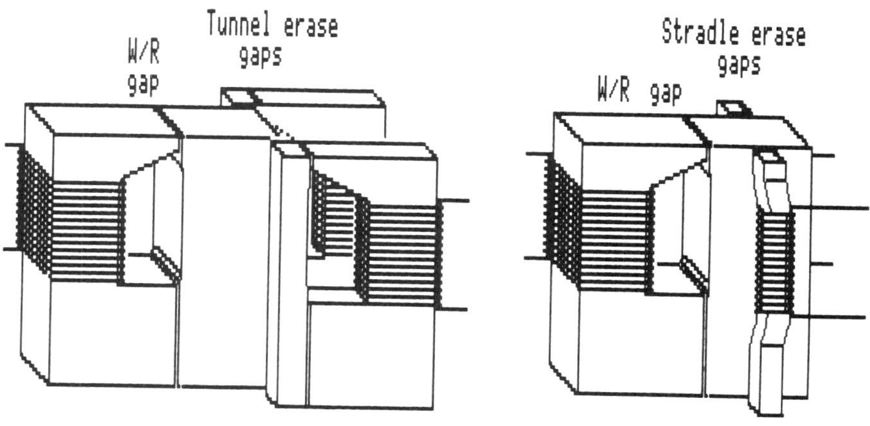

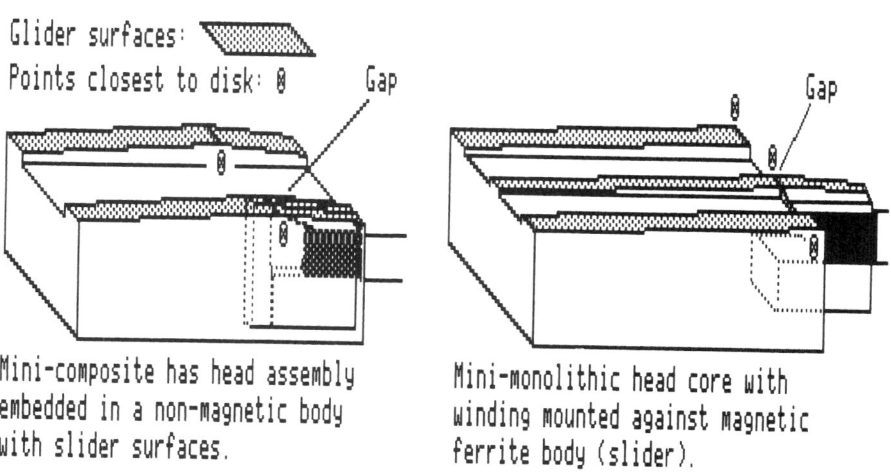

Fig. 7-2B. Example of magnetic disk loads.

heads, for instance the *magneto resistive heads*. A brief discussion of these will conclude the chapter.

MAGNETIC CIRCUITS

Magnetic heads in use today are ring-shaped cores of a flux conducting magnetic material, provided with a winding and a small gap in the core. This gap interrupts the magnetic circuit and is used in record heads to provide a concentrated recording field. In the reproduce head it collects the available flux from a recorded track, and the core links the flux through the winding where it produces a voltage.

The majority of all heads utilize a magnetic core as shown in Fig. 7-1. They may differ in size and geometry but are essentially alike in construction, whether for digital data, instrumentation, audio or video recording or playback.

Several *core shapes* are shown in Fig. 7-2 (without housings, adjacent tracks, or windings). The common core materials are laminated iron-nickel or ferrite, with occasional use of a special alloy for long wear life applications.

We can, for each of the cores shown, sketch a magnetic circuit from the flux path, while breaking down the more complex structure into simple geometric blocks for each of which we can calculate the *magnetic resistance*. This was done in Chapter 4 (see Fig. 4-22), and from an elementary circuit analysis we found expressions for the efficiency, record current required, and the self inductance of the winding:

$$\text{Efficiency } \eta = R_{fg}/\Sigma R_m \quad \textbf{(4.26)}$$
$$i_{write} = H_g L_g/n\eta \quad \textbf{(4.29a)}$$
$$\text{Inductance } L = n^2/\Sigma R_m \quad \textbf{(4.30)}$$

where

$R_{fg} = L_g/\mu_o D_g w$
L_g = gap length in meters
D_g = gap depth in meters
w = gap width (= track width) in meters
$\mu_o = 4\pi 10^{-7}$ Henries/meter
ΣR_m = sum of all magnetic resistances, tracing the circuit once.
H_g = Deep gap field required, as shown in Figs. 5-13 and 5-14.
n = number on turns in winding.

Optimize Efficiency

The magnetic circuit determines the efficiency of the head core, in addition to the value of L. A low value is achieved in a head core design where the flux path length through the core is short and the cross sectional area is large. So we want a small but wide core.

The permeability μ_r of the core material should be as high as possible, and traditionally, a designer will select an iron-nickel alloy, known by trade-names such as Permaloy or Mu-metal. This should assure the best efficiency.

A low write current or a high read voltage is obtained for a large number of turns, but it is limited due to restrictions on the inductance value. It must be small enough so the head impedance has a resonant maximum above the highest signal frequency (or bias frequency). And the time constant of head inductance and amplifier impedance must be smaller than the rise time of the pulse edges in pulse recording; there is possibly a desired time constant for each application.

Assume a prototype head has been built, the core is a laminated iron-nickel material, and testing is underway. This may be a bench test where a current carrying wire is placed in front of the head to simulate the playback of a constant flux recording. The current source is high impedance to assure a constant current from low to high frequencies. The induced voltage should ideally increase 6 dB/octave; but above a few kHz a slower rise is observed and we wish to know why.

Another measurement is made to verify the calculated inductance. It may be quite correct at low frequencies, but decreases toward higher frequencies, and the impedance bridge reveals that the resistive component of the winding resistance apparently increases quite dramatically.

EDDY CURRENT LOSSES

The effects of increased resistance values are named *core losses* and have their origin in circulating currents in the core. Fig. 7-3 shows how a magnetic field penetrates a sheet of conductive material. When the field changes value, as in an AC field, then circulating currents I_e are induced inside the sheet, in accordance with Lenz's Law.

This law is quite like Le Chatelier's law, which states that all systems at rest will react to an outside change by an action that opposes the change. Lenz's law applies specifically to the induced currents, making them have direction so they create a field that is in opposition to the field that creates them.

If the AC field varies fast enough (higher frequencies), then the induced fields may completely oppose the outside field, an effect used in shielding. The effectiveness of such shielding is not reduced by drilling a number of small holes in the sheet, since the circular currents still are operative. Hence, we get good shielding effect of conductive screens, such as Faraday shields. An example is the door of a microwave oven, where holes permit the user to look at the destruction of a good filet mignon, but the field cannot radiate out.

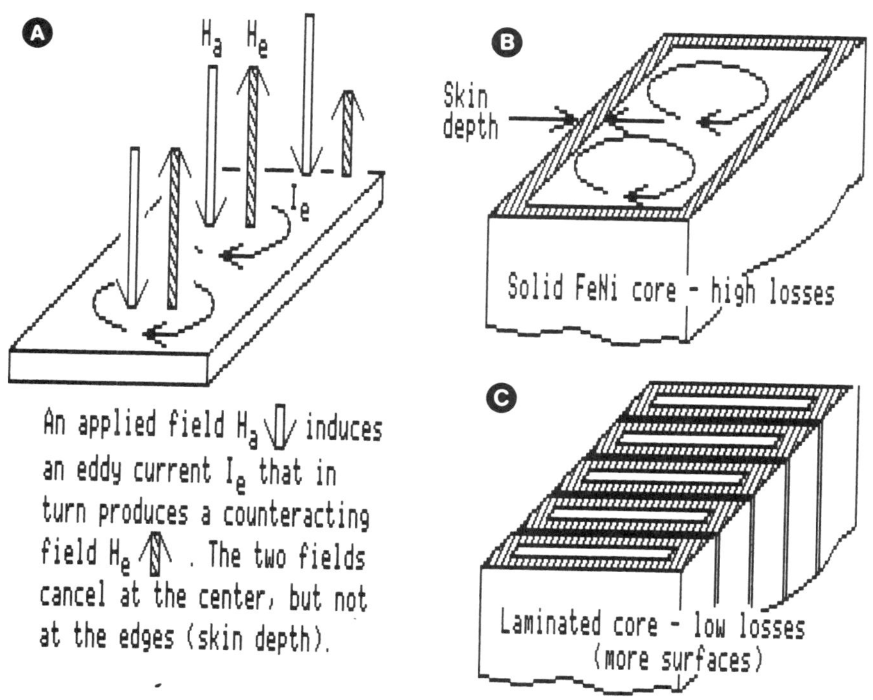

Fig. 7-3. Eddy currents and skin depth, (A). Eddy currents in sheet metal, (B). Surface flux in a bar at high frequencies, (C). The laminated core in a magnetic circuit (a head core) carries the high frequency flux on the lamination surfaces only, not in their center section; this increases the magnetic resistance.

When the eddy currents are strong enough, no magnetic field goes through the center section of the sheet. Since the eddy currents increase with frequency, we will, with increasing frequency, have the flux concentrated at the edges of the sheet. This will hold true when we next increase the thickness of the sheet so it, in the limit, becomes a rectangular bar. It is then logical to talk about a surface flux that only penetrates into the bar to a certain depth, the *skin depth*.

This skin depth is small for materials with high permeability μ_r and small resistivity σ; and it does, as we have seen, decrease with higher frequencies. We could therefore, in our expression for the magnetic resistance of the core (part of ΣR_m), substitute the area A with an effective area A′, that decreases in size with increasing signal frequency.

The skin depth can be calculated from (after Bozorth):

$$\delta = (\tfrac{1}{2}\pi) * \sqrt{\varrho/\mu_r F} \quad \text{cm} \tag{7.1}$$

where

ϱ = resistivity in μohm•cm
μ_r = relative permeability (at DC)
f = frequency in kHz.

This formula would allow the calculation of the area available for the flux, at each frequency we desire. The use of these values in the expressions for the efficiency and the inductance would indeed show a decrease in the value of both as we raise the frequency.

It would not provide an answer to the increase in resistance. Upon studying the literature (Bozorth, Peterson, and Wrathall, Lee) we learn that a complex value for the permeability must be used:

$$\mu = \mu' - j\mu'' \tag{7.2}$$

where j is the complex number $\sqrt{-1}$.

Use of this expression is in agreement with the real world and will give us correct results in our later analysis of the head impedance, losses, and the resistance type noise generated in the head.

Do note that the complex permeability is not a true material constant—it depends not only upon the material, but also its shape and the signal frequency. An expression for μ is found from the references and is given in Fig. 7-4.

The parameter K is:

$$K = 2\pi d_{cm}\sqrt{\mu_{rdc} f_k/\varrho} \tag{7.3}$$

where:

d_{cm} = lamination thickness in cm
μ_{rdc} = relative permeability at DC
f_k = frequency in kHz
ϱ = resistivity in μohmcm

The expressions in Fig. 7-4 are formidable, but rather easy to handle today with a small subroutine in a computer program. The expressions do simplify, at high frequencies, to:

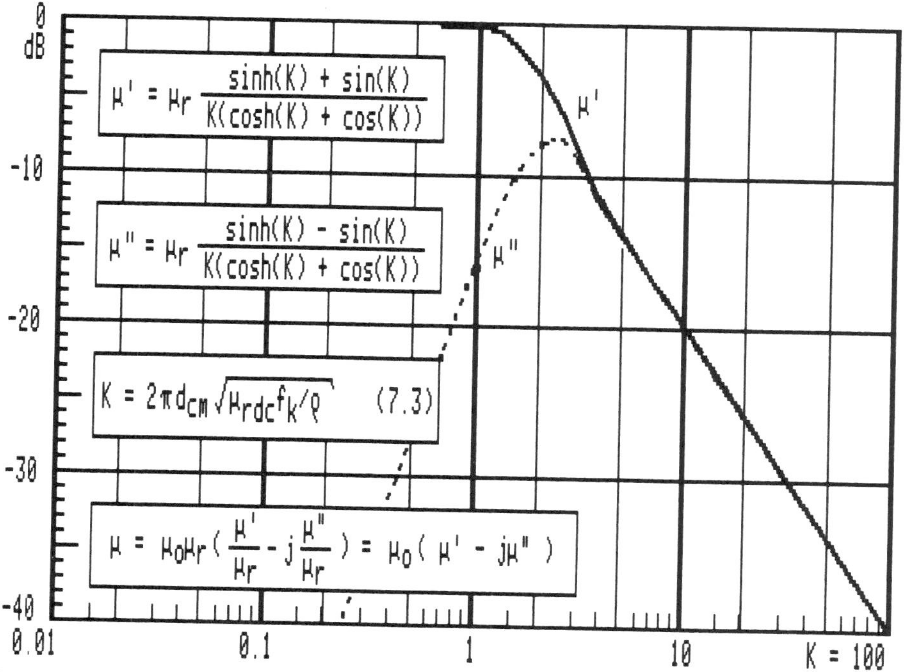

Fig. 7-4. Complex permeability spectrum. $\mu = \mu_o\mu_r\,(\mu'/\mu_r - j\,\mu''/\mu_r) = \mu o_o\,(\mu' - j\mu'')$. See text for value of K.

$$\boxed{\mu = \mu_o\mu_r * (1-j)/K.} \qquad \textbf{(7.4)}$$

The complex permeability therefore has a phase angle of 45 degrees at high frequencies, and decreases at a rate of 3 dB/octave (K is proportional to the square root of the frequency, see formula (7.3)).

Figure 7-4 shows that the permeability is mostly real and equal to μ_r up to a value of $K = 1$, after which it falls and becomes complex. When $K = 1$ the skin depth equals the lamination thickness, and the core has a permeability equal to the low frequency value. At higher frequencies the flux is carried by less than the full lamination thickness, which is equivalent to it having a smaller permeability.

There is a full 180 degrees phase shift between the surface flux and the flux at a depth of δ. The latter opposes the surface flux, as we explained earlier when using the Lenz Law to explain eddy current losses.

The significance of the complex permeability is that at high frequencies it reduces the inductance L and also appears as a loss at the head winding terminals. A toroidal inductor without an airgap has an impedance:

$$\begin{aligned}
Z &= R_{wdg} + j\omega L \\
&= R_{wdg} + j\omega n^2/\Sigma R_m \\
&= R_{wdg} + j\omega n^2\mu A_c/L_c \\
&= R_{wdg} + j\omega n^2(A_c/L_c) * (\mu' - j\mu'') \\
&= R_{wdg} + j\omega n^2\mu'(A_c/L_c) - (-1)\mu''\omega n^2(A_c/L_c) \\
&= R_{wdg} + \mu''\omega n^2(A_c/L_c) + j\omega L\mu'/\mu \qquad \textbf{(7.5)} \\
&= R_{wdg} + R' + j\omega L'.
\end{aligned}$$

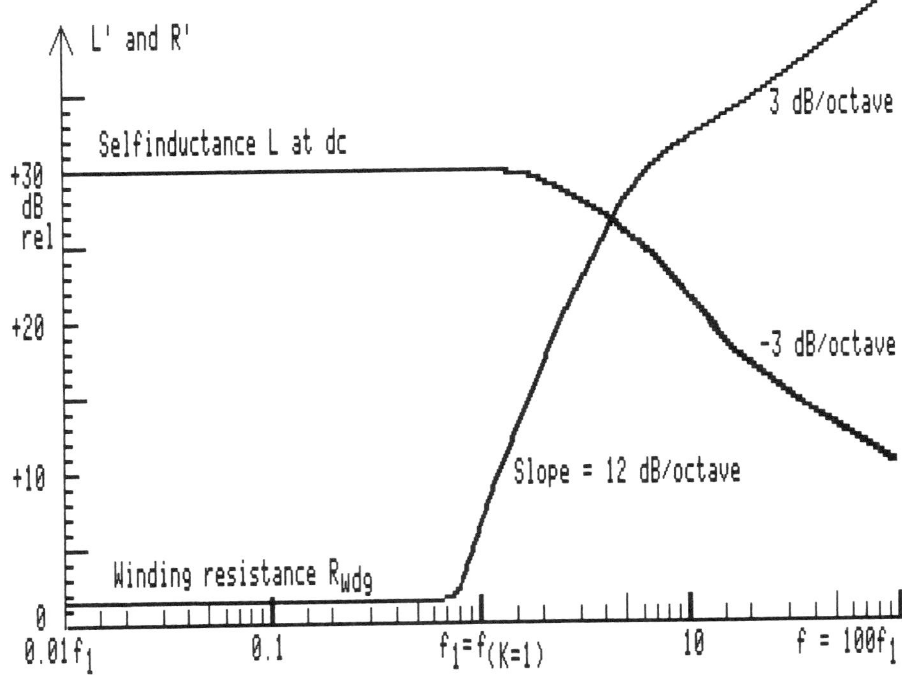

Fig. 7-5. Variations in L and R for inductor; (Toroid, no air gap).

(NOTE that no subscript is used for the permeabilities which means they represent the absolute values, i.e., $\mu = \mu_o * \mu_r$, $\mu' = \mu_o * \mu_r'$ etc.)

The net effect is an added resistance value R′ and an inductance value L′ that decreases as μ'/μ. The total magnetic resistance in a head assembly is due to core sections and air gaps; only the core sections include the μ-term to be substituted with $\mu' - j\mu''$. This complicates the computations.

The net effect is the same, possibly diluted by the presence of the air gaps: the inductance value decreases at higher frequencies (μ'/μ), and the head's impedance has an added resistive component that traces the curve for μ'' versus frequency, i.e., it rises 6 dB/octave for $K < 1$, and decreases 3 dB/octave for $K > 1$. The term ω is a steady 6 dB/octave rise, and the overall pattern of R′ is therefore an impressive 12 dB/octave rise when $K < 1$, and a modest 3 dB/octave rise when $K > 1$. See Fig. 7-5.

SELECTION OF CORE MATERIAL

We have discussed the magnetic circuit and the high frequency effects of eddy currents, and can now apply our gained knowledge toward selection of a suitable magnetic material. It should have the following properties:

For efficiency:	High permeability μ_r
For low hysteresis losses:	Low coercivity H_c
For low remanence:	Low coercivity H_c
For low eddy current losses:	High resistivity ϱ,
	Low permeability μ_r and thin lamination

Table 7-1. Properties of Soft Magnetic Materials (Average, from Manuf. Catalogs).

MATERIAL:	μ_r init	H_c Oe	B_{sat} G	σ μohmcm	ε *)	T_c °C	Composition:
4-79 Permalloy	20,000	0.05	8,700	55	13.2	460	79 Ni, 17 Fe, 4 Mo
HyMu 80	40,000	0.02	7,500	60	12.9	460	80 Ni, 15 Fe, Mo, Mn, Si
Tough Permalloy	40,000	0.02	5,000	90	-	280	Ni, Fe, Ti
Sendust Alfesil	10,000	0.06	10,000	90	15	500	85 Fe, 6 Al, 9 Si
Vacodur 16	8,000	0.06	8,000	145	15	350	Fe, Al
Metallic glass	20,000	0.01	5,500	130	12.5	250	Fe, Co
MnZn ferrite	4,000	0.1	5,500	10^7	11.1	170	
NiZn ferrite	3,000	0.2	3,000	10^{10}	9	100-200	*) 10^{-6}cm/cm/°C

To this we should add ease of machining or forming, and good mechanical wear characteristics (which is somewhat equivalent to a reasonable hardness).

The required high permeability for achieving a good efficiency at low frequencies overrides the conflicting low permeability for small eddy current losses. And it is resolved by using cores made from thin laminations, or from ferrites with high resistivities.

Table 7-1 is a chart of materials currently used for fabricating magnetic heads. We will now discuss their various merits.

METALLIC CORE MATERIALS

A high permeability is achieved when the walls between magnetic domains are easily moved, and when a few hindrances are in their way. From this we can conclude that the materials should be uniform, isotropic, and free from impurities.

We will also find that the core manufacturing processes may produce stresses from machining, bending , pressing (into epoxies), and from polishing and lapping of the pole tip interface. The material should be insensitive with regard to *magnetostriction*. Stress in a magnetostrictive material will produce an ordering of the domains and, therefore, a change in permeability. And a magnetic field will produce a similar ordering of the domains with resulting stress leading to expansion or contraction of the material. This property is utilized in magnetostrictive transducers of, for instance, sonar transducers and ultrasonic cleaning devices.

Iron-Nickel Alloys

A material with very low crystalline anisotropy and low magnetostriction is the iron-nickel alloy with about 80 percent Ni. The variation of λ_{100} and λ_{111} (magnetostriction in crystal direction (100) and (111)), and of the *crystal anisotropy* K, is shown in Fig. 7-6. They are all near zero around 80 percent Ni. That is the composition of 78 Permalloy, a popular soft magnetic material. Other material compositions have been analyzed and are described by Hall (see ref.).

When other elements are added to an Fe-Ni base, several useful alloys result. The addition of 4 to 5 percent molybdenum increases the initial permeability and about triples the electrical resistivity (so eddy current losses are lower). These alloys are called "4-79 Permalloy" and "Supermalloy", and the properties of 4-79 Permalloy are listed in Table 7-1.

Mu-Metal is obtained by the addition of 5 percent copper, and often 2 percent chromium. This material is easier to roll into thin sheets, and has by now expanded into a small family of alloys.

HyMu-80 has high μ_r, low losses, and is easy to photoetch into laminations. (Stamping of laminations for cores is really a thing of the past). *HyMu*-80, *Mark* II is a new alloy with excellent high frequency properties.

HyMu-800 alloys have smaller grains and also smaller domain sizes. It is available in grades A, B, and C. Grade A is manufactured in sheets down to 0.001 inch thickness and has the best electrical performance. Grade B has a minute (and carefully controlled) amount of abrasive particles embedded uniformly throughout the material, thus providing hardness for a high resistance to wear from the tape moving across the head. Grade C is employed mostly for deep drawn cans for shielding and for head housings.

The HyMu-800-B has a lower permeability than HyMu-800-A due to the added impurities, which are rolled-in aluminum oxides. While this provides for a tenfold increase in wear resistance, it also hinders domain wall movement. Its initial permeability is only 10,000 which compares with 70,000 for HyMu-800-A.

We can, at this point, temper our strive for the highest permeability since materials with this property are very sensitive to any form of stress and coldworking. This occurs in three steps of head manufacturing:

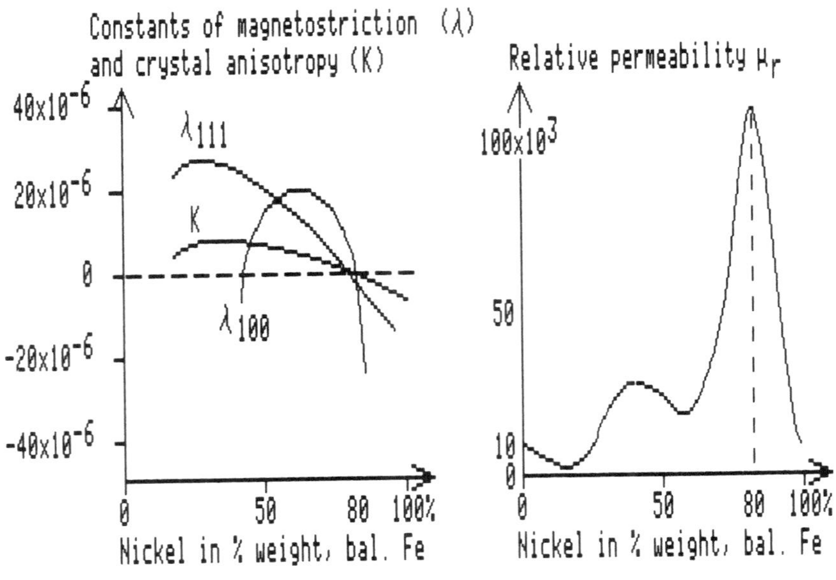

Fig. 7-6. Properties of permalloys and Hy-Mu alloys influence of nickel content on magnetostriction and permeability in Permalloys (after Bozorth).

(1) Stacking and gluing together of thin laminations to obtain a final thickness equal to the track width.

(2) Inserting and clamping the finished core into the head block (they are in general held in place with epoxies).

(3) Lapping of pole tip interfaces.

All three processes lead to a severe degradation of the permeability. It is found that a permeability of at best 2,000-5,000 is retained after stacking and gluing of laminations (ref. McKnight). The lacquers or epoxies used in bonding the laminations together (and providing for insulation) shrink during use, and result in compression stresses in the laminations.

Now, a material like HyMu-800-B has already been "down graded" by the addition of impurities (the hard abrasives), and therefore, is not degraded much further by the additional processing stresses. This is clearly evident in Fig. 7-7, which shows that there is little difference between HyMu-80 and HyMu-800-B after stacking and embedding of the laminations (ref. Bendson).

Another form of strengthening the material consists of adding titanium and niobium to the nickel-iron alloy. This results in an interstitial molecular structure. These materials are known as *Tough-permalloy*, and *Recovac* (ref. Miyazaki, Pfeifer, Radeloff).

Figure 7-7 provides another interesting piece of information. When we, in Chapter 10, go through the design and performance calculations of heads, we will find that the non-magnetic gaps in the magnetic circuits offset the benefit of a very high permeability at low frequencies. It doesn't matter much whether it is 5000 or 100,000. We will find that we should concentrate on selecting a thin lamination and a material that will keep up a reasonable permeability at high frequencies. Here we note, in Fig. 7-7, that HyMu-80 and HyMu-800-B behave almost identically after encapsulation. Also, lowering the permeability makes the K values small, and extends the head's operating range.

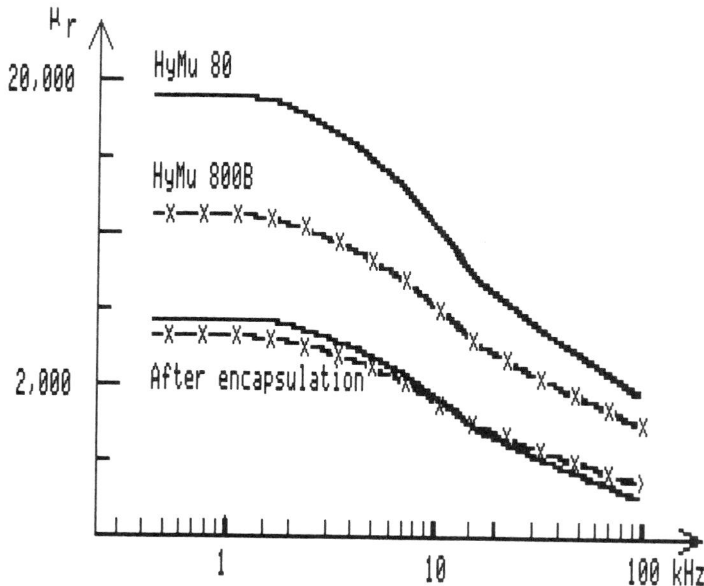

Fig. 7-7. Permeability μ_r is reduced when core is encapsuled. HyMu 800B is affected less than HyMu 80.

There are two additional materials that play a prominent role in modern magnetic heads, the iron-aluminum alloys, and the amorphous metallic glasses.

Iron-Aluminum Alloys

The first Fe-Al alloy was *Alfenol* (84 percent Fe, 16 percent Al), a hard alloy with modest permeability and high saturation density. Unfortunately, it is very difficult to work; rolling it into thin sheets, tend to make it buckle and therefore make the stacking of etched laminations quite difficult.

Sendust (85 percent Fe, 6 percent Al, 9 percent Si), also named *Alfesil,* is also a hard alloy, but very brittle and impossible to roll into sheets. *Spinalloy* is another name for this alloy. It has higher permeability than Alfenol and is often employed as tip plates on a regular permalloy, or better, ferrite cores, as shown in Fig. 7-2. The material has been tried in ordinary head fabrication (ref. Tsukagashi, Tsuya).

Vacodur has properties similar to Alfenol, and appears easier to process as evidenced by its use in heads by European manufacturers.

Why all these names for essentially the same material? AlfeNOL was first developed by Naval Ordnance Laboratory, SENDust came from the university of Sendai, SPINalloy is a product from Spin Physics, and VACodur is from Vacuumschmelze AG.

Metallic Glasses

Glassy metals are soft magnetic materials that are prepared in continuous ribbon form by rapid quenching directly from the melt (ref. Raskin and Davis, Liebermann). The alloy solidifies so fast that grains do not form; the structure is amorphous.

This brings about some unique characteristics, such as very high strength and low energy losses. The alloys that produce soft magnetic materials are made from iron and nickel plus nonmetallic substances such as boron and silicon. Trade names are *Metglass* (Allied Corp.) and *Vitrovac* (Vacuumschmelze). The ribbons are typically 50mm wide, and are cast at a rate of 30 meters per second. The cooling rate is about one million degrees centigrade per second.

The initial permeability for $Fe_{40}Ni_{40}B_{20}$ is 10,000, the losses are comparable to the FeNi alloys, and B_{sat} is 10 kGauss (ref. O'Handley and Narashmhan). Also, it is much more forgiving to handling than the NiFe alloys; it can be bent and twisted without appreciable loss in magnetic performance.

VERY THIN LAMINATIONS; THIN FILMS

We will now have a final discussion on the high frequency permeability in metals. We have learned that a high permeability requires easy domain wall movements and that a certain energy is associated with these movements, much like moving a spring loaded mass back and forth. An equation for the wall movement will, as a matter of fact, be identical to the equation of motion for a spring loaded mass.

The wall movements will decrease in amplitude above a certain resonance frequency, called the *wall relaxation frequency* f_w, and the permeability, thereafter, decreases. A detailed discussion is beyond the scope of this book, but we will quote the results from which we can determine the cut-off frequencies (ref. Boll):

$$\boxed{f_w = 0.84\, B_{rs}/\mu_{rdc} \quad \text{MHz}} \qquad \textbf{(7.6)}$$

where

B_{rs} = remanence after saturation in Gauss
μ_{rdc} = relative permeability at DC

There is also a resonance for magnetization rotation (see formula (7.8) under ferrites).

For Supermalloy we find f_w = 0.07 MHz, and for Mu-metal f_w = 0.2 MHz. In the practical case of a laminated core with final permeability below 5000 the frequencies are in the 1-2 MHz range.

The significance of this frequency is that it does not pay to reduce the lamination thickness beyond a certain thickness. (It should be emphasized, though, that f_w is a function of the material—it is not influenced by the lamination thickness.)

We first realized a decrease in effective permeability due to eddy current losses, and then we established a relationship between the lamination thickness, permeability, conductivity, and a frequency f, where eddy currents begin reducing the effective permeability. This relationship is shown in Fig. 7-8 by the sloped lines, shown for two values of the initial permeability, μ_r = 100,000 and μ_r = 5,000.

This relationship fails when the lamination thickness is below 3 μm (0.118 mils). The reason for this is the earlier mentioned wall resonance frequency f_s. If we eliminate the frequency f in our previous calculation, substituting the factor 0.84 then we find:

$$2\delta_{min} = \sqrt{\varrho/B_{rs}\mu_r} \qquad \textbf{(7.7)}$$

The operating frequency limitation from eddy currents and from the spin frequencies are shown together in Fig. 7-8 for the initial permeabilities of 100,000 and 5,000 for HyMu materials. Both curves show that the permeability will drop very fast if the lamination, in either case, is thinner than 3 μm (0.118 mil) (ref. Boll).

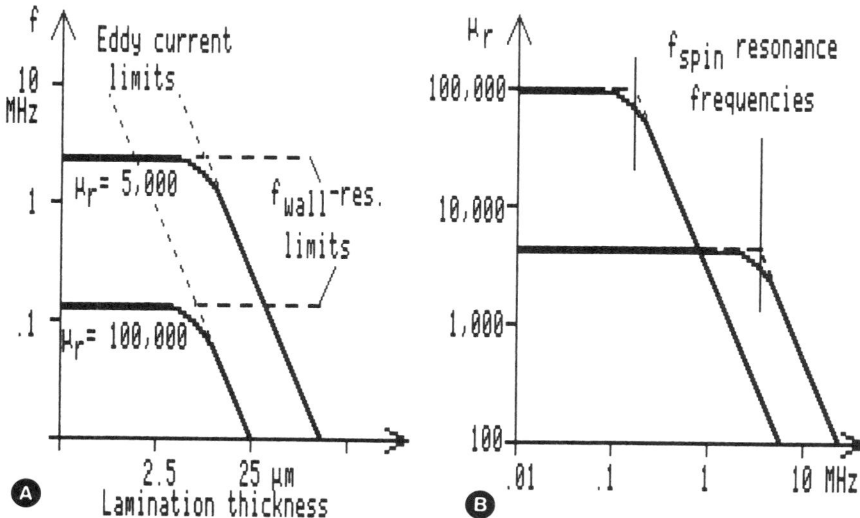

Fig. 7-8. Frequency limitations for laminated cores, (A). Operating frequency is limited by eddy currents (sloped lines), and by wall resonance frequency (horizontal lines), Bs = 8000 G, ∂ = 60 μohmcm; (B) Very high permeability causes early roll-off in permeability due to lower spin frequency, combined with eddy current losses, (A) (after Boll).

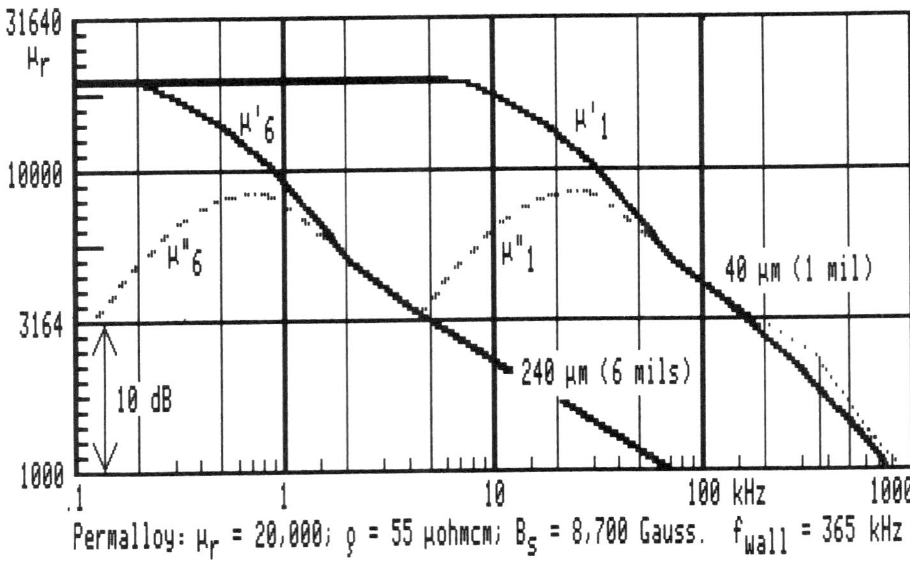

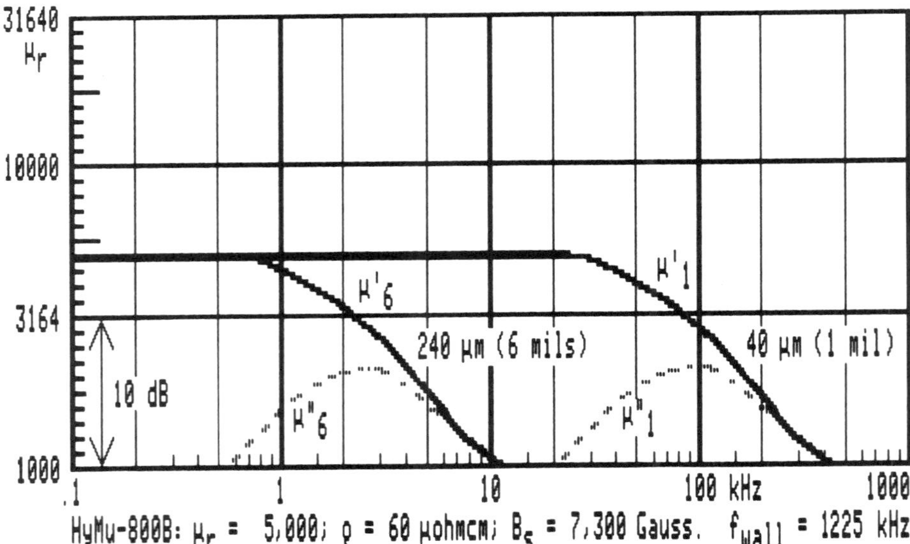

Fig. 7-9. Permeability spectra for permalloy and Hy-Mu 800-B.

The lamination with the very high permeability has a wall resonance frequency around 100 kHz and the permeability rolls off faster than eddy currents alone would cause (Fig. 7-8).

The essence of this discussion was to point out that a material with very high permeability is detrimental to the overall performance of a high frequency head. Also, it does not pay to make cores from laminations that are thinner than 3 μm = 0.118 mils. Now, such thin laminations are virtually impossible to work with in a piecewise assembled head, but they are easily deposited in thin film heads. And we have, from the above, learned that little, if anything, may be gained by deposition of laminations so thin that eddy current limitations are non-existent; the wall and magnetization rotation resonance frequency limitations remain.

We will, in later sections, calculate the efficiency of several heads, and will find that it is almost as high for materials with modest permeability (2-5000) as for very high permeability (100,000) materials.

The lower of the two curves in Fig. 7-8 does not show the early roll-off due to a low wall resonance frequency, and is, therefore, preferred over the very high permeability material. Figure 7-9 shows the permeability curves (or permeability spectrum) for two popular core materials, Permalloy and HyMu-800-B. The effects of eddy currents and wall resonance frequency roll offs are included.

Deposited Films

Very small head structures can be manufactured using the deposition and etch techniques developed by the semiconductor industry. The design of thin film heads are covered in Chapter 9.

There are limits to the smallness of the magnetic films, though. We already learned that problems occur when the thickness is reduced below 3 μm (formula (7.7)). It is generally found that the permeability decreases and the coercivity increases as the films are made thinner (ref. Feng). A double layer magnetic film appears to improve matters (ref. Herd).

It has been observed that distinct walls form in thin films (ref. Kryder, et al.). Figure 7-10 illustrates how the domains will form if the film is isotropic, i.e., its properties are independent of direction. The central wall will move from one side to the other during flux reversals, which

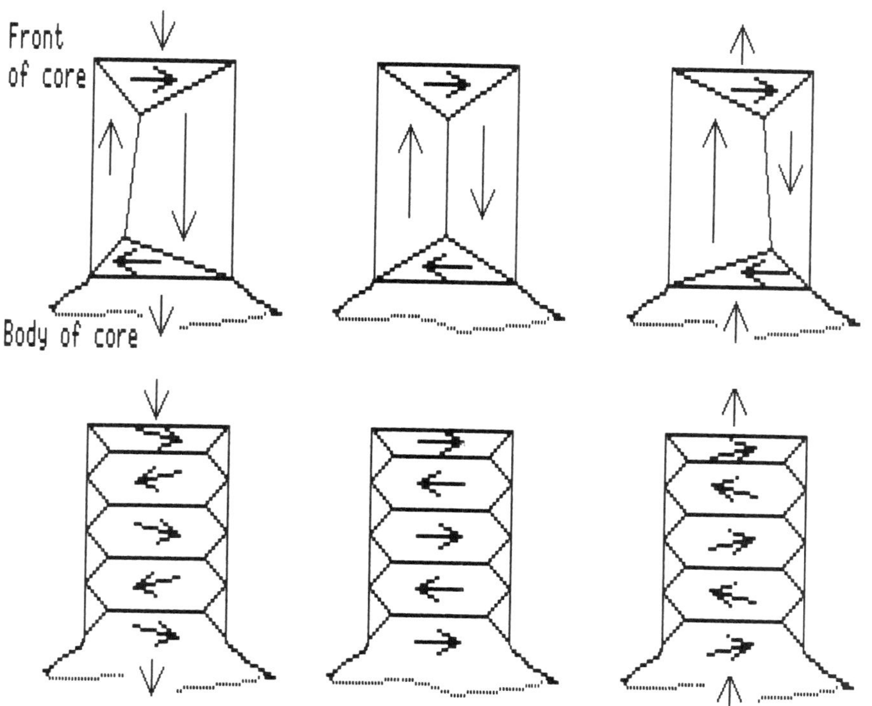

Fig. 7-10. Domain patterns in thin film head pole piece. Top: Isotropic material (after Wells and Savoy). Bottom: Anisotropic material (after Jones).

is a slow process. The film's permeability is improved by depositing an anisotropic film so the magnetization reversal takes place through rotation, which is a fast process (ref. Wells and Savoy, Jones).

FERRITE MATERIALS

Ferrites are magnetic ceramics that have come to play an important role in magnetic heads. Early after their discovery in the forties they were employed in erase heads because of low losses. Ferrite is a sintered material, and the first ferrites had a high porosity (5 percent at that time), which precluded their application for heads requiring precise gap dimensions.

These magnetically soft ferrites are generally formed as $MOFe_2O_3$, where MO is a bivalent metal oxide, such as Manganese (MnO) or Zinc (ZnO) mixed with the Fe_2O_3 powder. They are solid solutions and are sintered together. Their outstanding features are high resistivity with reasonably good magnetic properties. This means that they can operate with virtually no eddy current loss at high frequencies, and thus, be superior to laminated permalloy cores.

Efforts toward reducing the porosity succeeded in the sixties by hot pressing the ferrites during final sintering. The results were materials with 0.1 percent porosity or less. Earlier high density materials were also hot pressed, but at lower temperatures. These materials were primarily used for audio heads.

The details of ferrite fabrication are proprietary to the manufacturers, but the basic processing steps are similar:

(1) A powder of ferric oxide Fe_2O_3 is mixed with NiO or MnO (or NiO + ZiO or MnO + ZiO) in distilled water. They are thoroughly mixed in a ball mill to produce smaller particle sizes and a uniform mixture.

(2) After milling, the mixture is dried and pressed into loose blocks, which are presintered at 900-1100 C. This produces a partial formation of ferrite.

$$MO + Fe_2O_3 \rightarrow MOFe_2O_3$$

(3) This material is ground again to promote mixing of any unreacted oxides and to achieve further reduction in particle size, to about 0.5-1.0 μm. (Smaller sizes may be obtained by a method where an aqueous solution of Fe, Zi, and Ni hydrated sulphates are dehydrated in a spray drier at 250 degrees C. Decomposition at 800 degrees C in air produces particle sizes of 0.1-0.2 μm.)

(4) The powder is mixed with an organic binder and formed into its final shape which may be slabs about 1 by 2 by 3 inches, or 1″-2″ thick disks of 6″ diameter.

The final sintering has several process variables that to a great extent determine the final product: temperature, pressure, atmosphere, and time. Their combined effects upon grain size is shown in Fig. 7-11. This shows the proportionality to final sizes with temperature, pressure and time. A small grain size and freedom from pores provide a ferrite with higher resistance to wear and chipping. This conflicts with the desire to have a high permeability, which results from large grain sizes as shown in Fig. 7-12 (ref. Snelling).

The temperature cycle extends from one to many hours, and must be carefully controlled (ref. Withop). The atmosphere is even more critical, since it has a pronounced effect upon permeability and saturation induction.

The improved technique of applying pressure during the final sintering originated in ceramic tool manufacturing in Japan—a rather popular skill there. The pressure is in the order of 500 kg/cm^2 and may be applied in one or two directions, or may be isostatic. These hot pressed ferrites are successfully applied in heads (ref. Rigby, Kehr, and Meldrum).

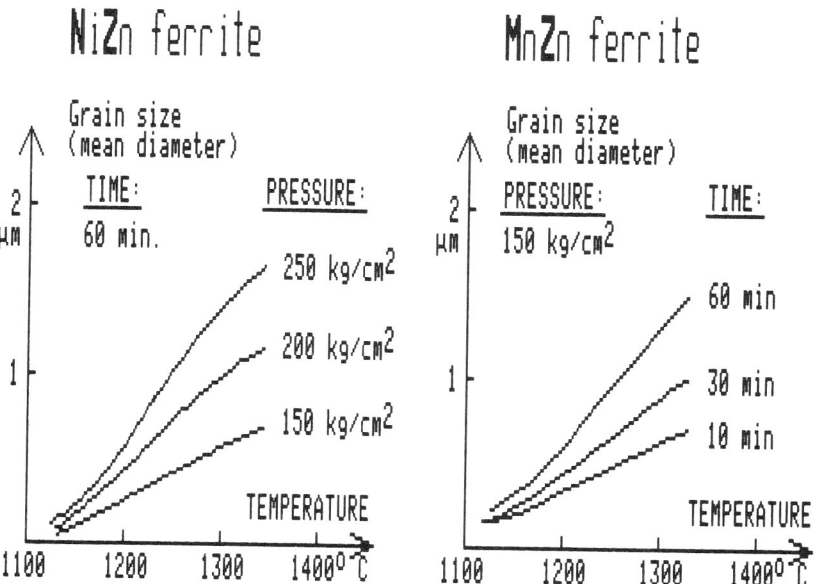

Fig. 7-11. Effects of pressure, time, and temperature on ferrite grain size (after Sugaya).

It is also possible to grow *single crystal ferrites* with magnetic properties similar to the poly-crystalline material. They often exhibit better wear characteristics, but are also fragile. The single crystal material is used for audio and video heads. One drawback of the single crystal material is anisotropy in regard to permeability and wear characteristics. The permeability varies as much as 2:1, and wear rates 3:1. Another factor is the limited range of ferrite compositions from which single crystals can be grown.

The magnetic properties of ferrites are heavily influenced by the starting powder composition, and the final sintering process. Manufacturers do not list the exact composition of their ferrites, calling them merely MnZn or NiZn types, with a letter/number specifying the various grades.

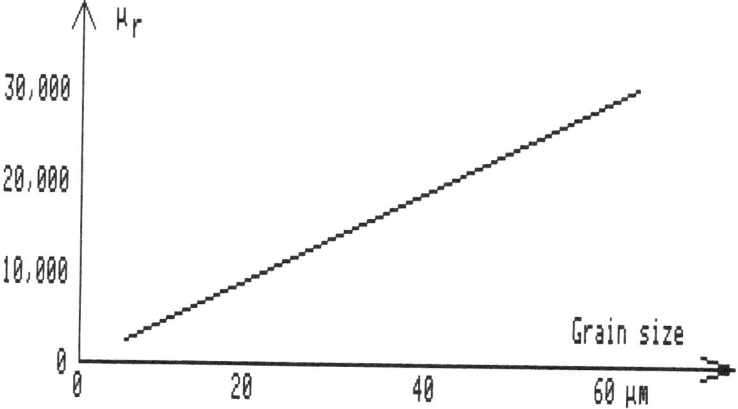

Fig. 7-12. Large grains result in high μ_r (after Snelling, Fig. 3.6).

The *MnZn ferrites* have initial permeabilities of 2000-5000, coercivities of less than 1.0 Oe, and resistivities of greater than 10^{+6} μohmcm (ref. Brissette, Takama).

NiZn ferrites are designed for very high frequency operations, with initial permeability of 100-200 and resistivity in the order of 10^{+10} μohmcm.

The properties of both ferrites are heavily influenced by the percent content of Zinc. As the nickel content is decreased from 50 to 30 percent of the MO content, then both saturation flux density and coercivity drop, while the permeability increases. (It is thus possible to tailor-make a ferrite.) Also, the higher the content of Zinc, the lower the Curie temperature.

Low Curie temperatures T_c are typical for ferrites; above this temperature the spontaneous magnetization disappears (Fig. 7-13). Composition plays a major role for the Curie temperature, intrinsic magnetization M, and the permeability μ_r; the latter is shown to have a very large variation with temperature. This, for instance, makes use impossible in video heads, where the frictional heat from the head-coating friction may exceed T_c.

The permeability of ferrites decreases toward higher frequencies. This is due to a slow-down in the rotation of the magnetization, or the spin's directions. Ferrites therefore exhibit a *permeability spectrum* with a roll-off at the spin frequency:

$$\boxed{f_s = 1.87\ B_{rs}/\mu_{rdc} \text{MHz}} \tag{7.8}$$

where

B_{rs} = remanence after saturation in Gauss
μ_{rdc} = relative permeability at DC

This spectrum is shown in Fig. 7-14 for three values of the permeability. A high spin frequency is obtained by sacrificing permeability, which will decrease the head's efficiency somewhat. The value of the remanent saturation flux density should be high, which also concurs with a desired high value for the material's application in a recording head.

Ferrites are always used as solid cores, and eddy currents may reduce the effective permeability for thick cores at high frequencies. The permeability spectrum for representative fer-

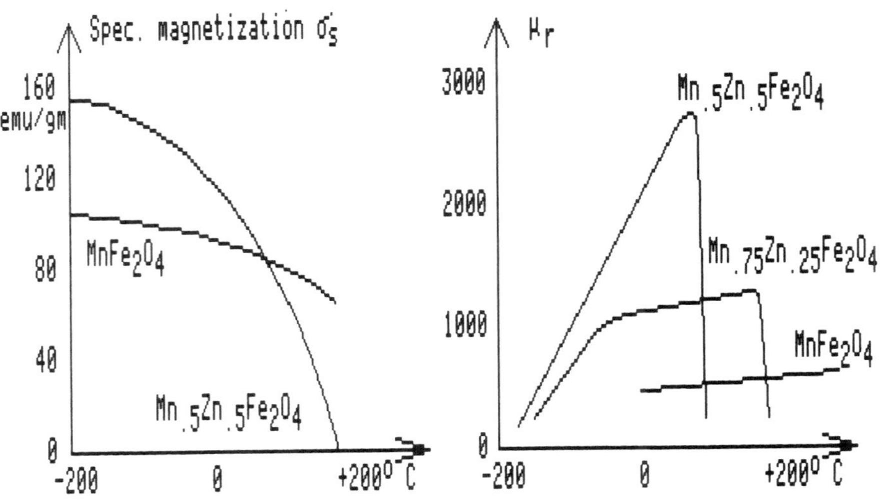

Fig. 7-13. Temperature dependence of magnetization and permeability.

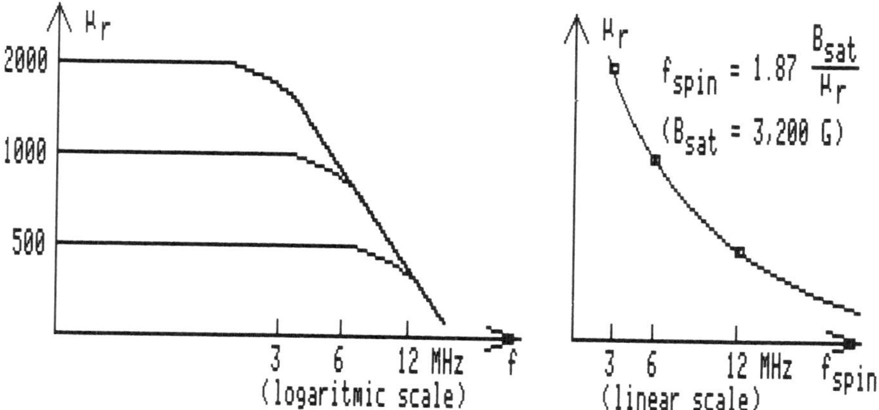

Fig. 7-14. Permeability spectrum for ferrites.

rites are shown in Fig. 7-15: manganese-zinc (MnZN) and nickel-zinc (NiZn). It is clear that eddy currents in MnZn can only be disregarded at low frequencies or when the core is very small. Modern high resistivity ferrites have to some extent eliminated this problem (ref. Brissette).

All curves shown in this chapter are guides only and accurate calculations of head performance requiring use of data from the materials data sheet. It is always best to make one's own measurements on a ring core sample of the ferrite (and as a sample that is free from machining stresses).

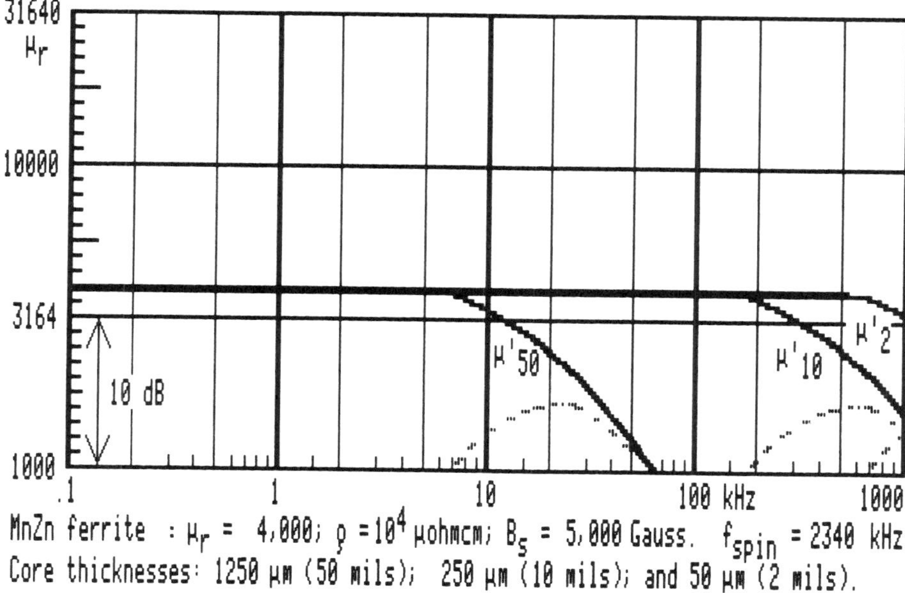

Fig. 7-15. Permeability spectrum for manganese-zinc and nickel zinc ferrite.

FLUX SENSING MATERIALS

We have so far considered the read signal from a tape or disk as the signal that is generated by flux changes in an inductor (the head core plus winding). There exist other means for reading the flux from a tape recording, such as the incorporation of a flux-sensitive material, or generation of second harmonics in a magnetic modulator.

Hall Effect

The semiconductor Indium-Antimonide will generate a voltage when magnetic flux lines pass through it; a DC-bias current is required, see Fig. 7-16. The voltage is proportional to the bias current J and a material constant K. The current is limited by the element's heat capacity and heat sink arrangement.

The element should be thin, and is in practice placed in one of the gaps in a conventional head structure. One practical layout (after Camras) is shown in Fig. 7-16B, where the element is placed in the front gap. The back gap is made large to concentrate the flux in the front gap.

It is correct that this *Hall element* will respond to a steady (DC) flux, but this is often projected to the erroneous statement that a Hall Head will read frequencies all the way down to DC. The core structure is still responsible for gathering the flux from the tape, and is limited in long wavelength response. An improved response may be obtained by deleting the back gap, but even this structure will gather less and less flux as the wavelength increases. Flux sensitive heads do not, therefore, respond well to DC during playback.

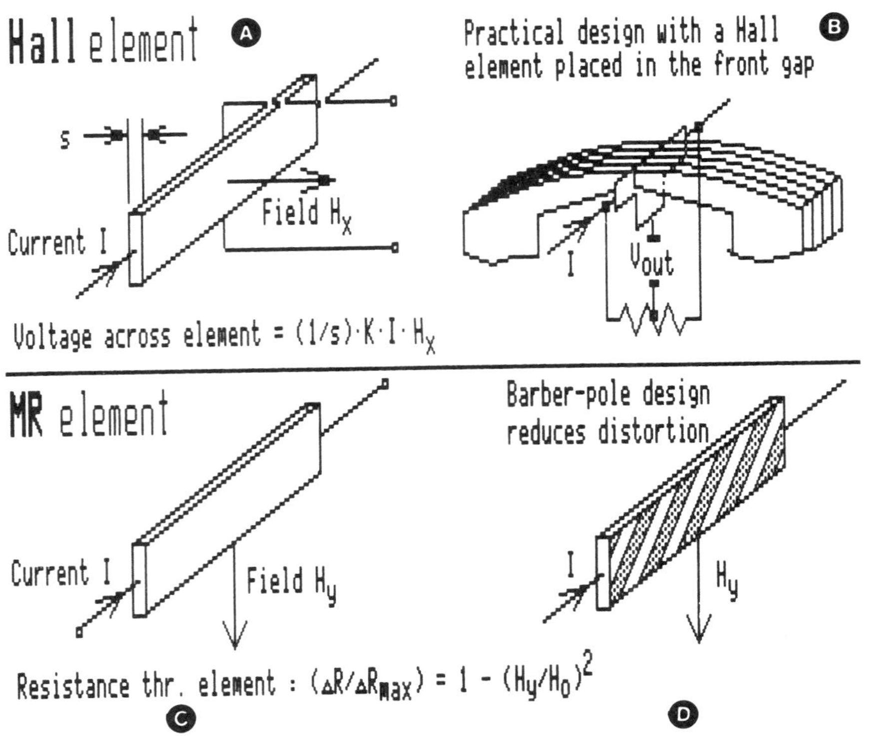

Fig. 7-16. Flux-sensitive heads, (A). Hall Effect, (B). Practical Hall element, (C). Magnetoresistive Effect, (D). Barber-Pole MRH.

Magnetoresistance Effect

The electrical resistivity of certain iron-nickel alloys will increase under the influence of an external magnetic field (ref. Thompson). The basic concept is shown in Fig. 7-16C.

The flux from a tape has a longitudinal and a perpendicular component. The latter, H_y, will change the element's resistance value as indicated, where H_o is the magnetic field that is proportional to the current density J in the element (ref. Druyvesteyn, et al.).

The response of ΔR to H_y is non-linear. It was found that this relationship could be improved by introducing an external bias field at an angle of 45 degrees (ref. Kuijk). Thus the *barberpole* head arrived, named after the striped pole seen outside barber shops. Other approaches (coupled magnetic strips) also achieve this linearity (ref. Jeffers).

The performance of these MRH assemblies (Magneto-Resistive-Heads) have for some reason not led to a large share of the market place for heads. A lot of activity is reported, and there appears to be one drawback: high noise level (spikes) from fluctuating heat generated by the head-to-tape friction. This should not matter in a flying head assembly.

From the standpoint of fabrication, the MRH appears just as attractive as the metallic thin film heads discussed earlier. Recently IBM has employed the MRH element in a magnetic bridge circuit, so the element is not in contact with the moving tape, and no frictional heat noise is produced (Fig. 10-29).

Other Flux Sensing Methods

Many other concepts for playback have been reported, tried, and used for special purposes. *Magnetic modulators* showed great promise, but were extremely sensitive to external noise; the SNR (signal-to-noise-ratio) was only 40 dB.

Wayne Johnson discovered a principle whereby the magnetic coating of the tape itself aced as a modulator. The SNR was about 80-90dB and response was flat all the way down to DC (a tape standing still would provide an output proportional to the coating magnetization, with proper polarity). The pursuit of this technique was given up when it was found that writing and reading a digitized version of the signal provided the same high SNR plus freedom from time displacement errors (see Chapter 27).

REFERENCES FOR CHAPTER 7

General

Jones Jr., R.E., and Mee, C.D., "Recording Heads," in *Magnetic Recording*, Vol. 1, Ed. by C.D. Mee and E.D. Daniel, McGraw-Hill, 1987, pp. 244-336.

Chen, C-W., *Magnetism and Metallurgy of Soft Magnetic Materials*, Dover, N.Y., 1986, 571 pages.

Bradley, F.N., *7 Materials for Magnetic Functions*, Hayden Book Co., 1971, 348 pages.

Roters, H., *Electromagnetic Devices*, John Wiley & Sons, 1955, 559 pages.

Katz, E.R., "Numerical Analysis of Ferrite Recording Heads with Complex Permeability," *IEEE Trans. Magn.*, Nov. 1980, Vol. MAG-16, No. 6, pp. 1404-1409.

Eddy Currents

Lee, E.W., "Eddy-Current Losses in Thin Ferromagnetic Sheets" *Proc. IEE*, C105 58/13 pp. 337-342.

Boll, R., "Metalliche Magnetwerkstoffe bei hohen Frequenzen," *Technische Informationsblatter*, Vacuumschmelze AG, 1960, 8 pages.

Boll, R., "Wirbelstron- und Spinrelaxationsverluste in duennen Metallbandern bei Frequenzen bis zu etwa 1 MHz," *Zeits. fur angw. Physik*, May 1960, Vol. 12, No. 8, pp. 212-223.

Boll, R., "Uber Bezichnungen dem Schaltkoeffizienten und der Grenzfrequenz ferromagnetischer Werkstoffe," *Zeits. fur angw. Physik*, Aug. 1960, Vol. 12, No. 8, pp. 364-370.

Deller, R., "Hochfrequenz- und Impulseigenschaften dunner weichmagnetischer Bander," *Zeits. fur Angew. Physik*, Mar. 1963, Vol. 15, No. 3, pp. 253-256.

Peterson, E., and Wrathall, L.R., "Eddy Currents in Composite Laminations," *JAES* (reprint), Dec. 1977, Vol. 25, No. 12, pp. 1026-1032; Original: *Proc. IRE*, Feb. 1936, Vol. 24, No. 2, pp. 275-286.

Metallic Materials

Boll., R., *Soft Magnetic Materials*, Siemens/Heyden, 1977, 349 pages.

Hall, R., "Single Crystal Anisotropy and Magnetostriction Constants of Several Ferromagnetic Materials Including NiFe, SiFe, AlFe, CoNi and CoFe," *Jour. Appl. Phys.*, 1959, Vol. 38, pp. 816-819.

Boll, R., "Dunne Bander aus Weissmagnetisden Werkstoffen und ihre Anwendungsformen," *Elektronik*, Oct 1961, Vol. 10, No. 10, pp. 293-298.

Miyazaki, T., "New Magnetic Alloys for Magnetic Recording Heads," *IEEE Trans. Magn.*, Sept. 1972, Vol. MAG-8, No. 3, pp. 501-502.

Kolotov, O., "Pulse Switching in Thin Magnetic Films," *IEEE Trans. Magn.*, Dec. 1974, Vol. MAG-10, No. 4, pp. 1023-1027.

Hubert, A., "Domain Wall Structures in Thin Magnetic Films," *IEEE Trans. Magn.*, Sept. 1975, Vol. MAG-11, No. 5, pp. 1285-1290.

Bendson, S., "New Wear-Resistant Permalloy for Optimum High-Frequency Permeability and Machinability," *JAES*, 1976, Vol. 24, pp. 562-567.

McKnight, J., "The Permeability of Laminations for Magnetic Recording and Reproducing Heads," *JAES*, Preprint no. 1265, May 1977.

Feng, J., "Permeability of Narrow Permalloy Stripes," *IEEE Trans. Magn.*, Sept 1977, Vol. MAG-13, No. 5, pp. 1521-1523.

Hempstead, R., "Unidirectional Anisotropy in Nickel-Iron Films by Exchange Coupling with Antiferromagnetic Films," *IEEE Trans. Magn.*, Sept. 1978, Vol. MAG-14, No. 5, pp. 521-523.

O'Handley, R., "Low-Field Magnetic Properties of Wide Fe-Ni Metallic Glass Strips," *IEEE Trans. Magn.*, Mar. 1979, Vol. MAG-15, No. 2, pp. 970-972.

Liebermann, H.H., "Manufacture of Amorphous Alloy Ribbons," *IEEE Trans. Magn.*, Nov. 1979, Vol. MAG-15, No. 6, pp. 1393-1397.

Jones, R., "Domain Effects in the Thin Film Head," *IEEE Trans. Magn.*, Nov. 1979, Vol. MAG-15, No. 6, pp. 1619-1621.

Herd, S.R., "Magnetization Reversal in Narrow Strips of NiFe Thin Films," *IEEE Trans. Magn.*, Nov. 1979, Vol. MAG-15, No. 6, pp. 1824-1826.

Kryder, M., "Magnetic Properties and Domain Structures in Narrow NiFe Stripes," *IEEE Trans. Magn.*, Jan. 1980, Vol. MAG-16, No. 1, pp. 99-103.

Preifer, F., and Radeloff, C., "Soft Magnetic Ni-Fe and Co-Fe Alloys-Some Physical and Metallurgical Aspects," *Jour. of Magnetism and Magnetic Materials*, Apr. 1980, Vol. 19, No. 1-3, pp. 190-207.

Wells, O. and Savoy, R., "Magnetic Domains in Thin-Film Recording Heads as Observed in the SEM by a Lock-In Technique," *IEEE Trans. Magn.*, May 1981, Vol. MAG-17, No. 3, pp. 1253-1261.

Tsuya, N., "Magnetic Recording Head Using Ribbon-Sendust," *IEEE Trans. Magn.*, Nov. 1981, Vol. MAG-17, No. 6, pp. 3111-3113.

Raskin, D., and Davis, L.A., "Metallic Glasses: A Magnetic Alternative," *IEEE Spectrum*, Nov. 1981, Vol. 18, No. 11, pp. 28-33.

Tsukagoshi. T., Ito, H., Ogasawara, K., Namiki, T., Uasuda, S., Masumoto, Y., Ota, H., and Yamamoto, T., "Ribbon Sendust Magnetic Tape Heads," *JAES*, Dec 1981, Vol. 29, No. 12, pp. 867-872.

Radeloff, C., "Metallische Magnetkopfwerkstoffe," *Sonderdruck aus NTG-Fachberichten*, 1982, 7 pages.

Engstrom, H., "Equivalent Circuit of a Thin Film Recording Head," *IEEE Trans. Magn.*, Sept. 1984, Vol. MAG-20, No. 5, pp. 842-844.

Hilzinger, H.R., "Application of Metallic Glasses to the Electronics Industry," *IEEE Trans. Magn.*, Sept. 1985, Vol. MAG-21, No. 5, pp. 2020-2022.

Ferrites (Ceramics)

Snelling, E.C., and Giles, A.D., *Ferrites for Inductors and Transformers*, Research Studies Press Ltd. (Wiley), 1983, 167 pages.

Withop, A., "Maganese-Zinc Ferrite Processing, Properties and Recording Performance," *IEEE Trans. Magn.*, Sept. 1978, Vol. MAG-14, No. 5, pp. 439-441.

Takama, E., "New Mn-Zn Ferrite Fabricated by Hot Isostatic Pressing," IEEE *Trans. Magn.*, Nov. 1979, Vol. MAG-15, No. 6, pp. 1858-1860.

Brissette, Leo, "MnZn Ferrites Extend Performance," *Electronic Design* 1982.

Roess, E., "Soft Magnetic Ferrites and Applications in Telecommunication and Power Converters," *IEEE Trans. Magn.*, Nov. 1982, Vol. MAG-18, No. 6, pp. 1529-1534.

Rigby, E.B., Kehr, W.D., and Meldrum, C.B., "Preparation of Coprecipitated NiZn Ferrite," *IEEE Trans. Magn.*, Sept. 1984, Vol. MAG-20, No. 5, pp. 1506-1508.

Flux Sensitive Materials

Thompson, D., "Thin Film Magnetoresistors in Memory, Storage, and Related Applications," *IEEE Trans. Magn.*, July 1975, Vol. MAG-11, No. 4, pp. 1039-1050.

Kuijk, K., "The Barber Pole, A Linear Magnetoresistive Head," *IEEE Trans. Magn.*, Sept. 1975, Vol. MAG-11, No. 5, pp. 1215-1217.

Jeffers, F., "Magnetoresistive Transducer with Canted Easy Axis," *IEEE Trans. Magn.*, Nov. 1979, Vol. MAG-15, No. 6, pp. 1628-1630.

Druyvesteyn, W., "Magnetoresistive Heads," *IEEE Trans. Magn.*, Nov. 1981, Vol. MAG-17, No. 6, pp. 2884-2889.

Fluitman, J.H.J., "Comparison of a Shielded 'One-Sided' Planar Hall-Transducer with an Mr-Head," *IEEE Trans. Magn.*, Nov. 1981, Vol. MAG-17, No. 6, pp. 2893-2895.

BIBLIOGRAPHY TO CHAPTER 7

Heck, C., *Magnetische Werkstoffe und ihre Technische Anwendung*, Dr. Alfred Huetig Verlag Heidelberg, 1975, 748 pages.

Kornei, O., "Structure and Performance of Magnetic Transducer Heads," *JAES*, July 1953, Vol. 1, No. 3.

Boll, R., and Hilzinger, H.R., "Comparison of Amorphous Materials, Ferrites, and Permalloys," *IEEE Trans. Magn.*, Sept. 1983, Vol. MAG-19, No. 5, pp. 1946-1951.

Metallic Materials

Hasegawa, R., *Glassy Metals: Magnetic, Chemical and Structural Properties,* CRC Press, 1983, 280 pages.

Umesaki, M., Magnetic Properties of Super Sendust Films,'' *IEEE Trans. Magn.*, Nov. 1982, Vol. MAG-18, No. 6. pp. 1182-1184.

Murata, Y., and Shirae, K., ''Susceptibility of Amorphous Magnetic Materials Over a Wide Frequency Range,'' *IEEE Trans. Magn.*, Sept. 1984, Vol. MAG-20, No. 5, pp. 1302-1304.

Major, R. and Martin, M., ''The Effect of Grain Orientation on the Initial Permeability of Mumetal Strip,'' *IEEE Trans. Magn.*, Mar. 1970, Vol. MAG-6, No. 1, pp. 101-105.

Preece, I. and Thomas, R., ''The Effects of Various Stresses on Mumetal: 77 Percent Ni-14 Percent Fe-5 Percent Cu-4 Percent Mo,'' *IEEE Trans. Magn.*, Sept. 1971, Vol. MAG-7, No. 3, pp. 554-556.

Ferrites (Ceramics)

Smit, J. and Wijn, P., *Ferrites,* Wiley, 1959.

Lax, B., and Button, K.J., *Microwave Ferrites and Ferrimagnetics*, McGraw-Hill 1962 752 pages.

Neel, L., ''Proprietes Magnetiques des Ferrites: Ferrimagnetisme et Antiferromagnetisme,'' *Ann. Phys.*, Paris, March 1948, pp. 137-98.

Sugaya, H., ''Newly Developed Hot-Pressed Ferrite Head,'' *IEEE Trans. Magn.*, Sept. 1968, Vol. MAG-4, No. 3, pp. 295-301.

Mizushima, M., ''Mn-Zn Single Crystal Ferrite as a Video-Head Material,'' *IEEE Trans. Magn.*, Sept. 1971, Vol. MAG-7, No. 3, pp. 342-345.

Monforte, F., ''Pressure Sintering of MnZn and NiZn Ferrites,'' *IEEE Trans. Magn.", Sept. 1971, Vol. MAG-7, No. 3, pp. 345-350.*

Lemke, J., "Ferrite Transducers," Ann. N.Y. Acad. Sci., Mar. 1972, Vol. 189, pp. 171-189.

Lagrange, A., ''Preparation and Properties of Hot-Pressed Ni-Zn Ferrites for Magnetic Head Application,'' *IEEE Trans. Magn.*, Sept. 1972, Vol. MAG-8, No. 3, pp. 494-497.

Kornei, O., ''Survey of Flux-Responsive Magnetic Reproducing Heads,'' *JAES,* Mar. 1954, Vol. 2, No. 3, pp. 86.

Chapter 8

Magnetic Fields From Heads

The function of the record head is to create a permanent magnetization pattern on a magnetic tape or disk moving past. The recording process is performed by the magnetic field from the gap in a magnetic core, energized by a current through its winding.

Upon playback, or read-out, the head performs a sensing function whereby its field pattern senses any magnetization moving past the gap in the head and converts the detected magnetization into flux through the core; this flux will in turn induce a voltage in the winding.

Investigation of the field in front of the magnetic head is therefore playing an increasingly important role in recording technology. This is spurred by the advent of perpendicular write/read modes, and the requirements for high density tracks. We wish to answer the following questions:

A. How does the field strength vary on the trailing side of the gap in the write head? It must decay fast for the best writing of short transitions, and short wavelengths.

B. What is the direction of the field on the trailing edge of the write gap? Longitudinal? Perpendicular? Or both?

C. How much wider than the head core is the field? This will ultimately set the limit for the smallest trackwidth, for writing as well as reading.

We need to explore the fields from various head configurations, and will in this chapter discuss several ways of doing this:

We will start by describing some past and present head designs, and their fields. Plots of these fields can be made by free-hand, aided by measurements or models. We will next develop a simple method for computing and plotting the fields from current carrying wires, and demonstrate how it can be used to model some of the head fields.

This development leads us gently into the mathematics of fields, paving the way to finite-difference and finite-element computations. A full description of these methods is beyond the scope of this book, so ample references are supplied.

With our knowledge of fields we can proceed to investigate how wide a track is laid down; it will always be wider than the core width dimension, due to stray fields on the sides, and it poses an upper limit to the number of tracks that can be written parallel to each other. This limitation will also apply to the read process. In this connection the reader is reminded that the design of magnetic fields with sharply defined fields is as difficult as designing electric circuits for use in salt water. Electric circuits enjoy a difference in conductivity between copper and air of more than 10^{20}, while magnetic circuits have magnetic conductances that only differ by 10^2 to 10^4.

We are also working with dimensions that preclude using the design techniques from antennas, where multiple elements aid in focusing the field. Any useful directivity will require an operating frequency of 10^{14} Hz, which happens to be near the frequency of light.

The contents of this chapter leaves the inventor-to-be of novel heads with the knowledge to cleverly apply materials and processes, such as the phasing of multiple pole pieces; that's why this chapter on fields has been made quite detailed.

FIELDS FROM MAGNETIC HEADS

The field lines from a gap in a ring core head are semicircular patterns outside the recording gap. This pattern has been studied extensively in the past years, since it, together with the tape magnetic characteristics, determines the recording resolution. The direct recording process without bias was introduced in Chapter 5, and will be further treated in Chapter 19 (Digital recording; write). It will then become clear that a desired feature of the recording field is that the field strength decreases rapidly with distance from its maximum value over the center of the gap.

Figure 8-1 illustrates the field patterns associated with the conventional ring core head (so-called, in spite of its rectangular shape, since its original form was a round core shape). We are first of all interested in the field in and around the front gap. The field into the coating, included that from both corners, does the recording. A picture of the field pattern can be developed by employing a couple of magnetic poles (cut from Mu-metal lamination stock), energized by a DC-current through a suitably located winding, see Fig. 8-2. Place a stiff piece of paper on top of the magnet, with iron powder evenly sprinkled on top of the card. Now gently tap the card, and the powder will move about to form the patterns shown in Fig. 8-3.

The field on the side and in back of the front gap dilutes the head's efficiency; the stray flux at the back gap has little bearing on the head's performance, and will not be discussed further.

The *leakage flux* around the coil decreases the head's efficiency when applied as a write head. This leakage is increased when shielding around the core, or between cores in a multitrack head, is employed.

Mathematical expressions for the various fields are often needed in calculations of recording performance. The most commonly used equations for the field in front of the write-gap is due to Karlqvist, and the expressions for the longitudinal component H_x and perpendicular component H_y are expressed in Fig. 4-24. In the past only the longitudinal (also called horizontal) component was considered, justified by the fact that the tape coating particles were longitudinally oriented and demagnetization losses were not too great.

Today's packing densities are very high, and the assumption of a purely longitudinal recording scheme is not valid. This matter was covered in Chapter 5, and much work has revealed the presence of strong perpendicular components in recordings.

There is, for the time being, no unified treatment of the recording process which expresses itself in a few useful formulas. All subsequent discussions in this book, therefore, will rely upon the reader's understanding of the interaction between a medium's switching field distribution and the write head field pattern. And we will use the curves of constant, total field strength, rather than those of constant longitudinal or perpendicular field components.

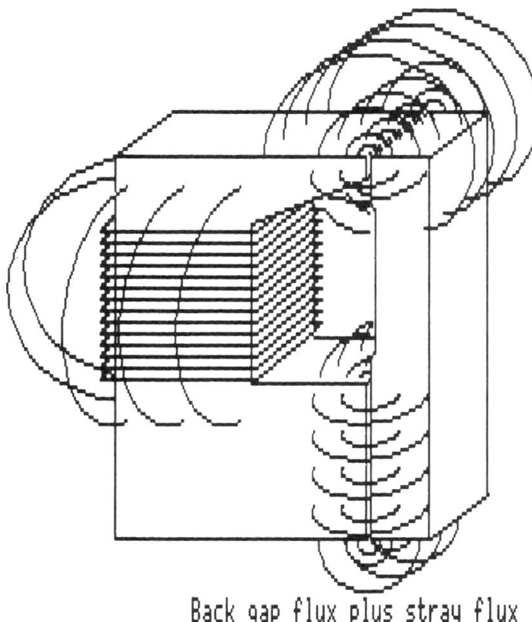

Fig. 8-1. Magnetic flux paths in and around a write head.

We will further assume a value of one for the coating's relative permeability; this appears justified in the write process, where the coating is magnetized to saturation.

HIGH GRADIENT HEADS

Improvements in the field gradient over that of the stray field from a gap are possible. Attempts to do this have not been great successes, mainly due to difficulties in the manufacturing processes. The *crossfield head*, created by Camras, uses superposition of two AC-bias fields with pole pieces and their strengths arranged in a manner that produces a higher gradient. This is illustrated in Fig. 8-4. This head has (to the author's knowledge) only been used by the early Roberts (Akai) and Tandberg recorders.

Another design utilizes the limitation of field penetration due to eddy currents. This head, called the *driven gap* record head, has the bias current running in a conductor wedged between two pole pieces of highly permeable materials. If the bias frequency is high enough (above 50 MHz) then the field is confined, as shown in Fig. 8-5. The gradient can be several times higher than that for a conventional head.

The driven gap head has been used to record data with a rate of 60 MBPS, utilizing a bias frequency of 150 MHz (ref. Krey). Multichannel applications were difficult due to lack of uniformity between tracks; this could be a problem of the past if modern micro-circuit techniques are used in the manufacture, whereby the conductor geometry is closely controlled.

The bias current is in the order of 2 A for a coating with H_c = 300 Oe. The difficulty in providing this current at UHF was readily solved by connecting the gap conductor as the termination in a quarter-wavelength coax cable.

A variety of the driven-gap head has been employed in studio recorders and duplicators (ref. K. Johnson). A conductive shim was wedged between the two front pole pieces, and shorted

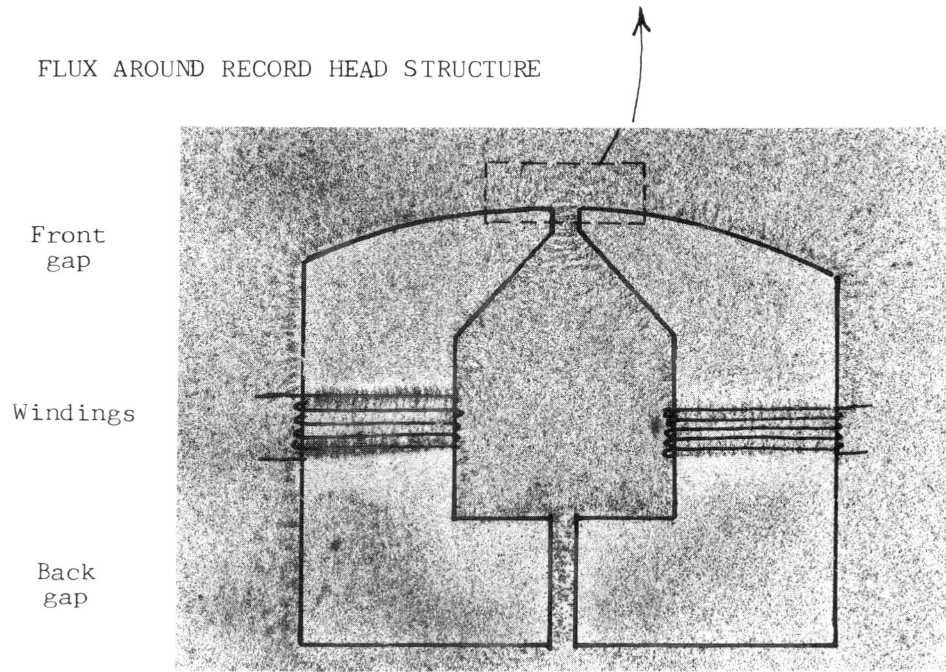

Fig. 8-2. Fluxlines made visible by iron powder (front gap).

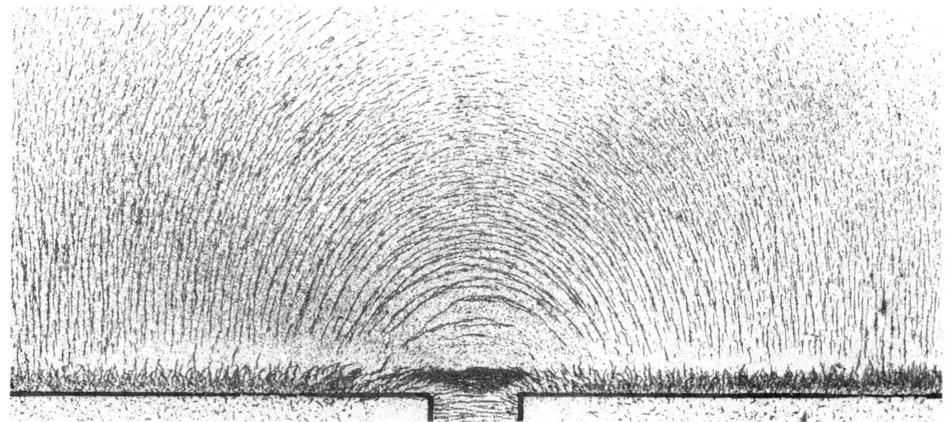

Fig. 8-3. Field lines made visible by iron powder.

to the head-housing block. A VHF bias applied to the ordinary winding would induce circulating currents in the shim, with such polarity that the total field was focused.

Figures 8-6, 8-7, and 8-8 show 3D-pictures of the field magnitudes in front of the standard ring core gap, the crossfield head, and the driven gap heads respectively. The superior field gradient of the latter is evident, as summarized in Fig. 8-9.

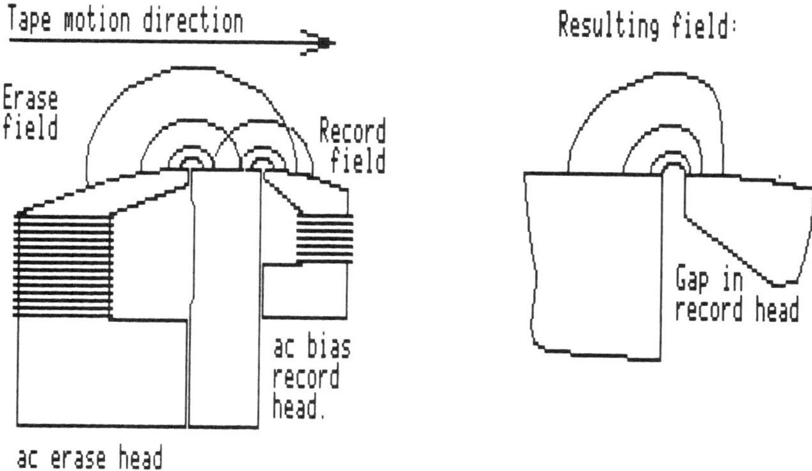

Fig. 8-4. Cross-field head (after Camras).

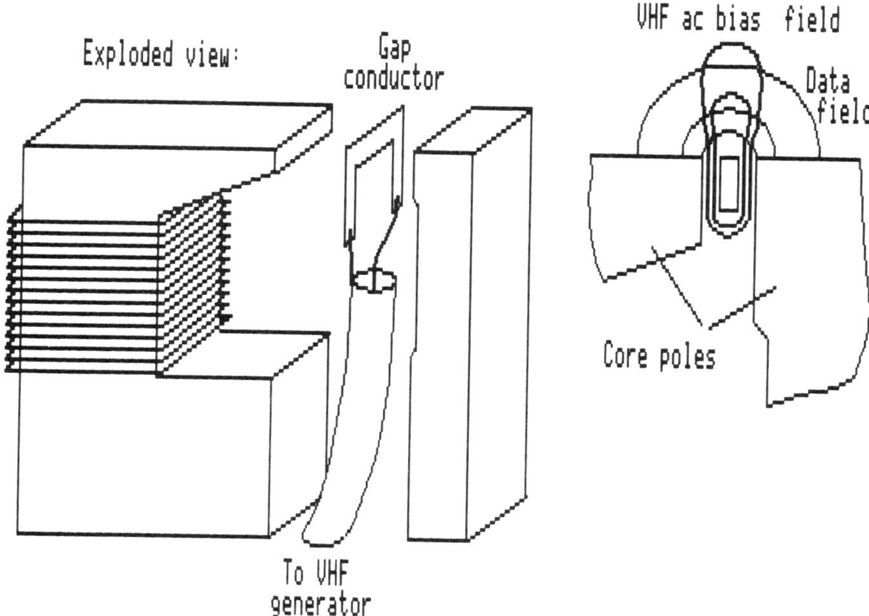

Fig. 8-5. Driven gap record head (after Johnson, Jorgensen).

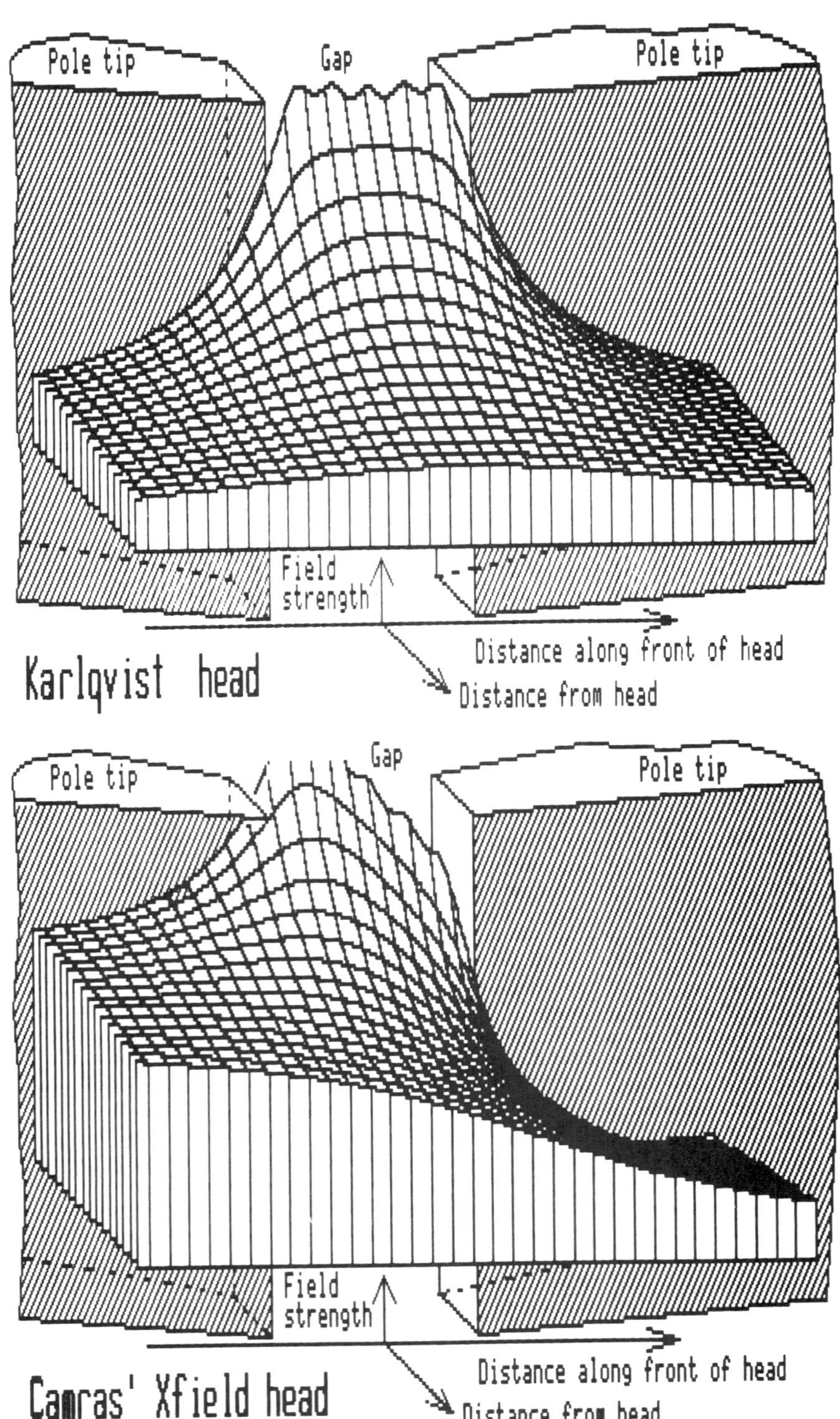

Fig. 8-6. Field strength in front of gap in a ring core head.

Fig. 8-7. Field strength in front of crossfield head gap.

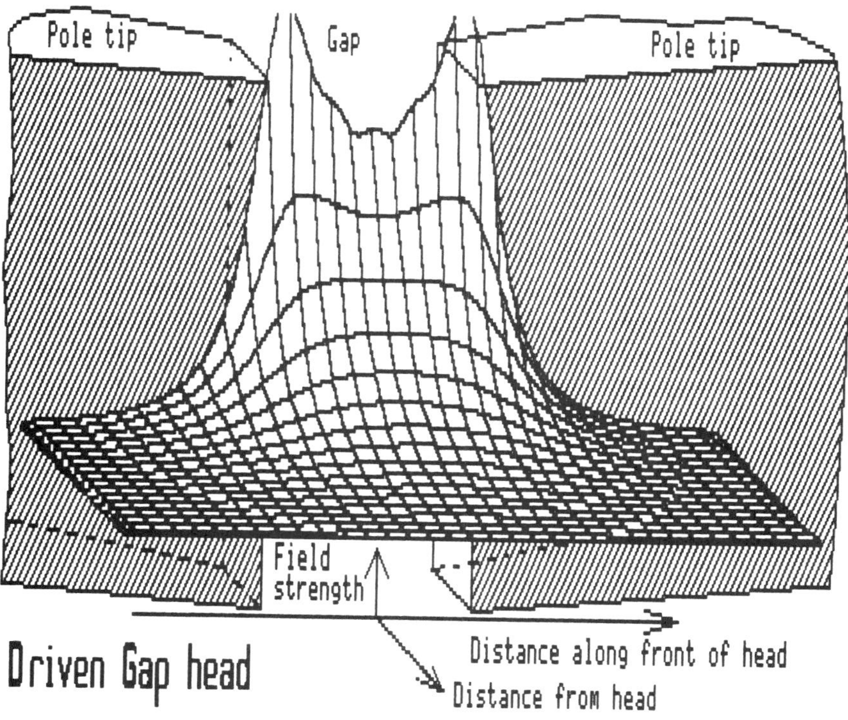

Fig. 8-8. Field strength in front of driven gap head.

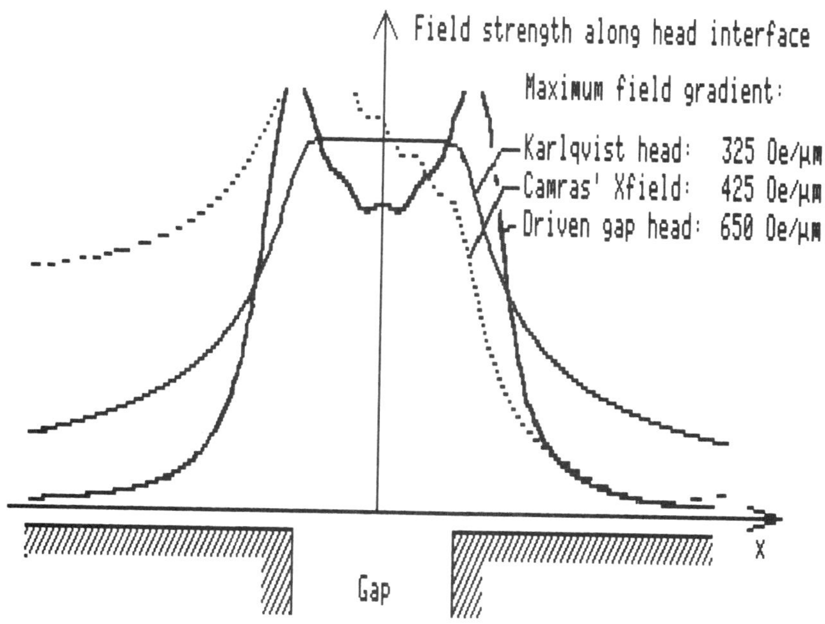

Fig. 8-9. Field strength in front of 3 heads, computed at a distance of ¼ gap length away.

The high gradient heads described above are, by virtue of their field patterns, also perpendicular recording heads (the H_y field being greater than the H_x field component). Experiments with heads having perpendicular fields continue in order to develop a more practical and efficient way to assist the perpendicular tape remanence. We shall return to a discussion thereof after we have learned more about computing and plotting head fields.

COMPUTATIONS AND PLOTS OF FIELDS

Iron Powder Patterns

A good qualitative method of establishing the shape of magnetic fields has been demonstrated in Fig. 8-3, using iron powder. A good source for the powder is the local hobby store: ask for Iron, Powdered Fe, item no. 721 from Perfect Parts Company, Baltimore, MD 21224.

How to Sketch Field Patterns by the Reluctance Method

Quite accurate field patterns may be drawn by following a few simple rules. It is recalled from high school physics that electric field lines in a simple capacitor (Fig. 8-10) run straight from plate to plate; and so it is for the magnetic field lines between two magnetic poles. Outside these uniform regions the field lines are curved, as shown.

The primary aids in drawing field lines are shown as broken lines: the potential lines. Here they are drawn in by free-hand, but they may be determined by the use of an electric model, as will be described later.

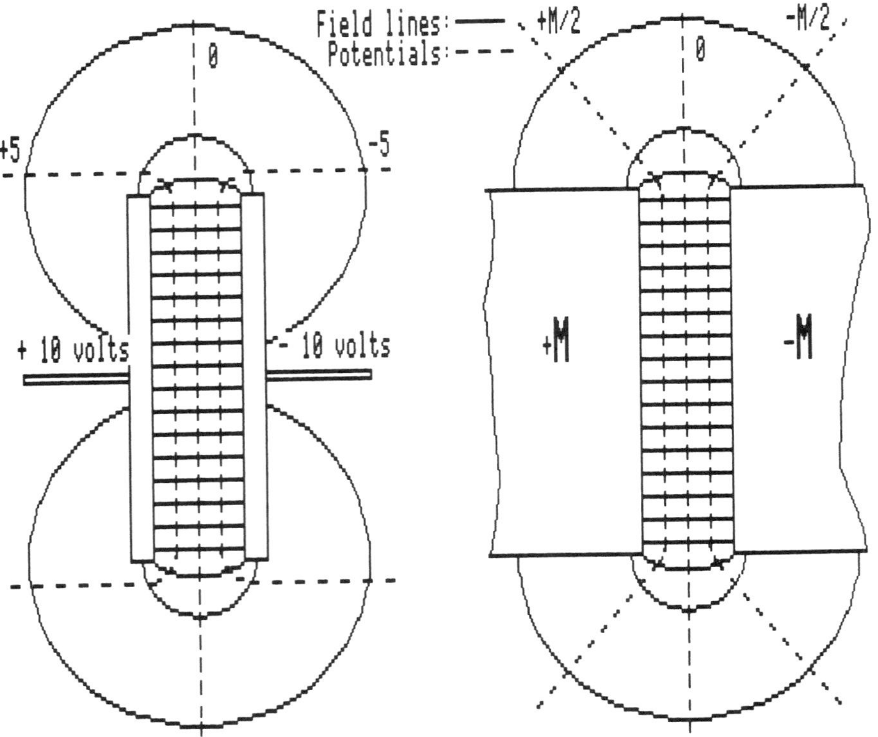

Fig. 8-10. Analogy between electric and magnetic fields.

If, for a moment, we look at the magnetic field lines we see that the distances between them increase as we move away from the gap between the poles. This simply means that the field strength decreases in the same proportion. The square pattern of field and potential lines between the poles divides the space up into *reluctance* elements, which all have the same value of reluctance (reluctance equals magnetic resistance).

Assume that the pole pieces extend w meters into the paper. The volume defined by the two sheaths following adjacent field lines, and of width w meters, is called a *flux tube*. It is a characteristic of the field line drawing that all flux tubes carry the same amount of flux; the magnetic potential difference equals 2M, and the flux in each tube equals $2M/\Sigma R_m$ (= $2M/4R_m$ in Fig. 8-10 since there are 4 cubes in series, each of magnetic resistance R_m). This is quite clear for the flux tubes in the uniform region between the poles.

Outside the poles the flux tubes become longer, but the distance between field lines increases so that the area of each tube also increases. This geometry change is such that the reluctance of each tube remains unchanged! This leads to the requirement that the elements dx, by, dy should remain square. That is not possible so the requirement is relaxed to the following rule:

1. Start the field drawing by using squares in a region where the field is uniform (as between the pole pieces in the front gap of a magnetic head).
2. Extend the potential lines into adjacent regions.
3. Draw the field lines so they form distorted square pillow patterns with the potential lines.
4. Step three must be done in such a way that the lines are perpendicular to each other where they intersect; this requirement overrides the prettiness of the pillow shape. A good check is to also draw a pillow's diagonals: they must be perpendicular to one another.

These rules are summarized in Fig. 8-11.

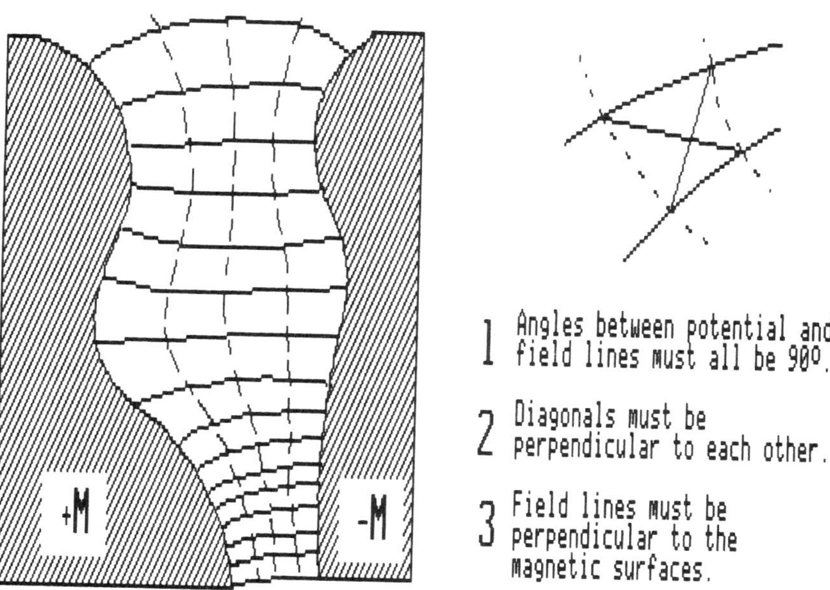

Fig. 8-11. Drawing of field liner orthogonal (perpendicular) to potential lines.

This graphic method does require a good estimate of the patterns of the potential lines, and that ability only comes with practice and experience. There is a simple method for the determination of points on potential lines, using the analogy that exists between electric and magnetic fields.

PLOT OF ELECTRIC POTENTIALS

This method is also known from high school physics, using so-called Teledeltos paper, an electrically conducting paper. The only problem is that such paper does not exist anymore—the author has searched in vain. An alternative is to use an electrolytic tank model (see references), but that tends to get a bit messy.

Another alternative is a different conductive paper, available from EDO Western Corporation, Salt Lake City, Utah 84115; part no. 22250-3. It is a black paper stock, coated with white wax, and is used in level recorders where a conductive stylus cuts through the wax and finds its proper position by sensing a voltage on the paper.

The procedure for making a model is now quite simple. Prepare a base by gluing 1⁄16″ artist cardboard onto one or more pieces of balsa wood. Next glue the conductive paper onto the cardboard using, for example, 3M Co.'s Super 77 Spray Adhesive.

Now use a soft pencil to trace the pole patterns onto the paper. Next scrape the wax overcoat away inside each pole pattern, exposing the black conductive paper in 3⁄32″ wide stripes. Use the flat side of a tool screwdriver so as not to cut or scratch the carbon paper.

The electrodes for the pole pieces themselves are best made from a 0.005″ flexible brass stock (from K and S Engineering Co., Chicago, IL 60638). Cut the pieces to exact size so they fit with the center lines of the scraped patterns.

A good contact is essential on the outside edges of the pole patterns and may be achieved by either beveling the edges of the brass pieces, or holding them down against the paper with thumbtacks. The author found better results by cutting a jagged edge with a scissor, see Fig. 8-12. (Conductive paint proved of little value due to cracking, i.e., open circuit.)

The model is now finished. Connect a battery (or power supply) to opposite pole electrodes, and a probe with a pinpoint as shown in Fig. 8-12. Trace the paper for points with the same voltages; this will give the pattern for one potential line. Continue on and trace points of different voltages to find different potential lines. It is a good idea to select voltages that are E/10 volts apart.

Finally trace the points of equal potential with a pencil, and the potential line plot is complete. To now add field lines to this plot, we must use the reluctance method. Since this method involves trial and error, use a copy machine and do the rest of the work on the obtained copy (i.e., starting with squares in an area where the potential lines are parallel and evenly spaced, working into the more difficult area). With training an accuracy of 5 to 10 percent for the finished field pattern plot is possible.

When outlining the pole pieces care must be taken to use the correct scaling. For magnetic heads a side view is generally used, and this may be scaled up without restrictions, provided the thickness of the core (track width) does not change. Changing the thickness as in microcomposite heads requires that the scaling correlates with the magnetic resistance of the various elements of the circuit. Round structures are particularly tricky, as for example the linear motors (voice coil motors) for head positioning in large disk drives. Nortronics has published a useful note on this topic (ref. Farquhar).

Examples of field plots are shown in Figs. 8-13 through 8-15. The short pole pieces in Figs. 8-13 and 8-14 illustrate a thin film head's poles, and they do not focus the field as one intuitively would think; they do provide a 10 to 15 percent improvement in the field gradient. The potential lines were plotted using the reluctance method, and compared with computed potentials (ref. Potter).

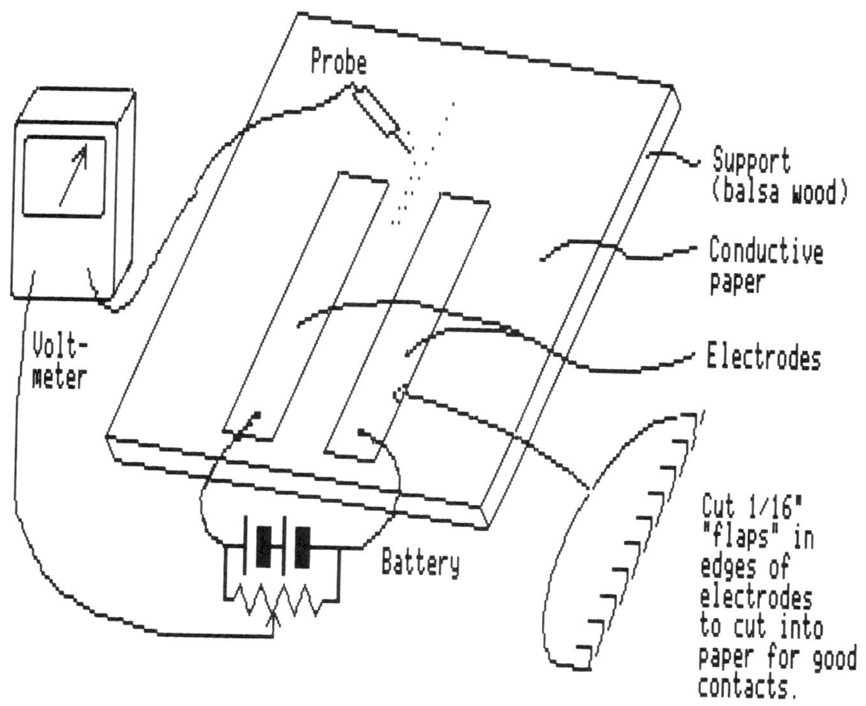

Fig. 8-12. Experimental plot of potential lines.

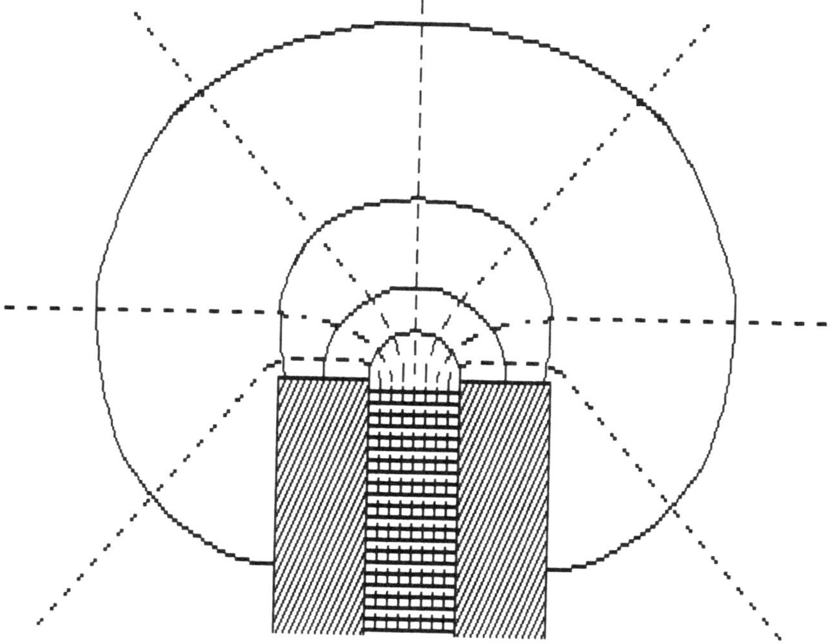

Fig. 8-13. Potential around thin film head pole pieces (after Potter, 1975). Field lines shown dotted.

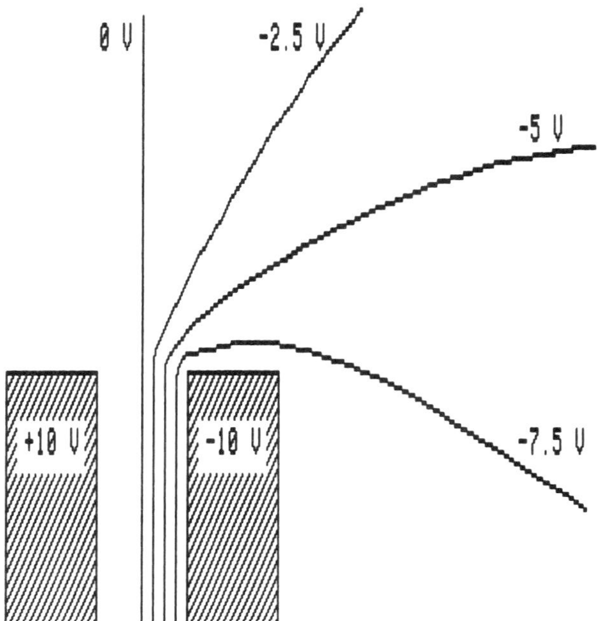

Fig. 8-14. Potential lines, short pole head (after Elabd, further readings).

Figure 8-15 shows the potential and the field lines from a pole protruding through a slot in another pole piece; the dimensions into the paper are at least an order of magnitude greater than the length of the slot. The potential lines allowed for a reluctance method plot of the field lines, as shown.

The next step toward accurate plots of field and potential lines is analytic computation. Interested readers have come across literature on the topic, and found mathematics of a rather advanced level. Often the intuitive feeling of "what goes on" is totally lost in these derivations. If the reader really wants to get into advanced field theory there is no way around the mathematics, but the next section will demonstrate that we can come a long way with simple fields from current carrying wires and from magnetic poles.

FIELDS FROM MAGNETS AND CURRENTS

No piece of magnetic material has ever been observed with *only* a north or a south pole (though that doesn't stop some physicists from seeking the elusive "magnetic monopole."However, separating magnets into north and south monopoles—in our minds—allows for calculations of magnetic fields that accurately match real devices. For the remainder of this section, we refer to monopoles as "poles."

The magnetic field external to a magnet, permanent or driven by a coil, originates at one pole, and ends on the other. The field emanating from a pole has a strength that decreases proportional to the distance squared (from Chapter 3):

$$H = \frac{p}{4\pi r^2}$$

where p is the pole strength in AM.

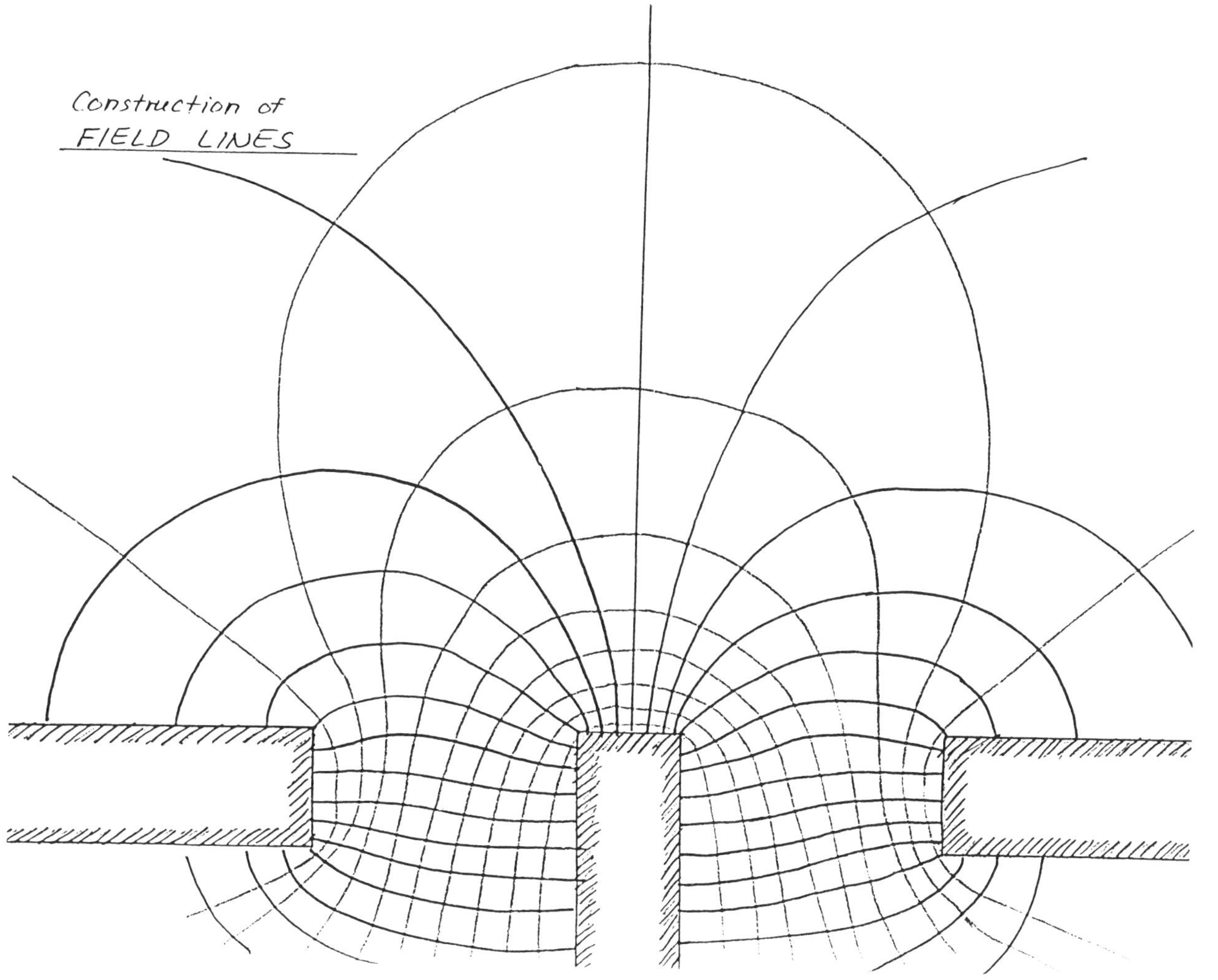

Fig. 8-15. Field lines from pole in slot (courtesy of Springer Technology, Inc.).

If we locate one pole at the origin of an X-Y coordinate system then we can calculate the field strength at point (x,y) by:

$$H = \frac{p}{4\pi(x^2 + y^2)}$$

We can also calculate the field strength for the general case, where the pole is located at (x_o, y_o):

$$H = \frac{p}{4\pi((x-x_o)^2 + (y-y_o)^2)}$$

Field lines radiate from the pole, like spokes from the center of a wheel. However, the points of equal magnetic potential are located on circles centered at the pole.

We can now calculate the **field strength from a pole**—and we must always account for the other, opposite pole, and must at any given point add its field. That means an additional expression for H, with a different (x_o, y_o).

We know that the **direction of the field** is outwards from the positive pole, and will now seek a more complete expression for the field using its components in the direction of the x- and y-axis.

We can write, with a pole located at the origin, for all points on the Y-axis:

$$H_x = 0.$$

$$H_y = \frac{p}{4\pi y^2} = \frac{p*y}{4\pi y^3}$$

Similarly, for all points on the X-axis:

$$H_x = \frac{p}{4\pi x^2} = \frac{p*x}{4\pi x^3}$$

$$H_y = 0$$

At all other points we can write:

$$H_x = \frac{p*(x-x_o)}{4\pi((x-x_o)^2 + (y-y_o)^2)^{3/2}}$$

and

$$H_y = \frac{p*(y-y_o)}{4\pi((x-x_o)^2 + (y-y_o)^2)^{3/2}}$$

and the total field strength is:

$$H = \sqrt{H_x^2 + H_y^2} \; \bullet$$

Figure 8-16 illustrates how the field direction is determined from the ratio of two components H_x and H_y, which is the slope, or direction of the tangent:

$$\text{Slope} = \frac{\text{rise}}{\text{run}}$$

$$= \frac{\text{y-component}}{\text{x-component}}$$

$$= \frac{H_y}{H_x}$$

$$= \frac{(y - y_o)}{(x - x_o)}$$

which, for multiple sources, should be calculated:

$$\text{Slope} = \frac{\Sigma H_y}{\Sigma H_x}$$

Example 8.1: A monopole of strength .01 AM, and is located at (0.001, 0.001) meters Find the strength of the field at point (0.002, −0.003). Express the answers in Oersteds

Answer: $H = .01/4\pi r^2 = .01/4\pi(1^2 + 4^2) * 10^{-6} = 46.8$ A/m = 0.588 Oe.

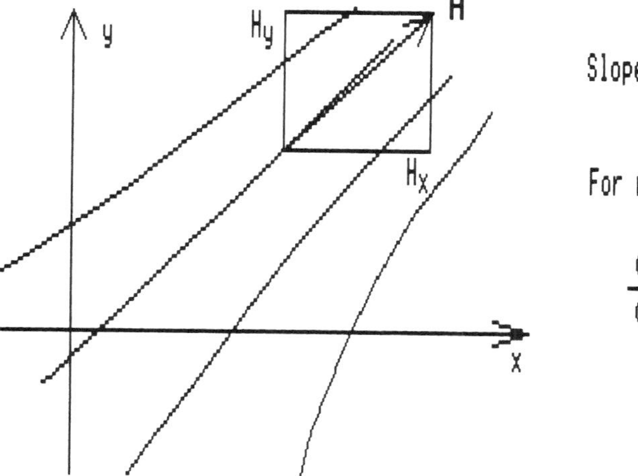

Fig. 8-16. Slope expression —> differential equation.

The magnetic field from a **current carrying wire** has a direction that is tangent to a circle which has the wire at its center. The direction is easily determined by the right hand rule:

Hold the wire with the right hand, so the thumb points in the direction of the current; then the field direction follows the direction of the other fingers.

The field strength is inversely proportional to the distance, and its value can be found from formula (3.8), see Fig. 8-17, where positive current means current flowing in the $-z$ direction:

a. For points on the Y-axis:

$$H_x = \frac{I}{2\pi y} = \frac{I*y}{2\pi y^2}$$

b. For points on the X-axis:

$$H_y = -\frac{I}{2\pi x} = -\frac{I*x}{2\pi x^2}$$

General expressions for H_x and H_y, at a point (x,y), for a wire located at (x_o, y_o) are:

$$H_x = \frac{I}{2\pi} * \frac{y - y_o}{(x - x_o)^2 + (y - y_o)^2}$$

and

$$H_y = \frac{I}{2\pi} * \frac{-(x - x_o)}{(x - x_o)^2 + (y - y_o)^2}$$

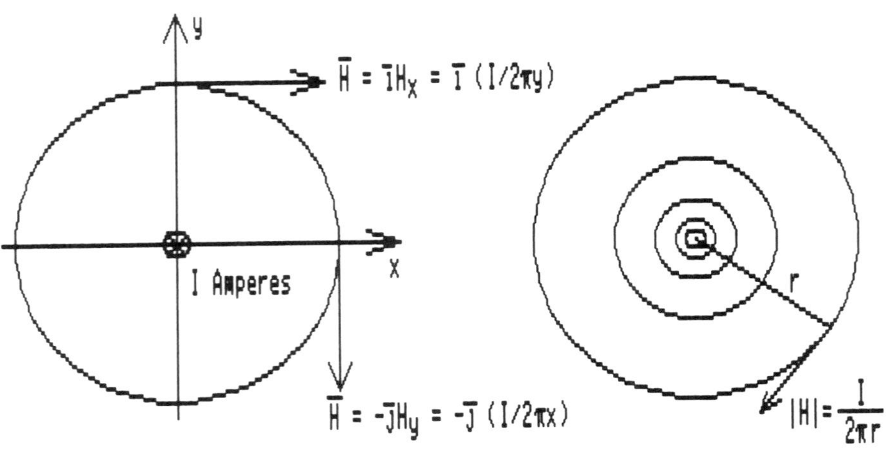

Fig. 8-17. Expressions for the magnetic field from a current carrying wire.

If we in these formulas use meters for the x and y values, and Amperes for the current, then the field strength values are in A/m; multiply by $10^3/4\pi \cong 80$ to find values in Oersteds.

We could find the same result by simply dividing the current I with the distance $R = \sqrt{(x^2 + y^2)}$ and with 2. And we know the field lines follow circular paths, so the direction of the field is readily known. It can be expressed as the slope:

$$\text{Slope} = \frac{\text{rise}}{\text{run}} = \frac{H_y}{H_x} = -\frac{x - x_o}{y - y_o}$$

Exercise 8.2: Find the field vector (strength and direction) from a single current carrying wire, located at (0,0) at points: (0,2), and (0, −1).

The current in the wire is 2π Amperes. Express the magnitude in Oe.

MULTIPLE FIELD SOURCES. A FIELD EQUATION

Things become more complicated for the field from two or more current carrying conductors. The field values from each are easily calculated at a given point, but they may not be added directly.

One must first add all the H_x components, then all the H_y components, and the square root of the sum of the squares will provide the true field magnitude. The direction of the field vector is, as before, given by:

$$\frac{dy}{dx} = \frac{\Sigma H_y}{\Sigma H_x} \qquad \textbf{(8.1)}$$

where H_x and H_y are the values of the x- and y-components of the field vector, from one or more conductors.

One could proceed to calculate the field strength and direction at a great number of points, and then interpolate to outline the points where the field strength is the same. If we connect these points together we obtain *field lines*, and several of these will provide a good illustration of the field pattern.

The expression dy/dx is used to find the direction of the field vector at any given point. Assume that this is done at a series of points, closely following each other, and in such a fashion that a new point lies a short distance away, in the direction of the field. Then we are actually tracing a *field line*.

It so happens that dy/dx is not just the field direction; it represents the *differential equation* for the field—from one or more conductors. For the field from a single wire, located at (0,0), we find, using the expressions for H_y and H_x from above:

$$\frac{dy}{dx} = \frac{-x}{y}$$

and the solutions to this differential equation are circles. Similarly, for a wire located (x_o, y_o) the differential equation is:

$$\frac{dy}{dx} = -\frac{x - x_o}{y - y_o}$$

Similarly, for several conductors:

$$\frac{dy}{dx} = - \frac{\Sigma H_y}{\Sigma H_x}$$

A solution may be found to such differential equations, but often with great difficulty. Only the field from a single wire has a simple solution, as can be found in any textbook on differential equations:

$$(y - y_o)^2 = - (x - x_o)^2 + c,$$

which are circles with radius $\sqrt{C}$, centered at (x_o,y_o).

The information we basically need about fields from various head configurations and from magnetization distributions in media, are:

- Field magnitude
- Field gradient
- Field pattern

Field magnitudes are readily calculated from the x and y components of **H**. Likewise, calculating the field strength at equally spaced points along a line will provide information about

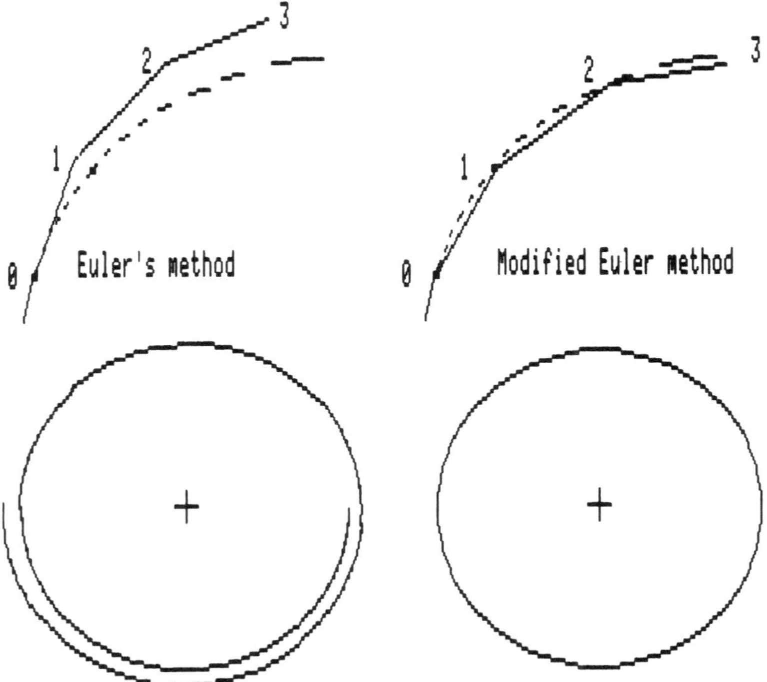

Fig. 8-18. Modified Euler's method is superior to the first order Euler method for tracing solution (field lines) to differential equations (see Appendix).

the gradient, along that line by merely listing (or better, plotting) the values for the differences between adjacent points.

The field patterns are readily plotted by a numerical solution of the differential equation; we already did that in the plot of the circle earlier. We can refine the technique by the following algorithm outline:

A. Compute the slope dy/dx at a point (X0,Y0).
B. Move a short distance Δ along that slope to a new point (X1,Y1).
C. Compute a new slope at that point.
D. Now move the distance Δ along the slope (from C) to point (X2,Y2).
E. Use (X2,Y2) as new start point (X0,Y0), and start again with A.

Repeat the loop A-E until the desired length of the field line has been plotted (for example, 500 steps, or if the line moves outside the plotting window or approaches the position of the field source.

Additional lines can be plotted with new start points, selected randomly within the plotting window, or in accordance with some rule.

This algorithm is known as Euler's method and is illustrated in Fig. 8-18. It has a built-in error that will make a circular pattern a growing spiral; this error is reduced by using the modified Euler's method, described in the Appendix. The reader will there find a small computer program to make field plots from experimental head structures, using a driven gap head as an example.

FURTHER MATHEMATICS ON FIELDS

We can formalize a mathematical expression for the field H by introducing the *unit vectors* **i** and **j**. **i** has a magnitude of 1.0 and always points in the X-direction, while **j** points in the y-direction, as we saw in Fig. 8-17.

Assume a wire is located at the origin of an X-Y coordinate system; the field strength can now be expressed in vector notation:

a. For points on the Y axis:

$$\mathbf{H} = H_x\mathbf{i} = (I/2\pi y)\mathbf{i} = (I*y/2\pi y^2)\mathbf{i}$$

b. For points on the X axis:

$$\mathbf{H} = H_y\mathbf{j} = -(I/2\pi x)\mathbf{j} = -(I*x/2\pi x^2)\mathbf{j}$$

The use of the bold letter **H** indicates that the quantity is a *vector* and thus has a direction in addition to its magnitude. For all other points (x,y) a general expression holds:

c. At any point (x,y):

$$\begin{aligned}\mathbf{H} &= \mathbf{i}H_x + \mathbf{j}H_y \\ &= \mathbf{i}(I*y/2\pi(x^2 + y^2)) \\ &\quad - \mathbf{j}(I*x/2\pi(x^2 + y^2)) \\ &= \frac{I}{2\pi} * \frac{1}{x^2 + y^2} * (\mathbf{i}y - \mathbf{j}x) \qquad \textbf{(8.2)}\end{aligned}$$

The vector component magnitudes are H_x and H_y, while **i** and **j** are unit vectors in the directions of the X and Y axes, respectively. General expressions for H_x and H_y, at a point (x,y), for a wire located at (x_o,y_o) are:

$$H_x = \frac{I}{2\pi} * \frac{y-y_o}{(x-x_o)^2 + (y-y_o)^2}$$

and

$$H_y = \frac{I}{2\pi} * \frac{-(x-x_o)}{(x-x_o)^2 + (y-y_o)^2}.$$

If, in these formulas, we use meters for the x and y values,and Ampere for the current, then the field strength values are in A/m; multiply by $10^3/4\pi \cong 80$ to find values in Oersteds.

We could find the same result by simply dividing the current I with the distance $R = \sqrt{(x^2 + y^2)}$ and with 2π. And we know the field lines follow circular paths, so the direction of the field is known. The matter becomes more complicated for the field from two or more current carrying conductors. The field values from each are easily calculated at a given point, but they may not be added directly.

One must first add all the H_x components, then all the H_y components, and the square root of the sum of the squares will provide the true field magnitude. And a little thought will inform us that the direction of the field vector is provided for by:

$$\frac{dy}{dx} = \frac{H_y}{H_x}$$

where H_x and H_y are the values of the x and y components of the field vector, from one or more conductors.

The differential equations we have arrived at are all solutions to a second-order, partial, differential equation, which is named Laplace's equation, in two dimensions:

$$\frac{\delta^2\phi}{\delta x^2} + \frac{\delta^2\phi}{\delta y^2} = 0 \text{ or } \nabla\ \phi^2 = 0 \tag{8.3}$$

where ϕ is the scalar magnetic potential from which we find $\mathbf{H} = -\nabla\phi$.

The mathematically interested reader is referred to the textbooks for the derivation of Laplace's equations, starting with Maxwell's equations.

An experienced magnetician will now start field investigations from Laplace's equation, which describes how a scalar function behaves in a region (such as above a head gap). Information about the function's values at all points at the edges of the region (surface of the head) must be provided, in order to find the solution for ϕ.

Closed-form analytical solutions exist only for simple geometries of magnetic circuits. There are two numerical methods for obtaining solutions: the finite-difference method and the finite-element method. The interested reader is referred to the literature listed at the end of this chapter.

Commercial programs are available for these computations. They have been of great use in examining thin film heads; examples of the division into elements of the space around pole pieces are shown in Fig. 8-19, and the computed flux patterns in Fig. 8-20 (Ref. Ansoft Corp., University Technology Development Center, 4516 Henry Street, Pittsburgh, PA 15213, USA.)

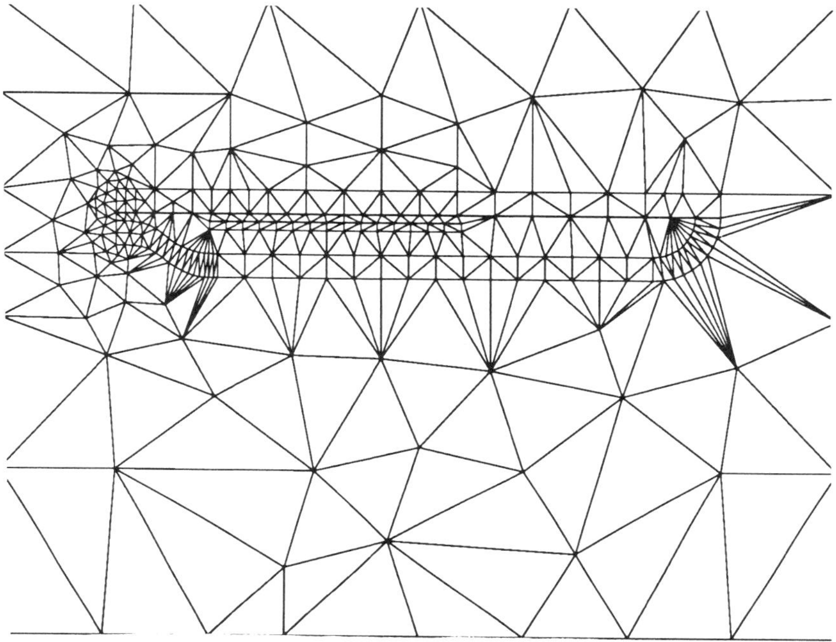

Fig. 8-19. Generated mesh for a finite element computation of field around a thin film head (courtesy Ansoft Corp.)

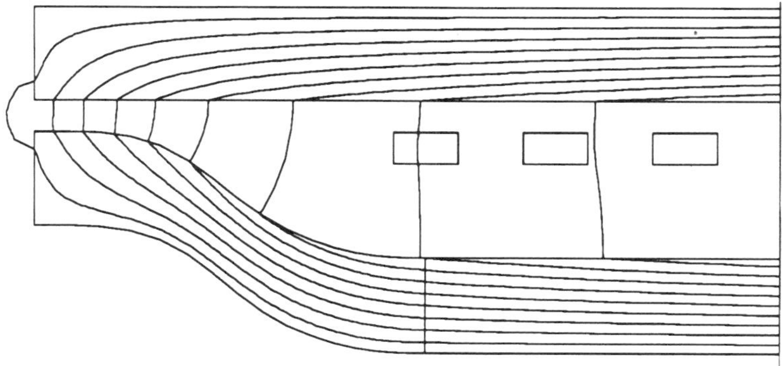

Fig. 8-20. Computed flux lines in the front of a thin film head (courtesy Ansoft Corp.).

PERPENDICULAR HEAD FIELDS

Since 1980 several individuals and groups have worked to develop a more practical and efficient write head to emphasize the perpendicular remanence. We know that this remanence pattern has little or no demagnetization and will therefore dominate at very short wavelength recordings. The results from work on perpendicular heads are shown in Figs. 8-21 through 8-23.

Iwasaki first promoted perpendicular recording in the late seventies, and started experimental work using a single pole head, also called the probe head. Now, since single poles do not exist, the counter pole was placed on the backside of the coating, see Fig. 8-21. This presents a very practical problem; the spacing between poles would have to be quite large to accommodate a disk or tape, and this would seriously impair the field gradient.

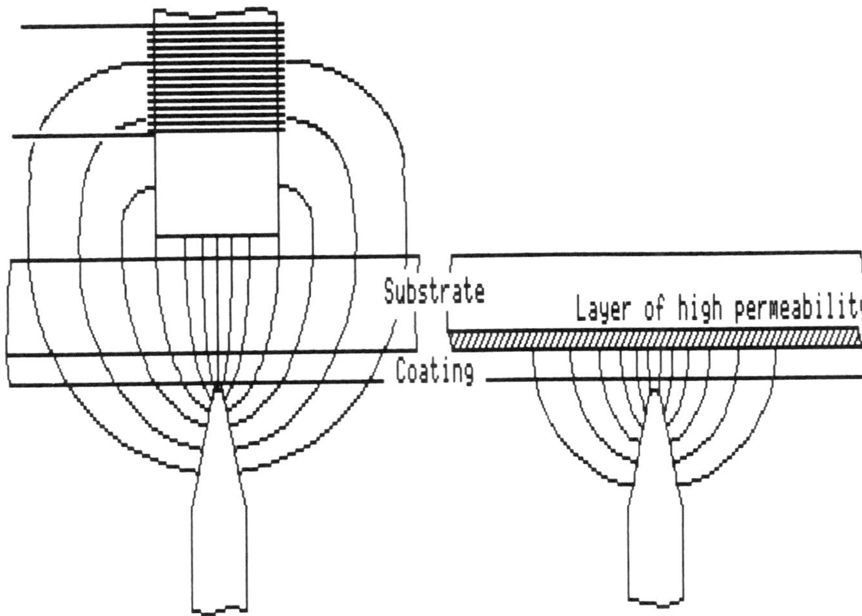

Fig. 8-21. First perpendicular heads (left, after Iwasaki).

A solution was quickly found by providing the coating with a high permeability underlayer such that the field lines from the pole would seek a path through the underlayer, and return some distance from the pole. Practical ways of doing this are illustrated in Fig. 8-22.

A standard ring core, or thin film head (see Chapter 10) will do quite well in recording onto a perpendicular coating. And there is quite a debate about the merits of the fields from

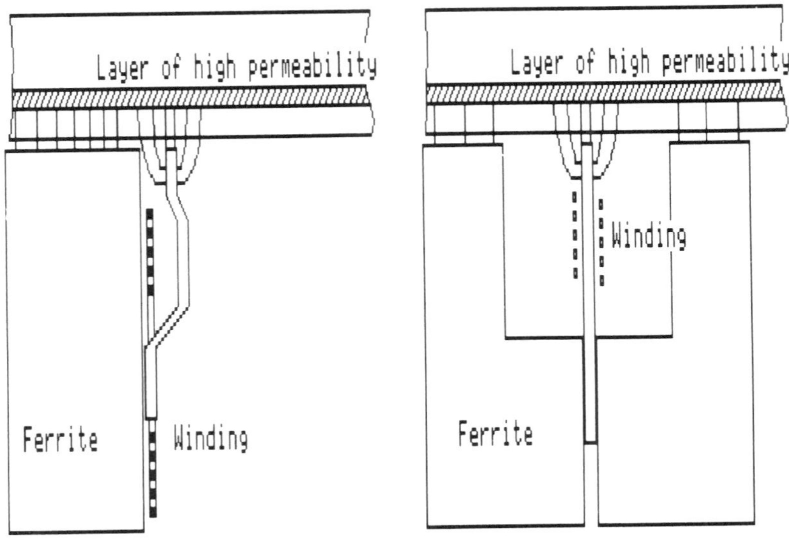

Fig. 8-22. Current perpendicular heads (after Vertimag).

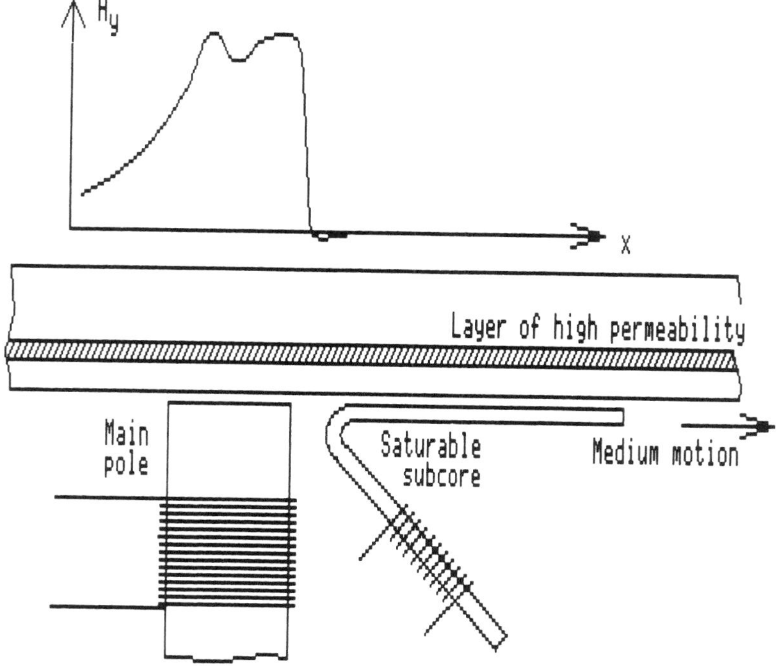

Fig. 8-23. High gradient perpendicular head (after C.S. Chi).

the perpendicular heads (Mallinson, Bertram, Minuhin, see ref.). Minuhin claims that the characteristics of the probe models are inferior to those of the ring core head; he also shows that the resolution of the probe head is reduced if a coating with soft magnetic underlayer is used. And thus the debate goes on.

A recent head by Chi (see ref.) is shown in Fig. 8-23. This design follows the thoughts behind the cross field and driven-gap head by shaping the field by an auxiliary pole. As of the writing of this chapter, Chi's head had not been reduced to practice. Progress in heads is, as stated in the chapter introduction, highly dependent upon the limits of materials and fabrication skill.

SIDE WRITE/READ EFFECTS (NARROW TRACKS)

The quest for increasing storage capacity of magnetic tapes and disks has naturally led to narrower tracks. We have so far disregarded any variations of the field strength in the z-direction, perpendicular to the plane of the paper. The results of a finite-element analysis by D.A. Lindholm of the field around a narrow track head are shown in Fig. 8-24, and the fields at the corner of a head core are shown in Fig. 8-25. The three dimensional views of the fields H_x, H_y, and H_z are very useful in assessing the wide writing effects. The recorded track will be slightly wider than the core.

Assume that the recorded track width equals the width in the coating where the field strength equals or exceeds the coercivity H_c of the medium. The final track width can then be determined from Fig. 8-26, showing the total write field which includes: the ordinarily assumed field configuration (top), the side fields, and the wedges connecting the two.

The amount of extra trackwidth ϵ on both sides has been analyzed and calculated by Y. Ichiyama. His results are shown in Fig. 8-27, giving all operating conditions plus a graph for

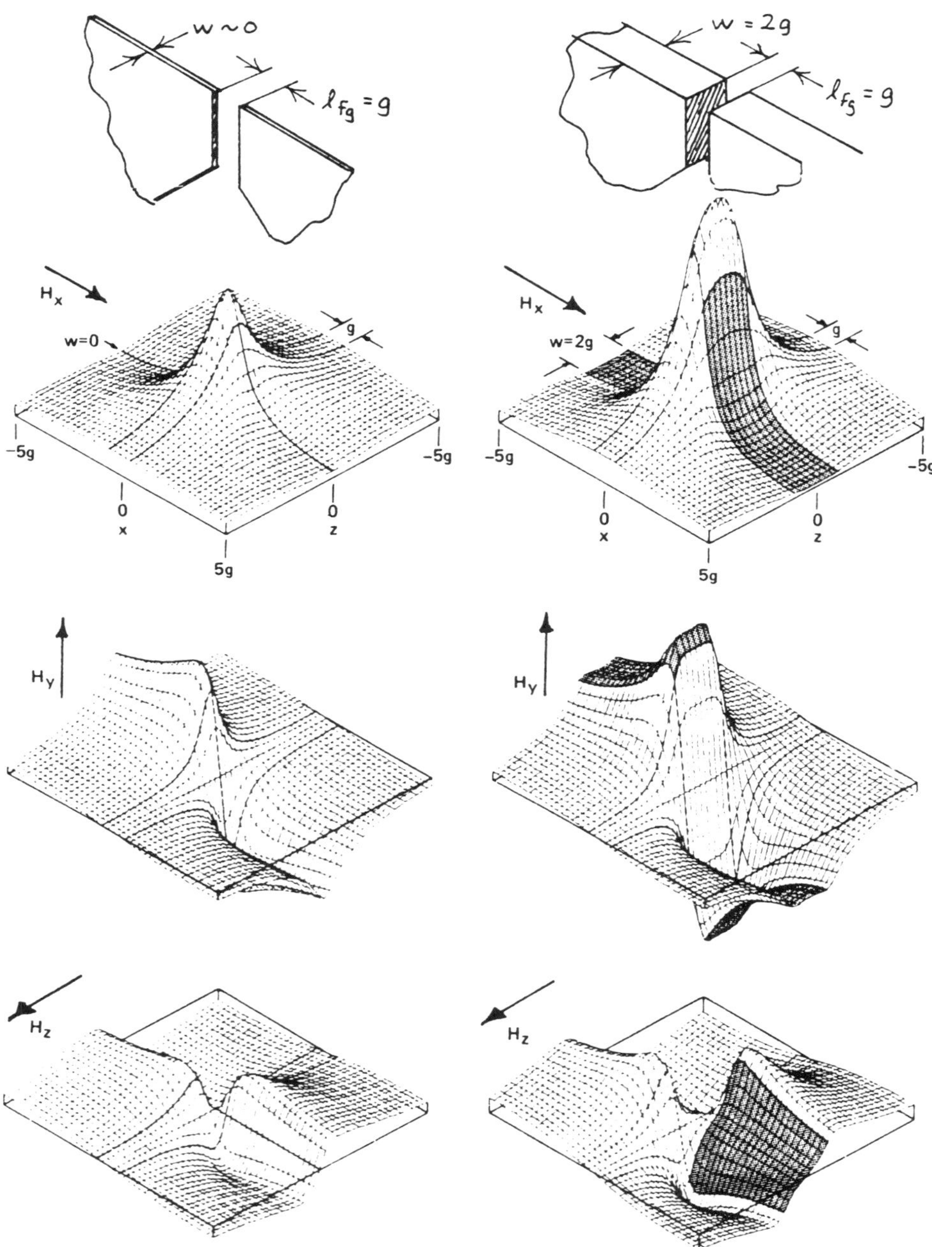

Fig. 8-24. Magnetic field components from narrow track magnetic heads (courtesy of Dennis Lindholm of Ampex Corp.).

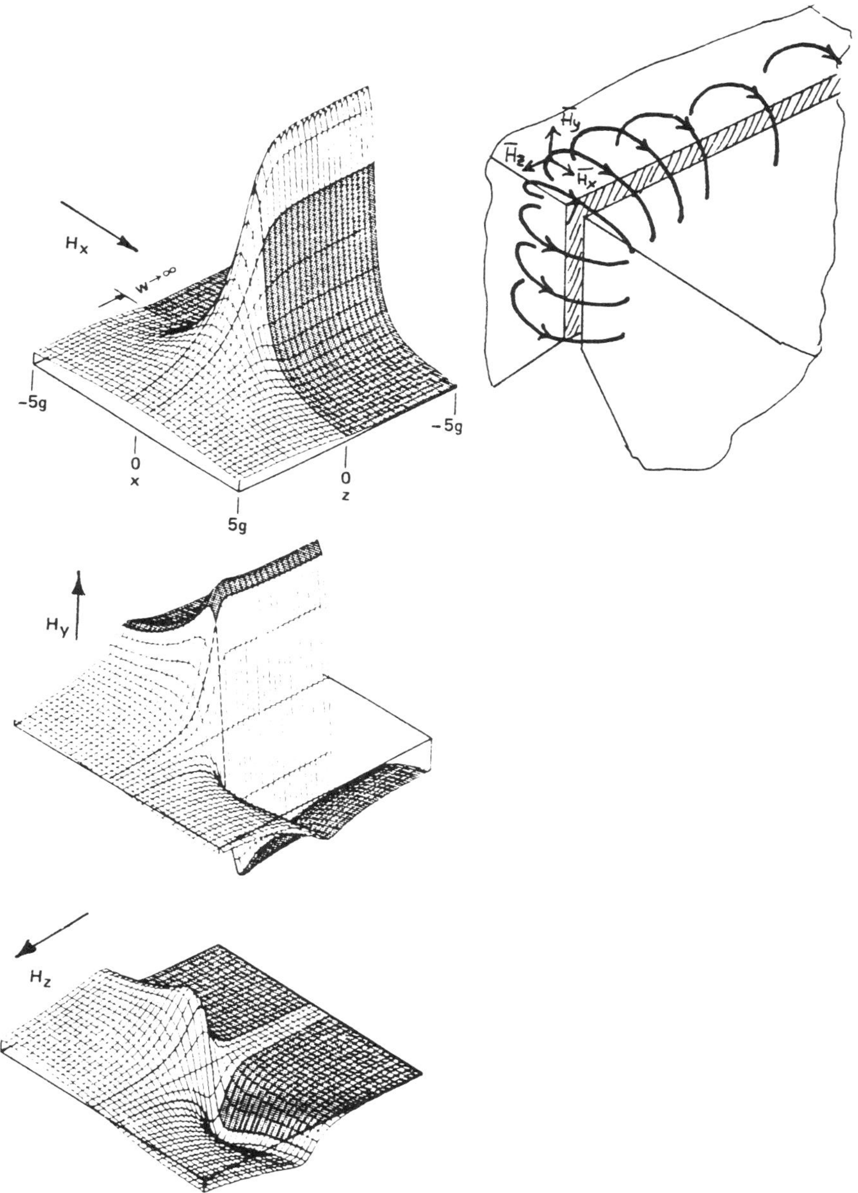

Fig. 8-25. Magnetic side fields at the corner of a magnetic head.

finding the answer. It covers also a tapered core, which has a wider track than a core with parallel sides. Straight cores are always preferable, but may not be used when the track width decreases below 100 μm (4 mils), since a certain amount of core body must remain just for core handling (and higher efficiency).

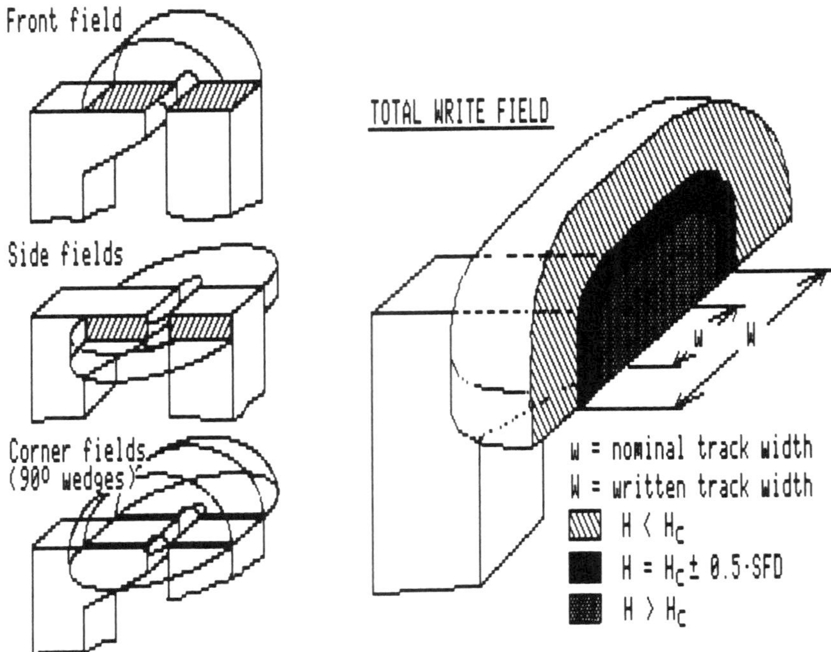

Fig. 8-26. Total front gap field.

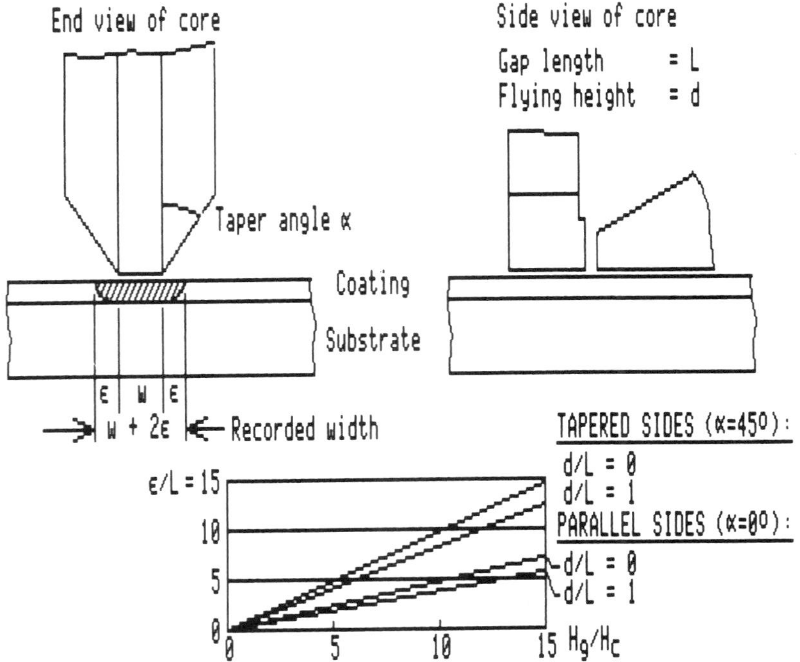

Fig. 8-27. Written track is 2ϵ wider than write core (after Ichiyama).

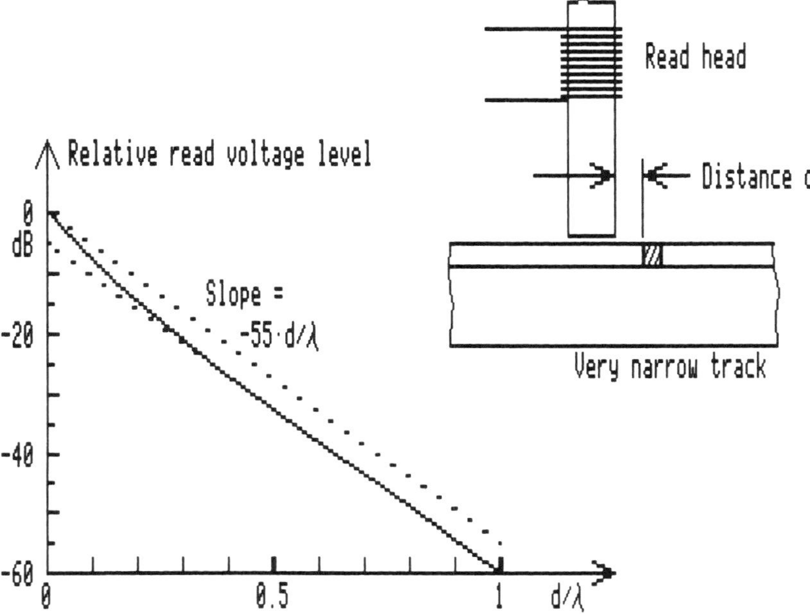

Fig. 8-28. Side reading (after van Herk, 1978).

Another effect of the side field is the associated side sensing function, which equals the head field divided by a unit current. This may lead to *cross talk* from adjacent tracks, and sets the limit for how many tracks can be placed parallel to each other. Each recording system has its own cross talk requirements, and the designer is referred to the papers listed in the references and Bibliography.

One valuable clue is found in work recently done by van Herk. An early assumption was that the cross talk, or *side sensitivity*, varied similar to the Wallace's spacing loss, $-55d/\lambda$ dB. This applied to a very narrow track running off the core, as shown in Fig. 8-28. Van Herk found that the loss actually approaches $55d/\lambda$ dB plus another 6 dB.

When recording surface area is at a premium , as for example in home video recorders, a clever solution applies. It was introduced by Sony, and is made by purposely misaligning the azimuth of two video heads, which are then placed 180 degrees apart on the head drum. The heads are offset in azimuth by plus and minus 7 degrees, and are synchronized upon playback to read their own tracks laid down earlier (see Fig. 6-15 in Chapter 6).

There is now a misalignment error of 14 degrees to the adjacent track; this causes a near complete loss in picking up cross talk, and the tracks can therefore be placed next to each other, without the use of a guard band.

ERASE HEADS

The principle of erasure of magnetic tapes was illustrated in Fig. 5-24A: the tape is exposed to a decaying alternating field in which the initial value of the field is great enough to saturate the tape (typically five times the coercivity). This procedure leaves the tape in a state of zero net magnetization.

The erasure is in practice implemented by passing the tape over a magnetic head with a strong AC field in the gap (erase head), or by placing the reel of tape in a device with a strong magnetic field that slowly decays to zero (bulk degaussers).

Heads with an alternating and decaying field can be made from *permanent magnets* in arrangements, as shown in Fig. 8-29. The wrap angle, where the tape moves away from the head, is critical and may need adjustment for a particular tape. The last field "seen" by the tape should be of such magnitude that it, combined with the preceding (and opposite) field magnitudes, leaves the tape with $B_r=0$. (The tape magnetization will follow diminishing minor loops.)

These permanent magnet erase heads require no power, but must be moved away from the tape when not used.

A record head with a fairly long gap and an AC current serves well as an erase head. It does not provide complete erasure of a high level recording. At the point where the field strength of the AC erase field has decayed to approximately the coercive force H_c of the tape there exists, in reality, a recording zone. Any foreign field that may be present will therefore be recorded onto the tape.

McKnight found that for a given erasing system operating at a given erasing current, and erasing a given wavelength, the signal is erased to a *certain percentage* of the original signal; this is almost independent of the original signal level.

Short wavelength recordings are more easily erased than those of long wavelengths.

Examples of erasures are shown in Fig. 8-30. The higher degree of erasure for shorter wavelengths appears to be related to reduced re-recording; the available flux from short wavelengths is small compared with the flux from long wavelengths, assuming identical recording levels.

It also appears likely that the fixed percentage of erasure in one erasure following another is related to the re-recording phenomenon. But there is no explanation for the increased difficulty in erasing a tape recorded with a higher bias level than another.

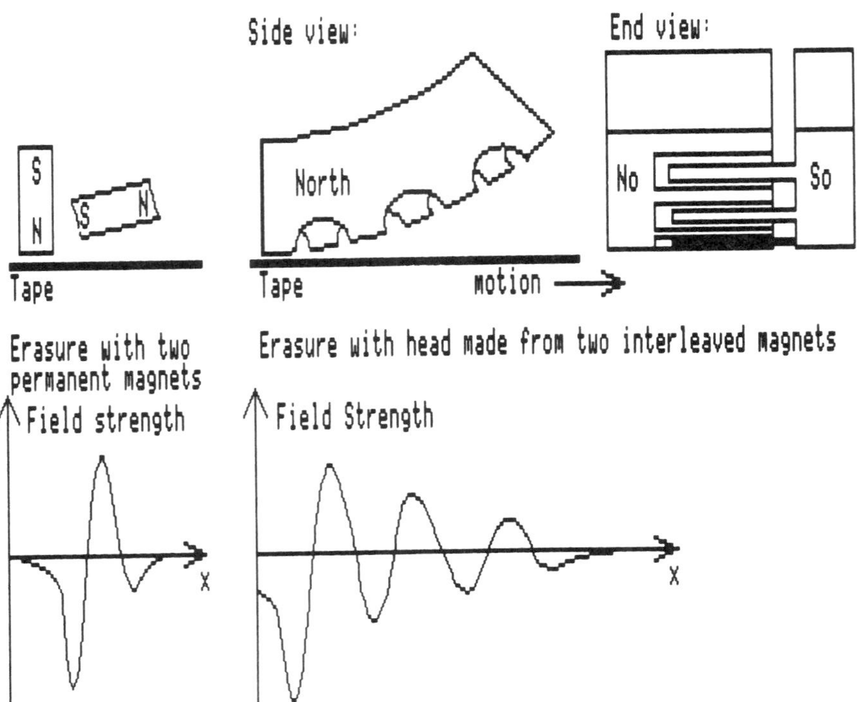

Fig. 8-29. Erasure with permanent magnets.

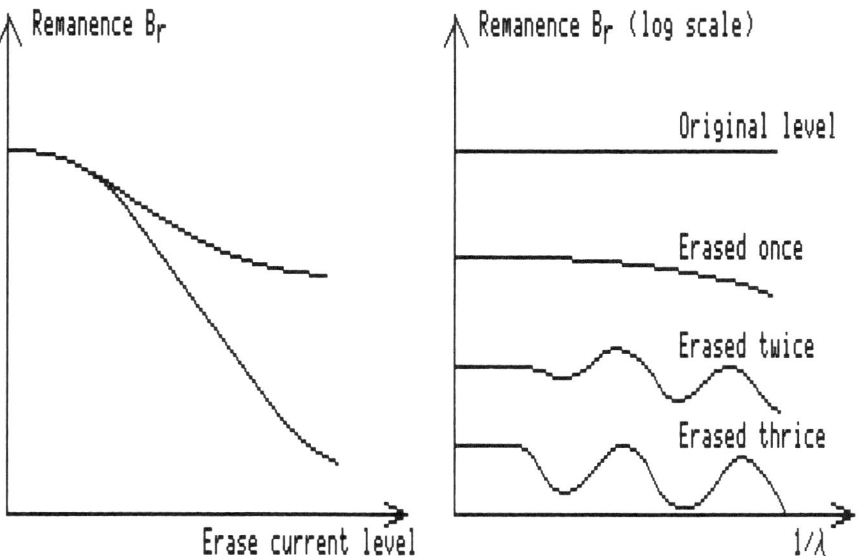

Fig. 8-30. Incomplete erasure as function of wavelength recorded (left), and number of passes over erase head (right) (after McKnight).

It appears that there are two sources for the field that creates the re-recording. Keep in mind that the data field strength in an AC-bias recording is much smaller than the level required for a non-bias recording. A fraction of an Oe will suffice, if added to the AC-bias field near the recording zone.

The erase head has a sensing function of its own, and will ''see'' magnetizations prior to their passage through the erase field. This sensing results in a flux through the core (as a read head), and will cause re-recording thereof. Bear in mind that any ring core type head acts as a read as well as a write head, at any time, as illustrated in Fig. 8-31. The two processes occur independent of each other, possibly with some coupling due to non-linearities in the head permeability due to high write levels.

The other source for the re-recording field occurs primarily in digital recordings, where difficulties arise in erasing while writing new information. Actually, the write field acts as an erase field for the old data. The situation is again like that shown in Fig. 8-31, but often aggravated by the fact that the bit length of the old data is quite close to the length of the write field (shown crosshatched). This sensing results in a flux through the core (as a read head), and will cause re-recording thereof. Bear in mind that any ring core type head acts as a read as well as a write head, at any time, as illustrated in Fig. 8-31. The two processes occur independent of each other, possibly with some coupling due to non-linearities in the head permeability due to high write levels.

Repeat erasures can be accomplished during a tape pass if two (or more) erase gaps are incorporated in the erase head. They must belong to separate core structures, spaced away from each other, as shown in Fig. 8-32. This is accomplished with a double center leg that isolates the re-recording fluxes to their respective head sections.

The frequency of the erase current must be so high that each tape section undergoes several hundred decaying field cycles as it leaves the center of the erase gap. If the tape speed is 7.5 IPS and the erase gap length is 40 mils (1mm), the frequency should be above 40 kHz.

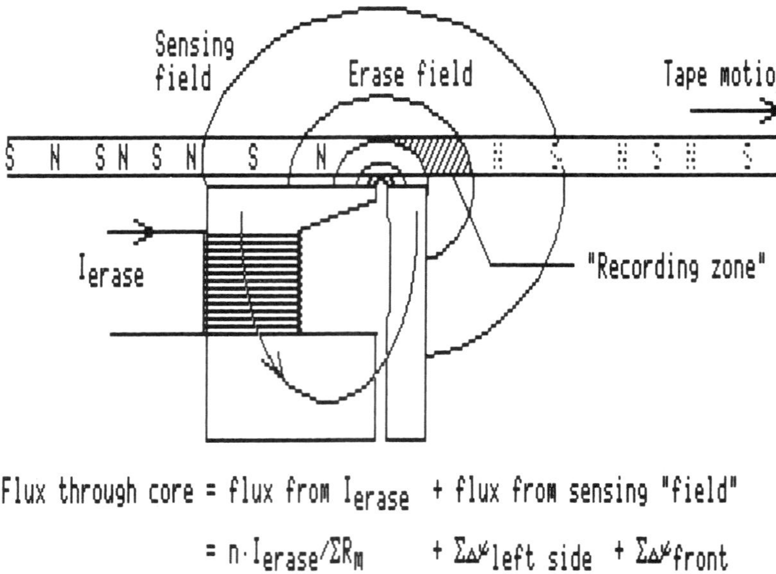

Fig. 8-31. Re-recording during erasure.

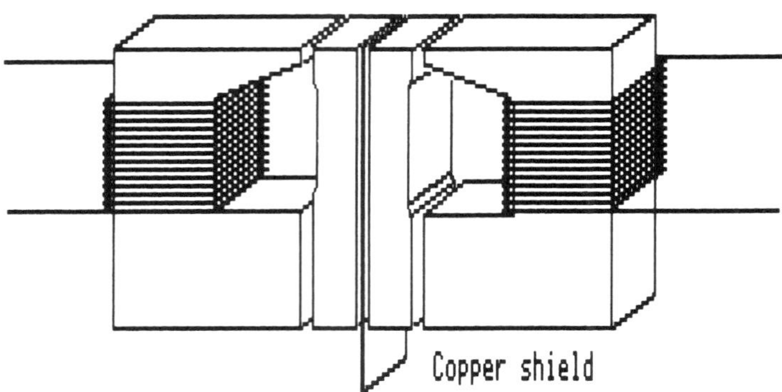

Fig. 8-32. Double gap erase head.

STRAY FLUX COMPUTATIONS

Magnetic circuits control the flux as poorly as an electric circuit immersed in salt water controls the current flow. While the difference in electric conductivity is infinite between air and conductive wires (at DC), the difference in magnetic conductivity is at most 100,000, often only 3-1000 and decreasing at higher frequencies because of eddy currents.

We should therefore include the *stray flux conductance* through the air, shunting the front as well as back gaps in a magnetic head structure. This is illustrated in Fig. 8-33, where all elements making up the front gap stray flux are identified and characterized.

When calculating the magnetic resistances of so many parallel elements it is easier to calculate the sum of the *permeances* (magnetic conductances), and then find the magnetic resistance from the reciprocal of the sum of permeances.

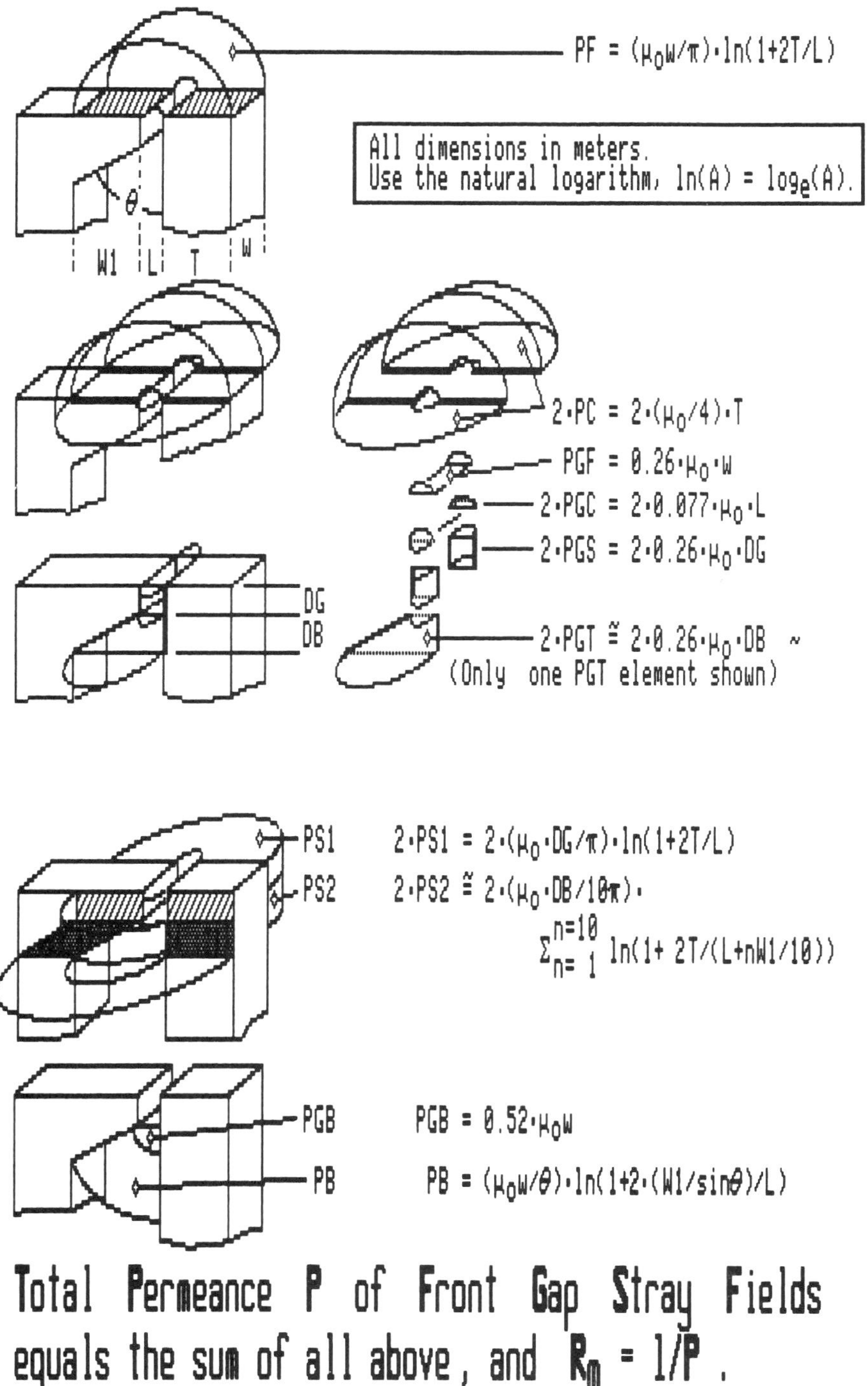

Fig. 8-33. Magnetic permeances (conductivities) around the gap in a magnetic head.

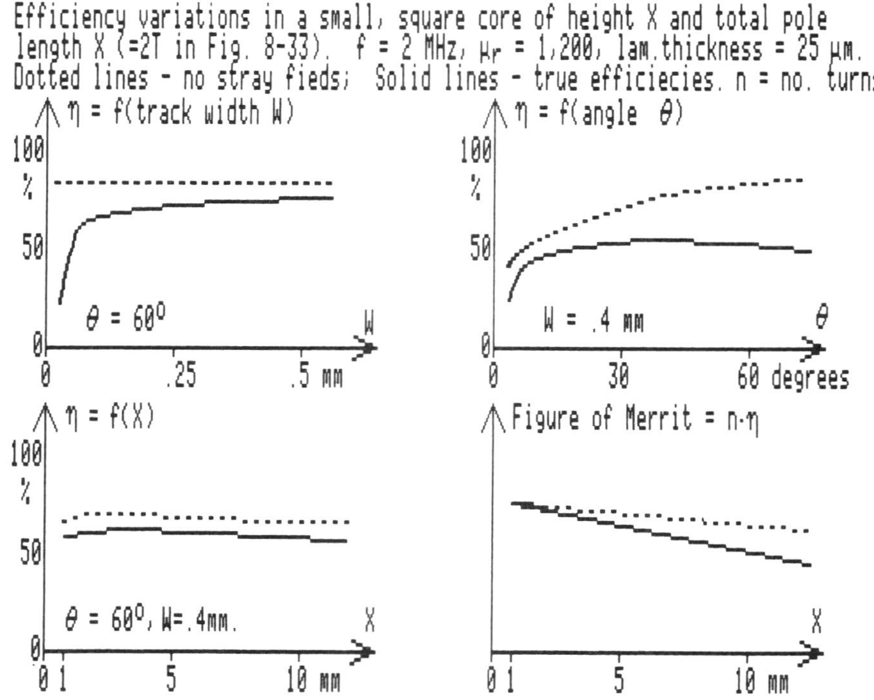

Fig. 8-34. Measures of head performance for varying magnetic circuit geometries.

The effect of stray fluxes cannot be neglected for narrow track cores, especially at high frequencies. The net effects of stray flux will become apparent in the calculations of head performance later in Chapter 10. Results are illustrated in Fig. 8-34, showing computed measures of head performance excluding stray fluxes (broken lines) and including them (solid lines). Note in particular how calculations for the variations in efficiency versus core thickness (track width) differ.

The formulas for the various permeances come from integrations of small flux tubes, summing their contributions to obtain the total permanence of the element. This is quite simple when, as an example,we deal with the permeance of the front gap of dimensions L long, w wide, and Dg deep. We divide the volume up into unit cubes with sides equal to 1 μm.

Magnetic resistance of a flux tube (also called a magnetic field cell, see Kraus, ref.) with cross section dx*dy is:

$$\begin{aligned} R_m &= \text{length}/\mu_o \bullet \text{area} \\ &= dx/\mu_o \bullet dy \bullet w \end{aligned} \tag{8.4}$$

For a cube with sides 1 μm we find:

$$R_{m\mu} = 10^{+13}/4_{\pi} \quad 1/\text{Hy} \tag{8.5}$$

and the permeance is $4\pi 10^{-13}$ Hy. We treat these elements as admittances, adding them when they are in parallel, and dividing by N when N elements are in series. The permeance of the gap is thus found by multiplying by w*Dg, and divide by L. A numerical example will clarify:

$$w = 125\ \mu m = 5 \text{ mils},$$
$$D_g = 50\ \mu m = 2 \text{ mils},$$
$$L = 2\ \mu m = 80 \mu in.$$

$$PG = (125*50/2)*4\pi 10^{-13} \text{ Hy}$$
$$= 3.93 * 10^{-9} \text{ Hy}$$

and hence

$$R_{gap} = 1/PG = 255 * 10^{+6} \text{ Hy}^{-1}$$
$$= 255\ \mu\text{Hy}^{-1}$$

The formulas in Fig. 8-33 provide for all stray flux around the gap in a magnetic head. We will again use these in Chapter 10.

LINKAGE AND LEAKAGE FLUX COMPUTATIONS

Flux Linkage

Figures 8-35 and 8-36 illustrate two impediments to a full coupling of the flux from the write current into the gap area of the write head.

First of all, not all of the field generated by the write head coil produces core flux. A few field lines from turns in the middle of the winding do not even get into the core. This inefficiency is termed *flux linkage.*

It is quite clear from the illustration on the right, where the field from a single conductor couples flux into a toroid shaped core, that flux linkage varies with the separation between

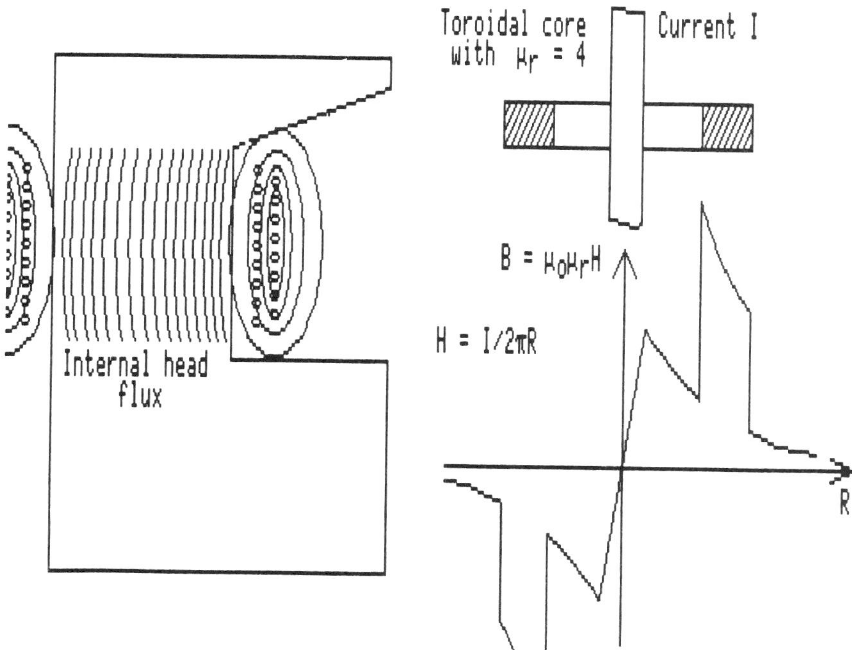

Fig. 8-35. Incomplete coupling of field into core (flux linkage, 100%).

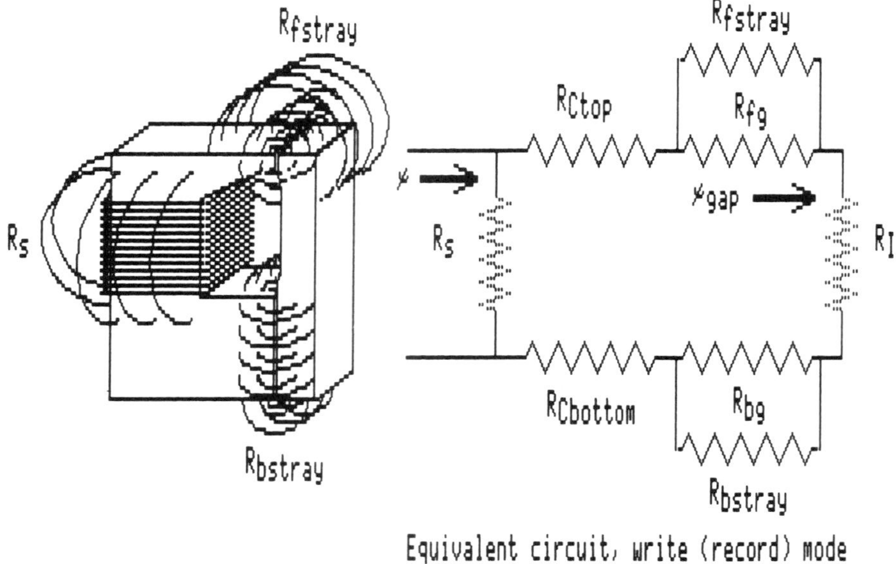

Fig. 8-36. Flux leakage around write coil.

conductor and core. Since the field strength is higher on the inside of the core than on the outside, and since H is inversely proportional to the distance from the wire center, the highest flux is achieved for a small diameter core.

When placing the winding onto the write head core, care must be taken in packing the windings as close to the core as possible in order to keep the flux linkage close to 100 percent. It is very difficult to arrive at a number for the flux linkage efficiency.

Flux Leakage

Flux leakage occurs when a portion of the flux strays from the core before reaching the front gap, as shown in Fig. 8-36. This causes a loss that can be expressed as a *coupling efficiency*:

$$\begin{aligned} \eta_{coupl} &= R_s/(R_s + R_{Ctop} + R_{fgs} + R_I + R_{bg} + R_{Cbottom}) \\ &= R_s/(R_s + \Sigma R_{ms}) \qquad \textbf{(8.6)} \end{aligned}$$

The stray flux at the front gap is included in R_{fgs}, as well as in ΣR_{ms}. Use the formula for PF in Fig. 8-33 to estimate the value of R_s (see Fig. 10-5 for a computed example).

When shields are placed near cores (as in multitrack heads) the leakage flux increases, as shown in Fig. 8-37. These shields serve two purposes. First, they reduce the transformer cross talk between adjacent cores. Second, two sets of write and read heads are often used, and the channels are interleaved, i.e., one set of heads handles all odd numbered tracks, the other all even numbered. Then shields serve the additional function of short circuiting the flux from the interleaved tracks moving past the read head.

An inside core with shields on both sides has higher leakage than an outside core; this will additionally cause track-to-track non-uniformity (sensitivity), and it is common practice to install shields outside the outer cores also.

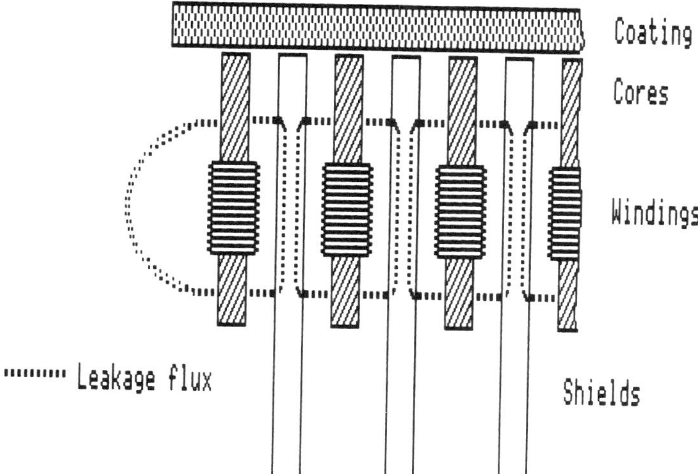

Fig. 8-37. Flux leakage to shield around write coil.

SHIELDING

The last topic that belongs in this chapter on head fields is the shielding of heads. This is more of an art than a scientifically studied topic—we know more about what does not work (ref. Jorgensen, Lopez) than we do about predicting what should work. It is apparent that *field strength levels of more than 5 Oe* affect AC-bias recordings as well as the writing of digital data; in the latter case the digital data field acts as an AC-bias source for the recording of the foreign field. The situation is aggravated when overwriting on old data.

During playback "hum" pick-up from power transformers and motors may occur, and only shielding of the head can reduce this pickup. Audio heads are often built into housings of soft magnetic materials, as shown in Fig. 8-38, top-right.

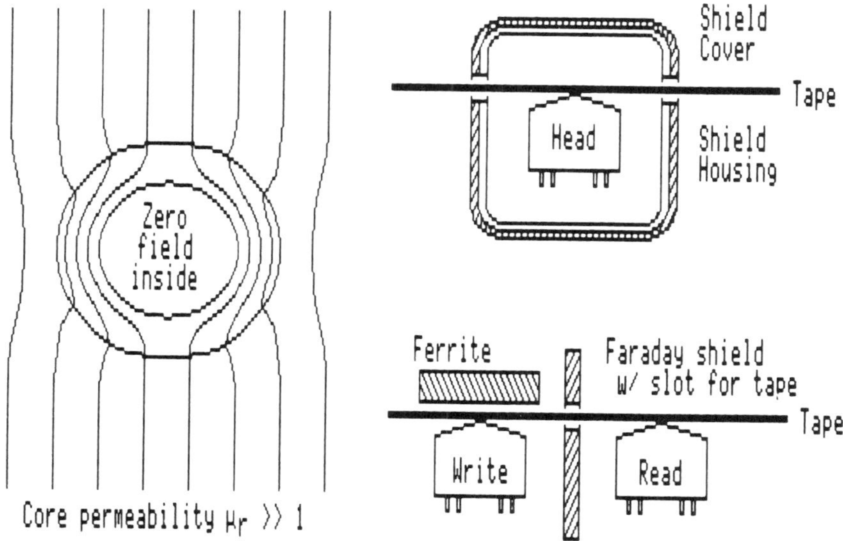

Fig. 8-38. Shielding of heads.

The magnetics involved with shielding is treated by Mager (see ref.), with some application to magnetic heads. A practical shield is shown in the figure, top right, where a cover is closed after the tape has been loaded. The housing itself is made from annealed nickel-iron or shielding material from PerfectionMica. A sandwiched layer of copper has been used in the past.

Deflection of fields (hum-bugging) is used as a last resort to prevent foreign flux from entering the heads. This is achieved by placing pieces of permalloy in the head's vicinity to deflect hum fields.

Another problem facing the equipment designer is radiation of fields from the heads to nearby electronic equipment and indeed from the write head to the read head. Insufficient attenuation of the radiated write field often causes a very poor signal-to-noise ratio when attempts are made to measure the read signal while writing.

A small slab of ferrite often helps (see Fig. 8-38, lower right). It is a good choice to use the type used for antennas in transistor radios. Further reduction is possible by installing a Faraday shield between the heads. It has a small slot open for the tape to pass through and opens up only for loading and unloading of tapes.

REFERENCES TO CHAPTER 8

Fields from Magnetic Ringcore Heads

Westmizze, W., "Studies on Magnetic Recording, II: Field Configuration Around the Gap and the Gap-Length Formula," *Philips Res. Rep.*, June 1953, Vol. 8, No. 3, pp. 161-183.

Karlqvist, O., "Calculation of the Magnetic Field in the Ferromagnetic Layer of a Magnetic Drum," *Trans. Roy Inst. Tech.*, June 1954, No. 86, pp. 3-27.

Fan, G.J., "A Study of the Playback Process of a Magnetic Ring Head," *IBM Jour.*, Oct. 1961, Vol. 5, pp. 321-325.

Potter, R., "Analytic Expression for the Fringe Field of Finite Pole-Tip Length Recording Heads," *IEEE Trans. Magn.*, Jan. 1975, Vol. MAG-11, No. 1, pp. 80-82.

Ichiyama, Y., "Reproducing Characteristics of Thin Film Heads," *IEEE Trans. Magn.*, Sept. 1975, Vol. MAG-11, No. 5, pp. 1203-1205.

Computations of Fields

Boast, W. B., *Vector Fields*, Harper and Row, New York., 1964, 610 pages.

Smith, G.D., *Numerical Solution of Partial Differential Equations: Finite Difference Methods*, Clarendon Press, Oxford, England, 1978, 304 pages.

Chari, M.V.K., and Silvester, P. P., *Finite Elements in Electrical and Magnetic Field Problems*, John Wiley & Sons, 1980, 219 pages.

Davies, A.J., *The Finite Element Method: A First Approach*, Clarendon Press, Oxford, England, 1980, 287 pages.

Artigue, M., and Gautheron, V., *Systemes Differentiels—Etude Graphique*, CEDIC, France, 1983, 180 pages.

Silvester, P.P., and Ferrari, R. L., *Finite Elements for Electrical Engineers*, Cambridge University Press, 1983, 209 pages.

Irons, B., and Shrive, N., *Finite Element Primer*, Halsted Press (Wiley), 1983, 157 pages.

Amort, D.L., "The Electrolytic Tank Analog," *Electro-Technology*, July 1962, 7 pages.

Brede, D. W., "Electrolytic Analog Magnetic Head Field Studies," *IEEE Trans. Audio*, Jan. 1964, Vol. AU-12, pp. 9-19.

Benedict, R.P., "Analog Simulation," *Electro-Techn.*, Dec. 1963, pp. 73-90.

Laubsch, H., Kisch, S., and Sladek, J., "Modelluntersuchingen zom Wiedergabevorgang Dynamischer Magnetspeicher im Elektrolytischen Trog," *Hochfreq. und Elektroakustik*, Jan 1970, Vol. 79, pp. 144-149.

Szczech, T.J., "Analytic Expressions for Field Components of Nonsymmetrical Finite Pole Tip Length Magnetic Head Based on Measurements on Large-Scale Model," *IEEE Trans. Magn.*, Sept. 1979, Vol. MAG-15, No. 5, pp. 1319-1323.

Baird, A., "High-Resolution Field Measurements Near Ferrite and Thin-Film Recording Heads," *IEEE Trans. Magn.*, Sept. 1980, Vol. MAG-16, No. 5, pp. 794-796.

Szczech, T.J., Steinback, M., and Jodeit, Jr., M., "Equations for the Field Components of a Perpendicular Magnetic Head," *IEEE Trans. Magn.*, Jan. 1982, Vol. MAG-18, No. 1, pp. 229-232.

Fayling, R., "Studies of the 'Magnetizing Region' of a Ring-Type Recording Head," *IEEE Trans. Magn.*, Nov. 1982, Vol. MAG-18, No. 6, pp. 1212-1214.

Hoyt, R.F., Heim, D. E., Best, J. S., Horng, C. T., and Horne, D. E., "Direct Measurement of Recording Head Fields Using A High Resolution Inductive Loop," *IBM Report RJ* 4117, Nov. 1983.

Ruigrok, J.J.M., "Analysis of Metal-In Gap Heads," *IEEE Trans. Magn.*, Sept. 1984, Vol. MAG-20, No. 5, pp. 872-874.

Elsbrock, J.B., and Balk, L. J. "Profiling of Micromagnetic Stray Fields in Front of Magnetic Recording Media and Heads by Means of a SEM," *IEEE Trans. Magn.*, Sept. 1984, Vol. MAG-20, No. 5, pp. 866-869.

Tortschanoff, T., "Survey of Numerical Methods in Field Calculations," *IEEE Trans. Magn.*, Sept. 1984, Vol. MAG-20, No.5, pp. 1912-1917.

Farquhar, A., "Effects of Scaling on Electronmagnetic Devices," Aug. 1985, Nortronics Note.

Harris, W., "Direction Fields," *Nibble Magazine*, Nov. 1985, Vol. 6, No. 11, pp. 100-103.

Dugas, M., Bonin, W., and Judy, J., "A Finite Element Analysis of the MSP Single Pole Perpendicular Recording Head," *IEEE Trans. Magn.*, Sept. 1985, Vol. MAG-21, No. 5, pp. 1554-1556.

Perpendicular Head Fields

Camras, M., "An X-Field Micro-Gap Head for High Density Magnetic Recording," *IEEE Trans. Audio*, May 1964, Vol. AU-12, pp. 41-52.

Johnson, W., and Jorgensen, F., "A New Analog Magnetic Recording Technique," *ITC Conf. Proc.*, Aug. 1966, Vol. 2, pp. 414-430.

Krey, K., and Lieberman, A., "Principles and Head Characteristics in VHF-Recording," *ITC Conf. Proc.*, Fall 1973, Vol. 9, pp. 49-59.

Lopez, O., "Interaction of a Ring Head and Double Layer Media—Field Calculations," *IEEE Trans. Magn.*, Nov. 1982, Vol. MAG-18, No. 6, pp. 1179-1181.

Minuhin, V., "Comparison of Sensitivity Functions for Ideal Probe and Ring-Type Heads," *IEEE Trans. Magn.*, May 1984, Vol. MAG-20, No. 3, pp. 488-494.

Mallinson, J.C., and Bertram, H.N., "On the Characteristics of Pole-Keeper Head Fields," *IEEE Trans. Magn.*, Sept. 1984, Vol. MAG-20, No. 5, pp. 721-724.

Luitjens, S.B., Smits, J.W., and Zieren, V., "The Write Field of a Ring Head for a Double Layer Perpendicular Medium," *IEEE Trans. Magn.*, Sept. 1984, Vol. MAG-20, No. 5, pp. 724-729.

Chi, C.S., and Szczech, T.J., "A Write Head Design with Improved Field Profile for Perpendicular Recording," *IEEE Trans. Magn.*, Sept. 1984, Vol. MAG-20, No. 5, pp. 836-838.

Shiiki, K., Shiroishi, Y., Shinagawa, K., Kumasake, N., and Kudo, M., "Probe Type Thin Film Head for Perpendicular Magnetic Recording," *IEEE Trans. Magn.*, Sept. 1984, Vol. MAG-20, No. 5, pp. 839-841.

Side Write/Read from Adjacent Track

Lindholm, D., "Magnetic Fields of Finite Track Width Heads," *IEEE Trans. Magn.*, Sept. 1977, Vol. MAG-13, No. 5, pp. 1460-1462.

Ichiyama, Y., "Analytic Expressions for the Side Fringe Field of Narrow Track Heads," *IEEE Trans. Magn.*, Sept. 1977, Vol MAG-13, No. 5, pp. 1688-1689.

van Herk, A., "Analytical Expressions for Side Fringing Response and Crosstalk with Finite Head and Track Widths," *IEEE Trans. Magn.*, Nov. 1977, Vol. MAG-13, No. 6, pp. 1764-1766.

van Herk, A., "Side-Fringing Response of Magnetic Reproducing Heads," *Jour. AES*, April 1978, Vol. 26, No. 4, pp. 209-211.

van Herk, A., "Measurement of Side-Write, -Erase, and -Read Behavior of Conventional Narrow Track Disk Heads," *IEEE Trans. Magn.*, Jan. 1980, Vol. MAG-16, No. 1, pp. 114-119.

Erase Heads

McKnight, J.G., "Erasure of Magnetic Tape," *Jour. of AES*, Oct 1963, Vol. 11, No. 10, pp. 223-233.

Katz, E.R., "Erase Profiles of Floppy Disk Heads," *IEEE Trans. Magn.*, July 1984, Vol. MAG-20, No. 4, pp. 528-541.

Lemke, J.U., "Re-recordings during Overwrite," unpublished, *THIC Meeting*, Jan. 1986.

Stray Flux Computations

Katz, E.R., "Finite Element Analysis of the Vertical Multi-Turn Thin-Film Head," *IEEE Trans. Magn.*, Sept. 1978 , Vol. MAG-14, No. 5, pp. 506-508.

Palmer, D.C., and McDaniel, T.W., "Spurious Signal Pickup from the Outer Rails of a Ferrite Recording Head," *IEEE Trans. Magn.*, Sept. 1984, Vol. MAG-20, No. 5, pp. 912-914.

Leakage Flux

Roters, H.C., *Electromagnetic Devices*, 1941, John Wiley & Sons, 561 pages.

Hammond, M.A., "Leakage Flux and Surface Polarity in Iron Ring Stampings," *IEEE Proceedings*, Jan. 1955, Vol. 43, No. 1, pp. 138-147.

Chen, T.H., "The Coupling Effect Between the Dual Elements of a Superimposed Head," *IEEE Trans. Magn.*, Nov. 1981, Vol. MAG-17, No. 6, pp. 2905-2907.

Shielding

Jorgensen, F., "The Influence of an Ambient Magnetic Field on Magnetic Tape Recorders," *ITC Conf. Proc.*, Fall 1975, Vol. 11, pp. 373-390.

Mager, A., "Magnetic Shields," *IEEE Trans. Magn.*, March 1970, Vol. MAG-6, No. 1, pp. 67-75.

Lopez, O., Lam, T., and Stromsta, R., "Effects of Magnetic Fields on Flexible Disk Drive Performance," *IEEE Trans. Magn.*, July 1981, Vol. MAG-17, No. 4, pp. 1417-1422.

BIBLIOGRAPHY TO CHAPTER 8

Reece, G., *Microcomputer Modelling by Finite Differences*, Halsted Press (John Wiley & Sons), 1986, 126 pages.

Lowther, D.A., and Silvester, P.P., *Computer-Aided Design in Magnetics*, Springer-Verlag, 1986, 324 pages.

The Electromagnetics Problem Solver, Research and Educ. Assoc., New York, 1983, 830+ pages.

Kraus, J.D., *Electromagnetics*, Mc-Graw-Hill, New York, 1984, 775 pages.

Plonus, M.A., *Applied Electro-Magnetics*, Mc-Graw-Hill, New York, 1978, 599 pages.

Armstrong, R.L., and King, J.D., *The Electromagnetic Interaction*, Prentice-Hall, New York, 1973, 493 pages.

Foster, K., and Anderson, R., *Electromagnetic Theory—Problems and Solutions*, St. Martin's Press, Great Britain, 1970, 212 pages.

Ferrari, R.L., *An Introduction to Electromagnetic Fields*, Van Nostrand Reinhold, 1975, 202 pages.

Parton, J.E., and Owen, S.J.T., *Applied Electromagnetics*, MacMillan Press Ltd., 1975, 258 pages.

Schey, H.M., *DIV, GRAD, CURL, and all that*, W.W. Norton, New York, 1973, 163 pages.

Polivanov, K., *The Theory of Electromagnetic Fields*, MIR Publ, Moscow (Imported Publ, Chicago), 1983, 271 pages.

Fields from Magnetic Ring Core Heads

Szczech, T., "Improved Field Equations for Ring Heads," *IEEE Trans. Magn.*, Sept. 1983, Vol. MAG-19, No. 5, pp. 1740-1744.

Otto, R., "Experimental Determination of Electrical and Magnetic Fields by Models" (In german), *Frequenza*, Oct. 1961, Vol. 15, No. 10, pp. 309-315.

Szczech, T.J., and Palmquist, K.E., "A 3-D Comparison of the Fields from Six Basic Head Configurations," *Intl. Conf. Video and Data 84*, April 1984, pp. 17-22.

El-Hakim, Y.A., "A Study of the Field Around Magnetic Heads of Trapezoidal Shape," *IEEE Trans. Audio*, March 1967, Vol. AU-15, No. 1, pp. 19-26.

Elabd, I., "A Study of the Field Around Magnetic Heads of Finite Length," *IEEE Trans. Audio*, Jan. 1963, Vol. AU-11, No. 1, pp. 21-27.

Perpendicular Head Fields

Minuhin, V.B., "Field of Probe Heads with Ideal Winding," *J. Appl. Phys.*, April 1985, Vol. 57, No. 1, pp. 4006-4009.

Lindholm, D., "Image Fields for Two-Dimensional Recording Heads," *IEEE Trans. Magn.*, Sept. 1977, Vol. MAG-13, No. 5, pp. 1463-1465.

Wade, R.H., "The Measurement of Magnetic Microfields," *IEEE Trans. Magn.*, Jan. 1976, Vol. MAG-12, No. 1, pp. 34-39.

Szczech, T.J., "Exact Solution for the Field of a Perpendicular Head," *IEEE Trans. Magn.*, Nov. 1981, Vol. MAG-17, No. 6, pp. 3117-3119.

Minuhin, V.B., "Characteristics of Ideal Probe Heads with Ideal Windings in the Presence of a Permeable Media Underlayer," *IEEE Trans. Magn.*, July 1985, Vol. MAG-21, No. 4, pp. 1289-1294.

Side Write/Read from Adjacent Track

McKnight, J., "The Fringing Response of Magnetic Reproducers at Long Wave-lengths," *Jour. of AES*, Feb. 1972, Vol. 20, No. 2, pp. 100-105.

van Herk, A., "Side Fringing Fields and Write and Read Crosstalk of Narrow Magnetic Recording Heads," *IEEE Trans. Magn.*, July 1977, Vol. MAG-13, No. 4, pp. 1021-1028.

Hoyt, R.F., and Sussner, H., "Precise Side Writing Measurements Using a Single Recording Head," *IEEE Trans. Magn.*, Sept. 1984, Vol. MAG-20, No. 5, pp. 909-911.

Chapter 9
Manufacture of Head Assemblies

A large variety of magnetic heads for recording and playback are available from a number of manufacturers. Figure 9-1 shows state-of-the-art heads in high packing density tape (IBM 3480) and disk (IBM 3380) drives. But that is just a fraction of the types, and many applications are generally ahead of standards. Figure 9-2 illustrates the evolution of magnetic heads for all applications, from the large, single-track ring core in the Magnetophone to very small heads (for disk heads see ref. Collier). The summary sheet ended up as Table 9-1, which gives the reader some idea about the variety of transducers available today. The preparation of a summary sheet for magnetic tape and disk heads turned out to be a most frustrating task because of the vast diversity in design and application.

A number of multitrack tape heads are shown in Fig. 9-3, and Fig. 9-4 shows a standard 14 channel head assembly manufactured to IRIG standards (Inter Range Instrumentation Group). The tracks are interleaved (i.e., all even numbered tracks are in one head stack and all odd numbered in the other). Full shields are used for maximum reduction of crosstalk.

Beneath the surface of these heads lies numerous hours of ultra precise manufacturing, machining, and assembly. The finished head must meet a set of rigid standards covering track widths, spacing dimensions, and other mechanical tolerances. An important mechanical tolerance is the maximum deviation of each track from being in-line (skew). Record or reproduce sensitivities must be within certain limits, for all tracks. Frequency response (efficiency, gap length, losses) must be uniform and alike for all tracks. And the assembly must withstand a hostile environment of tape abrasivity and often temperature extremes.

The temperature requirement dictates that core material, bonding epoxies and head housing materials have essentially the same coefficient of expansion versus temperature.

The fabrication processes for metal core heads and for ferrite core heads are pretty well established, and are described first in this chapter. Multitrack and single track all-ferrite heads

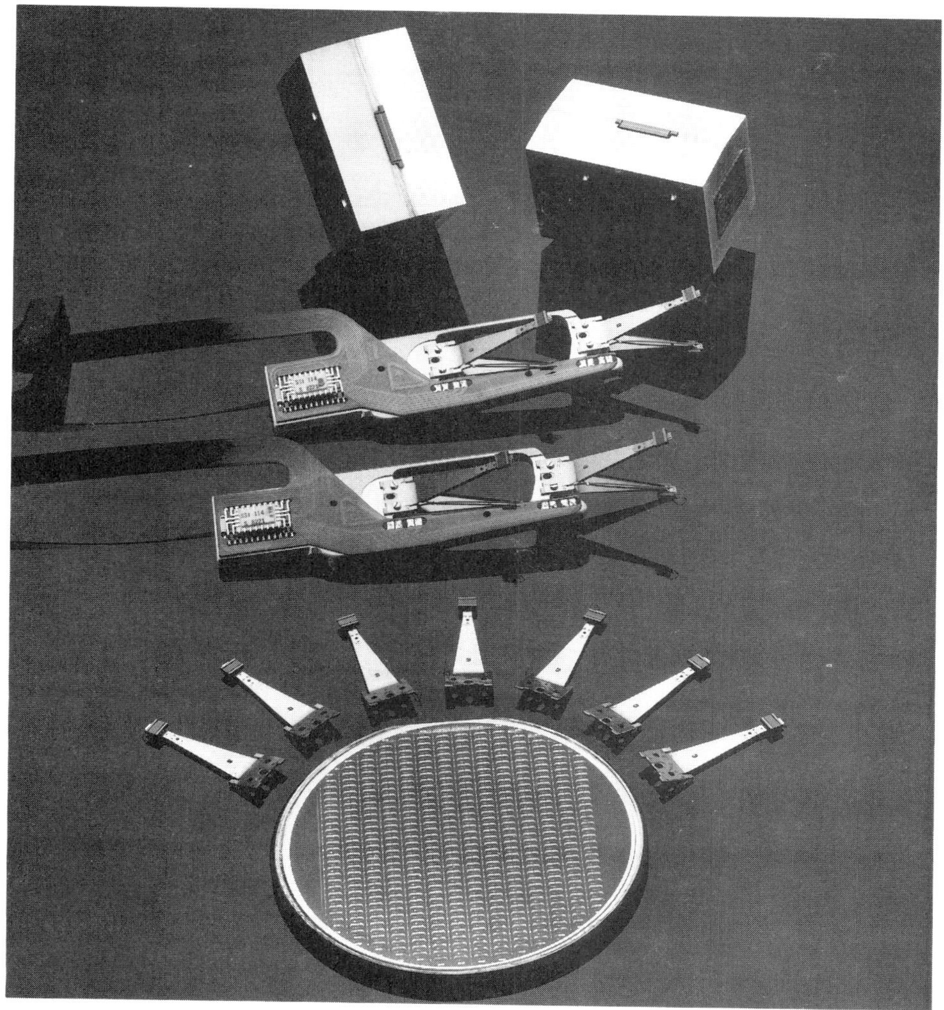

Fig. 9-1. Thin-film heads are shown here in three manufacturing stages. From bottom to top of the photograph are: a finished wafer on which thin-film sensors are deposited; thin-film disk head sliders (here mounted on flexures); complete thin-film disk head/arm assemblies; and thin-film tape heads (courtesy of Applied Magnetics, Inc.).

for diskettes and rigid disks are covered next, and the novel approach of using semiconductor fabrication technology in production of thin film heads is covered later.

The chapter ends with a discussion of specifications, some methods for the test and measurement of magnetic head performance, and a design test guide.

HEADS WITH LAMINATED METAL CORES AND SOLID FERRITE CORES

Many heads have cores for each track, that are made from laminated Ni-Fe alloys, or from solid ferrite. So-called hard-tipped heads have ferrite cores with wear resistant pole tips (Alfenol, Sendust, Spinalloy, Vacodur).

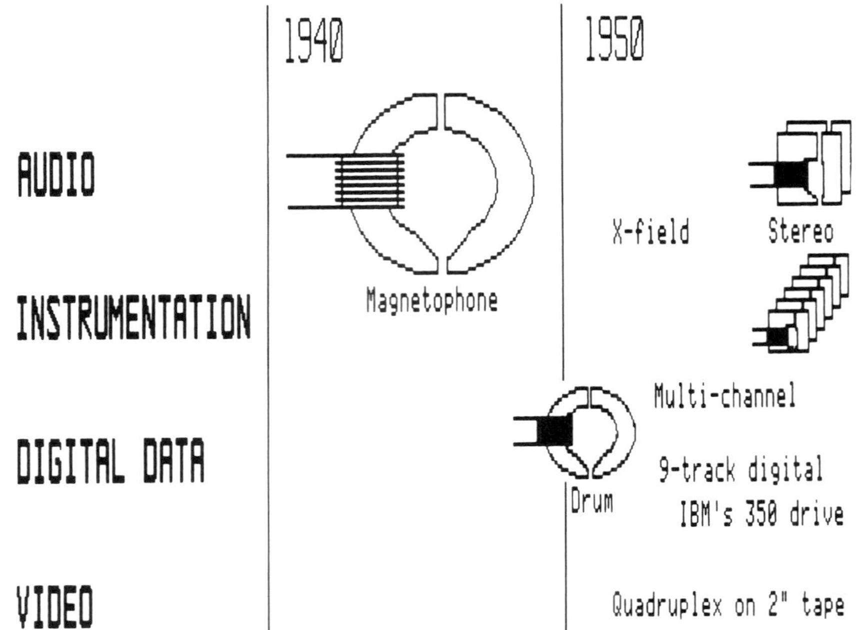

Fig. 9-2. Evolution of magnetic write/read heads.

Table 9-1. Average Values of Track Widths and Densities in Magnetic Recording.

APPLICATION/media:	Trackwidth	No. tracks	Upper freq./bit rate	BL μin
AUDIO cassette	.021"	4 parallel	20 kHz (bias 100)	46
INSTRUMENTATION	.025"	16-80 ""	4 MHz (bias 16)	15
VIDEO, studio	.004"	1 slanted	FM to 10 MHz	46
VHS	.0012"	1 """	FM to 4.2 "	27
Betamax	.0008"	1 """	FM to 5.5 "	25
8/mm	.0008"	1 """	FM to 5.4 "	14
DIGITAL tape cassette	.020"	4 parallel	75 KBPS	100
tape cartridge	.020"	8 ""	90 ""	100
reel-to-reel	.050"	9 ""	1250 ""	167
tape cartridge	.015"	16 ""	3000 ""	100
DIGITAL floppy	.014"	48 TPI	250 KBPS	167
floppy	.008"	96 "	250 ""	167
floppy	.006"	192 "	500 ""	100
Mini-disk	.0045"	140 "	250 ""	167
Cartridge	.0025"	300 TPI	1130 ""	56
"Hard disk"	.0015"	600 "	2-10,000 KBPS	100
Winchester	.0008"	1,000 "	2-15,000 ""	67
Magneto optical	.00007"	15,000 "	?	67

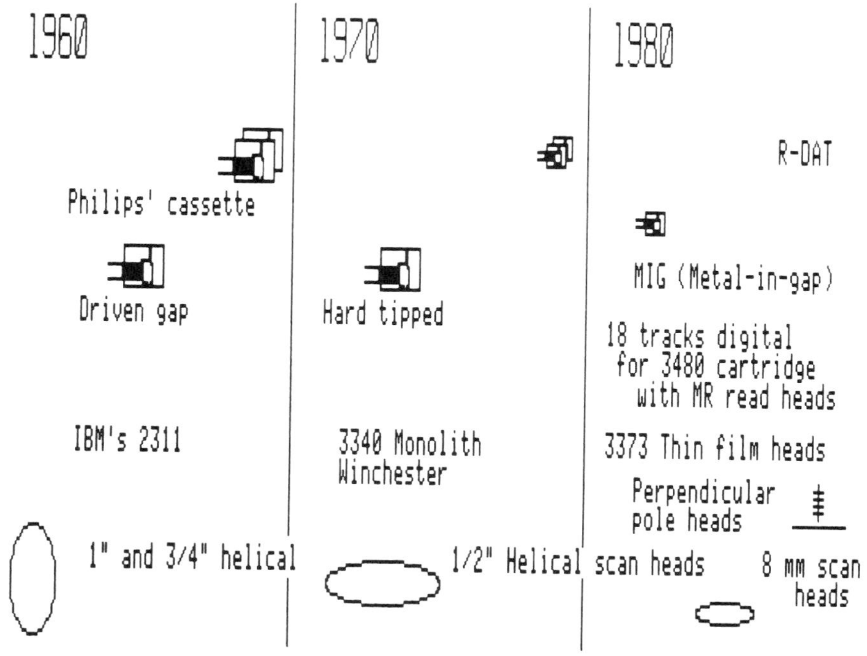

Fig. 9-2. Continued.

The flow chart in Fig. 9-5 shows the fabrication steps, from Mumetal stock to the finished head assembly. Large laminations are often stamped, annealed and stacked to form the core. Photochemical milling (or etching) is, however, a preferred technique nowadays (ref. Boll, 1965).

Any *coldworking* of a soft magnetic alloy will reduce its permeability drastically; there are no magnetic materials completely free from magnetostriction. If the sheets for etching are bent with a radius of only 10-20 cm, a large reduction in permeability takes place. And a pressure of:

$$981*10^3 \text{ N/m}^2 = 142 \text{ lbf/in}^2 = 10 \text{ kgf/cm}^2$$

will reduce the permeability of a well-annealed Mumetal by a factor of ten (ref. Boll, 1956). The manufacturer must be aware of these limitations when stacking the cores for bonding under pressure, and later during application of epoxies to keep them in place in the head housings (ref. Liechti). It is good practice to include a sample ring in each lamination sheet; this ring can be tested for permeability of the cores from its lot. A simple check of the inductance of a few turns will do; the derivation for formula (7.5) shows that the inductance value is directly proportional to μ_r.

The *coil windings* are either wound on a separate bobbin and slipped onto a core leg (when the core geometry permits it) or they are placed directly on the core, which then is provided with an insulation of oilpaper, teflon, or tape. One problem in the design of multitrack heads is finding room for the winding. Shielding between tracks must be provided for and should, in interleaved assemblies, have a thickness equal to the track width. This leaves only a distance equal to the track spacing for winding plus coil.

Grounding of the cores and the shields is essential for low noise operation. Not only does this reduce pickup of interfering electric fields, but it also prevents the buildup of electrostatic charges from the moving tape (a Van de Graaff generator). Some manufacturers mix graphite

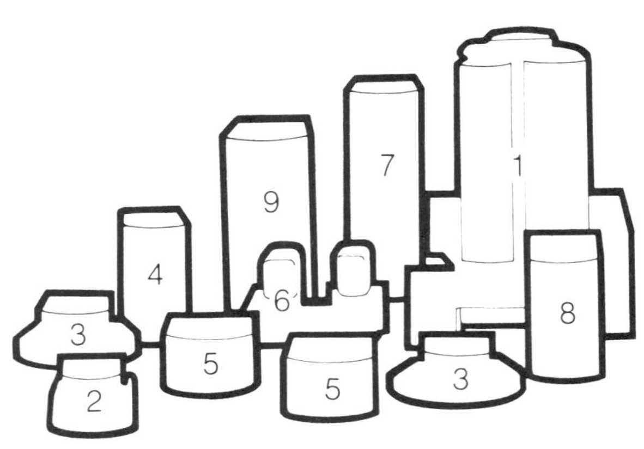

1. 3M—M-79 (24 track-2")
2. ATR—100 (Playback—2 track-¼")
3. Scully 280 (2 track-¼")
4. Scully (8 track-1")
5. Ampex AG 440 (2 track-¼")
6. Lockheed (Space Shuttle—2 track-¼")
7. Stephens (16 track-2" metal)
8. Stancil-Hoffman (28 track-1")
9. Ampex (16 track-2")

Fig. 9-3. Magnetic heads for reel-to-reel audio machines, high speed duplicators, voice loggers, tape certifiers and instrumentation (courtesy SAKI Magnetics, Inc.).

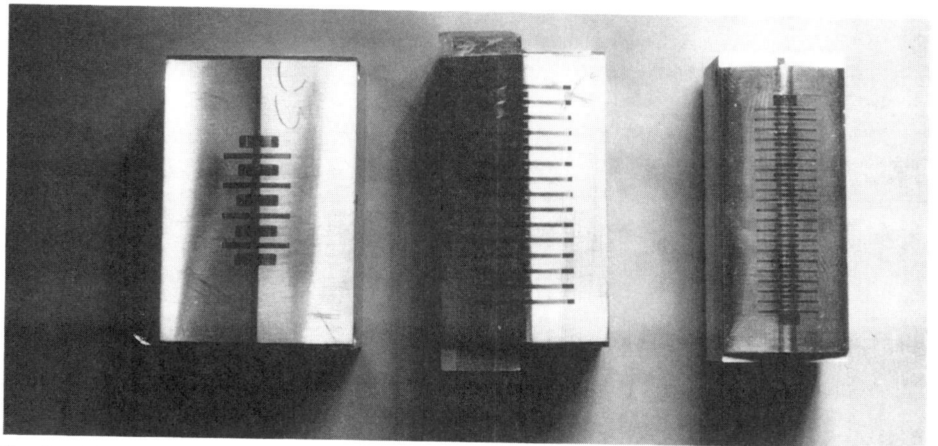

Fig. 9-4. Fourteen channel head assembly (2 × 7 tracks, interleaved) (courtesy of Omutec, Div. of Odetics).

or other conductive powder in the bonding epoxies to eliminate build-up of any electrostatic charge.

The next critical step in the process is the *lapping* of the head-halves. One four channel head-half is shown in Fig. 9-6. They should have been artificially aged (by way of temperature cycling) to be free of stresses prior to the finish lapping process. The finished surface is checked

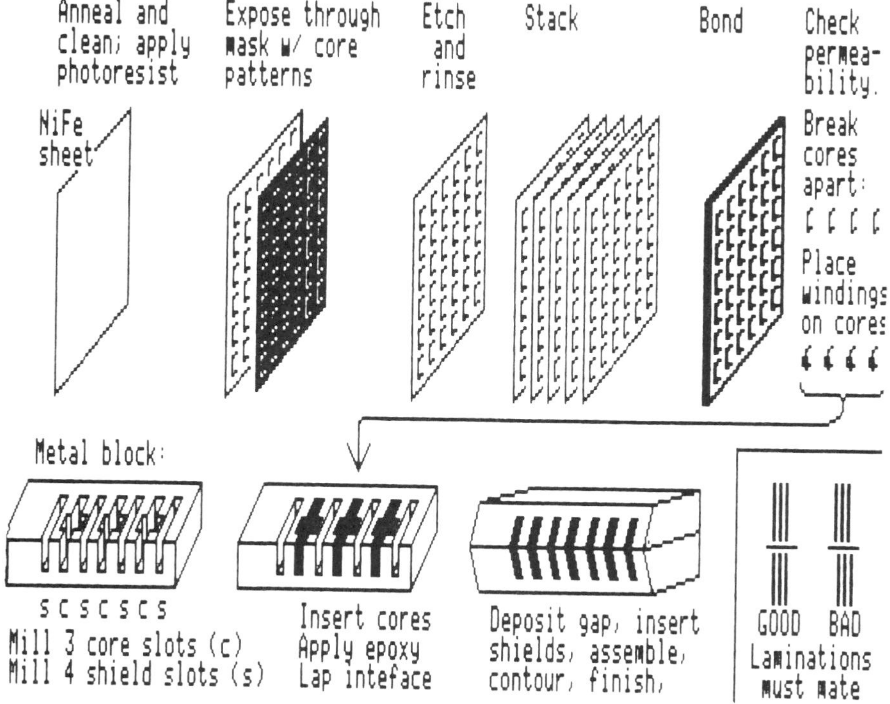

Fig. 9-5. Fabrication of multitrack head with laminated cores.

Fig. 9-6. Completed half-shells for 4 channel heads (courtesy of Omutec, Div. of Odetics).

with an optical flat placed against it, in monochromatic light, and should be within as little as one quarter of a lightband for narrow gap heads (one light band is, for example, in orange light equal to .3 μm (12 μin) since the wavelength of orange is 6000 angstrom = 0.6 μm = 24 μin).

During the grinding and final lapping of the mating surfaces of the head halves, great care must be exercised. The ultimate tensile strength of Mumetal is quoted to be $5.5*10^8$ N/m^2, and during grinding and lapping this force must occur in the minute areas where material is removed (ref. Rabinowicz).

The net result is that the outside surface layers of the interfacing poletips have very low permeability, which will appear as a gap that is longer than its physical dimensions. This is not to be confused with the fact that the magnetic gap of a head is 10-20 percent longer than its mechanical length; coldworking will add even more to these percentages. Only by careful and patient lapping methods can these layers be so thin as to be insignificant.

Gap spacers of various materials are used to control the length of the front gap in the head assembly. Cleaved Mica has been used extensively, while modern narrow gap heads have a silicon-monoxide layer for gap spacers. It is generally placed on a head-half by vacuum deposition techniques. This method can also be applied for spacers of conductive materials.

The halves are then mated and held together with screws or epoxy, or both. The laminations must meet so one set is a mirror image of the other; otherwise poor high frequency response results (remember the flux flows through the lamination surfaces at high frequencies, and should not flow into the middle of a mating lamination).

Finishing of the assembly is done by contouring the head surface by grinding and then lapping to its final dimensions. There are various (some proprietary) means of checking the grinding so a well-defined final gap depth is achieved. One method is a simple check of the inductance values of sample tracks.

For long life applications the gap must be about 5 mils deep (or provided with hard tips, see next section). Higher efficiencies are achieved in short gap heads for wideband applications by cutting the gap depth to 2 mils.

Overall shielding of the head assembly is a commonly used technique in audio recorders, where 60 or 50 Hz power line hum from the power transformer, and from motors, is a source of noise inside the signal frequency range (20-20,000 Hz). Shielding can be provided for by a magnetically soft shell around the head (see Fig. 8-38); the material may be drawn, annealed mumetal (ref. MuShield), or a Co-Netic material that can be cut and bent within wide ranges without losing its permeability (ref. Perfection Mica Corp).

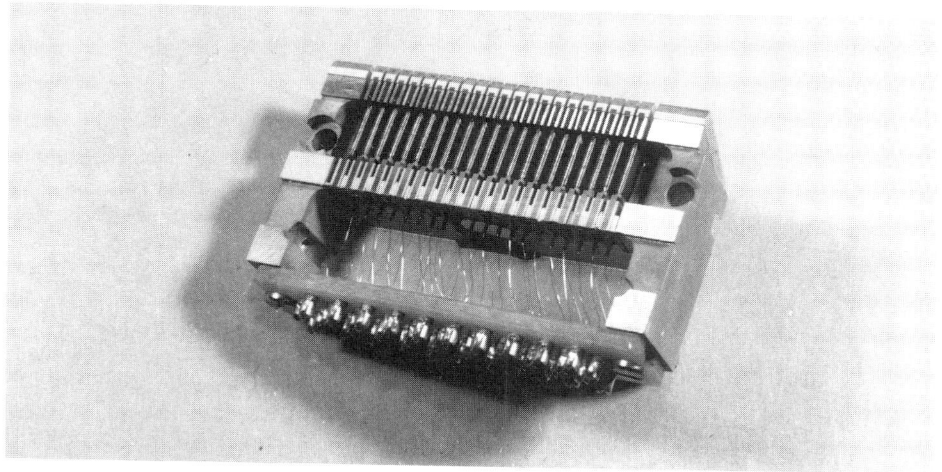

Fig. 9-7. Completed half shells for 8 channel heads (courtesy of Omutec, Div. of Odetics).

FERRITE HEADS WITH METAL TIP PLATES

Heads with Mumetal cores will in certain applications wear out too fast (see Chapter 18), and it has become standard practice to provide certain head types with *wear resistant poletips* (instrumentation, video, and some audio heads). The tip material is one of the hard iron-aluminum alloys of many names (for variation in composition): Alfesil, Sendust, Spinalloy, Vacodur, etc. The tips may be solid or laminated, and they are, as a step in the processing, placed in a tip plate.

The tip plate is placed against the head block and bonded to it so the tips make intimate contact with the core (Fig. 9-7). The magnetic reluctance in the gap between the tip and the core has a generally negligible effect on the net value of the head efficiency. But the hard tip must be laminated for good high frequency performance.

ALL FERRITE HEADS; MULTICHANNEL TAPE HEADS; SINGLE-CHANNEL DISK HEADS; METAL-IN-GAP HEADS

Ferrite materials offer a more straightforward manufacturing technique, with fewer steps. Single track as well as multitrack heads are made from ferrite bars ground to profiles, lapped, gapped, and bonded together with glass. The bonding process uses glasses with coefficients of expansion equal to the selected ferrites, and there are glasses with high as well as low melting temperatures. Occasionally both are used in the same head, a high temperature glass for the gap and a low temperature glass for bonding the head into the head housing.

Machining and grinding of ferrites requires experience and skill. Too fast and too deep a cut (high removal rate) can produce stresses that affect the magnetic properties. Final lapping operations are just as critical. When material is removed by the cutting and smearing action of the particles in the abrasives then tensile stresses in the surface results. During the motion of the scratching particle the material is deformed not only beneath the indenter, but also before, around and behind it (ref. Broese van Gruenau).

The result is a *dead layer* (called the "Beilby layer") where the stresses greatly reduce the material's permeability via the mechanism of magnetostriction. The thickness of this layer in HIP materials (hot-pressed ferrites) may be estimated by (ref. Wada):

$$\text{Thickness} \cong 4 * R_z + 50 \text{ angstrom} \qquad \mathbf{(9.1)}$$

where R_z equals the surface roughness in angstroms (mean value assumed, reference is not clear).

The stress inside the material, just below the surface, reaches values equal to the ultimate compressive strength of the material, and the stresses perpendicular to the direction of grinding are about 40 percent larger than those parallel to the direction of grinding (ref. Knowles).

The thickness of the dead layer can be determined by a straightforward experiment, illustrated in Fig. 9-8. A ferrite sample is provided in the form of a toroid. Wind a suitable number of turns (say 10) onto the core, and measure the low frequency inductance value L′. Next cut the toroid in half, and assemble (with a light pressure) to form a closed inductor again (Fig. 9-8, right). Now measure the inductance value L″, and find the thickness of the layer destroyed by the cutting plus the effective airgap due to surface roughnesses from formula (9.2).

Next lap and polish the surfaces to be flat with a lightband, and remeasure L″; it should be a higher value indicating a reduction in the thickness of the dead layer. Ultimately etch the surfaces with cold concentrated hydrochloric acid (HCl—use extreme care!), and it should be possible to achieve a very small value of BeL, the Beilby layer. The etching removed the stressed surface material quite uniformly, and with a ferrite with closed porosity no HCl penetrates below the surface (ref. Knowles).

Now apply your usual grinding method, and measure the resulting BeL; and repeat for the lapping and polishing processes. This method will aid in developing the optimum processing.

Another concern with ferrites is crystal pullout. Such voids in the gap edges will deteriorate the head's short wavelength resolution and these voids in the surface may collect dirt and dust and may cause "crashing" of flying heads, scratches in tape and generation of dropouts.

The NiZn ferrites are reportedly of higher stability and better intergrain bonding, and therefore easier to machine with high precision and little chipping. They do, on the other hand, appear to have higher magnetic losses from machining than do MnZn ferrites.

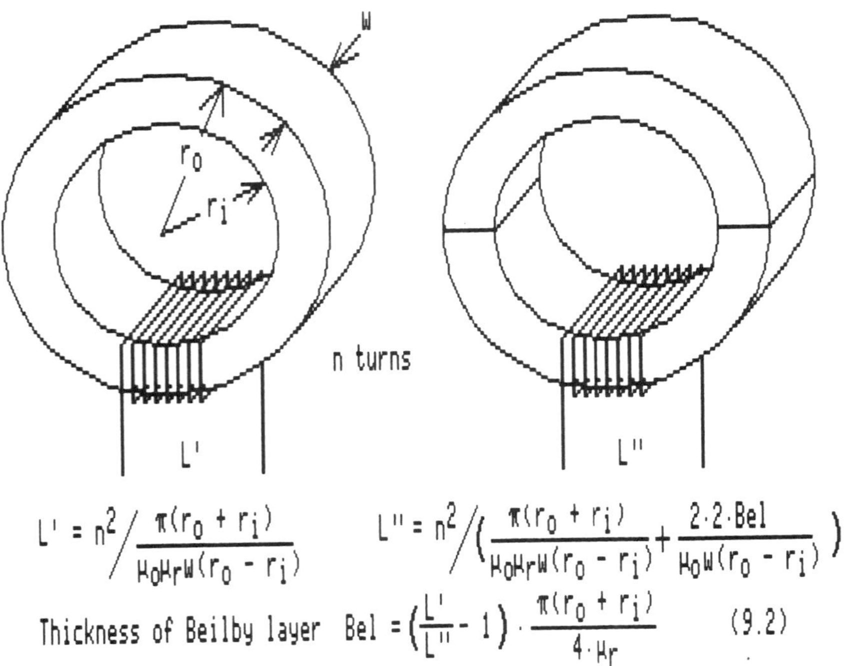

Fig. 9-8. Simple procedure for calculating effects of surface grinding.

Wear resistance of the two ferrite types are inadequately described in the literature and the conclusions tend to be contradictory.

A large variety of head configurations are made today, as the following descriptions and illustrations will show.

A *multitrack* head can be made from comb structure illustrated in Fig. 9-9. It has a disadvantage in that the tracks are sharing one or another leg of their magnetic circuits. The cross-talk performance can be calculated and measured and may be acceptable for digital recording, but not for direct recordings demanding a high signal-to-noise ratio. The joint portion of the magnetic circuits could be avoided by a comb-like back plate and removal of excess ferrite after the second bonding operation.

Single track heads for disk use have cores that are diced, ground, and sliced from ferrite as shown in Fig. 9-10. More complicated structures have been developed in the past few years. For diskette use, write/read heads with *tunnel-erase* and *straddle-erase* cores are shown in Fig. 9-11. Although the latter version trims the written track closer to the write/read gap, the first appears to have become industry standard (ref. Butsch); on track width, trimming, TPI etc., see Chapter 10.

Heads for rigid disk applications are shown in Fig. 9-12. They are required to fly over the surface of the disk at a distance of 0.3 - 0.4 μm, and operate at frequencies in the MHz range. The technique of flying a head over a moving surface is covered in Chapter 18, Head/Media Interface, and the design of very high frequency heads in Chapter 10.

Some ferrite core and core assemblies are shown in Fig. 9-13. Their assembly often involves the use of glass as a bonding agent, rather than epoxies that tend to change dimensions with time (humidity, absorption, etc.). Glasses match the ferrites quite well in terms of coeffi-

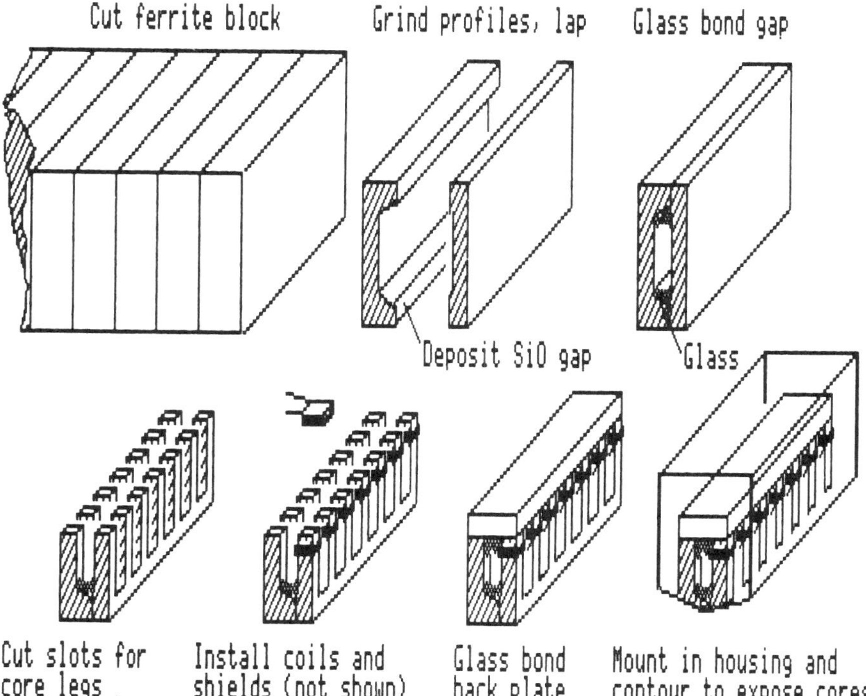

Fig. 9-9. Fabrication of multitrack ferrite head.

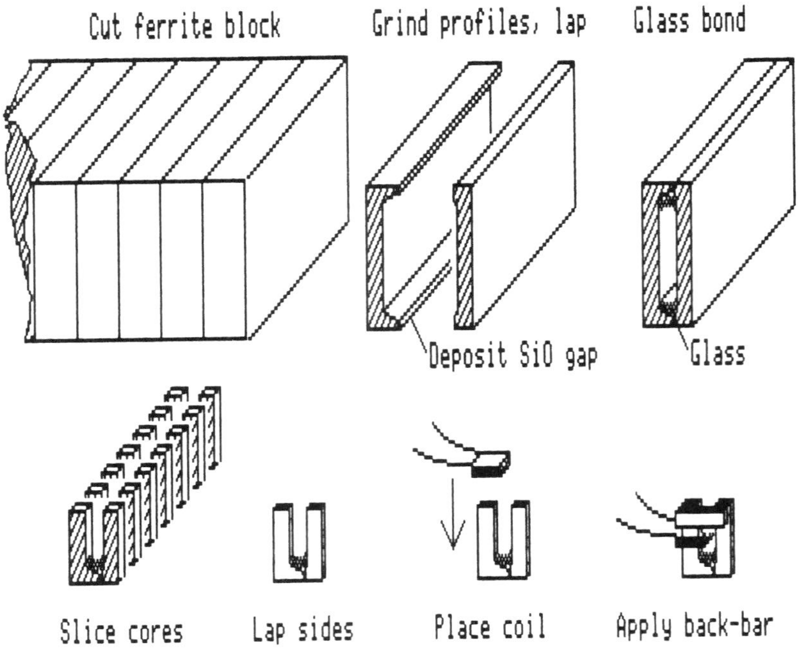

Fig. 9-10. Fabrication of single track flying head with ferrite core.

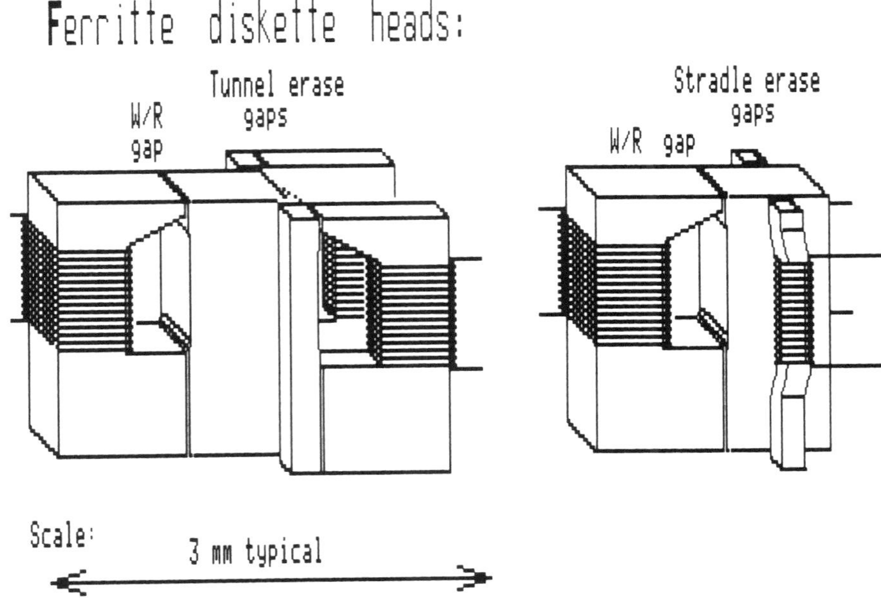

Fig. 9-11. Diskette write/read heads with track trimming erase cores.

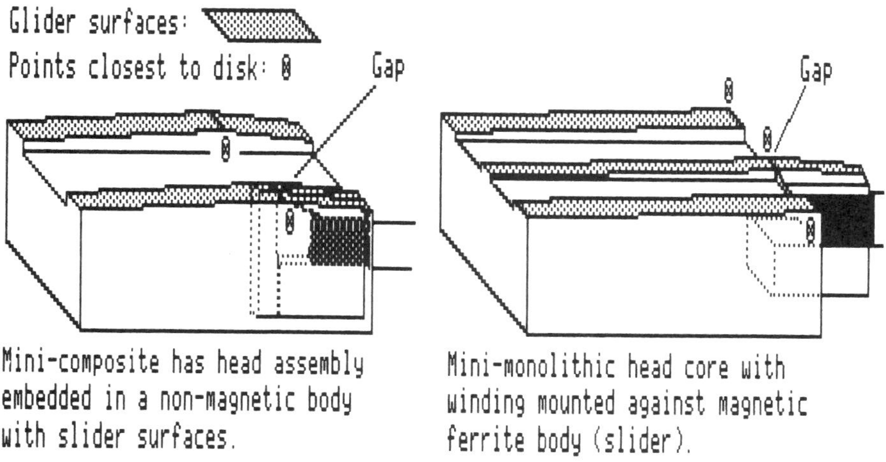

Fig. 9-12. Ferrite heads for rigid disks.

cient of expansion, and easy to apply and melt to form a meniscus holding for core halves together (see Figs. 9-14, 9-15). A high melting temperature glass is used in the gap region, while the glass used for holding the core in the slider (or housing) has a lower melting temperature.

The glasses react with the ferrite, and tend to diffuse into the surfaces and partly destroy the permeability. Experiments similar to the one shown in Fig. 9-8 will reveal the extent of diffusion. Certain glasses containing lead (for low temperature melting points) will interact with

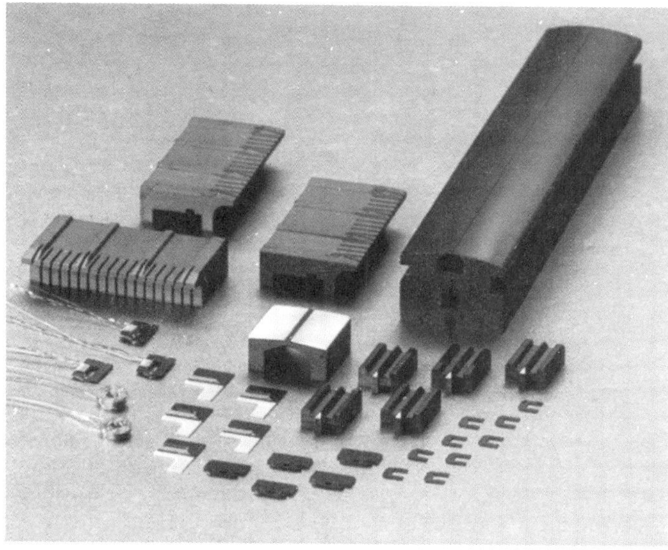

Fig. 9-13. Ferrite cores and assemblies (courtesy Tranetics, Inc.).

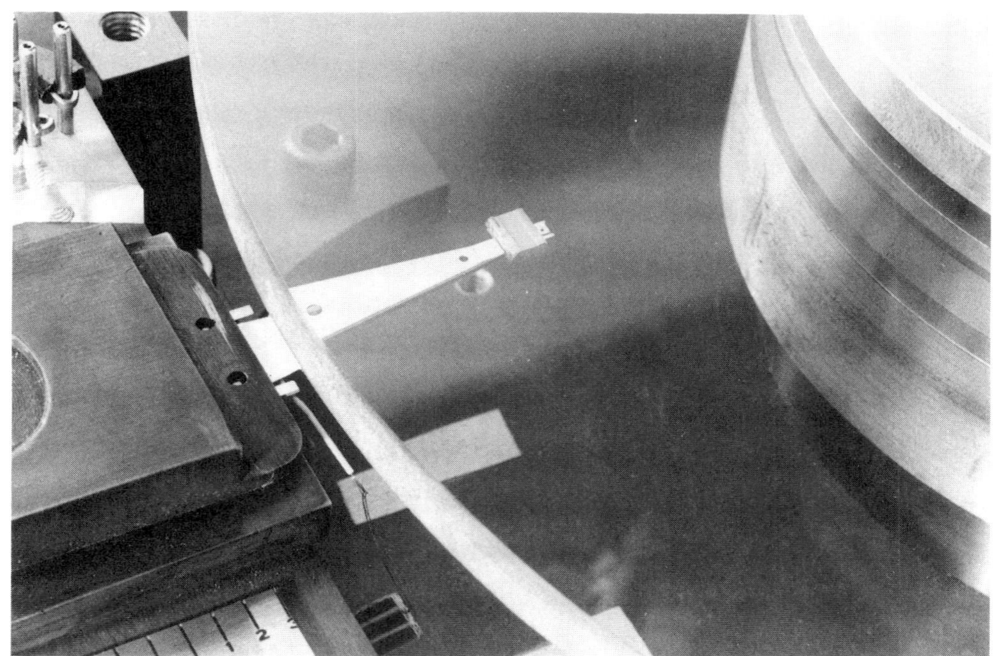

Fig. 9-14. Mini composite head (courtesy of Tranetics, Inc.).

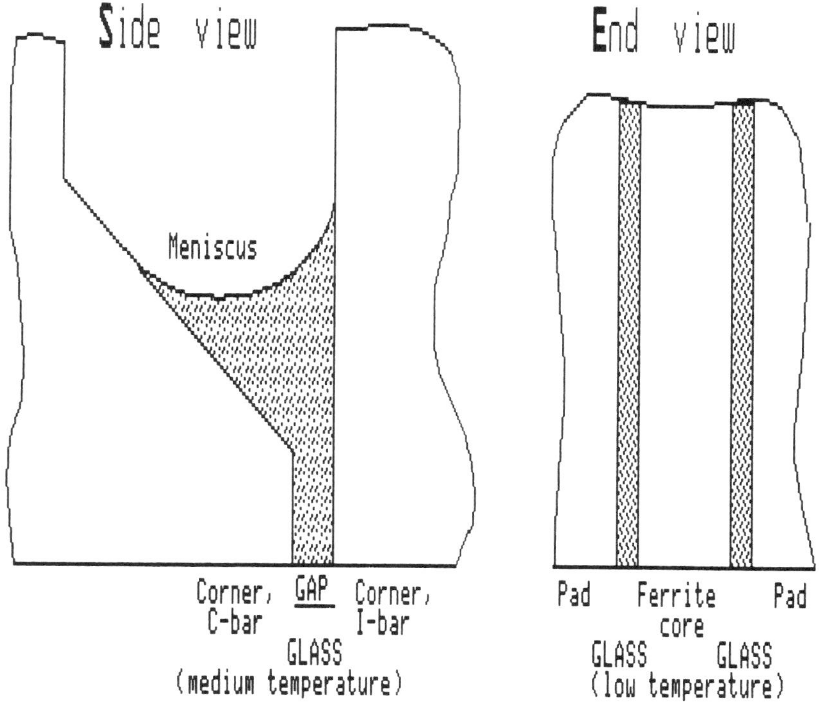

Fig. 9-15. Glass bonding of ferrites.

the atmosphere and form solid products, which may grow out from the junction between glass and ferrite (ref. Freitag). This appears within a few days, and if the corrosion products are long, whisker crystals they will eventually break off and lead to failure in otherwise well-filtered and sealed disk systems.

Another method of bonding is to deposit a thick-film of a non-magnetic ferrite between two magnetic ferrite disks previously prepared by hot-pressing. The materials are then bound by hot-pressing (ref. Chabrolle). Another method employs diffusion bonding using a ceramic between the two ferrites (ref. Rigby).

There is finally the method of using single crystals. They are employed exclusively in video recorder heads due to the high cost of fabricating them by the conventional Bridgeman method. By bringing a single crystal into contact with a polycrystal ferrite it is possible to grow the crystal into the polycrystal by heating in the solid state and crystallizing this polycrystal (ref. Tanji).

Recently heads have been made with metal-in-the-gap, as shown in Fig. 9-16. The purpose is to use the high flux density material Al-Fe for the pole piece, to avoid pole tip saturation problems (see later in Chapter 10). The metal is deposited onto the pole pieces prior to the deposition/insertion of a gap spacer. A discussion of deposition and annealing techniques is found in the reference listing (Coughlin).

A final word on an interesting phenomenon that may occur in ferrites. The propagation speed of an electromagnetic wave inside the material is much lower than the speed in air due to the material's dielectric constant and its permeability:

$$v = v_{air}/\sqrt{\epsilon_r \mu_r} \tag{9.3}$$

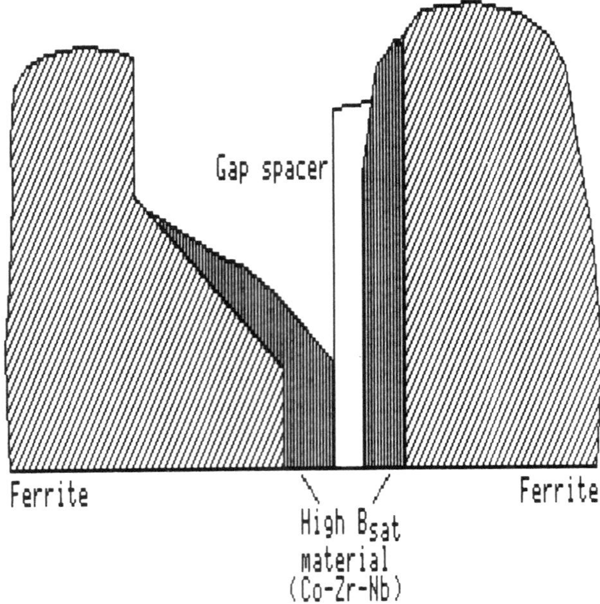

Fig. 9-16. Gap with metal deposited on pole pieces (MIG - metal-in-gap).

Resonances (standing waves) can occur say at 1 MHz when the smallest distance is in the order of millimeters (ref. Deschamps).

Furthermore, the velocity of the wave front (phase velocity v_c) is only:

$$v_c = \omega\delta \tag{9.4}$$

where $\omega = 2\pi f$ and δ is the skin depth (see formula (7.1)). The propagation speed of the wave front may relate to turn-on and turn-off phenomena in ferrite heads?

THIN FILM HEADS

We can, from the preceding section, see that the manufacturing of multichannel heads becomes more difficult as the track density and the upper frequency limit both increase. This is in particular a problem in disk files for computer storage. There are units that utilize only one head assembly with one data channel. The disk may have a couple of hundred concentric tracks and the head must therefore be moved back and forth to select a given track. This movement requires precision mechanical actuators (linear motors) and it takes time to settle down at a track.

In computers time is precious, and a head-per-track assembly could eliminate the time to change tracks. A partial solution would be the movement of a multi-channel head (shorter travel equals shorter time) or several interlaced heads.

A possible candidate for the head-per-track assembly is the *thin film head*, under development since the early seventies. The evolution of a ring core head into a single turn thin film head is illustrated in Fig. 9-17.

The modern version of a 9-turn thin film head is shown in Fig. 9-18. A large number of these heads are fabricated during a production cycle. They are deposited onto an insulating

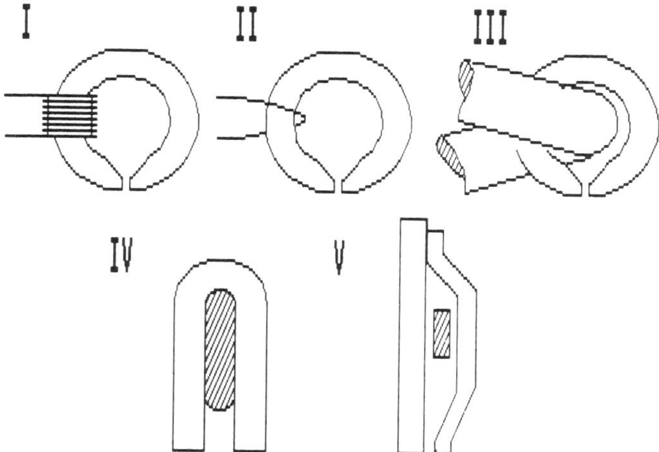

Fig. 9-17. Evolution of thin film heads.

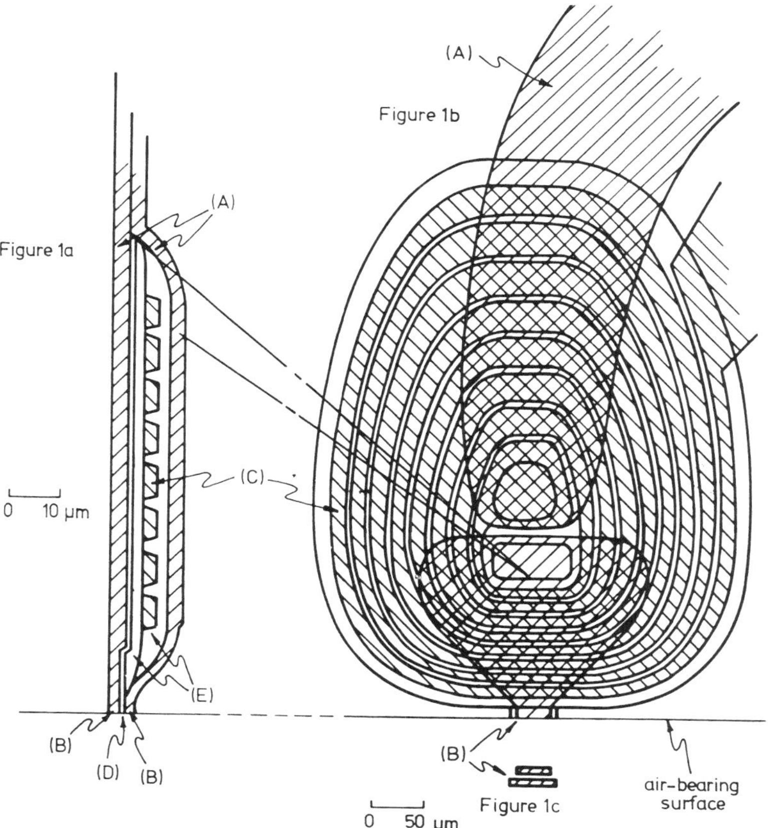

Fig. 9-18. 1a: A schematic cross section of the film head showing the magnetic layers (A), pole tips (B), conductor turns (C), gap layer (D), and insulation layers (E). 1b: A planar view of the film head. 1c: Pole-tip structure at the air-bearing surface (from *IBM Disk Storage Technology*, Feb. 1980)

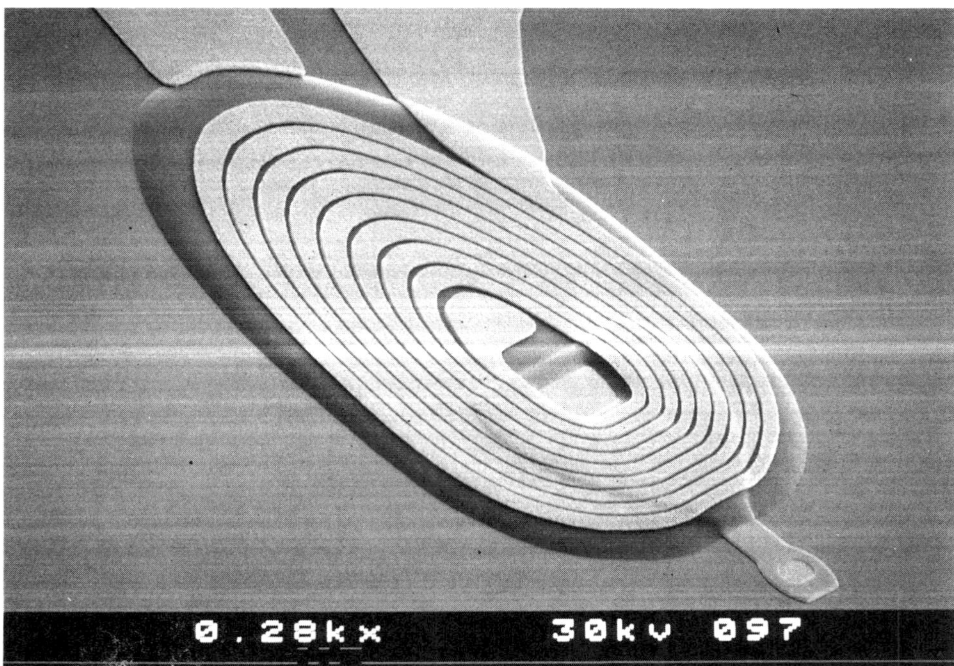

Fig. 9-19. Thin film head coil, electron microscope photo. (courtesy of Applied Magnetics, Inc.).

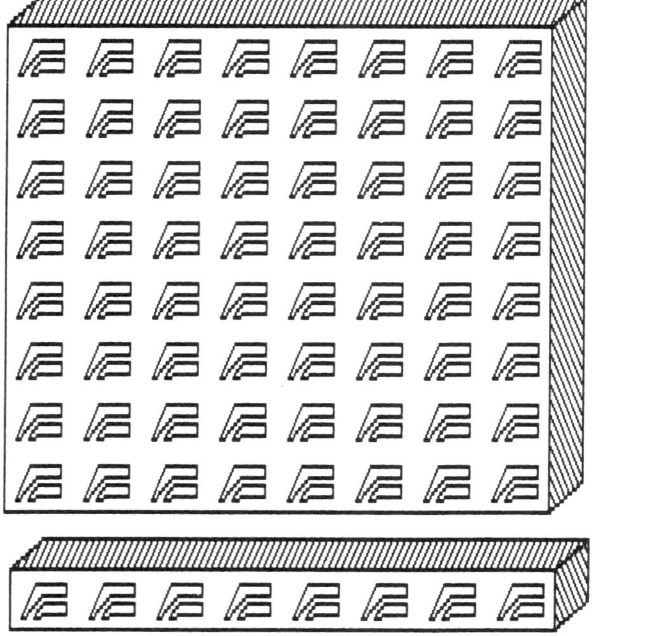

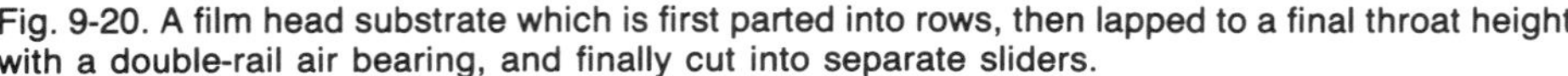

Fig. 9-20. A film head substrate which is first parted into rows, then lapped to a final throat height with a double-rail air bearing, and finally cut into separate sliders.

substrate (silicon) and consist of three layers: Permalloy, copper, and Permalloy. The copper has been shaped into a small 9-turn conductor (see Fig. 9-19), and the Permalloy sheets are joined at the center of this conductor. The front gap length is controlled by a thin gap layer, and there is no back gap.

The fabrication technology has been borrowed from the semiconductor industry fabrication of integrated circuits. The obvious advantages are: low cost, track-to-track uniformity and production repeatability. The substrate is a silicon wafer. In one example of fabrication a few hundred angstroms of Titanium is flashed onto the surface to provide good adhesion to the first layer of magnetic material, which is a Permalloy film, deposited at a 45° angle, which gives best magnetic results. Titanium is again flashed and now a copper layer is deposited. The process of etching the copper and Permalloy pattern is made by a photochemical milling. A final layer of Permalloy is deposited through a mask.

When the heads are finished they reside on a wafer, that subsequently is sliced into small sections which each end up as single (or dual) head sliders (Fig. 9-20). Several wafers, that each will contain thousands of heads, are processed together (see Fig. 9-21).

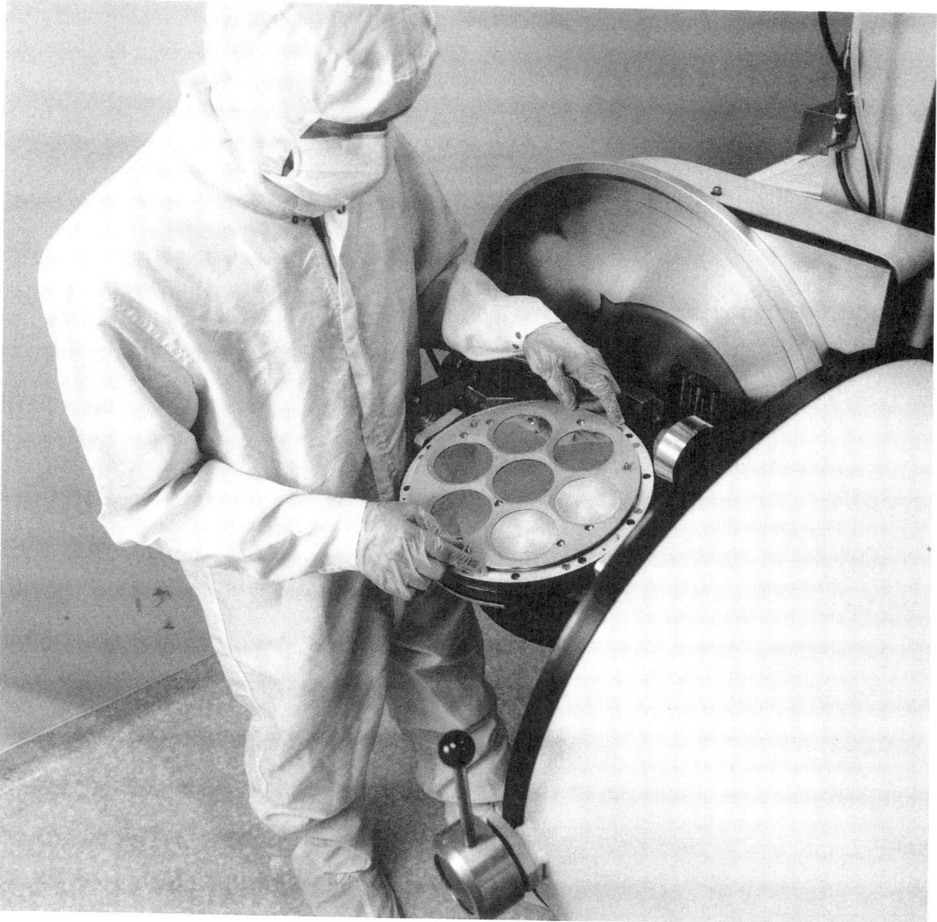

Fig. 9-21. Production ion milling system is used for precision momentum-transfer surface removal in processing thin-film sensors (courtesy of Applied Magnetics, Inc.).

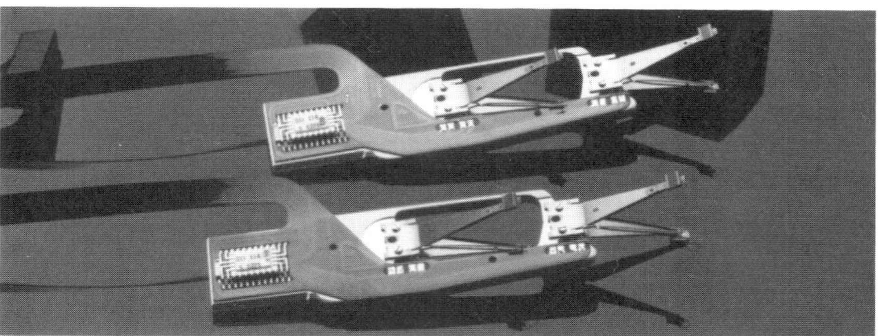

Fig. 9-22. Thin film head assembly (courtesy of Applied Magnetics, Inc.).

A final head assembly is shown in Fig. 9-22 for application in a rigid disk drive.

The exact manufacturing details are of course proprietary, but there are several good papers that can enlighten the interested reader. The concept of thin film heads, then called *integrated heads* was kicked off in 1971 by Lazzari (see ref.), and there have since been numerous papers on thin film heads. For initial guidance, the reader may benefit from these listed: the first single turn heads (ref. Chynoweth); head fabrication, circuitry and performance (ref. Taranto); fabrication by ion etching methods (ref. Nakanishi); multitrack heads (ref. Wakabayashi); an update review (ref. Lazarri, 1982); a planar construction for magnetic printing, or for multitrack read/write application (ref. Springer); deposition technique (ref. Wagner); IC process technology (ref. HP Journal).

PRODUCTION TOLERANCES

One of the major difficulties in the fabrication of multi-track heads is achieving satisfactory production yields. The processes are sequential and the final quality can rarely be tested until the assembly is completed. Head manufacturers have therefore implemented stringent quality control steps throughout production. Needless to say, a good portion of the fabrication is done in a clean room environment. Minute dust particles are known to have ruined production yields.

We recall the expression for a head's efficiency, which is a proportionality factor in its sensitivity to record or playback:

$$\eta = R_{fg}/\Sigma R_m \qquad \textbf{(4.26)}$$

where R_{fg} equals the reluctance of the front gap, in parallel with the stray fields around the front gap, and ΣR_m is the sum of the magnetic reluctances going once around the equivalent circuit of the head.

The gap reluctances are controlled by the spacings L_{fg} and L_{bg} between the two core halves, the depths of the gaps (front gap: d_{fg}, back gap: d_{bg}), and the track width w. The core reluctance is determined primarily by the relative permeability of the core material, and is subject to gross changes during the fabrication process (stamping or etching of metallic laminations, or machining of ferrites; pressure from insertion and epoxy bonding; and lapping operations).

The acceptable production tolerances can be determined if we differentiate the expression for η with respect to R_{fg} or $R_c = R_{c1} + R_{c2}$, the total core reluctance:

$$\begin{aligned}\delta\eta/\delta R_{fg} &= (\Sigma R_m - R_{fg}) / (\Sigma R_m)^2 \\ &= (1 - \eta) / \Sigma R_m\end{aligned}$$

or

$$\boxed{\delta\eta/\eta = (1 - \eta) * (\delta R_{fg}/R_{fg})} \quad \textbf{(9.5)}$$

$$\delta\eta/\delta R_c = -R_{fg}/\ (\Sigma R_m)^2$$

or

$$\boxed{\delta\eta/\eta = -R_c * (\delta R_c/R_c)\ /\ \Sigma R_m} \quad \textbf{(9.6)}$$

These last two equations enable the head design engineer to work out acceptable production tolerances.

Example 9.1: If the nominal value of η is 0.8 for a wideband instrumentation head with:

$$\begin{aligned} R_{fg} &= 30 \quad \mu H^{-1}, \\ R_c &= 2 \quad \mu H^{-1}, \\ \Sigma R_m &= 37.5 \quad \mu H^{-1}, \\ R_{bg} &= 3.5 \quad \mu H^{-1}, \end{aligned}$$

then we find, for an allowed + or -0.5 dB variation (= 5.9 percent) in η:

$$\delta\eta/\eta = \pm(1 - .8) * \delta R_{fg}/R_{fg}$$

or

$$\delta R_{fg}/R_{fg} <= 0.295,$$

or 30 percent variation in the gap reluctance. If we applied the tolerance to variations in the core reluctance we find:

$$\delta\eta/\eta = +- (2/37.5) * \delta R_c/R_c$$

or $$0.06 => +- (2/37.5) * \delta R_c/R_c$$

or $$\delta R_c/R_c <= 1.125,$$

or a 112 percent variation (in, say, the permeability of the core material).

If the gap length was changed to one third of the original length then R_{fg} would become 10 μH^{-1}, $R_{bg} = 1.2\ \mu H^{-1}$ and $\eta = 0.66$; this would be the case where the same core structure was used for a narrow gap high density recorder.

The permissible tolerances are now:

$$\begin{aligned} \delta R_{fg}/R_{fg} &<= 18 \text{ percent, or} \\ \delta R_c/R_c &<= 46 \text{ percent.} \end{aligned}$$

This example illustrates the difficulty in achieving a high yield with narrow gap, multitrack head assemblies. The shorter gap lengths must be much better controlled (from say 2.25 μm (90 μin) + $-$.7 μm (27 μin) to .75 μm (30 μin) + $-$ .135 μm (5.4 μin)!). And similar reasoning holds for the core permeability of nominal value of, say, 2,000, from a range of μ_r = 4,240 to 944 to a tighter range of μ_r = 2920 to 1370.

Here lies the reason why a head assembly with n tracks is not priced at n times the cost of a single track head, but rather n^q times, where $q>1$.

Accurate modelling of heads requires the inclusion of stray fields, and eddy current losses. Formulas (9.5) and (9.6) are instructive in making budgetary estimates of allowable tolerances in production, but a far more accurate insight is gained by making "what-if" questions to a computer program that models the head performance based upon all relevant input parameters.

The quality of the gap definition is also more demanding to produce in narrow gap heads. *Irregularities* in the gap line cause an additional loss for the write/read signals at short wavelengths (ref. Mallinson). Assume that the irregularities deviate from a straight line in a random fashion with variance σ^2 (The variance σ^2 is found from $\sigma^2 = \Sigma x^2/N - \overline{x}^{-2}$, where x are the measured gap lengths at N locations, and $\overline{x} = \Sigma x/N$ is the arithmetic mean value). This results in an additional loss in head output voltage:

$$\text{Irregularity loss} = A_t = 170 * (\sigma/\lambda)^2 \text{ dB} \quad \textbf{(9.7)}$$

where λ is the recorded wavelength. This result may be remembered by its similarity to Wallace's spacing loss formula $54.6*d/\lambda$ dB.

This loss is not the same as *gap scatter*, which refers to the potential difference in gap alignments in a multitrack head.

If the tape was recorded with a head having random irregularities with variance $\sigma_1{}^2$, and reproduced with a head with variance $\sigma_2{}^2$ then:

$$A_t = 170*((\sigma_1{}^2 + \sigma_2{}^2)/\lambda^2 \text{ dB} \quad \textbf{(9.8)}$$

Example 9.2: The deviations from a straight line at 10 locations along the gap were measured to be 12, 7, 3, 8, −9, 5, −14, 10, 4, and 7 μin. The variance σ^2 is 69, and the irregularity loss for $\lambda = 0.75$ μm (30 μin for a 4 MHz recording at v = 120 IPS) is then $170*69/30^2 =$ 13 dB. The standard deviation for this gap is 0.2 μm (8.3 μin).

SPECIFICATIONS

Table 9-2 is a comprehensive list of things not to forget on a specification drawing—it is advisable to check all of the items listed.

MEASUREMENTS AND TESTS

Techniques for measurements of the effective gap length, head core losses, self-capacitance and impedance are important aspects of magnetic recording.

Effective Gap Length

Since the magnetic gap lengths are always larger than the mechanical gap lengths, an error can be induced in a design by measuring the gap lengths under a microscope and using this dimension to establish the gap losses. The gap length can be determined accurately only by measuring the wavelength λ_o, where the induced voltage from the playback head goes through a null; the gap length is then calculated from the gap loss function. This results in:

$$\boxed{\text{Effective gap length} = \lambda_o} \quad \textbf{(9.9)}$$

Since this null normally is beyond the frequency range of the recorder (in a properly designed recorder by a factor of 2), the easiest measurement is undertaken by connecting the

Table 9-2. Comprehensive Listing of Magnetic Head Specifications.

DATA:

Digital:	Bit rate
	Upper frequency
Analog :	Upper data freq.
	Lower data freq.
	Bias frequency

MEDIUM:

Coating coercivity
Coating retentivity
Coating thickness
Coating surface smoothness
Head-medium speed
Head-medium spacing

CORE GEOMETRY:

Track width
Front gap length
Front gap depth
Core length
Core depth
Back gap length & depth

CORE MATERIAL:

Rel. permeability
Resistivity
Coercivity
Remanence
Lamination thickness

ELECTRICAL:

Self capacitance
Losses
Load impedance
Number of turns
Efficiency
Write sensitivity
Inductance = f(freq)
Resistance = g(freq)
Quality = q(freq)
Noise voltage = v(freq)
Output voltage
Signal-to-noise ratio
Crosstalk
Interface specifications

HEAD GEOMETRY:

Number of tracks
Track spacing
Gap scatter
Read-write gap spacing
Head surface contour
Provisions for mounting
Outline dimensions
Azimuth
Tilt

ENVIRONMENT:

Tape type
Tape tension
Temperature *)
Humidity *)
Shock and vibration
External magnetic fields
RFI
Adjacent head assembl.
*) Both oper. and store

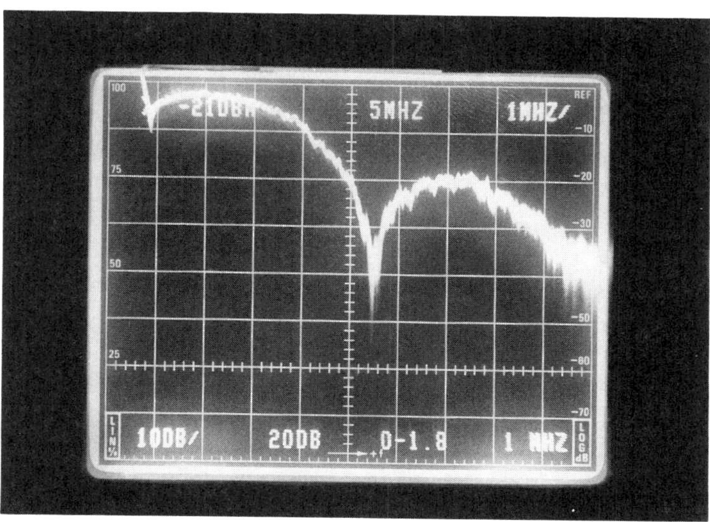

Fig. 9-23. Measurement of the effective gap length (Courtesy of Tranetics, Inc.).

head leads directly to a tone generator. (The higher frequencies required for this measurement makes the use of high-frequency bias questionable because of the generation of beat notes.) The level of the record current from the sine-wave generator should be of the same magnitude as the bias current normally used, which essentially means the tape is recorded to saturation. A series of tones are recorded and played back; then a curve can be plotted (Fig. 9-23). A formula for the effective gap length is provided. In the example shown, where the tape speed was 3¾ inches per second, the null at 16 kHz corresponds to an effective gap length of 6 μm = 235 μin.

Measurements of Core Losses

If the flux level through the core in a magnetic head (without losses) is held constant with frequency, the induced voltage will ideally rise 6 dB per octave. The constant flux can readily be provided as shown in Fig. 9-24 by a thin, straight wire placed in front of, and parallel with the gap. The departure from the straight 6 dB per octave line is evident, and is a measure of head losses.

The constant current is obtained by connecting the wire to a sine-wave generator in series with an induction-free resistance of a value equal to the recommended termination for the generator. Connecting a voltmeter across the resistor provides a method to keep the current constant, since most generators require slight readjustments of the output as the frequency is changed. Another voltmeter (high impedance) is connected to the head output terminals to measure the induced open-circuit voltage. Losses in the core reduce the induced voltage at high frequencies and these losses are represented by the distance between the 6 dB per oc-

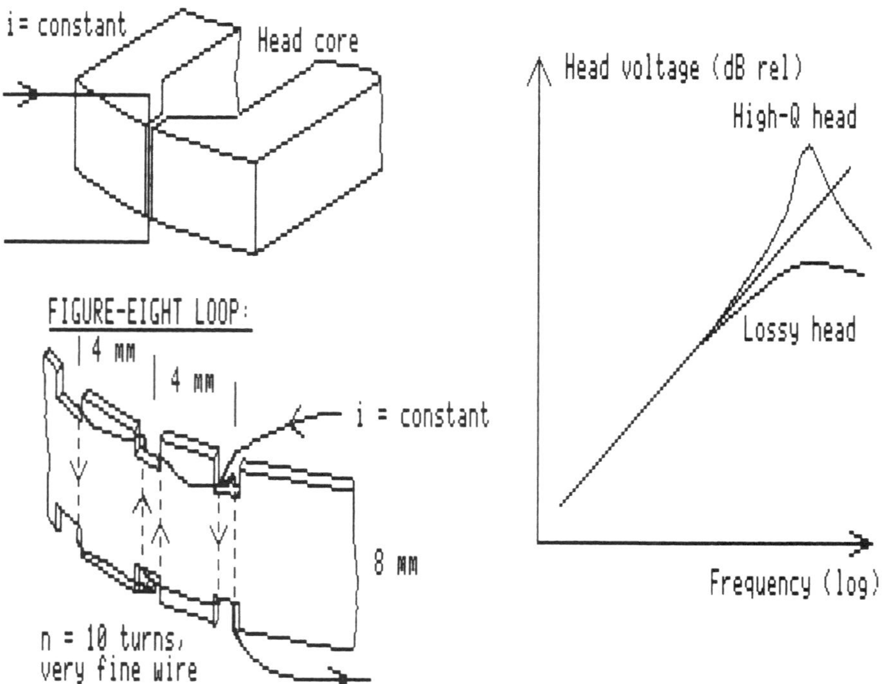

Fig. 9-24. Measurement of head core losses. Two signal injection techniques for headcore loss measurements and typical curves.

tave line and the measured curve in Fig. 9-24. The frequency range of the head will also be evident, since the head inductance and its self-capacitance will cause it to resonate at a certain frequency f_o.

Instead of using a straight wire, it may be advantageous to fabricate a small *figure-eight loop*, as shown in the same figure. It can be wound on a thin strip of plastic or celluloid, for example, and can easily be positioned in front of the head similar to the path followed by a magnetic tape. The loop should be pressed lightly against the head in such a position that the induced voltage is maximum; and if possible, tape it down so it will not move during the measurements.

Measurement of Head Impedance, Self-Capacitance, and Correction for the Inductance Value

The head impedance varies with frequency and needs to be plotted for the electronic circuit designer (write and read circuits). This is readily done using an impedance bridge; a simple test is outlined in Fig. 9-25, requiring only a sine wave generator, a frequency counter, and an oscilloscope.

The self-capacitance C_e is determined by substitution, as shown in Fig. 9-25 bottom.

The true inductance L_{true} is always smaller than the measured inductance L_{meas}. If the value of the tuning capacitor is C_1, then:

$$L_{true} = L_{meas}*C_1/(C_1 + C_e) \quad \textbf{(9.15)}$$

Measurements of output versus write current, of overwrite (erasure of old data), of resolution, and of noise are covered in Chapter 10.

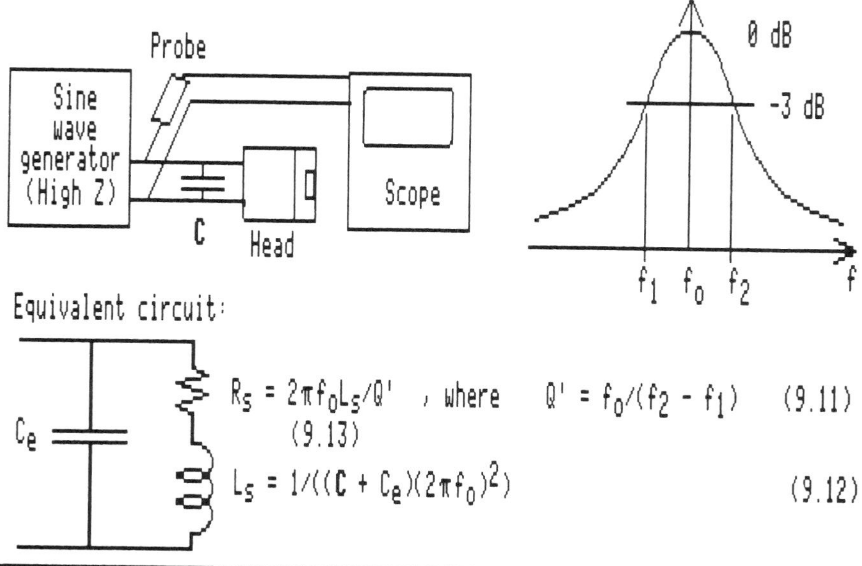

Fig. 9-25. Measurements of head impedance: Q′, L_a′, R_s′ and Ce.

Table 9-3. Design Test for Digital Recording (courtesy of Nortronics).

SYMPTOM	PROBABLE CAUSE	SOLUTION
Low Output	1. Poor tape contact	1. Check tape wrap angle and tension and/or face finish of head. Check pressure pad and look for tape wear pattern on face of head to see that gaps are being contacted.
	2. Excessive write current	2. Use write current = 150% of saturation current.
	3. Very low write current	3. Same as above.
	4. Tape speed too slow	4. Adjust speed to correct value.
	5. Poor tape (worn oxide)	5. Replace tape.
	6. Write and read coil connections reversed	6. Connect write head to writer and read head to read amplifier.
	7. Read coil open	7. Replace head.
	8. Open leads (poor connections)	8. Replace leads and/or clean connections.
	9. Read load excessive	9. A load impedance on read head of less than 10 Kohms/200 pf will reduce output somewhat depending on head impedance, frequency and load impedance.
	10. Write density too high	10. Use density for which output test is specified.
	11. Poor tape guiding	11. Adjust azimuth of head.
	12. Dirt on tape	12. Clean tape **and** head.
High Output	1. Tape speed too fast	1. Adjust speed to correct value.
	2. No load on read coil	2. Read head should have 10Kohm/200 pf for comparison with spec. Higher impedance will increase Eo a little.
	3. Resonant condition: Resonant Frequency = $\frac{1}{2\pi\sqrt{LC}}$ Where C is distributed C of head + leads + electronics	3. If correct density and tape speed results in approaching resonant frequency, then lower inductance head should be used.
	4. Write density is too low	4. Use density for which output test is specified.
High Write Current	1. Poor tape contact	1. Check tape wrap angle and tension and/or face finish of head. Check pressure pad and look for tape wear pattern on face of head to see that gaps are being contacted.
	2. Tape on backwards on reel to reel deck	2. Turn oxide side of tape towards the head.
	3. 0-Peak level used instead of peak-peak	3. Write current is specified in ma. peak-peak.
	4. Rise and fall time set improperly (too slow)	4. Set rise and fall times per spec.
	5. Shorted write coil	5. Replace head.
	6. Write density too low	6. Use density for which saturation test is specified.
	7. Dirt on tape	7. Clean tape **and** head.

Design Check List

This chapter concludes with the useful Table 9-3 of potential problems in the application of magnetic write/read heads.

REFERENCES, CHAPTER 9

Collier, D., Frank, P.D., and Aho, C.J., "Rigid Disk Heads Keep Pace with Growing Storage Needs," *Mini-Micro Systems*, Dec. 1984, Vol. 18, No. 12, pp. 127-133.

Laminated Core; Half-shell Construction

Boll, R., "Einbettung weichmagnetischer Werkstoffe in Kunststoffe," *Technische Informationsblatter M7*, Vacuunschmelze Aktiengesellschaft, Hanau, West Germany, 1956, 5 pages.

SYMPTOM	PROBABLE CAUSE	SOLUTION
Low Write Current	1. Open write coil	1. Replace head.
	2. Write and read coil connections reversed	2. Connect write head to writer and read head to read amplifier.
	3. Write density too high	3. Use density for which saturation test is specified.
	4. Defective writer or incorrect current measuring set-up	4. Check waveshape and amplitude of write current and current measuring set-up.
High Percentage of Crossfeed	1. High write current	1. Use write current = 150% of saturation current.
	2. Rise and fall times set too fast	2. Set rise and fall times per spec (20% of pulse length).
	3. Poor tape contact	3. Check tape wrap angle and tension and/or face finish of head. Check pressure pad and look for tape wear pattern on face of head to see that gaps are being contacted.
	4. Head not grounded	4. Make ground connection to case of head.
	5. Write and read coil connections reversed	5. Connect write head to writer and read head to read amplifier.
	6. Tape speed too slow	6. Adjust speed to correct value.
	7. No load on read coil	7. Loading head will reduce fast rise spikes due to instrumentation ground loops and crossfeed from fast rise.
	8. MU metal shield in cassette cartridge	8. Remove screening shield from cartridge.
	9. HY MU 800 or ferrite near face of head	9. Remove material from close proximity to head.
	10. Improper shielding on read/write cables	10. Separate read and write cables; read cable must be well shielded and shield grounded.
	11. Electro-magnetic radiation from motors, relays and electronics	11. Head must be shielded from strong magnetic fields.
Distorted Output	1. Shorted coil	1. Replace head.
	2. Shorted coil to ground	2. Replace head.
	3. Defective writer or incorrect current measuring set-up	3. Check waveshape and amplitude of write current and current measuring set-up.
	4. Defective read circuit	4. Check loading of read amplifier.
	5. Magnetic radiation into head	5. Head must be shielded from strong magnetic fields.
	6. Excessive secondary or ghost pulses	6. Check for excessive tape wrap angle (10° in and 10° out is sufficient) and for excessive write current (use 150% of saturation current).
	7. Low write current	7. Use 150% of saturation current.
	8. Incomplete erasure of old data	8. Use 150% of saturation current.
Low Pulse Resolution	1. High write current	1. Use write current = 150% of saturation current.
	2. Poor tape contact	2. Check tape wrap angle and tension and/or face finish of head. Check pressure pad and look for tape wear pattern on face of head to see that gaps are being contacted.

Boll, R., "Fotochemisch hergestellte Lamellen und Praezisionsteile aus weichmagnetischen Legierungen und anderen Sonderwerkstoffen," *Feinwerktechnik*, June 1965, Vol. 69, No. 6, pp. 241-246.

Rabinowicz, E., "Polishing," *Scientific American*, June 1968, Vol. 236, No. 6, pp. 91-99.

Staff, "Netic and Co-Netic Magnetic Shielding Manual," *Perfection Mica Co.*, 1322 Elston Ave., Chicago 22, Ill., 1980, 32 pages.

Staff, "MuShield Magnetic Shields and Shielding Materials," *MuShield Co. Brochure*, 1980, 8 pages.

Liechti, K.M., "Residual Stresses in Plastically Encapsulated Microelectronic Devices," *Experimental Mechanics*, Sept. 1985, Vol. 25, No. 3, pp. 226-231.

Ferrite Core Heads

Knowles, J., "The Origin of the Increase in Magnetic Loss Induced by Machining Ferrites," *IEEE Trans. Magn.*, Jan. 1975, Vol. MAG-11, No. 1, pp. 44-49.

Broese van Groenou, A., "Grinding of Ferrites, Some Mechanical and Magnetic Aspects," *IEEE Trans. Magn.*, Sept. 1975, Vol. MAG-11, No. 5, pp. 1446-1451.

Freitag, W., Mee, P., and Petersen, R., "Glass/Ferrite Interactions and Corrosion of Gap Glasses in Recording Heads," *IEEE Trans. Magn.*, Sept. 1980, Vol. MAG-16, No. 5, pp. 876-878.

Wada, T., "An Improvement of Ferrite Substrate," *IEEE Trans. Magn.*, Sept. 1980, Vol. MAG-16, No. 5, pp. 884-886.

Chabrolle, J., and Morell, A., "Magnetic Heads with Monolithic Gaps," *IEEE Trans. Magn.*, Nov. 1981, Vol. MAG-17, No. 6, pp. 3108-3110.

Butsch, O., "Refining head design in high-density minifloppies," *Mini-micro Systems*, Nov. 1981, Vol. 14, No. 11, pp. 221-224.

Girard, L., "Honeywell Solid Ferrite Heads," *THIC meeting*, Jan. 1982, 2 pages (unpublished).

Deschamps, R.G., "Mise en evidence des resonances electromagnetiques dimensionnelles dans des circuits magnetiques et en particulier dans les pots ferrites," *l'onde electrique*, May 1983, Vol. 63, No. 5, pp. 46-50.

Rigby, E.B., "Diffusion Bonding of NiZn Ferrite and Nonmagnetic Materials," *IEEE Trans. Magn.*, Sept. 1984, Vol. MAG-20, No. 5, pp. 1503-1505.

Tanji, S., Matsuzawa, S., Wakatsuki, N., and Soejima, S., "A Magnetic Head of Mn-Zn Ferrite Single Crystal Produced by Solid-Solid Reaction," *IEEE Trans. Magn.*, Sept. 1985, Vol. MAG-21, No. 5, pp. 1542-1544.

Coughlin, T.M., "Sendust Films for High Temperature Processing," *IEEE Trans. Magn.*, Sept. 1985, Vol. MAG-21, No. 5, pp. 1897-1899.

Thin Film Heads

"IC Process Technology," *HP Journal*, Aug. 1982, Vol. 33, No. 8, 36 pages.

Lazzari, J.P. and Melnick, I., "Integrated Magnetic Recording Heads," *IEEE Trans. Magn.*, Mar. 1971, Vol. MAG-7, No. 1, pp. 146-150.

Chynoweth, W., Jordan, J., and Kayser, W., "PEDRO: A Transducer-Per-Track Recording System with Batch-Fabricated Magnetic Film Read/Write Transducer," *Honeywell Computer Journal*, 1972, Vol. 7, No. 3, pp. 103-117.

Taranto, J., Stromsta, R., and Weir, R., "Application of Thin Film Head Technology to a High Performance Head/Track Disk File," *IEEE Trans. Magn.*, July 1978, Vol. MAG-14, No. 4, pp. 188-190.

Nakanishi, T., Kogure, K., Toshima, T., and Yanagisawa, K., "Floating Thin Film Head Fabricated by Ion Etching Method," *IEEE Trans. Magn.*, Sept. 1980, Vol. MAG-16, No. 5, pp. 785-787.

Lazzari, J.P., "Thin Film Heads," *Intl. Conf. Video and Data 1982, IERE Conf. Proc.*, No. 54, pp. 65-74.

Wagner, Udo, and Zilk, A., "Selective Microelectrodeposition of Ni-Fe Patterns," *IEEE Trans. Magn.*, May 1982, Vol. MAG-18, No. 3, pp. 877-879.

Wakabayashi, N., Abe, I., and Migairi, H., "A Thin Film Multi-Track Recording Head," *IEEE Trans. Magn.*, Nov. 1982, Vol. MAG-18, No. 6, pp. 1140-1142.

Jeffers, F., "Metal-In Gap Record Head," *IEEE Trans. Magn.*, Nov. 1982, Vol. MAG-18, No. 6, pp. 1146-1148.

Mallinson, J., "Gap Irregularity Effects in Tape Recording," *IEEE Trans. Magn.*, March 1969, Vol. MAG-5, No. 1, page 71.

Chapter 10

Design and Performance of Magnetic Heads

The performance of magnetic heads can be computed from the basic model of a resistance network, and the results can inform us about things like the write head's recording efficiency, current requirements and impedance. We can also get information about the possibility of signal distortion, and whether the head core may become permanently magnetized or not. Other items are pole tip saturation and potential heating of very small write heads.

For read heads we need to know about their sensitivity, efficiency and noise. For both head types susceptibility to external fields is important.

Keeping the book's title of a handbook in mind it would have been nice to have one large table that lists all head types and their performances. This proved to be completely impossible, and a different approach is used: The chapter will discuss the design and performance of several benchmark heads, designed to operate in ranges covering from audio frequencies up to 100 MHz. The information may then be applied to other heads, with proper interpolation.

The head designer may wish to learn from interdisciplinary engineering tasks; magnetic heads are inductive transducers and are related to other inductors and transformers, where books by Zinke and Grossner are recommended. Review papers on heads are also valuable; these include the classic paper by Kornei, and more recent ones by Nortronics and Applied Magnetics (Collier et al.).

Table 10-1 lists the 12 standard designs selected for analysis and description, in addition to a thin film head made from Ni-Fe material. The core materials were chosen from four different types to get a feel for their behavior:

Ni-Fe (Permalloy), 50 μm (2 mils) thick laminations
Ni-Fe (Permalloy), 25 μm (1 mils) thick laminations
Hard Ni-Fe (''Tough-malloy''), 15 μm (0.6 mils) thick laminations
MnZn Ferrite.

The materials properties are as listed in Table 7-1. Heads with AlFe(Si) poletips will behave as if the cores were made entirely from the ferrite used for the body of the core; the magnetic reluctance between the pole piece and the core is vanishing compared to the other head reluctances.

DESIGN PROCEDURE

The design of heads has clearly followed an evolutionary path where current designs are modified to new requirements, and so on. There are numerous considerations, often with conflicting demands on the design, and trade-offs are made.

A good starting point is to provide answers to all the items in Table 9-2, even if not applicable. The desired packing density for a given application will dictate the geometry of the track width and gap length, and the design can now proceed by adopting a core geometry that appears compatible therewith; this may be an existing core design.

The head's performance can now be computed in as much detail as desired and the design optimized. This chapter will show how, and illustrate the methods with computed results. This was done in part by a spreadsheet program, in part by Head/Media Design programs (see Ref. Jorgensen).

The computed performances are evaluated in light of the desired specifications, and changes may be made. The next step is the fabrication of a few engineering model heads, upon which measurements are made and compared with the computed results. Discrepancies should be appraised, and understood.

The modelling of a head rests on many formulas and certain approximations, in particular with regard to stray fields and eddy current losses. The first are computed from discrete approximations or by using finite element methods. The eddy currents cause a drop in permeability toward high frequencies (and the upsurge of an imaginary component μ''); these changes are readily computed, but neglect the fact that flux will run on the surface of the core, NOT in its interior, at high frequencies.

An example to keep in mind is the way two head halves with laminated cores match. If the core sections do not mate as mirror images of each other a poor performance will result; the flux coming up along a lamination surface will cross the gap and then bounce head-on into the misaligned lamination in the other core half and then spread out. The result is an irregular field in and around a gap in a laminated head—and this may be the way things are; we have no good way of observing this field irregularity.

The results from computations based upon head modelling may not agree completely with measurements on the prototype heads. Understanding the discrepancies and repairing the model is an important effort to provide for closer agreements between computations and measurements. The use of a model for trend analysis is nevertheless always excellent.

The sections in this chapter need not be read in sequence for the design of a head; rather scan through it, and use what is appropriate.

PRELIMINARY DESIGN

Select Upper Frequency; Determine Gap Length and Depth

The upper operating frequency f_u of a head is limited by the resonance between its inductance L and the sum of the capacitances of the coil (C_e), the cable, and the amplifier. The value of L is equal to the number of turns squared, divided by the magnetic impedance Z_m of the head structure, including the leakage field around the coil (Z_m is the complex value of the reluctance R_m that results at frequencies where μ_r becomes complex). This inductance was first introduced in formula 4.30.

For audio and instrumentation applications the selection of f_u is straightforward: for playback it must be higher than the highest data frequency, possibly by 50 percent to keep the self-resonance and phase shifts away. In the record mode f_u must equal or exceed the AC-bias operating frequency (60-100 kHz for audio, higher for instrumentation; typically 4-5 times the highest data frequency). The temptation to locate the resonance at the bias frequency to achieve a peaking effect should be avoided: First of all, heads do seldom have a quality factor Q exceeding two to five at bias frequencies (see Figs. 10-11 and -12); secondly, there will be variations in production values of L; the head impedance will also change with head wear.

In digital applications the data rate DR and encoding technique dictates f_u. FM-encoding results in f_u = DR, while MFM only requires $f_u = 0.5*DR$. This is for the fundamental frequency only, and should be multiplied with three if it is desired to include the third harmonic. It is assumed that the data rate DR is *after* error-correction coding. (Note: FM and MFM in digital recording are not what ordinarily is understood by frequency modulation; see Chapter 24 for details.)

With an established speed between the head and the tape/disk the resulting packing density in BPI and bit length BL can now be calculated as:

$$\boxed{\text{BPI} = 2*f_u/\text{speed (in IPS)}} \tag{10.1}$$

and

$$\text{BL} = 1{,}000{,}000/\text{BPI} \quad \mu\text{in.} \tag{10.2}$$

This number is used in setting the gap length:

$$L_{fg} = 0.6*\text{BL for high resolution} \tag{10.3a}$$
$$\quad = 0.9*\text{BL for high output.} \tag{10.3b}$$

The mechanical gap length should be specified shorter, by two factors:

A. Correct for the effective gap length (see formula (10.11)), i.e., subtract 15 percent.

B. Correct for dead layer (Beilby layer) on the surface of machined ferrites. Subtract four times the surface roughness per pole face (see formula (9.1)).

We can also establish the depth of the gap:

$$\begin{aligned} D_{gap} &= 4 \text{ mils (long life)} \\ &= 2 \text{ mils (high output)} \\ &= .2 \text{ mils (no contact).} \end{aligned} \tag{10.4}$$

Let us for a moment digress to the matter of units. We are currently in a transition going from μin, mils and inches to μm, mm and meters. Both units are in current use, with the following conversions:

$$\begin{aligned} 1\ \mu\text{m} &= 40\ \mu\text{in} \\ 1\ \mu\text{in} &= 25.4/1000\ \mu\text{m} \\ &= 25.4*10^{-9}\text{m} \\ 1\ \text{mil} &= 25.4\ \mu\text{m} \\ &= 25.4*10^{-6}\text{m} \end{aligned} \tag{10.5}$$

Table 10-1. Benchmark Head Configurations.

General Application:	AUDIO Playback	AUDIO Rec/Play	DIGITAL Write/Read	INSTRUMENTATION, HDDR Playback	Record	HDDR DIGITAL, VIDEO, HDDR FM, Analog and Digital Write/Read				
	Tapes, Cassettes, Cartridges, 5.25" & 3.5" Disks, Tapes					Rotating Head Machines, Rigid Disks				
Upper Frequency, MHz	0.02	0.1	0.2	2	10	10	10	10	100	
Core Material	50 µm MuMetal			15 µm MuMetal MnZn		25 µm MuMetal	MnZn	MnZn	MnZn	Thin films
Front Gap Length µm	1.25			0.75	1.75	0.375	0.375	1.25	0.7	0.7
Front Gap Depth µm	100			75		50		5	50	1.2
Track Width µm	625			250		25	25	25	25	15
Core Width µm	625			250		25	200	200	125	n/a
Core Size mm^2	6.25 x 6.25			4 x 4		3 x 3		2 x 2	2 x 2	1 x 1
Number of Turns	3000	535	225	50 #3, 26 #4	37 #5, 9 #6	45	5	12 , 12	8 , 8	9 or 18
Reference # in Figures: MuMetal	#0	#1	#2	#3	#5	#7				
Reference # in Figures: MnZn				#4	#6		#8	#9, #10	#11, #12)	(F)

Note: (9) is monolithic; (10) is micro-composite; (11) has μ_r = 4,000; (12) has μ_r = 1,000.

Select Trackwidth

Reduced trackwidth is one factor that has made possible the large amount of data that can be stored on magnetic tapes and disks. Only thirty years ago a good quality music recording required a speed of 15 IPS, with a track width equal to the tape width of 6.25mm (0.25 inches). Today a digital hi-fi recording can be contained within a video cassette, with a playing time of 4 to 8 hours.

Three considerations must be made when the track width is reduced. One, reduction of the track width itself reduces the signal-to-noise ratio by 3 dB for each halving of the width. Secondly, the amplitude can become modulated by mistracking, i.e., the track may not be moving past the read head with perfect registration, but may waver back and forth. Thirdly, the last action may also cause signals from an adjacent track to be picked up, which we classify as noise, since it is an unwanted signal.

The problem of mistracking with resulting signal amplitude modulation, and added noise from side reading is particularly troublesome when using a tape or disk made from PET-film (PolyEthylenTerephthalate, with tradenames such as Hostaphan, Mylar, Terylene; for properties, see Chapter 11 on Materials for tapes and disks). The dimensions of a PET film changes drastically with changes in temperature and/or humidity:

$$\text{Humidity change: } \Delta L = 1.1*10^{-5} \text{ m/m/percent RH}$$
$$\text{Temperature change: } \Delta L = 1.5*10^{-5} \text{ m/m/degree F.}$$

Example 10.1: A 5.25 inches diskette drive is to operate at temperatures ranging from 50 to 100 degrees F; add a 20 degrees temperature rise inside the drive. The humidity may vary over an 80 percent range. Estimate track width and track center spacing.

Answer: The radius of an outer track is 6.35 cm = 2.5 inches, and can change by $(50+20)*1.5*10^{-5}*6.25 = 0.0066$ cm, plus $80*1.1*10^{-5}*6.25 = 0.0056$ cm, totalling 0.0122 cm = 122 μm = $\pm$ 61 μm. We must design for worst case, and not use the mean value 61 μm, i.e., a track width of 122 μm = 4.8 mils could fail totally.

We must now add centering and head positioning tolerances, and find:

Maximum Error due to temperature change	66 μm
Maximum Error due to humidity changes	56 μm
Center hole, stamped to oversize	12 μm
Disk spindle, machined to undersize	12 μm
Head positioning	12 μm
Total, no compensation	158 μm
Compensation for temperature change	−33 μm
Compensation for humidity change	−28 μm
TOTAL CHANGE, peak-to-peak	97 μm

Without compensation the error would be 158 μm = 6.3 mils, calling for a track width of 0.32mm minimum. Temperature compensation is accomplished by building the drive transport and head arm assembly to match the PET film coefficients of expansion versus temperature.

The humidity is compensated partly by using a track index scale made from PET material, closely matching the disk. The main problem in this approach lies in the anisotropic residual stress gradients that exist in PET film; when exposed to temperature and/or humidity a circular track will distort into a pattern that grossly resembles a peanut shell (Ref. Greenberg, Brock).

From the above example we see that a track width of 320 μm can write a track that may be $\pm$ 80 μm off center line when later on read. This calls for a minimum width allocation of 320 + 2 * 80 = approximately 500 μm per track, and a minimum of 500 μm = 19.7 mils between track center lines. The resulting track density is 1000/19.7 = 50 TPI. The standard is 48 TPI, and a typical track width is 300 μm = 11.8 mils.

From the preceding discussion we can establish that the minimum track width and track center distance should be (pp = peak-to-peak):

$$\text{Track width} = 2 * (\text{TOTAL CHANGE})_{pp} \quad \textbf{(10.6a)}$$

$$\text{Track ctr. dist.} = 4 * (\text{TOTAL CHANGE})_{pp} \quad \textbf{(10.6b)}$$

Reductions in the read amplitude modulation can be made by trimming the written track with two narrow erase gaps, as was shown in Fig. 9-11. This also decreases the side-reading from the adjacent tracks, and it is possible to optimize the write/read and erase gaps for a given system. This should ultimately be done in a way that optimizes the ratio of signal to noise from adjacent and previously written tracks (ref. Edelman).

There is only one method for a drastic reduction of the tracking errors listed above, and that is a system where a prerecorded pattern is recorded on a disk during formatting. Dedicated bits are written between sectors and used in a servo system to keep the head on track (Ref. Kodak Data Technology, OEM manual 3.3 drive).

Select Core Size and Material

The choice of core size and material is a most frustrating task if the designer has little experience. Some guidance is necessary and is provided for in Table 10-1. The designs shown are typical of current heads listed in manufacturers catalogs and from recent papers on head performances (Trade journals, IEEE Transactions on Magnetics).

A decision is made for a start design, and computations will reveal how close the performance agrees with the specifications. The design is modified until the agreement is satisfactory (say within $\pm$ 20 percent, or $\pm$ 2 dB). Verifications of the design is carried out by making measurements on and with a few engineering models.

The relative permeability spectra for typical materials are shown in Fig. 10-1. Note that the relative permeability μ_r of MuMetal is listed as only 2,000, while manufacturers data sheet show much higher values. The value of 2,000 reflects the permeability that realistically exists in a head after finished fabrication (see Fig. 7-7).

WRITE HEAD EFFICIENCY

In Chapter 4 we arrived at an expression for the "deep-gap" field H_g in the front gap of a ring core head (formula 4.29). It contained multipliers for three efficiencies: coupling of the field from the coil into the core (Fig. 8-36, formula (8.6)), the ratio of the flux running through the gap to the total core flux, and finally the efficiency when the head operates in the read mode.

The field in front of the write gap depends upon the gap field strength; this matter was discussed in Chapter 5, and the designer can use Figs. 5-13 or 5-14 to determine the required field H_g. To do this he must know the desired write level, the coercivity of the magnetic media, and the dimensions (gap length, media thickness and spacing from the head surface).

We will now finalize the expression for the coupling efficiency so we can determine the required write current for a prescribed gap field.

The value of R_s, which was called R_{shunt} in Fig. 8-36, is difficult to calculate. Figure 10-2 illustrates the effect that the length of the coil has, using a short and a long coil.

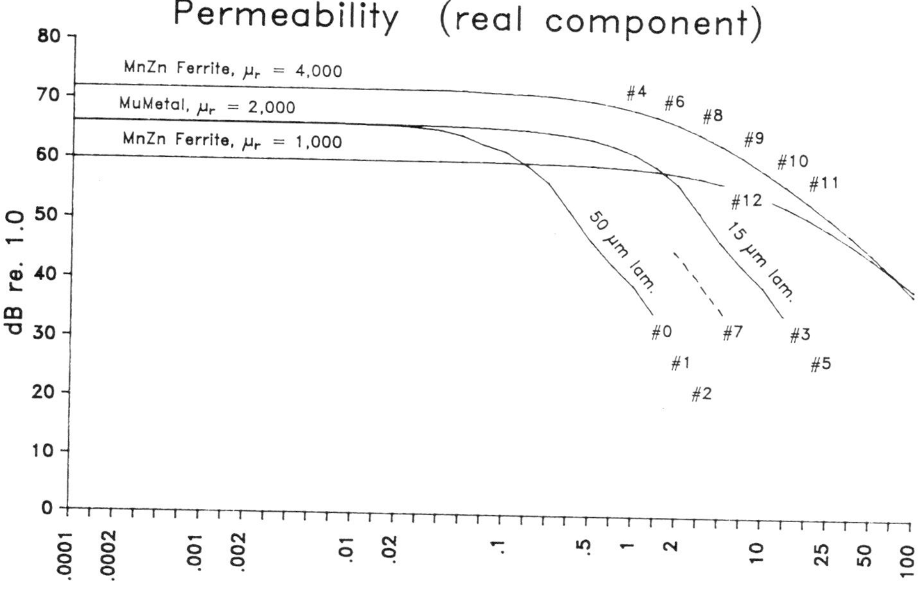

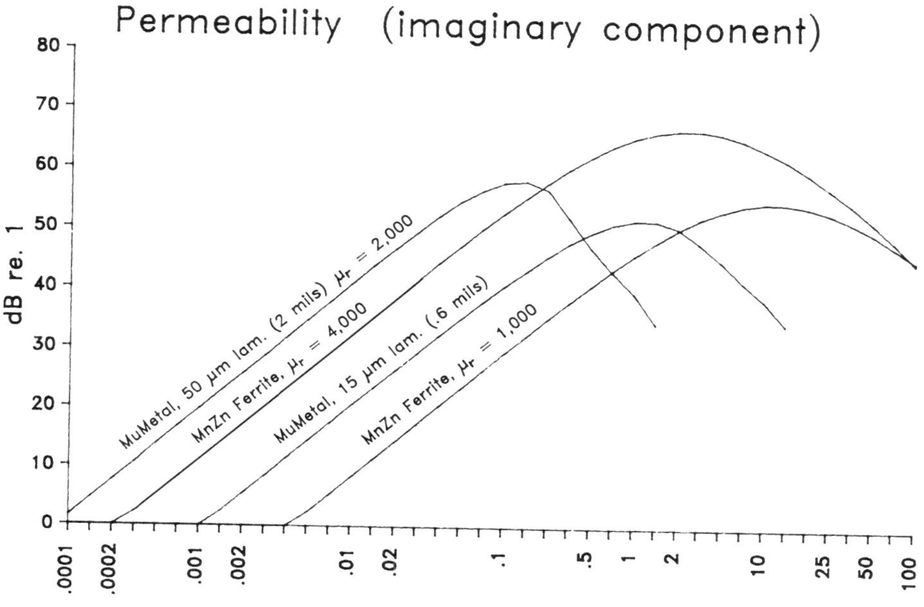

Fig. 10-1. Permeability spectra for MuMetal and MnZn ferrite. Top graphs show real component μ', bottom the imaginary component μ_r'' (numbers refer to benchmark heads listed in Table 10-1).

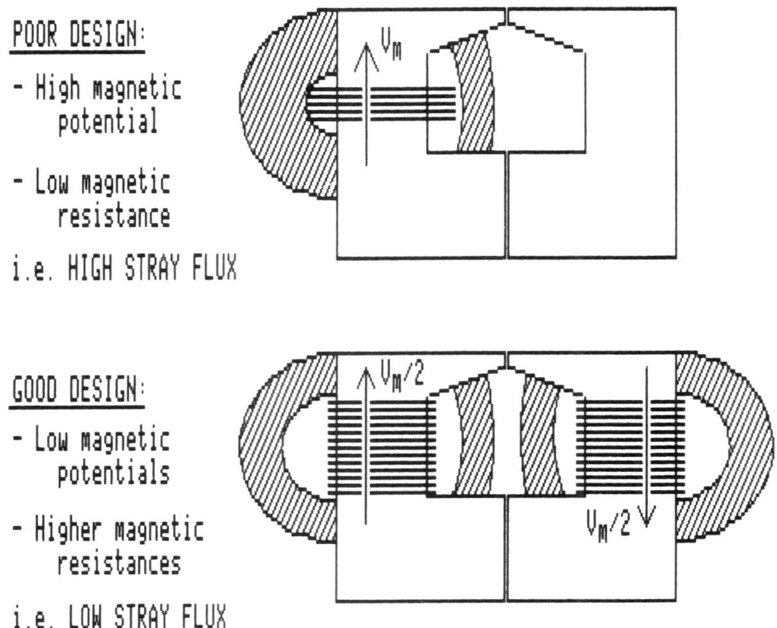

Fig. 10-2. Minimum flux leakage is achieved by using a long coil, split on both core halves.

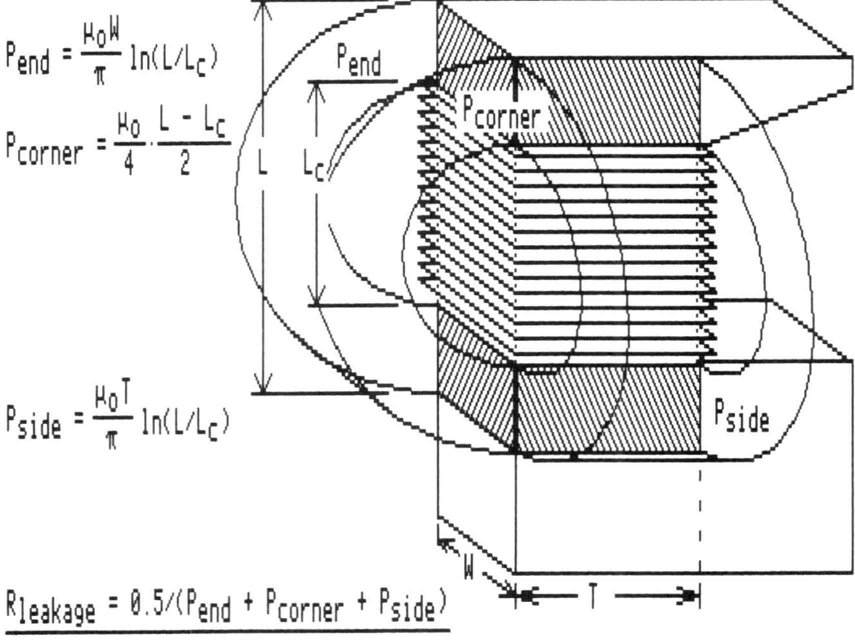

Fig. 10-3. Simplified algorithm for the flux leakage resistance R_s.

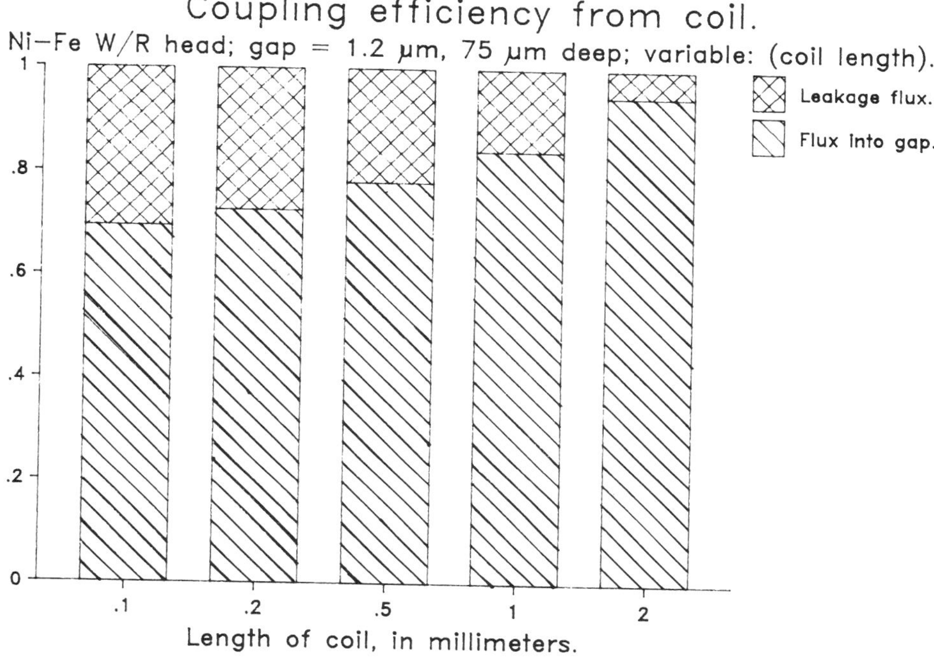

Fig. 10-4. Coupling efficiency improves with coil length.

Both coils develop a magnetomotive force n*i. The short coil makes a short distance in the core for the magnetic potential; that makes the effective value of R_s (R_{s1}) small, and the coupling efficiency small. On the other hand, a long coil that is split between the two core halves results in a large coupling efficiency.

Both cases assume that the coils are wound gradually from one end to the other so the potential is across the entire coil length. If wound back and forth and ending up with the last turn located on top of the first then the coils' electromagnetic length is quite short: the shunt reluctance will be small while, at the same time, the self-capacitance C_e will be large—both factors detrimental to an optimum design.

If we assume a gradual build-up of the magnetomotive force from one coil end to the other then a simple algorithm for the computation of R_s is shown in Fig. 10-3.

The analog circuit for a write head was shown in Fig. 8-36, where the shunt reluctance R_s around the coil was added. The inclusion of this element plus details of the stray fields around the gap are mandatory in computations of the performance of small size, high frequency heads. (Further details on R_s are necessary if the core is near magnetic shielding or housing.)

Figure 10-4 shows the changes in the coupling efficiency that take place when the coil is lengthened (the example was computed for head No. 8 in Table 10-1.)

Required Write Current

For the required write current we find:

$$i_{write} = H_g * L_g / (n * \eta_r * \eta_w * \eta_{coupl}) \qquad \textbf{(10.7)}$$

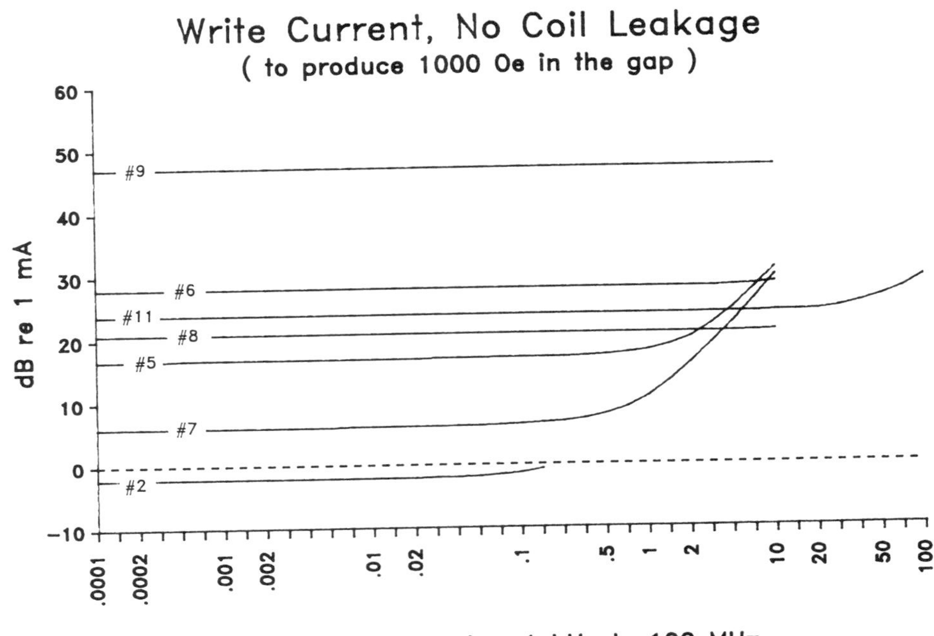

Fig. 10-5. Required write currents, no coil leakage (numbers refer to heads in Table 10-1).

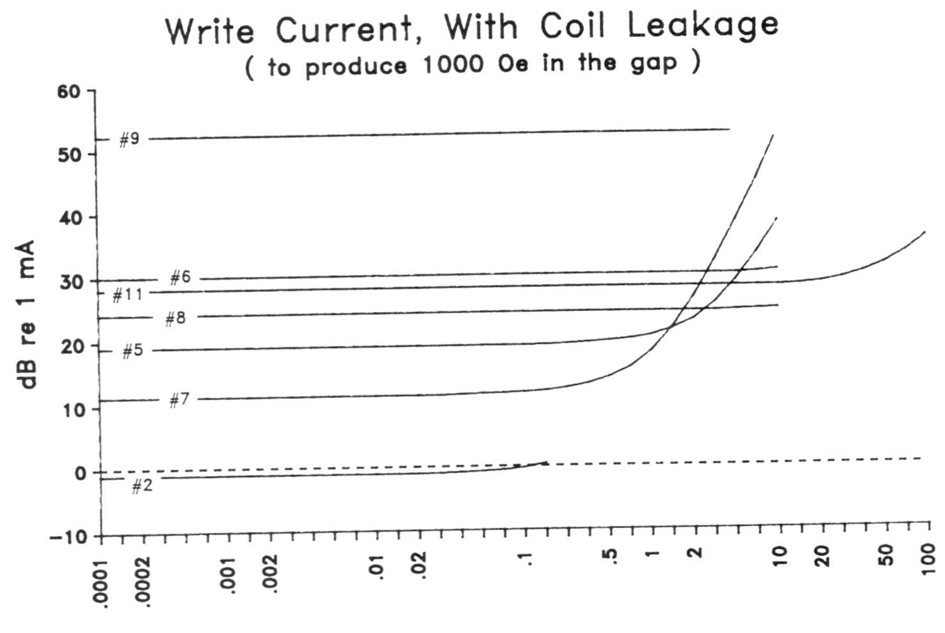

Fig. 10-6. Required write currents, with coil leakage (numbers refer to heads in Table 10-1).

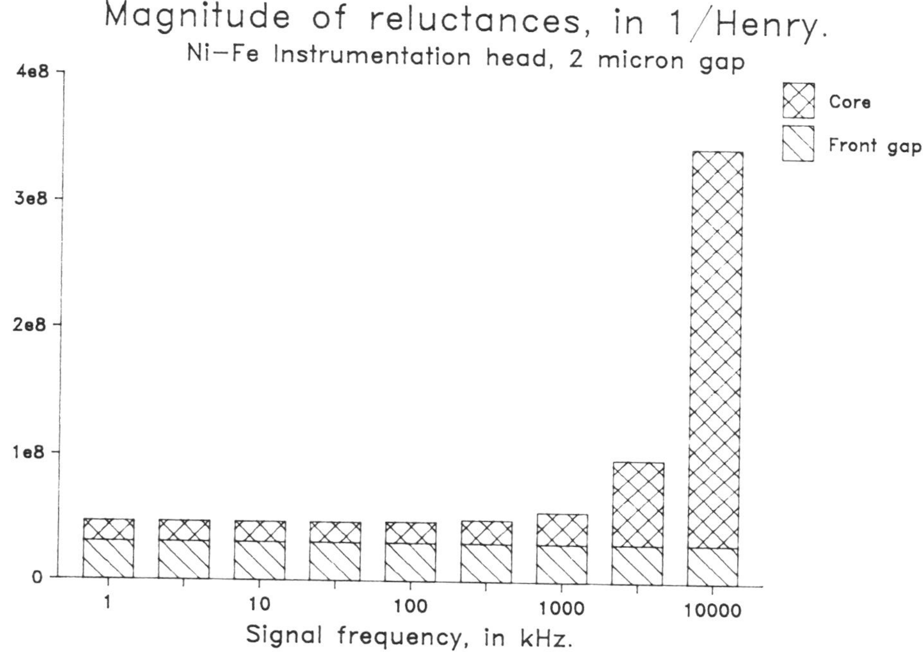

Fig. 10-7. Magnitudes of reluctances depend upon frequency, for Ni-Fe instrumentation heads with a 2 micron gap.

Expressions for the three efficiencies are: (4.26), (4.28) and (8.6). The magnetic resistance of the core elements are relatively easily determined, and the stray fluxes are determined by the formulas in Fig. 8-33.

A finer division of the core into more sections may be necessary in order to achieve sufficient accuracy; this will also preserve a feel for the magnitudes of the various elements (visible if programmed into the cells of a spreadsheet). This method follows the classic work described in Roter's book (See ref. Chapter 8, Leakage flux), and several papers cover the details, the first including crosstalk in a multitrack head assembly (Ref. Sansom, McKnight).

More recent papers use the full power of finite difference and finite element computations (Ref. Wood, Visser, Katz).

Computed results for the write currents for heads in Table 10-1 are shown in Fig. 10-5 for the ideal case of zero leakage flux; these graphs show the currents required to produce 1,000 Oe in the write head gap. More realistic results are shown in Fig. 10-6 where the computations have been repeated, now including the effect of the coil leakage. The required current levels are several dB higher at low frequencies, more at higher frequencies (R_s is independent of frequency, while the core elements' magnetic impedances increase with frequency, see Fig. 10-7).

READ (PLAYBACK) EFFICIENCY

The read efficiency was derived in Chapter 4, formula (4.26). Figure 10-8 shows the computed values, using R_{fgs} (which includes stray fields) and the total, complex value of ΣR_m. Phase shifts are associated with the roll-offs at high frequencies, and the read circuitry may use a conventional RLC-circuit equalizer to correct for the variations in magnitude and phase with frequency.

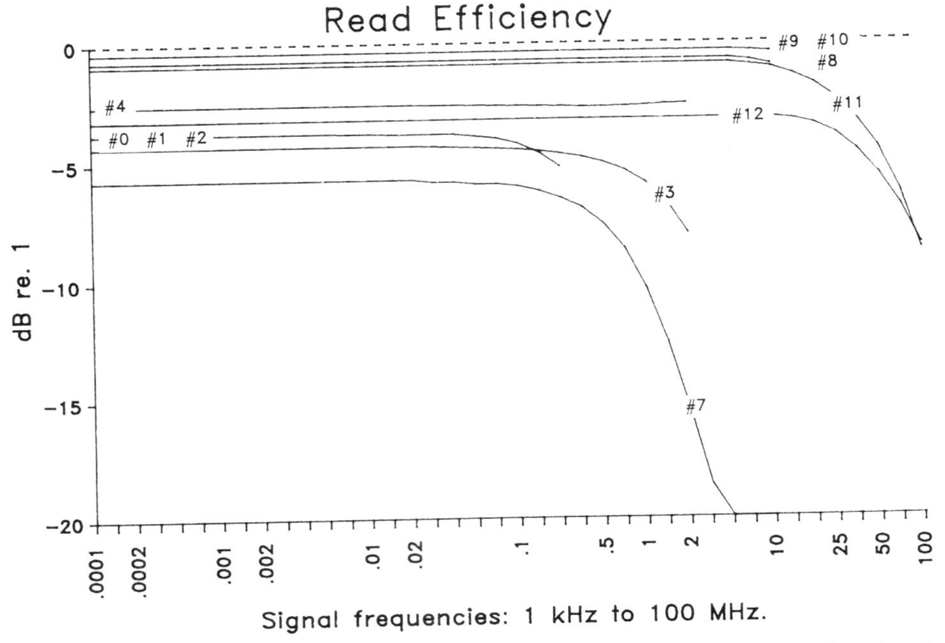

Fig. 10-8. Read efficiencies of magnetic heads as function of frequency (numbers refer to heads in Table 10-1).

The read voltage can be calculated by using the formula in Fig. 6-9.

HEAD IMPEDANCE

The head impedance is determined from:

$$\begin{aligned} Z &= R_{wdg} + j\omega L \\ &= R_{wdg} + j\omega n^2/\Sigma R_{ms} \end{aligned}$$

which is the same as used earlier for the impedance of a toroid, leading to formula (7.5). The core magnetic resistance ΣR_{ms} is now made up of the two core halves, the front and back gaps, stray fluxes and leakage fluxes (in parallel with the core impedances at the location of the coil(s)). Its value is complex, and varies with frequency. Computations are fairly straightforward, once the head core structure has been modeled, and the low frequency values of the core reluctances established.

The high frequency reluctances of the core are complex values since the permeability is complex: $\mu = \mu' - j\mu''$. The detail computations are as follows:

$$\begin{aligned} R_m &= L/\mu * A \quad \text{(at low frequencies)} \\ Z_m &= R_m \text{ at high frequencies} \\ &= L/(\mu' - j\mu'')*A \\ &= L*(\mu' + j\mu'')/(\mu' + j\mu'')*(\mu' - j\mu'')*A \\ &= L*(\mu' + j\mu'')/(\mu'^2 + \mu''^2)*A \\ &= (L/\mu*A)*\mu*(\mu' + j\mu'')/(\mu'^2 + \mu''^2) \\ &= R_m * (\mu\mu'/(\mu'^2 + \mu'' (\mu'^2 + \mu''^2 + \mu''^2)\ ) \qquad \textbf{(10.8)} \\ &= Z_{mreal} + jZ_{mimag} \end{aligned}$$

Z_m is computed for all elements in the head model, and added to give the total magnetic impedance. Let us designate that value $Z_M = Z_{Mreal} + jZ_{Mimag}$.

The head's impedance is, with a winding resistance of R_{wdg}:

$$\begin{aligned} Z &= R_{wdg} + j\omega L \\ &= R_{wdg} + j\omega n^2/Z_M \\ &= R_{wdg} + j\omega n^2/(Z_{Mreal} + jZ_{Mimag}) \\ &= R_{wdg} + j\omega n^2 * (Z_{Mreal} - jZ_{Mimag})/(Z_{Mreal}{}^2 + Z_{Mimag}{}^2) \\ &= R_{wdg} + \omega n^2 * Z_{Mimag}/(Z_{Mreal}{}^2 + Z_{Mimag}{}^2) + j\omega n^2 * Z_{Mreal}/(Z_{Mreal}{}^2 + Z_{Mimag}{}^2) \\ &= R_{wdg} + R' + j\omega L' \end{aligned} \tag{10.9}$$

where R′ is proportional to μ'' and L′ is proportional to μ'.

Computed values of L′ and R′ are shown in Figs. 10-9 and 10-10. R′ represents the eddy current losses. Notice the rapid increase in the value of R′ once it has sprung up from values below one ohm. It originates in Z_{mimag} which is proportional to μ'', which increases proportional to the frequency (see Fig. 7-4).

Z_{mimag} is further multiplied in formula (10.9) by $\omega = 2\pi f$. The result is a resistance value proportional to f^2, and it therefore increases 12 dB/octave ($12 \log_{10} 2^2 = 12$).

Heads No. 5 and 7 have MuMetal cores, and the reduced rise in R′ occurs when K approaches 1 (see Fig. 7-5 and formula 7.3).

The head impedance is now known and available for the circuit designer.

To the total impedance we must add the distributed capacitance of the winding C_e and the capacitances of the cabling and the amplifier impedance.

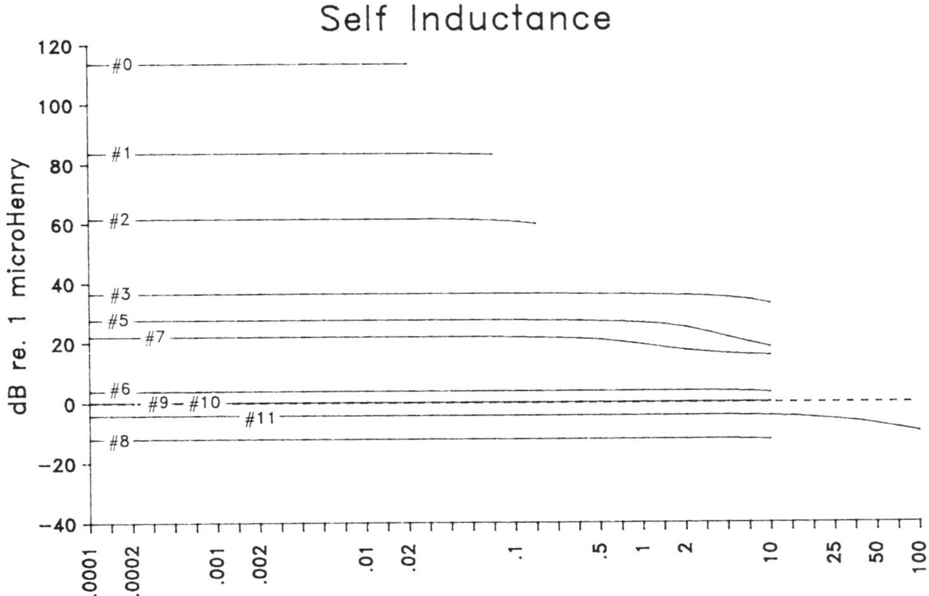

Fig. 10-9. Inductance L′ of magnetic heads (numbers refer to heads in Table 10-1).

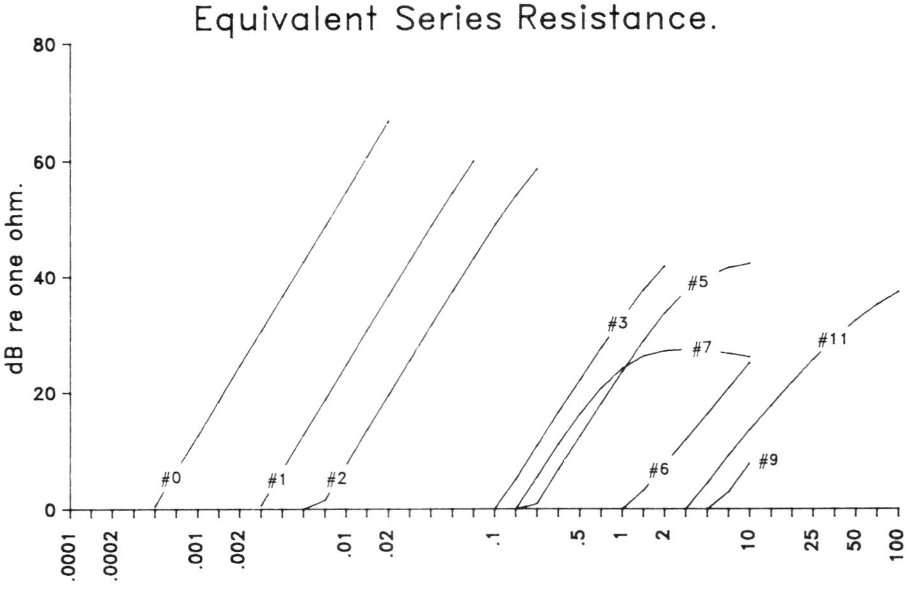

Fig. 10-10. Resistance R′ of magnetic heads (numbers refer to heads in Table 10-1).

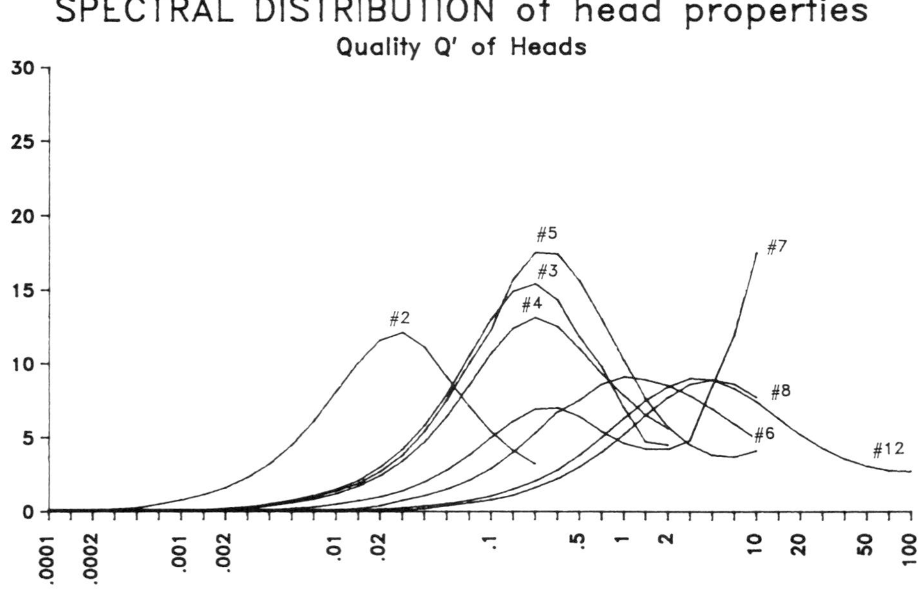

Fig. 10-11. Computed coil quality Q′ for magnetic heads (numbers refer to heads in Table 10-1); actual Q-values are lower.

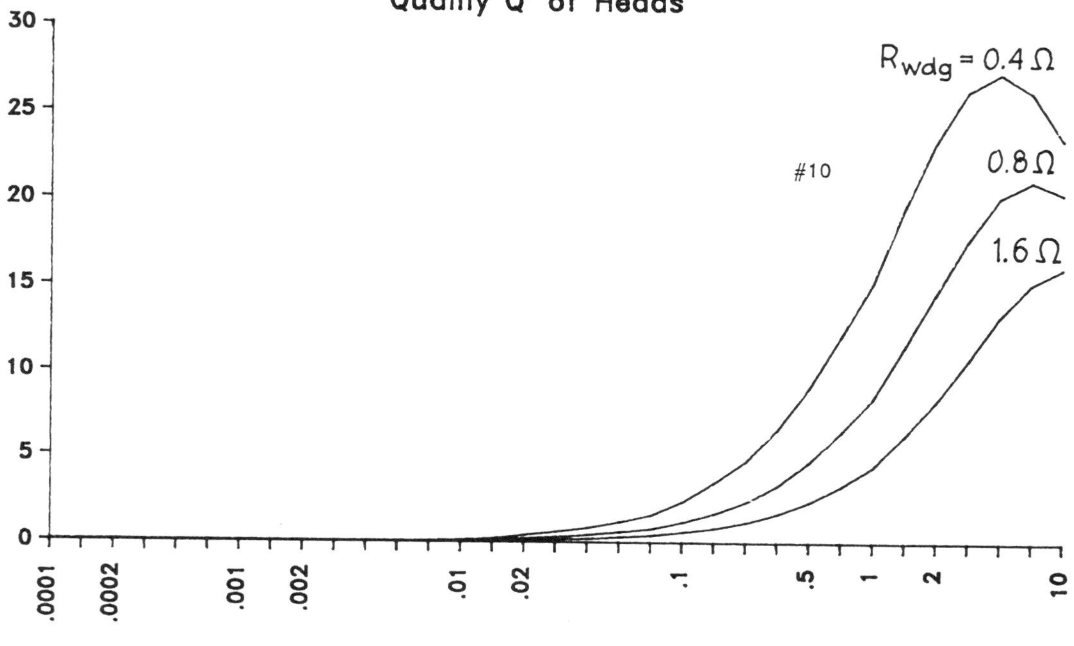

Fig. 10-12. Coil quality Q′ is affected at low frequencies by the coil wire resistance, at high frequencies by core losses.

The number of turns n is always selected as high as possible in order to minimize the required write current and to maximize the induced read head voltage.

When the number of turns has been determined then a wire size is selected that will fit into the available winding space. This normally results in a very fine wire (No. 40 or smaller) and R_{wdg} can now be calculated. Its value adds to the R′ values above, and the coil quality Q can be determined.

The ratio $\omega L'/(R_{wdg} + R')$ is the *coil quality Q′* of the head impedance. Figures 10-11 and 10-12 show the computed values of Q′ for most of the heads in Table 10-1. They are quite low values when compared with other types of RF-coils (radio frequency coils). They will be lower in the finished head due to coupled losses from shields (expand R_s to include shields or metal housing, if any), and dielectric losses (unpredictable magnitude).

Very low values of Q (less than three) will reduce the read voltage to lesser than the induced voltage, and simultaneously introduce a phase lag. For Q = 1 the loss is 3 dB and the phase angle −45°.

This loss is part of the often observed 5½ dB slope of the read head voltage versus frequency curve, at low frequencies; the slope should theoretically be 6 dB/octave.

Write Head Core Losses

The head core losses at the bias frequency and at the high level of bias current i_b are in the milliwatt range, and are calculated from:

$$P = i_b^2 * (R' + R_{wdg})$$

The losses are often just estimated as $V*i_b$, where V is the winding terminal voltage. Additional losses are hysteresis losses, which is covered later in this chapter under erase heads.

The two losses may together produce excessive heat in small core structures that are held imbedded in epoxies of low heat transfer capabilities.

Read Head Core Noise

There are two sources of noise in magnetic heads:

- Barkhausen noise, caused by jumps in the magnetic domain wall movements.
- Resistance noise. The thermal agitation in a resistor produces a noise voltage that is:

$$V_{noise} = \sqrt{4kTBR} \text{ volts} \tag{10.10}$$

where

K = Boltzmans constant
= $1.38 * 10^{-23}$ J/°K
T = Temperature in °Kelvin
= 273° + °C
B = Bandwidth in Hertz
R = Resistance value in ohms

The Barkhausen noise has been reported in connection with the switching of thin heads (ref. Fig. 7-10). Its contribution to the overall recorder noise level does otherwise remain undetected, although it is present in other devices (ref. Bittel). It appears to be one of the noise sources in magnetoresistive heads.

The noise voltage from the equivalent loss resistance R′, on the other hand, can be of sufficient level to impair the overall signal-to-noise of a tape recorder. This is particularly true of high-frequency wide-band recorders and video recorders. The computation of the noise voltage is straightforward once R′ has been determined.

The influences of μ' and μ'' on the head noise voltage was analyzed by Smaller (see ref.). He arrived at an expression for the narrow band signal-to-noise ratio (NB/SNR), defined as the tape signal output voltage divided by the per-cycle head noise voltage at any given frequency:

$$NB/SNR = \sqrt{\mu'^2 + \mu''^2 / \mu''}$$

which at very high frequencies, where $\mu' = \mu''$ becomes:

$$NB/SNR_{hf} \cong \sqrt{2\mu'}.$$

At lower frequencies, where $\mu' >> \mu''$, we can write:

$$NB/SNR_{lf} = \mu'/\mu''$$
$$= Q \text{ (with } R_{wdg} = 0)$$

from (10.9) and (10.8).

The SNR improves about 10 dB by using 2 mils lamination rather than 6 mils in audio heads. Further improvement is made by using ferrite cores (ref. Byers).

Some heads have been reported to be microphonic, i.e., they produce a crackling noise when a blank tape moves across their surface. These effects may relate to residual magnetostriction and/or domain wall pinnings (ref. Watanabe).

POLE TIP SATURATION

Modern recording materials with high coercivity in conjunction with short gap length heads have created a new problem for head designers. The field in the gap of the recording head must be higher than in the past, and the danger of saturating the core material becomes real. A brief example will illustrate the problem.

Example 10.2: A 1 μm gap length head is used for recording and playback in a cassette tape machine. Determine the record gap field when using a 5μm thick coating of a cobalt treated gamma ferric oxide tape with coercivity 600 Oe.

Answer: Assume an overbias setting that corresponds to a write level of 120 percent. From Fig. 5-14 we find that $\nu \cong 0.05$, i.e., $H_g = 600/0.05 = 12{,}000$ Oe. This corresponds to a gap flux density of 12,000 Gauss. A ferrite core cannot be used due to low saturation flux density, nor can Mumetal; only an Fe-Al alloy will handle the high flux level (see Table 7.1).

This example did happen a few years ago when high coercivity tapes were introduced into the audio cassette market, and similarly when a decision was made to make the 3.5 inch diskette from a 550 Oe material. In both cases a short read gap was required for adequate resolution; but using the same head for recording would result in saturation of the core pole tips during the record or write process.

A solution was simple in the audio field: adding a separate record head with longer gap. This increased the value of ν, and H_g could be lowered. The reader will recall that the length

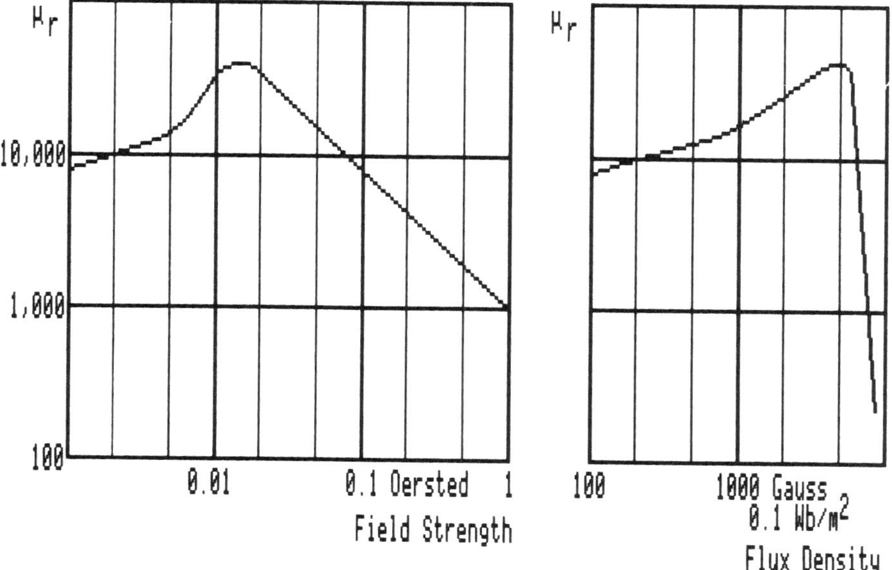

Fig. 10-13. Permeability μ_r varies with the magnetizing field H, or flux density B.

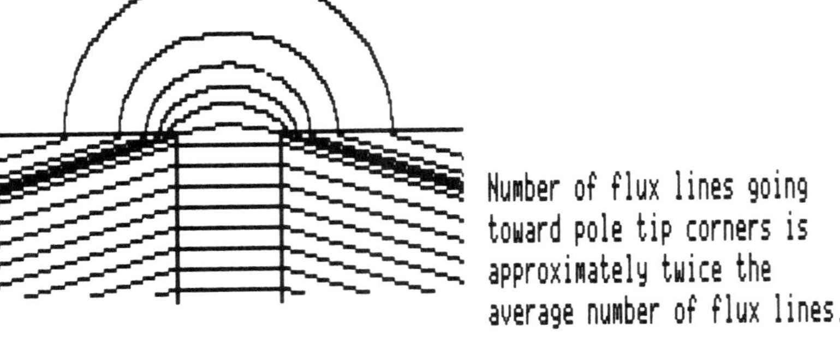

The corners of the pole tips start saturating when the head core flux approaches a value equal to 1/2 of J_{sat}.

The pole interfaces will saturate when the flux level reaches J_{sat}. Increasing the write current will not produce further flux in the gap (i.e. H_{gap} = constant = J_{sat}/μ_0).

Fig. 10-14. Flux patterns at the onset of pole tip saturation.

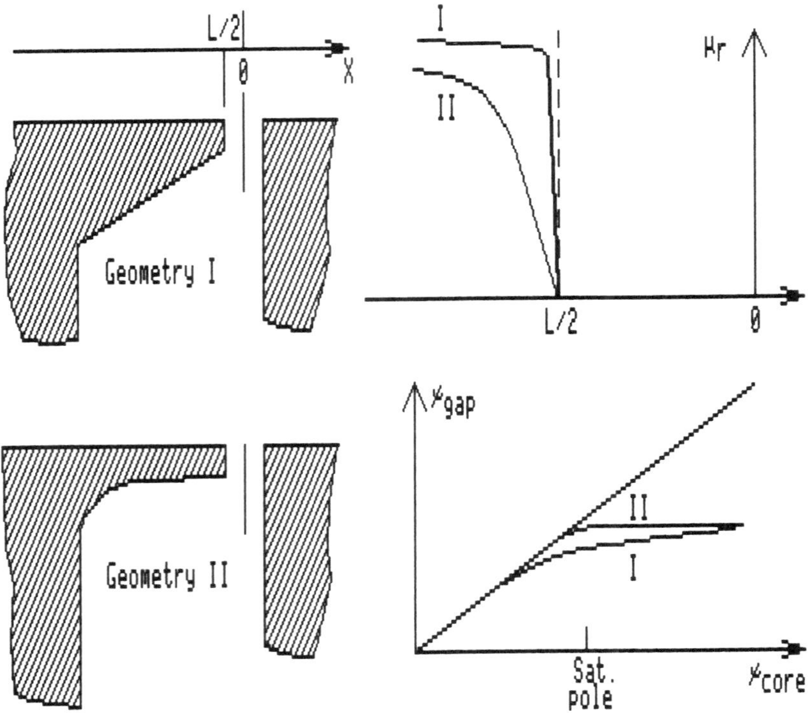

Fig. 10-15. Decrease in permeability near saturated poletip causes a smooth approach to saturation (curve I).

of the write gap has little influence upon the resolution, with the exception of situations where the recording is very shallow, i.e., ν is large.

The diskette problem was solved by making the coating thinner: c = 2 μm. That also increases the ν value. At the same time demagnetization of short bits is reduced. As a matter of fact, with packing densities of 5 KBPI (or more) the coating need only be c_{eff} = 0.44 * BL = 0.44 * 5 μm = 2.2 μm or less (ref. formula 8.8b).

The saturation process is complex: not the entire pole face saturates at first, as the gap field increases. It has been established that the corners saturate first, as shown in Fig. 10-14, and this occurs when the flux density in the gap approaches half the value of the material's saturation flux density (ref. Suzuki, Shibaya, Szczech).

The write resolution does not suffer much thereby (ref. Thornley, Fujiwara). This is partly so because the recording or writing takes place at some distance behind the trailing edge of the write gap.

When the gap flux is increased then the pole faces will at some point saturate (Fig. 10-14, bottom), and no further increase in gap flux is possible. Any attempt to drive more flux through the head will only result in an increase in the flux going through the stray flux elements around the gap, *not* through the gap.

Saturation does not happen like a switch was included in the equivalent circuit for the head. The transition is smoothed by the stray fluxes. This is emphasized by the fact that a head with an apex angle near 90 degrees will experience a gradual increase in the pole tip reluctance: The relative permeability μ_r is one at the pole tip surface when saturation starts.

The flux level inside the tip is at a level almost equal to saturation, and the permeability is much lower than it is for small flux levels. Hence the pole tip reluctance is high, and the

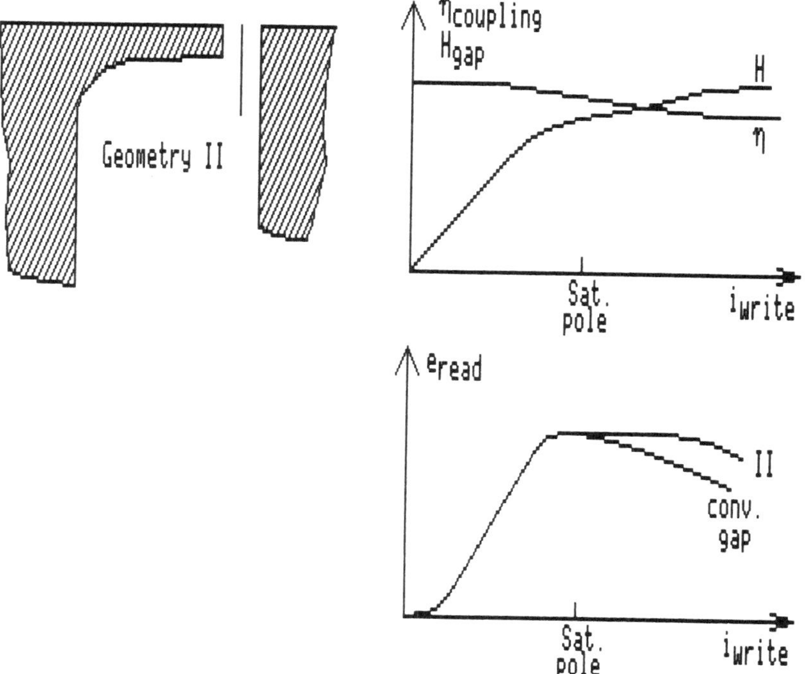

Fig. 10-16. Increase in core reluctance near saturated gap in a head with a 90 degree apex angle causes decrease in write efficiency and subsequently insensitivity to write current variations (after Tranetics, Inc.).

efficiency low. The gap flux will therefore level off gradually when saturation is approached, see curve I in Fig. 10-15. The effect is enhanced by the stray flux entering the pole piece behind the gap interface, where it may cause further saturation (ref. Valstyn).

This results in a lowering of the write efficiency (curve II in Fig. 10-16), and the gap field will remain almost constant over a range of drive currents, see Fig. 10-16. The net effect is a relaxation of the specification for the drive current.

The result is a write current response as shown in Fig. 10-16 (ref. Ed Packard, Tranetics Inc.).

Head with Metal-in-the-Gap (MIG)

The pole tip saturation problem can be overcome by using a write gap construction with deposited metal pole pieces, see Fig. 9-16. The improved performance of these heads are evident from Fig. 10-17 (ref. Jeffers).

Also the efficiency of a reproduce head can be improved by using a metal gap spacer, which at high frequencies increases the front gap reluctance due to eddy currents. Improvements of several dB have been reported (ref. McKnight).

The eddy currents are undoubtedly also at play in the MIG head, leading to higher read as well as write efficiency. A multigap effect appears to cause head bumps (waviness in amplitude versus frequency response) at high frequencies (ref. Ruigrok).

VARIATIONS IN PERMEABILITY (PRODUCTION TOLERANCES)

It was mentioned in Chapter 7, under Metallic Core Materials, that coldworking of the core material during head fabrication would significantly lower the relative permeability. The effect thereof upon the head's efficiency is shown in Figs. 10-18 through 10-20.

The read efficiency at low frequencies will become lower as the permeability drops in value.

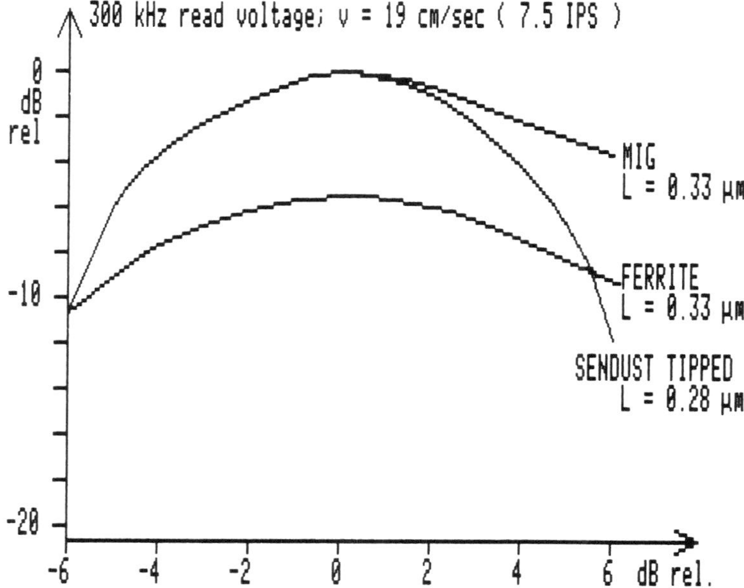

Fig. 10-17. Write performances for three head constructions (after Jeffers et al.).

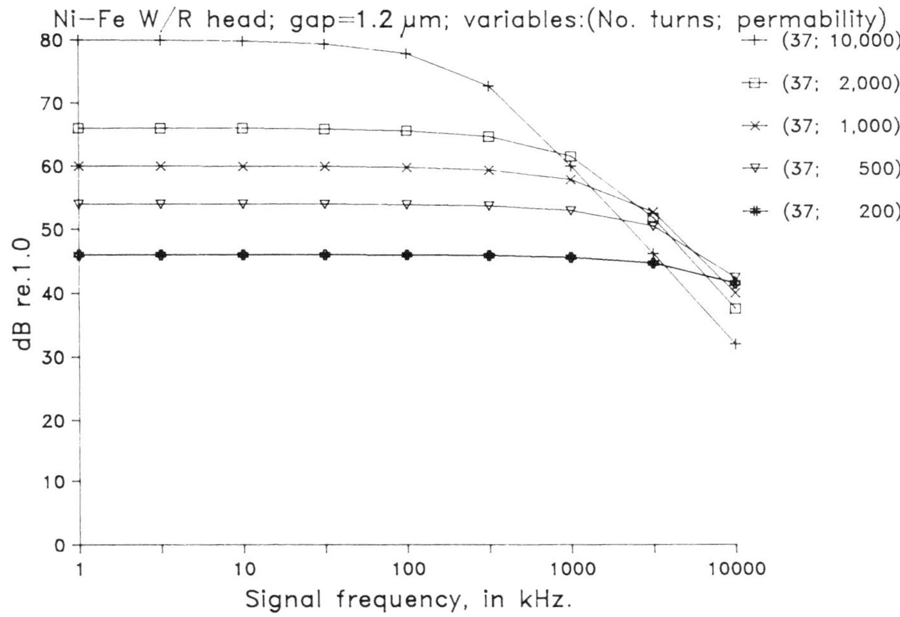

Fig. 10-18. Permeability μ_r' spectra for Ni-Fe Core, lamination thickness = 15 μm (0.6 mils).

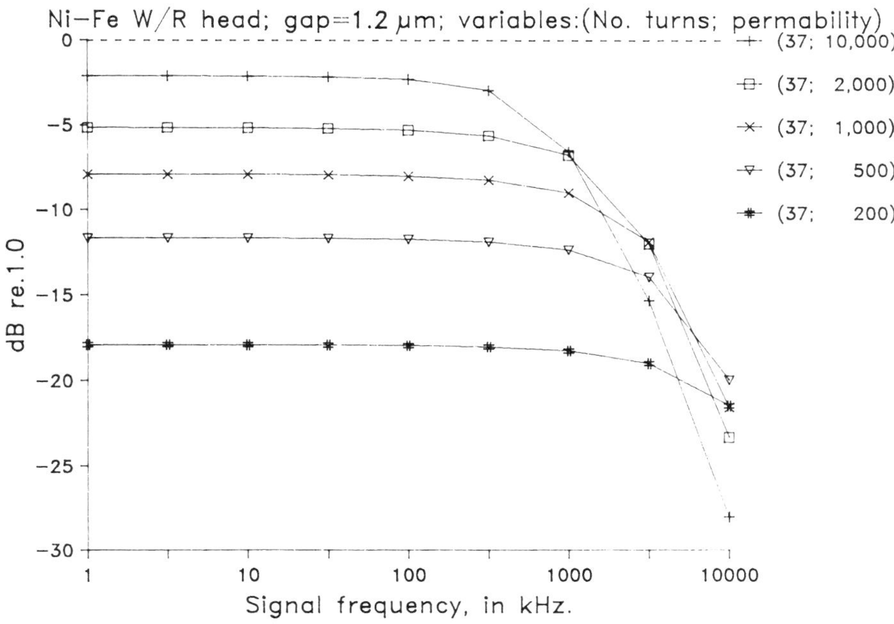

Fig. 10-19. Read efficiency as function of frequency, for five different values of the cores relative permeability (Ni-Fe core; head no. 5, gap length = 1.2 μ_m).

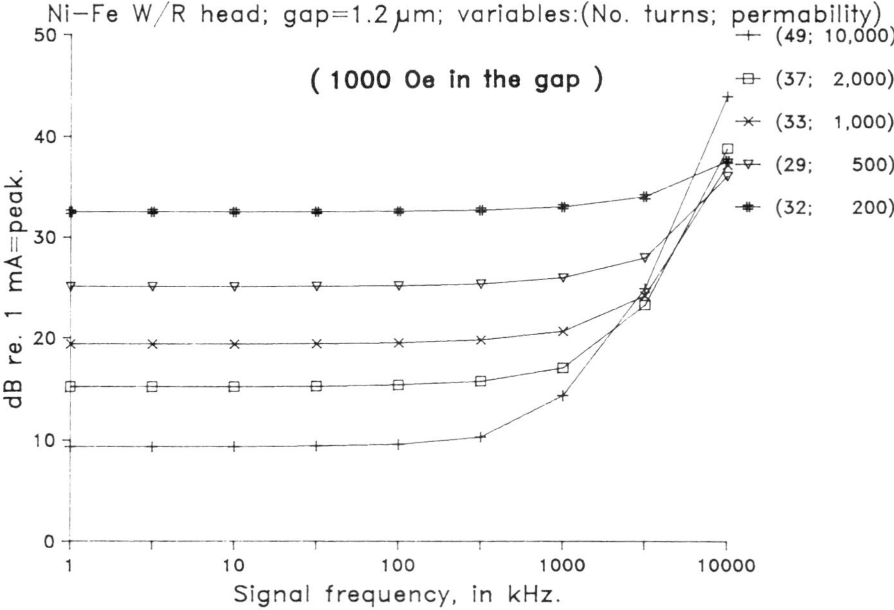

Fig. 10-20. Write current as function of frequency, for five different values of core permeability μ_r (Ni-Fe core; head no. 5, gap length = 1.2 μm).

This is not necessarily the case at high frequencies since the eddy currents are reduced when the permeability decreases (ref. formula (7.1) and (7.3)). The change in permeability spectra is shown in Fig. 10-18, and reflects in the efficiency curves in Fig. 10-19.

The write current requirements increase when the permeability decreases, see Fig. 10-20.

BENEFITS OF A LARGE BACK GAP

The low frequency efficiency of a magnetic head is highest when the back gap reluctance has a zero value. This is achieved by using a back gap with a large area and a short gap length. There are nevertheless a couple of good reasons for designing a long back gap into magnetic heads: In audio record heads the effects of non-linearities of the permeability are rendered "small," while the heads are less prone to become permanently magnetized. These topics are discussed next.

Distortion in Instrumentation and Audio Heads

The magnetic reluctance of the core is inversely proportional to the permeability μ_r which depends upon the strength of the magnetizing field H, or flux density B. The relationship between μ_r and H, and also μ_r and B, was shown in Fig. 10-13 for a typical metallic core material. The permeability increases with H (or B) up to a maximum value μ_{rmax}, whereafter it reduces dramatically as B approaches saturation.

This change in μ_r will change the values of the read efficiency η_r and therefore also the write current requirements. The increase in μ_r with flux density increases write efficiency; this means that the peak levels of a sine wave drive current will produce too high flux levels

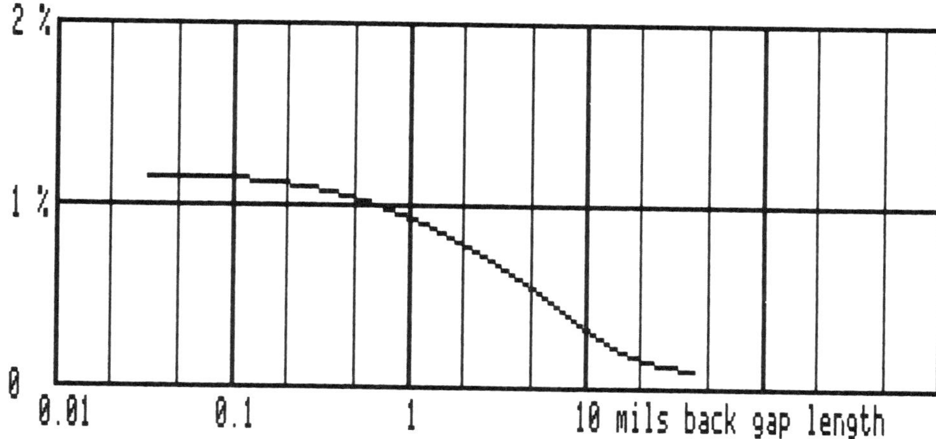

Fig. 10-21. Percent change in record efficiency for a 10 percent change in core permeability, as a function of the length of the back gap.

(peaking), and the recording will contain 3rd harmonic distortion components. (NOTE: This record head distortion is not the 3rd harmonic distortion measured on AC-biased recorders: it may be part of it—but should not be, with a properly designed record head.)

Attempts have been made to calculate this distortion using Bozorth's results for changes in μ_r at low field (Rayleigh loops). That approach is limited by:

- Approximations.
- Different induction levels exist in the various parts of the core ($B = \psi$/area, and the core area varies from tip to back gap).
- Eddy currents (skin depth) will affect the flux densities across the core cross sections.
- Superposition of data and bias signals in an AC-bias recording is difficult to handle.

A reasonable approach is instead to evaluate what percentage change occurs in the record current sensitivity when the permeability changes say 10 percent? The result of a series of calculations for various back gap lengths is shown in Fig. 10-21, and illustrates that a large back gap is beneficial in reducing the effect of changes in μ_r. (This technique is used in audio record heads.)

Permanent Magnetization of Heads

The core halves in a magnetic head form a magnetic circuit that is capable of maintaining a remanent magnetization B_{rs}'. If no gaps were present the level of remanence could equal the maximum remanence for the core material, B_{rsat}.

The presence of even very small air gaps causes demagnetization which lowers the level of remanence. In Chapter 4, example 4.1, we derived an expression for the maximum remanence in a magnetic head core with air gaps:

$$\boxed{B_{rs}' \cong \mu_o H_c / N} \qquad \textbf{(4.16)}$$

where $N \cong (L_{fg} + L_{bg})$/(Length of core plus gaps).

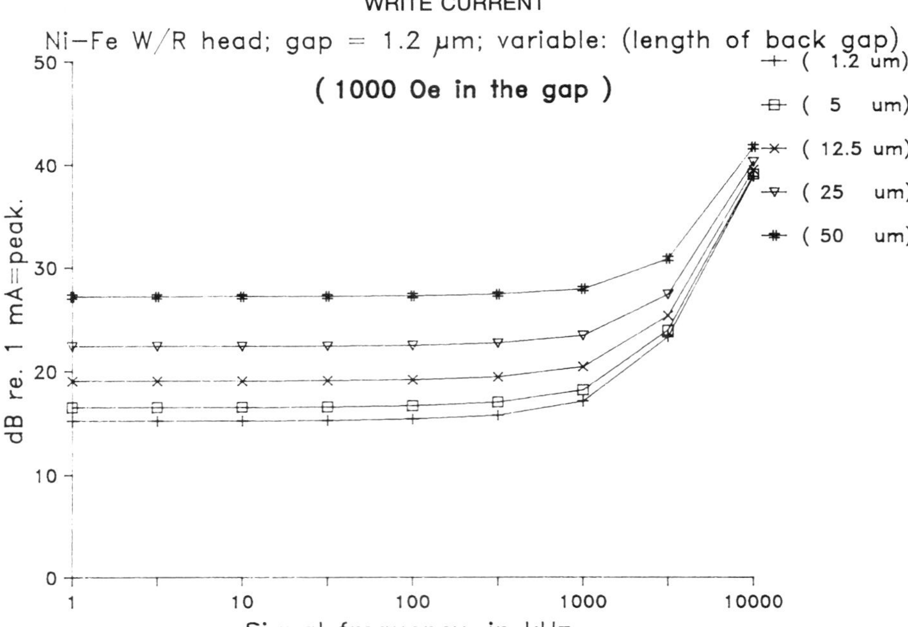

Fig. 10-22. Write current as function of frequency, for five different values of the back gap length (Head no. 5).

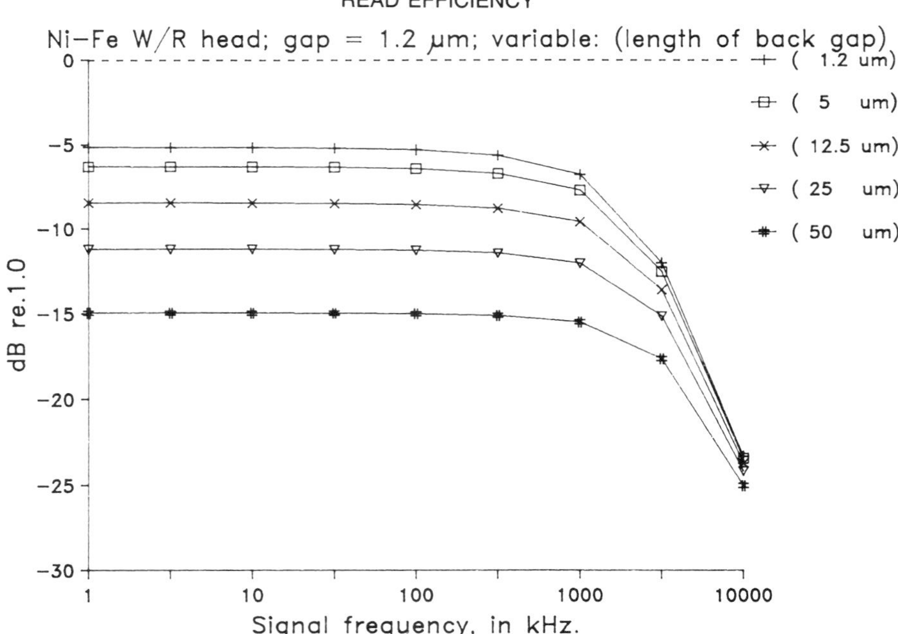

Fig. 10-23. Read efficiency as function of frequency, for five different values of the back gap length (Head no. 5).

When no back gap is present (highest efficiency) the demagnetization is small and current pulses (or turn-off current after saturation recording) may leave the core in a highly magnetized state. It may be high enough to cause partial erasure of a recorded track moving past the gap.

If the AC-current in biased recording is turned off properly (slow decay) then the head will automatically degauss.

Reproduce heads have shorter gaps than record heads, and are therefore more prone to become permanently magnetized.

A long back gap in any head has its price: reduced efficiency. This is shown in Fig. 10-22 for the write current requirements. A short back gap requires less current at low frequencies, but a larger amount of boost at high frequencies. A long back gap can just as well be used, requiring a current only 3 dB, since the write amplifier has to deliver the higher current at high frequencies.

The drop in read level for long back gaps may not be acceptable since the midband signal-to-noise ratio suffers thereby (Fig. 10-23).

EFFECTS OF CORE GEOMETRY

The wavelength resolution in magnetic recording is bounded by two mechanical dimensions: the length of the gap in the core will limit the short wavelength resolution while the overall core length will limit the long wavelength output.

The interactions between a head core and adjacent cores and recorded tracks are covered later in this chapter under Multitrack Heads.

Effect of Overall Core Size (Pole Lengths)

Low frequency recordings result in wavelengths that are comparable in size to the overall length of the head interface. This will result in a higher flux density through the body of the core (and hence winding), and the net effect is an increase in the head's efficiency at low frequencies. (Neither this phenomenon, nor the undulations listed later occur in record heads.) The equivalent diagram of the head core gradually changes for increasing wavelengths as shown in Fig. 10-24 top. The core reluctance is divided into 3 portions for clarity. The two ''sliders'' on the core tip reluctances move outwards with increasing wavelengths, hence increasing the efficiency.

This increase in output at long wavelengths is called *secondary-gap effect* (ref. Fritzsch.)

Another irregularity, *undulations*, occur in the same long wavelength range. This is easiest explained by examination of the flux through the entire core structure, as shown for λ = L and λ = L/2 in Fig. 10-24 bottom. The flux through the cores from the half wavelength magnets ''hanging'' over the edge of the core are in-phase with the main flux when λ = L, but out-of-phase when λ = L/2.

The total effect is a response function (called the *Spiegelfunktion*) that is a superposition of the secondary gap effect and the undulation. (A slang term is *Head Bumps*.)

The latter are lessened by avoiding tape-head contact at the corners of the core, shown in Fig. 10-25A.

The undulations are modified if the core is inside a shield can, which provides a flux return path for the overhanging magnets (Fig. 10-25B).

They can also be suppressed by using asymmetrical core halves (Fig. 10-25C). Some computer write/read heads are of this construction, and may exhibit phase distortion since the reproduce system is asymmetrical.

The undulation effects are so different from one head design to another (including the effect of trackwidth), that no general formula can be given for their magnitudes.

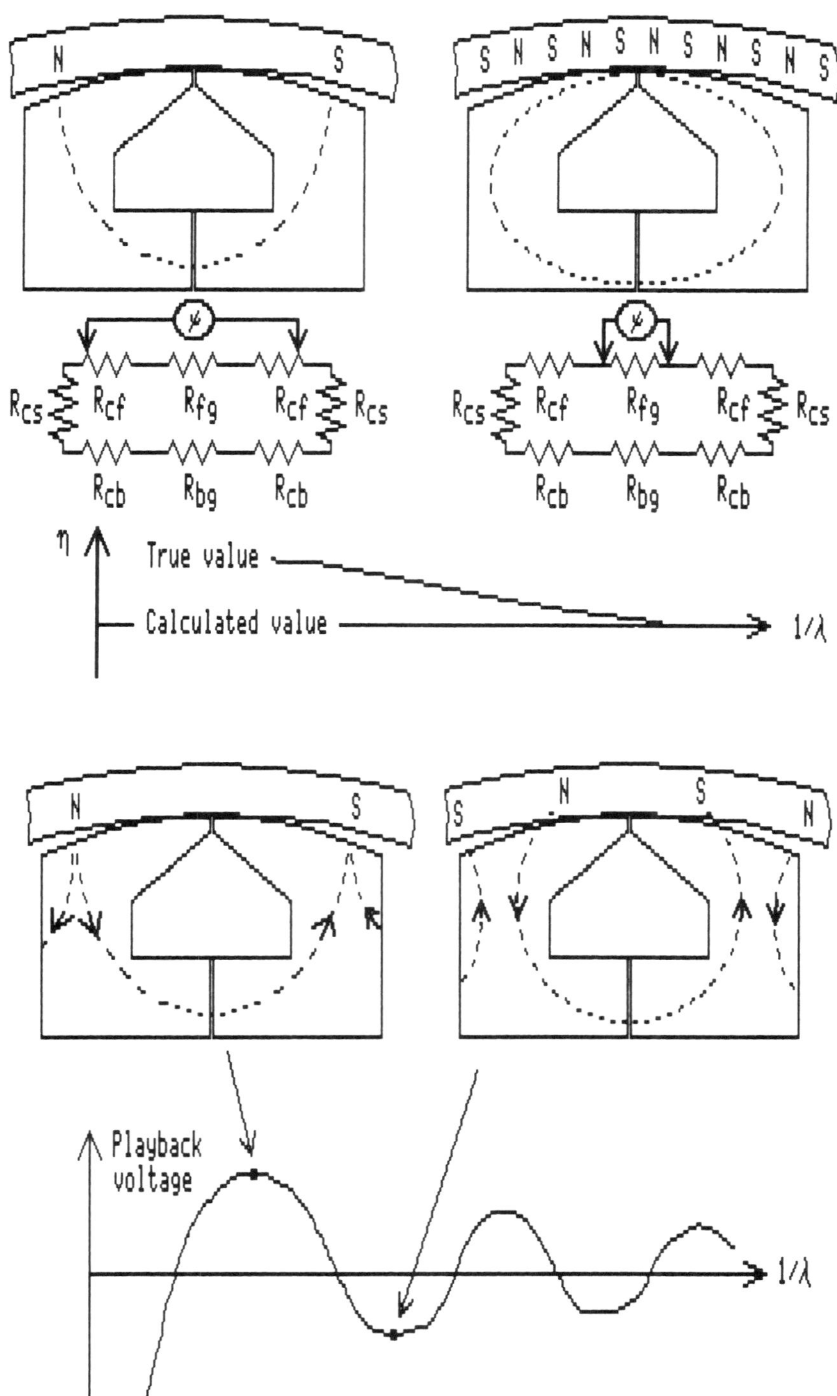

Fig. 10-24. Reproduce head core contour effects (after Fritzsch).

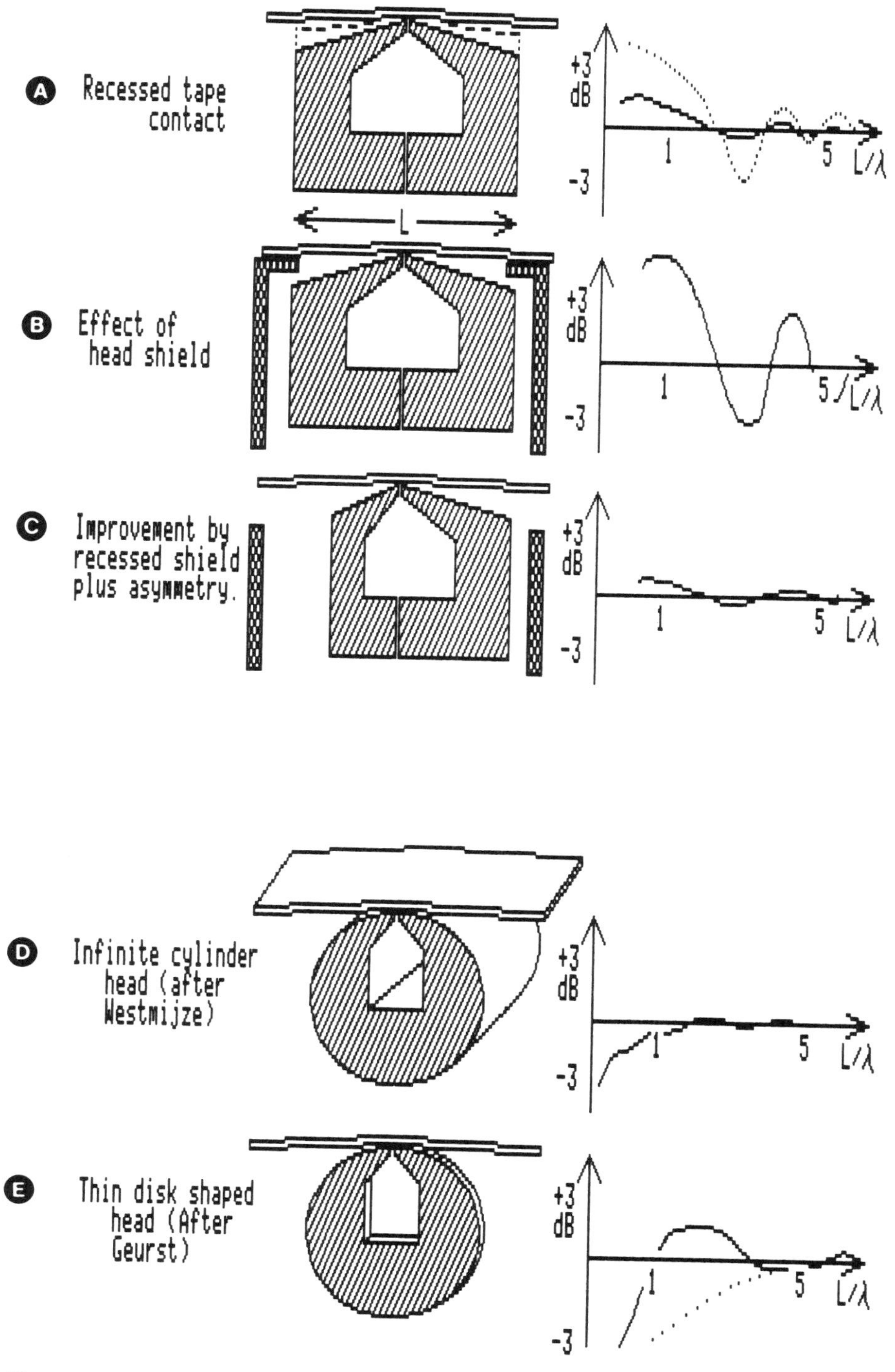

A Recessed tape contact

+3 dB, -3, 1, 5, L/λ, L

B Effect of head shield

+3 dB, -3, 1, 5, L/λ

C Improvement by recessed shield plus asymmetry.

+3 dB, -3, 1, 5, L/λ

D Infinite cylinder head (after Westmijze)

+3 dB, -3, 1, 5, L/λ

E Thin disk shaped head (After Geurst)

+3 dB, -3, 1, 5, L/λ

Fig. 10-25. Contour effects ("head bumps").

Finite Width of Core (Track Width)

A cylindrical head of infinite width was the base of Westmijze's studies. If the width is reduced toward zero then the core becomes a round disk, and this is a better approximation to the real world of narrow track heads.

Geurst investigated the behavior of such a narrow head. He found that there were much stronger fluctuations for the disk head, and that wavelengths as small as one tenth of the diameter were still detectable. The two heads, a long cylindrical and a short disk are shown in Fig. 10-25, D and E. The associated head bumps are shown for a wrap angle of 2 × 5° (solid lines) and 0° (broken lines).

There is obviously a higher degree of wavelength interference in the round disk head. The patterns of the fluctuations reflect the sensing field H_s (see Chapter 6). Since the field strength at sharp edges are quite high, strong fluctuations in H_s can also be expected, such as at the edges of the disk head and at the edges of the core where the tape enters or leaves, and also at the corners of shielding housing (Fig. 10-25B) (ref. Baker, Lindholm).

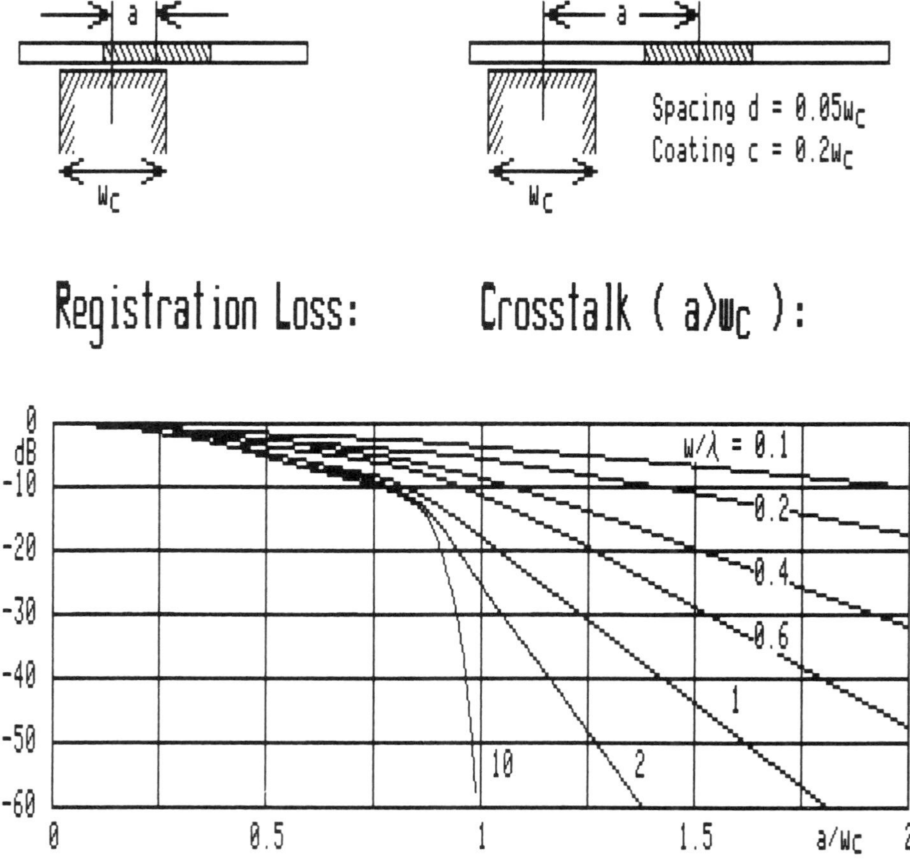

Fig. 10-26. Registration loss and crosstalk (after Lindholm, van Herk).

Side-Reading Effects (Cross-talk)

The sensing field exists in the entire space around the read gap, like shown in Figs. 8-24, -25 and -26. Then the question is: "What does the sensing field see from adjacent tracks on the tape or disk?". Any signals picked up in this way are named crosstalk, and are part of the overall recorder system noise.

Eldridge and Baaba investigated this crosstalk and found that it behaved quite like the well-known Wallace spacing loss when a head is spaced away from a media surface. They found a crosstalk figure of 65 $*$ d/λ dB, where d is the lateral distance from the side of the head to an adjacent track.

Lindholm and van Herk calculated the crosstalk using a three-dimensional field. Their results are comparable and are shown in Fig. 10-26. Two results can be deduced from these graphs:

- There exists a *registration loss* when the tracking between the reproduce head and the recorded track varies. This may occur when recording and playback is done on two different drives, and temperature/humidity conditions alter the tracking.
- Another loss is the *read crosstalk* from an adjacent track (when center-of-head to center-of-track distance a is greater than the track width. The crosstalk (equal to the reduced output from an adjacent track due to spacing d) was shown in Fig. 8-28.
- Another result was derived by van Herk for the increased output a playback head will produce when it reads a track that is wider than the core. This *side-fringing response* is strongest at long wavelengths (low frequencies) and must be considered when a multitrack tape reproducer is calibrated with full-track standard tapes (ref. Melis et al.).

EFFECTS OF COATING PERMEABILITY

Gap Length

The classical expression for the loss of signal, that increases with decreasing wavelength, is:

$$\text{Loss} = \sin(X) / X$$

where

$$X = \pi L/\lambda$$

When the wavelength λ equals the length L the output is zero.

The length L is not equal to the mechanical gap L_{fg} of the front gap in the reproduce head. The sin(X)/X formula was first derived for optical (motion picture) playback, where L was the slit width of the optical scanning head. In late 1940 Daniel and Axon found that the measured gap-length response did not agree with gap loss formula. The magnetic gap length appeared to be 1.15 times the mechanical length.

In 1953 Westmijze published his now famous study of the magnetic recording process, and verified that the semi-infinite head has an "effective" gap length that is about 10 percent longer than the mechanical gap. The tape relative permeability was assumed equal to one.

In 1974 Siakkou (see ref.) investigated the effect of tapes with isotropic permeability and found that L in our previous formula should be substituted with:

$$L \cong L_{fg} \sqrt{(1.5\mu_{rt} + 1)/(\mu_{rt} + 1)} \qquad \textbf{(10.11)}$$

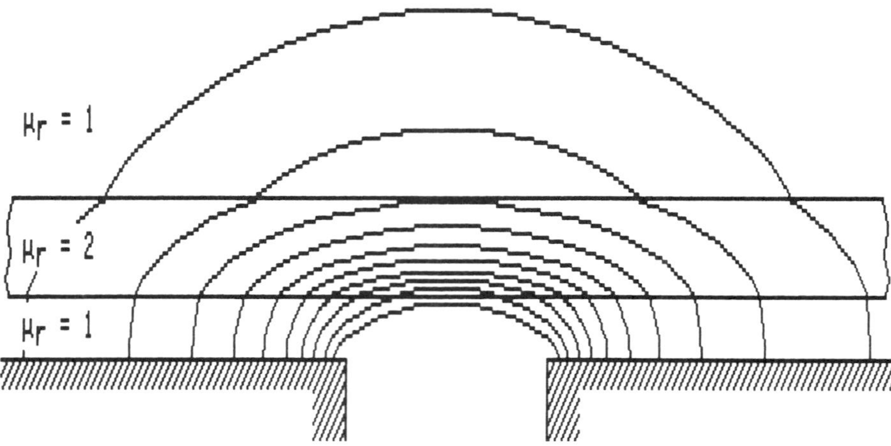

Fig. 10-27. Change in head field pattern due to media permeability μ_r = two.

When $\mu_{rt} = 1$ we have that $L = 1.12 * L_{fg}$, which agrees with Westmijze's work. For $\mu_{rt} = 2$ we find $L = 1.15\ L_{fg}$, and in the limit (for μ_{rt} very large) we find $L = 1.22 * L_{fg}$.

Spacing Loss; Coating Thickness Loss

The general playback theory assumes that the tape or disk coating has a relative permeability equal to one. If it is greater than one then the sensing field $H_s(x,y)$ in our previous formula (see Chapter 6) will be different from the sensing field derived from Karlqvist's expressions. The resulting flux ψ through the playback core will thus be modified and the result of the reciprocity calculations in Chapter 6 will likewise be modified.

Westmijze's evaluation was based upon a tape coating with uniform permeability in all directions (isotropic). He concluded that the reproduce flux was reduced at short wavelengths for μ_r. Siakkou arrived at essentially the same result; the mathematical derivation led to a modification of the factor c′ in the formula in Fig. 6-9, namely that c′ should be multiplied with μ_{rt}.

The result thereof is a coating thickness loss that starts at a frequency that is lowered by a factor of μ_{rt}; Fig. 10-27 illustrates the phenomenon. The sensing lines (corresponding to field lines) that sense the magnetization in the back of the coating "think" that the magnetization is further away, i.e., the coating is thicker.

Recent work by Niel Bertram considers a coating with *anisotropic permeability*; μ_{rx} in the direction of tape motion and μ_{ry} perpendicular to it. The numerical values for a well-oriented tape are typically $\mu_{rx} \cong 3$ to 5, and $\mu_{ry} \cong 1$ to 2; their values are determined as the tangent to the hysteresis loop where is crosses the B-axis, at B_{rsat}.

Bertram concluded that the anisotropic permeability reduces the coating thickness loss with 3-5 dB in AC-biased recordings, and about 2 dB in unbiased recording (digital), for a media with $\mu_{ry}/\mu_{rx} = 4$.

He also found that the multiplier 55.4 in the spacing loss formula should be 110, or double the traditional Wallace loss, for small values of the spacing (note slope in Fig. 8-28).

The increased output due to $\sqrt{\mu_{ry}/\mu_{rx}}$ and the increased loss due to $\sqrt{\mu_{rx}\mu_{ry}}$ may offset each other, while they are difficult to observe and separate in actual measurements using modern tapes with high anisotropy and small spacings due to surface finish.

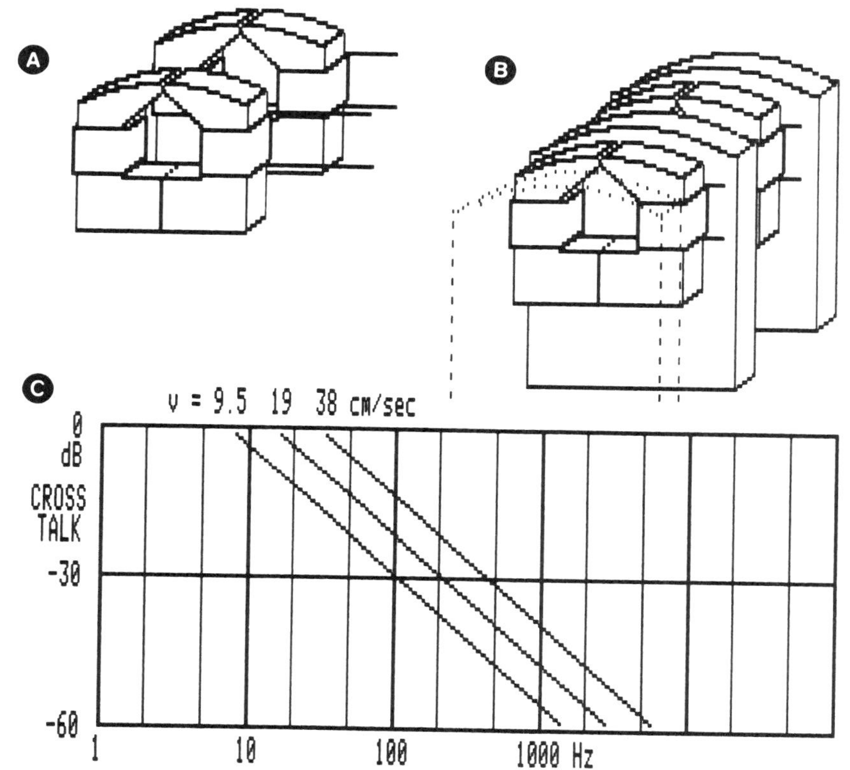

Fig. 10-28. Crosstalk in magnetic heads. A. Transformer coupling between channels via stray flux and capacitive coupling. B. Reduction of crosstalk by shielding. C. Typical crosstalk versus frequency for adjacent tracks in an IRIG head assembly.

CROSSTALK IN MULTITRACK HEADS

Transformer Coupling Between Cores

Several head cores placed close to each other will interfere by transformer and capacitive coupling (Fig. 10-28). This form of crosstalk can be greatly reduced by insertion of shields (MuMetal-Copper-MuMetal) between the cores. It is essential that full shields are used. If they are split (as shown by the broken line), then the gap line between them may not only behave as a secondary recording gap, but the shielding efficiency is lowered. The secondary recording may be on top on an "in-between track" when staggered head assemblies are used.

The level of crosstalk can be calculated from an analysis of the stray flux patterns between the structures shown in Fig. 10-28 (ref. Tanaka).

Thin film heads do not always offer an easy way of inserting shielding; reduction of crosstalk has been accomplished by insertion of feedback from a record track to the disturbed read track.

Crosstalk during recording results in permanent noise on recorded tracks.

Read-while-Write crosstalk is a noise source that can be reduced by individual shielding of the record and playback head assemblies. The placement of a ferrite block immediately in front of the record gaps will reduce this crosstalk.

Side-Reading Effects

During playback of interleaved recordings, the fringing field of the playback gap picks up flux from adjacent channels (in addition to transformer coupling in the head assembly itself). This effect is particularly pronounced at long wavelengths. It is good practice to make the width of the shield between tracks equal (or greater) than the width of the tracks that ride over them.

Figure 10-28 serves as a yardstick for predicting crosstalk. The sloped lines are smooth approximations of measured values of crosstalk in a 31-track system, using direct electronics, for adjacent tracks with a separation of 0.80mm (.030") center-to-center. They point out crosstalk as a problem when low frequencies are recorded and played back at high tape speeds.

Crosstalk can be reduced by using FM record techniques (see Chapter 27), and/or using a slow tape speed. Critical data may also be separated by recording them on tracks that are not located immediately next to each other. The use of AC-bias reduces the required data currents, and therefore also the read-while-write crossfeed (assuming the higher frequency AC-bias signal is filtered away).

THIN FILM HEADS

The potential and applications of thin film heads were first outlined by Lazzari at the Intermag Conference in 1973, following a thorough description of a single turn head by Chynoweth et al. in 1972 (see ref.).

Thin film heads operate much like conventional ringcore heads. The efficiencies are lower due to the low reluctance leakage path between the two metal planes, and this eventually limits the number of turns that is practical to use (ref. Katz, Miura et al.). The magnetic circuit must be carefully designed since it may saturate just behind the gap region (ref. Kelley et al.).

The finite pole lengths have a low-frequency roll-off that can be utilized in combination with the gap's short wavelength roll-off to form a bandpass filter.

The net wavelength response is a spatial equalizer, equivalent in effect to a pulse slimming delay line equalizer. Asymmetry of the pole lengths can be used to compensate the phase shift in the media magnetization (ref. Kakehi et al., Aoi et al., Chi, and Singh et al.).

A flat thin film head construction has potential for multitrack applications with freedom from crosstalk and RFI sensitivity (ref. Springer).

MAGNETORESISTIVE HEADS

MR heads were introduced to the reader in Chapter 7, see Fig. 7-16. Operation of MR heads is well covered in the literature (ref. Kornei, Cole et al., van Gestel et al., Collins et al., and Ruigrok). More recent techniques of implementing the MR head are covered by Simmons et al., Jeffers et al., and O'Connor et al.

These heads have not found widespread use, in spite of their simple construction. One possible reason is the noisy output that results when used in tape heads; a moving tape will contact the head at isolated asperities, and the frictional heat produces noise in the MR elements.

This problem is best resolved by placing the MR element away from the tape contact area, and conduct the tape flux to the element through a suitable magnetic network. This appears to be the technique used in the new IBM half inch tape cartridge drive.

A dual channel head with a magnetic bridge circuit has recently been proposed (ref. Vinal). It is intended for use with a perpendicular media with soft magnetic underlayer, see Fig. 10-29.

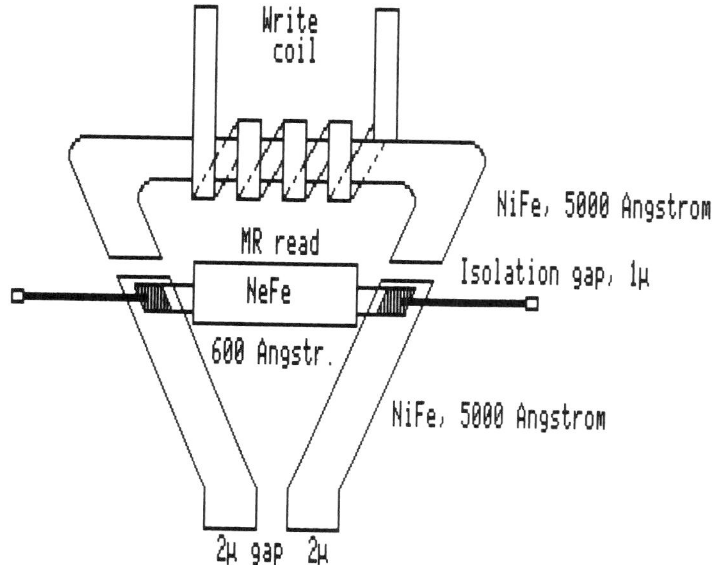

Fig. 10-29. Twin track read-write head with inductive write circuit, and magnetoresistive read element (after Vinal).

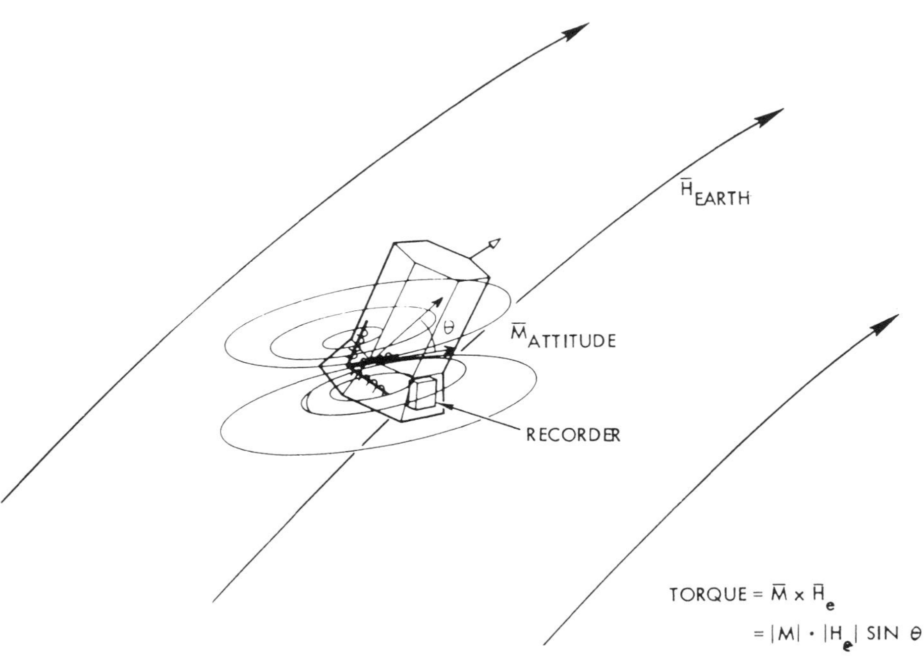

Fig. 10-30. Magnetic attitude correction for spacecraft.

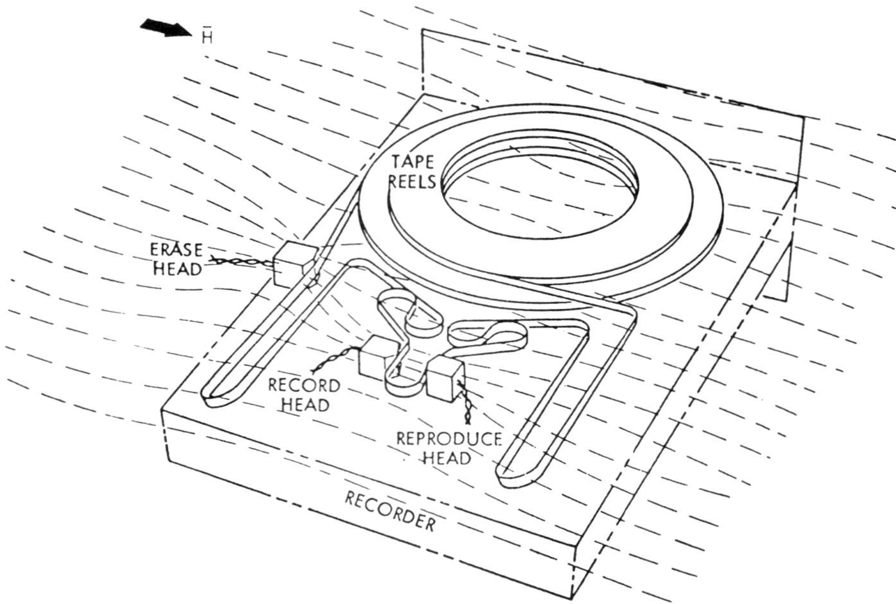

Fig. 10-31. Flux linkage from external field through magnetic tape unit.

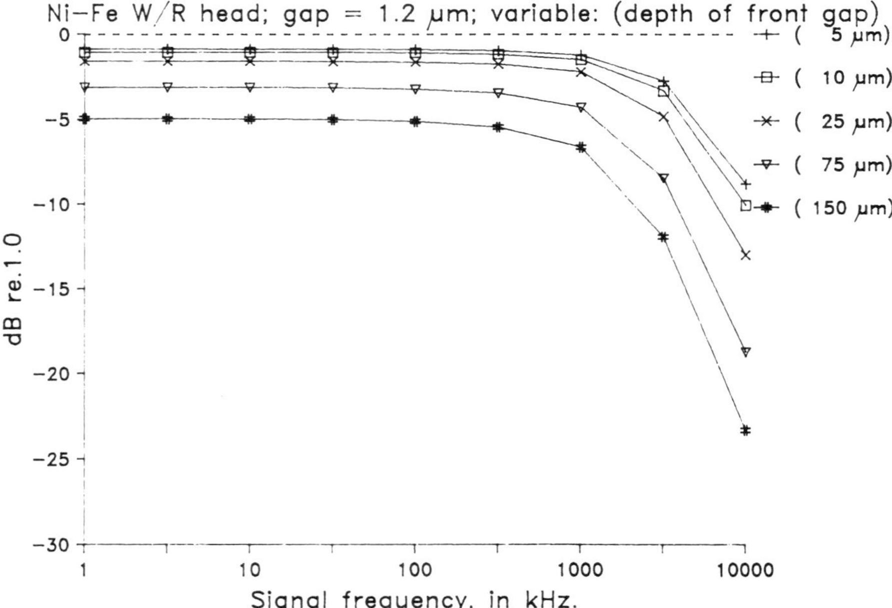

Fig. 10-32. Read efficiency (typical) of magnetic heads increases with decreasing front gap depth (after wear).

ERASE HEADS

The construction and performance of erase heads were covered in Chapter 8 (see Figs. 8-29 through -32). The text also treated the problem in writing new data on top of old data, i.e., supposedly erasing the old data with the field from the new write process.

Just how well this is done is measured by the *over-write* ratio, which is defined as the ratio of an original 1f (or half-band-edge signal) to that remaining after being overwritten by a 2f (band-edge signal. Note: FM digital recording is a form of frequency shift coding, using two frequencies spaced an octave apart; hence one signal is of frequency f, or 1f, and the other of double the frequency, or 2f). Over-write of 30 dB is considered adequate for proper performance (ref. Wachenschwanz et al.).

Measurements of over-write require knowledge of the original read level of the 1f signal. A better method appears to be the *Write-over* method that specifies the ratio of residual 1f signal to new 2f signal (ref. Ed Williams, Read-Write Corp.). A recommended method is to measure the residual spectral noise level to the 1f or 2f levels, in conjunction with the signal's *resolution*, i.e., the ratio of the 2f read voltage to the 1f read voltage; it should be no more than 80 percent (2 dB).

Core Losses

Two losses account for the power that is required to drive an erase head (and similarly, by smaller amounts, write heads for high coercivity media). These are eddy currents and *hysteresis* losses; the total power loss is:

$$\begin{aligned} P_{total} &= P_{eddy} + P_{hysteresis} \\ &= kf^2 + W_h f \text{ watts/m}^3 \end{aligned} \qquad \textbf{(10.12)}$$

where

W_h = area of hysteresis loop in watt*sec/m^3
k = material constant
f = frequency in Hz.

The eddy currents, kf^2, can also be written:

$$P_{eddy} = 10^{-4}\pi^2 d^2 B_o^2 f^2/6\,\sigma \text{ watts/m}^3 \qquad \textbf{(10.13)}$$

where

d = lamination thickness in cm
B_o = flux density amplitude in Gauss
f = frequency in Hz

σ = resistivity in $\mu\Omega$cm

Example 10.3: A MuMetal erase core operates at 50 kHz at a flux density level of 5000 Gauss. The core is made from 2 mils lamination; the total core volume is 1 cm^3. Find P_{total}.

Answer: From Table 7.1 we find the resistivity s = 60 $\mu\Omega$cm and H_c = 0.02 Oe = 1.6 A/m. The flux density level is 5000 Gauss = 0.5 Wb/m^2, so the total area of the rectangular BH-loop is W_h = 1.6*0.5*4 = 3.2 Ws/m^3. The lamination thickness is d = 0.0051 cm. The eddy current loss calculated to:

$$P_{eddy} = 4.3 * 10^6 \text{ watts/m}^3 * 10^{-6}\text{m}^3$$
$$= 4.3 \text{ watts}$$

and the hysteresis loss is:

$$P_{hysteresis} = 3.2 * 5 * 10^4 * 10^{-6}$$
$$= 1.6 \text{ watts.}$$

The total power loss in the MuMetal erase core is 5.9 watts.

The losses can be reduced significantly by using a NiZn ferrite core.

The head core losses cause a temperature rise in the core, and the material should therefore be connected to a heat sink through a heat conducting epoxy (or direct contact to metal parts, if possible). Ferrite materials must also have a reasonably high Curie temperature, and the design must not have flux levels that exceed the B_{sat} levels.

The heating of erase heads caused designers of early audio equipment to incorporate an interlock so the heads were inoperative unless the tape was moving and thus cooling the heads!

An analysis as shown in the example above should be carried out for most write heads that work at high write flux levels (ref. Monson).

A final note on the circuitry to drive erase heads: The current waveform must be absolutely symmetrical, i.e., free from even harmonic signals and DC-currents. Otherwise the result is erased tapes with high residual noise levels. Push-pull circuits are recommended. The basic oscillator may operate with the erase head in a tank circuit in its oscillator, while a frequency doubler delivers a synchronous bias signal to the record head bias drivers.

It is not advisable to use separate erase and bias oscillators without some form of synchronization; otherwise the result may be a set of recorded beat signals between the two oscillators.

RADIO FREQUENCY INTERFERENCE

Radiation from Write Heads

The write current signals into write heads contain VHF signals that can radiate from the heads; they are dipole antennas, although so small that their efficiency as antennas is very poor. If the radiation levels are troublesome shielding of the heads may become necessary (see Fig. 8-38).

Shielding of radio frequencies can be successful if an aluminum housing is used around the head structure, and by using a head cable that is a two-conductor over-all shield, properly grounded (i.e., ground the shield in one or both ends, see Ott's book for details).

A multi channel recorder with AC-bias can successfully suppress radiation by alternating the phase of the bias signals to a row of write cores.

Susceptibility to External DC-Fields

Shielding of heads for attenuation of outside fields may also be necessary. The earth field itself has caused many problems by linking through the write head and appearing as an unwanted DC-level mixed in with the write field. The results were (and are) noisy and distorted recordings.

A particularly grave situation may occur when heads are demagnetized with an AC-degausser; the result has often been a permed (DC magnetized) head, with subsequent noise and distortion problems.

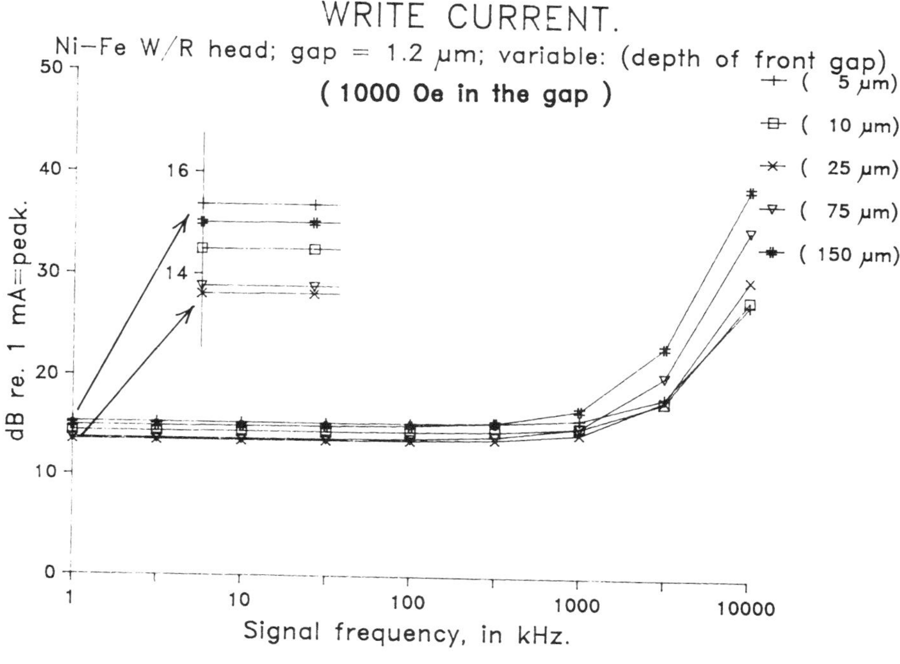

Fig. 10-33. Change in write current requirement after head wear (not necessarily typical).

Another example is a spacecraft recorder placed in the vicinity of three orthogonal (perdendicular to one another) electromagnets that has been proposed for spacecraft attitude control. Applying power to one or more of the magnets will bring about a torque, in relation to the Earth's field, and thus slowly turn the spacecraft. The power is generated from solar panels and therefore avoids using ordinary propellant fuel for attitude boosters.

Figure 10-30 shows the concept, and Fig. 10-31 illustrates how an external field can link through a recorder and be collected by the heads, with the ill results reported previously. This was modelled in a large coil (1 meter diameter), and the results revealed the following field strength limits versus performance impairment (ref. Jorgensen):

- H > 5 Oersteds: Write function affected.
- H > 25 Oersteds: Loss of write function.
- H > 30 Oersteds: Loss of read while write.
- H > 35 Oersteds: Total failure, any mode.

The 5 Oersted limitation is the same value as determined by Lopez and Stromsta (see ref.). This presents a major design problem in shielding of the field from magnets in permanent magnet position motors in disk drives. The actuator may have a voice coil motor with a field of several thousand Oe, located one to two inches away from the disk circumference; the field from the motor must at that point be less than 5 Oersteds = 400 A/m.

CHANGES IN HEAD PARAMETERS WITH WEAR

The front gap reluctance in heads increases when the heads wear. The read efficiency increases with wear, in particular at high frequencies. This is an interesting situation where an item gets better as it wears out. An operator should be alert to the subtle improvement in a recorder's amplitude versus frequency response; it may be just before the head wears out and the performance will nose dive. See Figs. 10-32, 10-33.

A read head will always increase its output as it wears, while a write head generally requires less drive current as it wears out. An earlier edition of this book showed some dramatic curves for these effects (Figs. 7-56 and 7-57, in that edition). The author has since improved the head modelling and found that no general rule applies for write heads; each head must be analyzed apart from others.

REFERENCES TO CHAPTER 10

General

Kornei, O., "Structure and Performance of Magnetic Transducer Heads," *JAES*, July 1953, Vol. 1, No. 3.

Staff, "Design Digest for Applying Magnetic Heads in Digital Recording," *Nortronics, Bulletin 9224*, 1976, 24 pages.

Collier, D., Frank, P.D., and Aho, C.J., "Rigid Disk Heads Keep Pace with Growing Storage Needs," *Mini-Micro Systems*, Dec. 1984, Vol. 18, No. 12, pp. 127-133.

Jorgensen, F., "Head/Media Design," design programs, 1986, *Danvik*, P.O.Box 30791, Santa Barbara, CA 93130, USA.

Zinke, O., and Seither, H., *Widerstaende, Kondensatoren, Spulen und ihre Werkstoffe*, Springer Verlag, 1982, 350 pages.

Grossner, N.R., *Transformers for Electronic Circuits*, McGraw-Hill, 1983, 467 pages.

Select Track Width

Greenberg, H.J., Stephens, R.L., and Talke, F.E., "Dimensional Stability of Floppy Disks," *IEEE Trans. Magn.*, Nov. 1977, Vol. MAG-13, No. 6, pp. 1397-1399.

Brock, G.W., "Instability of Flexible Magnetic Recording Media," *Symposium on Magn. Media Manuf., MMIS*, May 1983, paper No. A-2, 22 pages.

Staff, "Flexible 5¼ inch Drive, OEM Manual," *Data Technology Corp. (Eastman Kodak Co.)*, 1984, 28 pages.

Edelman, H., and Covault, M., "Design of Magnetic Recording Heads for High Track Densities," *IEEE Trans. Magn.*, Sept. 1985, Vol. MAG-21, No. 6, pp. 2583-2587.

Efficiency, Write Heads

McKnight, J., "Magnetic Design of Tape Recorder Heads," *JAES*, March 1979, Vol. 27, No. 3, pp. 106-120.

Sansom, D., "Recording Head Design Calculations," *IEEE Trans. Magn.*, Sept. 1976, Vol. MAG-12, No. 3, pp. 230-233.

Wood, R., Lindholm, D., and Haag, R., "On the Bandwidth of Magnetic Record/Reproduce Heads," *IEEE Trans. Magn.*, Sept. 1985, Vol. MAG-21, No. 5, pp. 1566-1568.

Visser, E.G., Van Rijn, L.R.M., and Maas, J.J.F., "An Improved Measurement of the Absolute Efficiency of Magnetic Heads by Saturating the Gap Field," *IEEE Trans. Magn.*, July 1985, Vol. MAG-21, No. 4, pp. 1283-1288.

Katz, E., "Numerical Analysis of Ferrite Recording Heads with Complex Permeability," *IEEE Trans. Magn.*, Nov. 1980, Vol. MAG-16, No. 6, pp. 1404-1409.

Head Core Noise

Smaller, P., "Reproduce System Noise in Wide-Band Magnetic Recording Systems," *IEEE Trans. Magn.*, Dec. 1965, Vol. MAG-1, No. 4, pp. 357-363.

Bittel, H., "Noise of Ferromagnetic Materials," *IEEE Trans. Magn.*, Sept. 1969, Vol. MAG-5, No. 3, pp. 359-365.

Byers, R.A., "Theoretical Signal-to-Noise Ratio for Magnetic Tape Heads," *IEEE Trans. Magn.*, June 1971, Vol. MAG-7, No. 2, pp. 254-259.

Watanabe, H., "Noise Analysis of Ferrite Head in Audio Tape Recording," *IEEE Trans. Magn.*, Sept. 1974, Vol. MAG-10, No. 3, pp. 903-906.

Pole Tip Saturation; Distortion

Suzuki, T., and Iwasaki, S., "An Analysis of Magnetic Recording Head Fields Using Vector Potential," *IEEE Trans. Magn.*, Sept. 1972, Vol. MAG-8, No. 3, pp. 536-537.

Shibaya, H., and Fukuda, I., "The Effect of the Bs of Recording Head Cores on the Magnetization of High Coercivity Media," *IEEE Trans. Magn.*, Sept. 1977, Vol. MAG-13, No. 3, pp. 1005-1008.

Szczech, T.J., Wollack, E.F., and Richards, D.B., "A Technique for Measuring Pole Tip Saturation of Low Inductance Heads," *IEEE Trans. Magn.*, July 1978, Vol. MAG-14, No. 4, pp. 197-200.

Thornley, R.F.M., and Bertram, H.N., "The Effect of Pole Tip Saturation on the Performance of a Recording Head," *IEEE Trans. Magn.*, Sept. 1978, Vol. MAG-14, No. 5, pp. 430-432.

Fujiwara, T., "Record Head Saturation in AC Bias Recording," *IEEE Trans. Magn.*, May 1979, Vol. MAG-15, No. 5, pp. 1046-1049.

Fujiwara, T., "Distortion Fluctuation Phenomena in Audio Magnetic Heads," *IEEE Trans. Magn.*, Jan. 1980, Vol. MAG-16, No. 1, pp. 111-114.

Jeffers, F., McClure, R.J., French, W.W., and Griffith, N.J., "Metal-In-Gap Record Head," *IEEE Trans. Magn.*, Nov. 1982, Vol. MAG-18, No. 6, pp. 1146-1148.

Valstyn, E.P., and Packard, E., "Optimization of Ferrite Heads for Thin Media," *IEEE Trans. Magn.*, Sept. 1986, Vol. MAG-22, No. 5, pp. 847-849.

Metallic Gap Materials

McKnight, J.G., "How the Magnetic Characteristics of a Magnetic Tape Head are Affected by Gap Length and a Conductive Spacer," *Jour. AES*, Mar. 1979, Vol. 27, No. 3, pp. 106-120.

Ruigrok, J.J.M., "Analysis of Metal-In-Gap Heads," *IEEE Trans. Magn.*, Sept. 1984, Vol. MAG-20, No. 5, pp. 872-874.

Effects of Core Geometry

Eldridge, D.F., and Baaba, A., "The Effect of Track Width in Magnetic Recording," *IRE Trans. Audio*, Feb. 1962, Vol. AU-9, No. 1, pp. 10-15.

Geurst, J.A., "Theoretical Analysis of the Influence of Track Width on the Harmonic Response of Magnetic Reproducing Heads," *Philips Res. Repts.*, 1965, Vol. 20, pp. 633-657.

Fritzsch, K., "Long-Wavelength Response of Magnetic Reproducing Heads," *IEEE Trans. Audio and Electroacoustics*, Dec. 1968, Vol. AU-16, No. 4, pp. 486-494.

Baker, B., "Long Wavelength Response of Shielded Elliptical Reproduce Heads," *IEEE Trans. Magn.*, Sept. 1977, Vol. MAG-13, No. 3, pp. 1009-1012.

Lindholm, D., "Secondary Gap Effect in Narrow and Wide Track Reproduce Heads," *IEEE Trans. Magn.*, Sept. 1980, Vol. MAG-16, No. 5, pp. 893-895.

Luitjens, S. and Van Herk, A., "A Discussion on the Crosstalk in Longitudinal and Perpendicular Recording," *IEEE Trans. Magn.*, Nov. 1982, Vol. MAG-18, No. 6, pp. 1804-1812.

Effects of Coating Permeability

Siakkou, M., "Influence of Coating Permeability in Thin-Film Pulse Recording," *IEEE Trans. Magn.*, Dec. 1969, Vol. MAG-5, No. 4, pp. 891-895.

Siakkou, M., "Playback of Magnetic Recordings with a Reproduce head with Finite Gap Length and Tapes of Various Permeabilities," (in German), *Zeitschrift elektr. Inform. und Energietichnik*, 1974, Vol. 4, pp. 311-316.

Bertram, H., "Anisotropic Reversible Permeability Effects in the Magnetic Reproduce Process," *IEEE Trans. Magn.*, May 1978, Vol. MAG-14, No. 3, pp. 111-118.

Multitrack Heads

Melis, J., and Nijholt, B., "A Comparison of Measured and Calculated Fringing Response of Multitrack Magnetic Reproducers," *Jour. AES*, Vol. 26, No. 4, April 1978, pp. 212-216.

Tanaka, K., "Some Considerations on Crosstalk in Multihead Magnetic Digital Recording," *IEEE Trans. Magn.*, Jan. 1984, Vol. MAG-20, No. 1, pp. 160-165.

Thin Film Heads

Chynoweth, W., Jordan, J., and Kayser, W., "PEDRO - A Transducer-Per-Track Recording System with Batch-Fabricated Magnetic Film Read/Write Transducers," *Honeywell Computer Journal*, April 1972, pp. 104-117.

Lazzari, J.P., "Integrated Magnetic Recording Heads Applications," *IEEE Trans. Magn.*, Sept. 1973, Vol. MAG-9, No. 3, pp. 322-326.

Katz, E., "Finite Element Analysis of the Vertical Multi-Turn Thin-Film Head," *IEEE Trans. Magn.*, Sept. 1978, Vol. MAG-14, No. 5, pp. 506-508.

Miura, Y., Kawakami, S., and Sakai, S., "An Analysis of the Write Performance on Thin Film Head," *IEEE Trans. Magn.*, Sept. 1978, Vol. MAG-14, No. 5, pp. 512-514.

Vinton Kelley, G., and Valstyn, E.P., "Numerical Analysis of Writing and Reading with Multiturn Film Heads," *IEEE Trans. Magn.*, Sept. 1980, Vol. MAG-16, No. 5, pp. 788-790.

Kakehi, A., Oshiki, M., Aikawa, T., Sasaki, M., and Kozai, T., "A Thin Film Head for High Density Recording," *IEEE Trans. Magn.*, Nov. 1982, Vol. MAG-18, No. 6, pp. 1131-1133.

Aoi, H., Saitoh, M., Tamura, T., Ohura, M., Tsuchiga H., and Hayashi, M., "Pole-Tip Design of High Density Recording Thin Film Heads," *IEEE Trans. Magn.*, Nov. 1982, Vol. MAG-18, No. 6, pp. 1137-1139.

Springer, G., and Jorgensen, F., "A Novel Magnetic Transducer for Magnetographic Printing," *IEEE Trans. Magn.*, Sept. 1985, Vol. MAG-21, No. 5, pp. 1548-1550.

Chi, C.S., "A Thin Film Head Design Technique to Restore Read Pulse Asymmetry," *IEEE Trans. Magn.*, Sept. 1985, Vol. MAG-21, No. 5, pp. 1569-1571.

Singh, A., and Bischoff, P., "Optimization of Thin Film Heads for Resolution, Peak Shift and Overwrite," *IEEE Trans. Magn.*, Sept. 1985, Vol. MAG-21, No. 5, pp. 1572-1574.

Flux Sensitive Read Heads

Kornei, O., "Survey of Flux-Responsive Magnetic Reproducing Heads," *JAES*, July 1954, Vol. 2, pp. 145-150.

Cole, R.W., Potter, R.I., Lin, C.C., Deckert, K.L., and Valstyn, E.P., "Numerical Analysis of the Shielded Magnetoresistive Head," *IBM Jour. Res. Develop.*, Nov. 1974, Vol. 18, pp. 551-555.

van Gestel, W.J., Gorter, F.W., and Kuijk, K.E., "Read-out of a Magnetic Tape by Magnetoresistance Effect," *Philips Tech. Rev.*, March 1977, Vol. 37, No. 2/3, pp. 42-50.

Collins, A.J., and Jones, R.M., "Review of Un-shielded Magnetoresistive Heads," *Intl. Conf. Video and Data 1979*, IERE Publ. No. 43, pp. 1-17.

Ruigrok, J.J.M., "Analytical Description of Thin-Film Yoke Magnetoresistive Heads," *Philips Jour. Res.*, 1981, Vol. 36, Nos. 4-5-6, pp. 289-310.

Simmons, R., Jackson, B., Covault, M., Wacken, C., and Rausch, J., "Design and Peak Shift Characterization of a Magnetoresistive Head Thin Film Media System," *IEEE Trans. Magn.*, Sept. 1983, Vol. MAG-19, No. 5, pp. 1737-1739.

Vinal, A.W., "Considerations for Applying Solid State Sensors to High Density Magnetic Disk Recording," *IEEE Trans. Magn.*, Sept. 1984, Vol. MAG-20, No. 5, pp. 681-686.

Jeffers, F., Freeman, J., Toussaint, R., Smith, N., Wachenschwanz, D., Shtrikman, S., and Woyle, D., "Soft-Adjacent-Layer Self-Biased Magnetorestive Heads in High Density Recording," *IEEE Trans. Magn.*, Sept. 1985, Vol. MAG-21, No. 5, pp. 1563-1565.

O'Connor, F.B. Shelledy, and Heim, D.E., "Mathematical Model of a Magnetostrictive Read Head for a Magnetic Tape Drive," *IEEE Trans. Magn.*, Sept. 1985, Vol. MAG-21, No. 5, pp. 1560-1562.

Erase Heads

Monson, J.E., Ash, K.P., Jones Jr., R.E., and Heim, D.E., "Self-Heating Effects in Thin-Film Heads," *IEEE Trans. Magn.*, Sept. 1984, Vol. MAG-20, No. 5, pp. 845-847.

Wachenschwanz, D., and Jeffers, F., "Overwrite as a Function of Record Gap Length," *IEEE Trans. Magn.*, Sept. 1985, Vol. MAG-21, No. 5, pp. 1380-1382.

Radio Frequency Interference

Jorgensen, F., "The Influence of an Ambient Magnetic Field on Magnetic Tape Recorders," *Intl. Telemetry Conf. Proc. 1975*, Vol. 11, pp. 372-390.

Lopez, O., and Stromsta, R., "Effects of Magnetic Fields on Flexible Disk Drive Performance," *IEEE Trans. Magn.*, July 1981, Vol. MAG-17, No. 4, pp. 1417-1422.

Part 4
Magnetic Tapes and Disks

Chapter 11
Tape and Disk Materials

The ultimate quality of a recording system is determined by the *recording media*. This is a most intriguing chemical product, in several forms and shapes, and one which is evaluated by its electromagnetic properties as well as mechanical uniformity, strength, and tolerances.

Valdemar Poulsen used piano wire for his first Telegraphon, but it was only a few years later when his associate, P. O. Pedersen, filed a patent on a metal ribbon with an electrochemically deposited magnetic coating. The later Stille recorder used a ¼″ wide steel tape. Experiments with magnetic particles coated onto a paper base, and with homogeneous, cast plastic tapes were tried in Germany in the late twenties.

The first coated plastic magnetic tape was conceived only 50 years ago by Fritz Pfleumer in Germany. It was fabricated by BASF in 1934 and it used a black iron oxide Fe_3O_4 in a binder coated onto a cellophane base. The Fe_3O_4 is an iron ferrite that has good magnetic properties for recording but with severe drawbacks in the form of instability and susceptibility to print-through. The latter is a copy effect whereby recordings from adjacent tape layers imprint on the layer between them (see Chapter 14).

The brown *iron oxide* γFe_2O_3 was soon introduced, and much refined remains the workhorse of the recording and storage industry today.

The *base film* was changed from paper to PVC (Polyvinylchloride) in 1944, and later to today's PET (polyethylene terephthalate), or polyester, or Mylar (tradename of duPont). We have also seen new particles emerge: chromium dioxide (CrO_2), cobalt doped or coated iron oxides, and recently iron powders (high energy tapes) and barium ferrite particles (perpendicular magnetization media). There are also particles with very high coercivities for credit cards and the like, where erasure resistant recordings are wanted.

Another recording material exists in plated disks. Platings on flexible films were tried but they were not successful. They were incompatible with in-contact recordings, where the metal had to be in moving contact with metal interfaced heads, and they were further very sensitive to mechanical damages (nicks, creases). Corrosion was also a problem. Efforts do continue to make it a viable product.

This chapter will first survey and discuss materials for both particulate and plated media. The magnetic properties, their measurements, demagnetization and noise are discussed in Chapter 12. Chapter 13 consists of a description of the manufacture of tapes as well as disks, while the performances of the end products are covered in Chapter 14.

PARTICULATE MEDIA COMPONENTS

It is customary in the media industry (tapes and disks) to classify the magnetic powders by their intrinsic properties, such as the *specific magnetization* σ_s (in EMU/cc), and the mean value H_c of intrinsic coercivities h_{ci}. The unit EMU is equal to 4π Maxwell-centimeter (or $4\pi*10^{-10}$ Weber-meter).

It should be noted that σ_s, which corresponds to the remanence, or retentivity, of the final coating is insensitive to the particles internal structure, while h_{ci} is structure sensitive.

Magnetization of a Powder Magnet

The powder pack (the media coating) will have a smaller saturation magnetization J_{sat} due to a volume occupancy, or *packing fraction p*, that is less than one, typically in the range of 0.35 to 0.40. And the coercivity H_c depends largely on the particle sizes and shapes, as we shall see. Both J_{sat} and H_c are, therefore, extrinsic properties of the coating. The intrinsic value σ_s is the absolute *maximum* magnetization inside a solid sample of the material (and therefore the magnetization in each domain, or single domain particle).

The level of magnetization in a *saturated, non-oriented* tape coating is:

$$\begin{aligned} B_{rsat} &= J_{rsat} \\ &= 0.5 * J_{sat} \\ &= 0.5 * p * \sigma_s \end{aligned}$$

where p is the packing fraction of particles in the coating ($0.0 < p < 1.0$). The theoretical maximum of a packing fraction of one is of course unrealistic; its maximum value is about 0.4. A powder of cubic particles has a packing fraction of approximately 0.17 when loosely packed, and 0.25 when compacted. It can increase from 0.35 to 0.40 when compressed under high pressure, and this is also the maximum for a tape coating. Higher values result in a weak coating, leading to oxide shedding, debris formation on heads and guides, and subsequent signal drop-outs.

The value of B_{rsat} can be increased by *orienting* the particles in the direction of the anticipated magnetization, in the past always longitudinal for audio, instrumentation and computer tapes, and transverse for two inch broadcast video tapes. The measure of orientation is the *squareness* S, which is defined as the ratio J_{rsat}/J_{sat}, both measured on the JH-loop. S ranges from 0.5 to a theoretical maximum of one.

Replacing the factor 0.5 with S, and converting to Wb/m^2 results in:

$$\begin{aligned} B_{rsat} &= S * p * \sigma_s \text{ EMU/cc} \\ &= S * p * \sigma_s * 4\pi 10^{-4} \text{ Wb/m}^2 \end{aligned}$$

It has become common practice to quote the *flux level* from a coating in nWb/m, and the end user can then multiply this figure with the recorded trackwidth in meters to obtain the flux from that track. Thus we arrive at the following formula (with mixed units for practical reasons):

$$\psi_{rsat} = 0.4\pi * S * p * \sigma_s * c \quad \text{nWb/m} \tag{11.1}$$

where:

$S = J_{rsat}/J_{sat}$ = the squareness
p = packing fraction (~0.35 – 0.45)
σ_s = specific magnetization in EMU/cc
c = coating thickness in μm

A *high output* media obtains its high flux level by maximizing the four items in formula (11.1). This technique has long prevailed in the audio and instrumentation field, where a very high signal-to-noise ratio is necessary (see Chapters 27 and 28).

Media for *data storage* aim at very high packing densities of bits along a track. We will examine which magnetic properties will enhance this.

A thin coating is desired when pulses of opposite polarities are recorded. Between bits there exist transition zones, and it is shown in Chapter 19 that the length Δx of this zone is inversely proportional to the coating thickness (for same levels of overwrite).

Demagnetization in all recordings is decreased when a thin coating is used, or when the recording is made only partially into the coating. The latter principle is used in wideband instrumentation recorders, used for analog data as well as in high density digital recording, HDDR (see Chapter 27). This does have the same effect as that of decreasing the coating thickness (less demagnetization). No mechanical difference can be observed between a thick and a thin coating except for the fact that a thin coat tape is more pliable, provides better head-to-tape contact, and affords longer playing time.

An increase in coercive force H_c will reduce the effect of demagnetization. The BH-loops for two coatings are shown in Fig. 11-1. They differ only in coercive force, and the advantage

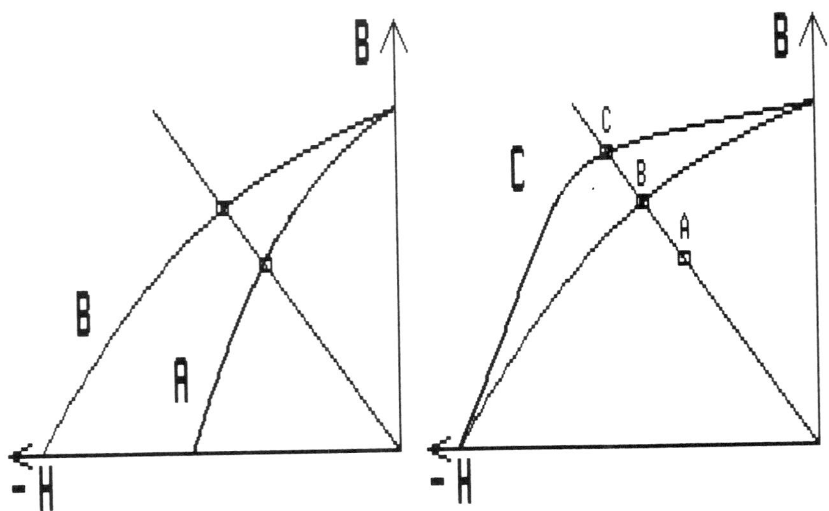

Fig. 11-1. Maximum remanence in magnetic tapes, (A). Low coercivity, (B). High coercivity, (C). High coercivity, good orientation. Demagnetization is high in A, small in B, non-existing in C (for the slope of 1/N shown).

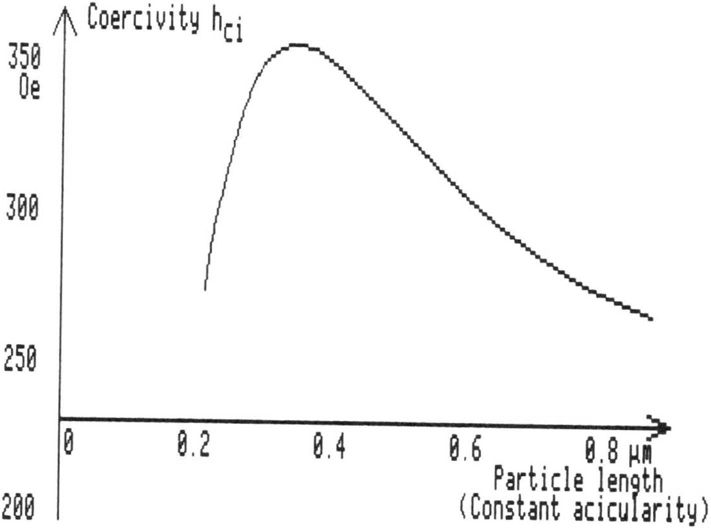

Fig. 11-2. Coercivity of Gamma ferric oxide particles. (After E.D. Daniel).

of the higher value of H_c is clearly evident by the higher values of both open-circuit remanence and short-circuit remanence (during contact with a high permeability reproduce head).

Final improvements can be achieved by orientation of the particles, which results in a more square loop (Fig. 11-1C). The analysis of pulse recording in Chapter 19 will further show that the steep slope of the sides of the hysteresis loop will enhance short transition zones; the slope corresponds to the switching field distribution. They are made steep not only by orientation of the particles, but also by using a magnetic powder material that has a narrow distribution range of single particle coercivities. The latter is also beneficial in anhysteretic ac-bias recordings.

The mean coercive force H_c of an assembly of particles depends upon the material as well as the particle size. Large particles will have smaller shape anisotropy and may each consist of more than one domain. The net result is low coercivity, as shown in Fig. 11-2.

On the contrary, very small, and in particular, nonacicular particles lead to non-magnetic behavior, also called superparamagnetic, below a critical size that happens to coincide closely with the thickness of the walls in that material.

The coating's coercivity is, in other words, controlled by the particle sizes. As the size is reduced, it is typically found that the coercivity increases, goes through a maximum, and then tends to go toward zero. The behavior changes simultaneously from multidomain to single-domain, when the particle size is below a critical diameter, which is not well defined.

Predictions of the critical sizes (and volumes and shapes) for single-domain and *superparamagnetic* transitions are difficult to make with any accuracy as the theory is not well enough understood.

Our knowledge of the single particle magnetization is limited, and gets worse when we put several of them together. The added complexity is due to interaction fields between the particles. The idealized view that a particle is a perfect ellipsoid in shape and consists of a single domain is shattered by the observation that they actually are of irregular shape, polycrystalline and with cubic anisotropy.

Further, a high proportion of particles are found to be multiple, that is they comprise several particles lying side by side; most particles in recording tapes are of this type (ref. Knowles).

The single particle coercivity can be estimated. For a needle shaped particle with predominant shape anisotropy and with a ratio of length to diameter greater than 10 we have:

$$h_{ci} = (N_d - N_l) * M_{sat} \quad \textbf{(11.2)}$$

where

N_d = demagnetization factor along the short axis d
N_l = demagnetization factor along the long axis l
M_{sat} = J_{sat}/μ_o

A spherical particle would thus have $H_c = 0$ since $N_d = N_l$ for l/d = one. Then the particle coercivity is controlled by its crystalline anisotropy. Such particles have been developed by Pathe in France and used in coatings named ''Isomax'' (Tradename of Spin Physics, Eastman Kodak Co.). The final coercivity is controlled by a cobalt treatment which brings about an equiaxis anisotropy, i.e., a coating made from these particles will magnetize equally well in the longitudinal and perpendicular mode. The commercial use of Isomax has been hindered by the classic cobalt problems: excessive sensitivity to temperature and stress.

Calculations of the total coating coercivity, based upon a single particle's coercivity, is also difficult because of *particle interactions*, as was shown in Fig. 3-25. It has been suggested that the coercivity is best represented by:

$$H_c = (h_{ci} - H_i)^2/h_{ci} \quad \textbf{(11.3)}$$

where h_{ci} is the coercivity of an individual particle and H_i is the interaction field applied to a given particle by its neighbors.

When *crystal anisotropy* prevails, then the coercivity is expected to be independent of the packing fraction p. For particles with *shape anisotropy*, the inverse is true and the coercivity H_c decreases with increasing p:

$$H_c = h_{ci} * (1 - p) \quad \textbf{(11.4)}$$

When p = 1, all particles are everywhere in contact; shape anisotropy is lost and the coercivity becomes zero—if other forms of anistropy are absent.

The ultimate behavior of coatings must be determined by extensive experimental measurements, while work continues on the theoretical aspects.

It was mentioned earlier that particles should have a narrow range of coercivities. Noting Fig. 11-2, this means a narrow range of particle sizes and shapes. One type of *distribution* that is found in practice is shown in Fig. 11-3 (ref. Daniel). This curve indicates that the particle volume follows a logarithmic distribution when plotted, not in terms of numbers of particles, but in terms of mass of particles.

Many of the conclusions we have arrived at concerning recording performance have been verified. Pulse width has been found to be proportional to H_c^{-x}, where x is between ½ and 1; the remanence had no influence on the pulse width (Ref. Bate).

The signal amplitude was, in three investigations, found to be proportional to $(H_c J_{sat})^{0.5}$, $H_c^{0.15} * J_{sat}^{0.85} * c^{0.15}$, or plainly $J_{sat}^{1.0}$ (ref. Bate et al.).

The pulse playback phenomenon called peak-shift (see Chapter 21) could not be traced to any particular properties of the magnetic media. This problem is complex and involves the position and phase of the ''in-depth of recording''.

We can, in summary, set goals for selection of a magnetic particle material:

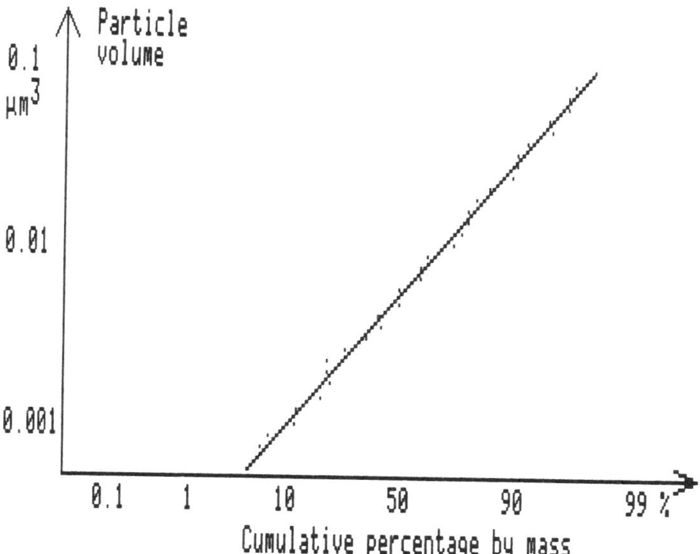

Fig. 11-3. Size distribution of ferric oxide particles. The cumulative percentage, by mass, of particles with less than a given volume is plotted on a probability scale versus particle volume on a logarithmic scale. The data was obtained from examination of electron photomicrographs and confirmed by sedimentation tests (after E.D. Daniel).

- The *coercivity shall be high*, consistent with compatibility and record (write) capabilities. A high coercivity provides maximum utilization of the material's magnetization and a high packing density. It also reduces the effects of demagnetization of recorded signals.
- The *specific magnetization shall be high* in order to achieve a high playback (read) signal.
- The *individual particles shall be small.* This leads to increased coercivity and reduced noise.
- The *particles shall, when feasible, be oriented*. This increases the overall signal output and improves resolution (steeper sides of the hysteresis loop).

The matter of selecting a *coating thickness* will be a trade-off decision. Audio tapes for studio use will employ a fairly thick coating (10 μm = 400 μin) and offset the short wavelength losses from demagnetization by running the tape at a higher than normally required speed (76 or 38 cm/sec (30 or 15 IPS) instead of 19 cm/sec (7½ IPS)).

In high density digital storage recording, the signal-to-noise ratio of a thick coating can easily be relaxed, and a thin coating is chosen for its better resolution. As a matter of fact, 75 percent of the output from a recording is contributed by a coating layer of thickness 0.22 $* \lambda$, or 0.44 $*$ BL (from Chapter 6)—and the coating should be no thicker.

OXIDE PARTICLES

Several oxide materials used for tape production are listed in Table 11-1. We will briefly discuss each of them.

Ferric Oxide, Fe_3O_4

This was the particle used earliest in magnetic tape production, in the thirties. It has a specific magnetization that essentially is the same as the other iron oxide materials, but the

Table 11-1. Recording Media Materials and Their Properties.

	Particle Length μm	Aspect Ratio	Magnetization Wb/m^2	EMU/cc	Coercivity kA/m	Oe	Curie temp. T_r	Spec. Surf. Area m^2/gr	SFD Δh_r	Comments:
γFe_2O_3	0.2 - 0.5	5:1 - 10:1	0.44	350	22 - 34	270 - 420	600	15 - 30	.20 - .30	Easily dispersed
Co-γFe_2O_3	0.2 - 0.5	6:1	0.48	380	32 - 52	400 - 650	520	25 - 30	.25 - .45	(epitaxial, not doped)
MP	0.1 - 0.5	3:1 - 10:1			28 - 56	350 - 700		15 - 35		New
CrO_2	0.1 - 0.5	10:1 - 20:1	0.50	400	36 - 56	450 - 700	120	10 - 20	.25 - .30	
Ba-ferrite	0.05-0.1 across, .02 thick		0.40	320	56 - 240	700 -3000	350	20 - 25	.10 - .30	New; very promissing
Fe	0.1 - 0.5	5:1 - 10:1	1.4	1100	56 - 176	700 -2200	770	20 - 60	.40 - .60	Difficult processing
Fe-Co		5:1			86	1,075	1,000			
SmCo5			2.5	2000	800	10,000			.50 - .90	
Co-Ni	film	n/a	1.5	1200	24 - 112	300 -1400				
CoP	film	n/a	0.82	650	24 - 80	300 -1000				

low value of H_c =90 Oe lowered the useable remanences to about 25 percent ot the saturation remanence (Fig. 11-1A). The tapes suffered from aftereffects (loss of level, in particular at short wavelengths) and heavy print through, and the material was soon superseded by gamma ferric oxide.

Gamme Ferric Oxide γFe_2O_3

This has been and is currently, by far, the most used particle. It is inexpensive, uniform, and there exists considerable experience in its manufacture and use.

These particles are considerably shorter than one micron (40μin) (see Fig. 11-4). They are acicular and have a length-to-width ratio of approximately six to one. The oxide particles are manufactured by first dissolving iron in an appropriate acid. The resulting particles, called alpha particles, have the elongated shape of the finished magnetic oxide, but are non-magnetic.

To continue in the manufacturing process, the particles are heated and the oxide is reduced to the Fe_3O_4 form in an atmosphere of hydrogen or natural gas. Then the particles are reoxidized under a controlled temperature and the final particles in the form of gamma ferric oxide are obtained:

	Red		Black		Brown	
(FeO)OH ----->	αFe_2O_3	--------->	Fe_3O_4	--------->	γFe_2O_3	**(11.5)**
Process temperature: 120		reduction 350-500		oxidation 250-330	degrees C	

Small amounts of Zn or Cr salt may be added to improve particle stability and dispersion properties.

The (FeO)OH and αFe_2O_3 powders were used as color pigmentation in the paint industry, and that is how some companies got into the tape business (I.G. Farben, now BASF, and C.K.Williams, now Pfizer Pigments, Inc.). This, of course, ties in with their knowledge of applying binders (paint) to the base films.

The shape and size of the final particles are largely determined by the preparation of the (FeO)OH-Goethite. Agglomeration and sintering of the particles occur at the high temperature steps of dehydration and reduction which are often combined into one step.

A small particle size is advantageous since the coating then has better short wavelength response (less demagnetization because of higher coercivity) and less noise (more particles per coating volume). Measured data on experimental tapes are shown in Fig. 11-5 (ref. Podolsky).

A new type of particle, named *NP (Non Polar)* has an ellipsoidal shape and therefore no internal demagnetization from an irregular shape (ref. Corradi et al). The morphology of these particles allows a much better packing and homogeneous distribution of the particles providing for a higher remanence and lower noise. It is also suggested that they may provide for a perpendicular recording surface, in addition to the Ba ferrite particles to be described further on. Figure 11-6 shows the uniformity and range of sizes for the NP particles.

Cobalt-Substituted (Doped) Iron Oxide

It has long been known that cobalt doped iron oxide particles can be fabricated with a large range of coercivities, H_c = 150 – 800 Oe. In these particles up to four percent of the ferric ions have been replaced by cobalt ions.

Fig. 11-4. γFe_2O_3 particles (courtesy of Pfizer Pigments, Inc.).

It has also been known that the cobalt doped particles in a tape suffer from temperature dependence of the coercivity and loss of remanence after storage at elevated temperatures. This is due to temperature sensitivity of the crystal anisotropy. A user might adjust storage conditions to live with these disadvantages, but will be faced with repeated signal losses each time the tape is played. This is due to a high magnetostriction constant and the signal would literally be erased after a sufficiently large number of passes (see Chapter 14).

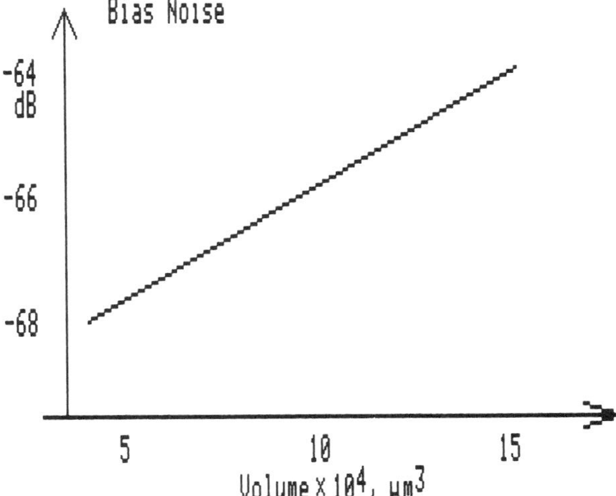

Fig. 11-5. Relationship of bias noise to $\gamma-Fe_2O_3$ calculated particle volume (after Podolsky).

Cobalt-Adsorbed (Surface) Iron Oxide

These particles are surface modified iron oxides, obtained by producing a reaction of the Fe_3O_4 or γFe_2O_3 particles, with a solution containing Co^{2+}. Their size and shape are therefore that of γFe_2O_3. The resulting material consists of particles with a thin surface layer of cobalt-iron-oxide. The depth and composition of the layer determine both the coercive force and its temperature dependence.

The process control over the final material coercivity is a great advantage. New tape products should not only be superior to their predecessors, but also compatible with existing equipment (ref. Umeki et al.)

Chromium dioxide was introduced early, and recording equipment was provided with a switch for normal (γFe_2O_3) and high (CrO_2) bias currents to the record head. The cobalt-adsorbed particles were introduced as *Avilyn* tapes, and the coercivity was adjusted to match the high bias setting.

Isotropic Iron Oxide

A new iron oxide particle named *Isomax* was introduced a few years ago by Spin-Physics. The particles are shaped like rice, and are essentially isotropic. By cobalt treatment it is possible to introduce anisotropy so a non-oriented coating has a high remanence for longitudinal as well as perpendicular magnetization. Thus Isomax delivers the best of both worlds.

It is unfortunate that a serious flaw exists, due to excessive thermal and magnetostrictive instabilities, that makes their use for ordinary tapes questionable. For rigid disks it should not matter. Efforts continue to correct these instabilities such as reducing the temperature coefficient of the coercivity to what is believed to be acceptable ($(1/H_c) \times (dH_c/dT) = -0.5 \times 10^{-3}$ per °C) (ref. Speliotis, 1986).

The JH-loop has the same squareness, whether measured in the plane of the coating (longitudinal) or perpendicular thereto.

Chromium Dioxide

Chromium dioxide was developed during DuPont's investigations into compounds formed at intense heats and pressures. It soon became evident that chromium dioxide possessed mag-

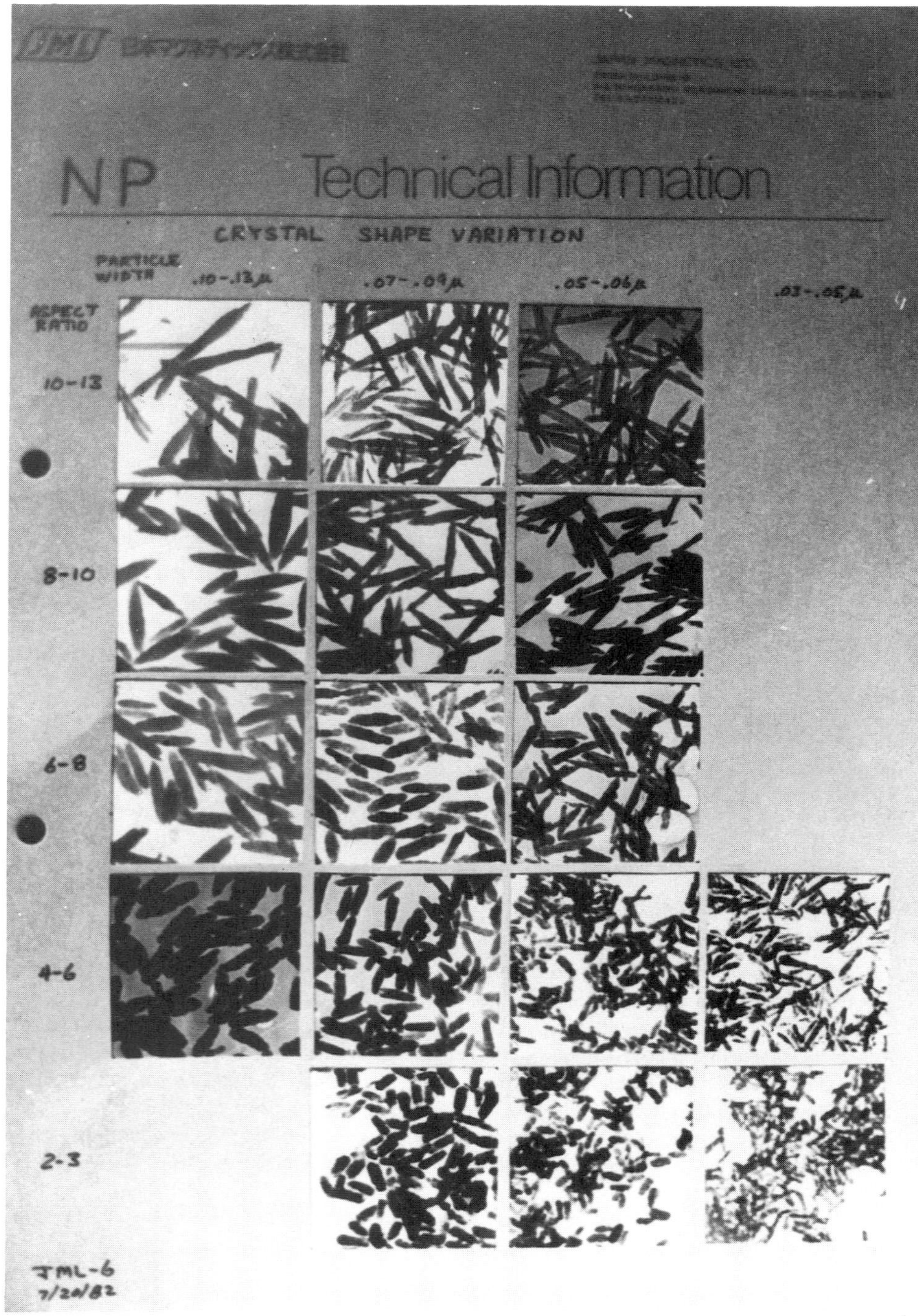

Fig. 11-6. NP–particles are available in a range of sizes and shapes (courtesy of Magnox, Inc.).

Fig. 11-7. CrO_2 particles (courtesy of Pfizer Pigments, Inc.).

netic properties superior to iron oxides and could be made in a needle shape well-suited to magnetic recording applications, being (at that time) 0.8 μm long and with a length to diameter ratio (aspect ratio) of ten to one. Commercial production of CrO_2 began in 1967.

One advantage is that the particles can be made of uniform size and free of dendrites. They are readily dispersed, easily oriented, and give excellent high density (short wavelength) recording performance. Coercivities range from 450 to 670 Oe, and the particle aspect ratio can be controlled over a wide range (ref. Chen et al.).

Their disadvantages are the cost (three times that of ferric oxides but about the same as co-absorbed ferric oxide) and at times a higher wear rate (see Chapter 18). Wear is often overrated and can be judged as a trade-off for better performance (1000 hours head life instead of 2000; that is still a lot of tape playing).

Modern CrO_2 particles are made in a range of sizes, obtained by doping with Antimontrioxide Sb_2O_3. A high doping level (approximately one percent) results in a higher H_c and smaller, more uniform particles (Fig. 11-7, ref. Braginskij). The popularity of CrO_2 is evident from the recent choice by IBM to use these particles for the new 3480 digital tape system.

METAL PARTICLES

Metal particles (pure Fe) have a magnetization that is almost three times that of ferric oxide, and they can be made smaller than the oxide particles without the onset of paramagnetism (Fig. 11-8).

The coercivity is high, which presents a two-fold problem: lack of compatibility with existing equipment and a potential danger of pole tip saturation in recording heads. On the other hand, a high coercivity is desirable for high density recordings.

The coating procedure for a metal particle tape presents its own set of problems. The particles are highly reactive with the atmosphere and will rust or even burn when exposed to air. They will need a protective coating to prevent contact with air.

Most modern tapes are made with polyurethane binders, which have high transmission rates for water vapor. A protective and permanent coating must therefore be developed for the Fe particles.

Indications are that some of the coating problems are being solved (ref. Chubacki et al.). The durability is approaching that of conventional tapes, and the SNR is about 10 dB above oxide tapes for the new 8mm video camera, and is similar for the 47mm disk used in electronic still cameras.

BARIUM FERRITE

Fine Ba-ferrite particles has been developed during the past few years, and show great promise as a particulate, perpendicular recording medium. The particles, about 0.08 μm (3.2 μin) in average diameter, are thin hexagonal platelets with easy magnetization axes normal to their plane (Fig. 11-9). The addition of Co and Ti makes the coercivity controllable over a wide range, without significant reduction in magnetization, see Fig. 11-10 (ref. Kubo et al.). This allows for the design and optimization of tapes for use with ferrite or metal head recording capability (ref. Isshiki et al.). Ferrite heads have tendencies to saturate when writing on high coercivity tapes or disks, see Chapter 10.

Recent data verify the excellent properties of Ba-ferrite, as shown in Fig. 11-11, illustrating packing densities approaching 80,000 fr/cm (200,000 FRPI) (ref. Fujiwara). The signal envelope has less amplitude modulation due to a more uniform particle dispersion in the coating, and the SNR is 6 dB higher than a low noise γFe_2O_3 tape.

PARTICLES WITH VERY HIGH COERCIVITY

A considerable effort was made during the seventies to develop rare-earth-cobalt materials for high-energy permanent magnets. The high coercivity of these materials is well suited for secure credit-card applications. Barium ferrites have also been used and require a recording head made from an iron-cobalt alloy. $BaFe_{12}O_9$ and $SmCo_5$ particle tapes have been compared and both were found suitable for erasure-resistant applications (ref. Fayling et al.)

Fig. 11-8. Fe particles (courtesy of Pfizer Pigments, Inc.).

THIN MAGNETIC FILMS

Metallic coatings on circular aluminum substrates are in widespread use in all hard disk memories. The coatings are thin (.2 – .5 μm = 8 – 20 μin) but the output is comparable or better than the thicker oxide coatings because the packing factor is one in films, and there is no intrinsic demagnetization.

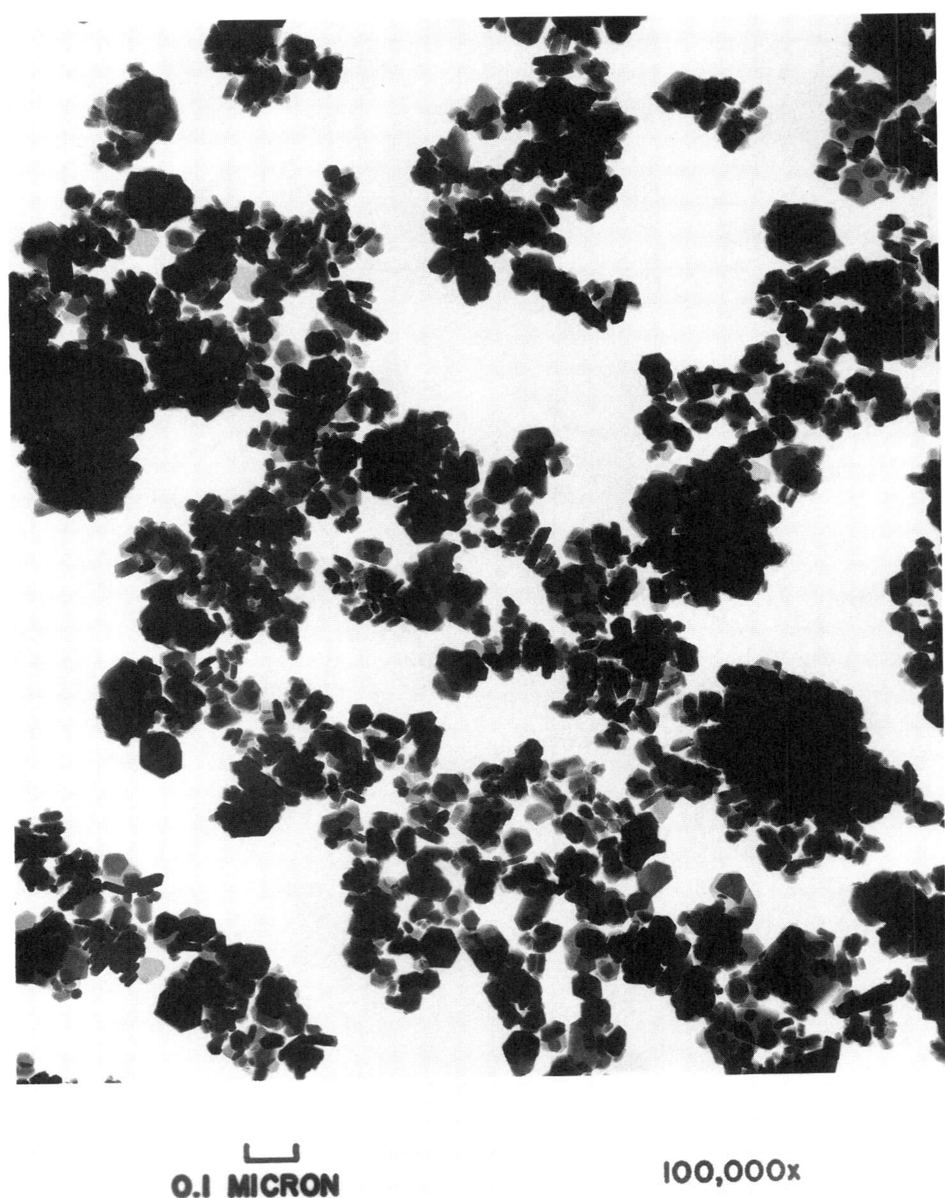

Fig. 11-9. Barium ferrite particles (courtesy Pfizer Pigments, Inc.).

Film coatings on flexible media remain in development in order to solve the problems of corrosion and excessive head wear. The wear is in particular a problem when thin film tapes are used in conjunction with metal heads, rather than ferrites. And all ferrite heads remain a bit of a mystery—one head manufacturer has excellent results with head life while another has problems with wear and crystal pull-outs. There is also the matter of using a stable, long-lasting lubricant, without which the contact between the head and the metal film would be like

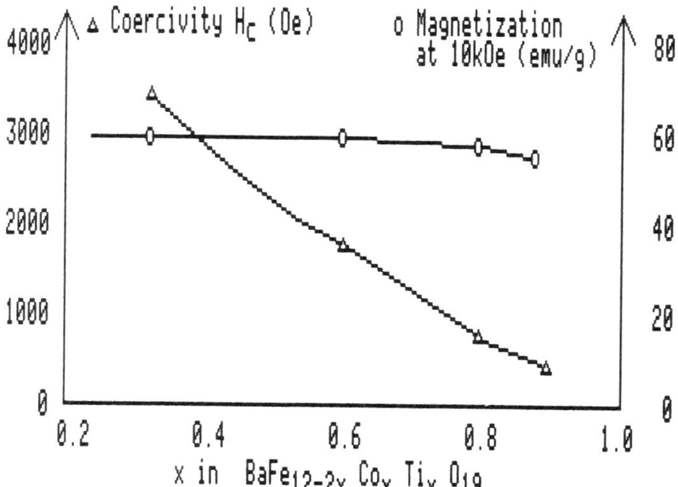

Fig. 11-10. Co and Ti substitution effect on coercivity and magnetization at 10 kOe for Ba ferrite particles (after Kubo et al.).

a bearing without grease. Recent efforts to manufacture such tapes are described in the paper by Feuerstein et al.

The films can be deposited by several methods, which will be covered in Chapter 13, and the range of materials has grown considerably during the past decade.

Some of the earliest coatings were made from *cobalt phosphorus* (CoP) or *cobalt-nickel* (CoNi) by vacuum deposition, chemical deposition, or electro deposition.

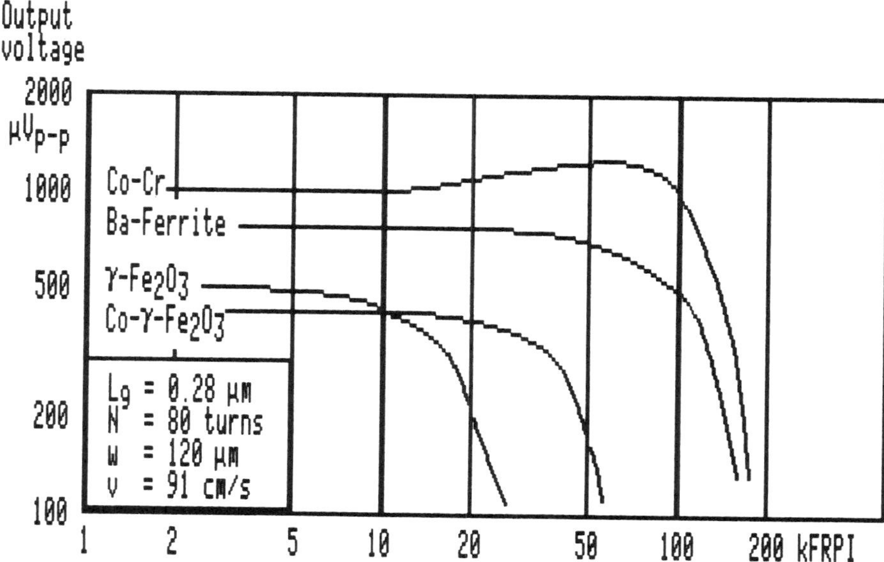

Fig. 11-11. Density response curve for barium ferrite flexible disk along with Co-Cr sputtered perpendicular, Co-γ-Fe_2O_3 longitudinal and $\gamma-Fe_2O_3$ longitudinal disks (after Fujiwara).

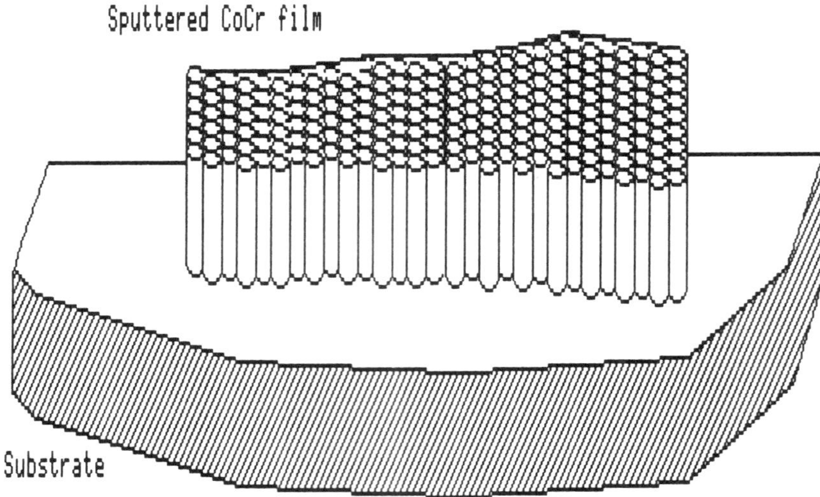

Fig. 11-12. Columnar structure of CoCr film.

More recent coatings are deposited by sputtering, and a fairly large number of materials have been evaluated, with many claims to superiority over particulate oxide coatings.

Ferrite films have been made with a coercivity of 40 kA/m (500 Oe), and remanences (B_{rs}) ranging from 0.05 to 0.26 Wb/m^2 (500 to 2600 Gauss) (ref. Ianagaki et al.). *Rare earth films* ($Fe_{1-x-y}Tb_xGd_y$) shows promise (ref. Desserre et al.) as do CoSm thin films (ref. Kullmann et al.).

Cobalt and *cobalt-chromium* are both good candidates for perpendicular thin film recording surfaces, made by electroplating (ref. Chen et al.) or by RF (radio frequency) sputtering (ref. Sagoi et al.). Both materials can be made in a columnar structure that enhances the perpendicular magnetization mode, see Fig. 11-12. This should enhance their ability to carry a high packing density, in contrast to some of the ordinary films that suffer from resolution limitation due to the formation of jaggies, (see Fig. 3-33, and the paper by Chen et al.).

FLEXIBLE MEDIA SUBSTRATES

Several methods have been employed to support the recording surface. Tapes and flexible disks use plastic films, in the early days a *cellulosetriacetate* material. Todays tape and diskette are coated on a PET film (polyethylene terephthalate) film, often called Mylar (a tradename of Du Pont). Rigid disks use an aluminum platter.

Cellulosetriacetate was abandoned because of its low tensile strength, causing a tape to break due to careless tape recorder operation. Nevertheless, the break was always clean and easy to splice; *Mylar* will stretch to about twice its normal length before breaking, ruining that portion, and making splicing a questionable operation.

The surface of PET (polyurethane terephthalate) is inferior to that of PVC. The PET is one of the many long-molecule materials, where cross-linking causes uneven surfaces, primarily due to lumping. Efforts continue to develop new base films by modifying the PET, and by mixing polymers—a very difficult process due to intermolecular interactions and difficulties in building a uniform molecular structure from the different long chains (ref. Braginskij).

High temperature application requires special films such as *polyimide* film, named Kapton (a tradename of Du Pont). Polyimides maintain the physical characteristics up to 400° C and do not support combustion (ref. Ochsner). Tapes in flight data recorders onboard commercial airliners are therefore made with a Kapton base film.

Polyimide substrates are candidates for flexible disks for perpendicular recording. Sputter metallizing of cobalt chromium is most effectively done at a temperature of approximately 200° C, where most other films would have failed completely.

A new film, named *Ryton* (tradename of Phillips Petroleum) is emerging (ref. Campbell). Its chemical name is polyphenylene sulfide (PPS), it has a tensile strength comparable to PET, but it's hygroscopic expansion coefficient is only 0.2×10^{-5} cm/cm/(percent RH)—only one fifth of the value for PET. Its tear strength is two to five times less than PET's and the thermal expansion coefficient is higher than for PET. (Author's note: Comparative numbers were not available at the time of writing. PPS is manufactured in a way similar to PET film, and both are biaxially stressed with properties that vary with direction away from the machine direction).

MANUFACTURE OF PET FILM

PET belong to a group of quite complicated polymer esters, made from terephthalic acid and ethylene glycol (ref. Mark). The molecular weight is between 15,000 and 30,000, and the raw material exists in the form of granules of a mean of size 4 by 4 by 2.5mm (ref. Braginskij).

The creation of a continuous film is started by thoroughly drying the granules so that the water content is below 0.01 percent, at a temperature between 150 and 180 degrees centigrades. Any excess humidity will result in film defects (small bubbles and blisters).

The granules are next fed into one end of a screw conveyor, where they are melted at a temperature of 280 to 290 degrees centigrade and form a liquid syrup-like substance (Fig. 11-13). This liquid is forced through an extrusion head to form a liquid sheet that immediately is cooled by contact with a cooling drum.

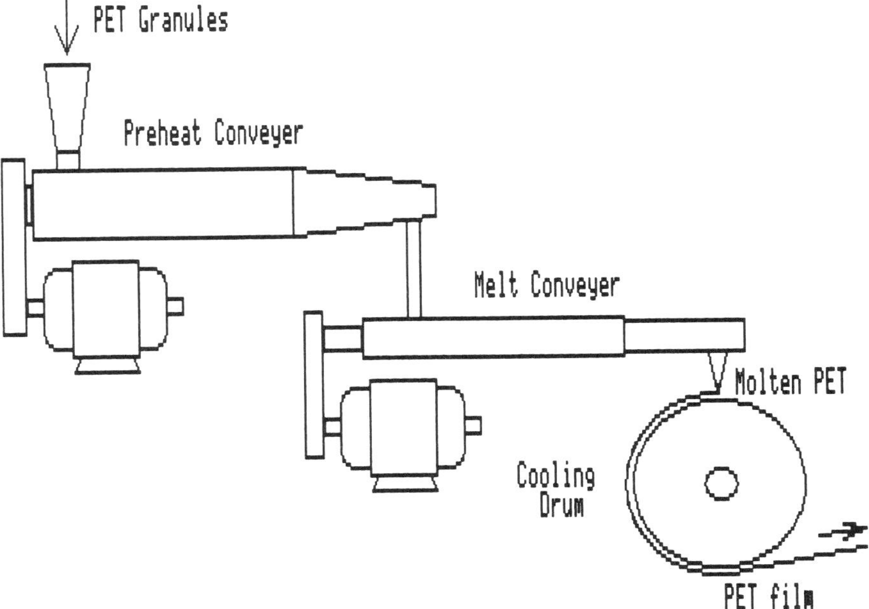

Fig. 11-13. Screw conveyers for fabrication of PET films.

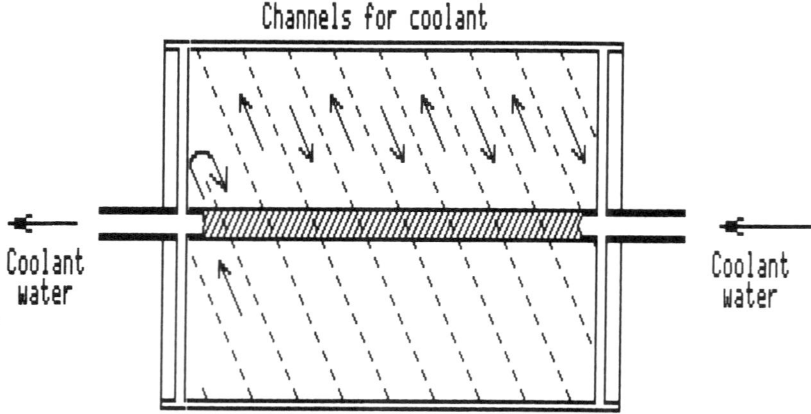

Fig. 11-14. Cooling drum for PET film.

It is important that the liquid sheet is extruded under a uniform pressure since pulsations result in thickness variations of the PET film. This can be accomplished by a double screw conveyor, or a single conveyor followed by a gear pump.

The PET may at the moment of extrusion crystallize into spherulites; this is prevented by rapid cooling when it touches a drum, that has a cooling water system built into its surface, see Fig.11-14. Good contact between the PET and the drum is required, and can be obtained by an air stream pushing onto the PET film, coating of the drum with Glycol or even the use of electrostatic adhesion (ref. Braginskij).

The fabricated film is next stretched in order to increase its strength. This must be done in the machine direction, which corresponds to the motion of the film, as well as in the direc-

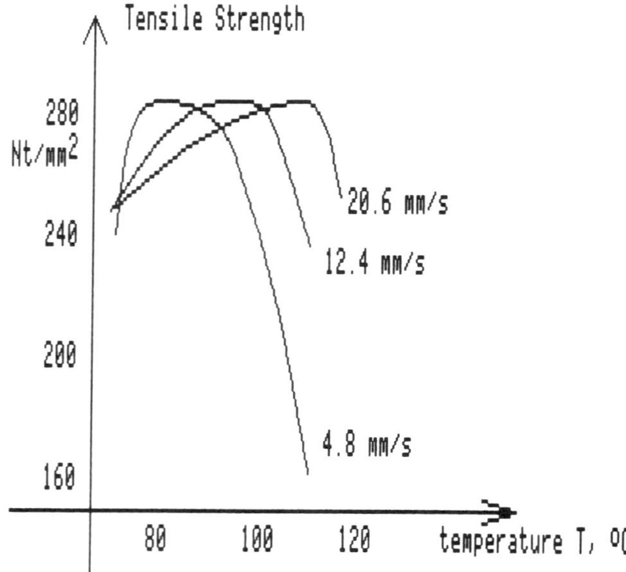

Fig. 11-15. The tensile strength of a PET film depends upon the velocity and temperature of the post-stretching process (after Braginskij).

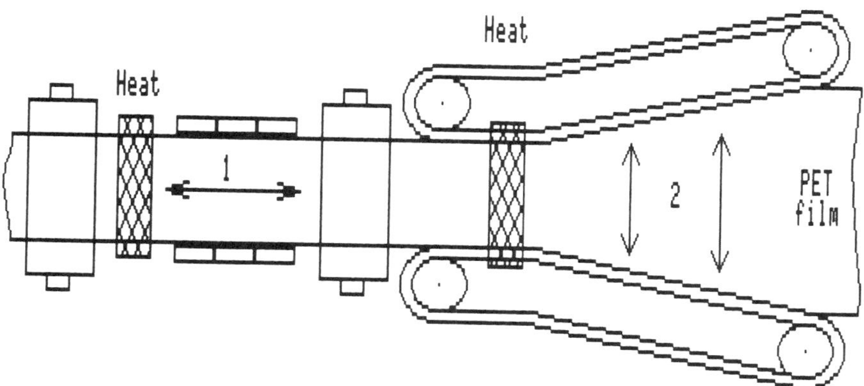

Fig. 11-16. Longitudinal (1) and transverse (2) stretching of PET film (after Braginskij).

tion of its width. If done in the machine direction only the result is an alignment of the molecules leading to strength in just that direction only. The temperature and the speed with which the film is stretched will influence its tensile strength, as shown in Fig. 11-15.

The process takes place in a piece of equipment where the film is heated, stretched lengthwise and then sideways, and finally fixed (cross-linking of the long molecules) and cooled (Fig.

Table 11-2. Properties of Coated PET Film. Values Are Typical; Use for Budgetary Estimates Only.

ANSI recommended HEAD/TAPE environment:	
NORMAL	+10 to +35 °C, 40 to 60 % RH
EXTENDED	0 to +45 °C, 25 to 95 % RH
Youngs modulus of elasticity, static	3,500 - 5,500 Nt/mm^2
Youngs modulus of elasticity, dynamic	5,000 - 8,000 " "
Youngs modulus of elasticity of binders, less than	500 " "
Yield point, at 0.2 % offset	50 " "
Tensile break strength, machine direction	275 " "
Tensile break strength, transverse direction	175 " "
Density	1.4 gr/cm^3
Mass per unit area of a 1 mil thick tape	3.6 mg/cm^2
Velocity of sound	2,000 m/sec
Coefficient of thermal expansion	$1.9 \times 10^{-5}/°C$
Coefficient of hygroscopic expansion	$1.1 \times 10^{-5}/\%RH$
Moisture absorption	< 0.6 %

$1\ Nt/mm^2 = 10^6$ Pa (Pascal) = 1 MPa = $10^7\ dynes/cm^2$ = 145.1 psi = 10.2 kgf/cm^2

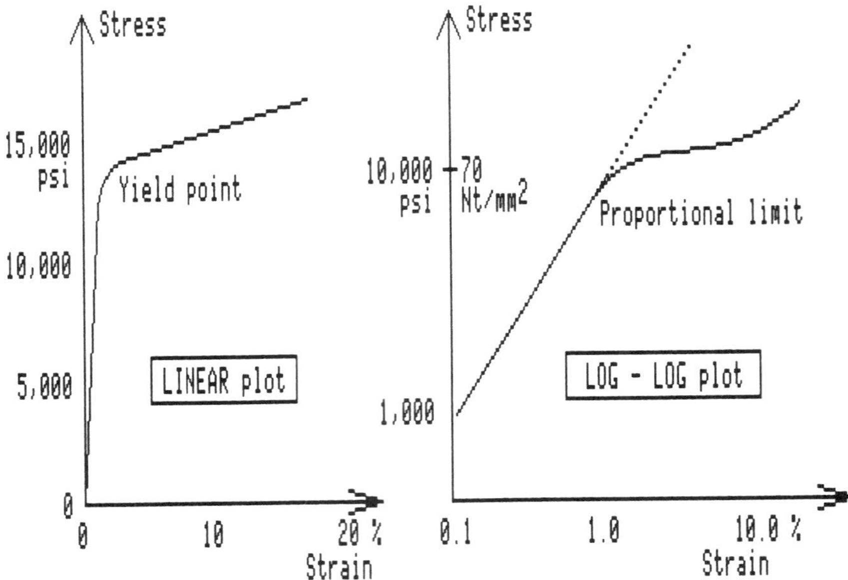

Fig. 11-17. Comparative stress–strain plotting (courtesy of E.F. Cuddihy, JPL, Pasadena, Ca).

11-16). This process is referred to as tensilizing and is done at a speed of 6-60 m/minute. The machinery is a rather large structure, from 30 to 50 meters in length, 3-6 meters wide and 2-3 meters tall.

The finished PET film that varies in width from 0.6 to 3 meters is trimmed from its edges and wound onto drums; care must be taken to avoid dust (which may originate from the slitting operation). Any small particles wound within a pack cause deformation of the film and will be the origin of potential dropouts.

Table 11-2 summarizes the properties of modern PET films. The values are averages of published data and should be used for budgetary design purposes only. Comparative stress-strain curves are shown in Fig. 11-17. In a tensilized (stressed) PET film there exist residual stresses that will deform and buckle the film if it is heated.

The long-chained molecule links are furthermore such that the films have properties that differ in directions away from the machine direction (length of web). The modulus and thermal expansion coefficient are affected, as shown in Table 11-3.

This leads to problems when diskettes are stamped out from a coated PET film. Changes in temperature cause different degrees of expansion in different directions, and a round track may end up being oval. Sufficient trackwidth must be used to avoid loss of contact between the head and the deformed track; this limits the TPI.

PET films are inherently very smooth which is desirable for a recording substrate, but leads to problems in the handling of the film (difficult to wind, blocking of surfaces). It is therefore general practice to roughen the surface by embossing, sand blasting, chemical etching, and so on. It is also possible to produce the rougher surface by altering the polymer by incorporating inorganic additives either grown or injected into the polymer prior to film extrusion. Silicon dioxides, titanium dioxide, calcium carbonate etc. are often used (ref. Heffelfinger).

After the biaxial stretching the result is a bumpy surface. The surface roughness is often graded by the use of a measurement of the arithmetic mean deviation from the center line, called the center line average (CLA) or Ra (see further in Chapter 14). PET for video tapes

Table 11-3. Angular Variation of Modulus and Thermal Expansion Coefficient for a Mylar "A" Polyester Film. (Courtesy of E.F. Cuddihy, JPL, Pasadena, Ca.).

Angular Position degrees	Youngs Modulus dynes/cm^2	Youngs Modulus MPa (Nt/mm^2)	Thermal Expansion Coefficient
0	4.67E10	4,670	1.68E-5 /C°
22	4.83E10	4,830	1.71E-5
45	6.03E10	6,030	0.91E-5
67	6.27E10	6,270	0.70E-5
90	5.82E10	5,820	1.55E-5
112	4.74E10	4,470	1.57E-5
135	4.40E10	4,400	2.88E-5
157	4.37E10	4,370	2.60E-5

Note 1: 0 degrees is machine direction.
Note 2: Published values for the thermal Expansion Coefficient is 1.70E-5 /C°.
Note 3: Thermal Expansion Coefficient in the thickness direction varied between 33E-5 to 59E-5 /C°.

for ½ inch VCR has Ra = 0.035 μm (= 1.4 μin), while Ra = 0.005 μm (= 0.2 μin) for higher density ME media (ref. Heffelfinger).

RIGID MEDIA SUBSTRATES

The definitive material for rigid disk substrates is aluminum, ranging in thicknesses from 6 to 1.5 mm; brass has been used but is a thing of the past.

Aluminum is produced by the electrolytic reduction of aluminum oxide dissolved in a molten cryolite bath. The result can be 99.9 percent pure aluminum, with iron and silicon impurity elements. A large cast ingot is the start of operations involving rolling and annealing to relieve stresses.

Pure aluminum is very soft so magnesium is therefore added (three to four percent, by weight). Magnesium dissolves in aluminum, providing a solid solution that provides structural strength. A small amount of manganese is added to absorb the iron impurities into intermetallic compounds. The manganese improves corrosion resistance, which is enhanced by the further addition of chromium (ref. Westerman).

Circular blankets are stamped or cut from the finished aluminum sheets. They must be perfectly flat which may be achieved by additional machining on a lathe, followed by a polishing operation. Some disks turned with a diamond tool have the smoothest surfaces.

Considerations have been given to the use of glass or molded plastic substrates, but experiments have not been very successful. Established disk drive designs cannot readily be switched to use a substrate that has different coefficient of expansions etc. from the original design. The potential for plastic substrates are in novel disk drive designs (ref. McLaughlin).

An interesting combination substrate/recording film has been introduced by 3M Co. in its stretched surface disk. The substrate, or main structural member, is a circular disk with a raised rim on the outside perimeter as well as one near the inside hole. A coated PET film is placed on both sides, stretched and then welded to both rims, very much like a drum skin (see Chapter 14 for further details).

BIBLIOGRAPHY TO CHAPTER 11

General

Koester, E., "Recording Media," in *Magnetic Recording*, Ed. by C.D. Mee and E.D. Daniel, McGraw-Hill, 1987, pp. 98-243.

Monson, J.E., "Recording Measurements," in *Magnetic Recording*, Ed. by C.D. Mee and E.D. Daniel, McGraw-Hill 1987, pp. 376-426.

Craig, D.J. (Editor), *Magnetic Oxides*, John Wiley and Sons, 1975, 2 volumes, total 798 pages.

Granum, F., and Nishimura, A., "Modern Developments in Magnetic Tape," *Intl. Conf. Video and Data 79*, IERE Conf. Proc. No. 43, July 1979, pp. 49-61.

Bate, G., *Recording Materials*. Chapter 7, Ferromagnetic Materials, Vol. 2, North-Holland, 1980, 126 pages.

Staff, *Magnetic Materials—A Glossary*, Pfizer Pigments Inc., 1981, 12 pages.

Substrates

Rodriquez, F., *Principles of Polymer Systems*, McGraw-Hill, 1970, 560 pages.

Moore, G.R., and Kline, D.E., *Properties and Processing of Polymers for Engineers*, Soc. of Plastic Engrs., Prentice-Hall, 1984, 209 pages.

REFERENCES TO CHAPTER 11

Particulate Media

Braginskij, G.I., and Timoteev, E.N., *Technologie der Magnetbandherstellung*, Akademie-Verlag, DDR - 1080 Berlin, Leipziger Strasse 3-4, 1981, 320 pages. Chapter 4.

Koester, E., "Recording Media", in *Magnetic Recording*, Ed. by C.D. Mee and E.D. Daniel, McGraw-Hill, 1987, pp. 98-243.

Bate, G., and Alstad, J.K., "A Critical Review of Magnetic Recording Materials," *IEEE Trans. Magn.*, Dec. 1969, Vol. MAG-5, No. 4, pp. 821-839.

Daniel, E.D., "Tape Noise in Audio Recording," *Jour. AES*, Mar. 1972, Vol. 20, No. 2, pp. 92-99.

Umeki, S., Saitoh, S., and Imaoka, Y., "A New High Coercive Particle for Recording Tape," *IEEE Trans. Magn.*, Sept. 1974, Vol. MAG-10, No. 2, pp. 655-657.

Fayling, R.E., and Bendson, S.A., "Magnetic Recording Properties of SmCo5," *IEEE Trans. Magn.*, Sept. 1978, Vol. MAG-14, No. 5, pp. 752-755.

Podolsky, G., "Relationship of Gamma-Fe_2O_3 Audio Tape Properties to Particle Size," *IEEE Trans. Magn.*, Nov. 1981, Vol. MAG-17, No. 6, pp. 3032-3034.

Knowles, J.E., "Magnetic Properties of Individual Acicular Particles," *IEEE Trans. Magn.*, Nov. 1981, Vol. MAG-17, No. 6, pp. 3008-3013.

Kubo, O., Ido, T., and Yokoyama, H., "Properties of Ba Ferrite Particles For Perpendicular Magnetic Recording Media," *IEEE Trans. Magn.*, Nov. 1982, Vol. MAG-18, No. 6, pp. 1122-1125.

Chen, H.Y., Hiller, K.M., Hudson, J.E., and Westenbroek, C.J.A., "Advances in Properties and Manufacturing of Chromium Dioxide," *IEEE Trans. Magn.*, Jan. 1984, Vol. MAG-20, No. 1, pp. 24-26.

Corradi, A.R., Andress, S.J., French, J.E., Bottoni, G., Candolfo, D., Cecchetti, A., and Masoli, F., "Magnetic Properties of New (NP) Hydrothermal Particles," *IEEE Trans.Magn.*, Jan. 1984, Vol. MAG-20, No. 1, pp. 33-38.

Chubacki, R., and Tamagawa, N., "Characteristics and Applications of Metal Tape," *IEEE Trans. Magn.*, Jan. 1984, Vol. MAG-20, No. 1, pp. 45-47.

Fujiwara, T., "Barium Ferrite Media for Perpendicular Recording," *IEEE Trans. Magn.*, Sept. 1985, Vol. MAG-21, No. 5, pp. 1480-1485.

Isshiki, M., Suzuki, T., Ito, T., Ido, T., and Fujiwara, T., "Relations between Coercivity and Recording Performances for Ba-Ferrite Particulate Perpendicular Media," *IEEE Trans. Magn.*, Sept. 1985, Vol. MAG-21, No. 5, pp. 1486-1488.

Speliotis, D.E., "Advanced Particulate Magnetic Media," *SMART Symposium*, May 1986, Paper No. PS-6, 33 pages.

Deposited Films

Inaqaki, N., Hattori, S., Ishii, Y., Terada, A., and Katsuraki, H., "Ferrite Thin Films for High Recording Density," *IEEE Trans. Magn.,* Nov. 1976, Vol. MAG-12, No. 6, pp. 785-787.

Chen T., and Martin, R.M. "The Physical Limits of High Density Recording in Metallic Magnetic Thin Film Media," *IEEE Trans. Magn.*, Nov. 1979, Vol. MAG-15. No. 6, pp. 1444-1446.

Desserre, J. and Jeanniot, D., "Rare Earth–Transition Metal Alloys: Another Way for Perpendicular Recording," *IEEE Trans. Magn.*, Nov. 1983, Vol. MAG-19, No. 6, pp. 1647-1649.

Feurestein, A., and Mayr, M., "High Vacuum Evaporation of Ferromagnetic Materials–A New Production Technology for Magnetic Tapes," *IEEE Trans. Magn.*, Jan. 1984. Vol. MAG-20, No. 1, pp. 51-56.

Sagoi, M., Hishikawa, R., and Suzuki, T., "Film Structure and Magnetic Properties for Co-Cr Sputtered Films," *IEEE Trans. Magn.*, Sept. 1984, Vol. MAG-20, No. 5, pp. 2019-2024.

Substrates

Braqinskij, G.I., and Timoteev, E.N., *Technologie der Magnetbandherstellung*. Akademie-Verlag, DDR - 1080 Berlin, Leipziger Strasse 3-4, 1981. 320 pages, Chapter 3.

Mark, H.F., *Giant Molecules*, Time-Life Books, 1968, 200 pages.

Mark, H.F., "Giant Molecules," *Scient. Amer.*, Sept. 1957, Vol. 197, No. 3, pp. 81-89.

Heffelfinger, C.J., "Improved Performance Polyester Films," *SMART Symposium*, May 1986 Paper No. WS 1-C-1, 26 pages.

Ochsner, J.P., "Polyimide Films for Recording Substrates," *SMART Symposium*, May 1986, Paper No. WS 1-D-2, 17 pages.

Campbell, R.W., "Biaxially Oriented Polu(Phenylene Sulfide) Film," *SMART Symposium*, May 1986. Paper No. WS 1-D-1, 16 pages.

Westerman, E.J., "Improvements and Problems in Aluminum Alloys for Magnetic Media Substrates," *SMART Symposium*, May 1986, Paper No. WS 1-A-1, 27 pages.

McLaughlin, H.J., "Problems in Non-Aluminum Substrates for Rigid Disk Media," *SMART Symposium*, May 1986, Paper No. WS 1-B-3, 22 pages.

Chapter 12

Magnetic Properties of Tapes and Disks

Magnetic recording surfaces are made from small particles or from deposited films. We discussed the basic magnetic properties of these materials in Chapter 11 and will now describe how they behave when finished into a coating for writing and reading information. We are particularly interested in those properties that relate to the performance of the write/read processes we examined in Chapters 4, 5, and 6.

We can briefly summarize how the shape and the magnitudes from the BH-, or better, JH-loop relate to performance:

- The current required for writing data onto a recording surface is proportional to the coating coercivity H_c.
- Long wavelength read output is proportional to the value of B_{rsat} (= J_{rsat}).
- Short wavelength read output is reduced by demagnetization; this loss can be reduced by using a coating with a high value of coercivity H_c.
- The read output in digital recording is 75 percent of its possible maximum when the coating thickness equals 0.44 * BL (bit length). A thicker coat produces only a small improvement in signal level, but increases demagnetization losses (which really is a function of the bit geometry) and increases peak shifts. The optimum coating thickness is therefore approximately 50 percent of the longest bit length (or 25 percent of the longest wavelength).
- Resolution (and packing density) is improved by having a BH- (JH-) loop with steep sides. This results if the particles have a low SFD (switching field distribution), and are well oriented in the direction of the magnetization.

The coating's hysteresis loop does therefore give us all essential information to predict the write/read performance. We will start the chapter by outlining two methods for the measurements of the magnetization curve, the BH-meter and the VSM (Vibrating Sample Magnetometer). Either instrument will serve as long as it is properly calibrated, but there have been discrepancies between results from the two instruments. This issue has recently been resolved, and the cause will be described (time effects in materials magnetism).

The BH-loop for a given material will vary in accordance with the particle loading, orientation, etc. Knowledge of the exact behavior is important for the optimum design of a media coating, as well as for production quality control. It is also desirable to know the magnetization loop of a single particle and experimental data are reported to illustrate this.

The chapter concludes with a review of the important issue of noise, from a particulate and also from thin film media.

MEASUREMENTS OF BH Loop

The coercivity H_c represents tape sensitivity, and is an important specification. It can be measured on a BH-meter or a VSM; some research groups also use torque-meters.

BH-Meter

A coated tape or disk sample, which may be several layers thick, is placed inside a large field coil in the BH-meter (Fig. 12-1) (ref. Newman). It is magnetized by the 60 Hz field H_a, and the change in flux ψ is sensed by the small coil around the sample. This coil also picks up the field H_a, but this component is cancelled by the voltage induced by H_a in a series connected balance coil.

The induced signal $d\psi/dt$ is integrated, amplified, and connected to the vertical deflection on an oscilloscope. A small resistor in the bottom of the field coil provides a voltage proportional to H_a, which provides the horizontal deflection. Thus the J-H loop is displayed on the

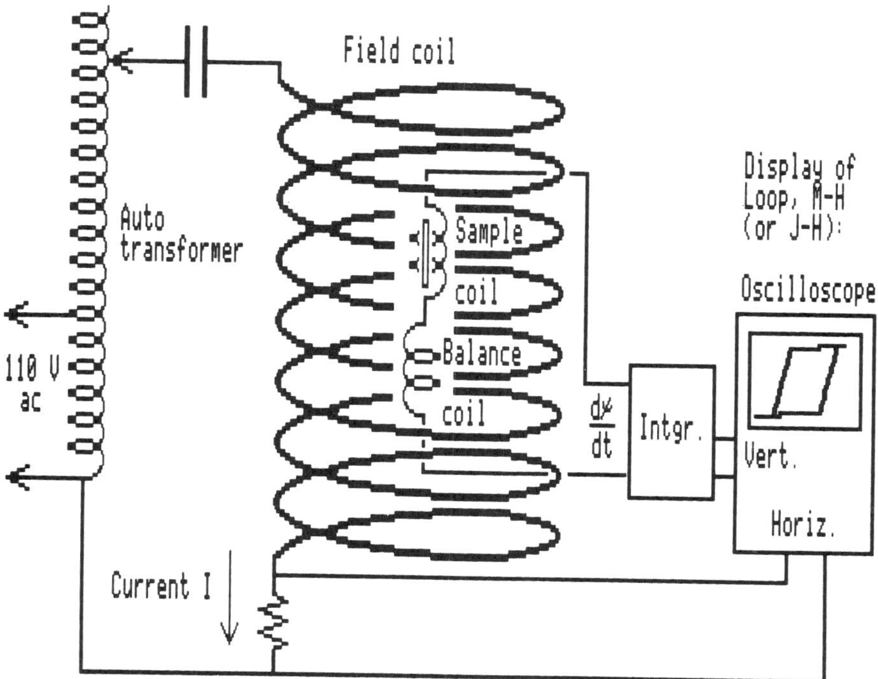

Fig. 12-1. BH-meter, simplified.

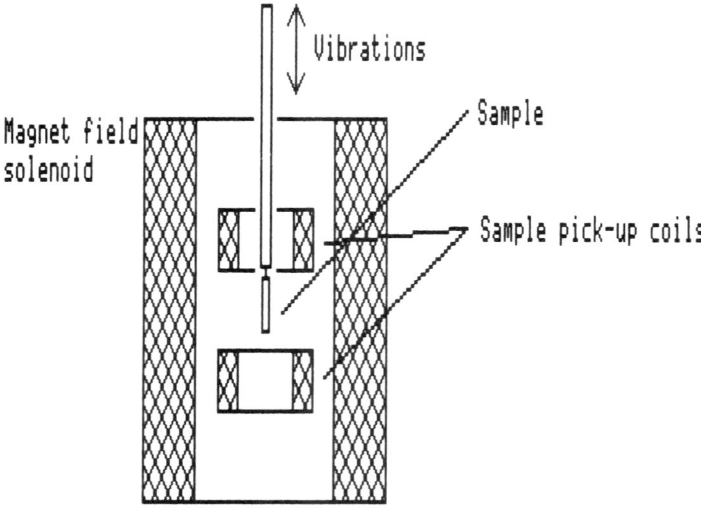

Fig. 12-2. VSM, simplified. In this configuration, field intensity is parallel to sample motion.

scope, and may be photographed or otherwise plotted for reading the magnitude of H_c, J_{rsat},SFD, etc.

Vibrating Sample Magnetometer

The VSM instrument (Fig. 12-2) operates by measuring the magnetization in a sample by mechanical oscillation of the sample in a gradient sensing field provided by two coils. The motion of the sample can be in the same direction of the field (ref. Newman), or perpendicular thereto (ref. Foner).

The sensing mechanism of the coils is much like the sensing field from a head; it is passive but senses magnetization. The sample's magnetization may be some remanent state from an earlier magnetization process, or it can be brought about by a magnetic DC-field (which itself it not sensed).

Symmetrical Read Head

There is finally a ''poor man's'' method to determine the absolute flux level which requires a symmetrical playback head, where the geometries of the front and back gaps are identical (Fig. 12-3). The core halves are made from high grade laminated mumetal in order to assure that the core reluctance is at least 1,000 times less than the gap reluctances at long wavelengths. The flux is determined from:

$$\begin{aligned} e &= n * \eta * d\psi/dt \\ &= n * 0.5 * \omega * \psi_m * \cos \omega t \end{aligned}$$

where:

$\omega = 2\pi f$
$\eta = 0.5$

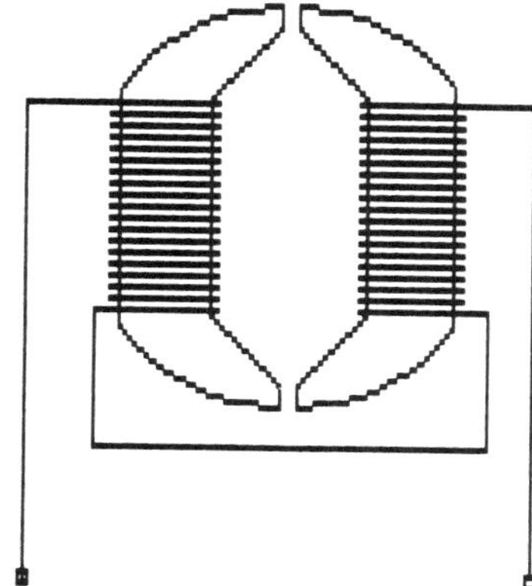

Fig. 12-3. Symmetrical read head.

Rewriting and dividing with the trackwidth w (in m) gives:

$$\psi m = e / (n\mu f w)$$

and additional division with the coating thickness c (in m) gives:

$$B_{rs} = e / (n\pi fcw) \qquad (\text{Wb/m}^2, \text{ volts}) \qquad \mathbf{(13.1)}$$

Sample Size

The samples are usually small pieces of tape, a few inches long for the BH-meter, and small squares for the VSM. Several layers may be used to provide increased signal output. The sample preparation is more difficult for disks, which actually have to be cut up so a square centimeter sample can be provided. An aluminum substrate will cause eddy current losses in the BH-meter, and results may be erroneous (ref. Newman). Errors will also prevail in the VSM if the substrate is conductive and the field gradient is not constant (ref. Kerchner et al.).

Measurements in the plane of the coating seldom require any correction for demagnetization. For a circular tape coating sample of diameter one cm and coating thickness 5μm = 0.0005 cm the diagonal demagnetization factor is zero (longitudinal or transverse), while it is almost one in the perpendicular direction.

Correction for Demagnetization in Sample

It is necessary to correct the BH-loops for the quite large demagnetization factor involved in perpendicular measurements. An approximation for the demagnetization factors, diagonally (N_x and N_y) and perpendicularly (N_z) are (ref. Bate; formulas changed to MKS values for N (nominal maximum = one)):

$$N_x = N_y = 1.13 * c / a \qquad \mathbf{(13.2)}$$
$$N_z = 1 - 2.26 * c / a \qquad \mathbf{(13.3)}$$

where:

c = sample thickness
a = sample diameter

Bate did find that the perpendicular properties for a longitudinally oriented coating could be determined from the transverse properties (diagonally, 90 degrees to the longitudinal direction); this eliminates the need for correction of demagnetization. However, this will not apply to a perpendicularly oriented coating without the cylindrical symmetry of the angular distribution.

COERCIVITY, REMANENT FLUX, AND SFD

We will now examine longitudinal, transverse, and perpendicular properties.

Longitudinal JH-loop

The level of magnetization in a tape or disk coating may be determined by formula (11.1), and it is proportional to σ_s = particle specific magnetization, which depends on the particle material used. It is also proportional to the particle loading, or volume fraction p, the coating thickness c, and the squareness ratio S. The latter three factors must therefore be included in a comparison of various tapes.

The largest variation will be found in the coating thickness which is 15 μm (600 μin) for high output audio tapes and early computer tapes, 6 to 3 μm (240 to 120 μin) for most tapes, and 3 to ½ μm (120 to 20 μin) for high resolution tapes and disks. ψ_{rsat}, which is proportional to the coating thickness, will vary widely without being an indicator of the tape's or disk's merit as long as it provides an adequate playback or read voltage.

Transverse JH-loop

The remanent magnetization of all tape and disk products includes the influence of particle orientation. Perfectly aligned, ideal particles should possess no remanence in directions perpendicular to the alignment direction. This corresponds to a squareness ratio of one, which is never achieved.

The net result is a transverse remanence that is somewhat less than the longitudinal remanence. It can be measured by properly oriented samples with a BH-meter or VSM.

Another factor is introduced to illustrate how good the orientation is: the orientation factor, which is simply the ratio between the saturation remanence in the longitudinal direction

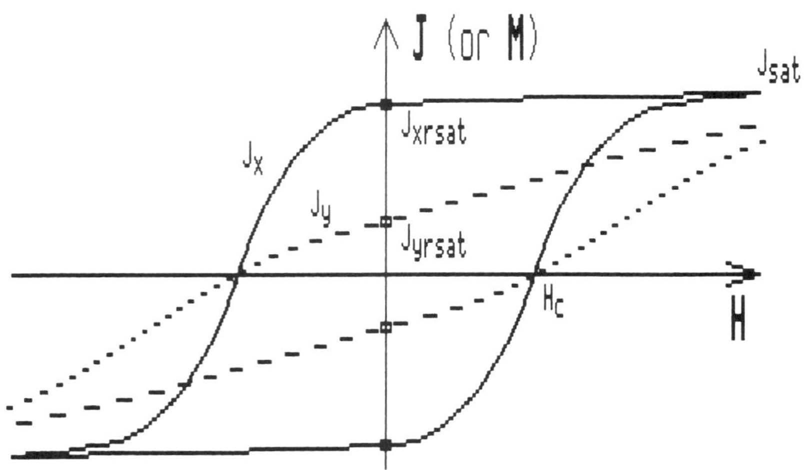

Fig. 12-4. JH-loops for a longitudinally oriented medium.

and the saturation remanence transverse thereto. Both are measured in the coating plane with perpendicular (sometimes called vertical) magnetization at ninety degrees to the coating plane.

The transverse remanence should equal the longitudinal remanence in flexible disk products to avoid cyclic patterns in the read signal. This ideal situation is found in a peripheral orientation, as in a rigid disk. An orientation in the coating plane is inherent in the extrusion of the wet coating onto the substrate for flexible disks, and any further alignment is discouraged by mechanical or electromagnetic agitation. This can, for instance, be achieved in the plane rotating orienter field. The squareness of a diskette cannot therefore exceed 0.7.

Figure 12-4 shows the magnetization loops for a longitudinally oriented coating; it is highly desirable for the designer of a write/read system to have both loops available from the media manufacturer so he can make a fair estimate of the written magnetization in a coating (see later this chapter, plus Chapters 19 and 20).

Perpendicular Properties

The curvature of the record head field lines will tend to generate a curved magnetization pattern in the coating, being perpendicular at the coating surface for in-contact recordings. The remanence pattern will be modified by demagnetization, and by the longitudinal alignment of the particles. The perpendicular remanence will be in the order of X percent of the longitudinal, where X is the ratio between perpendicular and longitudinal saturation remanence (at long wavelength).

Pure perpendicular remanence may prove advantageous at very high packing densities. Figure 12-5 shows the magnetization loops for a perpendicularly oriented coating.

Switching Field Distribution; Preisach's Diagram

Several methods are used for the measurement of the switching field distribution SFD, a quantity that was introduced in Chapter 5. It is a measure of a particulate coatings ability to store short transitions, i.e., a small SFD results in high resolution (for AC-bias recordings as well).

The recommended definition of SFD is the magnetizing field range wherein 50 percent of the coating's particles will switch magnetization. The SFD is then named Δh_r, and is derived from the JH-loop (or MH-loop) as shown in Fig. 12-6.

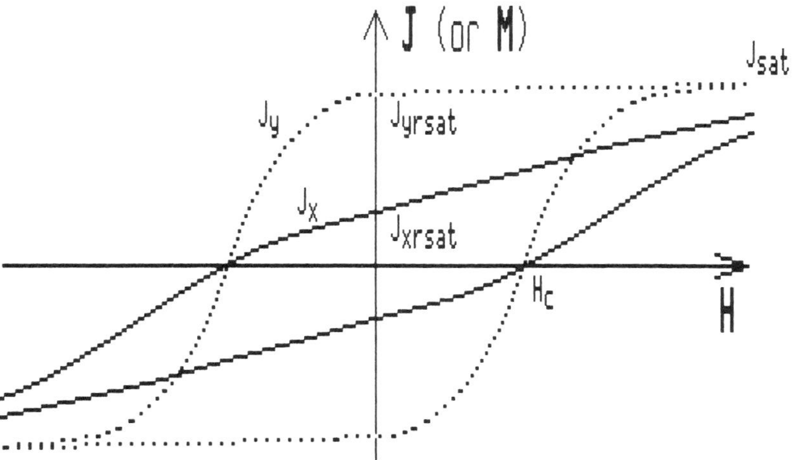

Fig. 12-5. JH-loops for a perpendicularly oriented medium.

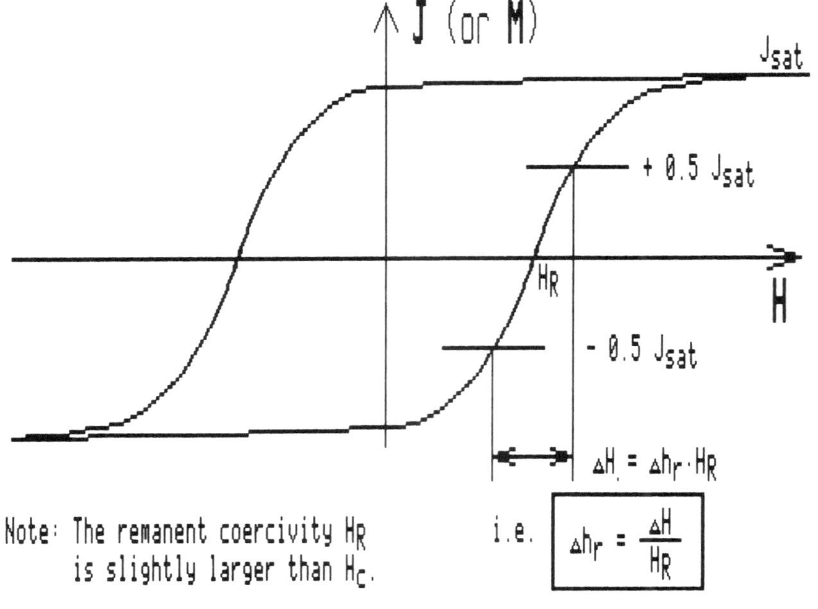

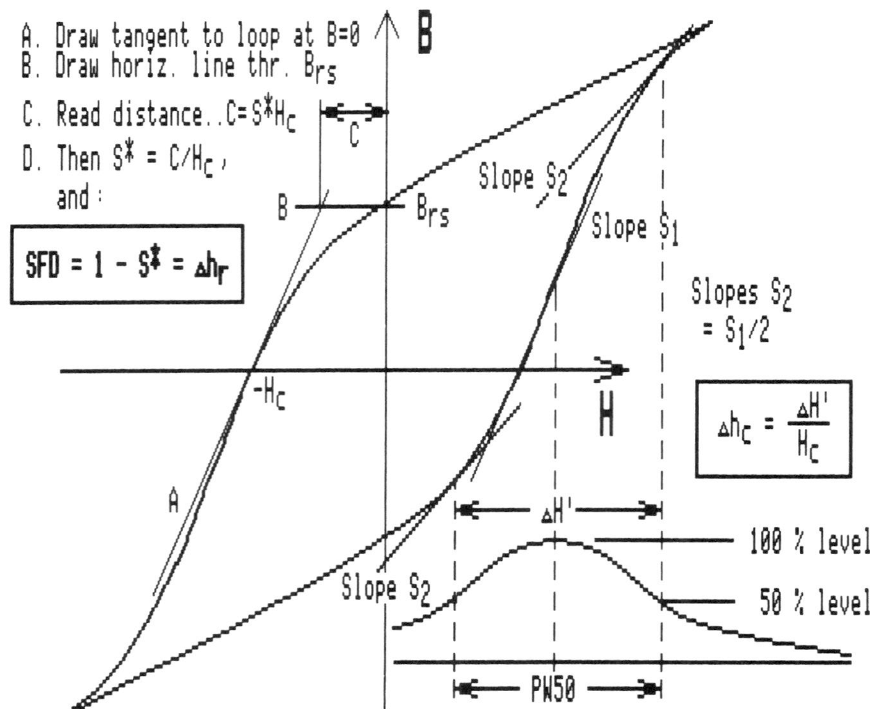

Fig. 12-6. Readings of SFD using the JH-loop, or the BH-loop.

Another way to determine the SFD is made by drawing the tangent to the BH-loop at B = 0. Next draw a line through point $(0,B_{rs})$, parallel to the H-axis. The distance from the intersection between the first drawn tangent and the B-axis equals $S^* \cdot H_c$. The SFD is then:

$$SFD = 1 - S^*$$

Values of $1 - S^*$ have been found to closely match the Δh_r values for the same coatings; it is therefore recommended as an easy method to determine the SFD (ref. Koester).

It has also been suggested that the BH-loop be differentiated, and the width of the differentiated curve measured at the level that is 50 percent of its maximum value. This method correlates with the PW50 measurements of read signals from a single transition; PW50 (in time, for example in μsec) equals the pulse width at the 50 percent amplitude points). The SFD measurement on the BH-loop results in a value named Δh_c; it does not correspond well with Δh_r.

A combined display of the SFD plus the interacting fields is possible by using a so-called Preisach diagram. A series of ac-bias measurements are needed to get the data points (see Chapter 19), and a three-dimensional plot of the distribution of the particles coercivities (h_{ci}) and their interacting fields (δH) will result in an illustration as shown in Fig. 12-7.

The coating sample is an audio tape. The two humps at low field levels are interesting: they tell us that the coating contains some large particles (low coercivities h_{ci}) that are subject to some large interacting fields.

A different view is obtained by letting the plotting program for Fig. 12-7 move the point of observation up over the distribution and look down. A tracing of constant altitude contours, plus the generation of shadows from a light source far to the right produces Fig. 12-8.

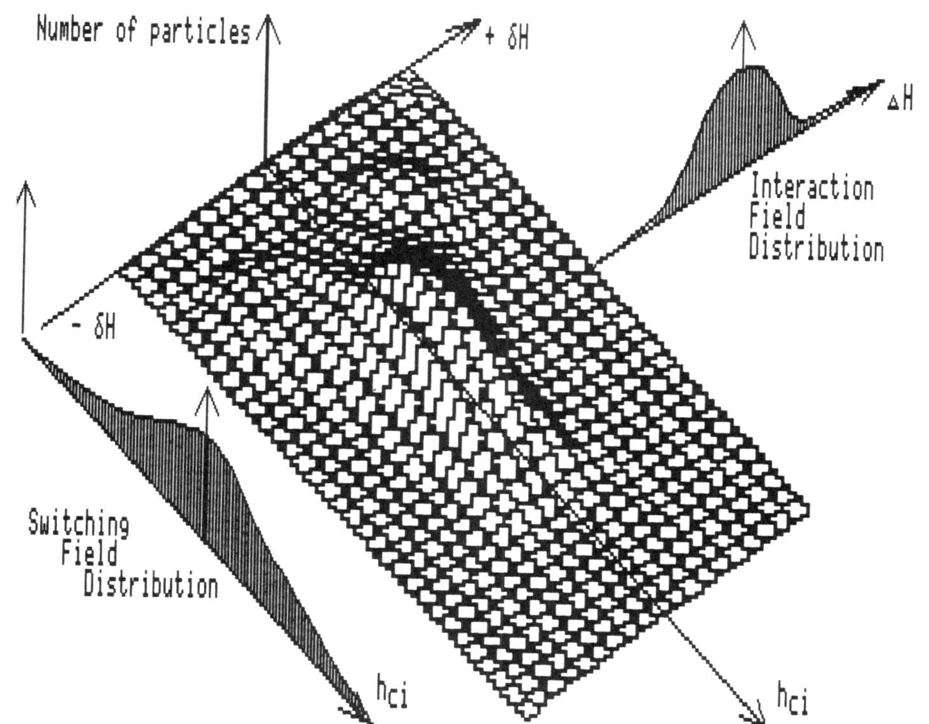

Fig. 12-7. Three dimensional distribution showing &Prob(h_{ci}) and &Prob(@H).

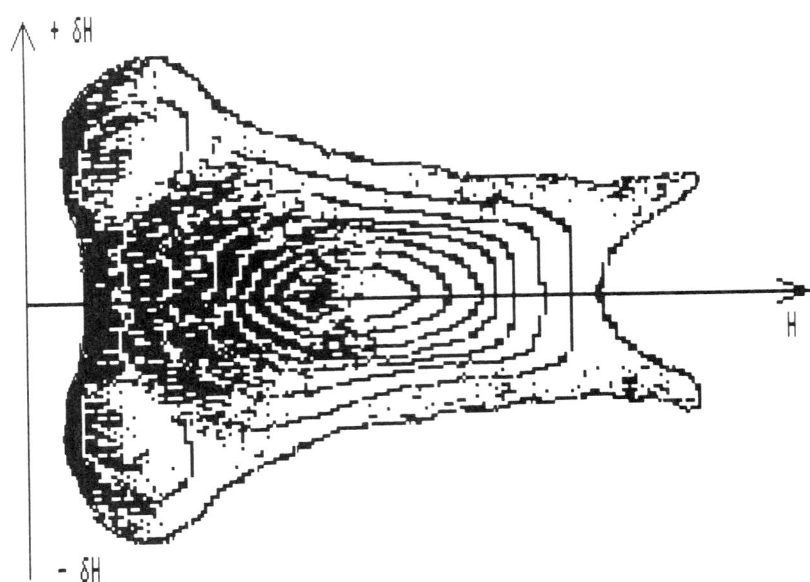

Fig. 12-8. A topographic view of Fig. 12-7, with shadow from a light source.

FACTORS INFLUENCING THE MAGNETIC PROPERTIES

Orientation

The most obvious factors that influence the magnetic behavior of a magnetic coating is the orientation of particles, or actually the preferred direction of magnetization. One example is shown in Fig. 12-9; the ratio between the extreme remanence values is the so-called orientation ratio. The change in coercivity is not as large as for the remanence.

This relatively small change in coercivity with field direction is important when studying the write process. The contour of the freezing zone at the trailing edge of a write gap will follow the field strength contour constructed by using the values versus field direction in Fig. 12-9.

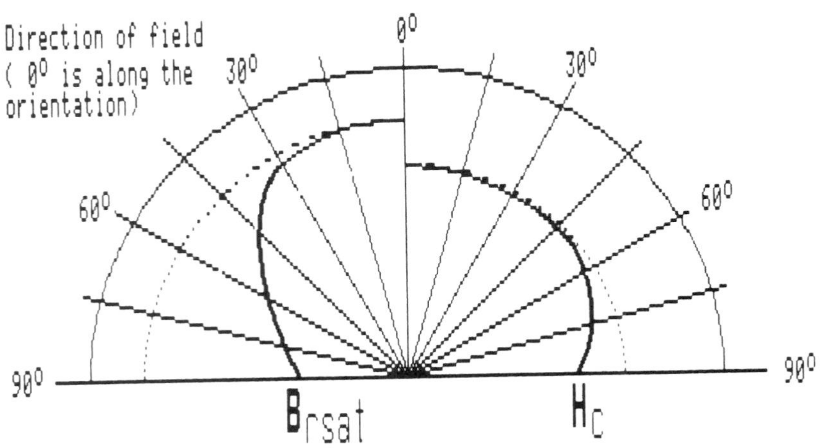

Fig. 12-9. Directional variations of remanence and coercivity (after Newman).

Fig. 12-10. Relationship of print-through to calculated gamma-Fe_2O_3 particle volume (after Podolsky).

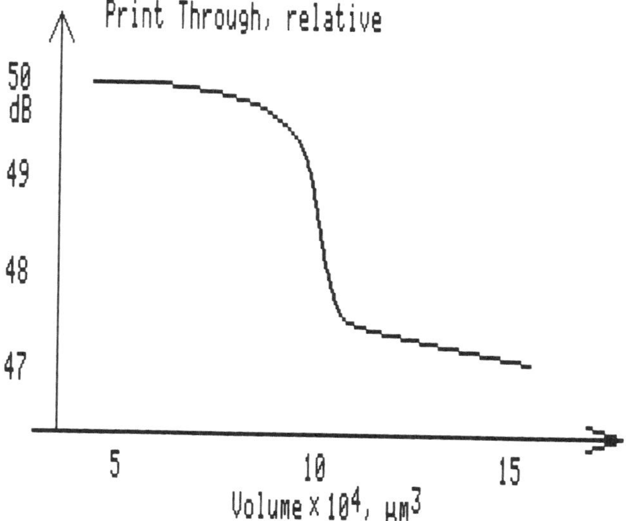

Particle Size

When the average volume of the particles in an assembly decreases we may expect that more particles will fall in the unstable region where their behavior is superparamagnetic. Figure 12-10 verifies this. Figure 12-11 shows the effect of particle size upon sensitivity and saturated output; both decrease for increasing particle volume at 15 kHz, while the opposite is true for the long wavelength output at one kHz (ref. Podolsky).

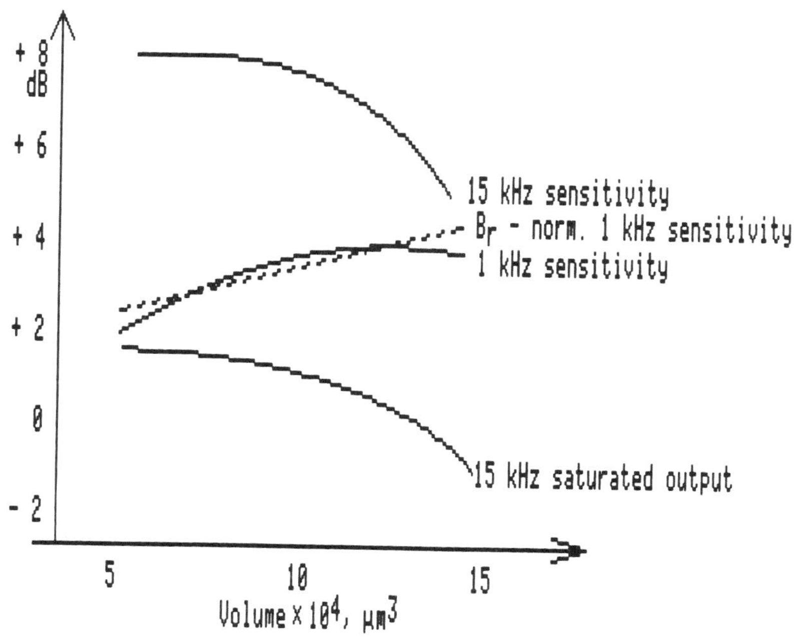

Fig. 12-11. Relationship of sensitivities at 1 kHz and 15 kHz saturated output at 15 kHz to calculated particle volume (after Podolsky).

Packing Factor P

The variation of coercivity with packing factor p is generally expressed as:

$$H_c(p) = h_{ci}(1 - p) \qquad \textbf{(12.4)}$$

where h_{ci} is the coercivity of an isolated particle. Equation (12.4) does not usually agree with experiments, and a novel computational approach (Monte Carlo treatment) combined with better understanding of the individual particles magnetization switching modes (see later in this chapter) provides better agreement (ref. Knowles, 1985).

Mixture of Two or More Particle Types

Makers of audio tapes have sought an improvement of overall response by making a double coated tape: First a thick coating of standard coercivity is laid down, whereafter it is overcoated with a high coercivity material. The bias current can now be adjusted for overbias of the thick undercoat; the record zone will still be quite short for recording of the short wavelengths in the top coat. And only the top coat magnetization contributes to the short wavelength output during read (remember: 75 percent flux comes from a thickness of 0.22 * wavelength).

The situation is illustrated in Fig. 12-12. This technique improves the frequency response, by providing the best of two worlds: a large signal output at long wavelengths due to the thick coating and a good short wavelength response due to a short recording zone.

The major shortcoming is a double transient response for a single transition. This caused audio signals from string instruments to have a muddled sound during the attack time (see Chapter 21 for computation). The same problems will occur for the so-called multimodal media, that also will have two (or more) recording zones, centered around the average value of the coercivities. The BH-loops for a bimodal and a trimodal media, where the particles are mixed within one coating layer, are shown in Fig. 12-13 (ref. Manly).

MEASUREMENTS ON SINGLE PARTICLES

When particles are suspended in a liquid one can observe their behavior through a microscope and for instance do experiments by turning them with a magnetic field (ref. Scholten).

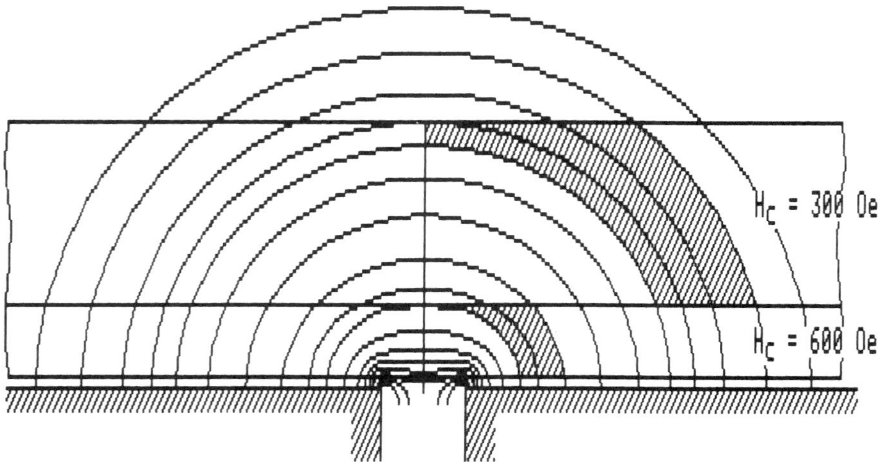

Fig. 12-12. Recording on a double coat medium provides optimum bias adjustment for long and short wavelengths. Frequency response is very good, but the transient response is very poor.

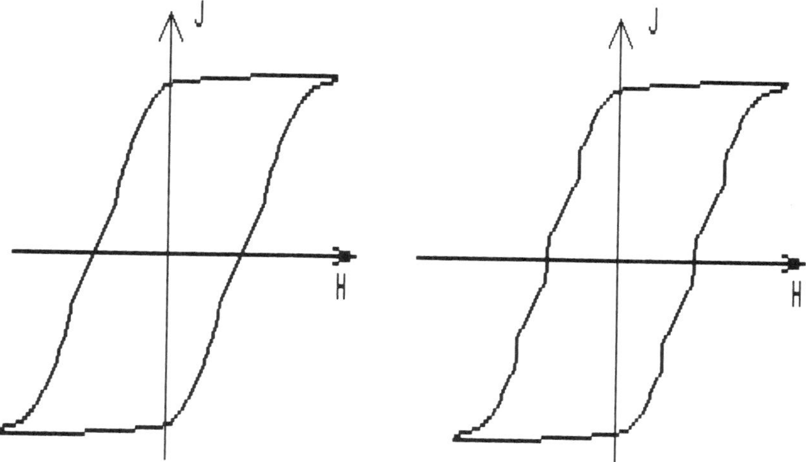

Fig. 12-13. JH-loops for a bimodal and a trimodal media (after Manly).

One approach has led to a fairly accurate way of finding the coercivities of these essentially single particles (ref. Knowles, 1980).

The particles are kept in a very dilute suspension, using a lacquer base with a viscosity of 100 P or more. This suspension was then sucked into a small glass tube, and an individual particle observed with the microscope, see Fig. 12-14. The particle was aligned by a continuously applied field of a few Oersted, and a large pulsed field was then applied in the opposite direction.

If this pulsed field was larger than the remanent coercive force H_r of the particle, then the magnetization was reversed and the particle subsequently rotated through 180°. By a sim-

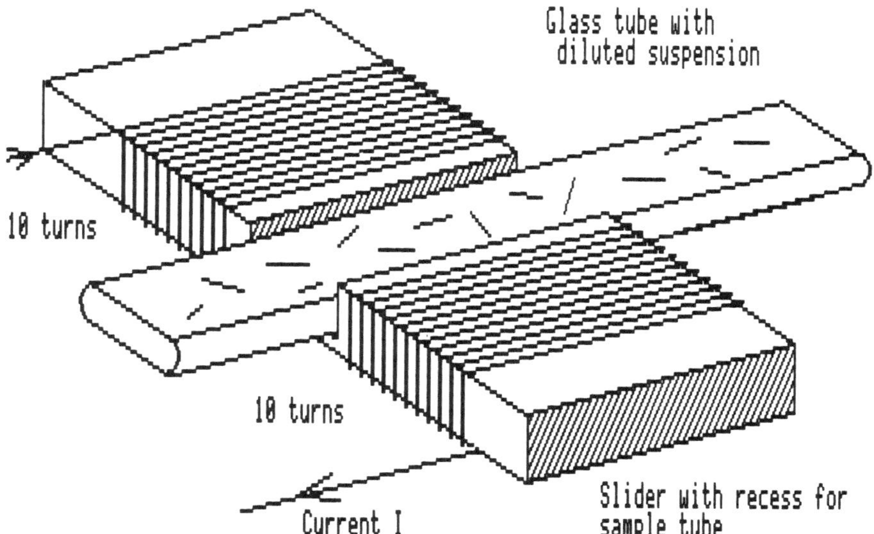

Fig. 12-14. Observation of rotation of single particles in a dilute suspension; remagnetization is done with an external, pulsed field.

ple extension of the method a remanent loop for the particle can be obtained. A histogram of many test results shows the coercivities for a typical tape sample; this curve corresponds to the distribution curve shown to the right in Fig. 3-30. These data permit the construction of the tapes overall magnetization loop, exactly as was done in Fig. 3-31.

Work continues in order to get a better idea about the magnetic behavior of particles, from the Stoner-Wohlfarth single domain model to the Jacob-Beans multi-domain particle with various magnetization modes—buckling, fanning, etc. (ref. Knowles, 1980, 1984).

AC-BIAS SENSITIVITY

AC-bias does, among other things, greatly increase the record current sensitivity. This can be expressed as an amplification factor, much like the alpha of a transistor, and is named the anhysteretic susceptibility χ_{ar}. It will be further characterized in Chapter 19, AC-bias Recording.

Kneller and Koester have found that χ_{ar}'s relation to the static magnetic parameters can be expressed as:

$$\chi_{ar} = p(1-p)M_s^2 \;/\; 2(1-m_r^2)H_r^2$$

where:

M_s = saturation magnetization (J_s/μ_o)
p = packing fraction of magnetic particles
m_r = remanence to saturation ratio (= squareness)
H_r = remanence coercivity (approx = H_c)

Typical values for χ_{ar} are in the range of 15 to 25.

EFFECTS OF TIME

The coercivity of a magnetic recording material is an important parameter; it determines the required field (and hence write current) of writing, and plays a role in the demagnetization process (such as transition broadening) during long term storage. It has been found that measurements that produce rapid changes of magnetization will yield a higher value of coercivity than those that operate on a longer time-scale (ref. Sharrock and McKinney).

The two types of measurements that were described earlier in this chapter will show such a discrepancy: The BH-meter operates at 60 (or 50 Hz) and the magnetization time is therefore less than 1/100 second. The VSM, in contrast, operates at a magnetization time of 10 seconds or more.

The difference in coercivity values is due to the thermally assisted nature of the magnetic switching process. Figure 12-15 shows the decay for two materials, top is cobalt treated γFe_2O_3, bottom ordinary γFe_2O_3. The formula shows also that the difference between short- and long-term coercivities become more pronounced with time.

Sharrock and McKinney give an example where they use 10^{-5} sec as an appropriate value of τ for the BH-meter, and 10 sec for the VSM: The values for H_c are 322 Oe for the BH-meter, 288 Oe with the VSM.

Signal amplitudes in magnetic tapes are also observed to decay with time; this decay is a function of the writing density, increasing with wavelength to 7.3 percent in 1 hour at 4 kfc/cm in one case (ref. Kloepper et al.).

Both phenomena are explained by time decays of magnetization, M, which are attributed to a thermal-fluctuation after-effect proposed by Neel (see ref.), involving irreversible switching of interacting particles.

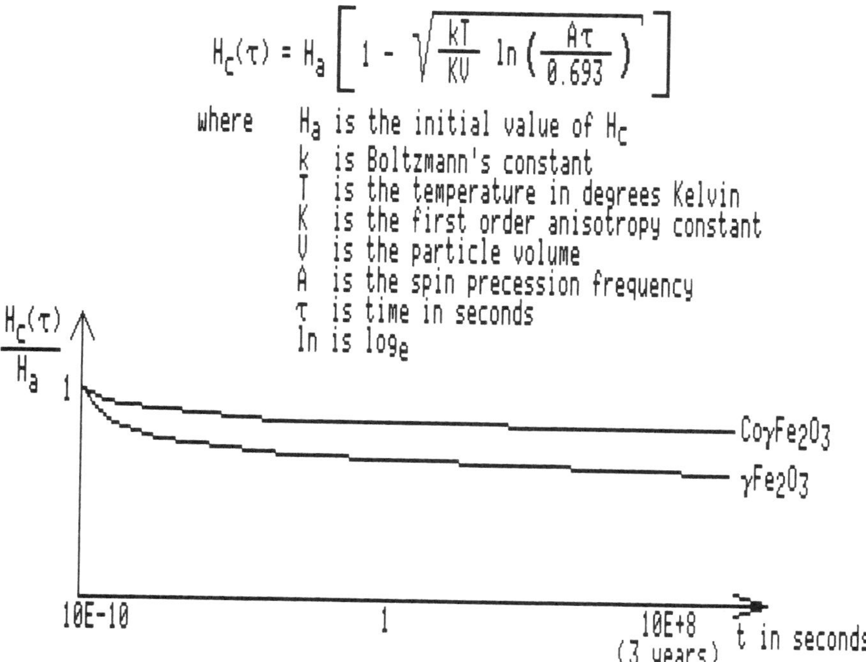

Fig. 12-15. Time dependent decay in the coercivity of particulate magnetic recording materials (after Sharrock).

The temperature does also play a role, as the presence of T in the formula for H_c in Fig. 12-15 shows; its exact role is also the target of recent investigations (ref. Oseroff).

Effects of Temperature

Magnetization in general will decrease with temperature, and eventually cease to be present at and above the material's Curie temperature. It will recur though when the temperature is lowered.

Figure 12-16 illustrates the variations in coercivities and remanences for modern magnetic particles (ref. Kubo, Fujiwara). These changes may cause errors in exact measurements of write- and read- experiments, and do also play a role in the exchange of recorded data (cross-play). An example will serve well to illustrate what might happen.

Example 12.1. Figure 12-17 shows the changed write (record) conditions that occur due to a change in temperature only, everything else remaining constant. The coating has a coercivity of H_c = 300 Oe at T = 0° C, and the writing is about 85 percent into the coating, and Δx equals $0.40L_g$.

At the higher temperature T = 50° C the write level has increased to 100 percent, and Δx has increased to $0.55L_g$. Therefore a lower resolution should result. This is confirmed upon read-out, but the higher level of magnetization is not realized into a higher flux since the remanence is also lowered at the higher temperature.

Read-out of the data at 0° C results in a higher flux level, but this may be masked by the longer Δx, i.e., smaller $d\psi/dt$!

Any write/read experiments should be done at a constant temperature, which means that several hours should be allowed for equipment warmup.

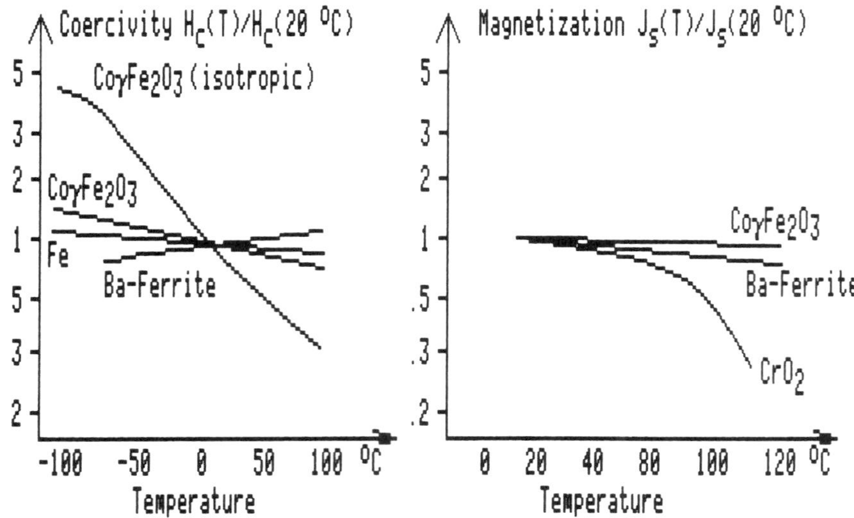

Fig. 12-16. Temperature dependence on coercivity and remanence for several magnetic particles (after Kubo, and Fujiwara).

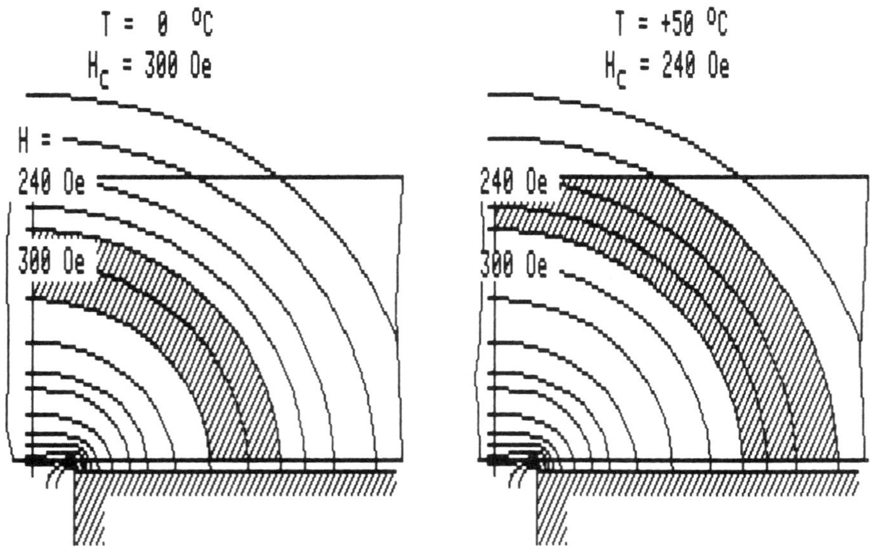

Fig. 12-17. Changes in write conditions due to changes in temperature; head field strength constant.

NOISE FROM PARTICULATE MEDIA

Noise in magnetic recording has several sources such as the electronic circuit's noise, head loss resistance noise, man-made noise (radio interference), and ultimately noise from the tape or disk coating itself.

Its origin is best understood if we at first envision a highly diluted coating where a few individual particles are well separated from each other. Each particle is a permanent magnet with a field strength that decreases as the inverse cube to square of the distance (near field).

The fields from these individual particles are sensed and summed by the reproduce head (Fig. 12-18). This random noise voltage will on a statistical basis increase proportional to the square root of their number n per volume unit:

$$E_{noise} = K * \sqrt{n}$$

where K is a proportionality factor.

A small number of particles will, when recorded, exhibit a sum remanence (in the direction of magnetization) that is proportional to the level of the recording field. Their number will, in an AC-bias recording (anhysteretic) be directly proportional to the record current (field) and their induced voltage is:

$$E_{signal} = K * n * i_w$$

where i_w is the write current amplitude, and K unchanged.

The ratio between the two voltages relates the signal-to-noise ratio:

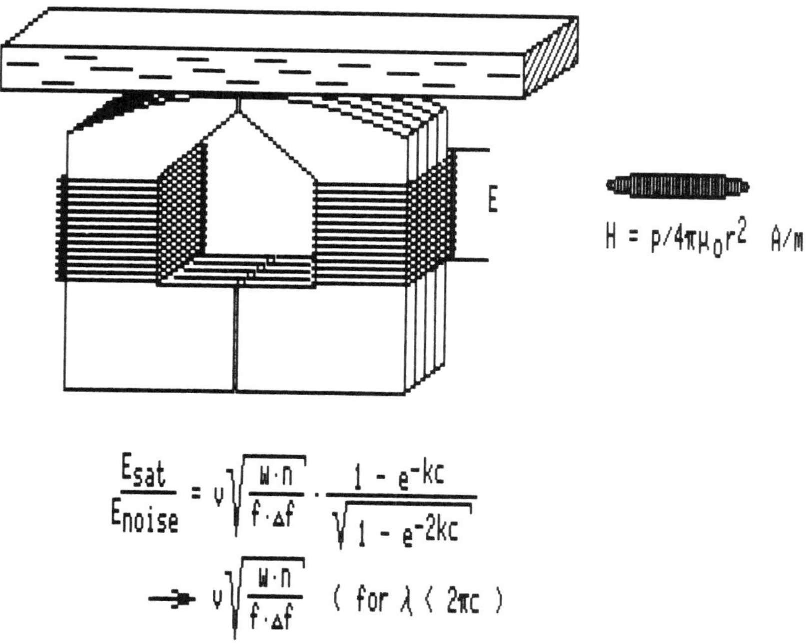

Fig. 12-18. Noise voltage generated by separate particles in a coating (after Mallinson):

v = media speed
w = track width
f = signal frequency
Δf = bandwidth observed
k = wavenumber = $2\pi f/v$
= $2\pi/\lambda$
c = coating thickness
n = N/V = particle density
(N no. particles in volume V)

$$\begin{aligned} SNR &= E_{signal}/E_{noise} \\ &= i_w * n / \sqrt{n} \\ &= i_w \sqrt{n} \end{aligned}$$

This simple relationship shows that the signal-to-noise ratio is proportional to the square root of the number of particles per volume unit. We derived this relation in a grossly simplified way and we should compare this result with a detailed analysis that resulted in the expression shown in Fig. 12-18 (ref. Mallinson).

These results show that the signal-to-noise ratio is proportional to tape velocity and the square roots of the track width and the particle density.

In the latter fact lies the reasoning for today's quieter tapes, as compared with early low density iron-oxide particle tapes. The new metal particle tapes have very small particle sizes (consequently higher density) and therefore low noise characteristics.

The noise of a tape recorder system is often measured with a ½ or ⅓ octave band-pass analyzer. This allows for a closer examination of the noise spectrum. The electronic's noise, for instance, is not flat, but boosted at both ends of the spectrum as a result of playback equalization. Erased tape noise is often slightly above the electronics' and will differ according to tape type.

AC-Bias Induced Noise

A noticeable rise in the playback noise level occurs when the tape has been recorded with AC-bias (and zero data record current).

This matter was investigated by Ragle and Smaller who suggested that the increased noise was a result of "recording of particle interaction fields under the influence of the bias." A reduction of the bias noise occurs only when the tape is separated from the head. When the tape is in contact with the highly permeable head core, magnetic images of the flux in the tape form in the head core material. The particle noise flux from coating and image combine, in the worst case, to double the mean square noise flux in the coating, giving an increase of 3 dB in the root mean square value.

This increase would be temporary if it were not for the presence of the bias field. Daniel has offered the theory that the effect of the bias is to cause the reinforced noise flux to be anhysteretically recorded on the coating as it moves away from the gap. (This is essentially the "recording" process earlier suggested).

An experiment verified the theory: a length of tape was cut from a reel, then interleaved with a portion of the remaining tape on the reel, so that coating-to-coating contact was made. The composite reel was bulk erased, the interleaved length of tape was removed, and the reel played back on a recorder: it was found that the noise spectrum was identical to the bias noise otherwise produced by the same tape. This confirmed the theory of rerecording of magnetic images of the noise.

DC Noise

The average number of particles seen by the playback head will vary slightly, as the tape moves by, and a random noise voltage is generated.

If, as an experiment, we feed a DC-current through the recording head we will observe a large increase in the noise voltage. This voltage is proportional to a varying magnetization along the tape and is related to a variation in the number of particles magnetized by the DC-current.

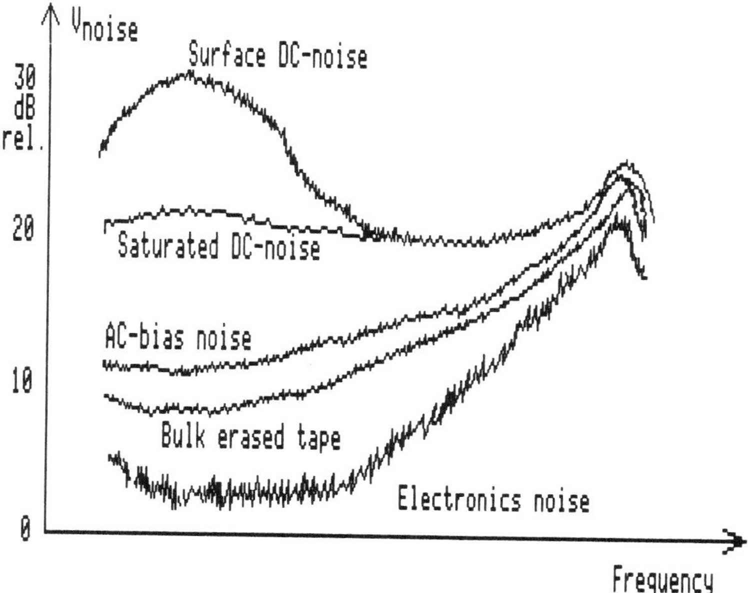

Fig. 12-19. Noise spectra measured on an audio recorder/reproducer (after Daniel).

When the DC-current is so strong that the coating is completely saturated, the noise voltage is directly proportional to the variation in the number of particles sensed by the playback head. Occasional voltage impulses will indicate the presence of particle agglomerates or voids.

If the DC-record current is reduced to a value somewhat below the saturation value, the noise level will in general increase at low frequencies. This noise reflects discrete projections (Drop-Ins) or asperities (Drop-Outs) in the tape surface. The noise spectrum of drop-outs, which cause the tape to form a tent with circumference d, is (after Daniel):

$$\Delta e = f^{1.5} * e^{-\pi df/v}$$

where Δe is the noise in a third octave bands df, f is the center frequency and v the tape velocity. From such measurements (see upper curve in Fig. 12-19) d has been determined to be about 7 mils.

A general conclusion to be drawn is that a uniform dispersion is essential in the preparation of low noise tape. Surface treatment after coating may result in improved short wavelength recording, but will have no effect on any imperfection already cast into the coating.

A note on standardization: the surface DC-noise can be measured with a DC-current that results in a maximum noise level. Or a value can be used that equals the effective level of a low frequency AC-current that results in a recorded level 10 dB below saturation (NAB, CCIR).

Modulation Noise

The playback of a non-distorted sine wave recording does not sound quite as good as its direct transmission to the loudspeaker. When the playback sine wave is displayed on an oscilloscope, it is also observed that its peaks cannot become sharply focused. This noise signal, riding on top of the recorded signal, is called modulation noise, and is similar in nature to the noise that arises when the record current is a DC-current (Fig. 12-20).

Fig. 12-20. Modulation Noise.

The signal-to-noise ratio in digital recording is the ratio of the peak pulse voltage output to the DC-record noise voltage. It is the noise that is present during the peak of a pulse that determines the bit-error-rate in a pulse peak detection circuit.

NOISE FROM THIN FILM MEDIA

Thin film media has taken on a very important role for the hard-disk drives for computer data storage. Thin films provide a higher read voltage than their oxide coated relative, with satisfactory SNR. The thin film disk would be the sole choice were it not for problems with head crashes, contamination, etc.

An ideal film would also be free from noise. The read head would sense alternating magnetizations, separated by perfectly straight walls. In reality the transitions are controlled by Neel-spikes, as earlier shown in Fig. 3-33.

See Fig. 12-21 for noise spectra; this illustration is about the only one published to date in the magnetic recording literature.

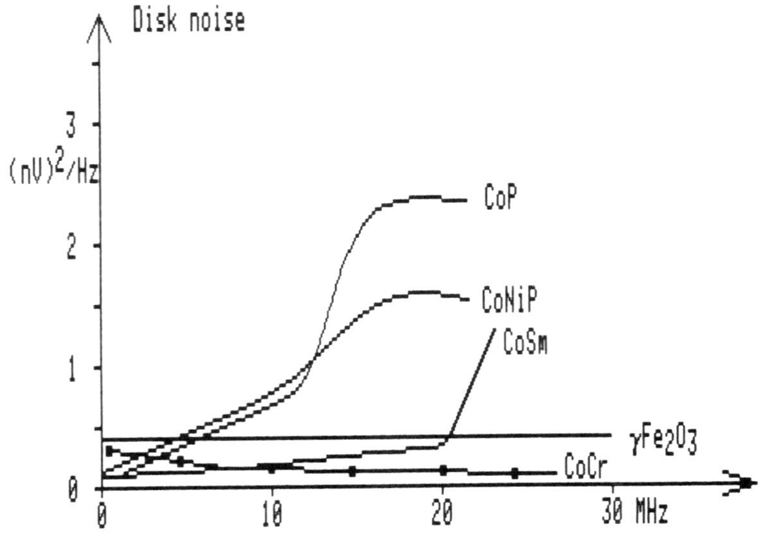

Fig. 12-21. Noise spectra representative of modern recording surfaces (after Belk et al.)

REFERENCES TO CHAPTER 12

Measurements of BH-loop

Foner, Simon, "Versatile and Sensitive Vibrating Sample Magnetometer," *Review Sci. Instr.*, July 1959, Vol. 30, No. 7, pp. 548-557

Newman, J.J., "Magnetic Measurements for Digital Magnetic Recording," *IEEE Trans. Magn.*, July 1978, Vol. MAG-14, No. 4, pp. 154-159.

Newman, J.J., "Correction to 'Magnet Measurements for Digital Magnetic Recording,'" *IEEE Trans. Magn.*, Nov. 1978, Vol. MAG-14, No. 6, pp. 1187-1188.

Kerchner, H.R., Sekula, S.T., and Thompson J.R., "Errors due to Field Nonuniformity in Vibrating Sample Magnetometry", *Rev. of Scien.Instr.*, Apr. 1984, Vol. 55, No. 4, pp. 533-536.

Bate, G., "The Cylindrical Symmetry of the Angular Distribution of Particles in Magnetic Tapes", *IEEE Trans. Magn.*, Sept. 1978, Vol. MAG-14, No. 5, pp. 869-870.

Coercivity, Remanence and SFD

Podolsky, G., "Relationship of γFe_2O_3 Audio Tape Properties to Particle Size," *IEEE Trans. Magn.*, Nov. 1981, Vol. MAG-17, No. 6, pp. 3032-3034.

Knowles, J.E., "Packing Factor and Coercivity in Tapes: A Monte Carlo Treatment," *IEEE Trans. Magn.*, No. 1985, Vol. MAG-21, No. 6, pp. 2576-2582.

Manly, W.A., "Multimodal Media for Magnetic Recording," *IEEE Trans. Magn.*, Nov. 1976, Vol. MAG-12, No. 6, pp. 764-766.

Kneller, E. and Koester, E., "Relation between Anhysteretic and Static Magnetic Tape Parameters," *IEEE Trans. Magn.*, Sept. 1977, Vol. MAG-13, No. 5, pp. 1388-1390.

Measurements on Single Particles

Scholten, P.C., "Magnetic Measurements on Particles in Suspension," *IEEE Trans. Magn.*, Nov. 1975, Vol. MAG-11, No. 5, pp. 1400-1403.

Knowles, J.E., "Magnetic Measurements on Single Acicular Particles of gammaFe_2O_3," *IEEE Trans. Magn.*, Jan. 1980, Vol. MAG-16, No. 1, pp. 62-67.

Knowles, J.E., "The Measurement of the Anisotropy Field of Single 'Tape' Particles", *IEEE Trans. Magn.*, Jan. 1984, Vol. MAG-20, No. 1, pp. 84-86.

Time and Temperature Effects

Sharrock, M.P., and McKinney, J.T., "Kinetic Effects in Coercivity Measurements," *IEEE Trans.Magn.*, Nov. 1981, Vol. MAG-17, No. 6, pp. 3020-3022.

Kloepper, R.M., Finkelstein, B., and Braunstein, D.P. "Time Decay of Magnetization in Particulate Media," *IEEE Trans. Magn.*, Sept. 1984, Vol. MAG-20, No. 5, pp. 757-759.

Sharrock, M.P., "Particle-Size Effects on the Switching Behavior of Uniaxial and Multiaxial Magnetic Recording Materials," *IEEE Trans. Magn.*, Sept. 1984, Vol. MAG-20, No. 5, pp. 745-756.

Neel, L., "Theorie du Trainage Magnetique des Substances Massives dans le Domaine de Rayleigh," as reported by Cullity, B.D. in *Introduction to Magnetic Materials* , Addison-Wesley, page 472, 1972.

Oseroff, S.B., Clark, D., Shultz, S., and Shtrikman, S., "Temperature Dependence of the Time Decay of the Magnetization in Particulate Media," *IEEE Trans. Magn.*, Sept. 1985, Vol. MAG-21, No. 5, pp. 1495-1496.

Kubo, O., Ido, T., and Yokoyama, H., "Properties of Ba Ferrite Particles for Perpendicular Magnetic Recording Media," *IEEE Trans. Magn.*, Nov. 1982, Vol. MAG-18, No. 6, pp. 1122-1124.

Fujiwara, T., "Barium Ferrite Media for Perpendicular Recording," *IEEE Trans. Magn.*, Sept. 1985, Vol. MAG-21, No. 5, pp. 1480-1485.

Noise from Particulate Media

Ragle, H.U., and Smaller, P., "An Investigation of High-Frequency Bias-Induced Tape Noise," *IEEE Trans. Magn.*, Mar. 1965, Vol. MAG-1, pp. 105-110.

Mallinson, J.C., "Maximum Signal-to-Noise Ratio of a Tape Recorder," *IEEE Trans. Magn.*, Sept. 1969, Vol. MAG-5, No. 3, pp. 182-186.

Daniel, E.D., "Tape Noise in Audio Recording," *Jour. AES,* Mar. 1972, Vol. 20, No. 2, pp.92-99.

Noise from Metallic Films

Belk, N.R., George, P.K., and Mowry, G.S., "Noise in High Performance Magnetic Recording Media," *IEEE Trans. Magn.*, Sept. 1985, Vol. MAG-21, No. 5, pp. 1350-1355.

Chapter 13

Manufacture of Magnetic Tapes and Disks

A large number of firms worldwide are engaged in manufacturing many types of magnetic tapes and disks. These products are all of similar construction, although there may be subtle differences between the different brands and different types of tapes.

The basic requirements for a magnetic tape are as follows: the coating shall be completely uniform and have a perfectly flat surface, which assures good contact with the recording and playback heads. The combination of the coating and film base shall be completely pliable and at the same time possess adequate mechanical strength, so stretching or breaking of the tape is prevented. And finally, it should be completely insensitive to storage, temperature changes, and humidity changes.

The vast majority of flexible media coatings are basically composed of a paint with magnetic particles instead of color pigmentation. These coatings are used for all diskettes (including the disk inside the 3½ diskette; diskette cartridges made by Iomega; and by Kodak in their 5-10 Mbyte disks), and for virtually all types of tapes—audio, video, computer, and instrumentation.

In the past rigid disk media also used a particulate coating, but solid metal films have been developed that provide higher outputs and better SNR (signal-to-noise ratio). Such disks are therefore found in all modern hard disk drives. Application of metal films in coating flexible tapes have shown great promise, but they are limited in practical applications due to corrosion and higher susceptibility to damage. A sophisticated and lasting lubrication is required, in particular where metal heads are used.

A modern magnetic tape consists of a 25 μm (= one mil) thick plastic film with a coating of magnetic particles in a binder, with a thickness of from 2.5 to 10 μm (= 0.1 to 0.4 mils); diskettes are made on a thicker base film, typically 75 μm (= three mils). Fig. 13-1 shows the cross-section of a magnetic tape as viewed through an electron microscope. The coating appears to consist of a binder material with a large number of magnetic particles. The binder

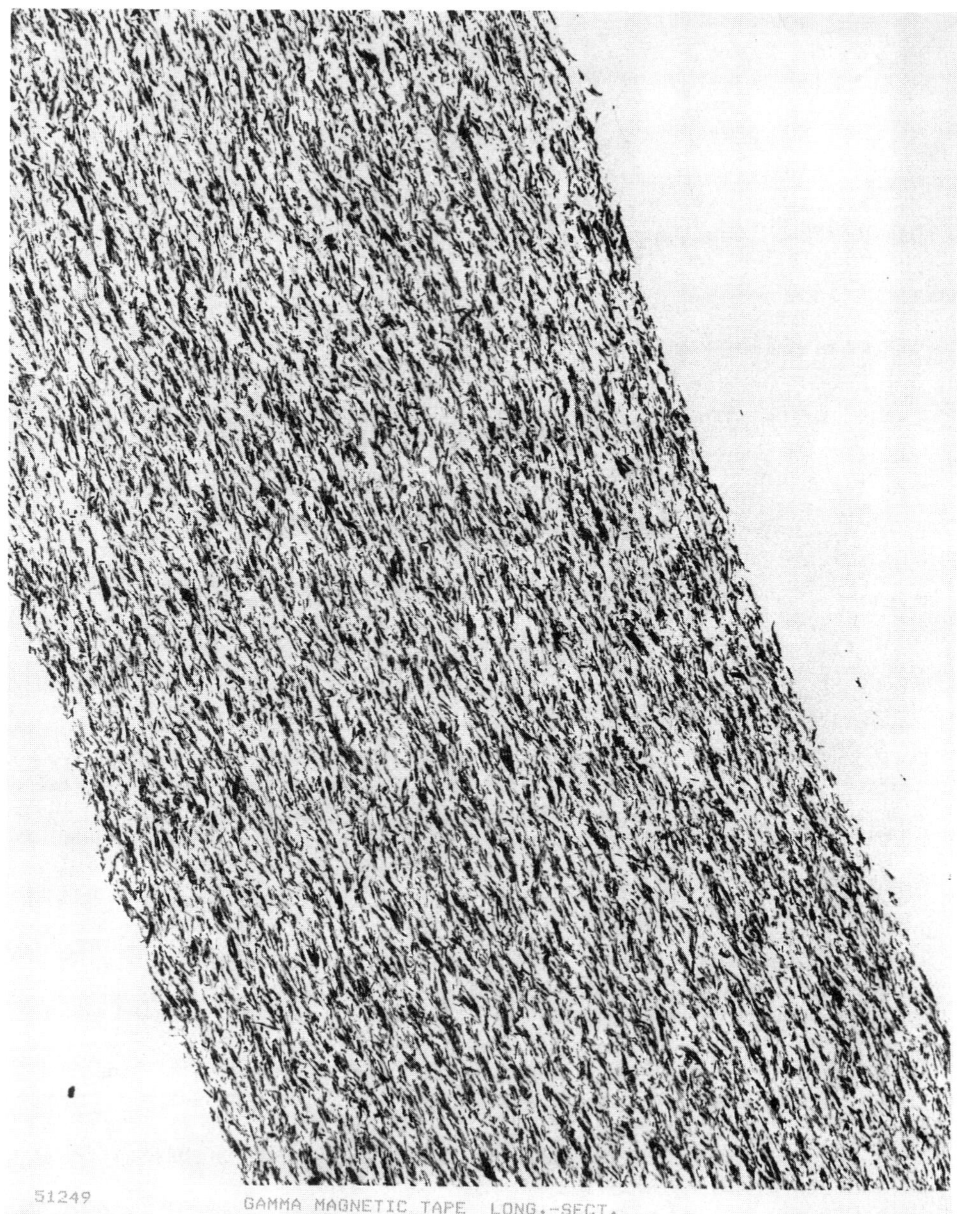

Fig. 13-1. SEM photos of cross section of a coating for a magnetic media (courtesy Memorex Corp.).

is, in essence, the cement that holds the magnetic particles together when the dispersion has been applied to the base and the solvents have dried.

The base film is commonly a polyethylene terephthalate (PET) material; fabrication and properties were covered in Chapter 11.

The materials and processes used in making flexible particulate media and solid metal film will be covered in this chapter. Although the detailed chemical formulations and processes are trade secrets, literature is very scarce. A general outline of magnetic tape and disk manufacturing is possible however. Considerations in the development of a magnetic tape coating are well summarized in papers by Sischka and by Naumann and Daniel.

This chapter is mainly describing the particulate media, which most users will actually handle. The plated rigid disks used in the so-called hard drives are inaccessible and proprietary, so the reader is referred to the scant literature on the making of these disks.

BINDER INGREDIENTS

Typical ingredients for the binder holding the magnetic particles are shown in Fig. 13-2. The chemical formulation of the binder material depends largely upon the material used for the base film since it must adhere very strongly to it. Since most tapes today have plastic bases, we find that the binder dispersions usually contain the following components.

Binder Material

A binder must satisfy several basic requirements. It must be totally dissolvable in a solvent so the magnetic particles can be uniformly dispersed, and remain so after the binder has solidified again as a coating. It must then also possess the necessary adhesion to the substrate and have strength and elasticity.

Early tapes were made on a base film of cellulose acetate with binders of polyvinyl chloride plastic and later cellulose nitrate of polyvinyl chloride. Both were closely related to the base film, and displayed excellent adhesion and flexibility, but were sensitive to light and heat.

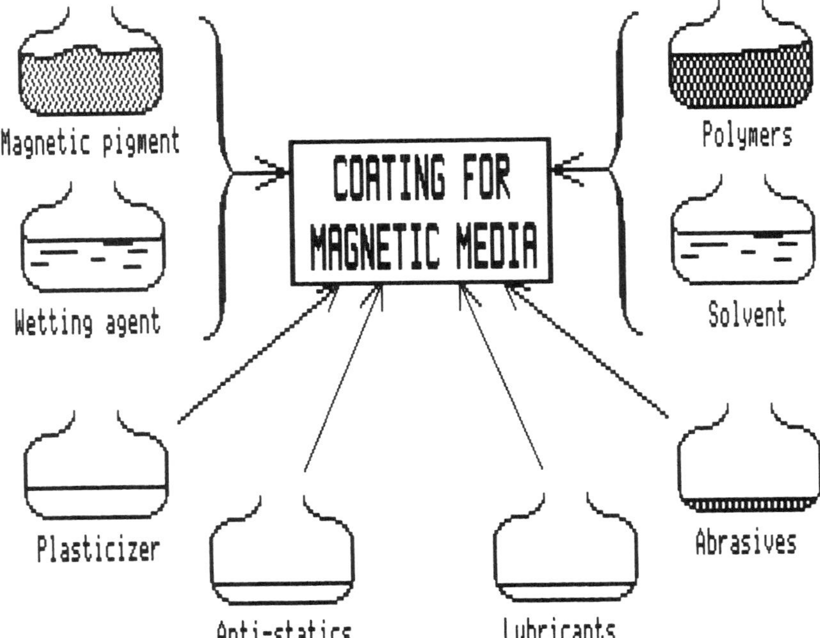

Fig. 13-2. Typical ingredients for a magnetic tape coating.

The most commonly used binders today for use on PET film are polyurethanes, which range from thermoplastic to thermoset, or polyvinyl, like "Saran", which is thermoplastic. A thermoplastic material can repeatedly be heated and thereby softened to take a new shape; this plays a role in the general plastic industry since scrap material can be re-used. A thermoset material is "fixed" during a heating cycle, and can no longer be made soft.

One important consideration in the selection of the binder material is the amount of solvent required to dissolve the material. All coating plants have strict requirements as to the recovery of solvents rather than letting them escape into the atmosphere where they contribute to air pollution. The recovery process is costly and energy consuming.

A remarkable solution would be a binder that merely required water rather than an organic solvent. This would require a binder that could later be treated to become non-water absorbent.

Solvents

A solvent plays an indispensable role in dissolving the binder material so it becomes a liquid into which the particles are dispersed and which can readily be applied to the base film. Two or more solvents are generally included in dispersions.

The process of dissolving the binder can most easily be explained as a breaking of the crosslinking and binding between the binder molecules. The cause is predominantly due to the thermal motion of solvent molecules that cause a weakening of the intermolecular forces. If these forces between the solvent molecules equal those between the binder molecules, then the latter becomes dissolved. But if the forces between the solvent molecules are greater than between solvent and binder molecules, then no solution takes place; this is also the case when the forces between the binder molecules are greater than between solvent and binder molecules.

The solution process is quite complicated and requires a good deal of experimentation in order to get the optimum formulation and process. This includes the drying process where the entire amount of solvent is removed from the binder. It must leave a stress free polymer structure where cross linking reoccurs in an orderly fashion, well bonded to the substrate.

Plasticizer (Softener)

Using solvent alone to prepare the binder would leave a very hard and brittle film on the substrate. Polymers are characterized by a *glass temperature* below which the molecular network is rigid with no allowance for translations or rotations between molecules. For nitrocellulose, for example, this temperature is 160 degrees C. A lowering to zero degrees C is achieved by adding 20 percent Rizinusoil (ref. Braginskij).

The plasticizer molecules must work their way into the polymer structure to be effective. The amount used may equal or exceed that of the binder, and the material must be such that it does not wander about and, for example, settle on the PET backing of a tape, which is possible in a wound reel of tape.

Wetting Agents

All pigments need to be surface treated prior to media production in order to improve their "wettability" which enhances their uniform dispersion in the binder. The particles used in the coating are magnetic and will therefore attract each other. They also tend to agglomerate into lumps; it is the purpose of the wetting agent to break up such agglomerates and provide for the best possible dispersion of the particles. Examples of agglomeration are shown in Fig. 8-7. Such agglomerates deteriorate the signal-to-noise ratio of a recording.

Small particles do also have electrically active surfaces, and the charges hold a film of air which in effect is bonded to the surface. This film of air must be replaced with a film of solvent containing the binder. A substantial amount of work is necessary to do this since the air film is firmly bonded to the pigment surface. This work is done by a milling device where the dry oxide is placed, and then small amounts of a wetting agent are added incrementally. The particles will begin to form balls that eventually coalesce into a continuous high viscosity mass.

The wetting agent is often lecithin, not the commercial liquid type used in the food industry but rather an oil free type now prepared for the media producers (ref. Lueck). Long chain fatty acid amines or polyglycerides are also used.

The wetting process is quite complex and beyond the scope of this book (see Braginskij pp.179-194). It is nonetheless interesting to note the very large area that the wetting agents must work on and that must be uniformly coated with the binder. An appreciation thereof is achieved if the reader considers the area of one cubic centimeter; it equals 6 square centimeters.

Now divide the cube into 8 smaller ones, this increases the area by 6 square centimeters to a total of 12. Divide once more, now into 8 × 8 cubes, and the area increases to 24 cm^2. If n is the number of times we divide the cube, then the area is 6×2^n cm^2, while the length of one cube edge is $1/2^n$ cm. Set n = 16 and you will find a cube with side length = 0.156 μm which is about the average size of magnetic particles used. The corresponding area has grown to 39.2 square meters! Compare this with the number in the column for the specific surface area of particles in Table 11-1.

Anti Static Agents

Magnetic tapes are inherently good insulators and may attain very high electric potentials that will only slowly bleed off. This necessitates tape transports that have grounded guides and heads, or they may act as Van deGraaff generators. Electrostatic charges on a tape not only cause noise when they discharge through a head assembly, but will most certainly attract airborne dust particles that cause drop-outs.

Carbon is added to dispersions in order to reduce the electrical resistance and thereby prevent build-up of electrostatic charges. The carbon can be in the form of graphite or soot. Both will weaken the mechanical strength of the coating and yet must occupy a fair amount of the binder volume to be effective (10 to 25 weight percent of the binder, or 2 to 6 weight percent of the total coating).

This binder weakening can be avoided if the carbon is applied as a back coat. Polyethylene and polypropylene are both excellent binders for soot. The back coat technique brings two benefits; any differential expansion (and hence cupping) of the tape is compensated and the danger of layer-to-layer slippage on the reels is greatly reduced. This reduces the danger of cinching.

Carbon powder originates in the soot from gas flames burning with insufficient oxygen supply. The carbon particles are recovered by placing a water cooled baffle in the flame and later scraping off the soot. Another method obtains the graphite from a thermal decomposition of gas in a preheated chamber; the end product is acetylene graphite if acetylene is decomposed.

Graphite is not an amorphous structure as earlier believed; modern electron microscopes have revealed a micro crystalline structure, held together by the Van der Waal forces and chemical bindings. The conductivity of polymers is affected when the distance between the graphite particles decreases below 10 nm (0.01 μm = 0.4 μin). This occurs when the graphite particles occupy 10 to 15 percent of the polymer volume, or two to three percent of the total binder volume. It is also found that the conductivity is highest for small graphite particles.

Graphite is named Carbon Black in the media industry. The particles are considerably finer than the magnetic particles (ref. Burgess):

Carbon Black:	Particle diameter:	Surface area:
(fineness level)	μm	m^2/gm
Low	0.09	24
Medium	0.03	88
High	0.01	259

Processing techniques and associated parameters are described in Burgess' paper, and in Braginskij, pages 194-201.

Lubricants

To overcome stickiness and scrape-flutter problems a lubricant is generally added to the binder material. Great care is exerted in the type selection and the amount applied, since it may transfer to the backside of the base material. This would result in poor friction characteristics between the capstan and the rubber puck. Lubricants should be selected and applied so they become an integral part of the polymer structure.

There are many candidates for the lubricant. In modern tapes and disks with very smooth surfaces the lubrication problem is two-fold. The primary function is to lubricate the binder-head interface, the latter being ferrite/ceramics or metal; this problem is traditional. The next function is to prevent, if possible, the build-up of binder polymer molecules onto the head surface (''clear varnish'', see Chapter 30). The frictional properties of a polymer-to-polymer interface invites stick-slip behavior into the media motion, and is a degenerative mode where the polymer build-up increases until spacing losses disrupt operations.

The lubricants are traditionally added to the coating mix and become part of the binder system, acting at times also as a plasticizer. These materials are fatty acids (ref. Mihalik).

Another approach places the lubricant on the surface of the dried coating. The most used lubricants for this method are the perfluoro-polyethers (ref. Bagatta, 1984). These materials do not interact with most plastic materials and are practically insoluble in most organic solvents. They are stable up to 260 °C, and are very efficient in reducing the dynamic coefficient of friction and the wear rate of tested polymer materials in sliding contact against head surfaces (tested by Fulmer Research, England; ref. Bagatta, 1986).

The method of application to the surfaces vary: for rigid disks a spray system is generally used, while for flexible disks sometimes the dip coating is preferred. For tapes an additional roll coating machine is used.

Abrasive Agents

Only recently were abrasive powders introduced into the binders to remove any polymer build-up on the heads. The material is typically aluminum oxide, and is included in a very sparse amount to avoid excessive head wear. (In some audio cassettes a few inches of the leaders are mildly abrasive in order to remove any build-up on the heads; head cleaning as such with a Q-tip has been eliminated.)

Properly dispersed alumina particles provide a gentle cleansing action with little wear or damage to the head. The average particle size is 0.4 to 0.6 μm (=16 – 24 μin). The variance is controlled to eliminate very large particles that could cause damage, a scratch in the head surface, or a head crash in a flying head application. The limit for the maximum particle size varies from five to less than one micron (200 to 40 μin).

The amounts of alumina used vary from one producer to another; general guidelines are (ref. Crowe et al.):

Rigid disks	0 - 1 percent
Flexible disks	0.5 - 2 percent
Computer tapes	0.5 - 2 percent
Video/audio tapes	1 - 3 percent

Other Additives

A host of other additives for binders are cited in the literature, each contributing to an improvement in one or another characteristic of a magnetic media. An example is oleic acid, a fatty acid still used in some applications. It serves as an effective plasticizer, but may form a thin film on the media surface that most likely is accompanied by stick-slip friction (ref. Mihalik).

Other additives modify the basic binder polymer. Crosslinking (thermoset) is enhanced by isocyanate, and the result is less layer-to-layer adhesion, better scratch resistance, better cohesive strength, and more resistance to solvents than polymers which are cross-linked.

The entire formulation issue is complex and again reflects the trade-off decision processes in magnetic recording as a whole. The reader is referred to papers by Williams and Markusch (polyurethane coatings), Mihalik (additives), and Brown, Ansel, Laskin and Schmid (new E-Beam curable coatings).

DISPERSION OF THE MAGNETIC PARTICLES

The dispersion process distributes the oxide particles uniformly throughout the binder. There must be a sufficient amount of binder available in order to thoroughly disperse the particles, and the process must be one of true dispersion, not forcible milling. The particle agglomerates must be broken up by the shearing forces in the agitated binder liquid, not by contact or collisions between agglomerates.

The destruction of an agglomerate is illustrated in Fig. 13-3, simplified. The two spheres will move as shown under the influence of the shear force τ, produced by agitation of the binder liquid. The parameter K is given by (after Braginskij, p.208):

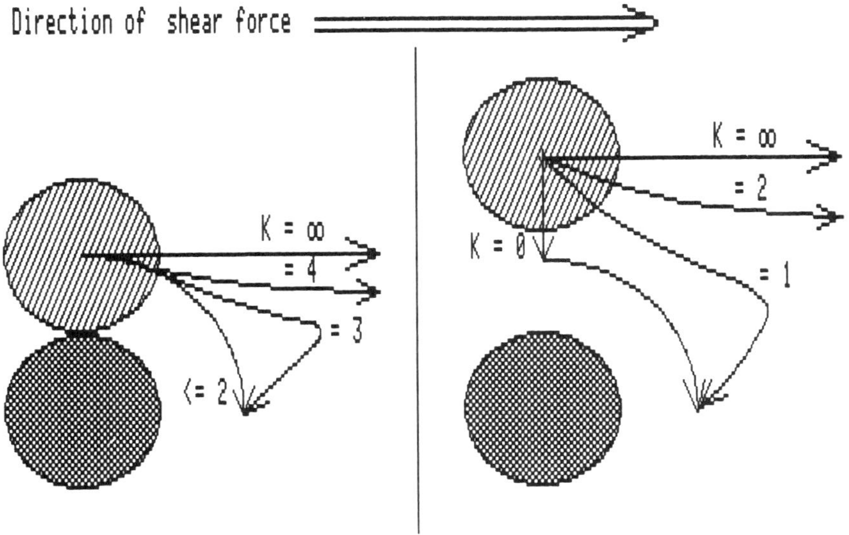

Fig. 13-3. Shear forces break down an agglomerate structure (after Braginskij).

$$K = 6\pi R\tau/F_a \tag{13.1}$$

where:

R = radius of spheres
τ = shear force
F_a = force between spheres

The force between the spheres is predominantly magnetic and varies inversely proportional to the distance squared for the far field and to the distance cubed in the near field. Hence this force is large for very small particles such as pure iron (which in addition has higher magnetization). From Fig. 13-3 it is clear that K must exceed a value of three for particles in contact, and less for particles not in contact. (The exact number is irrelevant in this discussion; the formula was introduced to characterize the action of breaking up agglomerates.)

Formula (13-1) shows that a certain shear force must be exceeded, or no dispersion will take place. It is further advantageous to continuously alter the direction of the shear forces, thereby increasing the probability of dispersing all agglomerates.

The Dissolver

A traditional dissolver is shown in Fig. 13-4. It is used to produce the shear that will dissolve the solid binder material parts into the solvent, and often to disperse the magnetic pigments. The rotating blade(s) and speed plus the dimensions of the container are carefully designed to produce a laminar flow as shown. The rotational speed of the dissolver disk must reach a

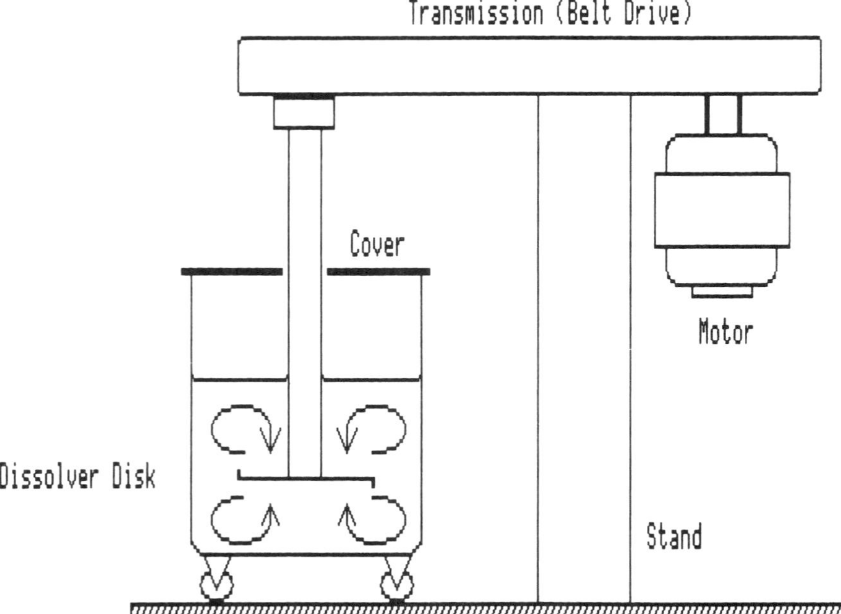

Fig. 13-4. Traditional dissolver with stirrer (dissolver disk).

certain speed so the necessary shear forces are produced; otherwise the machine will merely act as a stirring or milling device for the solvent/binder mix.

The process can be monitored by observing the energy consumed by the motor drive. It will increase rapidly during the first few minutes and decrease during the next ten or so minutes. This means that the dissolving process is completed within 15 to 20 minutes.

The geometry of the rotating blade is of course of paramount importance for the process. Its efficiency is improved by providing its perimeter with small teeth-like structures.

The Kneader

The viscosity range is limited in the dissolver; it will in the limit only move the material in the immediate vicinity of the stirrer. Many modern formulas call for a high viscosity which can be handled a so-called kneader. It is also becoming popular in the pre-wetting process of the magnetic pigments. They are loaded dry into the kneader and the wetting agent is added incrementally. Small balls start forming and they eventually coalesce into one large ball, which becomes the premix to which binder and the other additives are added (ref. Missbach, Lueck).

The Ball Mill

The oldest and most common dispersion method is a ball mill, which is simply a jar or a bell that is partially filled with a dispersing media of either metallic or ceramic balls, cylinders, rods, or a random shape such as sand. The time of the milling process may last from a few hours to several days, and it is dependent upon the chemical composition of the binder and the particles used. Too long of a milling time has been found to break down particles, so the process is therefore aimed at shorter milling cycles. A horizontal mill is shown in Fig. 13-5.

The horizontal ball mill consists of a slowly rotating metal or porcelain drum, provided with a cooling jacket. The drum is partially filled with the milling media and their motions are greatly influenced by the rpm of the drum, from that of an avalanche to a waterfall, as shown in Fig. 13-6.

The size of the balls range from a couple of millimeters in diameter down to slightly less than one millimeter. The optimum size depends upon the operating conditions of the mills, rpm, mix viscosity, etc. It is also important that the dispersion takes place by generating liquid shear forces in the small volumes that exist near the points where the balls either touch or are very close to each other. The action must not develop into one where too many particles break up into smaller particles that are superparamagnetic.

A vertical mill is shown simplified in Fig. 13-7. It completely encloses the mix, which is advantageous, and the shear forces are produced by the rotating disk(s). These are available with different cross sections, and different geometries such as a worm or a set of eccentric rings. The latter can now become a horizontal mill again.

Filtration

Lumps, aggregates, agglomerates, and down right foreign particles cannot be tolerated in the mix for the coating process. The mix must pass through one or more filters. Excessively large particles are prevented from passage, either at the filter surface or deeper down in its structure.

Sintered metal filters were once used but have now been replaced by pleated polypropylene filters, with an absolute rating in the 2.5 to 5 micron range. The polypropylene yarn is graded in size as well as closeness in structure, as shown in Fig. 13-8. This provides the deep filter action.

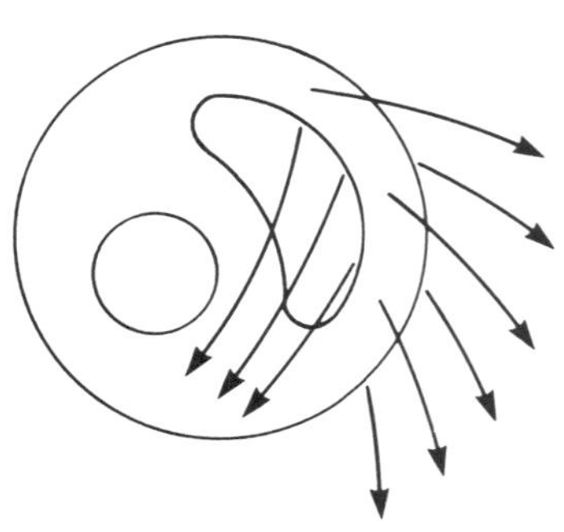

Fig. 13-5. 20 Liters Horizontal Media Mill. The media agitation is done by eccentrically mounted disks on a rotating shaft (courtesy of Netzsch Incorporated).

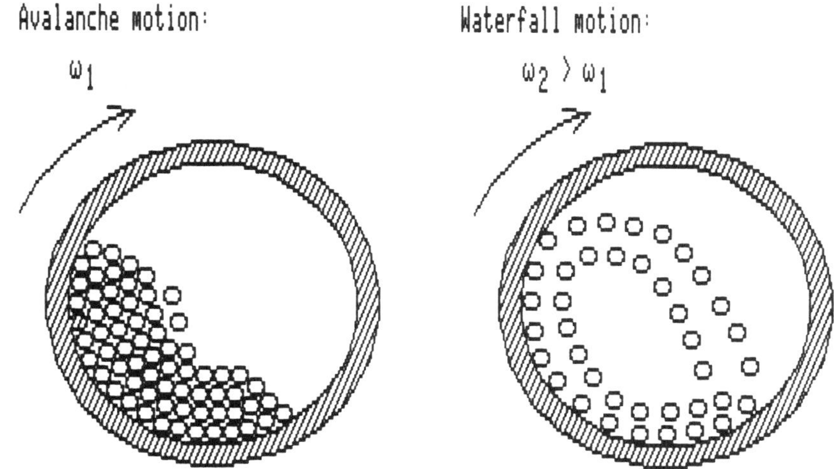

Fig. 13-6. Ball mill, horizontal.

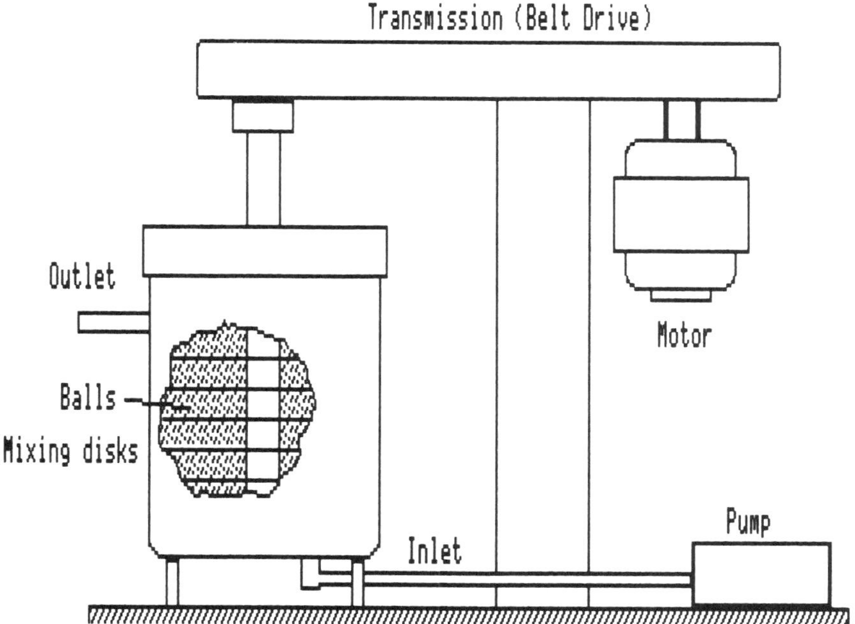

Fig. 13-7. Ball mill, vertical.

COATING OF FLEXIBLE, PARTICULATE MEDIA

After the milling process, the material is fed to a coater which applies the coating onto the PET film. A complete coating line is shown in Fig. 13-14, while the coating process is shown simplified in Figs. 13-9 and 13-10. The coating machinery is really a giant tape transport. Its function is to feed a cleaned and stress relieved PET film at a constant speed to a station where

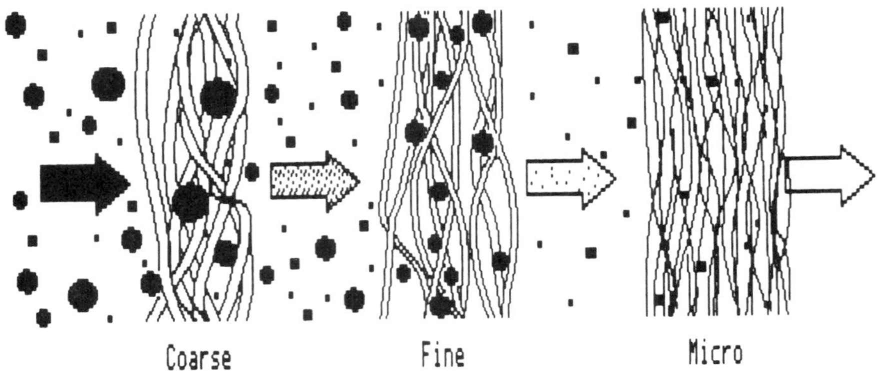

Fig. 13-8. Graded filter, using polypropylene yarn (after Missbach).

the wet coating is applied. A following station orients the magnetic particles in the coating and the coating must then be dried.

The drying is done in one or more zones in an air ventilated duct, and further heating of the coating causes it to cure into a hard yet pliable surface. The surface finish is at this point fairly rough, and therefore dull to the eye; this will cause spacing losses later on, and the surface is improved by treatment in a calendar station. Here the coated web loops through several nips existing between heated, polished steel-to-cotton rollers, pushed together with a very high force. The result is a high gloss finish coating. Care must be exercised so that the coating does not become over-stressed, which can cause a number of particles to break into the smaller, superparamagnetic particles.

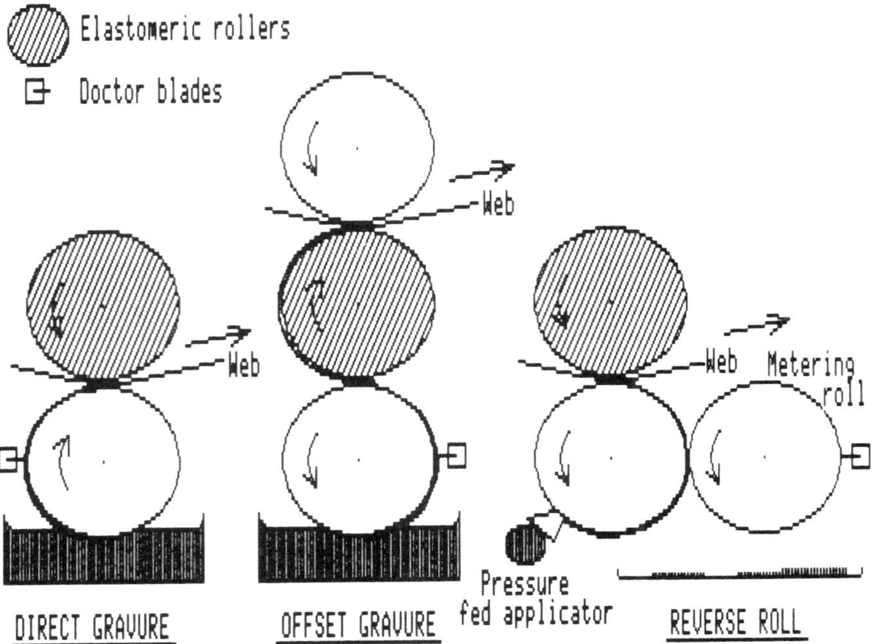

Fig. 13-9. Nipped coating processes.

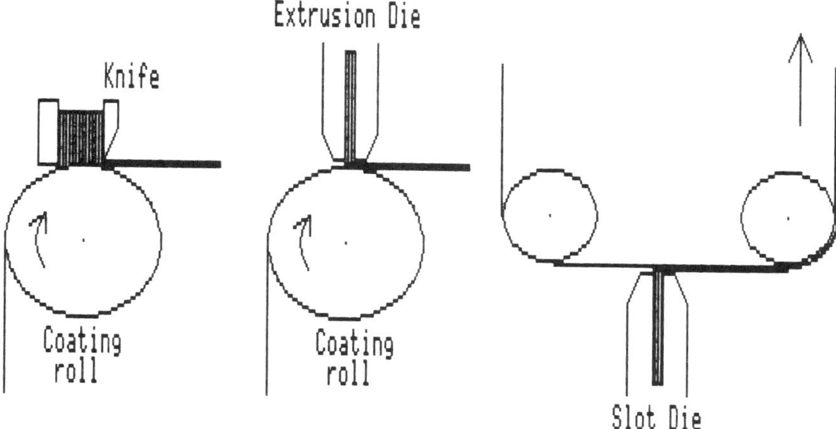

Fig. 13-10. Flow coating processes.

The finished web is wound onto a roll that may proceed to "slitting" where it is cut into the planned tape widths, or it may go into storage for further aging (crosslinking). Another destination for the web may be back to the front-end of the coating line to be coated on the backside, either with carbon-black for a high quality video or instrumentation tape, or with another magnetic coat so it may later be stamped out into double-sided diskettes.

The design and construction of such a coating line have much in common with paper production lines and instrumentation tape transports. Web guidance, tension controls, servo sensors, and servo controls operate within 10 milliseconds in a modern coating line—that is a servo system with a 50 Hz bandwidth (ref. Landskroener, Brakinskij).

Methods of Coating

Several coating processes are used and are roughly divided into two categories as shown in Figs. 13-9 and 13-10. The *direct gravure* process utilizes a gravure roll (a cylinder with etched cells) that picks up the coating from a pan or a pressure fed applicator. Excess material is wiped off with a doctor blade before the coating material is transferred to the PET film in the nip between the gravure roll and the backing roll that has a resilient surface.

The pattern of the cells in the surface of the gravure roll is imprinted on the PET film so the coating film is discontinuous (split). This requires a post-coating smoothing that for instance can be done by a PET sheet that wipes the coated surface. An *offset gravure coater* tends to reduce the split film patterns. Further improvements are possible by using *reverse gravure systems*, where the applicator roll moves opposite the backing roll.

A very smooth coated surface can be made with the *reverse roll coater* that has numerous configurations, one shown in Fig. 13-9. The wet coating is applied through a pressure fed applicator, and excess material removed with a metering roll, which in turn is cleaned with a doctor blade. The coat is transferred, or "pealed off" in the nip between the coating roll and the backing roll.

The other coating process category includes the so-called unnipped systems which apply the coating in a more direct fashion to the PET film. The *Knife coater* is the veteran, in use since the forties. The gap between the knife blade and the film determines the coating thickness, and is set by adjusting the backing roll to knife distance. This will necessarily include the film thickness—and the variations thereof; the latter variations can be eliminated from the metering process by the *extrusion die* or the *slot die* methods.

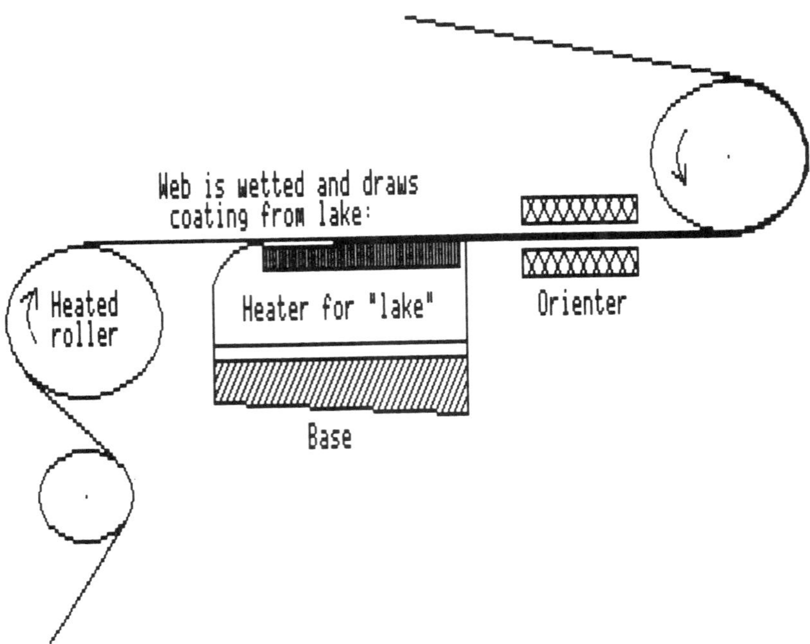

Fig. 13-11. Special Coater for Thin-Coat Particulate Media (after Garey and Lueck).

Two novel methods have recently been disclosed. One is for very thin coatings, less than one micron thick. This is very difficult to achieve, nearly impossible, with any of the methods discussed above. Without proprietary details the method is shown in Fig. 13-11. This is an artist's rendering showing how the film surface drags onto itself a thin layer (by wetting) from the lake, or pool, of coating mix (ref. Garey and Lueck).

The other method pertains to the making of a *perpendicularly oriented particulate media*. It has been tried in the past to orient the coating particles with a magnetic field perpendicular to the coating, so that the particles stand on end. The outcome, shown in Fig. 13-12, results in a coating surface with a roughness like one would see flying closely over a redwood forest. Abrasion would be a rather predominant problem.

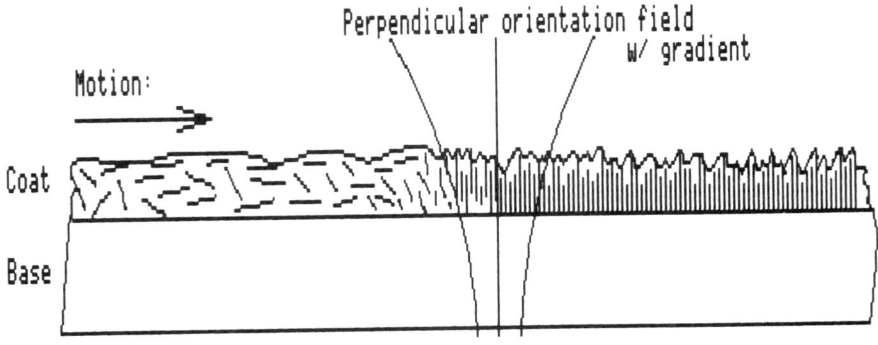

Fig. 13-12. Coating with perpendicularly oriented particles has excessive surface roughness. Gradient in orienter field pulls particles down toward and against base film.

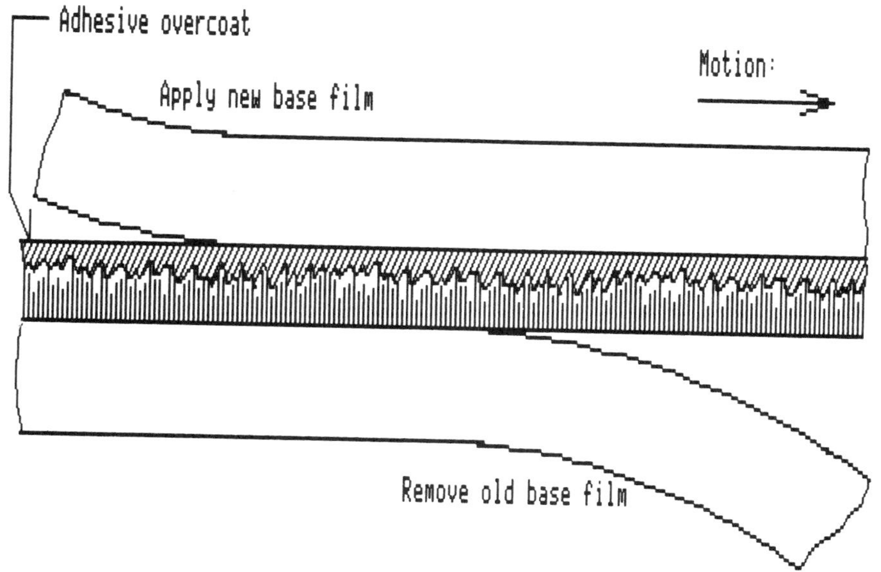

Fig. 13-13. An overcoat is applied to the tape from Fig. 13-12, and is adhered to the new base film; the old base is peeled off. This transfer of the coating provides a new, perpendicular oriented and smooth recording surface (after Speliotis).

Imagine that the orienting field has a gradient that pulls all the particles toward the PET surface, and that the coating is a binder with poor adhesion to the film. It is then possible to apply an adhesive to the coat, press it against another PET film and peel off the coating. The result is shown in Fig. 13-13, a transferred coating with all particles perpendicular and perfectly flush with the surface (ref. Speliotis).

ORIENTATION OF PARTICLES

Immediately after coating, while the binder is still wet, the coated web undergoes orientation in a magnetic field of 1000 to 5000 Oe strength, whereby the magnetic particles are aligned. This provides (as mentioned earlier) for increased output at long wavelengths. The effectiveness of orientation is generally referred to as the squareness ratio, which is the ratio between the remanent saturation magnetization J_{rsat} and the saturation magnetization J_{sat}. For randomly-oriented particles, the squareness ratio is 0.5, and for ideally and perfectly oriented particles the ratio is equal to 1.0, providing for a potential 6 dB increase in long wavelength output. Practical values for the squareness ratio normally fall around 0.70 to 0.90, which gives an increase of 3 to 5 dB.

Most tapes have the magnetic field applied in the direction of tape travel. One exception is the two-inch wide tape used in transverse scanning video recorders, and another is the web used for diskettes. In the latter a uniform distribution of the particle orientation is made over 360 degrees by a proprietary, rotating magnetic field. This results in a stable signal amplitude when the finished disk rotates through one revolution.

The mechanism of rotating particles in the viscous coat has been examined (ref. Newman and Yarbrough, 1969) and applied to actual coating processes (ref. Newman, 1978). This work plus the analysis of orienting magnet design (ref. Bate et al.) set guidelines for the design of orienting magnets.

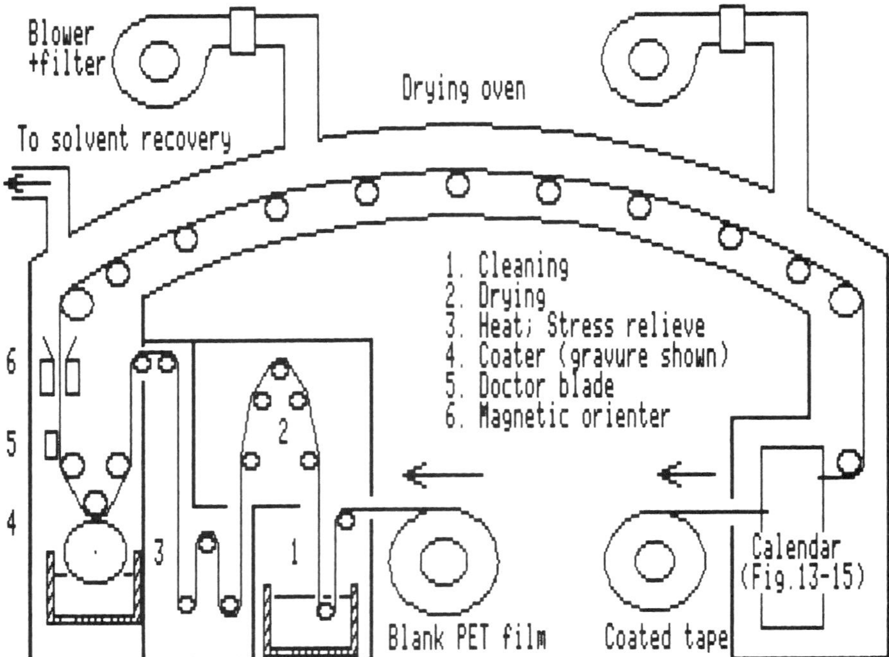

Fig. 13-14. Drying oven for thermoset magnetic coatings.

DRYING AND CURING OF THE COAT

After coating and orienting, the tape enters a drying oven, which normally is over a hundred feet long, where all the solvents are evaporated and removed with the aid of heat and airflow. The drying oven is typically designed as shown in Fig. 13-14, divided into zones. The solvents are removed (and recovered), and the degree of cure of the coating is a function of time as well as temperature; together they dictate the speed of the web.

There is a rapid removal of solvents in the first zone, yet the speed of this process must be limited or a crust will form on top of the otherwise wet coating; this will hinder the ongoing extraction of solvents from the interior of the coating, and a non-uniform coating structure results.

In the following zone the coating starts to solidify and further solvent removal now causes a complicated, slightly porous coating to form. The coating is simultaneously cured by heat, causing crosslinking between the polymer molecules.

A total removal of the solvents is not always possible within the time frame of the normal drying process. It is advantageous to store and age the coated web for a period of anywhere from one day to one month, a time period which also would allow for complete curing of the binder.

Drying by Microwaves, UV- and E-beam Radiation

An in-depth heating of the wet coating is possible by exposing it to a very high frequency electromagnetic field. This is effective when the waves are in the microwave (0.001 to 1 m wavelength) to infrared range (780 nm to 0.001 m). The inside of the coating heats up and the solvents move to the surface, evaporate and are removed by the air stream. The combination of RF-heating and a hot air stream can shorten the drying oven significantly.

Reducing the signal wavelength into the ultraviolet band causes a photopolymerization of certain polymers. No solvents are used, the polymerization process takes place during a fraction of a second, and the long drying oven has in effect been eliminated, as has any need for storage/curing.

Polymerization can also be accomplished by directing a stream of electrons toward the particular polymers, mixed with the magnetic pigments. The electron stream, or E-beam, is generated in a high voltage device, and requires only ⅕ the energy of UV-curing source. The apparatus is more complicated since the E-beam reaction with the polymers must take place in vacuum, or an inert atmosphere; and x- and gamma-rays must also be contained. The technique is new, and the paper by Brown et al. is highly recommended (detailed with many references).

CALENDERING, FINISH AND SLITTING OF TAPES

The web proceeds to the final treatment named *calendering,* which compacts the binder and smooths its surface. The machinery for this process is another loan from the paper industry; one configuration is shown in Fig. 13-15.

Further finishing treatment depends on the intended use of the tape. A burnishing process is very much like the first few passes over a head and will remove any high points on the coating surface. This procedure is common for all high packing density tapes.

The web is finally slit into the finished widths, which may be from one-sixth of an inch for audio uses up to 2 inches for video uses. This final slitting process must be carefully controlled since any width variation will cause skew (improper tracking of the tape). Also, the debris materials from the slitting action must be completely removed (with vacuum), or they will cause further problems when they become attached to or embedded in the tape coating.

The slitting action is very rough on the knives. They are essentially cutting into a plastic containing abrasive particles, which shortens the lifetime of the blades and changes their dimensions which will be reflected in the tolerances that can be achieved.

The slitting machine is normally separate from the coating line. It has its own web feed and guidance mechanism, friction drive for the web and the slit tapes, and tape-up hubs for

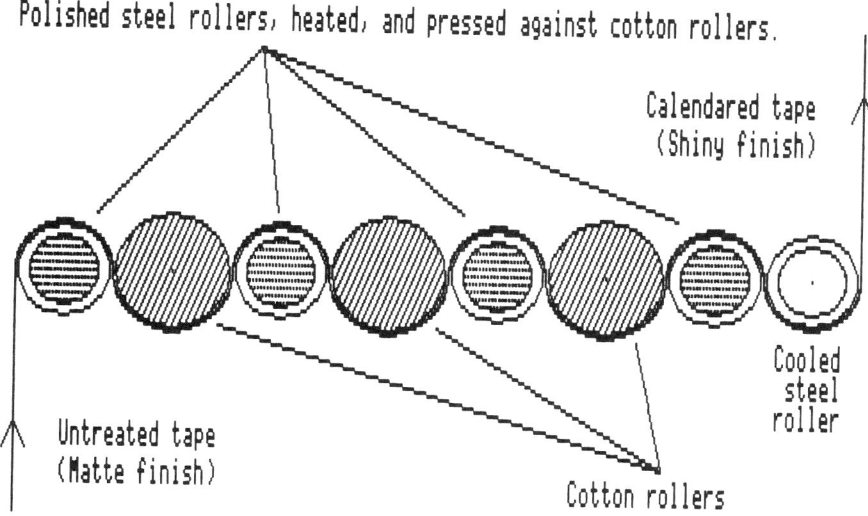

Fig. 13-15. Calender machinery.

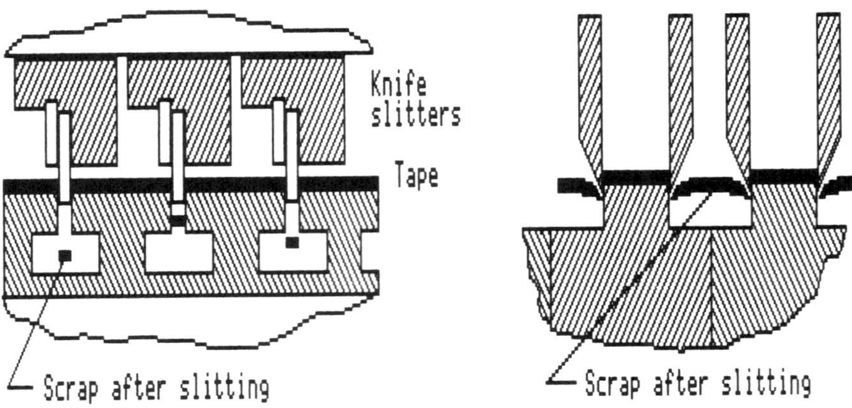

Fig. 13-16. Symmetrical knife slitters, utilizing less than 100 percent of the coated web.

the tapes. The operation on the web is a mix of knife/scissor operations, and typical slitters are shown in Figs. 13-16 and 13-17. Each version has its own advantages and disadvantages in terms of slitting tolerances, quality of cut, lifetime before sharpening/regrinding etc.

The speed of the slitting machine ranges from 80 to 200 m/min (= 52.5 to 131 IPS). Any faster speed would lead to problems with windage when winding the slit tapes onto the reel hubs; air drawn with the tape will take time to spread away when the tape is laid down on the reel pack, and a loose pack could result. This would make subsequent handling of the wound ''pancakes'' very difficult, and it is often practice to use a pack wheel for each reel hub to firmly lay down the slit tape.

The speed of the circumference of the rotating knives is slightly higher than the web speed to assure a clean cut; too high a speed would only cause excessive wear and shorten the life of the knives; a lesser speed would result in a very poor edge.

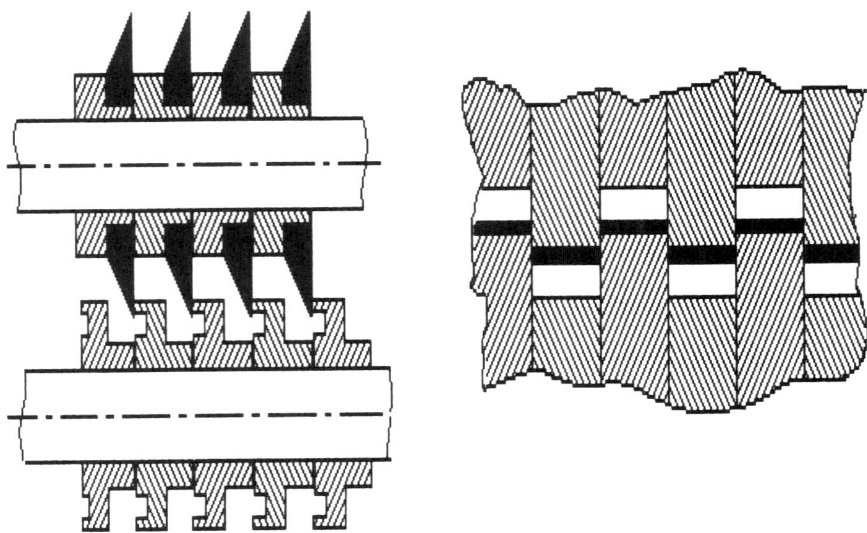

Fig. 13-17. Asymmetrical knife slitter, and symmetrical scissor slitter.

Fig. 13-18. Data tape cartridge.

Fig. 13-19. 8mm video cassette.

The slitting operation results in reels of tape, commonly called pancakes, which often need further processing. During the winding onto the pancakes the slit tape may pass over a cleaning blade, and then a tissue plus vacuum system remove all dirt and dust. The finished tape product may stay on the hubs that are provided with flanges, then sealed and packaged. The pancakes may also go on to a rewinder station for the lading of tapes into data, video or audio cassettes and cartridges. Figs. 13-18 and 13-19 show the packaging into a 450 feet data cartridge and into an 8/mm video cassette.

STAMPING, FINISHING, AND PACKAGING OF FLEXIBLE DISKS

Flexible disks with particle coatings are manufactured using the same principles as magnetic tapes. There are a couple of differences, though. The particles are not oriented in the machine direction, but as randomly as possible in the plane of the coating. And the total thickness of flexible disks are typically 75 μm (= 3 mils).

While calendering is a satisfactory surface treatment for many tapes one finds that flexible disks must undergo a polishing, or buffing, process before their surface roughness is small enough to minimize spacing losses and reduce head wear. This is done by clamping a disk against a rotating but firm support and placing the disk into contact with a buffing wheel.

The disk is placed in a protective jacket with an inside liner that protects both sides of the disk while collecting dust and foreign particles that inadvertently enter the jacket. Fig. 13-20

Fig. 13-20. 3.5 inch diskettes.

shows the construction of diskettes; the inside (fiber) liner can be chosen from a large number of materials such as rayon, polypropylene, polyesters, and copolymers (ref. Ostrowski).

COATING OF PARTICULATE RIGID DISK MEDIA

The base for rigid disks is most often a high quality aluminum substrate of thickness ⅛″ to ¼″. It is highly polished and then cleaned by dipping in acid, alkaline, and water, and then scrubbed by an ultrasonic process.

The coating is applied by either spraying or spinning, or both. The centrifugal forces on the wet coating cause it to spread out in a spiral fashion. The speed is highest at the periphery of the disk and the final coating tends to be rather thick there. This is, of course, advantageous for high density recording, but limits the useful area of a disk to a 2″-wide outer band.

The coated disks are then placed in ovens where the coating is hardened and bonded to the aluminum disk. A final step is polishing of the disk surface, using proprietary methods. The final coating thickness for high density disks is about 0.5 to 1.25 μm (20 to 50 μin).

Each disk is physically tested for coating defects such as voids, improper dispersion of oxide, wavy or sloping edges. Some surfaces are checked for hardness and smoothness. The disk is then placed into a single disk tester to check magnetic properties (output, noise) and potential defects (drop-outs, drop-ins). Finally, a head is made to fly at one half the regular height and this burns off any protruding chunks of oxide. This is called *burnishing*.

METAL COATING OF FLEXIBLE MEDIA

Coating of a continuous magnetic film on a flexible tape has been under development for the past two decades. Satisfactory performance has been achieved for the write/read signals, but the mechanical problems are only slowly being resolved.

Some of the first problems were in the area of exposing the base film to high deposition temperatures, followed by differential expansion between the metal film and the base, causing cupping. Contact problems are also evident when the metal film is run over a metal head. Ferrite heads have been shown to be a partial solution, and surface lubrication helps, but abrasion characteristics remain worse than for polymer coated tapes.

There is finally the environmental stability question of corrosion. This will always be more difficult for tapes than for rigid disks since the tapes are subject to bending shear and other stresses as they move through a transport. The stresses alternate from compressive to expansive and may in time cause cracks in the metal surface; the result is a severe, permanent drop-out, plus noise.

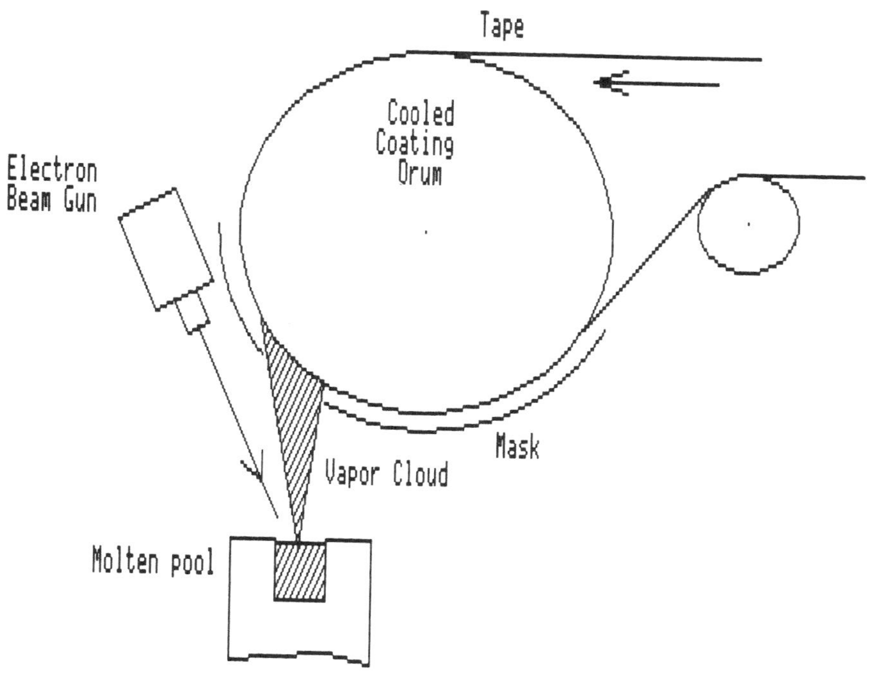

Fig. 13-21. Experimental coater for metal evaporated tapes (after Feuerstein et al.).

Figure 13-21 shows one of the more recent experimental coating devices that produce a product that is beginning to become commercially acceptable (ref. Feuerstein et al.).

COATING OF RIGID DISKS

Metallic coatings of CoP and CoNiP on aluminum substrates are used to achieve a thin coating without sacrificing other parameters. The thickness ranges from 0.1 to 0.25 μm (4 to 10 μin). A thin oxide coating would have very low magnetization, while a metallic film of comparable thickness would have magnetic moments comparable to that of thick oxide coatings. The volume factor is 1.0 and there is no internal demagnetization. It is further possible to obtain a wide range of coercivities (4 to 120 kA/m = 50 to 1500 Oe) in metal films.

The films can be deposited by several methods, such as electrochemical deposition, electrolytic deposition, and vacuum evaporation. The plated media have been in use for the past several decades, and has reached a rather sophisticated level of perfection in current production (ref. Suganuma et al., Garrison).

Composite materials cannot be deposited by evaporation due to the different vapor temperature of the ingredients. Instead, use is made of RF-sputtering techniques where the molecules to be deposited are literally knocked off the target surface and subsequently travel away to become deposited on the substrate, i.e., the disk (ref. Judy). Fig. 13-22 illustrates the process. Production of RF-sputtered disks has recently been described by Moore et al., and Drennan et al.

Ion-beam sputtering can be used to produce a film of CoCr that has the easy axis (the preferred direction of magnetization, i.e., anisotropy) perpendicular to the plane of the substrate and, therefore, enhances perpendicular recordings (ref. Gill et al.).

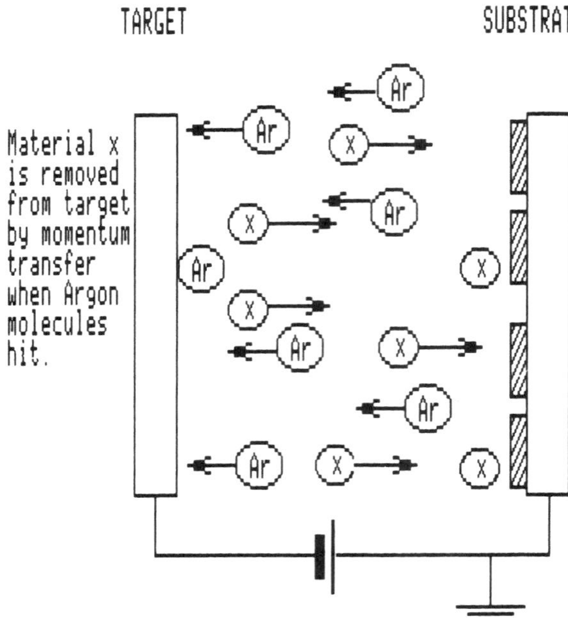

Fig. 13-22. Sputtering process, simplified.

RF-sputtering methods can also be used for the deposition of non-conducting oxide and ferrite films. The work is still experimental but the interested reader can find several examples in recent issues of *IEEE Transactions of Magnetics*.

QUALITY CONTROL METHODS

Substrates need to be tested prior to production, and several methods are useful for PET films (thickness, surface roughness, strength—see ref. Persoon, and Holloway), and for aluminum disks (ref. Morrison and Brar).

In-process measurements have been evaluated and some have been introduced into production as a means of quality control. Most of the measurement techniques are described in the next chapter, and we will look briefly at some of the sample preparations.

The liquid coating material can be sampled in a small capillary tube. A solid dispersion is obtained by immersing the sample into liquid nitrogen, and orientation is obtained by applying a field of 160 kA/m (= 2,000 Oe) during the solidification. Measurements are next made on the Vibrating Sample Magnetometer (VSM) and can be utilized to examine and follow the effect of milling, solvent changes, viscocity, stability etc. (ref. Fisher et al., further developed by Sumiya et al.). Measuring instruments and techniques have been developed that allow in-process testing, from the dry magnetic particles to the finished, calendered product (ref. Steinberg).

REFERENCES TO CHAPTER 13

Binder Ingredients; Dispersion

Sischka F.J., "Development of an Advanced Fe_2O_3 Cassette Tape Coating," *Jour. AES*, Oct. 1973, Vol. 21, No. 10, pp. 789-808.

Mihalik, R.S., "Binder/Additive Relationships," *SYMPOSIUM Mag. Media 83*, May 1983, paper no. E-2, 37 pages.

Brown, W.H., Ansel, R.E., Laskin, L., and Schmid, S.R., "Radiation Curable Coatings," *SYMPOSIUM Mag. Media 83*, May 1983, paper no. D-3, 69 pages.

Lueck, L.B., "The Wetting of Magnetic Pigment," *SYMPOSIUM Mag. Media 83*, May 1983, paper no. D-1, 21 pages.

Missbach, F.S., "Premilling, Milling, and Filtration Methology in Magnetic Media Dispersion Preparation," *SYMPOSIUM Mag. Media 83*, May 1983, paper no. B-1, 46 pages.

Bagatta, U., Corradi, A.R., Flabbi, L., and Salvioli, L., "Lubrication of Tapes with Fluorocarbon (Fomblin) Oils," *IEEE Trans. Magn.* Jan. 1984, Vol. MAG-20, No. 1, pp. 16-18.

Crowe, J.T., and Arvidson, D.B. Jr., "The Role of Abrasive Additions in Magnetic Media," *SMART Symposium*, May 1986, paper no. E-4, 13 pages.

Burgess, K.A., "Carbon Black: Effect on Electrical Conductivity in Polymer Systems," *SMART Symposium*, May 1986, paper no. E-3, 32 pages.

Williams, J.L., and Markusch, P.H., "Polyurethane Coatings," *SMART Symposium*, May 1986, paper no. D-2, 34 pages.

Bagatta, P.U., "Concepts of Advanced Lubricant Systems For All Forms of Magnetic Media," *SMART Symposium*, May 1986, paper no. WS 3-C-1, 39 pages.

Coating of Particulate Media

Landskroener, P.A., "Conventional Coating Methods," *SYMPOSIUM Mag. Media 83*, May 1983, paper no. B-2, 38 pages.

Garey, R.R., and Lueck, L.B., "Considerations in Thin-Layer Particulate Media Coating Processes", *SMART Symposium*, May 1986, paper no. WS 2-D-1, 12 pages.

Orientation

Newman, J.J., and Yarbrough, R., "Theory of the Motions of a Fine Magnetic Particle in a Newtonian Fluid", *IEEE Trans. Magn.*, Sept. 1969, Vol. MAG-5, No. 3, pp. 320-324.

Newman, J.J., "Orientation of Magnetic Particle Assemblies," *IEEE Trans. Magn.*, Sept. 1978, Vol. MAG-14, No. 5, pp. 866-868.

Bate, G., and Dunn, L.P., "On the Design of Magnets for the Orientation of Particles in Tapes," *IEEE Trans. Magn.*, Sept. 1980, Vol. MAG-16, No. 5, pp. 1123-1125.

Plated Media

Feuerstein, A., Lammermann, H., Mayr, M., and Ranke, H., "Production Coating Processes for Metal Thin Film Video Tape-State of the Art," *Intl. Conf. Video & Data 84*, April 1984, IERE publ. No. 59, pp. 45-52.

Suganuma, Y., Tanaka, H., Yanagisawa, M., Goto, F., and Hatano, S., "Production Process and High Density Recording Characteristics of Plated Disks," *IEEE Trans. Magn.*, Nov. 1982, Vol. MAG-18, No. 6, pp. 1215-1221.

Garrison, M.C., "Plated Media Manufacturing Methods," *SYMPOSIUM Mag. Media 83*, May 1983, paper no. MMS-G, 23 pages.

Gill, H.S., and Yamashita, T., "The Growth Characteristics of Ion-beam Sputtered CoCr Films on Isolation Layers," *IEEE Trans. Magn.*, Sept. 1984, Vol. MAG-20, No. 5, pp. 776-778.

Moore Jr, G.E., Seymour, R.S., and Bloomquist, D.R., "Manufacturing Thin-Film Discs," *HP Journal*, Nov. 1985, Vol. 36, No. 11, pp. 34-35.

Drennan, G.A., Lawson, R.J., and Jacobson, M.B., "In-Line Sputtering Deposition System for Thin-Film Disc Fabrication," *HP Journal*, Nov. 1985, Vol. 36, No. 11, pp. 21-25.

Judy, J., "Sputtered Perpendicularly Oriented Metal Coatings", *SYMPOSIUM Mag. Media 83*, May 1983, paper no. F-1, 39 pages.

Johnson, Jr., C.E., "The Current Status and Future Directions of Perpendicular Magnetic Recording", *SMART Symposium*, May 1986, paper No. PS-4, 16 pages.

Quality Control in Production

Morrison, J.R., and Brar, A.S., "Disk Substrate Requirements for Future High Areal Density", *SYMPOSIUM Mag. Media 83*, May 1983, paper no. A-3, 45 pages.

Fisher, R.D., Davis, L.P., and Cutler, R.A., "Magnetic Characteristics of Gamma-Fe_2O_3 Dispersions," *IEEE Trans. Magn.*, Nov. 1982, Vol. MAG-18, No. 6, pp. 1098-1110.

Steinberg, G., "Testing Dispersions and Magnetic Coatings", *SYMPOSIUM Mag. Media 83*, May 1983, paper no. TMM-2, 43 pages.

Persoon, A.H., "Characterization of Parameters for Developing Quality Control Programs for Magnetic Media," *SYMPOSIUM Mag. Media 83*, May 1983, paper no. TMM-1, 18 pages.

Sumiya, K., Hirayama, N., Hayama, F., and Matsumoto, T., "Determination of Dispersibility and Stability of Magnetic Paint by Rotation - Vibration Method," *IEEE Trans. Magn.*, Sept. 1984, Vol. MAG-20, No.5, pp. 745-747.

Chapter 14

Properties of Magnetic Disks and Tapes

In selecting a magnetic media, several factors should be considered:

- Field of application
- Frequency response; Data rate
- Storage capacity, or playing time
- Quality

The physical dimensions of disks, or diskettes, are dictated by the equipment, and the user's choice is pretty much a matter of pricing. He can pay for assurance of good durable data by purchasing brand name products of proven quality, or save money while running a risk of getting stung by buying non-brand name disks.

The playing time and the required frequency response both dictate the length of tape needed, which should be rounded up to the nearest standard configuration. By knowing the recorder's capabilities, the tape speed is established by the required frequency response and the tape length and the reel configuration are easily found from Table 14-1.

The field of application will roughly divide the tapes into four groups: audio, instrumentation, computer, and video.

Audio tapes are generally 6.25mm (.25 inch) or 4mm (.156 inch) wide tapes used by the sound recording industry, broadcast stations, and in homes. (One-half, one-inch and two-inch wide tapes are also used in sound recording studios, where several microphone outputs are recorded on individual tracks for later mixing). The 6.25mm wide tapes are typically wound on 7 or 10-½ inch reels for professional use and eight-track cartridges for entertainment. The 7-inch reels contain 1200 feet (one and one-half mil base thickness), 1800 feet (1 mil), or 2400 feet (one-half mil tensilized polyester), and are generally called out as standard, long-play, and extra long-play. The 1800-feet version seems to be the most popular, considering economy

Table 14-1. Tape Lengths, Speeds and Playing Time.

Length in feet	TAPE SPEED IN IPS 120	60	30	15	7-1/2	3-3/4	1-7/8	15/16
9600	16"	32"	1' 4"	2' 8"	4'15"	8'30"	17'	34'
7200	12"	24"	48"	1'36"	3'12"	6'24"	12'48"	25'36"
4800	8"	16"	32"	1' 4"	2' 8"	4'15"	8'30"	17'
3600	6"	12"	24"	48"	1'36"	3'12"	6'24"	12'48"
2400	4"	8"	16"	32"	1' 4"	2' 8"	4'15"	8'30"
1800	3"	6"	12"	24"	48"	1'36"	3'12"	6'24"
1200	2"	4"	8"	16"	32"	1' 4"	2' 8"	4'15"
900	1:30"	3"	6"	12"	24"	48"	1'36"	3'12"
600	1:00"	2"	4"	8"	16"	32"	1' 4"	2' 8"
450	0:45"	1:30"	3"	6"	12"	24"	48"	1'36"
300	0:30"	1:00"	2"	4"	8"	16"	32"	1' 4"
225	0:22"	0:45"	1:30"	3"	6"	12"	24"	48"
150	0:15"	0:30"	1:00"	2"	4"	8"	16"	32"

TAPE TYPE and c:	Reel size: Cassette	3"	5"	7"	10.5"	14"
Standard - 1.5 mils	300'	150'	600'	1200'	2400'	4800'
Long Play - 1.0 mils	450'	225'	900'	1800'	3600'	7200'
Extra Play - 0.5 mils	600'	300'	1200'	2400'	4800'	9600'

and playing time. The 4mm wide tapes are used in the popular cassettes with standard playing times of 2 × 15, 2 × 30 and 2 × 45 minutes.

The quality of tapes on the audio market varies widely and it is generally recommended that the user buy brand-name tapes. The so-called white-box tapes may be rejects. These tapes are likely to be high in abrasion and, in addition, quite likely to be rejects for one or another reason.

Variations in the frequency response of widely recognized tapes are minor, but may require different bias settings (and possibly equalization). It is, therefore, a good rule to stay with a given brand and type of tape once it has been selected. These comments are applicable for video tapes as well.

The ultimate test is performed by the recording and playing back of suitable program material and comparing the input with the played back signal. There should be no difference in level or frequency response. But this test does require, to be fair, that bias and equalization be properly adjusted. This test is generally referred to as an A-B test.

Instrumentation and *computer tapes* are high-quality products, designed and manufactured to meet a stringent set of requirements. Quantitative numbers are now attached to slitting tolerance, cupping, layer-to-layer adhesion, abrasion, wear characteristics, sensitivity, response, etc. While such specifications also exist for the manufacture of audio and video tapes, these are normally purchased on their established merits, where instrumentation and computer tapes are purchased against a set of specifications which from time to time are checked by the user. Instrumentation tapes are widely used in the aerospace industry and if a tape for some reason fails during an experiment, it can mean the loss of literally hundreds of thousands of dollars worth of scientific data. Thus, the requirements are more stringent.

Standards for tapes, reel, cartridges, cassettes etc. are available from EIA and ANSI; the interested reader can write to these organizations and request their catalogs and then order the needed standards. Address listings are found at the end of Chapter 28.

MECHANICAL PROPERTIES OF SUBSTRATES

Tapes

The base film for magnetic tapes and floppy disks today is almost exclusively PET or polyester (an oriented polyethylene terephthalate). In the early days *cellulose acetate* was used, but was later succeeded by PVC (polyvinyl chloride). The PCV film was not as strong as PET, nor would it withstand temperatures approaching 70°C. It did have a better surface than the first PET films, and the acetate films had the advantage of breaking clean rather than stretching under heavy loads, such as occurred during the start and stop operations in early recorders. A clean break could be spliced while a recording on a stretched tape was lost.

PET combines the strength required of a thin base film with the limpness needed for intimate head-to-tape contact.

Although polyester base materials are twice as expensive as any of the other two earlier base films, it is employed exclusively in the manufacture of precision magnetic tape for audio, computers, video, and instrumentation recorders. The strength of the PET film can be increased further by pre-stretching it. The stretching orients the long chain molecules in the film in the direction of the stretch. The disadvantage of such pre-stretched or tensilized polyester films

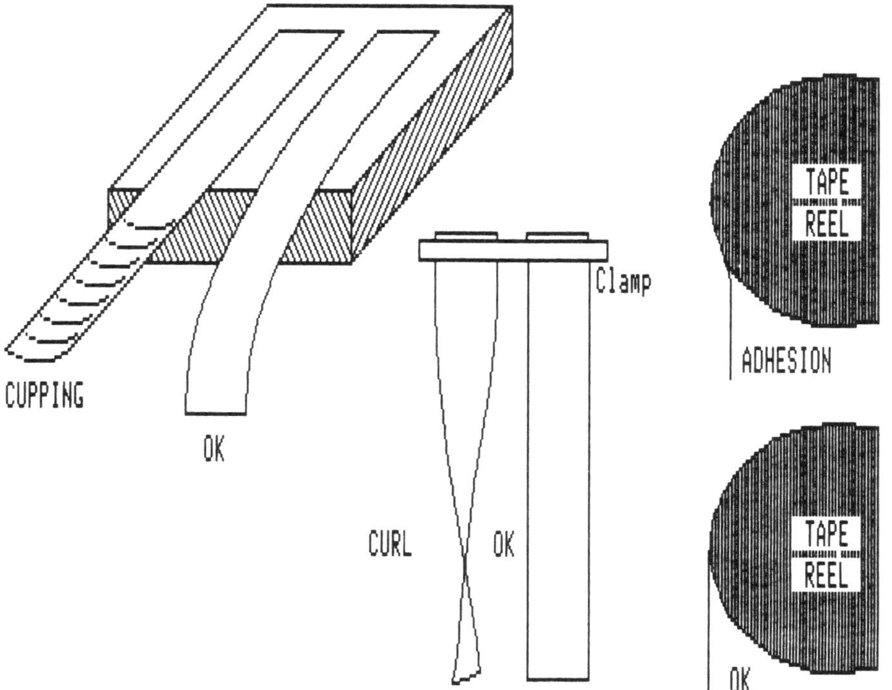

Fig. 14-1. Sketches showing the effects of cupping, curl, and layer-to-layer adhesion for a magnetic tape.

is that they have a memory; if they are re-heated beyond a certain point, they will shrink back close to their original size with consequent distortion of the recorded material.

Other types of distortion of the finished tape product are cupping, curbing and layer-to-layer adhesion. *Cupping* is illustrated in Fig. 14-1. Two tape samples are placed on a plane supporting surface. The cupped tape will stand straight out, while the better tape will bend in a smooth arc. Cupping is generally found in inferior tape and is due to improper manufacturing or differences between coefficients of thermal or hygroscopic expansions of coating and base film.

Curling results in a twist of a free hanging length of tape. This is again due to residual stress gradients in the PET film, possibly aggravated by a bad slitting process. Curl is detected by unwinding a few feet of tape, holding the reel up and looking down along the free-hanging length of tape. It should be a flat sheet, as illustrated in Fig. 14-1.

Layer-to-layer adhesion is detected by observing how easily the tape unwinds. Hold the reel vertically and slowly turn it to unwind the tape; it should unwind freely and smoothly. Any jerky motion caused by sticking to the tape pack will cause flutter in many recorders. This defect is particularly noticeable after prolonged storage of a tape.

The dimensional stability of PET films was in the past neglected due to the overshadowing problem of slitting tolerances. Tapes are slit to a final width with a tolerance of +.000″ to −.0008″ in the very best case. This 1 mil uncertainty sets an ultimate limit for the number of tracks that can be successfully recorded on one machine and played back on another; in reality differences in temperature and humidity cause additional dimensional changes.

Flexible Disks

A similar tolerance applies to floppy disks for the center hole. We must add to it the dimensional changes that take place in a PET disk. These are stress relaxations of built-in stresses created during the manufacturing process, plus temperature and humidity changes. How much and when they occur is difficult to predict. Furthermore, PET film properties vary across the web with the orientation of the polymers.

All the dimensional changes due to temperature, humidity, time and centrifugal forces may change an initially circular track into an elliptic or other non-circular track. This will obviously limit the number of tracks per inch radius of the disk.

These changes were discussed in the section on trackwidth selection in Chapter 10. There are methods for the determination of dimensional changes in a disk such as recording a pair of adjacent tracks followed by playback and comparison of the data (ref. Behr and Osborn); users of such methods must be aware of errors involved that may be substantial (ref. Izraelev).

Rigid Disks

The requirements to the aluminum blanks were covered in Chapter 11. The surface characteristics can be evaluated by a profilometer, or optically which does not disturb the surface. The surface properties of the substrate will reflect through the magnetic overcoat, particulate or metallic. This is not just in the form of asperities or the overall smoothness, but affects the magnetic properties of film as well (ref. Thompson and Mee).

MECHANICAL PROPERTIES OF COATINGS

Surface Properties

The magnetic coating on a tape is fabricated from magnetic particles dispersed in a suitable binder. The overall coating thickness and the surface finish of the coating bear a significant

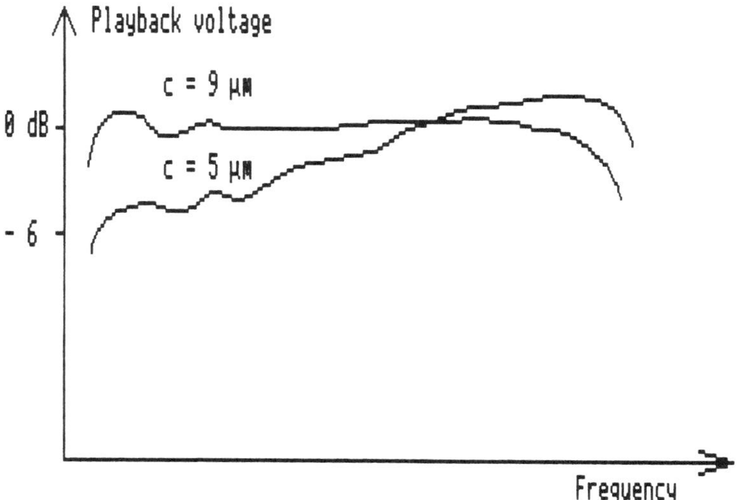

Fig. 14-2. Amplitude versus frequency responses for two tapes with different coating thicknesses, each with bias optimized at long wavelengths.

influence on the quality of recordings. For optimum recordings, different coating thicknesses require different amounts of high-frequency bias currents and a recorder is therefore tuned up for a particular tape that should be the only tape used for future recordings. The difference between a thick coating and thin coating is illustrated in Fig. 14-2. The thicker coating (''high-output'') gives a higher output at low frequencies but at the price of low output at high frequencies (overbias).

Digital storage response is measured by resolution which is the ratio between the outputs at the frequencies 2f and 1f, expressed in percentage (Fig. 14-9). Here, like above, the outcome depends upon the media thickness, write current, packing density etc., and no resolution rating can be assigned to a media without qualifications.

The surface characteristics have a significant bearing on the high-frequency performance of a magnetic tape. Any spacing between the magnetic heads and the tape surface causes a reduction of high-frequency (short wavelengths) response, and a rough tape surface does, in effect, act as a spacing between the magnetic coating and the head. In the manufacture of magnetic tape it is therefore common practice to surface-treat the tape to obtain a finished product that has as smooth a surface as practical.

Surface flatness can be evaluated by means of a profilometer. This instrument moves a very small diamond needle across the tape surface and any motions of the needle (like a phonograph pickup) are amplified and the output waveform plotted on a paper chart (Fig. 14-3A). Although the diamond stylus may deform the tape surface under the high unit area pressure, it is still a useful instrument in comparing various tapes. Photographs of typical tape surfaces are shown in Figs. 18-13 and 18-14.)

Another method of gauging the surface flatness of a recording media results when a MR-head is placed in contact with the moving surface since localized heat from friction between the surface asperities and the MR-element results in a voltage output (ref. Talke et al.).

A superior technique has been employed in recent years by using laser interferometry (ref. Wahl et al., Perry et al, Robinson et al.). The instrumentation is shown simplified in Fig. 14-3B, and two scans of a tape surface are shown in Fig. 14-4 (ref. Robinson). A laser beam, frequency shifted by an acousto-optic cell, is focused on a moving surface and reflected. Sur-

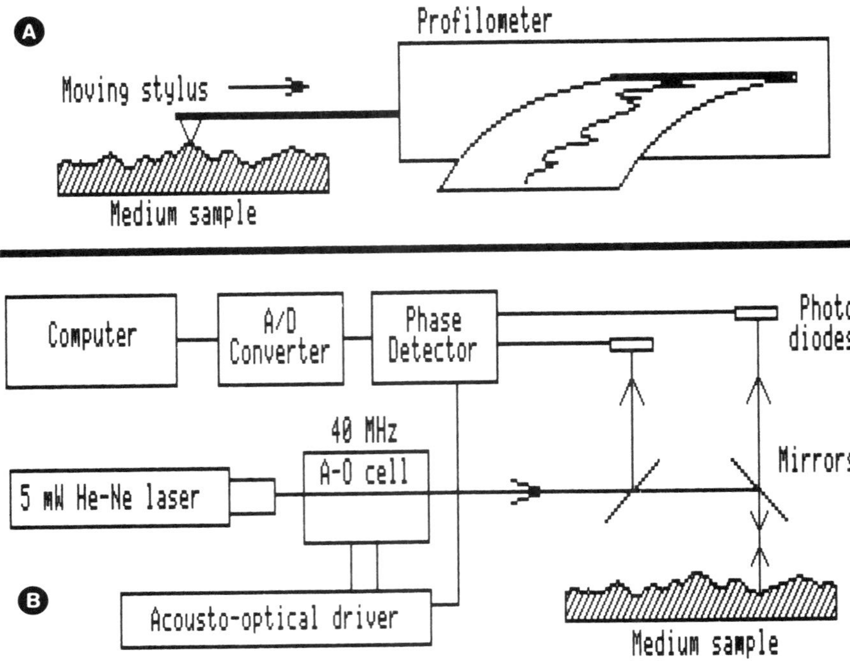

Fig. 14-3. Measurements of surface roughness. A: stylus instrument (Profilometer). B: light interference meter.

face measurements of the sample are derived from the phase shifts of the carrier frequency created by interference.

This measuring technique has allowed for collection of some very interesting data pertaining to the SNR of video tapes (ref. Robinson et al., Wierenga et al.). The RF-signal amplitude varied inverse proportional to the surface roughness, which is to be expected from the spacing loss rule. This in turn affected the SNR of the chrominance signal in the ½-inch VTRs with color-under, see Fig. 14-5. The chrominance signal is discussed in Chapter 29.

An ideally smooth tape does have an apparent disadvantage by causing an almost molecular adhesion to the head surfaces. This phenomenon is chemical-mechanical in nature. The effect demonstrates itself in two ways: gap smear or "varnish." *Gap smear*, which is particularly noticeable in microgap heads, will cause the core lamination material to flow across the gap and in essence short circuit it magnetically. The *varnish* phenomenon is a microscopic build-up of a clear film on the magnetic heads which may cause separations between the tape surface and the head surface. This varnish may come from resins in the coating material and should be eliminated by abrasive coating additives, such as alumina powder.

A tape must further exhibit good temperature stability when used in video recording. In a transverse-scan video recorder, the relative speed between the head and the tape is in the order of 500 to 1500 IPS and the localized heat generated by friction can be very high and cause a breakdown of the coating. This is particularly pronounced in the helical scan video recorders with still-frame capability; that is, when the tape motion is stopped while the scanning head rotates and thereby produces a single-frame picture of the video information. This generates locally concentrated temperatures and special binder formulations are therefore used for video tapes.

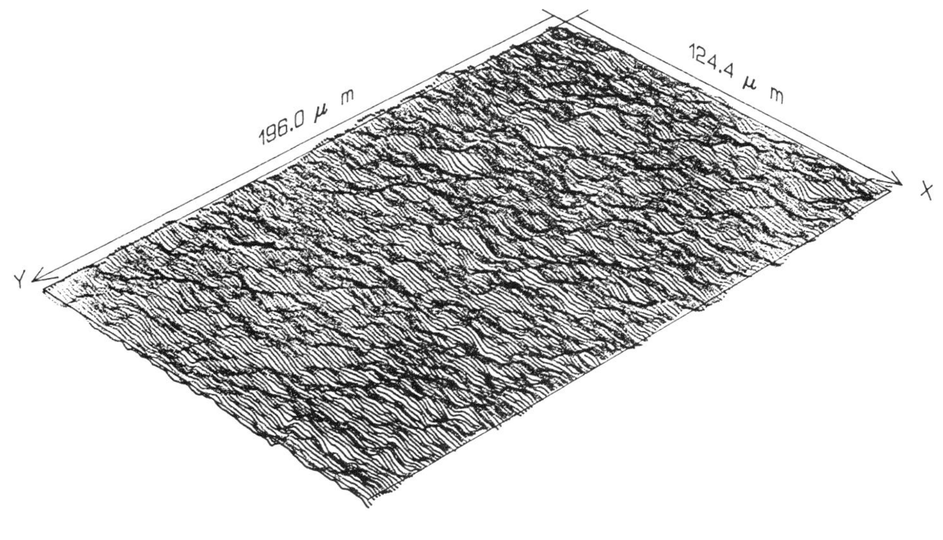

DEPTH SCALE = 200.0 × x,y dimensions

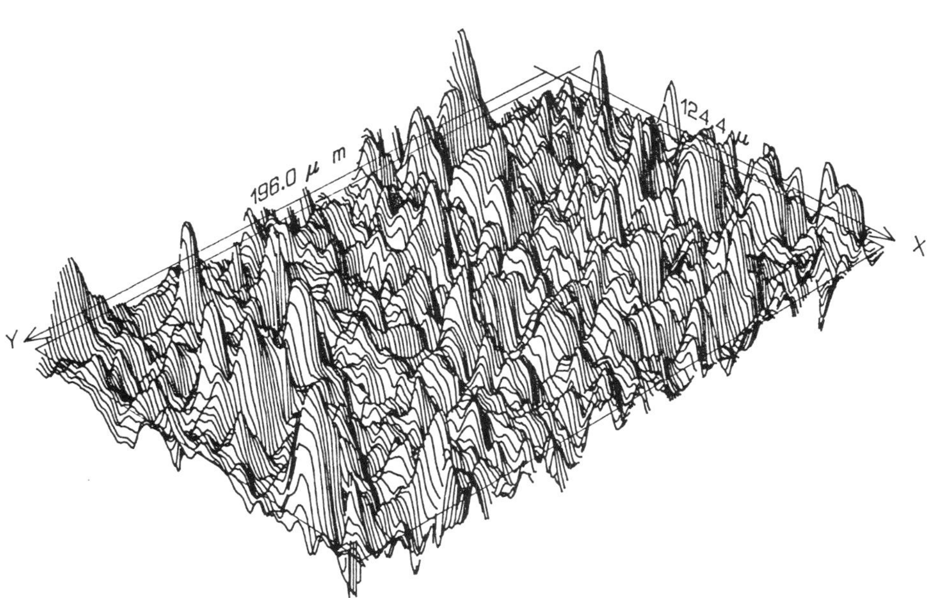

Fig. 14-4. Examination of recording surface by laser interferometry (after Robinson et al.).

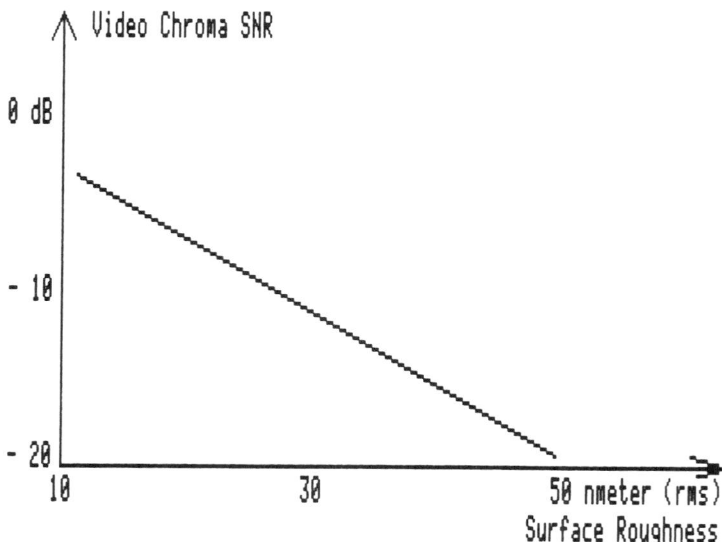

Fig. 14-5. Video chrominance SNR for a color-under system, as a function of the tape surface roughness (after G.M. Robinson et al.).

In recent years a second coating is often applied to the backside of many varieties of tapes. This coat is a binder containing carbon and serves to lower the resistivity of the tape and thereby prevent the annoying pop noises associated with electrostatic discharges. This coat is sometimes called *carbon black*.

The surface finish of this coat is matte, and it provides the additional advantage of increasing friction among tape layers and hence reducing the danger of tape cinching and slipping during fast starts and stops in the winding mode. It may also be beneficial due to the stronger grip of the tape between capstan and pressure puck.

There is finally the question of lubrication of disk and tape surfaces. Fluorocarbon appears to be the preferred material for disks (ref. Au-Yeung), while other, organic lubricants are used for tapes (Wright and Tobin, Nakamura et al.).

DURABILITY, ABRASIVITY

Certain applications require a coating binder that will endure a large number of passes over a head. Examples are computer tape drives, in-contact floppy disks, and satellite telemetry recorders. Still-framing in video recorders adds the additional environment of frictional heat, which decomposes most binders.

It is in general required that binders shall meet the following requirements:

Computer drive tapes:	>20,000 passes
Floppy disks:	> $3*10^6$ revolutions
Video tapes:	> 1 hour still framing

An often highlighted feature of helical-scan video recorders is their still-frame capability. In other words, with the tape motion stopped, the rotating head assembly will scan a particular portion of the recorded material and display it on the TV screen. A very high temperature is generated in the contact area between the rapidly moving head and the stopped tape (possibly several hundred degrees C). This action may wear out the tape or cause the formation of debris on the video head (called head clogging). Debris is a general term used for oxide and/or binder buildup on magnetic heads and tape guides. Some tapes have a still-frame capability of only a few seconds and are obviously rated poor, while others will last in excess of one hour. The still-frame test is consequently used as a figure of merit for a video tape.

Still-frame life of a tape is strongly affected by the surface micro roughness (ref. Tomago et al.); CoNi films can be optimized in this respect for their use in the 8mm cameras. Wear and durability of disks have been investigated by Tereda et al., Hempstead, Garrison and Smallen et al. (see references).

Test for Abrasivity

The abrasivity of a magnetic tape is a complex condition resulting from interplay between all of the following conditions:

Tape particles and binder formulation
Tape surface
Tape tension
Tape speed
Head material(s)
Humidity
Temperature

There are three cases to consider:

No abrasion. This condition often results in the build-up of binder material onto the head surface, mixed with dust particles, etc. The film thus formed is in general referred to as varnish, and may not be easily removed. It causes a spacing between the tape and the head surfaces and therefore a signal loss at short wavelengths.

Normal abrasion. Here the tape offers a small amount of abrasion, enough to keep the head surface free from build-up of varnish.

Excessive abrasion. This has the same desirable result as listed above, but causes rapid head wear.

The reader is referred to Chapter 18 under Results of Tape Wear Testing for further details.

ELECTRO-MAGNETIC PROPERTIES

The magnetic recording surfaces of tapes or disks are subject to a series of measurements to determine how well they may perform in a recording system. The chore has become almost overwhelming with a fairly large number of tests made in order to fully characterize a tape.

The list of test data should be compared against some standard values so the tapes or disks can be graded, and a proper selection made. There is today no complete set of standards and the choice in the end is made for brand A's tape because it has certain critical properties that are superior to brand B's tape, and because it has a known performance history.

This situation is not the fault of the standards' groups, but it is really a result of the rapid developments among media manufacturers. The past decade has seen more new and better products coming on the market than during the entire history of magnetic recording.

In order for the end user to have some kind of a yardstick there are test and alignment tapes on the market. They are designed for the purpose of bringing the reproduce head, amplifier and equalization within some defined tolerances. But the record side (bias, levels, equalization) must be set up using the selected tape. (Some test tapes have a blank reference section at the beginning or the end for this purpose but it is not likely to match the tape the use plans to operate on the recorder).

Normal Record Level

An example will bring light on this subject: the definition of a normal recording level. This level will first of all depend on the amount of bias used in the recording (which in general is defined by a standard or operating practice). The recording level is next adjusted to produce a one percent or a three percent third harmonic distortion. Let us choose the last, at a signal frequency of 1 kHz.

When we follow this procedure for a group of modern *audio tapes*, for example we will find that the absolute flux levels on the tapes may range from 200 to 1000 nWB/m. This has been of some concern in the *standardization of audio recording levels*. Practical reasons dictate that interchangeable tapes within the broadcast industry should all have identical program levels. This will eliminate the need for gain adjustments from program to program.

Deutscher Rundfunk in Germany has in the past used *peak levels* of 320 nWB/m for full track tapes and 510 nWB/m for stereo tapes. Peak levels are measured with peak reading instruments and displayed by LED devices or oscilloscopes. New standards define a *normal reference level* as 320 nWB/m at tape speeds of 19 cm/sec and higher, 250 nWB/m at 9.5 cm/sec and lower.

The normal reference level is set by first playing a prerecorded standard reference tape while adjusting the VU-meter gain control for a reading of 0 VU. Next a recording is made, on the selected tape and with the bias properly adjusted. The record gain is now adjusted so a normal input voltage (for instance 0 dBm) results in a level on the tape that indicates 0 VU on the playback VU-meter.

The recorded level depends on the amount of bias current. Figure 14-6 shows the relationship between output level and bias current; each frequency has an optimum bias current and a compromise must be made. A high bias current produces a clean, in-depth recording with low harmonic distortion, while a small bias current results in excessive high frequency response with high distortion levels.

A useful guide on distortion is that the *one percent distortion level (at a long wavelength) is about 10 dB below tape saturation*, and the five percent distortion level is 5 dB below tape saturation. This is *universal* for all tapes, γFe_2O_3, Co-Fe_2O_3, CrO_2, and Fe, *provided* the record signal corresponds to a recorded wavelength of 5 mils (125 μm) and that bias is adjusted for maximum signal output at that wavelength.

The standards for the various systems will be covered by a future IEC publication, No. 94 (interested readers can write IEC and/or ITA for further information, see end of Chapter 28 for address). Current practice is spelled out below, following the procedures spelled out on manufacturers tape specification sheets:

Audio Recording:

1. Adjust the bias level so a maximum playback signal results when recording a low level tone of 1,000 Hz at 19 cm/sec and higher speeds, 333 Hz at 9.5 cm/sec and lower speeds (Level at −20 dB re normal level). It is quite difficult to repeat this adjustment due to the flat maximum of the low frequency tone, as shown in Fig. 14-6. It is better to use a higher frequency (shorter wavelength) and establish how much overbias is required to match the maximum output at the low frequency. One manufacturer recommends 0.5 dB signal drop when using a frequency of 4,000 Hz, another 4 dB drop at 10,000 Hz.

2. Now adjust the signal record level to produce a 0 VU reading on the playback meter. This is the normal reference level (as we described a few paragraphs earlier in this text).

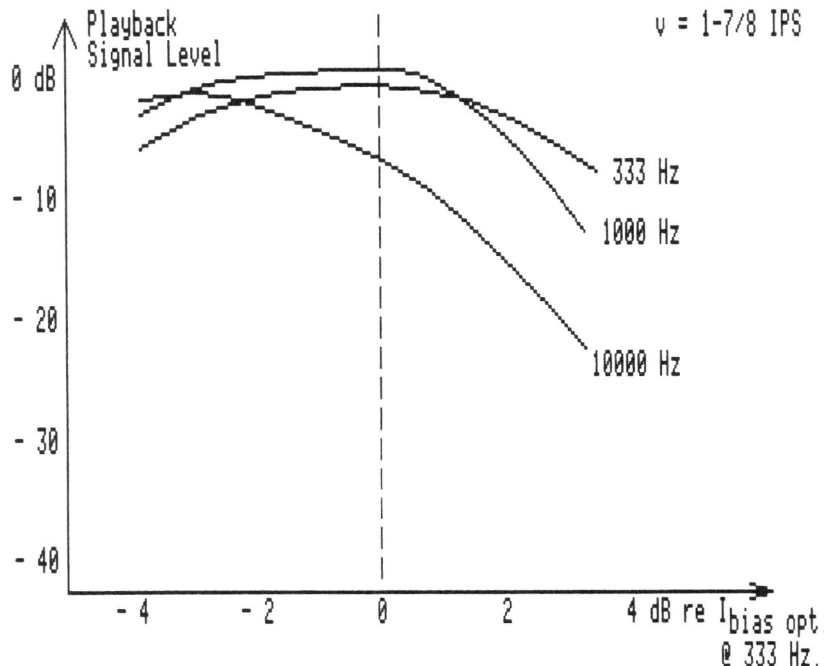

Fig. 14-6. Amplitudes of three test frequencies, plotted as function of the AC-bias current level.

3. Measure h_3, the third harmonic distortion level; it should be below 1 percent.

4. Appraise the tape by measuring the maximum output levels it can produce (MOL). This is generally done at two frequencies:

MOL_{333} is the *maximum level output* when the record level has been adjusted to produce three percent harmonic distortion on the tape (some manufacturers use five percent). MOL_{333} versus i_{bias} is shown in Fig. 14-7 for a typical tape.

MOL_{10} is the *difference in output between the normal reference level and the tape level at saturation, recorded at 10 kHz* (some use 12.5 kHz). A typical curve is shown in Fig. 14-7.

A different procedure is used in *instrumentation recording,* where the record level is adjusted to reach a tape magnetization level that has either a one percent or three percent third harmonic distortion level. Either could be used as the normal reference level similarly to that used in a studio recorder, but the level will not necessarily produce the standard flux level of 320 nWB/m (a thick coating may result in 1,000 nWB/m while a thin coating will produce only 200 nWB/m). For further details consult with IRIG Doc. 118-79, Vol. III.

High Density Digital Recording may or may not use AC-bias. An appraisal of tape is made by evaluating its qualities as a wideband magnetic instrumentation tape, combined with HDDR error data.

Direct Digital Recording requires a record current that will produce a field that is greater than the coercivity H_c of the coating. The write current is normalized to a value i_{s95} that produces 95 percent of the maximum read level, and is then adjusted upward to $2.1 \bullet i_{s95}$ for low bit density recordings (< 1600 PBI) or $1.8 \bullet i_{s95}$ for high bit density (> 1600 BPI). It may be even less for higher density recordings (5,000 to 60,000 BPI).

Updating of digital recordings is made by overwriting on top of old data, and a typical criteria is that old data must be reduced below −30 dB, referenced to the new data, see Fig. 14-8.

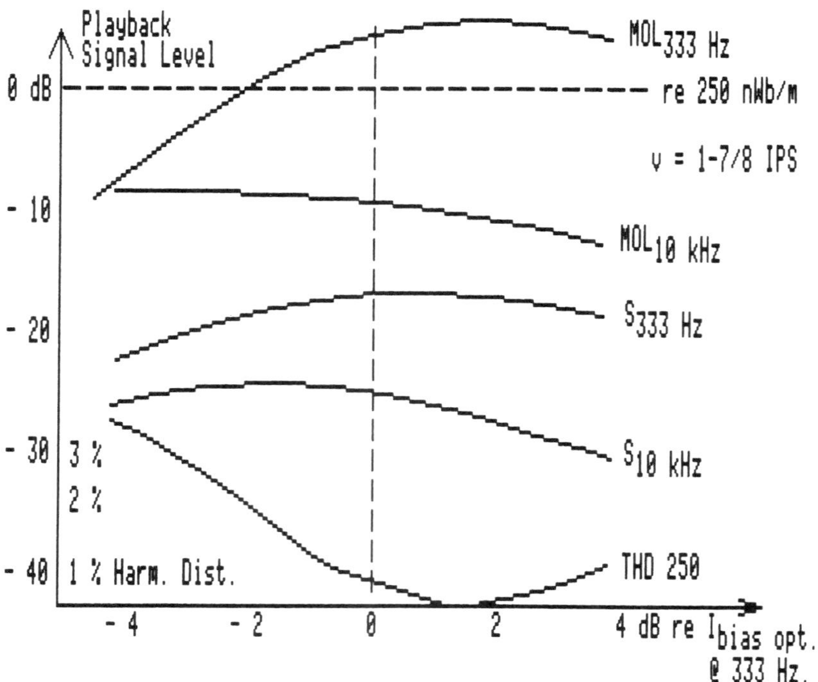

Fig. 14-7. Variations in MOL, Sensitivities S and distortion with AC-bias current level.

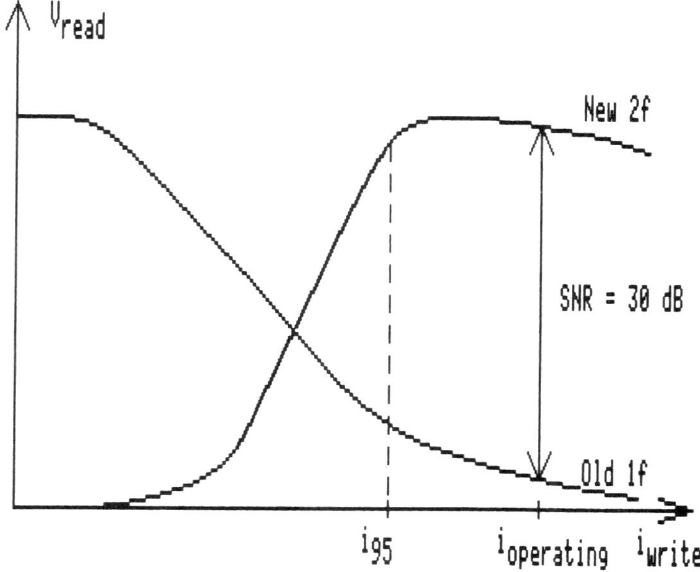

Fig. 14-8. Amplitude of 2f signal and overwrite of old 1f signal, as functions of write current.

This results in a method for adjusting the *normal operating level in digital recording*:

1. Write data at 1f.
2. Overwrite the 1f data with a new 2f signal, and increase the write current until the old 1f signal is 30 dB below the new 2f data (or better: until the SNR is better than 30 dB). Repeat step 1 with the new write current, and verify step 2; if necessary, readjust the write current, and repeat steps 1 and 2.
3. Use the write current from step 2 to write 1f and 2f data. Measure the corresponding peak-to-peak voltages V_{1f} and V_{2f}. The resolution can now be calculated: Resolution = 100 * V_{2f}/V_{1f} percent.
4. When adjusting i_{write} for a disk system do steps 1, 2 and 3 at an inner and an outer track.

Video Recording has the simplest procedure of all: Adjust the record current for maximum RF-signal output during playback. At this point evaluate the SNR of the tape.

Resolution, Amplitude Versus Frequency Response

A media for direct *digital recording* should retain the shortest possible transition zones while providing an adequate signal output. This requires both a high value for the remanence B_{rs}, the coercivity H_c and furthermore a thin coating for small demagnetization (BL > coating thickness). The reduced output from a thin coat can be compensated for by increasing the remanence. This results in a small degradation in resolution and a thin metallic coating may provide the optimum future coating for disks.

A narrow range of individual particle coercivities is also desirable, providing shorter transition zones at flux reversals in digital recording and providing a shorter recording zone in AC-bias recording. This coercivity range is commonly referred to as *switching field distribution*, SFD. It is the width in Oersteds of the derivative of the J-H curve in the switching region around H_c. See Fig. 12-6 for details.

A measure for the response of a *digital write/read system* is *resolution*, defined as the ratio (in percent) between the 2f and the 1f read voltages, see Fig. 14-9.

The frequency response of *AC-bias recordings* depends on several factors:

Switching Field Distribution SFD
Recorded Thickness c′
Surface Smoothness d′
Bias Setting.

The recorded thickness c′ enters into the formula for coating thickness loss:

$$A = 20 \log ((1 - e^{-2\pi c'/\lambda})/2\pi c'/\lambda) \text{ dB}$$

This loss results in a voltage versus frequency response curve that increases at a rate of 6 dB/octave at low frequencies and becomes level above a frequency f_c that corresponds to a recorded wavelength $\lambda = 2\pi c'$, (Fig. 14-10A).

We will therefore find different crossover frequencies for *different coating thicknesses*, assuming they all are normal biased, i.e., $c' = c$ (Fig. 14-10B).

The voltage versus frequency response is equalized in the reproduce amplifier by means of an integrator with a shelf having a response that corresponds to one of the curves in Fig. 28-1. The question is: which one should be standardized? If the tapes are recorded with a constant current then the equalizer should be set to one of the crossover frequencies listed in Table 14-2. This will assure a flat voltage versus frequency response during playback.

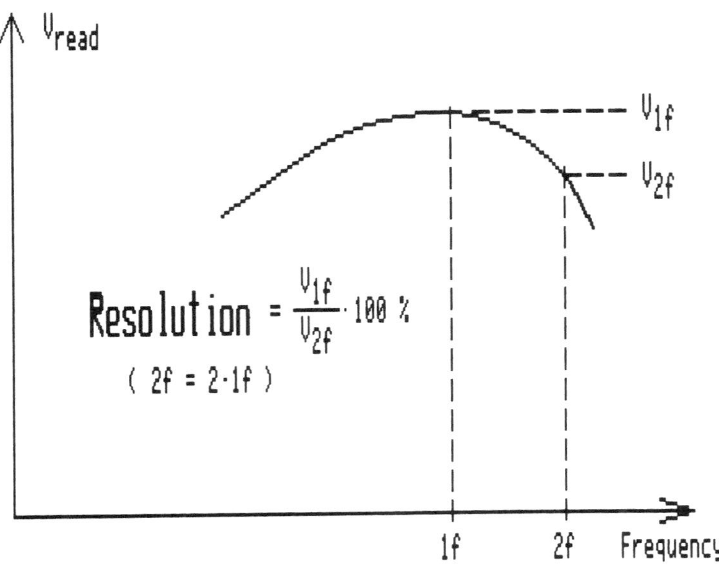

Fig. 14-9. Amplitude versus frequency in direct recording of square wave (digital signal).

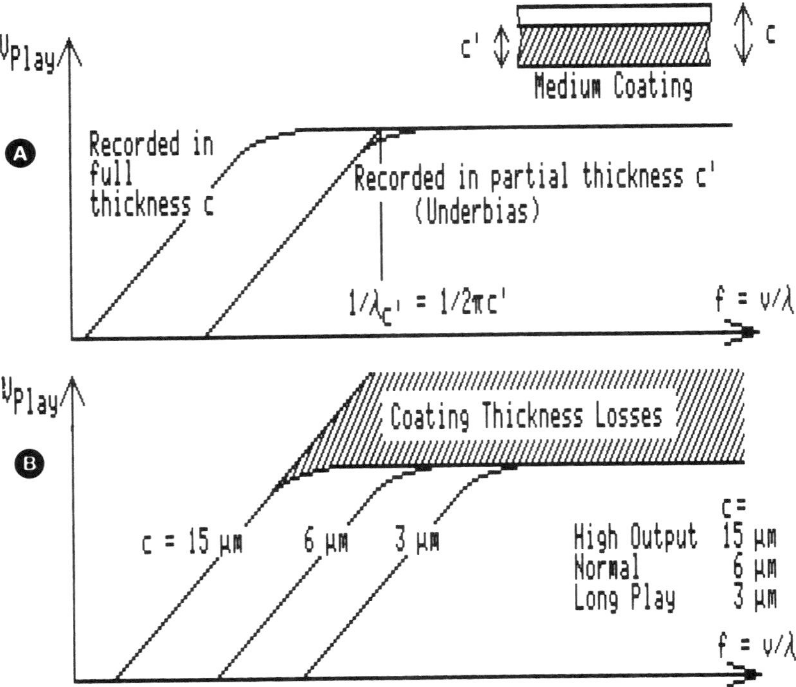

Fig. 14-10. Coating thickness losses. A: Effect of bias setting (recorded depth). B: Effect of coating thickness, bias optimized for each coating.

Table 14-2. Tape Speed, Coating Thickness and Time Constant.

Application:	Speed IPS	c= µm	f_c Hz	$\tau_c = 1/2\pi f_c$ µsec	IEC/DIN µsec	NAB µsec
Studio	30	10	11,935	13	35	18
Professional	15	5	11,935	13	35	50
Home A	7-1/2	5	5,970	27	50-70	50
Home B	3-3/4	5	2,985	54	90	90
Cassette	1-7/8	2.5	2,985	54	120	90

Matters are now made complicated by the fact that the studio recorder uses over-bias, the home recorder normal bias and the cassette recorder slight under-bias. Furthermore, it is common practice to boost the record current to achieve a uniform record level versus frequency curve. This is permissible for speech and music programs that have little energy at high frequencies, but it is not a recommended practice in instrumentation recorders.

The amount of *boost* has been standardized by NAB in the USA and DIN, CCIR and IEC in Europe. The standards specify that the record current is boosted above a certain frequency in order to provide a recorded flux that is constant down to a wavelength corresponding to a defined τ_c.

The selected standards for τ_c are listed in Table 14-2. A proper wavelength response (or frequency response) of a tape requires that it is biased properly and that the record current equalization is adjusted to provide a flat response through a playback system that has been equalized to the chosen standard.

Audio and Instrumentation Tapes, Test of Response

A-B comparisons of tapes are in practice often carried out by readjusting only the bias level and not changing the record boost. The relative changes in playback output are therefore a measure for the tapes ranking.

Tapes can be categorized into four types:

I Normal gamma-Fe_2O_3 for normal bias; t = 120 µS
II CrO_2, Co-treated gamma; t = 70 µS
III Ferrichrome, double coat
IV Metal (Fe) (has more head room); t = 70 µS

A tape can be tested for amplitude versus frequency response by first adjusting the recorder, using a standard alignment tape to adjust the playback equalizers, then the bias and record equalizers for a flat response, using a known tape, call it the reference. Then load the unknown tape to be compared, leaving all control settings. If the new response falls within a given envelope (say plus/minus 3 dB) then the new tape is compatible with the reference.

Next adjust bias in accord with the rules for setting up the reference tape, and re-measure the response; if the record current needs additional boost then the new tapes response is poorer than the reference; and vice versa. The amount of boost or reduction in record equalization is a direct measure of the difference between the tape's frequency responses.

The *recommended bias setting* is to first adjust the audio signal level at 1 kHz to 10 dB below MOL. Increase the bias level until the signal level upon playback reaches a maximum, and then decreases again. This procedure results in an overbiased condition; it is recommended to adjust the bias so the signal level drops the following amount:

0.25 dB at 19.6 cm/sec (7.5 IPS)
0.75 dB at 39 cm/sec (15 IPS)
1.25 dB at 78 cm/sec (30 IPS)

It is important to measure amplitude versus frequency response at a level that is well below the normal reference level, or the one percent 3rd harmonic distortion level. −20 or −30 dB below the reference level will provide the correct response, without tape overload at short wavelengths (see Fig. 14-11). This overload is partly inherent in the write process, partly in the pre-equalization employed.

It is not always necessary to measure the entire response in order to evaluate and compare tapes.

Standard measures have been defined by the *sensitivities* S_{333} Hz and S_{10} kHz. Both are measured as the dB difference in output between the standard reference level less 20 dB (i.e., 25 nWB/m) and the corresponding levels measured at 333 Hz and 10 kHz, recorded 20 dB below normal.

Relative frequency response is defined as the difference in dB between the S-values, against those of the reference tape.

A figure of merit of a tape's short wavelength response was introduced in Chapter 5, the *Switching Field Distribution*, abbreviated SFD. The switching field distribution is proportional

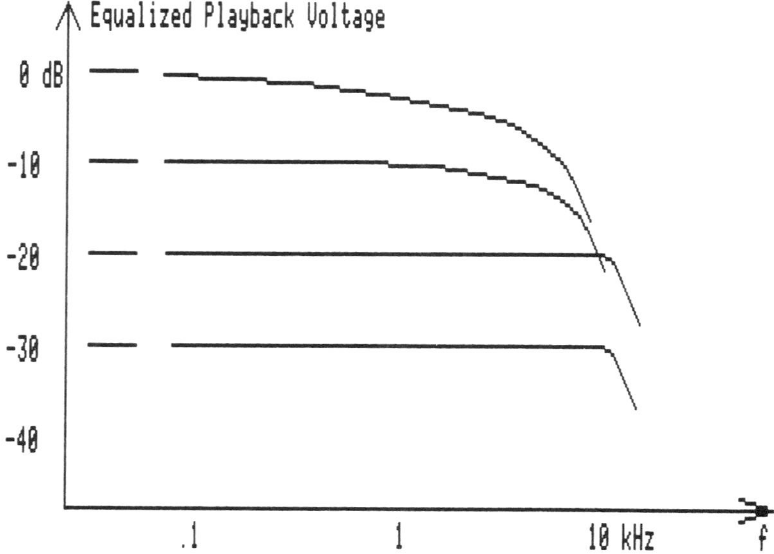

Fig. 14-11. Amplitude versus frequency response in audio recording must be made at low levels (−20 to −30 dB below normal reference level) in order to avoid overload.

to the steepness of the slope of the BH-loop and a high value is synonymous with good, short wavelength characteristics.

It should in closing be mentioned that the record head field characteristics will influence the resulting response, as well as the reproduce head losses: AB-tests are valid only when carried out on the same recorder, and all bias and equalization settings specified.

Long Wavelength Uniformity; Drop-Ins

Amplitude variations are common in the analog reproduce voltage and they are, at long wavelengths, directly related to variations in coating thickness. This applies to all methods of recording and reading, including digital.

Diskettes may have once-around amplitude variations which are the result of incomplete randomizing of the particles orientation in the plane of the coating. Variations in output are called *modulation*, and should be within ±10 percent.

Amplitude variations in tapes may vary in a cyclic fashion for every few feet of tape, in which case they originate from an eccentric roller in the coater during tape manufacturing. These variations are small, typically ± 0.5 dB, down to ± 0.1 dB over a few feet of tape.

Slow changes can occur from the beginning to the end of a reel of tape and are traceable to minute changes in the coating slurry during a coating run. These variations are in the order of 0.5 to 1.5 dB. Level changes from one reel to another are in the same range (0.5 to 1.5 dB).

Changes of very short duration may occur due to particle agglomeration or voids. The first are named *Drop-Ins*, the latter *Drop-Outs*. Both are detected quite easily by recording the tape with a DC current (to saturation). A flawless tape has perfectly constant DC-flux and therefore zero read voltage. Defects will appear as voltage pulses $d\psi/dt$ because ψ changes.

The long wavelength level variations can be annoying in instrumentation recording applications but are of seemingly little consequence in audio work due to the insensitivity of the ear to level changes below ± 1 dB.

Measurements are made simply by connecting a strip chart recorder to the recorder output. *Sensitivity uniformity* is defined as the difference in dB between the maximum and minimum playback levels of the sample tape upon which a recording has been made over the entire length under the same circumstances as sensitivity measurements.

Short Wavelength Uniformity

Changes in short wavelength response occur quite frequently in tape production due to variations in particle dispersion, coating viscosity and drying rates. All contribute to a surface that after callendering and other surface treatment techniques results in a few microinches of waviness.

The uniformity of the short wavelength output is easily measured by connecting a strip chart recorder to the recorder output. Variations are ± 1 dB for an exceptionally good tape recorded with a wavelength of 2.5 μm.

The surface regularity can be tested by the measurement of DC noise after the tape has been recorded with a DC record current that was adjusted for maximum noise output. The noise voltage output represents surface noise rather than coating dispersion noise.

DROP-OUTS

Drop-outs are the causes of momentary signal loss in the write/read process, leading to errors in digital systems, and degraded performance of audio, and in particular video, playback signals.

The definition of a drop-out varies according to the application. A signal drop exceeding 50 percent of the normal value is a drop-out for low packing density digital recording (< 1600

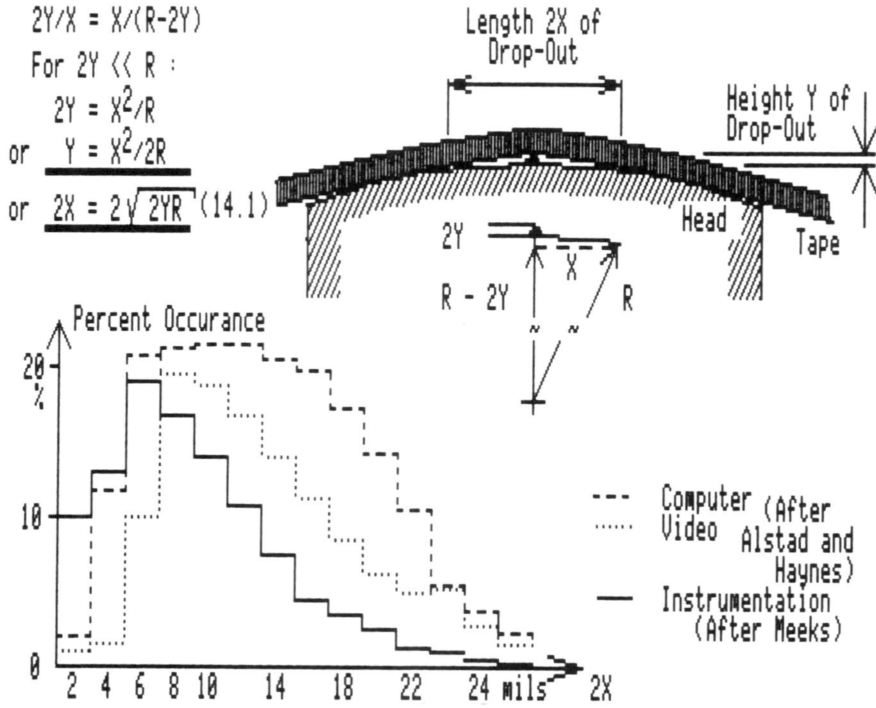

Fig. 14-12. Top: A drop-out forms a tent with height Y; bottom: probability distribution for width of drop-outs.

BPI), while a drop of only 35 percent may classify a drop-out in high density recording (> 1600 BPI). The length, and hence duration, of a drop-out is significant in selection and construction of error detecting and correcting codes.

A dust particle or a nodule in the coated surface will lift the tape away from the heads. During recording, this will move the tape away from the otherwise properly adjusted write field and reduce the recorded flux. During playback, it will reduce the signal by the normal spacing loss equal to $54.6\ d/\lambda$ dB, where d is the spacing between the tape and head surfaces, and λ is the recorded wavelength, also equal to 2 bit lengths.

Drop-outs are primarily caused by foreign particles that raise the tape surface in a tent like fashion, see Fig. 14-12 top. In the early days of tape recording drop-outs were often imperfections of all sorts in the coated surface (ref. Radocy, Kramer, van Keuren). Nowadays media can be manufactured virtually drop-out free, and high-grade computer tapes are certified to have no more than one error per each 100 feet, or even to be error free. Note: One error, or drop-out, may comprise numerous bits.

We can estimate the length of the drop-out, if we know the height Y of the particle between the tape and the head, the latter made with a contact radius of R. This situation is illustrated in Fig. 14-12 top, and we find by comparing a couple of triangle ratios that the drop-out length 2X is proportional to the squareroot of the height Y multiplied by the radius R (ref. Baker).

Effects of Environment Upon Drop-Out Activity

There is a remarkable correlation between much of the experimental data collected relating to drop-outs. Errors in digital recording systems are characterized by burst errors, i.e., the signal may disappear for a duration lasting hundreds of bits, which is disastrous without the proper error control coding.

Observations of the effective drop-out length on a tape has resulted in a mean value of 0.15-0.25mm (6-10 mils), which translates into asperity heights of 2-3 μm (80-120μ"). This mean value has been found by several independent observers, see Fig. 14-12 (ref. Alstad and Haynes, Meeks).

They are, for the most part, not fixed errors in the surface of the media. Some investigators of drop-outs would, when they detected an error, stop the transport and go back to the error location. In many cases there was nothing to be observed: the error was a dirt particle that moved about.

Another revealing experiment disclosed what is now known as the Wilmot effect: The drop-out from a tape stored on a reel with solid flanges is an order of magnitude less than tape stored on a reel with the standard openings (Ref. Perry et al.). A detailed analysis of the drop-out count from beginning to end of a tape is shown in Fig. 14-13: The tape portion that was exposed through the flange openings has the highest error count.

It is easy to envision the air flow that must enter from the sides and along a tape when it is lifted away from the reel pack. The air contains dust particles, and many of them will settle down on the tape due to possible electrostatic attraction. Some may even become permanently embedded into the coating surface when wound with the tape onto the take-up reel.

The only cure for tape and diskette drop-outs is therefore to have the best possible clean room condition around the drive installations. Smoking is of course out of the question.

The new IBM 3340 tape cartridge and drive protects the tape from its environment and should therefore be one order of magnitude better in performance than the comparable reel-to-reel tape drives.

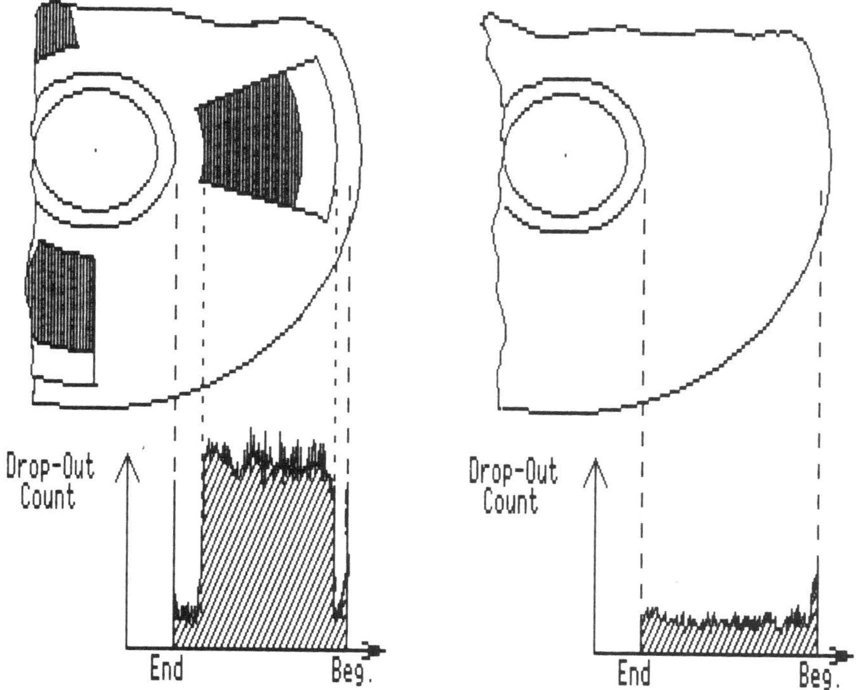

Fig. 14-13. Drop-out count along two tapes, the left wound on a reel with openings in flanges, the right with closed flanges.

The effect of drop-out has been studied and described in several papers listed in the references: digital systems (Jack-Kee and Middleton, Nunnelley, and Maedigger et al.), video tapes (Lee and Papin), and audio (Pichler and Pavuza).

Dead Layer and Velour Effect

Many users have observed that a medium may exhibit a difference in short wavelength output when operated in the forward motion versus reverse motion. This is attributed to the orientation of particles in the coating since they, during the coating and following magnetic alignment process, end up laying at a very small angle to the coating surface.

The remanence that exists in the very surface of a recorded media is perpendicular (ref. Bate and Dunn). Some of the particles do also appear at an almost right angle to the nearly perpendicular field lines at the interface between the coating and a recording head. Therefore, they may not be influenced strongly enough by the recording field; taking an average, one finds less remanence in the coating surface. This is in general referred to as the dead layer, and is related to the velour effect.

Figure 14-14 illustrates the differences in the interrelation between the head field and the particles in the surface of the coating. A weak recording results in A, where the particles are perpendicular to the field, whereas in B, a stronger recording results, where more particles agree with the field.

STORAGE STABILITY

All magnetic particles exhibit a certain instability that, among other things, can lead to *print-through*, or transfer of signals recorded on one layer of tape onto the adjacent layers. This manifests itself as a pre-echo and a post-echo of the recording and can be rather annoying in music recordings.

The instability that causes print-through is a time and temperature dependent property. Another instability will cause permanent *loss of signal*, in various amounts, and it is a property of the coating particles magnetostriction.

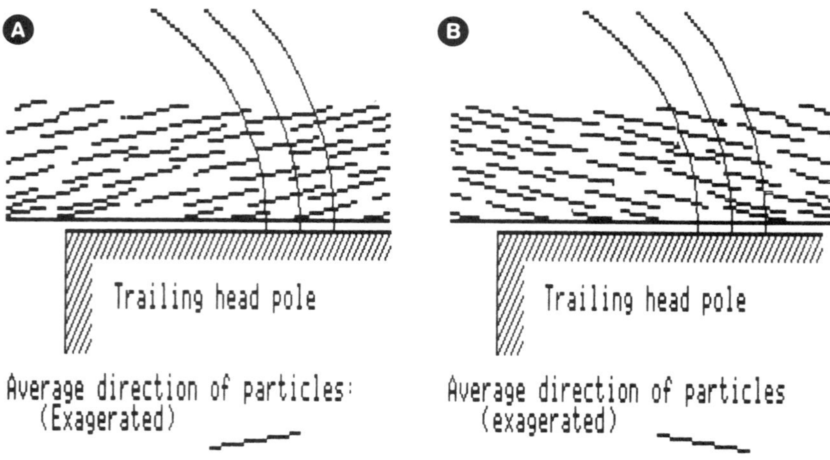

Fig. 14-14. Velour Effect, or "Dead" layer. A: Weak surface recording results when the particles in the coating are perpendicular to the field lines. B: Strong surface recording results when the particles are not perpendicular to the field lines.

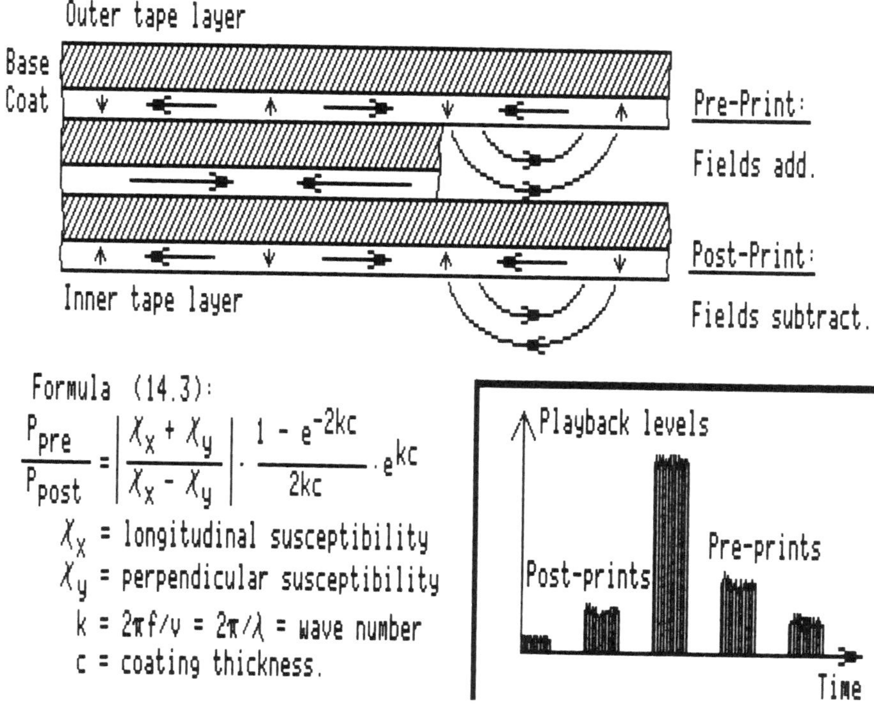

Fig. 14-15. Printing effect in stored magnetic tape. A: Pre- and post-print in a tape wound with the backing side out (normal reel-to-reel configuration). B: Playback levels of printed signals (after Daniel).

Archival stability of recordings are covered in Chapter 30, under Storage of Media, Archival. Time effects upon the magnetization was covered in Chapter 12 (see Fig. 12-15).

Print-Through

Some magnetic tapes suffer from a flaw called *print-through*, which is found in thin base tapes, and in particular those with small particle coatings. When the tape is wound onto a reel, adjacent coating layers are separated by a distance equal to the thickness of the base material. Flux lines from one layer will reach adjacent layers and under the influence of time, temperature, and external fields, cause a weak "recording" on the adjacent layers. This printed signal can be very annoying in audio recording, while it is seldom observed in video and instrumentation applications. A quiet pause before a loud orchestra opening may, by print-through, contain a faint prelude of the opening. This calls for caution in using thin-based tapes.

A distribution of particle sizes in a typical coating shown in Fig. 3-30 indicated the presence of some very small particles. γFe_2O_3 particles have small coercivities below a length of about 0.3 μm and eventually become super-paramagnetic below a length of 0.2 μm. If we assume a length-to-width ratio of 5 then the corresponding volumes are 0.001 μm^3 and 0.0003 μm^3. From Fig. 11-3 we read that the analyzed sample contains 4 percent particles that are smaller than 0.001 μm^3 and less than 0.5 percent smaller than 0.0003 μm^3.

The direction of magnetization in these particles can be reversed by the thermal energy kT, where k is Boltzmann's constant and T the absolute temperature. The probability of this occurring within a given time period can be expressed by a time constant where:

$$1/\tau = e^{-uJ_s h_{ci}/2kT} \quad \textbf{(14.2)}$$

where u is the particle volume, J_s is the magnetization and h_{ci} is the coercivity of the particle.

A particle characterized by a time constant τ will be susceptible to a magnetization reversal during the time τ. This time is relatively short (hours, days) for very small particles and is further shortened by an increase in T (τ for normal size particles is well in excess of 50 years).

In the absence of an applied field, the probabilities of the particle ending up magnetized in one direction or the other are equal. In the presence of a field, such as that ascending from an adjacent layer of magnetized tape, the particle will tend to end up magnetized in the direction of the field, even when this is very small. The total printed magnetization would then be obtained by integrating over all particles τ within the range 0 to t (from time of recording = 0 to time of playback = t).

If this time t is comparable to τ for the smallest particles (below 0.0003 μm), of which there were 0.5 percent, the print-through could be as much as 20 log(0.5/100) = – 46 dB relative to the normal record level.

Longer storage times, combined with elevated temperatures, may then include the larger particles and the print-through would be worse, equal to 20 log(4/100) = – 28 dB.

Print-throughs will be present on both of the tape layers adjacent to the recording. As the tape is played back the first signal heard would be a *pre-print*, then the original signal and finally a *post-print* (there may be additional, weaker pre- and post-prints).

Pre- and post-prints are different in magnitudes. This phenomenon arises because the printed magnetization has both longitudinal and vertical components and these, on playback, are additive for one set of prints and subtractive for the other set. The situation is shown in Fig. 14-15.

Daniel has expressed the ratio of the magnitudes of the prints as shown in the illustration.

The maximum print-through occurs when the wavelength of the recorded signal is approximately equal to the total tape thickness c + b. Under these conditions equation (14.3) reduces to:

$$P_{pre}/P_{post} \cong 1.3\ (\chi_x + \chi_y)/(\chi_x - \chi_y) \quad \textbf{(14.4)}$$

The multiplier will be 1 instead of 1.3, at shorter wavelengths.

When the tape is isotropic, the $\chi_x = \chi_y$, and the preprints will vanish, while the postprints will double. If the anisotropy is predominantly longitudinal, then $\chi_y = 0$, and the prints are equal.

In modern tapes $\chi_y = 0.25\ \chi_x$, resulting in:

$$P_{pre}/P_{post} = 2.2, \text{ or } 6.8 \text{ dB.}$$

A practical consequence is that European studio and broadcast stations wind their tapes with the oxide out. This makes the pre-print the more annoying print, weaker than the post-print. The same practice is followed in audio cassettes.

Maximum print-through occurs when a frequency of 1200 Hz is recorded at 15 IPS on 2 mils thick tape. This is unfortunately in the region where the ear is most sensitive. The situation is better for audio cassette tapes where the speed is only 1-⅞ IPS and the total thickness about 1 mil, resulting in a maximum print-through at a frequency of only 300 Hz, where the ear is less sensitive.

More recent work on print-through has been done by Tochihara et al., Stafford, and Corradi et al.

Reduction of Print-Through

Print reduction is possible by moving the tape past a very weak permanent magnet (ref. Radocy), or by feeding a very small amount of DC or AC current through the record head during playback. This is sufficient to erase the unstable particles that were responsible for the print effect.

One method of adjusting this print-erase current is to measure the output of a prerecorded 15 kHz pilot tone, well before the actual recording on the tape. Adjust the erase current for about 1 dB erasure of the pilot tone; this will assure print erasure while leaving the recording intact.

Measurement of Print-Through

A tape's print-through characteristics can be measured in the following way: record a very short section (a few cm) with a 1,000 Hz tone, or a worst-case signal (square wave at $f = v/2\pi(c+b)$) at maximum recording level (6 dB below saturation). Rewind immediately and store the tape for 24 hours at 30°C; (some standards call for 20°C. That will produce less print-through.)

Now play back the tape and record the output signal on a strip chart recorder. This will provide a record of the levels of pre-prints, original signal and post-prints. Typical print-through values for audio cassette tapes are shown in Fig. 14-16.

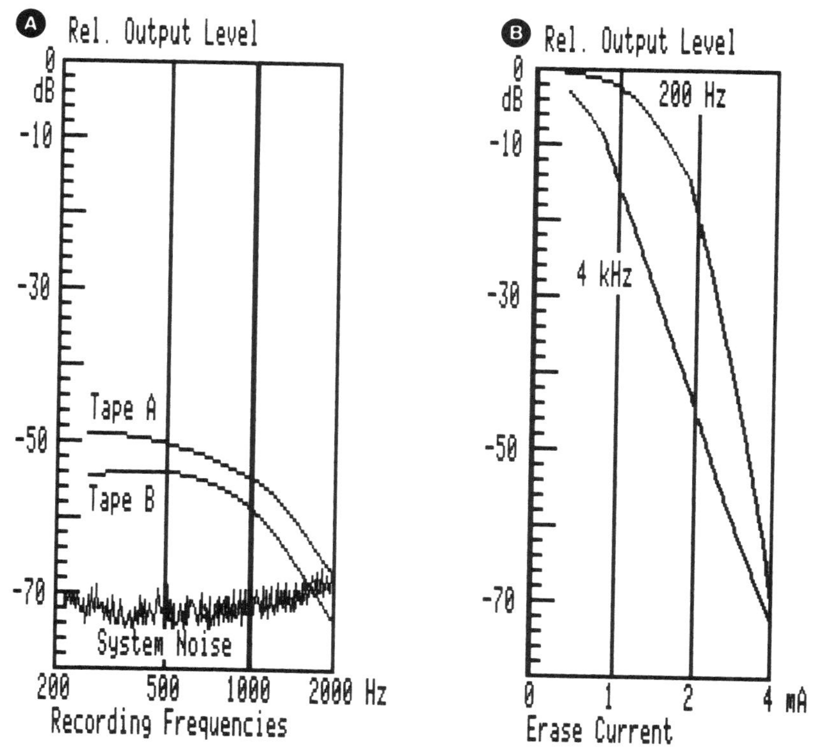

Fig. 14-16. A. Typical print-through levels versus frequency; B. Typical erasure versus current level.

Such records will assist in classifying tapes for storage and archival use since the amount of printing is in direct relation to the tapes instability over prolonged periods of time and possibly elevated temperature.

ERASEABILITY

No single number can represent a tape's eraseability. We may define a method of measurement but the result would be applicable only to a particular set of circumstances: type of erase head or bulk degausser or erasure by over-writing with new data in digital recording.

A tape becomes more difficult to erase after long-time storage, possibly affected by elevated storage temperature. A recording made with a high bias level is more difficult to erase than a tape recorded at a low bias level; it is also observed that high frequencies (short wavelengths) are easier erased than low frequencies, see Fig. 14-16.

Investigations into the erasure process are surprisingly few. The problems of insufficient erasure with erase heads were covered in Chapter 8, and it is apparent that bulk degaussers do not completely erase tapes. The author found only four papers on the topic. Manly examined the erasure process and the strange phenomenon that signals may reoccur after erasure and storage. Fayling continued this work and found that there were directional differences in the bulk erasure process, which requires very large fields to accomplish the erasure.

The problems are aggravated for users of high energy tapes (ref. Burke and Sanders). Overwriting old data with new is used in digital systems, and presents problems of its own. A recent investigation (ref. Christensen and Finkelstein) ranks the methods of erasure as follows (from poor to good): DC-erasure, two-gap DC-erasure (bringing B_{rs} to zero prior to rewriting data), AC-ferrite head erasure and finally bulk erasure. Only small differences were found between CrO_2 and Co-treated γFe_2O_3 tapes.

Measurement of erasure in audio and instrumentation is made by recording a 1,000 Hz signal at the standard recording level plus 5 dB with standard bias current. Then a portion of the recording is erased.

Signal-to-Erase-Noise Ratio is defined as the difference in dB between the playback output level of the recorded portion and that of the erased portion.

STRESS DEMAGNETIZATION

When a tape loops through a recorder it is deformed several times as it passes guides and heads. This induces stresses in the coating, and consequently on the particles. These forces will cause demagnetization of particles that are magnetostrictive.

Magnetostriction is a material property whereby mechanical forces cause an alignment of the domains in magnetostrictive material. A magnetization will, vice versa, cause an elongation or contraction of the material.

The cobalt-doped iron oxide particles possess a rather large magnetostriction (magnetostriction constant λ). Their *stress demagnetization* will appear as a signal loss that increases with the number of passes. Measured data confirms this, see Fig. 14-17 (ref. Daniel and Naumann).

These measurements were accelerated by running a prerecorded tape loop through a transport while observing the playback signal during each pass. The losses were found to be proportional to the number of passes and the recorded frequency.

Such losses are significant in applications where many repeated passes of a tape are anticipated, as in the analysis of signals in instrumentation recording, and these should be accounted for.

Stress demagnetization has been studied by Hoshi et al, and Izawa. Measurements on Co-Fe_2O_3 are reported by Flanders, on thin film disks by Terada et al, and on videotapes by Woodward.

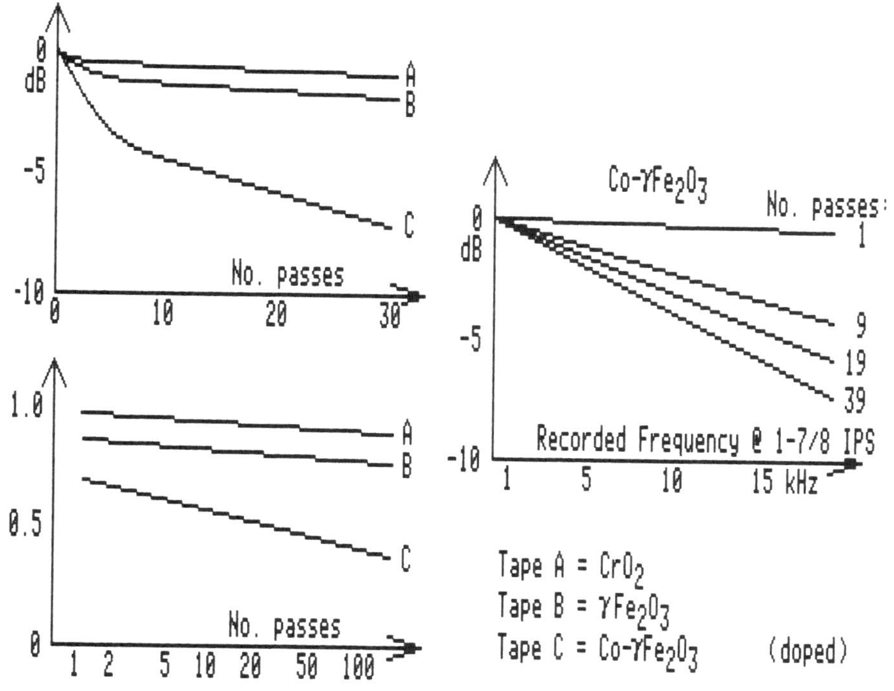

Fig. 14-17. Stress demagnetization and re-coil losses (see Fig. 4-13; repeated readouts may gradually reduce B_{rs}', in particular at short wavelengths, where the internal changes in field strength are significant) (after Naumann and Daniel).

BIBLIOGRAPHY TO CHAPTER 14

Scholz, C., *Handbuch der Magnetbandspeichertechnik*, Carl Hanser Verlag, Munich/Vienna, 1980, 392 pages.

Masscho, R., *Technique du Magnetophone*, Editions Radio, Paris, 1980, 328 pages.

REFERENCES TO CHAPTER 14

Mechanical Properties of Substrates

Behr, M.I., and Osborn, J.K., "Technique for Measuring Dynamically the Dimensional Stability of A Flexible Disk," *IEEE Trans. Magn.*, Nov. 1981, Vol. 17, No.6, pp. 2748-2750.

Izraelev, V., "On Determination of Thermal and Hygroscopic Expansion Coefficients of Pet Floppy Disk Substrates by the Recording Method," *IEEE Trans. Magn.*, Sept. 1983, Vol. MAG-19, No. 5, pp. 2253-2256.

Thompson, J.A., and Mee, P.B., "A Statistically Designed Investigation of Substrate Effects on CoCr Thin Films for Perpendicular Magnetic Recording," *IEEE Trans. Magn.*, Sept. 1984, Vol. MAG-20, No. 5, pp. 785-787.

Mechanical Properties of Coatings

Wright, C.D., and Tobin, H.G., "Surface Lubrication of Magnetic Tape," *ITC Conf. Proc.*, August 1975, pp. 336-344.

Talke, F.E., Teng, R.C., and Morrish, A.H., "Surface Defect Studies of Flexible Media using Magnetoresistive Sensors," *IEEE Trans. Magn.*, Sept. 1975, Vol. MAG-11, No. 5, pp. 1188-1190.

Wahl, F., Ho, S., and Wong, K., "A Hybrid Optical-Digital Analysis Processing for Surface Inspection," *IBM Jour. Res. Devel.*, July 1983, Vol. 27, No. 4, pp. 376-385.

Au-Yeung, V., "FTIR Determination of Fluorocarbon Lubricant Film Thicknesses on Magnetic Disk Media," *IEEE Trans. Magn.*, Sept. 1983, Vol. MAG-19, No. 5, pp. 1662-1664.

Perry, D.M., Moran, P.J., and Robinson, G.M., "Three Dimensional Surface Metrology of Magnetic Recording Materials Through Direct Phase Detecting Microscopic Interferometry," *Intl. Conf. Video and Data 84*, April 1984, IERE Publ. No. 59, pp. 57-62

Nakamura, K., Momono, K., Ota, Y., Itoh, A., and Hayashi, C., "Organic Lubricant Evaporation Method, A New Lubricated Surface Treatment for Thin Film Magnetic Recording Media," *IEEE Trans. Magn.*, Sept. 1984, Vol. MAG-20, No. 5, pp. 833-835.

Robinson, G.M., Moran, P.J., Peterson, R.W., and Englund, C.D., "Applications of Interferometric Measurements of Surface Topography of Moving Magnetic Recording Materials," *IEEE Trans. Magn.*, Sept. 1984, Vol. MAG-20, No. 5, pp. 915-917.

Wierenga, P.E., Winsum, J.A.v., and Linden, J.H.M.v.d., "Roughness and Recording Properties of Particulate Tapes: A Quantitative Study," *IEEE Trans. Magn.*, Sept. 1985, Vol. MAG-21, No. 5, pp. 1383-1385.

Robinson, G.M., Englund, C.D., and Cambronne, R.D., "Relationship of Surface Roughness of Video Tape to its Magnetic Performance," *IEEE Trans. Magn.*, Sept. 1985, Vol. MAG-21, No. 5, pp. 1386-1388.

Durability

Tomago, A., Suzuki, T., and Kunieda, T., "Effects of the Surface Configuration of Magnetic Layers on Durability Against Mechanical Stress," *IEEE Trans. Magn.*, Sept. 1985, Vol. MAG-21, No. 5, pp. 1524-1526.

Terada, A., Ishii, O., and Ohta, S., "Wear Resistance and Signal-to-Noise Ratio in the gamma-Fe_2O_3 Thin Film Disks," *IEEE Trans. Magn.*, Sept. 1985, Vol. MAG-21, No. 5, pp. 1520-1523.

Storage Stability

Hempstead, R.D., "Magnetic and Corrosion Properties of Nickel-Iron-Rhodium Films," *IEEE Trans. Magn.*, Nov. 1979, Vol. MAG-15, No. 6, pp. 1570-1571.

Garrison, M.C., "Affects of Absorbed Films on Galvanic Corrosion in Metallic Thin Film Media," *IEEE Trans. Magn.*, Nov. 1983, Vol. MAG-19, No. 5, pp. 1683-1685.

Smallen, M., Mee, P.B., Ahmad, A., and Freitag, W., "Observations on Electrochemical and Environmental Corrosion Tests for Cobalt Alloy Disk Media," *IEEE Trans. Magn.*, Sept. 1985, Vol. MAG-21, No. 5, pp. 1530-1532.

Drop-Outs

Radocy, F., "Defects in Magnetic Recording Tape: Their Cause and Cure," *Jour. AES*, Jan. 1955, Vol. 3, No. 1, page 31.

Kramer, A., "Locating Defects in Magnetic Recording Tape," *Jour. AES*, March 1955, Vol. 3, No. 3, page 143.

Van Keuren, W., "An Examination of Dropouts Occurring in the Magnetic Recording and Reproducing Process," *Jour. AES*, Jan. 1970, Vol. 18, No. 1, page 2.

Bate, G., and Dunn, L.P., "The Remanent State of Recorded Tapes," *IBM Jour. Res. Devel.*, Oct. 1974, Vol. 18, pp. 563-569.

Baker. B.R., "A Dropout Model for a Digital Tape Recorder," *IEEE Trans. Magn.*, July 1977, Vol. MAG-13, No. 5, pp. 1196-1199.

Alstad, J.K., and Haynes, M.K., "Asperity Heights on Magnetic Tape Derived from Measured Signal Dropout Lengths," *IEEE Trans. Magn.*, Sept. 1978, Vol. MAG-14, No. 5, pp. 749-751.

Meeks, L.A., "Characterization of Instrumentation Tape Signal Dropouts for Appropriate Error Correction Strategies for High Density Digital Recording Systems," *Intl. Conf. Video and Data 79*, July 1979, No. 43, pp. 199-215.

Perry, M.A., Blackwell, F., and Harris, R., "The Importance of Dropout Measurements in Ensuring Good Short Wavelength Recording on Magnetic Tape," *Intl. Conf. Video and Data 82*, 1982, IERE Publ. No. 54, pp. 23-41.

Jack-Kee, T., and Middleton, B.K., "Drop-Outs and Their Effects on Error Rates in a Digital Magnetic Tape Recording System," *Intl. Conf. Video and Data 82*, 1982, IERE Publ. No. 54, pp. 43-49.

Lee, T.D., and Papin, P.A., "Analysis of Dropouts In Video Tapes," *IEEE Trans. Magn.*, Nov. 1982, Vol. MAG-18, No. 6, pp. 1092-1094.

Nunnelley, L.L., "Determination of Defect Length in Disk Coatings by Autocorrelation," *IEEE Trans. Magn.*, Jan. 1984, Vol. MAG-20, No.1, pp. 93-95.

Maediger, C., Voelz, H., Wiollaschek, H.K., "Analysis of Signal Statistics and Drop-Out Behavior of Magnetic Tapes," *IEEE Trans. Magn.*, Aug. 1984, Vol. MAG-20, No. 5, pp. 765-767.

Pichler, H., and Pavuza, F., "Criteria for the Selection of Audio Tapes for Analog and Digital Recording According to Their Drop-Out Characteristic," *Audio Eng. Soc.* Preprint, Oct. 1984.

Print-Through

Daniel, E.D., "Accidental Printing in Magnetic Recording," *BBC Quarterly*, Oct. 1951, Vol. 5, pp. 241-256.

Radocy, F., "A New Device for the Reduction of Print-Through," *J. Audio Eng. Soc.*, March 1959, Vol. 7, No. 3, page 129.

Tochihara, S., Imaoka, Y., and Namikawa, M., "Accidental Printing Effect of Magnetic Recording Tapes Using Ultrafine Partiles of Acicular Gamma-Fe_2O_3," *IEEE Trans. Magn.*, Dec. 1970, Vol. MAG-6, No. 4, pp. 808-811.

Stafford, M.K., "A Print-Through Constant for Magnetic Particles," *IEEE Trans. Magn.*, Sept. 1976, Vol. MAG-12, No. 5, pp. 583-584.

Corradi, A.R., Andress, S.J., Dinitto, C.A., Bottoni, D., Candolfo, G., Cecchetti, A., and Masoli, F., "Print-Through, Eraseability, Playback Losses: Different Phenomena from the Same Roots," *IEEE Trans. Magn.*, Sept. 1984, Vol. MAG-20, No. 5, pp. 760-762.

Eraseability

Manly, W.A., "Erasure of Signals on Magnetic Recording Media," *IEEE Trans. Magn.*, Nov. 1976, Vol. MAG-12, No. 6, pp. 758-760.

Fayling, R.E., "Anisotropic Erasure and Demagnetization Characteristics of Recording Tapes Comprising Particles with Uniaxial Magnetocrystalline Anisotropy," *IEEE Trans. Magn.*, Sept. 1977, Vol. MAG-13, No. 5, pp. 1391-1393.

Burke, E.R., and Sanders, D.R., "The Erasure of High Energy Tapes," *IEEE Trans. Magn.*, Sept. 1985, Vol. MAG-21, No. 5, pp. 1374-1376.

Christensen, E.R., and Finkelstein, B.I., "Erasure Methods for High-Density Recording," *IEEE Trans. Magn.*, Sept. 1985, Vol. MAG-21, No. 5, pp. 1377-1379.

Stress Demagnetization

Daniel, E.D., and Naumann, K.E., "Audio Cassette Chromium Dioxide Tape," *Jour. AES*, Nov. 1971, Vol. 19, No. 10, pp. 822-828.

Flanders, P.J., "Magnetostriction and Stress-Induced Playback Loss in Magnetic Tapes," *IEEE Trans. Magn.*, May 1979, Vol. MAG-15, No. 3, pp. 1065-1067.

Woodward, J.G., "Stress Demagnetization in Videotapes," *IEEE Trans. Magn.*, Nov. 1982, Vol. MAG-18, No. 6, pp. 1812-1818.

Terada, A., Ishii, O., and Kobayashi, K., "Pressure-Induced Signal Loss in Fe_3O_4 and Gamma-Fe_2O_3 Thin Film Disks," *IEEE Trans. Magn.*, Jan. 1983, Vol. MAG-19, No. 1, pp. 12-20.

Flanders, P.J., "Elastic Stress-Induced Coercive Field Changes in Ni-Co-P Films Used in a Rotating Disk," *IEEE Trans. Magn.*, Sept. 1983, Vol. MAG-19, No. 5, pp. 1680-1682.

Izawa, F., "Theoretical Study on Stress-Induced Demagnetization in Magnetic Recording Media," *IEEE Trans. Magn.*, July 1984, Vol. MAG-20, No. 4, pp. 523-528.

Hoshi, Y., Matsuoka, M., Naoe, M., and Yamanaka, S., "Demagnetization of Co-Cr Films Induced by Stress and Heat," *IEEE Trans. Magn.*, Sept. 1984, Vol. MAG-20, No. 5, pp. 797-799.

Part 5

Tape Transports and Disk Drives

Chapter 15

Tape Transports

The drive apparatus required to move the magnetic medium past the write and read heads are conceptually simple mechanisms, tailored to move the medium at a constant speed. There are only four basic configurations: A long ribbon (tape) wound onto a spool, a short, wide ribbon (card) and a circular, rotating sheet that is thin and flexible (diskette) or thick and rigid (disk).

The tape or disk drives are precision mechanisms that move the medium past the head assembly at a perfectly constant speed, with a minimum of interference to the electromagnetic signal transfer. The interference may be time variations (constancy of medium speed, flutter, TDE), or amplitude and phase variations (head-medium spacing, drop-outs, tracking, coating thickness uniformity).

This chapter will cover the topics that pertain to tape drives:

- Basic tape transport
- Capstan and pinch roller
- Reeling and handling of tapes
- Vibrations in free tape spans
- Rotating head assemblies
- Cassettes and cartridges

Disk drives are covered in Chapter 16.

Tape units are vital components in data measurement and computer systems. Their demise has been predicted numerous times, and they have each time survived—lately in the form of numerous back-up drives, and in the new IBM 3480 drive. An introduction to computer tape drives is available to the reader in the papers by Harris et al., and by Wildman. Application of instrumentation recorder technology to high rate digital data is given by Zoeller.

DIGITAL TAPE DRIVES

Open Reel-to-Reel Transports

Tape drives employ cassettes, cartridges or reels with tapes of different widths and lengths. Table 15-1 lists typical tape configurations and their utilization in a number of mini-computer and computer drives.

The tape transports must satisfy several requirements. Very fast starts and stops are necessary and movement of the tape must be in a perfectly straight path. The last requirement is particularly critical in parallel track applications with high bit densities; here the bits from the tracks need to be read simultaneously. Tape skew will often necessitate the use of de-skew buffers to correct for time-displacement errors between tracks.

The fast start-stop requirements are evident from Fig. 15-1, where a tape is started from stand-still to a speed of 200 IPS over a length equal to half of the IRG (Inter Record Gap). The calculations show that a linear acceleration of $9*10^5$ in/sec^2 is required during less than 1.5 msec.

Such high accelerations were not possible in early digital tape drives. It was common to bring the tape into motion by pressing it against a rotating shaft called a capstan by means of a puck. Both a forward and a reverse capstan could be used, see Fig. 15-2. Stop of tape motion was accomplished by releasing the puck and simultaneously clamping the tape between the puck and a fixed post.

This system was improved by introducing a perforated capstan wherein a vacuum would attract the tape and move it with the capstan. To stop the tape, air pressure would force the tape against a fixed surface near the capstan.

In the sixties very low inertia motors with printed circuit armatures were developed. These motors could meet the tape acceleration requirements and with a programmed servo control of the speed the designer could also achieve good tape speed stability within the allotted acceleration time. The tape is typically wrapped around the capstan over 120 to 180 degrees, and follows its motions precisely.

Table 15-1. Tape Configurations.

Tape Width: mm	mils	APPLICATIONS:
2	80	Future R-DAT ?
4	156	R-DAT (Digital Audio with rotating head assembly).
6.25	250	Audio Cassettes, Digital Cartridges and Cassettes.
8	320	Home Video, High Capacity Digital Archival.
12.7	500	Studio Audio, VHS and BETA Video, Computer Storage Reel-to-Reel and Cartridge (IBM 3480), Instrumentation.
19	750	Professional Video.
25.4	1000	Studio Audio, Broadcast Video, Instrumentation.
50.8	2000	Studio Audio, Old Broadcast Video, Instrumentation.

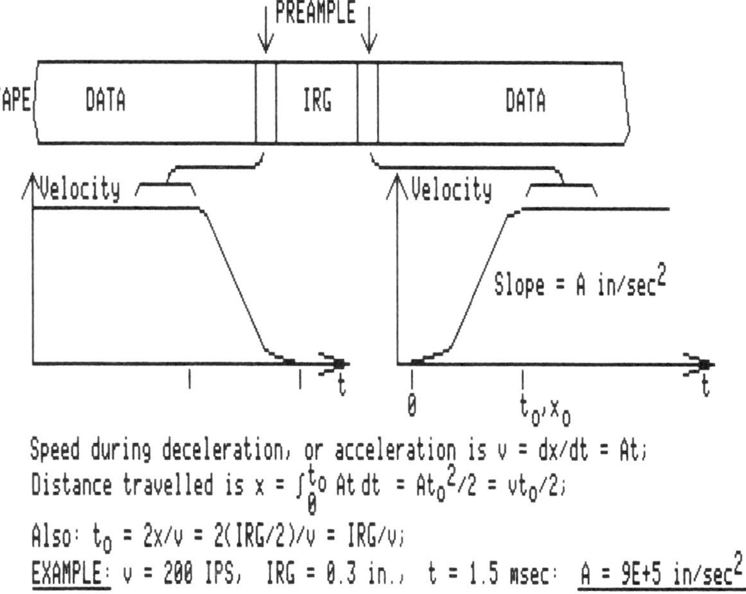

Fig. 15-1. Dynamics in fast start and stop mechanisms.

Another improvement in tape handling was the introduction of vacuum columns instead of tension arms to absorb the tape slack. Reel speed is servo controlled, which allows for control of the torque applied to the reel. When a reel of tape is subjected to excessive torque, it may happen that the tape pack shifts, causing cinching and subsequent damage to the tape. The tension with which the tape is wound on the reel contributes to the forces that prevent layers from slipping. These forces will decrease during storage because tape is a visco-elastic material.

The best recommendation for tape storage is to occasionally rewind the tape while possibly also cleaning it for debris (see Chapter 30). It is also suggested that the maximum acceleration that is applied to a reel should be kept below 10,000 rads/sec^2.

There are today a great number of tape drives that do not operate in the start-stop mode; they are named streamers and they move long lengths of tape in a write or read mode to copy databases or restore them. They may be operated in a start/stop mode, but at much lower speeds than the ordinary digital tape transports.

Design considerations for streaming tape transports are outlined in a paper by Cutler, and modern half-inch streamers are described by Curtis et al., Dong et al., Ruska et al., and Becker et al..

The proper operation of a digital tape drive can be tested by measuring the timing jitter on the bit transitions just prior to detection.

Cartridges; Cassettes

The amount of data that many mini and micro computers need to store in a backup archive, disks or tapes, will often require less tape than normally found on a ½″ tape reel. A smaller tape unit has thus emerged, the quarter inch tape cartridge, and in some cases an enhanced audio type cassette. The standard cartridge is shown in Fig. 15-3.

Currently two cartridges are being offered, the DC-300XL cartridge holds 450 feet of tape, while the DC-600A holds 600 feet and uses higher coercivity media (see Chapter 26 for track

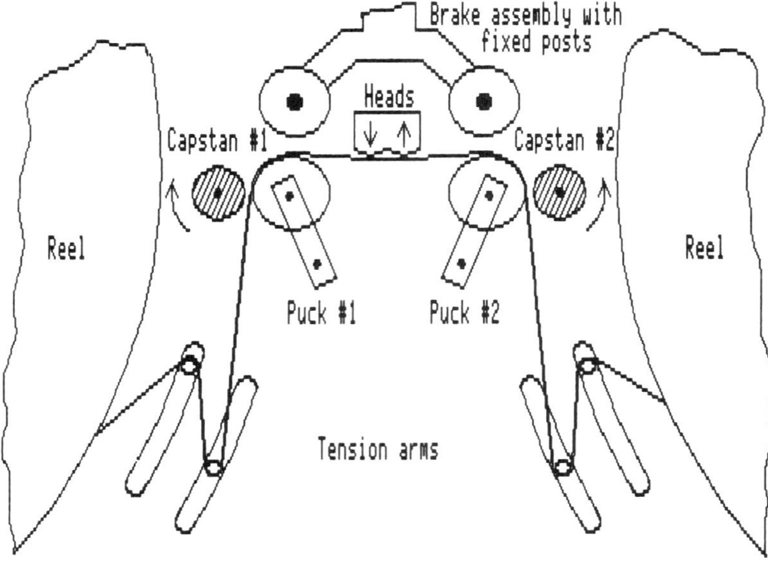

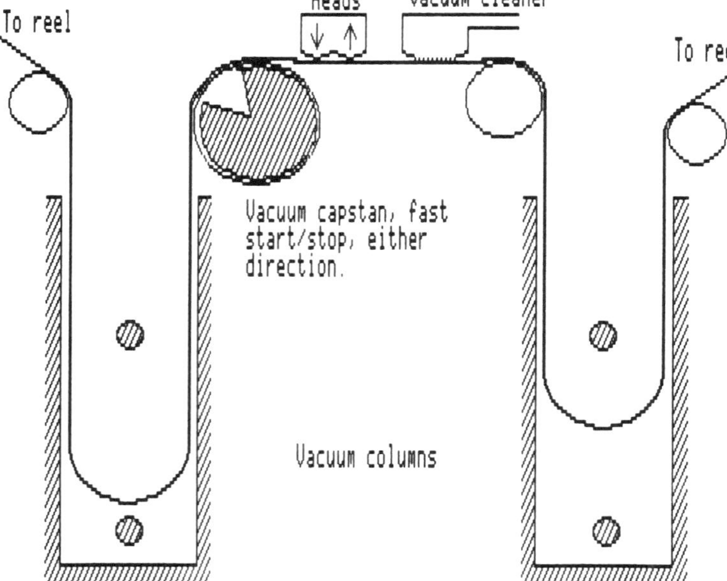

Fig. 15-2. Capstan arrangements in half-inch computer tape drives.

configurations). The housing consists of an aluminum plate where a number of rollers and guides are installed. The aluminum provides the rigidity and precision; the rest of the housing is molded clear plastic.

The tape is placed on two hubs without flanges, and is driven in one or the other direction by an endless PET belt, that in turn is powered from a rubber coated drive roller on the drive motor shaft. The head assembly is pushed into contact with the tape through an opening in the plastic housing; the opening is covered by a hinged door when the cartridge is not inserted into a drive.

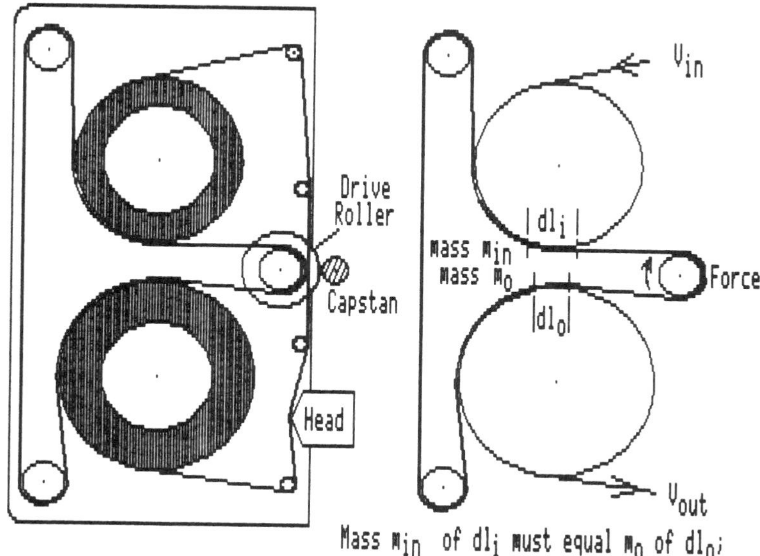

Fig. 15-3. The ¼-inch tape cartridge. The only points of interface with the tape drive is the write/read head and the drive roller. The PET belt drive generates the needed tension by a mass-flow concept.

A tension of about 57 grams (two ounces) is generated by mass flow principles, illustrated in the right portion of Fig. 15-3. The belt is pulled in on the top side of the capstan roller, and pushed out on the bottom side. Imagine that the belt is made of a very soft and compliant rubber; the ingoing thickness must therefore be thinner than the outgoing thickness when energy is supplied to tape motion and frictional losses.

Since the mass of the belt is constant we can then deduce that the length dl_o is shorter than dl_i, and that the speeds of the tape taken up and fed out therefore are different: $V_{in} > V_{out}$. The difference between the two tape speeds is compensated for by a slight elongation of the free tape span, and that is how the tension of about two ounces is generated over the heads, for tape motion in either direction (ref. Burdorf).

One drawback is a varying tension profile from beginning to end of tapes. The heads must have a special profile with additional contact points (outriggers) to assure a minimum change in wrap angle around the heads contact points.

The design concepts of a standard cartridge drive is covered by papers by Valliant and by Gennetten, while the miniature cartridge is described by Nordman et al.. State-of-the-art reports on cartridge tape drives have been prepared by Makmann and by Wright.

In mid-1984 IBM introduced the model 3480 ½-inch cartridge drive, see Fig. 15-4. The cartridge measured 1 × 4 × 5 inches, and houses a single reel; the drive employs an 18-channel thin-film head for 18-track recording on chromium-dioxide tape, storing about 200 Mbytes at 38 kbpi density. Its present use is limited to mainframes, primarily those with 3780-class disk subsystems.

A group of tape drive manufacturers has formed a Working Group for Half Inch Tape Cartridge Drives, and has endorsed the IBM cartridge format, but not the head design nor recording format. The group plans to use the cartridge in lower cost drives for other than mainframe uses (ref. Wright).

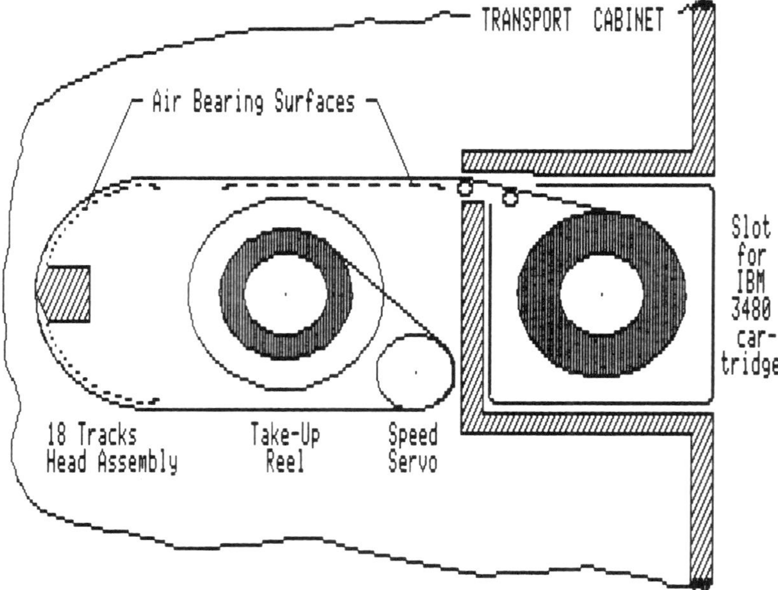

Fig. 15-4. The IBM-3480 tape cartridge transport, simplified.

A last format being used for storing data is the Philips audio cassette and a miniature version thereof (ref. Manildi, Buskirk et al.). This appears to be a shrinking market that lost out to the cartridges.

Finally, some readers may be interested in learning about the small magnetic strips, hand-pulled for playback, used for storing small programs, data etc. The design of the unit used in the HP-75 portable computer is described by Hoecker et al.

Tape Guidance; Skew

Bytes are written in parallel in 7 and 9 (and now also 18) track systems and they should be read back simultaneously. This situation never exists in practice because of tolerances in the manufacture of multitrack heads and in the slitting of tapes. The nominal width of a computer tape is 0.498 •0.002 inches and it may have a longitudinal curvature of up to ⅛″ over 36 inches. The curvature may, among other factors, cause the tape to ''weave'' through the drive and it could, over a distance of say 3 inches, depart as much as ±0.002 inches from the ideal, straight path. This translates into an angle of:

$$\beta \cong \tan\beta$$
$$= \pm 0.002/3$$
$$= \pm 0.00067$$

resulting in an edge-to-edge displacement of $0.5'' * \tan\beta = 335\ \mu\text{in}$. To this error from dynamic skew should be added the gap scatter and head tilt (from one drive to another) and the total maximum error could be as high as 1 mil between edge tracks.

An error of 1 mil is almost equal to one bit length at the conservative packing density of 800 BPI and the simultaneous read-out of a byte would be impossible. The problem is solved by storing the bit stream from each track in individual buffers, which are clocked in synchronism.

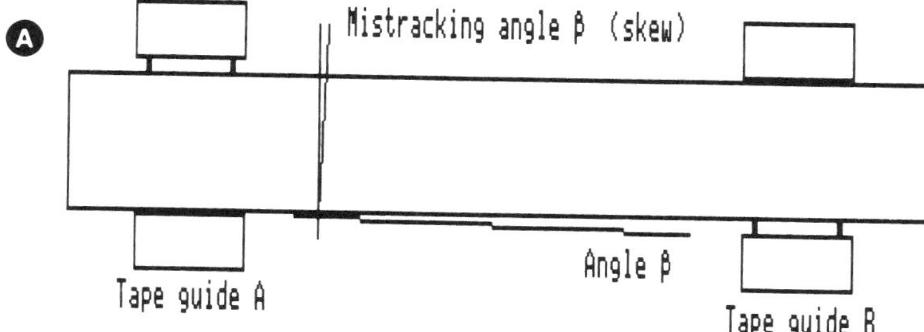

Fig. 15-5. A: Tape skew; B: De-skew buffers.

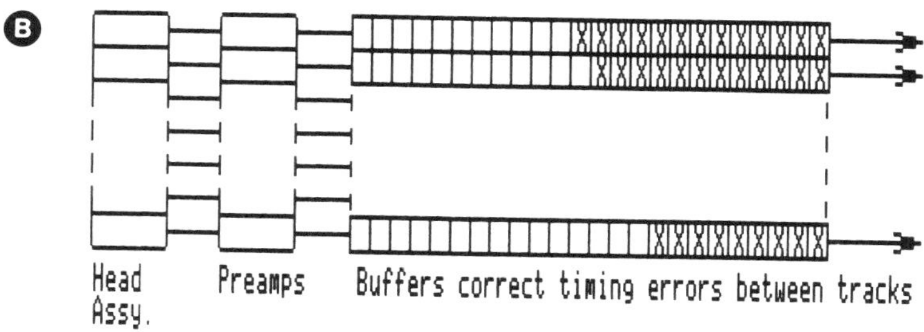

The skew changes quite slowly and the timing of the de-skew buffers is initiated by a known bit pattern stored in front of a block of records. Only the edge tracks need to be read for the pattern to be recognized by logic in the controller. The buffers are normally kept half-full (on the average) and time displacement errors between tracks are completely removed (Fig. 15-5).

ANALOG TAPE DRIVES

This section is devoted to a discussion of the audio, instrumentation and high-density-digital tape drive mechanisms. There are hundreds of different designs in operation today, but they are all quite similar when the dressings and trimmings are removed. The physical layout may differ but the transports all have several common design denominators, such as capstan drive, reeling system and tape guidance.

Most important is the striving for perfection in achieving a constant tape speed. Speed variations called flutter are easily detected by the human ear as a vibrato in music, or by sensitive detection circuits in instrumentation and digital equipment. Flutter implies that recorded events are not played back with the original timings between them and a measure for this is defined in the time displacement error (TDE).

The tape configuration (reels, cassettes, etc.) and the area of application dictate the overall tape transport layout, which will be discussed first. The major portion of this section is then devoted to the capstan drive and reeling systems and the concept of electro-mechanical analogies are introduced as a valuable aid in design work.

Tape Configurations

The ranges of tape widths and tape speeds have expanded several times from the original ¼ inch wide tape operating at 15 IPS on the German Magnetophone. No single transport can

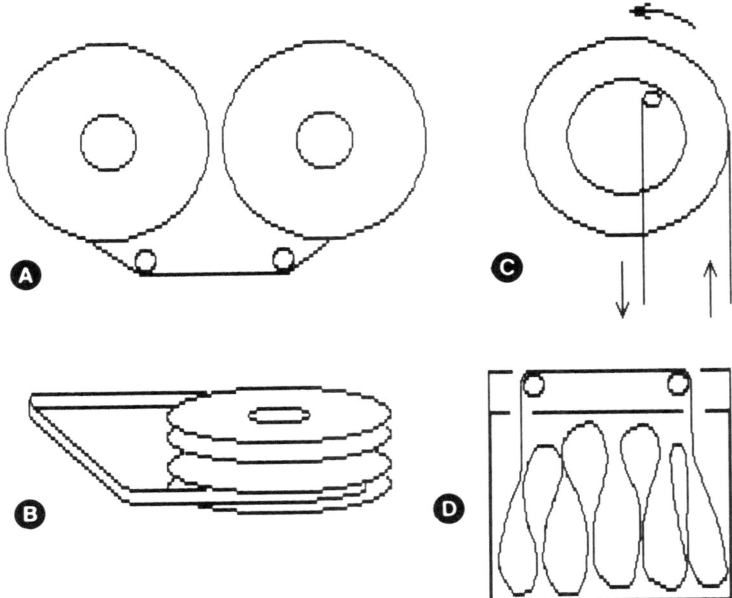

Fig. 15-6. Reel-to-reel arrangements: A. side-by-side; B: stacked; C: Endless loop cartridge; D: Endless loop bin storage.

handle all of the tape widths shown in Table 15-1. Their designs center around one tape width and a speed that can be selected from one or two adjacent speeds 4.75 cm/sec (1⅞ IPS), 9.5 cm/sec (3¾ IPS), 19 cm/sec (7½ IPS) etc.; instrumentation recorders offer up to eight standard speeds.

Layout of a Tape Transport

The physical layout of a magnetic tape transport is dictated by its application. Figure 15-6 shows four different layouts used for analog transports. Figure 15-6A illustrates the reel-to-reel concept which is so commonly found in home-type recorders (including cassette) and in many instrumentation recorders. Figure 15-6B shows a compact version where one reel is located concentrically (or coaxially) above the other, a feature which saves space. A cartridge design (Fig. 15-6C), used in many automobile installations and broadcast stations, is shown together with an endless loop contained in a bin (Fig. 15-6D). The latter concept is often used where a constant monitoring of data is required without the need for prolonged storage; one such application is in flight recorders where only the last few minutes of cockpit conversation must be stored.

Capstan Drives

The most important element in a magnetic tape transport is the capstan—a precision drive shaft against which the tape is pressed, often by a pinch roller. The speed of the capstan determines the tape speed. The diameter of the capstan shaft varies from about 1.5mm (1/16 of an inch) for slow-speed cassette recorders to about 75mm (3 inch) for high-speed, precision instrumentation recorders.

There are essentially three basic capstan drive systems (Fig. 15-7). The oldest and still a much used concept (Fig. 15-7A) is the open-loop drive where the capstan pulls the tape over

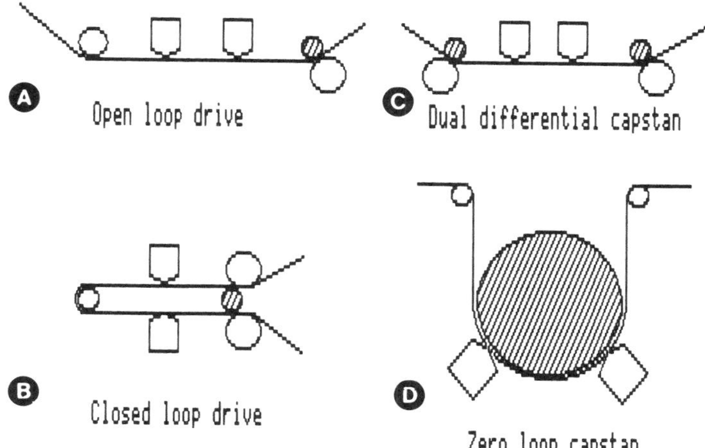

Fig. 15-7. Basic capstan drive configurations.

the head with tension provided by hold-back torque or a felt-pad arrangement. These transports often have an idler which provides isolation from any tape vibrations from the supply reel. In this regard, the closed-loop drive (Fig. 15-7B) is superior in isolating any speed variations from the reeling system and this concept is found in several instrumentation recorders.

The best drive system is shown in Fig. 15-7C, where two capstans with associated pinch rollers are used. One capstan rotates at a slightly higher speed than the other, thereby providing tension on the heads. Both capstans could rotate at the same speed with the diameter of one capstan being slightly larger than the other. This differential drive is in wide use in instrumentation recorders, and recently in studio and home recorders, in reel-to-reel as well as cassette recorders.

Figure 15-7D illustrates yet another drive concept, where the tape contact with the large diameter capstan is held by a vacuum. This method is referred to as a zero-loop drive and has the advantage that the tape is supported all the way past the heads, which eliminates scrape flutter.

Capstan Motors

The capstan shaft is driven by either an AC or a DC motor. In inexpensive home entertainment recorders the drive motor is quite commonly a squirrel-cage configuration which drives the capstan through a belt system. With such an indirectly driven system, the capstan shaft usually has a flywheel on it to smooth out any motor cocking or other speed variation. More expensive units utilize a hysteresis-synchronous motor in which the speed corresponds to the power line frequency. Again, a flywheel is used on the capstan shaft, or with higher-grade synchronous motors the capstan shaft may be the motor shaft itself.

Construction details of each motor are illustrated in Fig.15-8. The stator, contained within the motor housing, is the same for either motor. When the stator winding is connected to a multi-phase power line source, a rotating magnetic field is generated. The rotor on the squirrel-cage motor is constructed from stacked, nonconducting lamination with short-circuiting connections between the two end stators. The rotating magnetic field from the stator generates currents through the short-circuiting rods, and these currents in turn generate a magnetic field with a polarity opposite to that of the stator field. The stator currents, therefore, magnetize the rotor, which tries to follow the rotating magnetic field from the stator. When the rotor

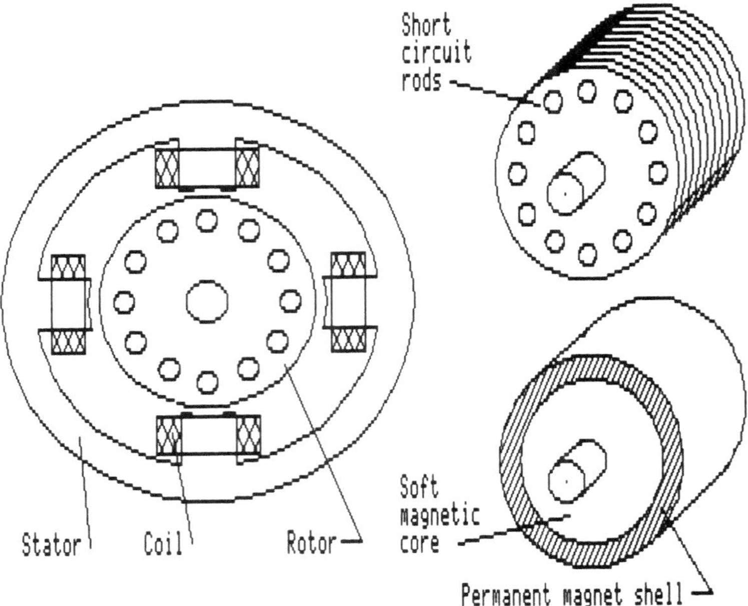

Fig. 15-8. Construction of squirrel-cage motor. The right hand drawing shows the armature in a hysteresis-synchronous motor.

approaches the speed of the field from the stator, the induced currents weaken, and if we assume that the rotor followed the stator field exactly, there would be no current induced and consequently no magnets would be formed in the rotor. Therefore, it would slow down. The squirrel-cage motor does therefore have a certain amount of slip. That is, it does not have a perfectly constant speed; the speed depends upon the load. This slip can vary from two to ten percent. That's why a squirrel-cage motor is used only in inexpensive recorders.

The hysteresis-synchronous motor has a composite rotor with an outer shell of a hard magnetic material and an inner core of soft magnetic material. The starting torque is higher because of the rotor's permanent magnetization due to the remanence in the hard magnetic material. The speed of the hysteresis synchronous motor follows the power line frequency exactly, since any slip would mean a constant remagnetization of the rotor. Figure 15-9 shows the difference in the speed vs. torque curves for the two types of motors. It is evident that the hysteresis-synchronous motor is superior, both with regard to constant speed and to a higher starting torque.

As mentioned above, most recorders are provided with two and sometimes three speed ranges. This is achieved in two ways: one, by changing the pulley ratio between the drive motor and the capstan shaft flywheel, which usually means mechanically shifting the drive belt (often a seamless cloth belt). Such a change can obviously only be made while the capstan motor is running. The second, and more practical solution, is obtained by using a two- or three- speed motor, with two or three times the number of poles on the stator. The speed selection is made by merely coupling all or only half of the poles to the power source.

In the instrumentation recorders, the speeds commonly used range from 120 IPS down through 60, 30, 15, 7½, 3¾ and even down to 1⅞ IPS. To make speed changes by means of a pulley and belt in this wide speed range is quite impractical. The choice of a drive system for such a demanding system, therefore, falls on a servo-controlled system, where the capstan drive motor is a DC type with a suitable tachometer mounted on the drive-motor shaft. The

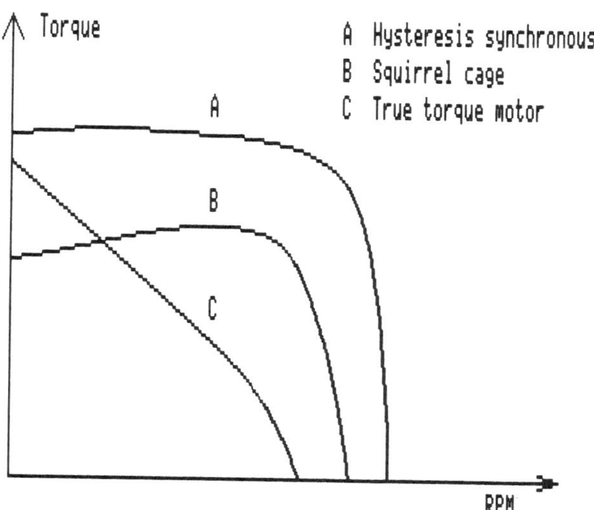

Fig. 15-9. Motor torque-versus-speed curves: A: hysteresis-synchronous; B: squirrel-cage; C: true torque motor.

output frequency of the tachometer is compared with a control reference frequency, and any frequency deviation results in an error signal which corrects the motor speed. Here, again, the drive motor can be coupled to the capstan shaft by means of a belt drive or the capstan shaft can be the motor shaft itself.

Servo Controlled Speed

In earlier instrumentation recorders, speed control was attained by recording a 17-kHz control frequency which was amplitude-modulated at a rate of 60 Hz. During playback, this signal was demodulated and fed into a phase-comparison circuit, which fed a precision 60-Hz motor-control oscillator whose output after amplification drove the synchronous capstan motor. This technique is referred to as speed control since the servo system operated through a high-inertia drive system with flywheels and belt drive and the maximum error rate that could be followed was around 1 Hz.

Instrumentation recordings are often made under adverse conditions (vibrations, etc.) and the inherent wow and flutter both from the recorder and the external vibrations in turn affect the frequency of the recorded tones. This will have an effect on any analysis work that is performed on data from the tape. Any change of speed causes a change of signal frequency, which is translated to a change in timing. Therefore, it is vital that this error, also called time-base error, be reduced. This requirement has led to a new family of instrumentation recorders which have a drive system as illustrated in Fig. 15-10.

The capstan drive motor is a low inertia, high-peak-torque motor, driven by a power amplifier controlled by a servo amplifier. When the tapes are recorded originally, one track is recorded with a control tone with a frequency ranging from 200 kHz down to 6.25 kHz, all dependent upon the speed selected. Upon playback this signal is amplified and fed to a phase detector and compared with a crystal-controlled control frequency. Any speed deviation results in a phase error signal which in turn controls the servo amplifier and causes the tape drive to speed up or slow down. The inertia of the capstan motor, and possibly the coupling elements, restrict the time for making speed corrections.

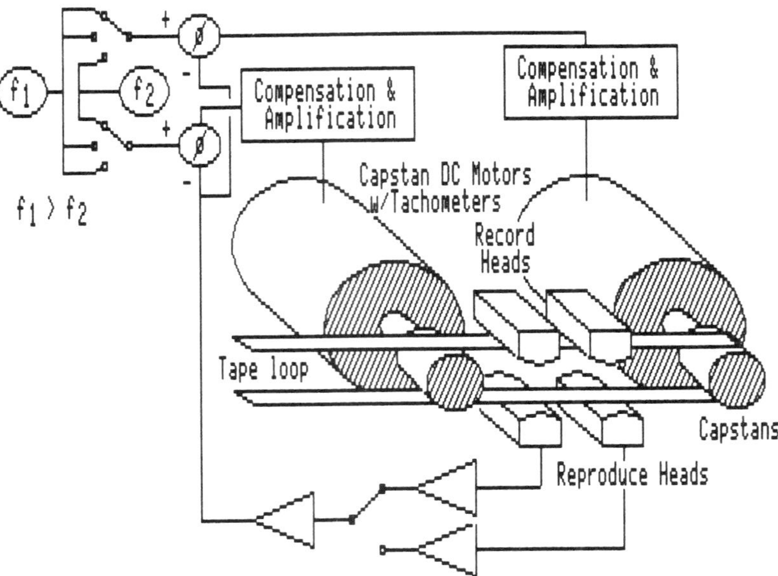

Fig. 15-10. Dual-motor servo control, shown in the playback mode.

A measure of merit is saturation acceleration (equal to the peak torque of the drive motor divided by the inertia). A low inertia drive with high saturation acceleration is far more capable of correcting speed variations than any of the more conventional transports with their heavier motors and flywheels. A numerical evaluation of tape speed performance is given in Fig. 15-12.

The degree to which a servo can suppress a flutter component is a function of the open loop gain of the servo at the frequency of the particular flutter term. If the velocity of the capstan is considered to be the same as the velocity of the tape (a valid assumption when the distance between head and capstan is short) the transfer function relating the tape velocity ω_o (in rads/sec) to the reference frequency ω_{ref} is (see Fig. 15-11, top):

$$\omega_o/\omega_{ref} = G(s)/(1 + G(s)H(s)) \qquad \textbf{(15.1)}$$

where:

$G(s)$ = transfer function for the forward path
$H(s)$ = transfer function for the feedback path

It can be shown that the relationship between output F_o and input flutter F_i is:

$$\boxed{F_o/F_i = 1/(1 + G(s)\,H(s))} \qquad \textbf{(15.2)}$$

From this equation it is seen that the output flutter is attenuated by the factor $(1 + G(s)H(s))$. Consequently, to obtain the maximum attentuation it is necessary to make $G(s)H(s)$ as large as possible and to provide this correction over as wide a bandwidth as can be achieved.

The complete transfer function becomes unwieldy when the individual transfer functions are substituted into our equation. A number of meaningful conclusions can, however, be reached

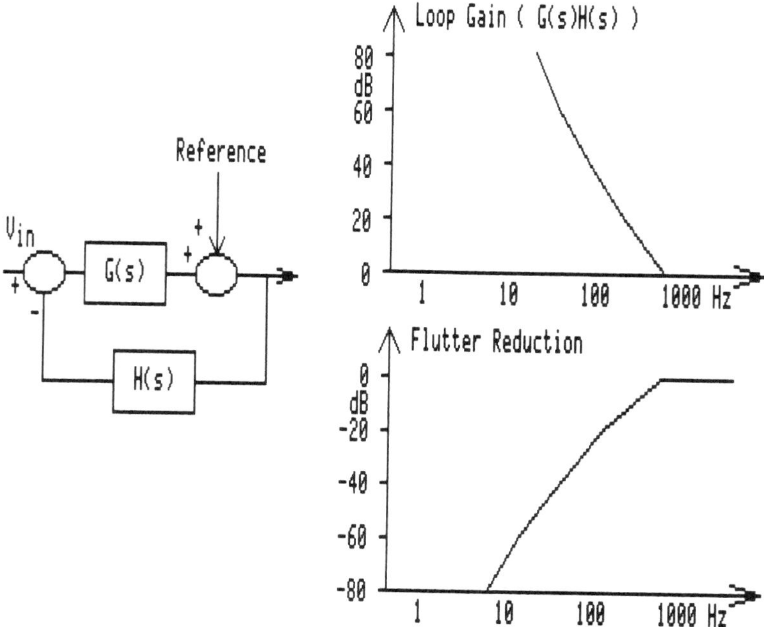

Fig. 15-11. Characteristics of a typical capstan servo.

by considering the general form of the expressions. Then, graphical techniques can be employed to evaluate servo performance of the actual transfer function.

Our previous equation exhibits unity in the frequency range where G(s) H(s) > > 1 (Fig. 15-11). Since G(s)H(s) has two poles at zero frequency (the compensation and feedback each introduce 1/s terms) the DC gain is theoretically infinite and $\omega_o/\omega_{ref} = 1$. Consequently, there can be no static error between the shaft velocities and the reference.

Also, from our last equation it can be seen that the useful bandwidth of the servo occupies the frequency range for which G(s)H(s) = > 1. Consequently, the servo will provide correction only for flutter that occurs within this bandwidth.

The degree of correction is in practice limited by several terms in G(s), arising from mechanical resonances in the system. The tape itself is elastic with a spring constant K that enters into the servo loop. Speed corrections propagate with the speed of sound in PET film (2,000 meters/second) and cause a minor delay. The capstan shaft and the motor itself have rotational resonances (ref. Diamond, Alexander et al., Hu). The head stacks may vibrate as cantilevers if not rigidly mounted. Most troublesome is still the tape since all flutter components are inside the servo loop and a closer examination of tape flutter is worthwhile (ref. Law).

Example 15.1: Figure 15-12 contains a few calculations for a typical example of the numbers pertaining to flutter reduction and the dynamics involved in correcting tape speeds.

If we assume a disturbing flutter (as it exists on the recorded tape) of 1 percent p-p at 15 Hz at the speed of 120 IPS, that corresponds to a TBE of ± 53 μsec, and the actual tape displacement error at the reproduce head is ± 0.000636 inches. If we assume an ideal dampened idler with infinite inertia (ref. Wente), then the capstan must correct these relatively instantaneous displacement errors by increasing or relaxing the tension in the tape. The slope of the stress-strain curve for the tape is 750,000 psi/in/in around $\sigma = 1250$ psi (corresponding to 10 oz tension in the ½", 1 mil tape).

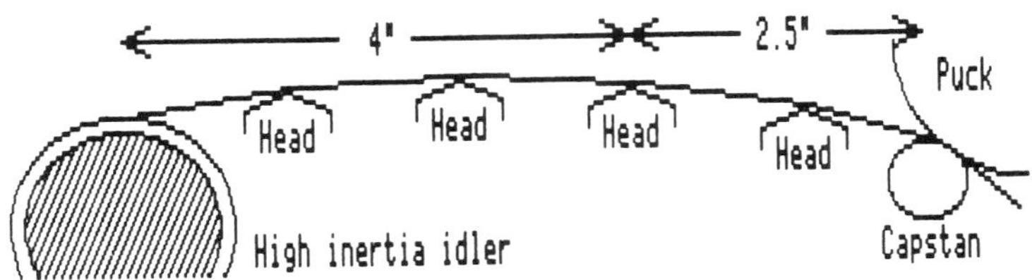

Flutter to be maximum 1 % p-p = ± 0.5 % peak at 15 Hz. The resulting TBE (Time-Base-Error) equals [Flutter/100]/2πf = 0.005/2π15 = 53 μSec peak.

Permissible displacement error at reproduce head = 53E-6 * 120" = 6.4 mils, which at the capstan corresponds to 6.4E-3 * (4/(4+2.5)) = 10.0E-3" = 10 mils.

The required change reflects in an elongation, or shortening, of the tape span equal to error/length = 10E-3 / 6.5 = 1.6E-3 in/in. The slope of the tapes stress-strain curve at T = 10 oz (1250 psi) is 750,000 psi/in/in.

The dimension change requires a force of 1.6E-3 (in/in) * 750,000 (psi/in/in) = 1,194 psi = 9.5 oz.

The tension in the tape may therfore vary from 0.5 to 19.5 oz, and the required acceleration is: dv/dt = d(120 + 0.3 sinωt)/dt = 0.3ω cos ωt, with a peak value of 0.3 * 2π15 = 28.2 inches/sec^2.

Fig. 15-12. Sample calculations for a single capstan servo system.

When the calculations are carried through, we find that the tension variations range from 19.5 oz to 0.5 oz. Had the flutter been larger, a tape slack condition would have occurred during each flutter cycle (negative tension).

This capstan configuration (open loop) would obviously not provide a good servo system. The dual capstan system provides an advantage: it will assist in pushing (or holding back) the tape, which moves as a platform under constant tension.

An instantaneous time displacement error of 53 μsec will still require a displacement correction of 6 mils at the reproduce head. But the speed of sound in tape is around 2,000 m/sec = 78,800 inches/sec, and if each capstan corrects for one-half of the length of the tape platform, the theoretical tape limit for error correction in a 4″ long tape is $\frac{1}{2} \times 4 \times \frac{1}{18800} = 25 \times 10^{-6}$ sec = 25 μsec, or an upper frequency limit of 20 kHz.

Capstans and Pinch Rollers

The objective of the capstan drive is to meter the tape motion at a constant speed and to generate or maintain a constant tape tension over the magnetic heads. The very basic capstan drive is a rotating shaft moving the tape at a constant speed.

The capstan shaft was in the past often driven from a two-speed hysteresis synchronous motor via a belt, or the motor shaft itself could serve as the capstan. This required a good motor, with freedom from hunting (ref. Clurman). Modern recorders use low inertia DC-motors with a tachometer and a servo system for speed control, or brushless DC-motors.

The tape is pulled by friction. In order to maintain a reasonable tension over the heads (approximately 20 ounces per inch tape width), it is necessary to clamp the tape to the capstan. An early practice and still in widespread use today, is the method of pressing the tape against the capstan with a compliant pinch roller. This forms a nip where the tape is clamped. The tape speed is now determined by the pinch roller. (The area on the pinch roller and the coefficient

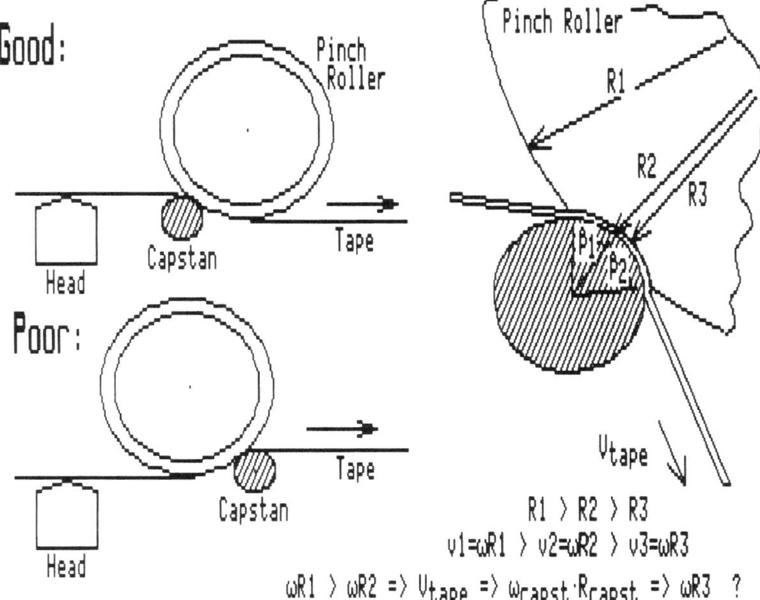

Fig. 15-13. Capstan-rubber-puck drive.

of friction between the roller and the tape is greater than between the capstan and the tape). See Fig. 15-14.

The true tape speed may not be equal to the surface speed of the capstan (Fig. 15-13). The pressure causes an indentation in the compliant pinch roller and the effective radius of the roller is reduced. The tape enters and follows the pinch roller during angle β_1 with a certain speed, and leaves the roller during angle β_2. An answer to the complexity of the capstan/pinch roller question may come some day from detailed analysis (ref. Durieu).

The surface speed of the pinch roller in the nip area is somewhere between v_2 and v_3, and so is the tape speed. The tape leaves during the angle β_2 and there the surface speed of the pinch roller is greater than the tape speed v. Therefore, the tape will slip and the relative

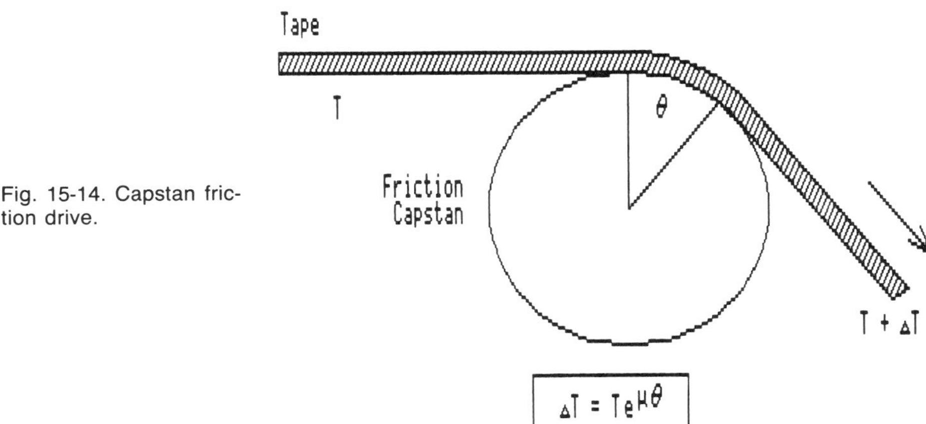

Fig. 15-14. Capstan friction drive.

speed difference will be so small that stick-slip friction occurs. This would also be true if the tape contacted the pinch-roller prior to the capstan (see Fig. 15-13, left, lower)—a bad situation which generates high flutter in the head area.

The tape and the pinch roller are both elastic materials, and slight variations in the take-up and hold-back tension from the reeling system will cause deviations in the tape speed. The tendency is for the tape to slow down toward the end of a reel. The implications may be a noticeable change in pitch if the recorded material is edited.

A few manufacturers specify the maximum speed deviation that may occur as slip and it is normally within one percent. The effect of slip upon editing can be minimized by using large tape reels (10½″ rather than 7″ diameter) and recording on less than half of the tape length. The change in pitch from sections spliced together may then not be noticeable.

Much can be said against the application of pressure felt-pads that press the tape against a tape guide post for tension control. But, they will, when properly maintained and adjusted, provide constant tension and hence assure constant tape speed throughout a reel.

REELING AND WINDING OF TAPES

While the capstan drive system plays a major role in maintaining speed accuracy, equal attention must be given to the reeling system. The supply reel must feed the tape to the head area in a perfectly smooth motion, while the takeup reel winds the tape onto the reel under constant tension and with a resulting smooth tape pack.

During winding or re-winding, when the capstan puck is disengaged, the tape should move with the highest practical speed from one reel to the other. When the winding or rewinding mode ends, both reels should stop in such a fashion that the tape will neither be overstretched nor will it throw a loop. This again puts a demand on the design of the braking mechanism for the reel system.

The best reeling is obtained by the use of two separate reel motors, which should be torque motors that provide for an almost constant tape tension (ref. Holcomb, Hittle). Such motors have an inverse speed-torque curve, as shown in Fig. 15-9, curve C. When the reel is empty and the speed is high, the torque is low and a given tape tension is provided. Next when the reel is filling up, the speed slows down and the torque increases, but the radius of tape accumulating on the takeup reel is increasing and the tension remains constant from the empty to the full reel (ref. Hartmann). The almost constant tension is desirable since tapes stored for a long period under excessive tension, or a highly varying tension through the reel, may become permanently damaged.

Winding of a tape at high speeds does not always happen with a smoothly flowing motion. Minute fluctuations in the reel motor torque (cocking) may cause the tape to flap back and forth, and the pack on the takeup reel will thus be bad. If the transport designer is faced with these problems he may find help by studying a similar phenomenon that occurs in band saws and other line structures (ref. Swope, Mote, Thurman et al, Mote et al.). The designer will also benefit from a recent paper on the handling of wide webs (ref. Pfeiffer).

Constant tension during record or play modes is important for the absolute tape speed, in particular with open-loop transport drives. The speed is affected by the supply reel tension and, in poorly designed recorders, the absolute speed of the tape can change several percent from the beginning to the end of a recording. If played back on the same recorder, the same tension variation exists and the playback speed is equal to the record speed. But if the tape is edited, and especially if a portion is moved from the beginning to the end or vice versa, there will be a noticeable change in the frequency or pitch of the recorded material.

The braking system, which (as earlier mentioned) works in conjunction with the reeling motors, must be carefully designed to avoid tape damage. A popular braking system is shown in Fig. 15-15, right. Each brake consists of a flat steel band with a felt coating. The braking

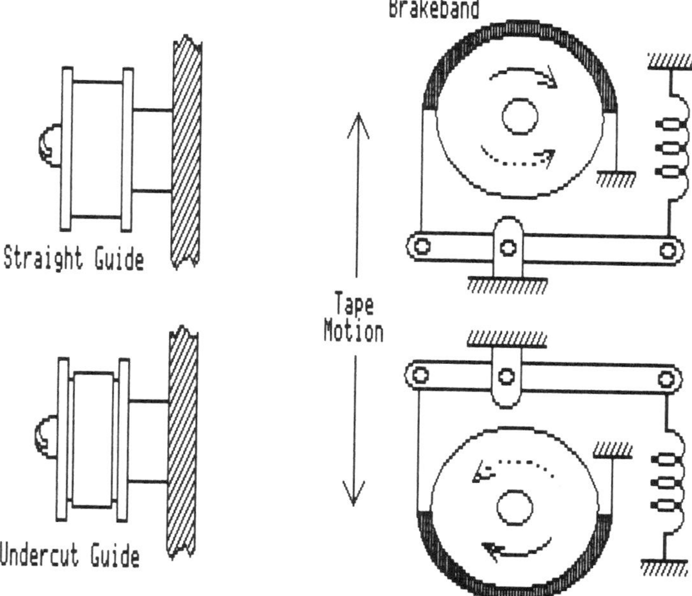

Fig. 15-15. Tape guides (left), and a typical reel braking system.

action is differential. In other words, if the tape is traveling from top to bottom when the brakes are engaged, the top wheel will receive a slightly greater braking action than the bottom wheel, and vice versa. This braking system is disengaged during playing or winding either by means of a lever or by solenoids. The use of solenoids is a more attractive solution since any power failure to the recorder will cause the brakes to automatically engage.

Spring loading for the brakes must also be given consideration during the design of a recorder. Most home recorders are designed for the use of 7-inch reels, but quite often such recorders are used with, say, a 7-inch reel for takeup and a 3-inch reel for supply (in the exchange of tape letters, for example). Here the dynamics of the braking action are different from that originally designed for two reels of the same size and the operator should take great care in starting and stopping a recorder with different reel sizes. The same problem exists in instrumentation recorders where the two commonly used reel sizes are 10½ inch and 14 inch diameter reels. The larger reel has far more inertia than the smaller reel. Therefore, some of these recorders are equipped with a switch that selects the proper braking action for the various reel sizes.

Dynamic braking is commonplace in instrumentation recorders and in better audio recorders. Proper current levels are fed to the reel motors, possibly servo controlled by sensing arms in the tape path, bringing the tape motion to a gentle stop. It is important that the tape path from the supply reel to the takeup reel be in one perfectly flat plane. This requires a rigid transport deck (which often is a casting) and also carefully mounted and adjusted tape guides. The tape guides must provide perfect edge guidance of the tape. Normally, two types of fixed guides are used, as shown in Fig. 15-15, left. The design shown at the bottom is undercut at the inside flange edges and is made out of one piece of metal. The undercut design is not very desirable, since the edges on the tape-supporting surface eventually cause creases in the tape with a resulting deformation of the tape itself. The design shown above it consists of three round pieces that are screwed together, thus presenting a perfect guiding edge and overall support of the tape surface.

Rotating guides, with or without flanges, are also employed. They may be flat or crowned rollers, mounted on precision roller bearings; the reader is referred to the papers listed for further details (ref. Eshelman et al, Paroby et al., Clurman).

Flanged guides in the head area are supposed to guide the tape without skew, or deviation, from the straight path. The clearance through the guide is manufactured to the exact specified maximum width of the tape. The actual tape width may be less, by as much as 2 mils, since the tolerance of state-of-the-art tape slitting is •mil. This may cause considerable ''wandering'' of the tape as it runs through the transport, with a maximum deviation of 2 mils over the length of the path between two guides. When the tape is recorded and played on the same transport chances are that the tape will wander in the same pattern. But if is recorded on one machine and played back on another a full deviation of 4 mils may occur over a span of say 4 inches.

This is equal to 0.001″ per inch, or a misalignment of α = atan(0.001) = 0.001 rads = 0.0057 degrees. The signal loss can then be calculated from the alignment loss formula in Chapter 6. A 2 MHz signal at 120 IPS could suffer a worst-case attenuation of 14.4 dB if the tape clears the guides in opposite ways during record and playback, while a realistic worst case would be only about 7 dB. A more annoying outcome is the amplitude variation associated with an uneven tape passing through the guide.

A similar skew problem occurs when the tape path is changed from one plane to another. This happens in recorders with coaxial reels. Sufficient lengths must be allowed for the tape to make its twist, so in no place is it subjected to a stress exceeding 3000 psi. Higher stresses may cause permanent tape damage, often in the form of stressed edges.

Some useful features may, in conclusion, be incorporated with the guides on a transport. These are: tape-break and end-of-tape sensors, footage counters and tape cleaners (vacuum, possibly with ionization).

Head Mounting

In magnetic recorders with low packing densities, it is quite common to find that the heads are mounted on a fixed base plate, which in turn is mounted on the capstan drive precision plate. Where the packing densities (i.e., wavelengths per length of tape) exceed 5,000 Hz per linear cm of tape, it becomes necessary to mount the reproduce head on an alignment plate so its azimuth can be adjusted to be exactly equal to that of the record head. The record heads are normally fix-mounted with a certain tolerance allowed for the azimuth which is obtained by careful grinding of the recording head mounting surface in the final step of its production.

Reproduce heads must be carefully shielded with a MuMetal housing or another suitable metal in order to minimize hum pickup from the drive motors, transformers, and any other external interference source. In wideband recorders that operate in the range of up to 4 MHz, additional shielding becomes necessary in order to avoid pickup of signals in the midwave radio-frequency band. Powerful broadcast stations in this range may be picked up by a sensitive reproduce head and appear as noise in the recorder output. In the design of the head mounting, attention should also be paid to the fact that signals from the record head may radiate into the reproduce head during recording, which makes alignment of the recorder's electronics difficult.

DESIGN FOR LOW FLUTTER

In order to keep flutter components low, the magnetic heads should be placed at a point where the tape speed is smoothest and this is normally as close to the capstan as possible. Attention should be paid here to the fact that the magnetic tape itself is an elastic medium and that longitudinal oscillations (similar to those of a violin string) occur. Whenever the tape runs over a fixed guide or magnetic head, it rubs against these fixed elements, resulting in oscillations (scrape flutter, see Chapter 17). Guide and head surfaces must be as smooth as possible, to minimize flutter and the wrap angle around the heads should be a minimum, normally in the order of 3 to 5 degrees. If the heads are located between the idler and the capstan, scrape flutter will be reduced. A rotating high-inertia guide will likewise serve as a grounding point for the longitudinal oscillations and such guides are frequently incorporated into precision tape drives.

Flutter and time displacement errors must be minimized in a tape drive, both for the sake of listening pleasure and for data integrity. It is true that modern digital recorders will remove flutter by the use of data buffers on the read side, with sizes proportional to the largest speed variations that must be removed. This fact should not cause the transport designer to relax his design efforts.

In all longitudinal tape oscillations there is a natural resonance frequency, which by a rule of thumb is 100,000 Hz per centimeter. So, if we have a free tape span from the in-going idler to the capstan of say 12 cm, the resonant frequency will be at 8,300 Hz. If, on the other hand, the tape span is 20 cm, the resonant frequency will be at 5,000 Hz. Thus, a short tape span is important. In the design of a precision tape recorder, it is quite useful to apply an equivalent electrical diagram of the electro-mechanical components in the recorder, as shown in Figs. 15-16, 15-17.

A valuable tool is available to the mechanical designer of tape drive mechanisms: The electro-mechanical analog. This approach is advantageous because the electrical engineers have developed a set of convenient symbols for circuit elements which can be used to represent a mechanical system with masses, springs, forces, etc.. The circuit diagram can be examined and its behavior treated; the results can then be applied to predict the mechanical systems behavior, and then corrections can possibly be applied.

Excellent tutorial examples are found in two now classic papers by Wolf and by Jacoby, supported by a classic textbook on the topic (ref. Olson). A simplified layout of the mechanical

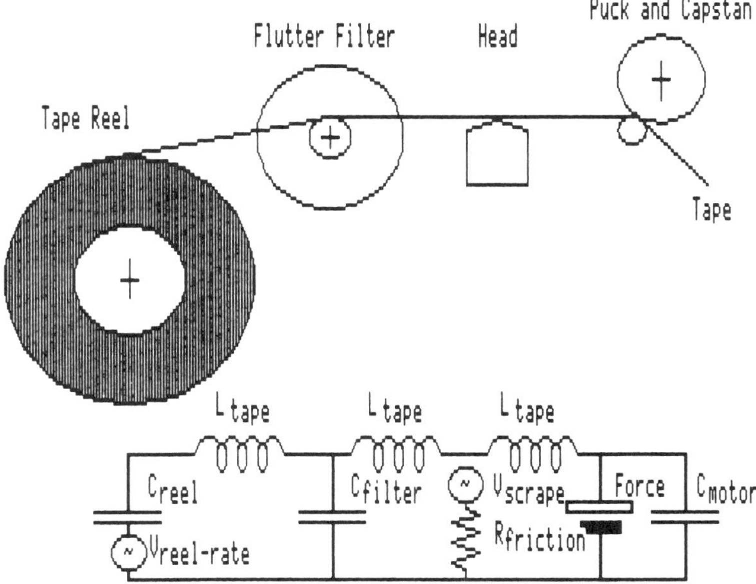

Fig. 15-16. Fundamental electromechanical analogies.

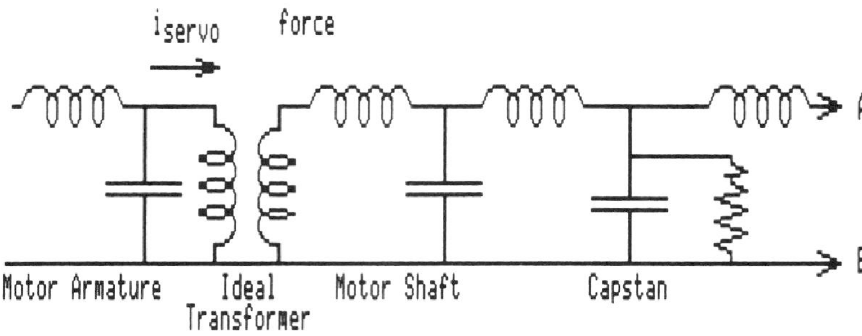

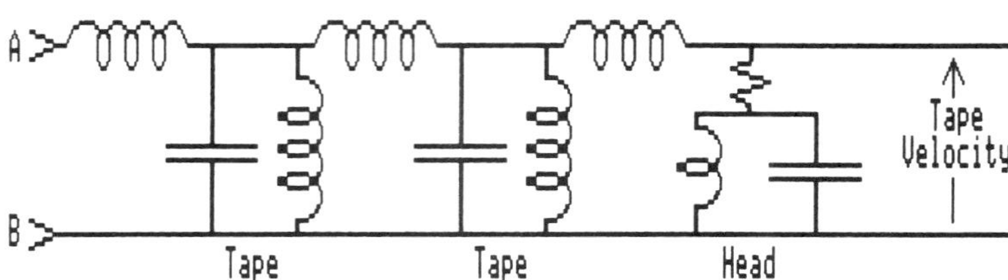

Fig. 15-17. Detailed diagram showing the elements in a tape drive, from feed-reel to capstan (after Wolf).

components is easily transformed into an electrical admittance diagram, which then is converted into an impedance diagram. Now all the tools of modern electrical system engineering can be used: LaPlace transform, impulse response, etc. The systems performance can next be traced back to the individual circuit components, or mechanical counterparts, and needed design changes can be made.

CONCLUDING COMMENTS

There are some papers of interest for mechanical designers that deserve attention:

Recorders with Rotating Head Assemblies

A full chapter in Robinson's book on *Videotape Recording* addresses the problem of geometrical errors in tracking on quadraplex and helical machines. One problem is mistracking in a helical transport which occurs when a tape's dimensions are changed due to temperature, humidity or excessive (or too little) tension. The situation is shown in Fig. 15-18. It is normally corrected by tension control on the video machine. Other papers address the geometry problem (ref. Wada (two papers), Fujiwara et al.(two papers)).

The problem of head contouring and contact are treated by Hahn and Talke et al. The new portable 8mm recorders are also covered (ref. Yamaguchi, Kirino et al.)

Audio Tape Cassettes and Cartridges

The basic Philips cassette is described in a paper by Hanson, and the effects of the plastic's dimensions etc. by Robertson. This brings up the topic of plastics in general, and some readers may be interested in a new book written on this topic for electronics applications (ref. Goosey). Wear evaluation of plastics may also be of interest (ref. Bayer).

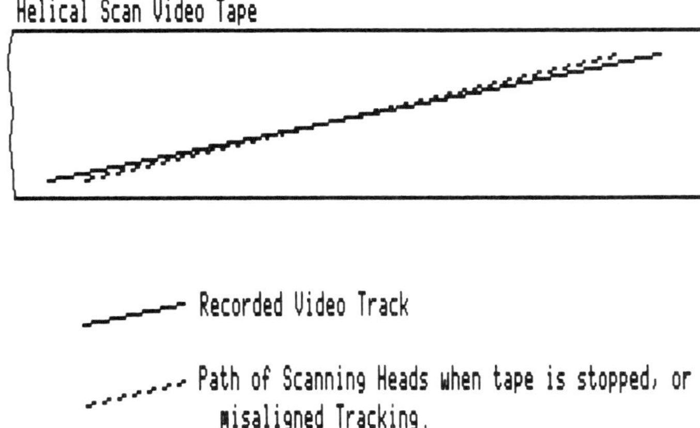

Fig. 15-18. Error in scanning of a helical video tape (see text).

The quarter inch tape cartridge is often thought of as a playback system for car use only. Broadcast quality has nevertheless been achieved (ref. Moris).

REFERENCES TO CHAPTER 15

Introduction

Volz, H.(Editor), and Schneider, C. *Grundlagen der magnetischen Signalspeicherung II: Magnetbander und Grundlagen der Transportwerke*, Akademie-Verlag, Berlin, January 1970, 133 pages.

Wildman, M., "Mechanical Limitations in Magnetic Recording," *IEEE Trans. Magn.*, Sept. 1974, Vol. MAG-10, No. 3, pp. 509-514.

Harris, J.P., Phillips, W.B., Wells, J.F., and Winger, W.D. "Innovations in the Design of Magnetic Tape Subsystems," *IBM Jour. Res. Devel.* Sept. 1981, Vol. 25, No. 5, pp. 691-699.

Zoeller, M., "An Ultra-High Data Rate Mass Storage System," *IEEE Trans. Magn,* July 1981, Vol. MAG-17, No. 4, pp. 1426-1431.

Digital Tape Drives; Streamers

Mackintosh, N.D., "A Margin Analyser for Disk and Tape Drives," *IEEE Trans. Magn.*, Nov. 1981, Vol. Mag-17, No. 6, pp. 3349-3351.

Cutler, D.S., "Design Considerations for Streaming Tape Transports," *Intl. Conf. Video and Data 82,* April 1982, IERE Publ. No. 54, pp.113-125.

Curtis, H.L., and Turley, R.T., "Development of a High-Performance, Half-Inch Tape Drive," *Hewlett Packard Journal,* March 1985, Vol. 36, No. 3, pp. 11-16.

Dong, J.W., Van Maren, D.J., and Emmerich, R.D. "Streaming Tape Drive Hardware Design," *Hewlett Packard Journal,* March 1985, Vol. 36, No. 3, pp. 25-29.

Ruska, D.W., Russon, V.K., Culp, B.W., Richards, A.J., and Ruf, J.A. "Firmware for a Streaming Tape Drive," *Hewlett Packard Journal,* March 1985, Vol. 36, No. 3, pp. 29-31.

Becker, J.C., DiTommasok, D.A.f, and Mortensen, S.J. "Low-Cost, Highly Reliable Tape Backup for Winchester Disc Drives," *Hewlett Packard Journal*, March 1985, Vol. 36, No. 3, pp. 34-35.

Digital Cartridges, Cassettes

Staff, *Streaming,* Archive Corp., Los Angeles, Calif., June 1982, 170 pages.

Burdorf, D.L., "New Concepts in High Reliability Tape Recorders for Spaceborne Data Storage," *ITC Conf. Proc.,* 1966, Vol. 2, pp.441-454.

Manildi, A.B., "Digital Mini-Cassette," *IEEE Trans. Magn.,* July 1978, Vol. MAG-14, No. 4, pp. 167-170.

Nordman, R.G., Smith, R.L., and Witkin, L.A. "New CRT Terminal Has Magnetic Tape Storage for Expanded Capability," *Hewlett-Packard Journal,* May 1976, Vol. 27, No. 9, pp. 2-14.

Valliant, W., "The Evolution of the High Density ¼ in. Cartridge Tape Drive," *IEEE Trans. Magn.,* July 1981, MAG-17, No. 4, pp. 1438-1441.

Buskirk, W.A., Gilson, C.W. and Shelley, D.J., "Compact Digital Cassette Drive for Low-Cost Mass Storage," *Hewlett Packard Journal,* May 1983, Vol. 34, No. 5, pg. 17-24.

Hoecker, K.R., Schwartz, J.R., Young, F.A., and Johnson, D.R., "Handpulled Magnetic Card, Mass Storage System for a Portable Computer," *Hewlett Packard Journal,* June 1983, Vol. 34, No. 6, pg. 15-23.

Gennetten, K.D., "Cartridge Tape Data Integrity Ensured at Five Levels," *Hewlett Packard Journal,* March 1985, Vol. 36, No. 3, pp. 39-43.

Makmann, M.T., "Standards Pave the Way for Tape Cartridges," *Computer Design,* May 1985 pp. 133-139.

Wright, M., "Cartridge Tape Drives - Special Report," *EDN,* Oct. 1985, Vol. 30, No. 24, pp. 109-118.

Tape Skew

Hu, P.Y., Hollman, W., and Argumendo, A.J., "Tension Gradient Measurement of Magnetic Tape," *IEEE Trans. Magn.,* Sept. 1984, Vol. MAG-20, No.5, pp. 921-923.

Rosback, Th.J., "A Digital Audio Time-Base Corrector for Linear Magnetic Recording," *Jour. SMPTE,* Nov. 1985, Vol. 94, No. 11, pp. 1180-1184.

Analog Tape Drives

Peshel, R.L., "The Application of Wow and Flutter Compensation Techniques to FM Magnetic Recording Systems," *IRE Natl. Conv. Rec.,* Pt. 7, 1975, pp. 95-110.

Jacoby, G.V., "The Design of a Magnetic-Recording-Tape Transport for Very-High Timing Accuracy," *IEEE Trans. Comm. and Control,* Jan. 1963, pp. 491-499.

Wolf, W., "Electromechanical Analogs of the Filter Systems Used in Sound Recording Transports," *IEEE Trans. Audio and Electroacoustics,* June 1966, Vol. AU-14, No. 2, pp. 66-85.

Olson, H.F., *Solutions of Engineering Problems by Dynamical Analogies,* Van Nostrand, 1966, 277 pages.

Servo Controlled Tape Speed

Wente, E.C., and Mueller, A.H., "Internally Damped Rollers," *Jour. SMPTE,* Oct. 1941, Vol. 50, No. 10, pp. 406-417.

Diamond, A., "Inertially Damped Servomotors: Performance Analysis," *Electro-Technology,* July 1965, pp. 28-32.

Alexander, Alan G., and Ling, Cheng C., "Calculating the Natural Frequencies of Rotating Machine Elements," *Machine Design,* Feb. 1967, pp. 137-140.

Hu, P.Y., "Vibration Characteristics of the Printed-Circuit Motor," *Jour. of Engr. for Industry,* May 1974, pp. 541-546.

Law, E., "Frequency Response of Tape Transport Servo Systems," *ITC Conf. Proc.*, 1976, Vol. 12, pp. 196-201.

Capstan and Pinch Roller

Clurman, S., "On Hunting in Hysteresis Motors and New Damping Techniques," *IEEE Trans. Magn.*, Sept. 1971, MAG-7, No. 3, pp.512-517.

Durieu, J., and Petit, M., "A 2-D Solution of the Contact Problem in the Capstan /Tape /Roller Mechanism of Magnetic Recorders," *Computer Meth. in Appl. Mech. and Engr.*, April 1984, Vol. 43, No. 1, pp. 21-36.

Reeling and Winding of Tape

Holcomb, A.L., "Film-Spool Drive with Torque Motors," *J. SMPTE*, Jan. 1952, Vol. 58, No. 1, pp. 28-35.

Hittle, C.E., "Automatic Torque Controller for Torque Motors," *Jour. SMPTE.*, July 1952, Vol. 59, No. 7, pp. 28-31.

Hartmann, G., "Zur Bremsvorgang bei Magnettongeraeten," *Elektronische Rundschau*, Feb. 1958, Vol. 2, No.2, pp.45-49.

Swope, R.D., and Ames, W.F., "Vibrations of a Moving Threadline," *Jour. Franklin Inst.*, Dec. 1963, Vol. 275, No. 12, pp. 36-55.

Mote, C.D., "A Study of Band Saw Vibrations," *Jour. Franklin Inst.*, Dec. 1965, Vol. 279, No. 12, pp. 430-444.

Thurman, A.L., and Mote Jr., C.D., "Free, Periodic, Nonlinear Oscillation of an Axially Moving Strip," *Jour. Appl. Mech.*, March 1969, pp.83-91.

Isabeau, John G., and Vogel, Charles A. "The High Speed Tape Transport," *ITC Conf. Proc.*, 1969, Vol. 5, pp. 532-551.

Eshleman, R.L., Meyers, A.P., Davidson, W.A., Gortowski, R.C., and Anderson, M.E., *Feasability Model of a High Reliability Five-Year Tape Transport,* IITRI Proj. No. E6225, NASA Contract No. NAS5-21692, 1973, 3 Volumes.

Paroby, W., and DiSilvestre, R., "Tape Tracking and Handling for Magnetic Tape Recorders," *Intl. Telemetry Conf. Proc.*, 1975, Vol. 11, pp. 345-357.

Pfeiffer, J. David, "Web Guidance Concepts and Applications," *Tappi,* Dec. 1977, Vol. 60, No. 12, pp. 53-60.

Clurman, S., "A Simple Tape Wrap Around A Guide: Some Complexities," *IEEE Trans. Magn.*, Nov. 1981, MAG-17, No. 6, pp. 2754-2756.

Mote, C.D. Jr., and Wu, W.Z., "Vibration Coupling in Continuous Belt and Band Systems," *Jour. Sound and Vibr.*, Sept. 1985, Vol. 102, No. 1, pp. 1-10.

Rotating Head Assemblies

Robinson, J.F., *Videotape Recording,* Focal Press, 1981, 362 pages.

Wada, Y., "Track Straightness in Helical Scan VTR," *IERE Proc. on Video and Data Recording Conf.*, 1973, pp. 51-60.

Fujiwara, Y., Eguchi, T., and Ike, K., "Tape Selection and Mechanical Considerations for the 4:2:2 DVTR," *SMPTE Jour.*, Aug. 1974, Vol. 93, No. 9, pp. 818-829.

Wada, Y., "Track Straightness in Helical Scan Video Tape Recorders," *Jour. SMPTE*, Dec. 1975, Vol. 84, No. 12, pp. 954-958.

Hahn, F.W., "Design Considerations for a Helical-scan Recording Head Contour," *Intl. Conf. on Video and Data*, April 1976, pp. 261-266.

Talke, F.E., and Tseng, R.C., "Submicron Transducer Spacing in Rotating Head/Tape Interface," *IEEE Trans. Magn.*, Nov. 1976, Vol. MAG-12, No. 6, pp. 725-727.

Fujiwara, Y., Eguchi, T., and Ike, K., "Tape Selection and Mechanical Considerations for the 4:2:2 DVTR," *Jour. SMPTE*, Sept. 1984, Vol. 93, No. 9, pp. 818-829.

Yamaguchi, T., "A New Tape Path Analysis for all VCR Tape Transport Systems," *IEEE Trans. Cons. Elec.*, Aug. 1985, Vol. 31, No. 3, pp. 398-404.

Kirino, T., Tominaga, T., Kasai, S., Ogihara, H., Kawamura, T., and Inatsu, M., "Progress Report on Recent Developments on One Manufacturer's ¼-*in. ENG Recorder,*" *Jour. SMPTE*, Jan. 1986, Vol. 95, No. 1, pp. 20-24.

Audio Cassettes; Cartridges

Goosey, M.T. (Editor), *Plastics for Electronics,* Elsevier Appl. Sci. Publ. Ltd. (Essex), 1985, 380 pages.

Hanson, E.R., "A Tape Cassette Standard," *Jour. AES*, Oct. 1968, Vol. 16, No. 4, pp. 430-435.

Robertson, H., "Focus on Shell Mechanics," *Audio*, Sept. 1981, pp. 31-37.

Moris, A.H., "A New Radio-Broadcast Cartridge Capable of Professional Open-Reel Performance," *Jour. AES*, Nov. 1983, Vol. 31, No. 11, pp. 810-816.

Bayer, R.G., and Payne, N.G. "Wear Evaluation of Molded Plastics," *Lubrication Eng.*, Feb. 1984, Vol. 41, No. 5, pp. 290-293.

Chapter 16
Disk Drives

Disk drives (rotating memories) provide almost instantaneous access to a very large number of data records; this is in contrast to tape drives where the tape first must be wound to a location where the data are recorded.

These rotating memories were born in 1957 in form of the IBM 350 disk drive, consisting of 100 fixed recording surfaces of 24 inches diameter. There were 20 tracks in parallel per inch radius (20 TPI), and the bit density was 100 BPI so the areal packing density was 2,000 bits per sq.in. The data rate was 8.8 kBPS (ref. Harker et al.).

The 8 inches diameter flexible disk was introduced in 1973, also by IBM, as a device for loading control microprograms and diagnostics. Since then the applications have broadened to serve as a media for distribution and exchange of data, programs, microcode, diagnostic procedures, and other digital information, and as a removable disk in system files (ref. Engh).

Today a double, flexible 5-¼ inch disk can store 21.4 megabytes (= 21.4*8 = 171.2 megabits) of data, operating with densities of 600 TPI and 18,000 BPI, which corresponds to an areal density of 10,800,000 bits per sq. in.; the data rate is 500 kBPS (ref. Iomega Corp., Beta 20 drive). Rigid disk drives appear to be settling at the 5¼ and 3½ inches diameter sizes (ref. Killmon).

The speed of developments has been dazzling at times, and is the result of progress in materials, processes, servo designs and precision manufacturing rather than any novel write/read techniques (ref. Mulvany and Thompson). Two books are listed for the interested reader (one by the staff at IBM, and one by Teja) and two magazines are referenced to stay current.

Data on a disk are stored in circular tracks, that are divided into sectors (see Fig. 16-1). The disks range in diameter from 35.6cm (14 inches) down to 5.1 cm (two inches), rotating at speeds from 3600 rpm to 300 rpm. Figure 16-2 summarizes the disk formats used today. The organization of data is a logistic problem that is outlined in Chapter 26, and the scope of

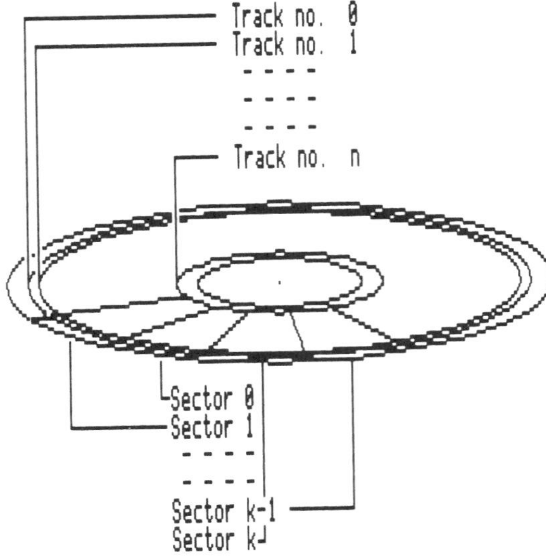

Fig. 16-1. Data are stored in sectors along tracks on disks.

this chapter is limited to a discussion of the required rotation of the media plus the mechanisms for positioning the head over the desired track.

We will first treat the flexible and rigid media drives, and then describe linear and rotary actuators for moving the heads. At low track densities it is relatively easy to position the head; at higher densities it may be necessary to use a track seeking technique, also called a closed-loop servo. We will finally address the problems associated with vibrations, shipments of disk drives, etc.

The overall performance of a disk drive is measured in terms of its *seek time*, which is the average time it takes for a head to move from one track to another, including the time to settle over the track plus the *latency*, which is the time for one revolution (some may quote the time for one-half revolution as the average latency). One should also consider the bit rates for the data, although that number is pretty well fixed by the interface standards (see Chapter 26).

FLEXIBLE DISK DRIVES

The evolution of the 8″ IBM diskette is well described in the earlier referenced papers, including a brief description of one of the microfloppies two years later. Many design details and requirements are similar to a tape drive, such as signal losses, friction and wear. Different are the vibration modes, which now correspond to those of a circular membrane, clamped at the center and free at the edges.

A high areal packing density on flexible disk can be achieved by applying a large number of concentric tracks, measured in tracks-per-inch (TPI). This design effort is hampered by the mechanical tolerance build-up for the exact centering of a diskette, from drive to drive, in combination with the changes in disk dimensions with changes in temperature and in humidity. These matters were discussed in connection with selection of track widths in the beginning of Chapter 10.

The flexible disk formats are shown in Fig. 16-2. The disks themselves are circular PET film disks, typically of thickness 75μm (3 mils), and packaged like shown in Fig. 13-20. The 8 and 5¼ inches formats are housed inside a PVC jacket, while the 3½ and 2 inch formats are enclosed in a rigid plastic housing. The 5½ inch formats for high packing densities are further enclosed in a rigid structure, or inside a cartridge.

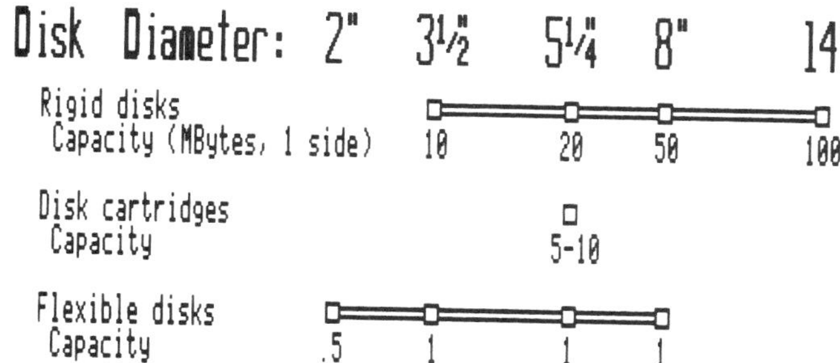

Fig. 16-2. Popular disk sizes.

The rotary motion of the disk is achieved by clamping the disk at the location of the center hole. This may be done like shown in Fig. 16-3, where the disk ends up seated against a spindle, centered and clamped by a cone shaped clamp. This clamping is done in conjunction with closing the door to the disk drive. The centering of the diskette may not always be perfect upon the first clamping due to friction; it is recommended to center the disk manually, as best as can be done, and then to double clamp, i.e., close, then open and close the drive gate one extra time after the disk has been inserted.

The clamping of the 3½ inch diskettes is done by a magnet that is brought up into contact with the center metal piece, and a pin on the drive spindle interlocks with the positioning hole in the center piece. This assures repeatable positioning of the tracks' center, drive to drive.

Clamping and centering of disks in the disk cartridges require both precision and strong clamping since they operate at 1,500 - 1,800 rpm, and dynamic balancing is important.

The rotation of the spindle can be provided by AC as well as by DC motors, through a belt or direct drive where the motor shaft is the spindle shaft. AC motors are simple and reliable, and run at speeds controlled by the line frequency. The latter is a disadvantage since four models must be provided for in order to accommodate any mix of the voltages 110 or 220 V, and 50 or 60 Hz line frequencies.

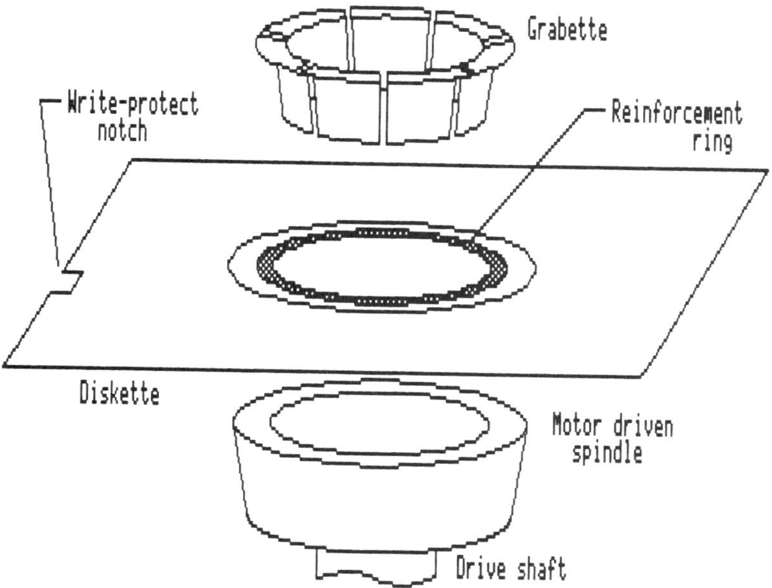

Fig. 16-3. Center clamping of diskette.

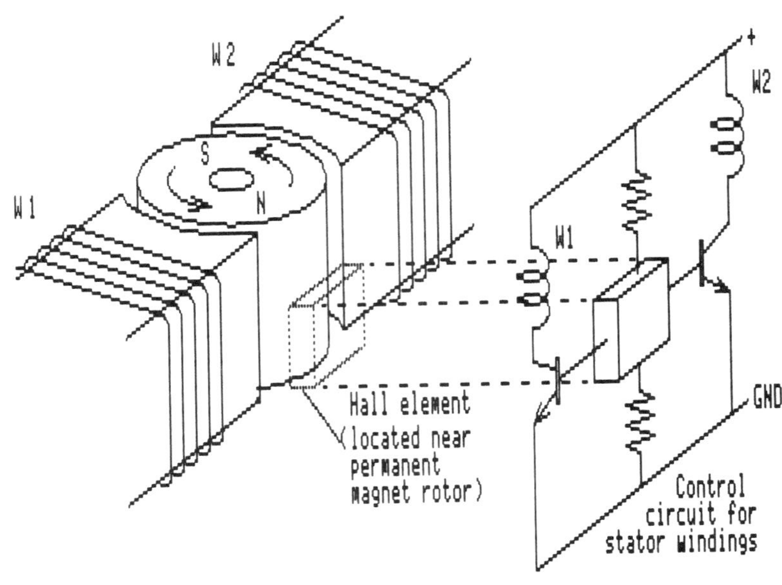

Fig. 16-4. Basic principle of a two-phase brushless DC motor using one Hall element.

DC motors used today are brush-free, and require a DC power supply plus speed control circuits. These motors can easily be made in a flat package, well suited for the shallow disk drives in use today. They are also much simpler to turn on and off as is required for each write or read cycle.

The principle of a brushless DC motor is shown in Fig. 16-4. The stator windings W1 and W2 are controlled by a hall element which in turn is switched by the rotating permanent rotor. This provides for the required commutation in a DC motor.

Tracking and maximum trackwidths are discussed by Resnik, while the first 3½ inch disk drive is described by Katoh et al. The rapid development in the field is evident by papers describing the achievement of 1.6 MBytes on a mini disk (ref. Adachi and Yano), growing to 34 MBytes on a 5¼ inch Co-Cr perpendicular media disk (ref. Okuwaki et al.). The latest papers are on high coercivity gamma ferric oxide on a 3 inch disk with 2 MBytes capacity (ref. Matsuda et al.), and the new Barium Ferrite perpendicular media on a 3½ inch disk with 4 MBytes capacity (ref. Imamura et al.).

RIGID DISK DRIVES

Rigid disk drives operate at much higher speeds (2,400 to 6000 rpm) and track densities (400 to 1,000) than do flexible diskettes. This places higher demands upon mechanical rigidity and tight tolerances; the disk, or disks, may be inside a removable container (diskpak), or they are permanently installed on a spindle inside the drive. The latter design allows for the maximum TPI number.

A rigid disk drive is in a constant run mode, in contrast with flexible disk drives that only rotate the media during write and read cycles. The heads in a rigid disk drive are flying over the surface, except for start and stop where they take off or land on designated areas on the disk; some designs do not have landing areas, the heads are lifted away from the disks when they are not rotating.

Disk surface finish, flatness and dynamic stability are usually considered to be the most critical mechanical parameters in a disk drive. However, there are also the aerodynamic problems, of mechanical as well as thermodynamic nature, generated by the air flow around the spinning disk(s).

Any kind of large-scale unsteady flow or pressure fluctuation is a random excitation source for the vibration of disk and positioner-arm assembly. These relative motions between disk and head introduce amplitude variations and off-track errors. Severe disk flutter causes data losses that are due to slider-disk interference (magnetostrictive impact losses) or even head crashes.

A non-uniform flow will also generate heat, and thermal expansion results. The heat generated in an eight platter 14 inch disk drive can be as high as 50 watts.

The amount of disk flutter and actuator arm vibration depends upon several factors (ref. Lennemann):

- The flutter may increase by a factor of five when two disks are used instead of one. For the increase may be up to 20 more disks.
- The form and dimensions of the disk drive housing will heavily influence the disk flutter.
- Steady air flow reduces disk flutter by an order of magnitude, and aids in cooling the disk(s).
- Proper form of the actuator arm and head assembly is important: A spoiler between disks prevents the development of unsteady air flow.

Many drives use 8 inch disks which produce far less air turbulence than the 14 inch disks due. The result is less disk flutter, arm vibration and heat generated. Also 10 inch disks are advantageous in this respect (ref. Mizoshita and Matsuo).

Fairly large motors are used to drive the precision spindle holding the disk(s). A high starting torque is needed in order to overcome the static friction from often many heads resting on the disks (in landing zones). Hence, brushless DC-motors are used, often mounted inside the spindle (Fig. 16-5), or underneath the base plate casting (Fig. 16-6). Preloaded and sealed ball-bearings are used to support the spindle shaft in both ends; some drives actually have spindle shafts that are supported fully in both ends to avoid any cantilever effects.

An enlightening design outline for the IBM 3340 disk drive is found in Mulvany's paper (1974), and a discussion of disk file manufacturing in papers by Mulvany and Thompson (1981),

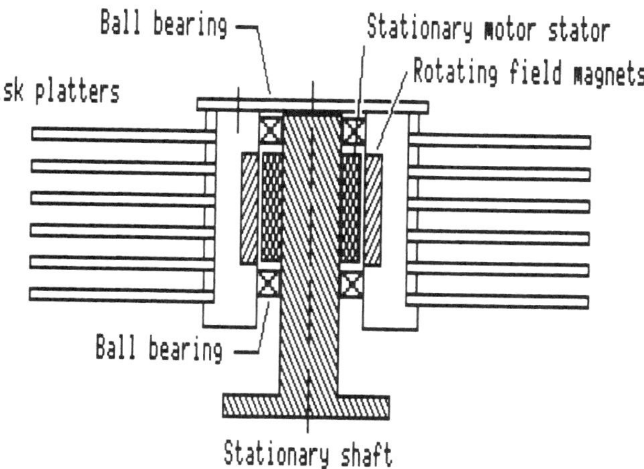

Fig. 16-5. Drive motor mounted inside hub in eight platter disk drive.

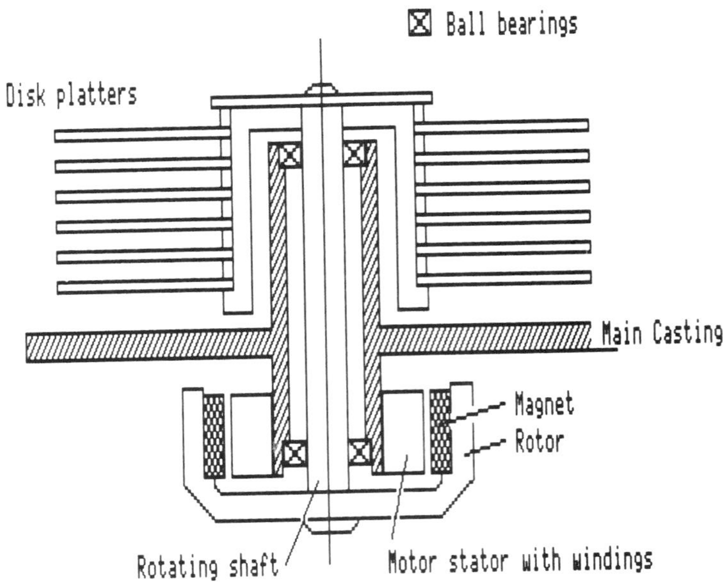

Fig. 16-6. Drive motor mounted under main casting in disk drive.

and Smith (1984). Rigid disk testing is covered by Opfer et al. (1985). A brief technology survey was written by Kaneko and Koshimoto (1982), and more extensively by McLeod (2nd edition, 1986).

Design of a rigid disk drive includes considerations for shock and vibration. Already mentioned are the landing zones for heads, or a mechanism that moves the heads away from the disk when it is not rotating at normal speed. Another measure to be taken is a clamp that prevents the disk(s) from rotating during transport of the disk drive.

Shock absorbers will, if properly designed, isolate the drive from bumps and vibrations. It is important to use the proper compliance with consideration for allowable sway space. It is worthwhile to consult suppliers of shock mounts plus applicable textbooks (Thomson, Nashif et al.). A disk drive should be able to withstand 40 Gs when non-operating (such as in shipment), and 10 Gs while operating (ref. Rahimi).

For the ultimate design assurance one does well in following the guidelines and spirit of the military specification MIL-E-5400. (Caution: strict adherence causes mental break-down in designers.)

ACTUATOR (HEAD POSITIONER)

The magnetic head must be positioned in contact with the correct track before writing or reading. This is accomplished by mounting the head on an arm that moved radially to the section of the disk allocated to the write/read operation.

There are two simple methods for moving the arm, or actuator: The lead screw and the capstan-band, both operated by a stepper motor. The lead screw was the earliest used method, shown in Fig. 16-7. The screw is powered by a stepper motor, which rotates the screw in small equal steps, which are translated into a linear movement of the head assembly.

The capstan-band system is shown in Fig. 16-8. The metal band is actually a spring that is fastened to the capstan on the stepper motor shaft. The compliance of the band preloads the positioner arm such that backlash is minimized, and repeat positioning of the head improved.

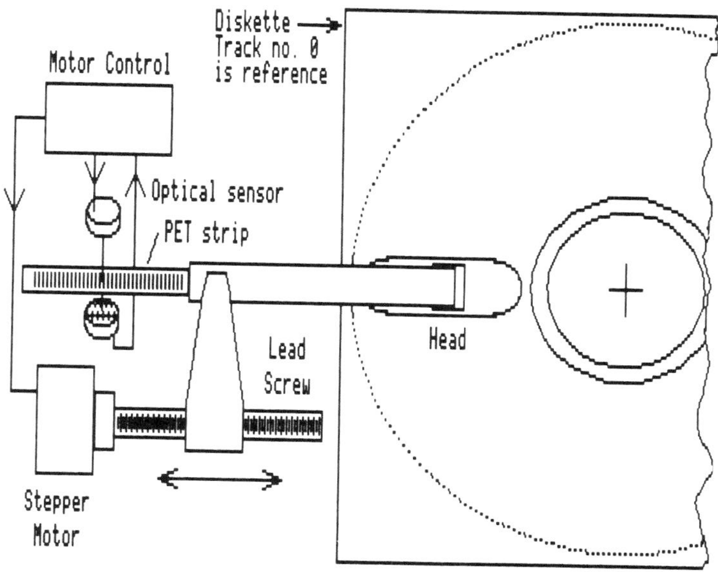

Fig. 16-7. Head positioning by stepper motor and lead screw.

The stepper motors are inexpensive, operates with steps of typically 1.8 degrees, with a 3 percent accuracy (ref. Kenjo).

In rigid disk drives many head assemblies and support arms are mounted on one common carriage which is moved as one unit. This is shown in Fig. 16-9 where a four-platter disk drive is shown in the top portion, and the head carriage is moved back and forth by a linear voice coil motor, identical to the "motor" in a loudspeaker.

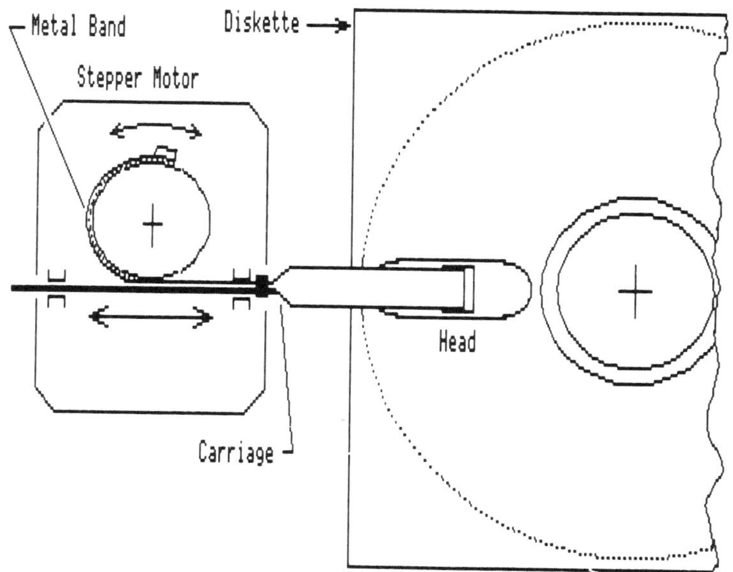

Fig. 16-8. Head positioning by stepper motor and capstan/band system.

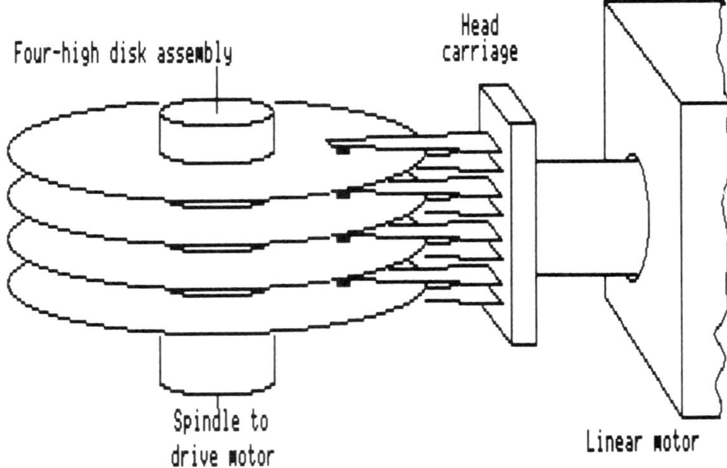

Fig. 16-9. Multiple head positioning by a linear motor and head carriage.

A single disk with four heads is shown in Fig. 16-10; here the heads only have to travel half the distance, which means shorter access time, and write/read operations can be optimized for the inner (low speed) and outer (high speed) heads individually.

The linear actuator, which is very much like a loudspeaker coil/magnet mechanism, was developed in the early seventies (Brown and Ma, Inoue et al.). The coil inductance is proportional to the square of the number of turns, and it must be low to allow for a rapid current rise so the actuator quickly shifts the heads to new track positions. On the other hand, a certain number of coil turns are necessary to achieve the desired force f = BLI, where L if the total wire length in the field B, and I is the drive current. The inductance value can be reduced

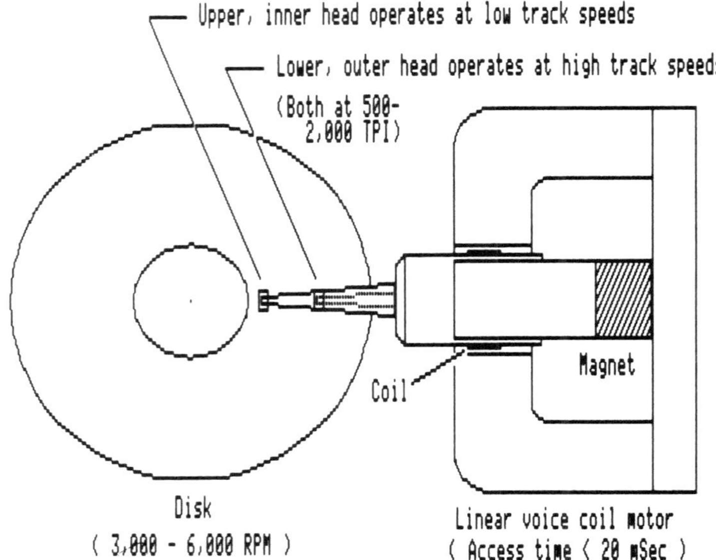

Fig. 16-10. Twin-head positioning, using a voice-coil motor.

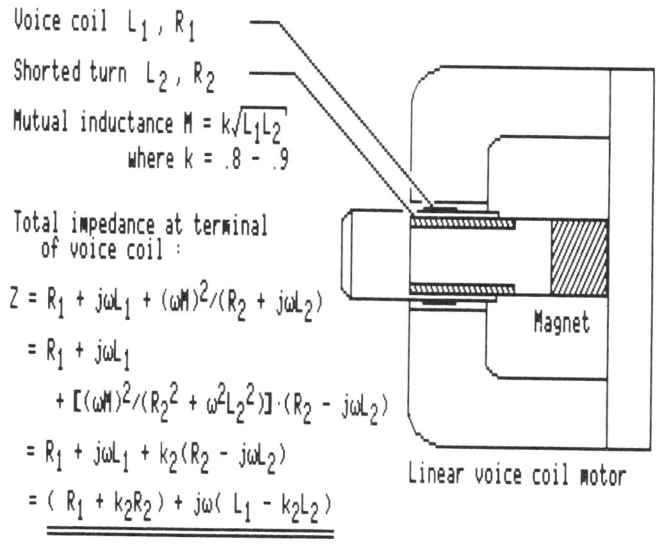

Fig. 16-11. Shorted turn in a voice-coil motor reduces the coil time constant.

by a clever design using a shorted turn inside the coil, mounted on one of the magnet poles (Fig. 16-11).

The shorted turn will act like a secondary winding on a transformer with the motor coil being the primary winding. Its action is easiest realized if we consider the value of the shorted turn as an overcoupled impedance from the secondary. This impedance can be expressed as: $(\omega M)^2/(r_2 + j\omega L_2)$, where $M = k\sqrt{L_1L_2}$ is the mutual coupling.

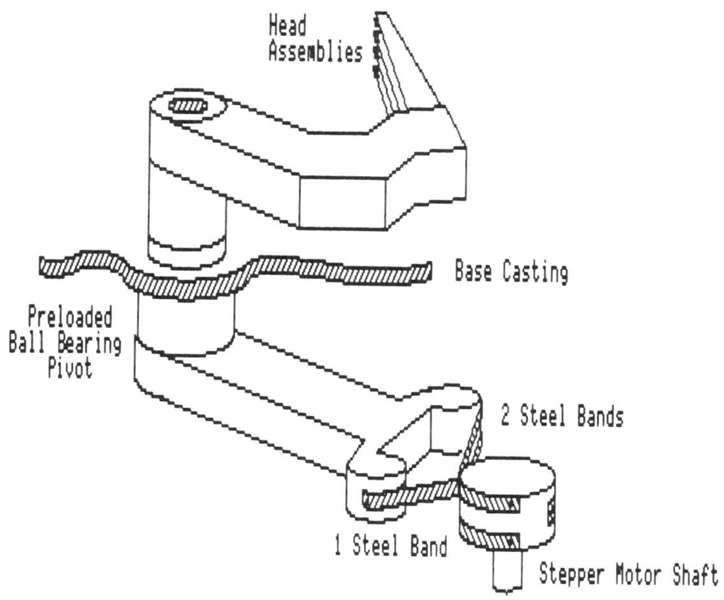

Fig. 16-12. Pivoting a head arm's motion around a preloaded ball bearing with capstan/band.

k is the coefficient of coupling, L_1 is the coil inductance and L_2 is the inductance of the shorted turn; R_2 is the resistance of the shorted turn.

The net result is a large reduction in the value of the primary inductance, and hence improvement in the rise time of the current. The shorted turn will absorb a fair amount of power and must have good thermal conductance to the motor structure, serving as a heat sink. Improvements in the response time range from 15 to 40 mS down to 10 to 15 mS (ref. Wagner 1982 and 1983).

The linear motor can be shaped in various ways. One approach results in a dual-path actuator that has more evenly distributed flux lines and lesser stray flux (ref. Dong). Another design replaces the large carriage with the head mounted in a comb fashion with individual actuators that are very flat and therefore fits between disks; the use of individual actuators results in increased system performance (ref. Scranton et al.). The design procedure for the servo signal electronics is covered in a recent paper (ref. Bell et al.), and observation of slider motion is important in that connection (ref. Miu et al.).

Rotary head positioners allows for more compact designs, and are found in several disk drives today, see Fig. 16-12 (ref. Heath, Hearn, Winfrey et al.).

The topic of this section and the following is vast, and the interested reader is referred to a 2-Days Course, or the notes therefrom by Oswald, Wasson and Wagner. A book by Pillay gives a good introduction to motors.

SERVO DESIGN

Floppy Disk Drives

The location of any recorded information is stored in a small directory on a disk, and will inform the disk operating system about the exact sector and track number where the recorded data are. It then becomes a matter for the head positioner to find the track.

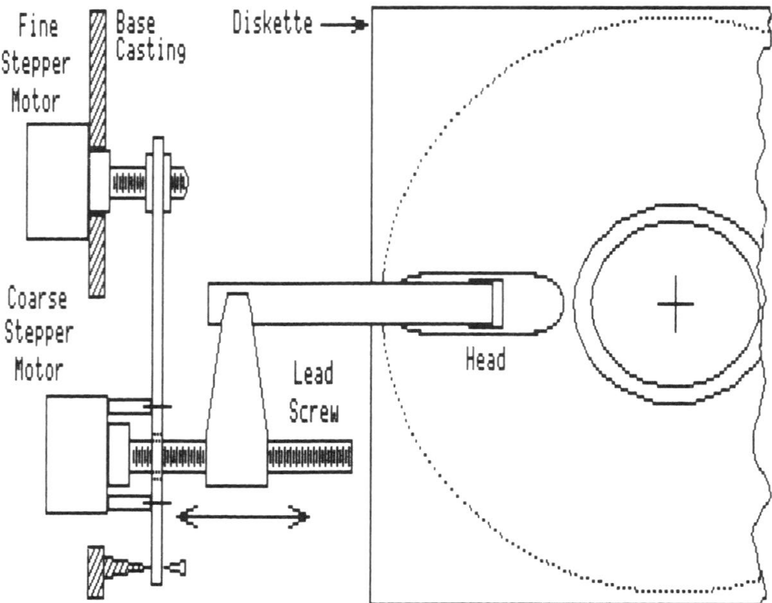

Fig. 16-13. Single burst servo signals between sectors allows for first order correction of head position.

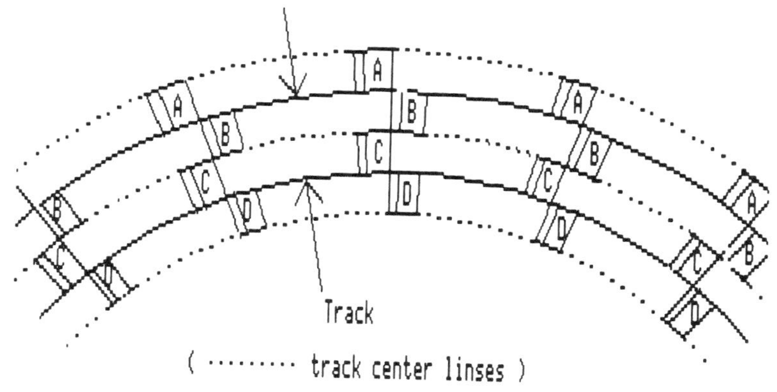

Fig. 16-14. Servo bursts are recorded on each side of the tracks (courtesy H. Thomson).

Ideally the head just moves to that track. This works well at low track densities (35 to 48 TPI or less), but not at higher. Mechanical tolerances build up, and temperature and humidity changes track size and dimension. The latter effects can be offset by a control scheme, where an optical sensor follows the head arms movements, and informs a microprocessor to advance or retard the head position. The optical sensor is made from PET, the same as the diskette, and this compensates for temperature and humidity changes.

The method is shown in Fig. 16-7, and has been used by Amlyn Corp. to achieve 170 TPI. The system recalibrates itself every 30 seconds by moving the head to a reference track for adjustment. The system is an open loop servo, and thus of limited use.

Drivetec used a different approach with a 192-TPI drive. Two stepper motors position the head; the coarse control motor simply steps at 1/192 inch, the other is a closed loop servo motor for fine adjustments, see Fig. 16-13.

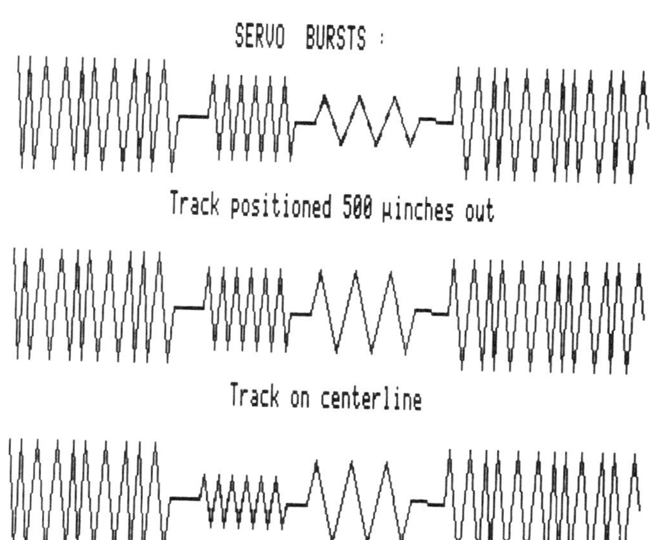

Fig. 16-15. Off- and On-track read signals from the servo bursts (see text).

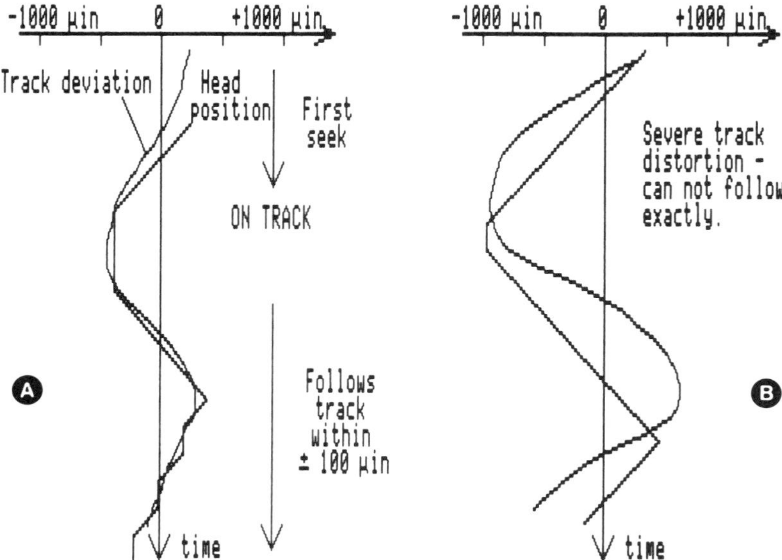

Fig. 16-16. Track following in the Drivetec servo system.

The media is preformatted with an index mark on each track and a servo burst signals between every sector of every track (each track has 32 sectors). The two servo burst signals, A and B, each contain an ID and two bytes of FF (1010 1010 1010 1010), and they are recorded by two heads, offset as shown by the recorded A-B patterns along tracks in Fig. 16-14.

During read the head alternately picks up A and B, and when these two signals are of equal amplitude then the head is on-track, see Fig. 16-15, center. If the head is off center then one or the other signal is stronger, indicating that the head is off. The fine stepper motor repositions the head by one step of 100μin per sector until the two signals are of equal amplitude (ref. Thompson); this makes the head track very well for normal track geometry distortions, and reasonably well for severe distortions, see Fig. 16-16 A and B.

Rigid Disk Drives

Track Accessing. Rigid disk drives demand faster and more precise positioning of the head. The access time should be in the 10 to 30 mS range. To position a head accurately, full forward power of the actuator is applied for as long as possible. Then reverse power is applied until the actuator comes to zero velocity which, if the control system functions properly, will position the head precisely on target.

This is what servo engineers call a "Bang Bang" control, see Fig. 16-17 (ref. Ananthanarayanan). The servo system needs to keep track of where the head is, and this is done in a difference counter which contains the number of tracks to the target, and which is updated as tracks are crossed (ref. Oswald, 1974 and 1980).

Track Following. When the head has been positioned over the right track the next task for the disk controller's servo is to keep it on track, irrespective of mechanical vibrations and aerodynamic disturbances, as well as changes due to varying temperature. At high track densities a direct-position feedback from the head itself is necessary to correct for any mistracking.

A servo signal must therefore exist on the track to be followed. In early disk drives with many platters it was practice to dedicate one recording surface to servo tracks. Since all heads

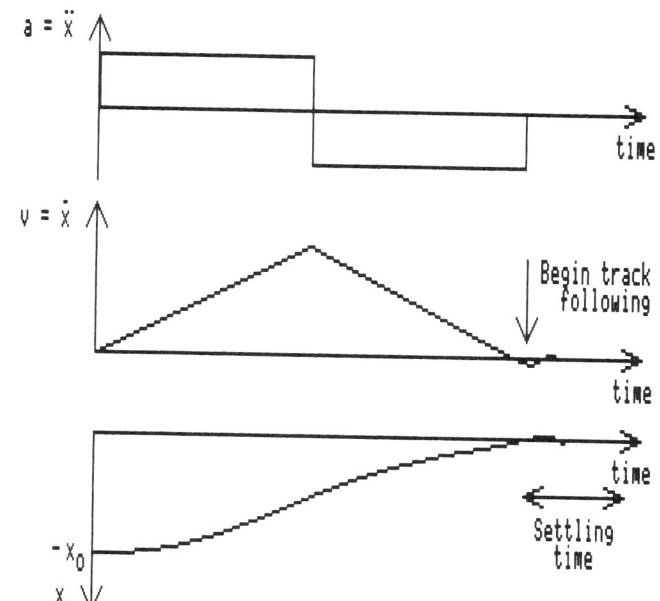

Fig. 16-17. Bang-Bang servo control for actuator motors.

are mounted on a common carriage they will all be positioned over their proper tracks (all these associated tracks making up one so-called cylinder). In modern high TPI systems the mechanical tolerances for doing this becomes unbearable, and the information from the disk surface with the servo information may be used as a detailed high-frequency servo error signal. The DC and low frequency correction signals comes from the data head currently reading information (ref. Commander and Taylor).

This information is written between sectors as a tribit signal, or as a null pattern, shown in Fig. 16-18. In this pattern, magnetic transitions of opposite polarity appear simultaneously on either side of the track center line. When the head is on track, these two transitions result in equal pulses of opposite polarity, ideally cancelling each other.

In an off track position cancellation does not take place, and a read voltage of one or the opposite phase occurs, in relation to a reference pattern. This will enable the servo to move the head back on track since the voltage amplitude is indicative of the amount of off-track position, and the phase will provide the direction to move the head in.

The signals and patterns used in these *embedded* servos vary from system to system (ref. McKnight, Ragle et al.). The discontinuous nature of embedded servo signals between sectors can be avoided by using so-called *buried* servo signals. These are initially written onto a disk with a long-gap recording head, so the entire thickness of the coating has a continuous servo track for each concentric data track; a short-gap then erases the surface portion to stabilize the servo amplitude.

Short-gap read-write heads then write and read high frequency data signals without interference. Using frequency-separation filters and an AC-biased pulse write method, an inductive head reads servos while simultaneously writing data (ref. Haynes, Hansen).

The design of servo mechanisms and associated electronics requires an analysis of the open loop response of the actuator plus motor system (ref. Samuels, Edwards), and careful design of filters (ref. Sidman). The servo signal itself should be analyzed and optimized (ref. Siegel and Marcus, Cooper).

The open loop system response will exhibit several electro-mechanical resonances, typically starting at a few hundred Hz and continuing up to several kHz. The phase response is

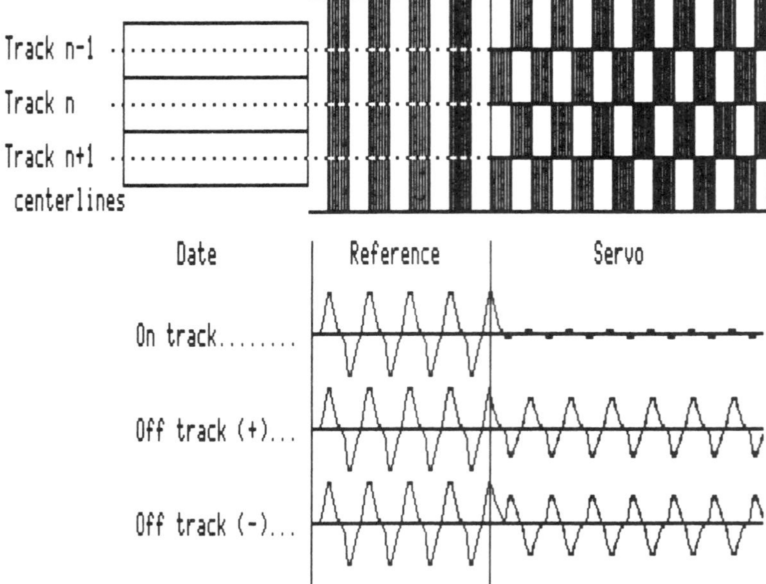

Fig. 16-18. Servo bursts are recorded between sectors; Amplitude and phase of servo signals provide correction signal to the motor drive electronics.

normally unpredictable at and above the lowest resonance, and it behooves the designer to identify and possibly increase that resonance frequency in order to make the servo as effective as possible (increase its bandwidth). Chapter 17 will discuss some of the encountered resonances in detail.

TESTING OF DISK DRIVES

The data written on disks are in serial format, and the write and read processes must be synchronized. There will be speed variations between the input and output data due to four timing functions: The speed at which the device generating the write signal operates, the rotational speed of the disk at the time of writing, the rotational speed at the time of reading, and finally the speed of operation of the device that is trying to interpret the recorded information.

The best way to synchronize the data is to carry the timing information along with the data stream in the form of clock pulses, that are separate from the data pulses. In single density encoding all data bits are stored in defined bit cells. Every bit cell contains a timing bit that is called the *clock* bit; the leading edge of the clock bit determines the start of each bit cell. If there is a data bit in the bit cell it will occur at the center of the bit cell. A detection circuit will normally identify this pulse as a "1", while the absence of a pulse is a "0".

Figure 16-19 shows an example; the 2 microsecond and 4 microsecond time intervals correspond to 500,000 and 250,000 bits per second respectively. Typical clock and data pulses would be approximately 200 nanoseconds wide at these bit rates. A bit stream that contains only clock pulses would represent the 1f, or lower frequency of operation of this particular system. If all data were 1's then the bit stream would represent the 2f, or higher frequency of operation. There can be only 1f or 2f frequency signals and the encoding scheme is therefore called *binary frequency modulation*.

The existence of a clock bit at the start of each bit cell makes data recovery straightforward. The recovery electronics lock onto the leading edge of each clock pulse, and a *window*

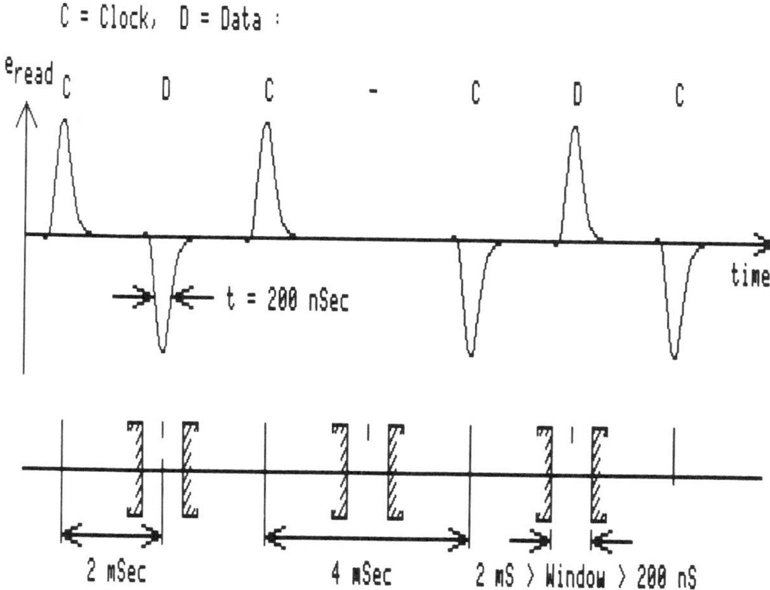

Fig. 16-19. Read-out of data signals must occur during the short duration called the window.

shortly thereafter opened to look for the presence (a "1") or absence (a "2") of a transition. The window length for data recovery is typically 50 percent of the bit cell, or 2 microseconds.

Single density, or FM, encoding is easy to implement, and reliable because there is a clock bit in every bit cell. The price is that half the storage space has been allocated to timing bits. That situation is improved in MFM, or *double-density encoding,* where all clock timing transitions are eliminated, except between zeros (see further in Chapter 24 on codes).

The detection circuits window must now be timed from a free running oscillator, that is adjusted to be operating at the correct rate, which in this case is 250,000 windows per second. Whenever a timing pulse occurs the oscillator is re-timed; the circuit for this type oscillator is named a phase-locked oscillator. It can also be an oscillator that operates at twice the speed, and therefore can use all transitions (one's or clock) to adjust its phase; the window then operates at half the oscillator rate.

Drift in disk speed (typically specified within a couple of percent) and other timing variations (peak-shifts, see chapters 19 and 21) will shift the transitions to be detected back and forth within the window, and may even fall completely outside. In that case an error occurs, and their number must be limited to something like only one in 10 billion. These events would be rare, and would not provide much information about such things as the drive speed stability, and the entire disk systems stability.

A test has evolved to examine the shifts of transitions within a window, the so-called *Phase Margin* tests (see Figs. 16-20). It basically consists of dividing the detection window up into smaller time windows, and during reading of data examine the timing of the leading edges, and log them into the appropriate timing windows. This will eventually provide a histogram of the system's timing accuracy, and allows characterization by Maximum, Minimum, Mean and Variance numbers.

Instruments are available for testing disk drives (ref. Lerma, Baldwin, Rosenblatt), preparation of test disks (ref. Mackintosch, 1982), and discussions of test results (Albert, Mackintosch (1984), Ruoff). Figure 16-21 shows two examples of margin testing on two drives before and after inclusion of a pulse slimming equalizer.

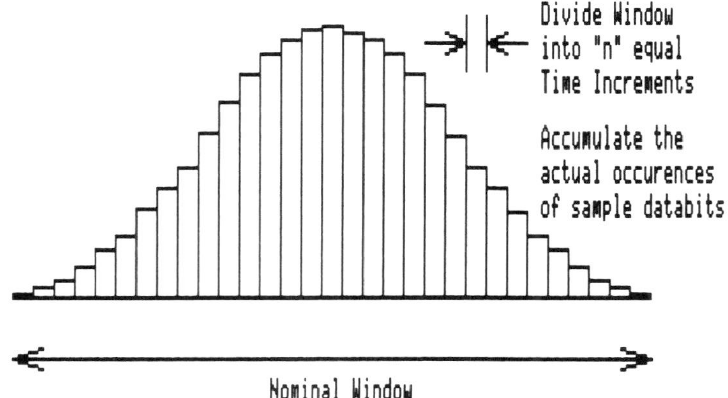

Fig. 16-20. Unsteady media motion in tape or disk drives cause jitter in the data stream. Margin analysis examines drives for amount of jitter.

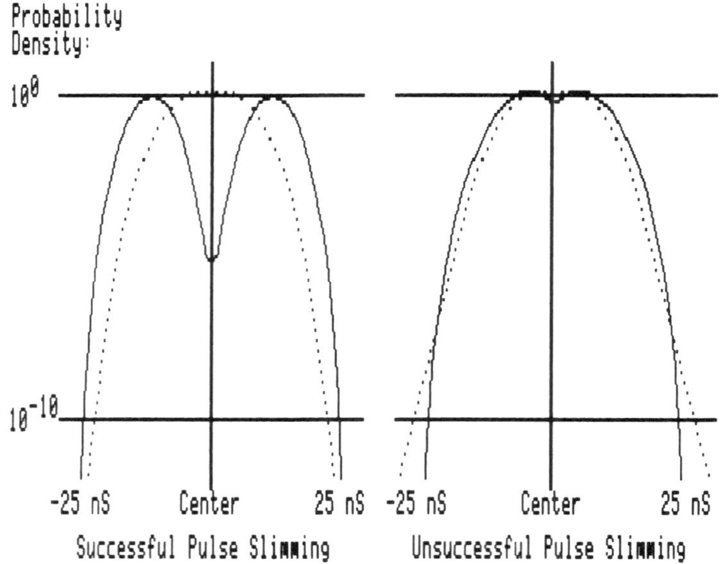

Fig. 16-21. Pulse slimming is a trade-off technique whereby peak shift is reduced but signal-to-noise ratio is degraded (see text) (after Mackintosh).

The peak shift distribution (solid line) is well defined, and the slope indicates a reasonable signal-to-noise ratio. Pulse-slimming will reduce the peak shifts, but reduced the signal-to-noise ratio. The dashed line plot to the left shows how the peak shift has been reduced by a factor of four, but the slope of the fall-off is now less steep, indicating that the signal-to-noise ratio has been degraded. Applying the same technique to a different drive system (Fig. 16-21, right) again reduced the peak shift, but the overall result is a loss in margin at levels below the 10^{-8} probability level.

REFERENCES TO CHAPTER 16

The journals *Mini-Micro Systems* (Cahners Publ. Co., Boston, MA) and *Computer Design* (PennWell Publ. Co., Littleton, MA) do regularly feature papers on novel designs, and special reports on tape, disk and diskette drives.

Introduction

IBM Disk Storage Technology, IBM Corporation, 5600 Cottle Road, San Jose, CA 95193, 1980, 97 pages.

Teja, E.R., *The Designer's Guide to Disk Drives*, Reston Publ. Co., Inc. (Prentice-Hall), June 1984, 127 pages.

Harker, J.M., Brede, D.W., Pattison, R.E., Santana, G.R., and Taft, L.G., "A Quarter Century of Disk File Innovation," *IBM Jour. Res. Devel.*, September 1981, Vol. 25, No. 5, pp. 677-689.

Mulvany, R.B., and Thompson, L.H., "Innovations in Disk File Manufacturing," *IBM Jour. Res. Devel.*, September 1981, Vol. 25, No. 5, pp. 711-723.

Engh, J.T., "The IBM Diskette and Diskette Drive," *IBM Jour. Res. Devel.*, September 1981, Vol. 25, No. 5, pp. 701-710.

Killmon, P., "Shrinkage Continues Unabated - Winchesters Settle in at 3.5 In.," *Computer Design*, January 1985, Vol. 24, No. 1, pp. 29-40.

Flexible Disk Drives

Pearson, R.T., "The Development of the Flexible-Disk Magnetic Recorder," *Proc. IRE*, January 1961, Vol.49, No.1, pp. 164-174.

Resnik, D., "Drive Mechanism Design Reduces Errors in Mini-Floppies," *Computer Design*, March 1980, Vol. 19, No. 3, pp. 160-168.

Katoh, Y., Nakayama, M., Tanaka, Y., and Takahashi, K., "Development Of A New Compact Floppy Disk System," *IEEE Trans. Magn.*, November 1981, Vol. MAG-17, No. 6, pp. 2742-2744.

Adachi, O., and Yano, N., "High Density Mini Flexible Disk Drive System," *Review Electrical Comm. Lab.*, February 1984, Vol. 32, No. 2, pp. 245-251.

Okuwaki, T., Kugiya, F., Kumusaka, N., Yoshida, K., Tsumita, N., andTanura, T., "5.25 Inch Floppy Disk Drive using Perpendicular Magnetic Recording," *IEEE Trans. Magn.*, September 1985, Vol. MAG-21, No. 5, pp. 1365-1367.

Matsuda, N., Ryohnai, H., and Mohri, N., "3 Inch High Recording Density Floppy Disk," *IEEE Trans. Magn.*, Sept. 1986, Vol. MAG-22, No. 5, pp. 1176-1178.

Imamura, M., Ito, Y., Fujiki, M., Hasegawa, T., Kubota, H., and Fujiwara, T., "Barium Ferrite Perpendicular Recording Flexible Disk Drive," *IEEE Trans. Magn.*, Sept. 1986, Vol. MAG-22, No. 5, pp. 1185-87.

Rigid Disk Drives

McLeod, J., *Winchester Disks in Microcomputers*, Elsevier Intl. Bull., 256 Banbury Road, Oxford, OX2 7DH, England, 2nd edition, 1986, approx. 200 pages.

Thomson, W.T., *Theory of Vibration with Applications*, Prentice-Hall, 1981, 493 pages.

Nashif, A.D., Jones, D.I.G., and Henderson, J.P., *Vibration Damping*, John Wiley & Sons, 1985, 453 pages.

Mulvany, R., "Engineering Design of a Disk Storage Facility with Data Modules," *IBM Jour. Res. Devel.*, Nov. 1974, pp. 489-505.

Lennemann, E., "Aerodynamic Aspects of Disk Files," *IBM Jour. Res. Devel.*, Nov. 1974, Vol.18, pp. 480-487.

Mizoshita, Y., and Matsuo, N., "Mechanical and Servo Design of a 10 inch Disk Drive," *IEEE Trans. Magn.*, July 1981, Vol. MAG-17, No. 4, pp. 1387-1391.

Mulvany, R.B., and Thompson, L.H., "Innovations in Disk File Engineering," *IBM Jour. Res. Develop.*, Sept. 1981, Vol. 25, No. 5, pp. 711-723.

Kaneko, R. and Koshimoto, Y., "Technology in Compact and High Recording Density Disk Storage," *IEEE Trans. Magn.*, November 1982, MAG-18, No. 6, pp. 1221-1226.

Smith, J.H., "Mechanical Design of a Large Disc Drive," *Hewlett-Packard Journ.*, January 1984, Vol. 35, No. 1, pp. 20-22.

Rahimi, A., "Designing Hard Disk Drives to Take Abuse," *Computer Design*, October 1984, Vol. 23, No. 10, pp. 141-148.

Opfer, J.E., Spenner, B.F., Natarajan, B.R., Baugh, R.A., Murdock, E.S., Morehouse, C.C., and Bromley, D.J., "Thin-Film Memory Disc Development," *Hewlett Packard Journal*, November 1985, Vol. 36, No. 11, pp. 4-10.

Actuator, or Head Positioner

Oswald, R.K., Wasson, K. and Wagner, J.A., *The Disk File Linear Actuator: An Introduction to the Magnetics and Control*, Wasson Assoc., P.O. Box 26800, San Jose, CA 95159, June 1981, 177 pages.

Pillai, S.K., *A First Course on Electrical Drives*, Wiley Eastern Ltd., Bombay, 1983, 208 pages.

Kenjo, T., *Stepping Motors and their microprocessor controls*, Clarendon Press, Oxford, 1984, 244 pages.

Brown, C.J. and Ma, J.T., "Time-Optimal Control of a Moving-Coil Linear Actuator," *IBM Jour. Res. Devel.*, Sept. 1968, Vol. 12, No. 9, pp. 372-379.

Inoue, Y., Sata, Y. and Hashizume, K., "New Linear Motion Actuator for Head Positioning," *Fujitsu Scien. and Tech. Jour.*, March 1974, pp. 95-118.

Heath, J.S., "Design of a Swinging Arm Actuator for a Disk File," *IBM Jour. Res. Devel.*, June 1976, Vol. 20, pp. 389-397.

Hearn, A.R., "Actuator for an Eight-Inch Disk File," *Disk Storage Technology*, IBM, Feb. 1980, pp. 83-88.

Winfrey, R., Riggle, C.M., Bernett, F., Read, J. and Svendsen, P. "Design of a High Performance Rotary Positioner for a Magnetic Disk Memory," *IEEE Trans. Magn.*, September 1981, Vol. MAG-17, No. 4, pp. 1392-1395.

Wagner, J.A., "The Shorted Turn in the Linear Actuator of a High Performance Disk Drive," *IEEE Trans. Magn.*, November 1982, Vol. MAG-18, No. 6, pp. 1770-1773.

Wagner, J.A., "The Actuator in High Performance Disk Drives: Design Rules for Minimum Access Time," *IEEE Trans. Magn.*, September 1983, Vol. MAG-19, No.5, pp. 1686-1688.

Dong, A., "Dual-Path Electromagnetic Actuator for a High Performance Magnetic Disk Drive," *IEEE Trans. Magn.*, September 1983, Vol. MAG-19, No.5, pp. 1689-1691.

Scranton, R.A., Worthington, T.K., Hunter, D.W., Thompson, D.A., Thompson, M.O., and Cocke, J., "A Novel, High Performance, Low Mass Disk Head Actuator," *IEEE Trans. Magn.*, September 1983, MAG-19, No.5, pp. 1692-1694.

Bell, F., Johnson, E.W., Whitaker, R.K. and Wilcox, R.V., "Head Positioning in a Large Disc Drive," *Hewlett-Packard Journ.*, January 1984, Vol. 35, No. 1, pp. 14-20.

Scranton, R.A., Thompson, D.A. and Hunter, D.W., "The Access Time Myth," *Intl. Conf. Video and Data Rec.*, 1984, IERE Publ. No. 59, pp. 145-149.

Miu, D.K., Bouchard, G., Bogy, D.B. and Talke, F.E., "Dynamic Response of a Winchester-Type Slider Measured by Laser Doppler Interferometry," *IEEE Trans. Magn.*, September 1984, Vol. MAG-20, No. 5, pp. 927-929.

Servo Designs

Oswald, R., "Design of a Disk File Head-Positioning Servo," *IBM Jour. Res. Devel.*, November 1974, Vol. 18, pp. 506-512.

Sidman, M.D., "An Adaptive Filter for the Processing of Position Error Signal in Disk Drives," *IEEE Trans. Magn.*, July 1978, Vol. MAG-14, No. 4, pp. 185-187.

Samuels, F.A., "Beyond the Second Order System in Track Following Servos," *IEEE Trans. Magn.*, July 1978, Vol. MAG-14, No. 4, pp. 178-181.

McKnight, B., "A Track Locating Servo System Utilizing the Data Heads as Absolute Position Transducers," *IEEE Trans. Magn.*, July 1978, MAG-14, No. 4, pp. 182-184.

Ragle, H., Blessum, N., and Stomsta, R. "Position Sensing for High Track Density Recording," *IEEE Trans. Magn.*, September 1978, Vol. MAG-14, No. 5, pp. 327-329.

Oswald, R.K., "The IBM 3370 Head-Positioning Control System," *Disk Storage Technology*, IBM, Feb. 1980, pp. 41-44.

Commander, R.D., and Taylor, J.R., "Servo Design for an Eight-Inch Disk File," *Disk Storage Technology*, IBM, Feb. 1980, pp. 89-97.

Haynes, M., "Magnetic Recording Techniques For Buried Servos," *IEEE Trans. Magn.*, November 1981, MAG-17, No. 6, pp. 2730-2734.

Hansen, N., "A Head-Positioning System Using Buried Servos," *IEEE Trans. Magn.*, November 1981, MAG-17, No. 6, pp. 2735-2738.

Ananthanarayanan, K.S., "Third-Order Theory and Bang-Bang Control of Voice Coil Actuators," *IEEE Trans. Magn.*, May 1982, Vol. MAG-18, No. 3, pp. 888-893.

Thompson, H., "Half-Height Minifloppy Stores 3.3 M Bytes," *Mini-Micro Systems*, July 1983, Vol. 16, No. 7, pp. 219-226.

Nakanishi, H., and Mizukami, M., "High Track Density Head Positioning Using Sector Servos," *IEEE Trans. Magn.*, September 1983, Vol. MAG-19, No. 5, pp. 1698-1700.

Edwards, S.A., "High-Capacity Disc Drive Servomechanism Design," *Hewlett-Packard Journ.* Jan. 1984, Vol. 35, No. 1, pp. 23-27.

Siegel, P., and Marcus, B., "Worst Case Code Patterns for Magnetic Buried Servos," *IEEE Trans. Magn.*, September 1984, Vol. MAG-20, No. 5, pp. 906-908.

Cooper, E.S., "A Disk File Servomechanism Immunized to Media Defects," *IEEE Trans. Magn.*, November 1985, Vol. MAG-21, No. 6, pp. 2592-2594.

Miscellaneous

Mackintosh, N.D., and Miyata, J.J., "A Standard Disk for Calibrating Head-Disk Interference Measuring Equipment," *IEEE Trans. Magn.*, Sept. 1982, Vol. MAG-18, No. 6, pp. 1230-1232.

Albert, M.S., "Testing Disk Drives from Head to Toe," *Electronics*, August 1983, Vol. 56, No. 17, pp. 147-151.

Mackintosh, N.D., "Evaluate Disk-Drive Performance with Margin Analysis," *Computer Design*, January 1984, Vol. 23, pp. 81-88.

Lerma, J., "Analyzer Integrates Disk Drives With Controllers, Systems," *Digital Design*, Apr. 1985, Vol. 14, No. 4, pp. 132-134.

Ruoff, J., "How to Test Winchester Disk Drives," *Computer Design,* May 1984, pp. 81-86.

Baldwin, D., "Disk Drive Analyzer Speeds Testing with Innovative Graphics," *El. Design*, May 1985, Vol. 33, No. 11, pp. 50-51.

Rosenblatt, A., "Analyzer for Checking Floppy-Disk Drives Draws on Video-Game Techniques," *Elec. Prod. Mag.,* December 1985, Vol. 28, No. 13, pp. 42-46.

Chapter 17
Vibrations in Tapes and Disks

A linearly moving tape, or a rotating disk will vibrate at a number of frequencies; this will manifest itself as flutter and timing errors in data streams (window margins, peak-shifts), and as excessive wear of both heads and media. The vibration modes for tapes and disks will be treated separately.

There are many sources for vibrations:

- Friction between heads and medium.
- Nearby vibrations (motors, ball bearings, etc.).
- Airborne noise that the media picks up just as a microphone membrane would.

The latter has been verified when listening to the detected output of flutter meters; the signal contains airborne sound from nearby conversation. Consider the fact that early ribbon microphones were constructed by mounting a flat conductive ribbon between two magnet poles, as shown in Fig. 17-1. A voltage will be induced in the ribbon when it moves back and forth in the flux lines (ref. Beranek, pressure-gradient microphones).

The affects of vibrations are an annoyance while listening to music from an audio recorder. This is illustrated in Fig. 17-2 where the human annoyance threshold for frequency modulation of tones is shown, together with the threshold for amplitude variations (ref. Stott and Axon, Sakai, McKnight (1972)). It is also possible to identify a recorder from measurements of its flutter (McKnight 1976). Data from instrumentation recorders will suffer accuracy, as several papers in the references describe (ref. Prager, Kashin et al., Minukhin, Slepov, McKnight, and Ratz).

The reader may waive all this aside by referring to the flutter elimination that a digital system easily provides by reading and dumping the digitized bits into a buffer, and then clock the bits out in a perfect time sequence; it is only necessary to keep the buffer about half full

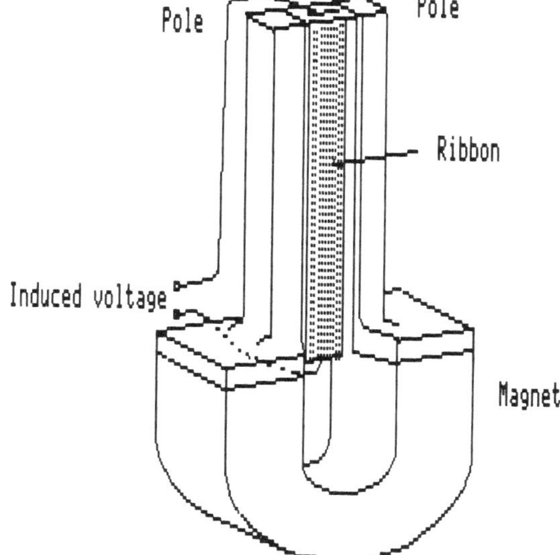

Fig. 17-1. Ribbon microphone (after Beranek).

of bits most of the time, and regulate the playback speed to assure this operation. The transport designer should still strive for the best possible tape motion; that leads to a buffer size smaller than a poorly designed transport would require.

This buffering is not required in a self-clocking digital read/write system for data and data bases, where the speed accuracy of the read mechanism easily can be a couple of percents off. Now speed variations show up in deterioration of the window margins, as discussed in Chapter 16, and again we find it beneficial to design a mechanism that will move the media at a perfectly constant and smooth speed; vibrations will affect the latter.

This chapter will therefore focus on vibrations that take place in the media, and that are determined by the dimensions and material properties of the media. Vibrations at large are

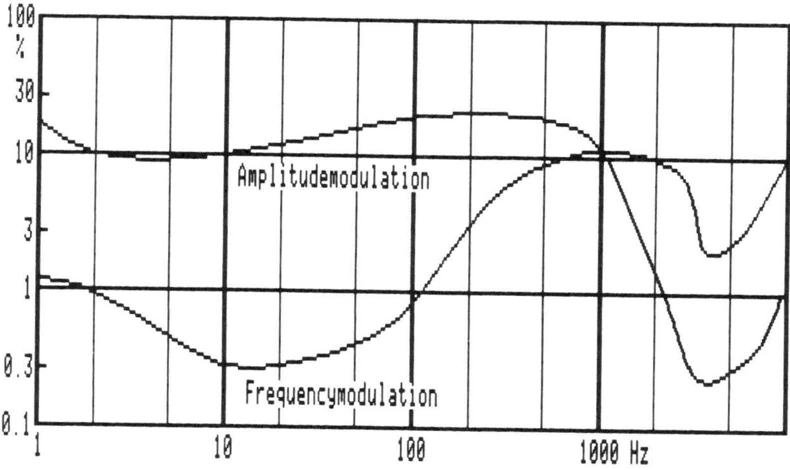

Fig. 17-2. The human ear's perceptibility limits to amplitude and frequency modulation of tones as function of the modulation frequency (after Stott and Axon).

already well treated in several textbooks (ref. den Hartog, Thomson, Lalanne et al.; the latter has several interesting computer programs written in Basic).

VIBRATIONS GENERATED BY FRICTION

The frictional behavior of two surfaces sliding against each other is a complicated science where a large amount of knowledge comes from observations and measurements. A first approximation for the vibration generating characteristics of friction is to equate it with a white noise generator that produces vibration components at all frequencies.

Friction cannot occur without some abrasion and wear, see end of Chapter 18. The force in a dry frictional process follows the general law that the force is proportional to the coefficient of friction multiplied with the force perpendicular to the surface, and independent of speed:

$$F = \mu P \tag{17.1}$$

This situation is shown in Fig. 17-3 top, left.

When a lubricant is added, the law no longer holds true. The drag force is now proportional to the perpendicular pressure as well as to the velocity, (Fig. 17-3 top, middle). When the velocity is very low there may be isolated contact points where the lubricant film is broken and the frictional forces increase drastically (Fig. 17-3 top, right).

Tape and disk coatings contain minute amounts of lubricants. This results in a composite picture of the frictional forces versus tape velocity as shown in Fig. 17-3 bottom. The dynamic coefficient of friction is lower than the static coefficient.

This results in a small region at very low speeds where the slope of the curve for the frictional force versus tape velocity is negative, causing a "stick-slip" motion. An interesting study of the irregular tape motion is found in a paper by Ristow, while Kalfayan et al. have many practical measurements and observations to add to our knowledge.

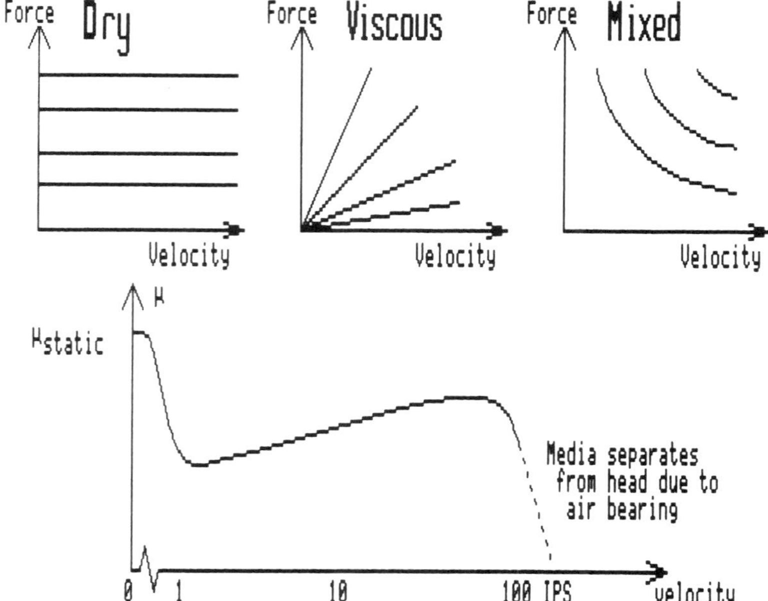

Fig. 17-3. Frictional forces versus velocity in a sliding contact.

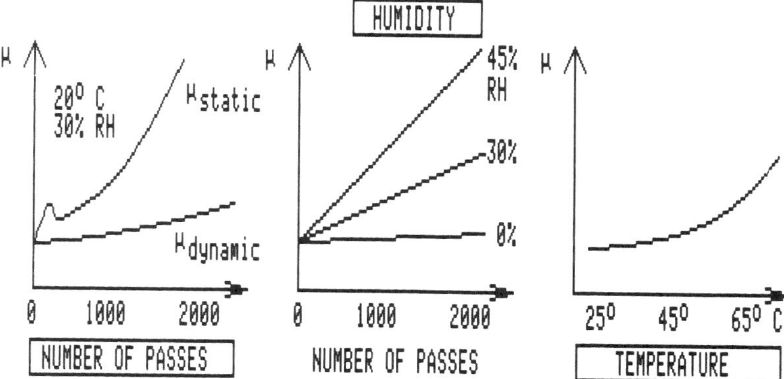

Fig. 17-4. Changes in coefficients of friction.

When the speed approaches 100 IPS (or 9.1 km/hour = 6 MPH) enough air is drawn along with the media surface to reduce the coefficient of friction to zero. At higher speeds an air foil bearing is formed, causing separation between the tape and the head.

Mean values of the coefficient of friction are listed in Chapter 18, in an ambient environment. They will vary with binder formulation, and they all show a generally upward trend after many passes, as illustrated in Fig. 17-4 left. This increase is understood by realizing that high points on the coating are worn down, providing for a larger contact area.

The increase due to rising temperature and humidity is not well understood, although it correlates with the fact that abrasion also increases with humidity which in turn requires higher pulling forces.

Stick-Slip Friction

Tapes have been found to move with a jerky motion at very low speeds due to a static coefficient of friction that is larger than the dynamic coefficient of friction. This is normally (and justly so) disregarded in high speed transports, but it has occasionally been brought into light in discussions regarding possible tape slippage over capstans (such slips may occur at the low capstan-to-tape speeds (ideally a relative stand-still), and any stick-slip activity would sort of pluck the resonance frequencies of the tape and thus cause high flutter).

If a tape moves along at a speed v_1 it would sense less friction if the speed moved up to v_2. And if there is sufficient tension and compliance it will do just that for a short moment, then advance to get too far ahead, then stop—and then go through v_1 and v_2 again. This jerky motion is stick-slip, and can be expected to occur at tape speeds less than the critical speed v_c, shown in Fig. 17-5.

The critical speed where stick-slip occurs can be calculated, using formula (17.2) in Fig. 17-5. Steinhorst (see ref.) introduced the dynamic value of Young's modulus of elasticity, which uses Poisson's ratio for its calculation. Poisson's ratio is defined as the ratio of lateral unit deformation to the unit longitudinal deformation when a material is strained; pulling or pushing a rubber-eraser illustrates Poisson's ratio.

The critical velocity is typically in the range of a few cm/sec; it is not exact since the coefficient of friction changes with several parameters, as we saw in Fig. 17-4.

VIBRATIONS IN TAPES

Flutter is quite often introduced into a recording by either eccentric reels, an improper hold-down mechanism for the reels, or by cogging in the reel motors. These flutter compo-

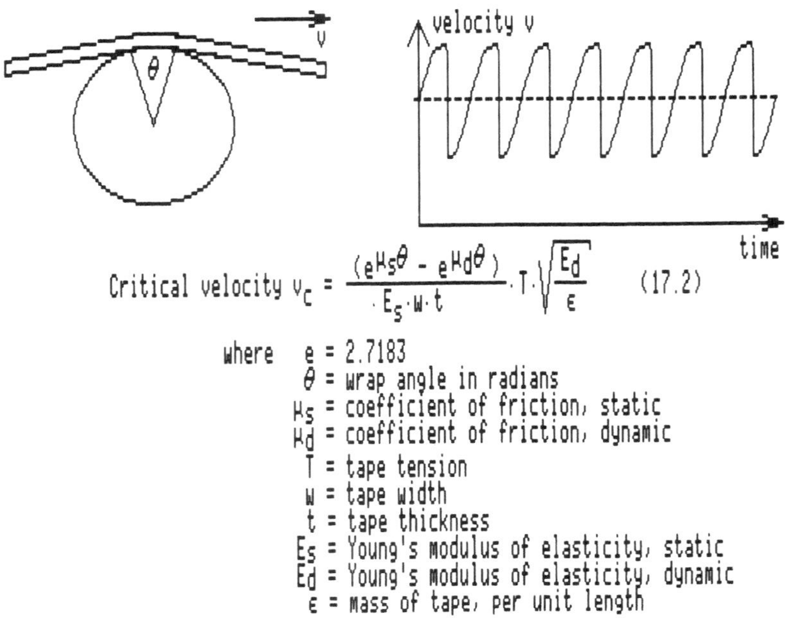

Fig. 17-5. Stick-slip friction (after Steinhorst).

nents are particularly noticeable when the reel is almost empty and they can be avoided only by a careful mechanical design. Several tape recorder manufacturers incorporate sensing arms in the tape path. The arms have transducers that detect any tension variations and can therefore compensate for reel tension variations or reel motor cogging by servo-controlling the torque on the reel motors.

Flutter from the capstan area is more often caused by a soiled pinch-roller or worn bearings in the pinch roller or capstan assembly. Another (although rare) cause of flutter may be an eccentric capstan. These drive shafts are ground to a high degree of precision, but may inadvertently have been bent or otherwise damaged.

Longitudinal Resonances

One contributor to flutter is the resonance frequency for the longitudinal oscillations in a free tape span. This has long been recognized as one cause of flutter and minimized by shortening all free tape sections (capstan to head, head to head, etc.) (ref. Hadady).

The length of the tape is determined by the length between any two adjacent points of support, i.e. capstan(s), rollers, guide and heads. The heads are generators of frictional forces, but are also grounding points insofar as the mechanical vibrations are concerned. And in many cases one or more high inertia rollers are placed in the tape path to break it up into shorter stretches since the resonance frequency is inversely proportional to the length, see formula (17.3) in Fig. 17-6.

Note that the frequency f_n is independent of the tape tension. The amplitude of oscillation is limited by the Q of the mass-compliance system. The example shows that the resonance frequency for a 5 cm long piece of 1 mil thick tape is 23.6 kHz; a lower value of E_d = 5.3 $*10^{10}$ dynes/cm^2 would result in 19.5 kHz. A rule of thumb is therefore that the fundamental frequency of the longitudinal resonance is about 20 kHz per 5 cm tape, or **100 kHz per**

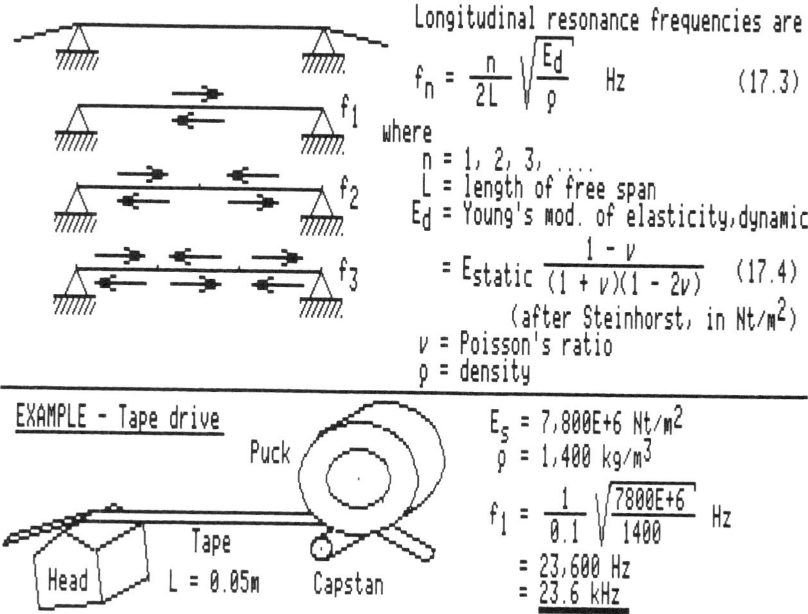

Fig. 17-6. Longitudinal resonance frequencies in a string (or free span of tape).

cm length of free tape; for any given length L simply divide 100 kHz by L (in cm) to find the resonance frequency (the harmonics are rarely of any practical influence).

Design of servo systems for speed control of tape drives will encounter this frequency as the ultimate limitation for the servo bandwidth. Any corrections produced by the capstan servo will reach the read head with the speed of sound in the tape. That speed is related to the fundamental resonance frequency f_1 by $v = 2L * f_1$.

Support of the tape with a soft, endless rubber belt has been found to reduce the vibration amplitude (ref. Steinhorst). This is illustrated in Fig. 17-7 that shows a ¼ inch tape moving past a magnetic head. Recording of a 2.7 kHz sine wave resulted in the playback signals shown in Fig. 17-8.

Modulation noise occurs due to flutter during playback. The use of a co-belt clearly reduces the noise, or flutter. The very compliant rubber belt adds to the tape mass per unit length, and will therefore reduce the flutter frequencies; its presence does also significantly lower the Q of the mechanical system.

Transverse Resonances

Another set of vibrations, seemingly more troublesome than the longitudinal mode, are the low frequency transverse resonances.

The formula for the transverse vibration resonances for a string is shown in Fig. 17-9, top (17.5), and for a membrane, bottom (17.6). There exist a host of resonance frequencies, and the short spans of tapes in tape recorders should be regarded as membranes—not as strings.

A membrane is a plane sheet whose stiffness is negligible compared to the restoring forces due to tension. Analysis of the motions of a membrane is more complicated than that of the motions of the corresponding one-dimensional system, the flexible string. The membrane has more freedom in the way it vibrates than the string has.

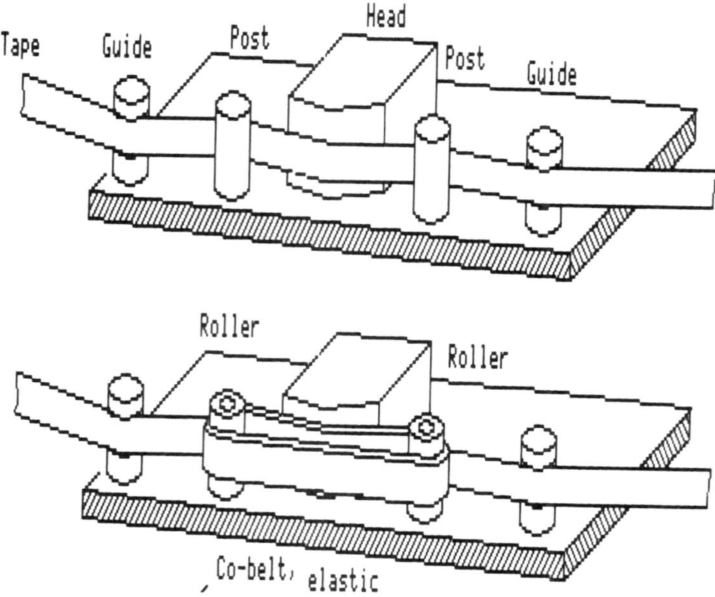

Fig. 17-7. Tape drive without and with a compliant co-belt (after Steinhorst).

It is useful to treat the tape span between two "ground" surfaces (guide, head, capstan) as a membrane. The allowed transverse oscillations of a square membrane can be calculated (ref. Morse), and verified by measurements. The author believes that these vibrations are the cause of so-called scrape-flutter.

The resonance frequencies can be calculated from formula (17.6) in Fig. 17-9, and there are actually a double infinity of solutions (m,n); a few of the vibration modes are shown in Fig. 17-10, and the computed resonance frequency spectra shown in Fig. 17-11.

A given tape span needs to be fitted to formula (17.6); "a" is the length of the tape (being clamped or passing over a head), but "b" is not equal to the width of the tape since the edges are free. A value of "b" can, however, be fitted from appropriate resonance measurements

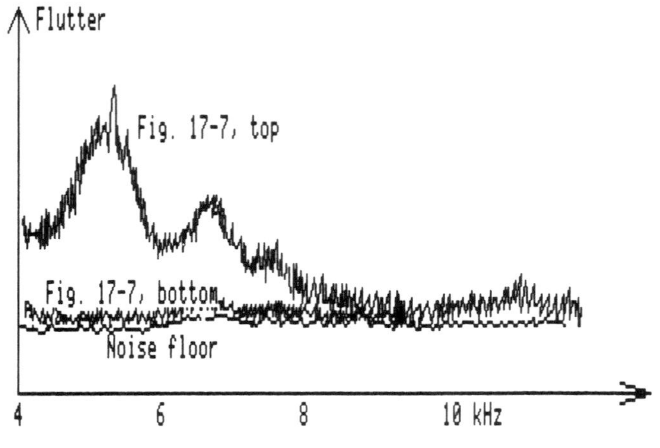

Fig. 17-8. Reduction of flutter with a co-belt drive (after Steinhorst).

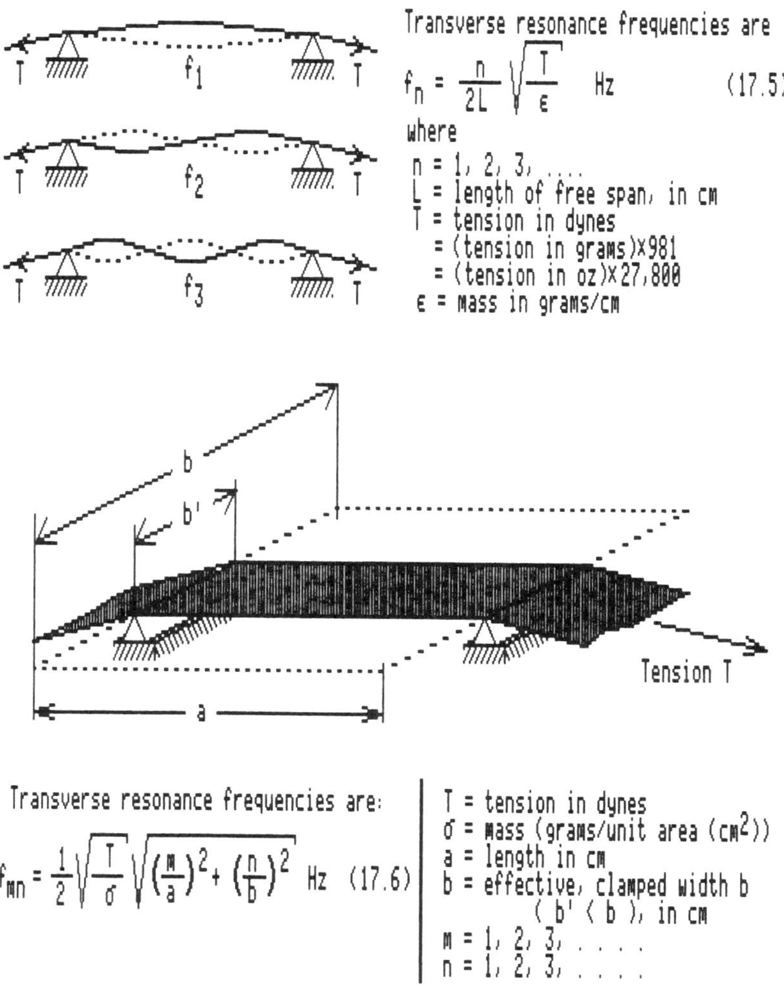

Transverse resonance frequencies are:

$$f_{mn} = \frac{1}{2}\sqrt{\frac{T}{\sigma}}\sqrt{\left(\frac{m}{a}\right)^2 + \left(\frac{n}{b}\right)^2}\ \text{Hz} \quad (17.6)$$

T = tension in dynes
σ = mass (grams/unit area (cm^2))
a = length in cm
b = effective, clamped width b
(b' < b), in cm
m = 1, 2, 3, ...
n = 1, 2, 3, ...

Fig. 17-9. Transverse vibrations in a string and in a membrane.

and was found to vary from several times the width of the tape for very light tension to slightly larger than the width for high tensions.

The spectra offers an explanation for the broad spectrum for **scrape flutter**. We see that the band of frequencies shift toward higher values for higher tape tension. And the individual frequencies will simultaneously move closer together (as a result of a smaller effective value of "b").

Note that these frequencies are tension dependent (T). The tape tension in any recorder varies from one instant to the next due to any or all of the following causes:

- Mechanical run-outs of rotating parts.
- Changes in coefficient of friction.
- Changes in tape cross section.
- Changes in tape elastic modulus.

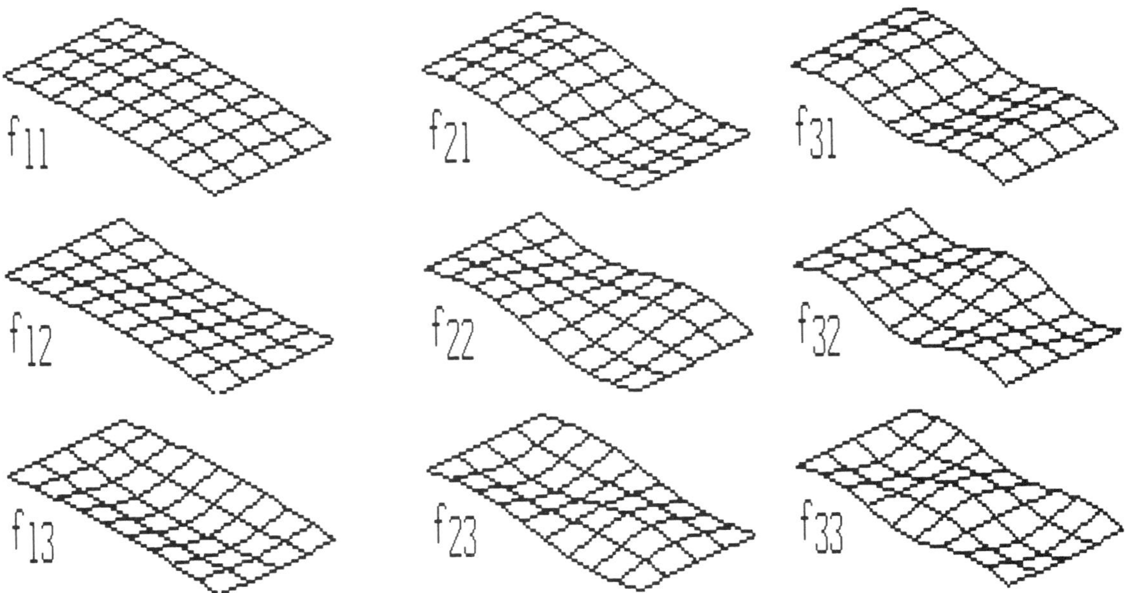

Fig. 17-10. Resonance frequencies in a rectangular membrane, clamped at the ends, free on the sides.

A further variation in tape tension may be generated by a servo controlled capstan. Large tension changes will in particular occur in a servo drive when it corrects for a large amount of flutter that was caused by vibrations of a recorder during recording. Tension variations from 0 to many ounces would result in a resonance spectrum extending almost down to DC, and this would display a pattern looking like grass on an oscilloscope connected to the flutter meter.

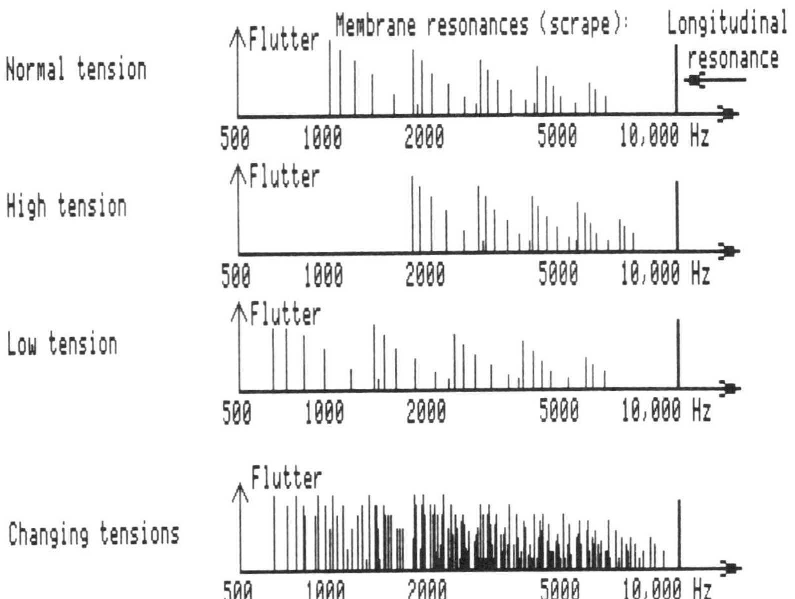

Fig. 17-11. Resonance frequencies in a 1.5 inch long piece of magnetic tape.

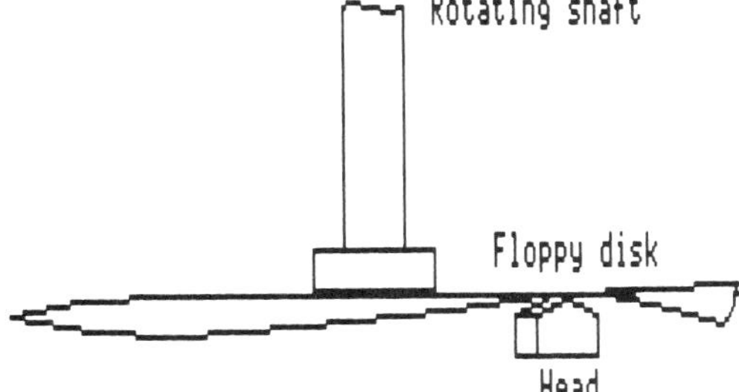

Fig. 17-12. Flexible PET disk conforming to a rigid head shape (after Pearson).

Reduction of the oscillation amplitudes is not possible by conventional means of adding a dampening layer; it would have to be prohibitively thick (Schwarzl, Dietzel (2)). A possible solution may be the installation of a low friction, pebbled surface parallel to, and immediately next to the tape; this would deter transverse vibrations (author).

VIBRATIONS IN DISKS

Disk drives employ disks that are clamped at their centers, and free on the outside; the flexible disks are best treated as circular membranes, while the rigid disks are similar to rigid, circular plates. Early investigations of the flexible disk drive were made by Pearson; fluttering of the flexible disk could be severe (Fig. 17-12), unless supported by a back plate air bearing (Fig. 17-13). The latter principle is used in the Iomega cartridge drive.

Resonance frequencies of the thicker, rigid disk platters can be computed as shown in Fig. 17-14. It is an approximation since the formula is for a disk simply supported at its center, not an annular support. The reader is referred to the work by Leissa, or Priola and Sitjia for details on annular clamped disks. The allowed resonance frequencies for a 3.5 inch aluminum disk are computed, and are: 490, 825, 2,355, 4,213,—etc. Hz.

The vibration flexures are shown in the bottom of Fig. 17-14. Note that the existence of an allowed frequency does not mean that the disk will start vibrating at that, or at any of the other allowed frequencies. But it may, and will, if excited.

The allowed frequencies in plate and membrane vibrations do not follow the common pattern of harmonics. They are the solutions to the wave equation in solids, and their values are often determined by nulls of Bessel functions; these functions are encountered later in the book in connection with frequency modulation, see Fig. 27-11, top.

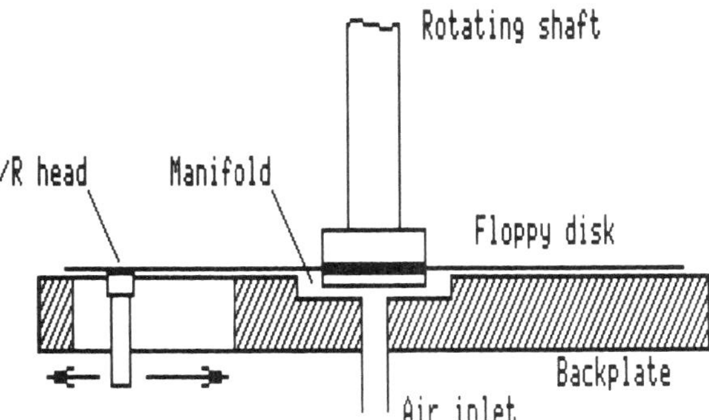

Fig. 17-13. Flexible disk supported by an air bearing backplate (after Pearson).

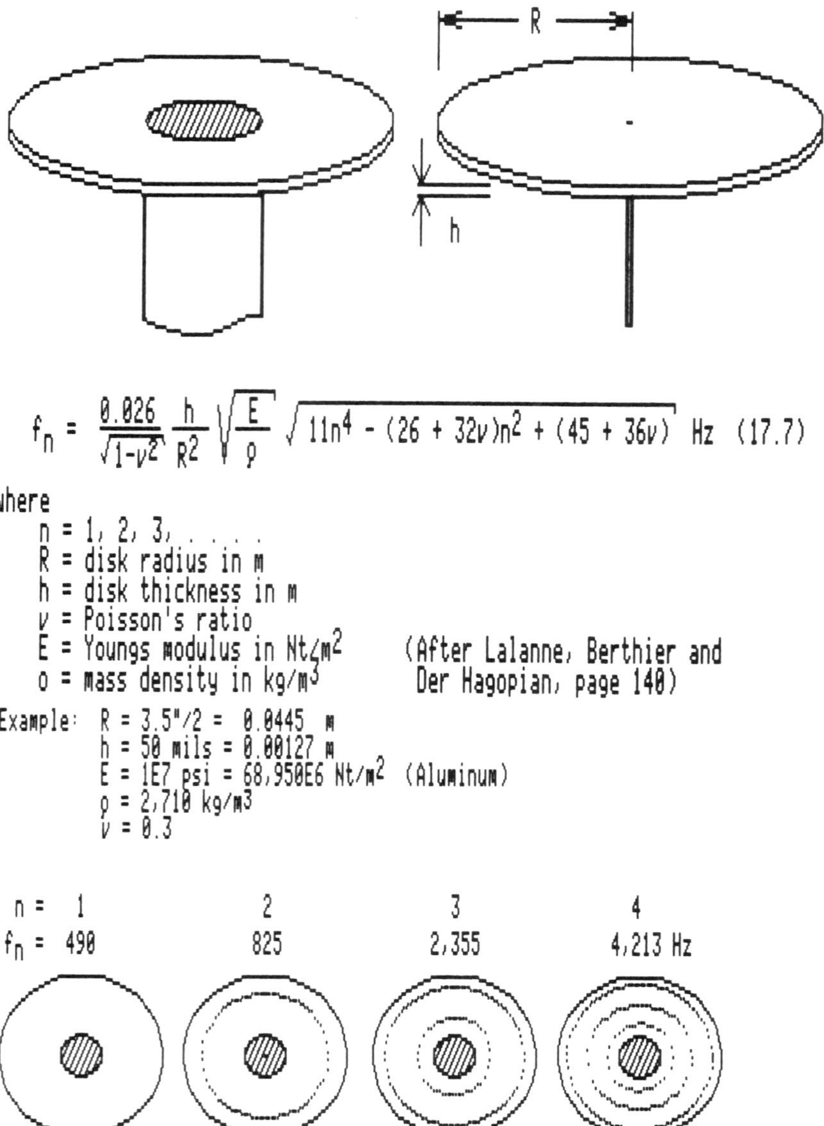

Fig. 17-14. Resonance frequencies in a circular membrane, clamped at the center, free at the edges (annular membrane).

The use of soap films to demonstrate allowed frequencies in thin membranes are shown on pages 186-187 in a book by French. A simpler experiment is shown in Fig. 17-15, where a 5¼ inch disk (jacket removed) is mounted (by press fit) onto a small paper cylinder glued to the center of a small loudspeaker. Excitation of the disk is done by connecting the loudspeaker to the output terminal on a sinewave oscillator. Pre-sprinkle the disk with an even, thin layer of ordinary salt (or sugar), and vary the signal frequencies. At certain frequencies patterns will develop. (Here: circular at 7,000 Hertz, peaks and valleys at 1600 Hertz.)

Fig. 17-15. Observation of resonance frequencies in a 5¼ inch diskette.

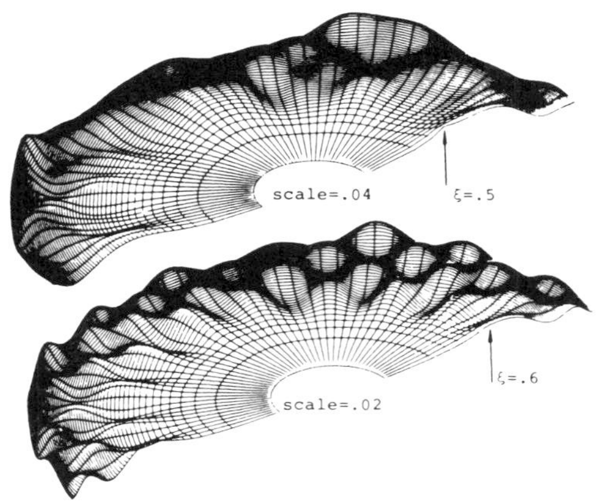

Fig. 17-16. Vibrations of a circular disk, finite element computation (after Benson).

Modern investigations use finite element analysis to study vibration modes (ref. Good and Lowery, Benson), see Fig. 17-16. Further information about vibration in membranes, rectangular as well as circular, can be found in the book listed (Morse, Baldock and Bridgeman, and French), while a good introduction to Bessel function is a book by Bowman.

Other aspects of vibrations in disk drives are covered by Richards, Ohta et al., and Yamaguchi et al..

MEASUREMENTS OF VIBRATIONS AND FLUTTER; MARGIN ANALYSIS

Several instruments are available for the measurement of flutter. A flutter meter is a constant-frequency generator whose output is coupled to the record amplifier while a frequency discriminator demodulates the output from the playback amplifier. The output of the discriminator is fed to a meter which can be reading either RMS or peak-to-peak. The output is available for observation on an oscilloscope.

Flutter during later playback of a tape may give a higher flutter reading than measured while recording. It also may be found that some low flutter frequency components will cancel each other so the playback flutter oscillates between a low value and a high value.

It is important to assure a high signal-to-noise ratio in flutter measurements. It is advisable to insert a band-pass filter during playback to eliminate low-frequency and high-frequency noise. Any noise will affect the discriminator output and result in a reading that appears as flutter. Also, adjust the record level for saturation of the tape during recording.

The performance of disk drives is measured with an instrument that measures the timing accuracy in the data stream. The technique is referred to as margin analysis, and was discussed in Chapter 16.

Direct observation of tape or disk vibrations is possible in several ways. The illustrations in this chapter generally show exaggerated amplitudes of vibrations that seldom are found in real drives. The simplest instrument that has been used to observe tape or disk vibrations is simply a stereo phono cartridge of a type that has very low equivalent mass (f. ex. Ortofon with 0.6 to 0.9 mg effective needle mass). Transverse or longitudinal vibrations are measured by connecting the coils in series or in parallel; this method was used to verify the computations shown in Fig. 17-10 (ref. author, not published).

A secondary indication of disk vibrations is hidden in the motions of heads and heads plus arm rest; methods of observation are a piezoelectric transducer mounted on the slider arm (ref. Yeack-Scranton), high speed motion picture film (ref. Millman et al.), and a two-beam interferometer (ref. Best et al.).

Direct observation of disk vibrations have been made with a laser doppler vibrometer (ref. Bouchard and Talke, Wiezin et al.), monomode fibre optic interferometer (ref. Lewin et al.), and electrostatic capacitance probes (ADE Corp., Newton, Mass 02166).

REFERENCES TO CHAPTER 17

Introduction

Den Hartog, J.P., *Mechanical Vibrations*, Dover Publ., 1984, 436 pages.

Lalanne, M., Berthier, P., and Der Hagopian, J., *Mechanical Vibrations for Engineers*, John Wiley and Sons, 1983, 266 pages.

Thomson, W.T., *Theory of Vibration with Applications*, Prentice-Hall, 1981, 493 pages.

Stott, A., and Axon, P.E., "The Subjective Discrimination of Pitch and Amplitude Fluctuations in Recording Systems," *Jour. AES,* May 1957, Vol. 5, No. 3, p. 142-155.

Sakai, H., "Perceptibility of Wow and Flutter," *Jour. AES,* March 1970, Vol. 18, No. 3, pp. 290.

McKnight, J.G., "Development of a Standard Measurement to Predict Subjective Flutter," *IEEE Trans. Audio Electroacoustics,* March 1972, Vol. AU-20, pp. 78-80.

McKnight, J.G., and Weiss, M.R., "Flutter Analysis for Identifying Tape Recorders," *Jour. AES,* Nov. 1976, Vol. 24, No. 11, pp. 728-734.

Timing Errors

Prager, R.H., "Time Errors in Magnetic Tape Recording," *Jour. AES,* June 1959, Vol. 7, no. 2, pp. 81-88.

Kashin, F.A., Nemeni, T.M., and Slepov, N.N., "Errors in Magnetic Recording Caused by Tape-Speed Instability," *Radio Engineering,* Oct. 1968, Vol. 23, No. 10, pp. 98-101.

Minukhin, V.B., "Signal Distortions in a Magnetic Recording Channel in the Presence of Oscillations of a Magnetic Tape Speed," *Telecom. and Radio Eng.*, Sept. 1969, Vol. 25/26, No. 9, pp. 101-107.

Slepov, N.N., "Effect of Tape-Velocity Fluctuations on the Spectra of Single-Sided Pulse-Width Modulation of the Second Kind," Part 1 and Part 2, *Radio Engineering,* Dec. 1970, Vol. 25, No. 12, pp. 116-121, and Jan. 1971, Vol. 26, No. 1, pp. 128-133.

McKnight, J.G., "Time Base Distortion in Continuous Recording Systems: Its Terminology, Measurement, Causes and Effects," *Jour. AES,* Jan. 1962, Vol. 10, pp. 44-48.

Ratz, A.G., "The Effect of Tape Transport Flutter on Spectrum and Correlation Analysis," *IEEE Trans. Space Electr. and Telemetry,* Dec. 1964, SET-10, pp. 129-134.

White, J., "Dynamic Simulation of the Zero-Load Slider Bearing," *IEEE Trans. Magn.,* Sept. 1983, Vol. MAG-19, No. 5, pp. 1668-1670.

Stick-Slip Friction

Ristow, J., "Die elektrische Nachbildung des Stick-Slip-Vorgaenges," I and II, *Hochfrequenz und Elektroakustik,* Dec. 1985, Vol. 74, No. 12, pp. 191-198 and Vol. 75, No. 1, pp. 1-9.

Kalfayan, S., "Study of the Frictional and Stick-Slip Behavior of Magnetic Recording Tapes," *Jet Prop. Lab. Tech. Report No. 32-1548,* April 1972, 19 pages.

Vibrations in Tapes

Morse, P.M., *Vibration and Sound*, McGraw-Hill, 1936, 351 pages.

Baldock, G.R., and Bridgeman, T., *The Mathematical Theory of Wave Motion*, Halsted Press, Div. of John Wiley and Sons, 1981, 261 pages.

French, A.P., *Vibrations and Waves*, M.I.T. Physics Series, W.W. Norton & Co, 1971, 316 pages.

Leissa, A.W., *Vibration of Plates*, NASA SP-160, NTIS No. N7018461, 1969, 353 pages.

Bowman, Frank, *Introduction to Bessel Functions*, Dover Publ., 1958, 135 pages.

Schwarzl, F., "Forced Bending and Extensional Vibrations of a Two-Layer Compound Viscoelastic Beam," *Acoustica,* March 1958, Vol. 8, No. 3, pp. 164-172.

Steinhorst, W., "Theoretische Betrachtungen zum elastisched Verhalten von Tonbandern", *Feinwerktechnik,* 1966, Heft 3, pp. 114-119.

Steinhorst, W., "Elastische Longitudinalschwingungen in Tonbandern speziell bei Stick-Slip-Anregung," *Feinwerktechink,* 1966, Heft 4, pp. 172-184.

Dietzel, R., "Zur Bestimmung des Verlustfaktors von eingezwaengten daempfungsbelaegen auf duennen Blechen," *Hochfreq. und Elektro Akustik,* 1967, Vol. 76, pp. 151-162.

Dietzel, R., "Vergleichende Untersuchung uber den Verlustfaktor einfacher undeingezwangter Dampfungsbelage auf dunnen Blechen," *Hochfreq. und Elektro Akustik,* 1967, Vol. 76, pp. 189-197.

Hadady, R. and Bentley, R., "Low Flutter High Environment Recorders," *ITC Conf.* 1967, Vol. 3, pp. 637-659.

Warburton, G.B., and Edney, S.L., "Vibrations of Rectangular Plates with Elastically Restrained Edges," *Jour. of Sound and Vibration,* July 1984, Vol. 95, No. 4.

Caldersmith, "Vibrations of Orthotropic Rectangular Plates," *Acoustica*, Oct. 1984, Vol. 56, No. 2, pp. 144-152.

Vibrations in Disks

Pearson, R.T., "The Development of the Flexible-Disk Magnetic Recorder," *Proc. IRE,* Jan. 1961, Vol. 49, No. 1, pp. 164-174.

Greenberg, H.J., "Flexible Disk - Read/Write Head Interface," *IEEE Trans. Magn.*, Sept. 1978, Vol. MAG-14, No. 5, pp. 336-338.

Richards, D., "The Relationship between Disk Surface Acceleration and Head-To-Disk Interaction," *IEEE Trans. Magn.*, July 1978, Vol. MAG-14, No. 4, pp. 194-196.

Zajaczkowski, J., "Stability of Transverse Vibrations of a Circular Plate Subject to Periodically Varying Torque," *Jour. Sound and Vibr.,* July 1983, Vol. 89, No. 2, pp. 273-286.

Benson, R.C., "Observations on Steady-State Solution of an Extremely Flexible Spinning Disk with a Transverse Load," *Jour. Appl. Mech.,* Sept. 1983, Vol. 50, No. 3, pp. 525-530.

Priolo, P., and Sitjia, C., "Efficiency of Annular Finite Elements for Flexural Vibrations of Thick Disks," *Jour. Sound and Vibr.,* Jan. 1984, Vol. 92, No. 1, pp. 21-32.

Lennemann, E., "Aerodynamic Aspects of Disk Files," *IBM Jour. Res. Devel.,* Nov. 1984, Vol. 18, pp. 480-488.

Good, J.K., and Lowery, R.L., "The Finite Element Modelling of the Free Vibration of a Read/Write Head Floppy Disk System," *Jour. Vibr., Acoustics, Stress and Reliab. in Design (ASME),* May 1985, Vol. 107, No. 3, pp. 329-333.

Ohta, N., Naruse, J., and Hirata, T., "Vibration Reduction of Magnetic in Disk Drive Mechanism," *Bull. of the JSME,* July 1985, Vol. 28, No. 241, pp. 1489-1496.

Yamaguchi, Y., Takahashi, K., Fujita, H., and Kuwahara, K., "Flow induced Vibration of Magnetic Head Suspension in Hard Disk Drive," *IEEE Trans. Magn.,* Sept. 1986, Vol. MAG-22, No. 5, pp. 1022-1024.

Bouchard, G., and Talke, F.E., "Non-Repeatable Flutter of Magnetic Recording Disks," *IEEE Trans. Magn.,* Sept. 1986, Vol. MAG-22, No. 5, pp. 1019-1021.

Measurements of Vibrations in Disks

Wiezien, R.W., Miu, D.K., and Kibens, V., "Characterization of rotating flexible disks using a laser Doppler vibrometer," *Optical Engr.,* July/Aug. 1984, Vol. 23, No. 4, pp. 436-442.
Lewin, A.C., Kersey, A.D., and Jackson, D.A., "Non-contact Surface Vibration Analysis Using a Monomod Fibre Optic Interferometer Incorporating an Open Air Path," *Jour. Phys., E (Scient. Instr.),* July 1985, Vol. 18, No. 7, pp. 604-609.
Millman, S.E., Hoyt, R.F., Horne, D.E., and Beye, B., "Motion Pictures of In-Situ Air Bearing Dynamics," *IEEE Trans. Magn.,* Sept. 1986, Vol. MAG-22, No. 5, pp. 1031-1033.
Best, G.L., Horne, D.E., Chiou, A., and Sussner, H., "Precise Optical Measurement of Slider Dynamics," *IEEE Trans. Magn.*, Sept. 1986, Vol. MAG-22, No. 5, pp. 1017-1018.
Yeack-Scranton, C.E., "Novel Piezoelectric Transducers to Monitor Head-Disk Interactions," *IEEE Trans. Magn.,* Sept. 1986, Vol. MAG-22, No. 5, pp. 1011-1016.

Electromechanical Analogies

Beranek, L.L., *Acoustics*, McGraw-Hill, New York, January 1976, 481 pages.
Hittle, C.E., "Twin-Drum Film Drive Filter System for Magnetic Recorder-Reproducer," *J. SMPTE,* April 1952, Vol. 58, pp. 323-328.
Bixler, O.C., "Mechanical Components for Handling Magnetic Recording Tape," *Trans. IRE,* January 1954, Vol. AU-2, pp. 15-23.
McKnight, J.G., "Mechanical Damping in Tape Transports," *J. AES,* April 1964, Vol. 12, pp. 140-146.
Wolf, W., "Electromechanical Analogs of the Filter Systems Used in Sound Recording Transports," *IEEE Trans. AUDIO and ELACOUST.,* June 1966, Vol. AU-14, No. 2, pp. 66-85.
Cole, Kevin A., "Application of Transport Dynamic Model to the Head/Tape Interface," *Presented at THIC Meeting,* Jan. 1986, 21 pages.

Chapter 18

Head/Media Interface

This chapter will describe and discuss the following topics:

- Signal losses.
- Friction.
- Flying heads.
- Wear of Magnetic Heads.
- Wear Mechanisms.
- Test Methods.
- Head Wear.
- Changes in Head Performance, after wear.

The head/medium interface is the kernel in any tape or disk system: Here the signal is laid down as a magnetic pattern, and later picked up during reproduction.

High storage capacity of a recording system is achieved by recording many bits per linear inch of tape along a track (named BPI - bits-per-inch), and many tracks side-by-side (named TPI - tracks-per-inch). A good short bit length performance requires a minimum spacing between the head and medium surface; zero spacing would be preferable in order to minimize signal losses.

SIGNAL LOSSES

Several signal losses occur during the magnetic write and read processes. The read voltage can be analyzed, and its mathematical expression will disclose three such losses (see Chapter 6):

Coating thickness loss : A $= (1 - e^{-2\pi c'/\lambda})/(2\pi c'/\lambda)$
Spacing loss : B $= e^{-2\pi d/\lambda}$
Read gap loss : C $= (\sin x)/x$ where $x = \pi L_g/\lambda$

The reader will note that nowhere in these expressions do we find the symbol f, used for frequency. The losses are dependent upon dimensions only—i.e., the recorded thickness c′, the spacing d between the head and the medium surface, and the gap length L_g. The recorded wavelength λ is equal to two bit lengths, and relates to a recorded frequency f as $\lambda = v/f$, where v is the head-to-track speed in m/s (IPS).

Only a portion of a recorded coating produces a flux. This is evident from the loss term called *coating thickness loss*. The formula for the loss can be reduced to as a simple rule: 75 percent of the flux that contributes to the read head voltage comes from an effective coating thickness c_{eff} that equals 0.44 times the bit-length:

$$c_{eff} = .44 \bullet BL. \tag{18.1}$$

If the bit-length BL = 3 μm (120 μin), then the coating need not be thicker than 0.44 3 = 1.3 μm (53 μin). It would be detrimental to make it thicker since a thicker coating leads to another type of loss, called demagnetization losses (see Chapter 4).

A distance between the head and the medium will cause a *spacing loss* that can be expressed as:

$$\text{Loss} = 55 \bullet d/\lambda \text{ decibels} \tag{18.2}$$

where d is the effective head/medium spacing, and λ is the recorded wavelength (equal to two bit lengths). This formula was found by taking 20 times $\log_{10}$ of the spacing loss factor, a convention often used by EE's to keep numbers reasonable, and in accord with the way many things in nature change (for example: objective hearing levels in relation to sound pressure levels).

The waveforms of recorded bits are basically rectangular patterns, that can be decomposed into a fundamental signal with a number of odd harmonics. The latter provide for the sharp transition times and the square corners in the waveform. They are of shorter wavelengths than the fundamental, and therefore have greater spacing losses. The net result of the spacing loss is a signal attenuation combined with decreased resolution.

In-contact operations are therefore desirable, such as in tape and diskette (floppy) drives. It requires a tape coating having a perfectly flat surface; this can cause problems: binder debris may buildup on the head and encourage stick-slip motion of the tape or disk. The result is time base errors and potential rapid head and/or media wear.

High data rates require a high velocity between head and medium, and in-contact operation is not always feasible. Rigid disk drives, currently of the Winchester or Whitney class, have heads hovering over the rotating disk surfaces at distances of 0.4 to 0.2 μm (16 to 8 μin).

Cleanliness and freedom from airborne dust particles are important in both systems. Many particles are large enough to make scratches in head or medium surfaces, and to cause head crashes. The latter is a catastrophic failure; Winchester/Whitney drives are therefore sealed units to keep dust particles out.

The *gap loss* leads to a simple relation between the required gap length and the bit-length: A long gap is desired for high head efficiency; but the signal output is zero when the gap length equals two bit lengths. The gap length is therefore made equal to or less than one bit length:

$$\text{Gap length} \leq \text{Bit length.} \tag{18.3}$$

An excellent review of the head-media signal transfer is presented in a paper by Yeh.

FRICTION

The head/media interface presents an environment of its own. There are frictional forces and wear processes. The head may be in contact with the medium or spaced away by an air film.

The engineer or technician that seeks solutions to problems regarding head wear or tape/disk abrasivity will benefit from studying the interdisciplinary topic of *tribology*. This term covers all sorts of mechanical and chemical interface problems between two contacting surfaces, at stand-still or sliding past one another. The books by Rabinowitz and by Halling are both recommended for an introduction to tribology.

It is common for tape and flexible diskette coatings to contain minute amounts of lubricant. This tends to make the drag force proportional to the contact pressure as well as to the velocity. When the velocity is very low there may be isolated contact points where the lubricant film is broken and the frictional forces increase drastically. This results in a coefficient of friction versus head/medium speed that starts at a high value at zero speed, decreases to a minimum around 2.5 cm/s (1 IPS), then rises slowly to a maximum in the vicinity of 100 cm/s (40 IPS), and then decreases to zero at speeds near 200 cm/s (80 IPS). At higher speeds a hydrodynamic air bearing is formed between the tape and the head. A characteristic curve for the coefficient of friction versus speed is shown in Fig. 18-1. The negative slope at velocities less than 1 IPS gives rise to a ''stick-slip'' motion (ref. Ristow, Kalfayan, and Mukhopadhyay).

Friction is needed in the capstan nip area to grip a tape and move it through a tape transport. Typical values for the coefficient of friction of an oxide surface are, at a speed of 25 cm/s (10 IPS) (at 18° C, 40 percent RH):

μ_{static} against rubber $\cong$ 0.8
$\mu_{dynamic}$ against rubber $\cong$ 0.35
μ_{static} against aluminum = 0.1 − 0.2
$\mu_{dynamic}$ against aluminum = 0.05 − 0.1

The values vary with binder formulation, and they all show an upward trend after many passes, and with increasing humidity.

Graphite is often used as a solid lubricant since it offers the additional benefit of lower electrical resistance. A tape may act like a Van de Graaff generator, which discharges the accumulated electrostatic charge at the head. This can result in noise in the read signal, or erratic frictional behavior (ref. Schnurmann).

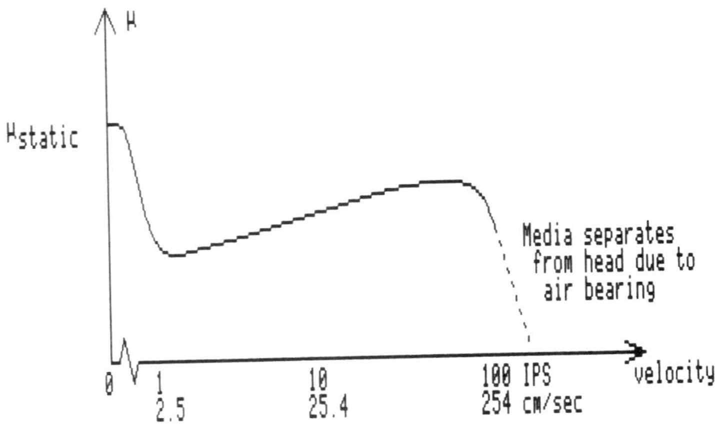

Fig. 18-1. Typical variations in coefficient of friction at the head/media interface.

Media surfaces contain hard particles that are highly abrasive: γFe_2O_3 or CrO_2. The resulting wear of the magnetic heads presents a costly maintenance problem, and the selection of a low-wear tape is a difficult task. Experience is still the master in this matter, although attempts constantly are made in order to find a quick and reliable test for the abrasion of a medium (see later).

It is possible to compute and predict the contour of a head, after wear (ref. Clurman), and to optimize video head contouring (ref. Hahn).

The wear of heads presents more of a maintenance problem than the mere replacement of worn heads. The electrical characteristics of the head change with wear: the reduced gap depth increases the efficiency. This means that less write current is required, and the read voltage from a worn head is higher than that from a new head. The increase in write efficiency will cause a changed magnetization pattern in the coating, and hence a changed read voltage wave form. The electronics must either be tolerant to these changes, or the system must be calibrated from time to time.

Foreign dust particles and media surface imperfections can cause scratches in heads and/or drop-outs (A drop-out is the temporary loss of signal). Examinations of media defects have in the past disclosed imbedded dirt particles, resulting in further demands upon cleanliness in the "wet end" of media coating facilities. As a result, modern tapes and disks are essentially drop-out free; drop-outs are introduced by the users!

Measurements of drop-out activities in many recorder installations under varying operating conditions have revealed that loose dust particles are serious generators of drop-outs. The use of reels with solid flanges, instead of flanges with spoke openings, netted an order of magnitude reduction in drop-out count (the "Wilmot" effect, see Chapter 14).

Video recorders, using rotating head assemblies, operate with very high head-to-media speeds, in the range of 3 to 25 m/s (120-1000 IPS), or 11 to 90 km/hour! It is generally recognized that an air film forms between a head and a medium surface when the speed is greater than 2 – 3 m/s. This means that a spacing loss will occur in most video recorders if measures were not taken to force the head approximately 50 μm (2 mils) into the tape on the units with highest speed, lesser at lower speed (tip penetration). That explains why still framing causes severe wear and deterioration of the tape. The associated temperature rise can be measured (ref. Sakai), and head materials selected that will remain magnetic at that temperature (i.e., have sufficiently high Curie temperature).

THE FLEXIBLE DISK/HEAD INTERFACE

The heads for diskette drives are single channel structures imbedded in a housing that mounts on a movable arm, together forming a so-called slider assembly. The head surface is rounded and the heads are named button heads.

The single-sided disk application is straightforward, as shown in Fig. 18-2, top. The double sided disks require a head on both sides. This is best done as shown in Fig. 18-2 center, where the top head is mounted on a lever that is part of the arm system. In early designs both heads were mounted on springs. That caused numerous failures due to vibrations in the head/diskette mechanical system.

The vibration problems were aggravated by flexures of the disk itself: it would undulate 2 to 3mm above and below its centerplane as it revolved, much like a warped phonograph record on a turntable (ref. Klevesahl and Stromsta).

The vibration problem is made worse by the dimple in the disk formed by the head. This contact problem has been analyzed in the past (Charbonnier, Greenberg, Adams, and Licari and King). It is unfortunate that all the work was done by spinning a disk in close proximity of a rigid back plate, with the head protruding through a hole in the plate. The plate served

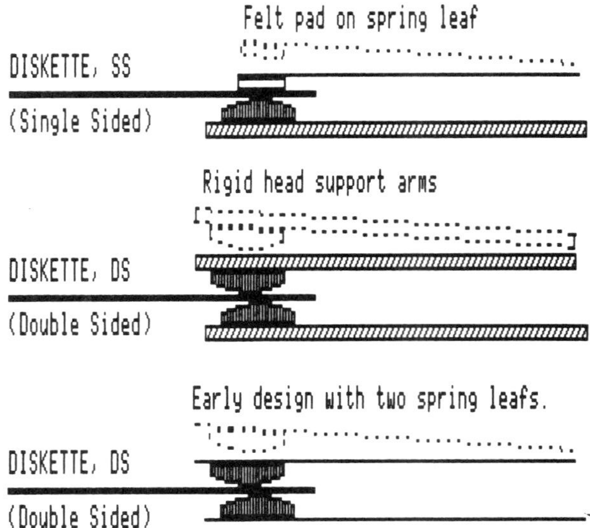

Fig. 18-2. Heads for single-sided (SS) and double-sided (DS) diskettes.

as a stabilizing platform for the disk. This method eliminated the real disk vibrations that otherwise occur when they are rotating in their jackets only (see Chapter 17 under Vibrations).

THE FLYING HEAD INTERFACE

Modern computers are insatiable with regard to the speed with which they wish to write and read data, in bit rates measured in bits-per-second (BPS). This will, in turn, require higher head-to-media speeds $v = \lambda \cdot f = BL \cdot BR$, where BL is the bit length, and BR the bit rate. The required speeds are above the 2-3 m/s threshold for the formation of an air film, and spacing losses will occur. Head-medium contact could be re-established by brute force (as done in the video recorders), but this would in a disk file quickly lead to excessive heat generation, and destruction of the head and disk surfaces.

The head assembly is therefore designed to fly over the disk surface, originally at a distance of 3 μm (120 μin). And the rpm of the disk is chosen high enough to allow for bit lengths that are long enough so spacing losses are small. In this fashion the rigid disks were born, at first using disks of diameter = 35.6 cm (14″).

The air bearings have been studied and described in several papers in the IBM Journal of Research and Development, November 1974 issue. A foil-bearing theory for the tape-head interface is described (ref. Stahl), together with measurements (ref. Vogel). A like treatment is made for the disk-head interface (Tseng), along with a technique for the measurement of flying height (ref. Fleischer).

A frequency domain analysis for the dynamics of a slider bearing is important for the design and appraisal of the entire disk/head/arm system (ref. Ono).

The flying attitude of the air lubricated head platform with sliders has only recently been analyzed. The structure is shown in Fig. 18-3, and definitions of certain key parameters are shown in Fig. 18-4.

The sliders behavior during the first couple of milliseconds after start-up is illustrated in Fig. 18-5. The vibrations do settle rather quickly, and the flying height is thereafter determined by the speed, the load force and the width of the slider bearing surfaces (see Fig. 18-6).

Any fluttering of the disk will excite head vibrations. An analysis which relies upon linear superposition is questionable due to a very complex behavior of the air (gas) film between the

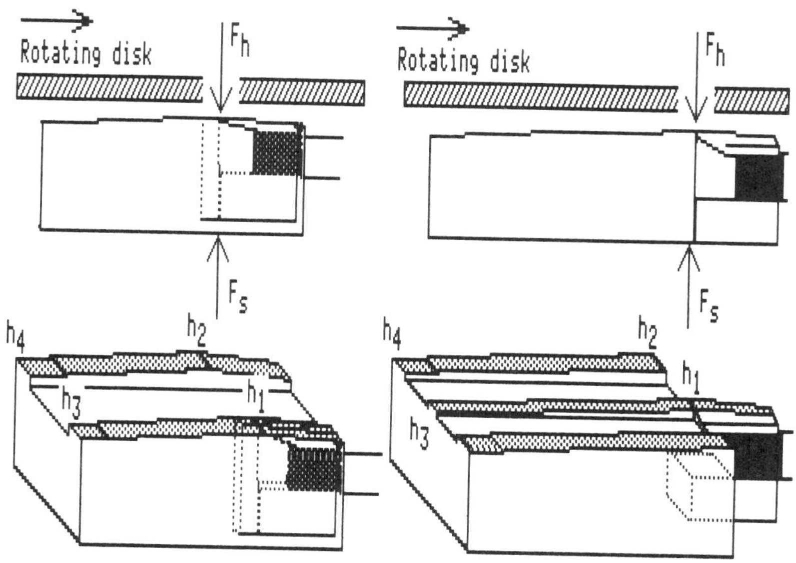

Fig. 18-3. Side view of a flying head, loaded with the slider arm force Fs and lifted by the air bearing force Fh. The head core is located at h1.

sliders and the disk surface. The usual modelling of the gas film is not valid for spacings less than 0.5 μm: In this operating region, the head clearance dimension is only a few molecular collisions.

Experimental observations must therefore provide us with most of our knowledge. A typical test set-up is shown in Fig. 18-7, where a head is loaded against a rotating glass disk. By shining light onto the head surface a set of fringe patterns develops, allowing for the determination of flight height and flying attitudes (ref. Ragle et al., Mizoshita et al.). The technique

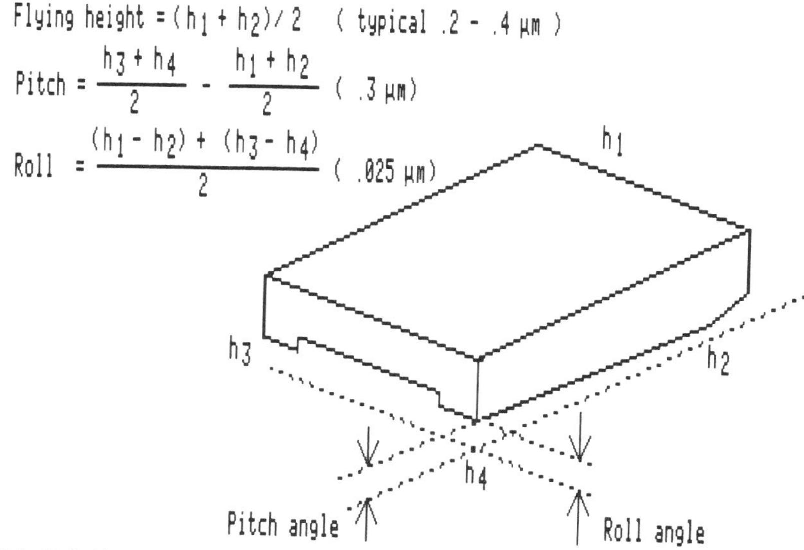

Fig. 18-4. Definitions of Flying Height, Pitch and Roll.

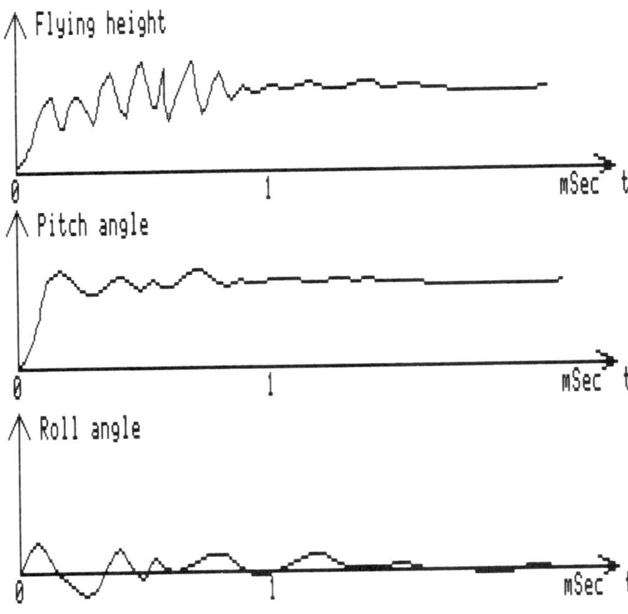

Fig. 18-5. Changes in the flying height, pitch angle, and roll angle during the first 2 milliseconds after disk start-up.

has been extended to Laser doppler interferometry which allows for the use of an ordinary coated disk (ref. Miu et al.).

These observations gain further value when information about the slider arm vibrations are added. Finite element analyses are reported in several papers (ref. Tagawa et al., Mizoshita et al.).

Additional information is available from a miniature accelerometer, mounted on the slider arm. It measures the noise generated by acoustic emission (AE), which are high frequency

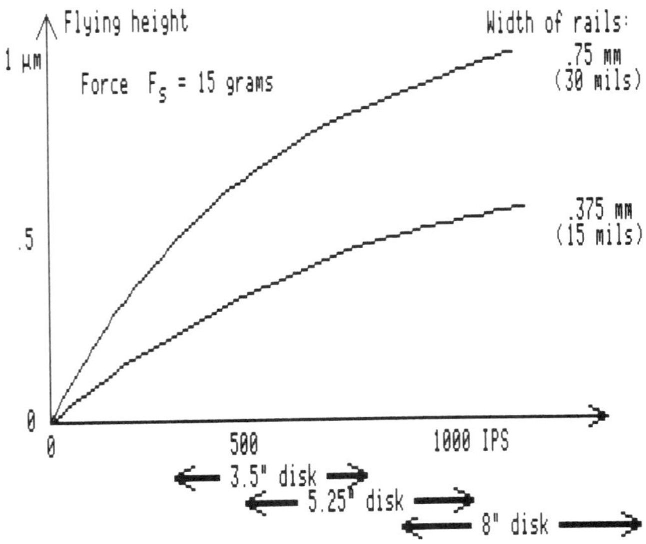

Fig. 18-6. Flying height as a function of speed along track; shown for two slider widths, 30 mils and 15 mils (courtesy Tranetics, Inc.).

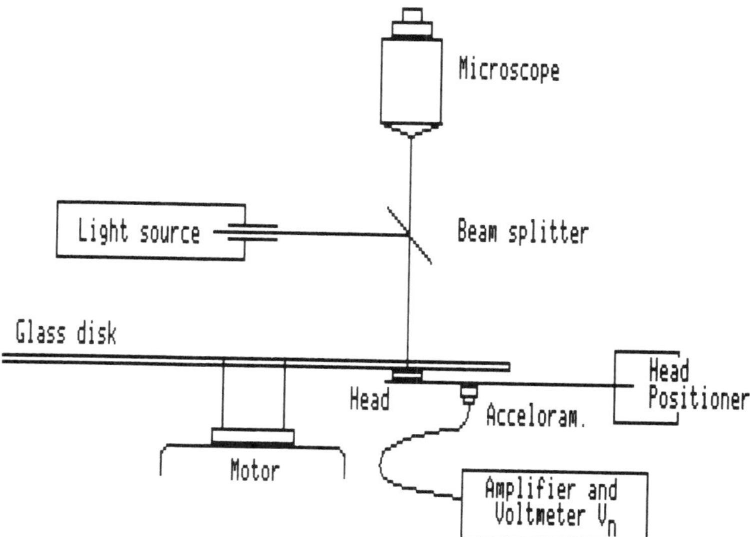

Fig. 18-7. Observation of flying attitude by means of fringe rings, and accelerometer pick-up of high frequency vibrations in slide arm, caused by head/disk interface friction.

elastic waves generated by the deformation or destruction of solids. AE is generated by the slider-to-disk contact (ref. Kita et al.).

The AE rms voltage and slider flying height are shown in Fig. 18-8. There are four distinct ranges: In range I the whole flat part of the slider touches the disk surface continuously. In range II the leading edge of the slider begins to fly, and in range III light measurements indicate that the slider flies without contact; obviously there are occasional collisions between the slider and projections of the disk surface.

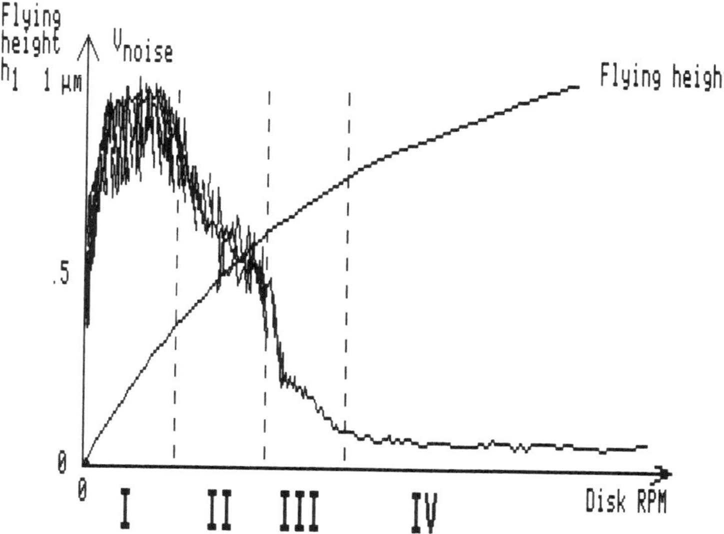

Fig. 18-8. Correlation between flying height and high frequency noise pickup (after Kita et al.).

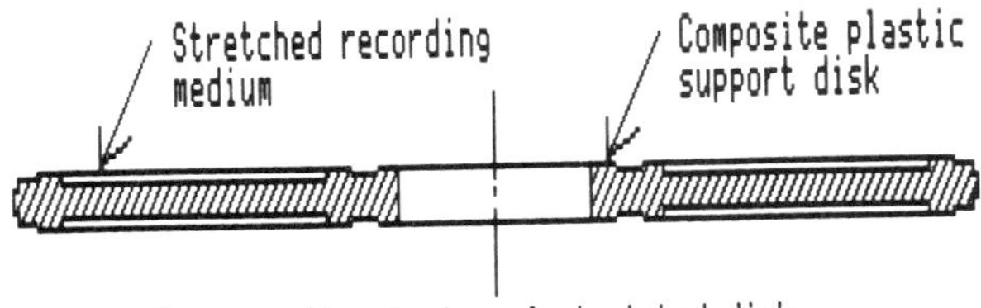

Fig. 18-9. Stretched membrane disk (3MCo.). Head is actually flying over surface (after Knudsen).

The fearsome head crash problems have been analyzed in two papers (ref. Kawakubo et al., Kumaron et al.). There is no easy cure: assure utter cleanliness, air filtration, perfect disk surfaces without large projections, good parking zones for the head; and avoid pressure from the marketing department.

A couple of applications use a flexible media that rotates at a high enough RPM so an air film is formed, and the head assembly is flying. One is the cartridge from IOMEGA, that has a flexible disk well protected within a plastic housing. It therefore functions as a low cost removable disk, with a storage capacity approaching rigid disks.

A similar approach is found in 3M Co's stretched membrane disk, see Fig. 18-9 (ref. Knudsen). This concept reduces the variables in temperature and humidity changes, and a higher number of parallel tracks-per-inch is therefore possible.

WEAR

Excessive wear of magnetic heads and/or disks is a problem that often occurs when a new recording medium or a novel medium drive is introduced, or when operating conditions are changed. Experience in the field of abrasion and wear does not always allow accurate prediction of performance, and each system must be evaluated separately.

Any vibrations in the head/medium interface area will aggravate wear, as will high humidity and high temperature. The chemistry of the coating binder appears crucial, and it is now common practice to add minute amounts of an abrasive powder (alumina) to the binder in order to abrade the head free from any debris formation.

This section will briefly review the fundamental wear theories, and analyze how well they relate to the wear of magnetic heads and media. Various test methods are described, and results from these should be useful in future developments of media and/or drives.

The Nature of Wear

The technology of wear is part of *tribology*, the science and technology of interacting surfaces in relative motion. It is without the exact laws of other sciences, and the guidelines set forth in this chapter are just that: Suggestions for an optimum design, based upon past data and experiences.

There are four generally accepted modes of wear:

A. Adhesive Wear. This process involves the joining, or welding of small interfering projections in mating surfaces, where material then is torn away from the weaker material by continuing motion.

The contacting surfaces touch each other only at high points, where plastic deformations take place, as shown in Fig. 18-10. The real area of contact between a polymeric magnetic medium and a rigid surface has recently been investigated and appears to be elastic in nature (ref. Bhushan).

Adhesion will clearly occur in cases where head and medium have been in prolonged stationary contact. Spacecraft recorders have this problem if left unattended; the tape drive must be exercised regularly.

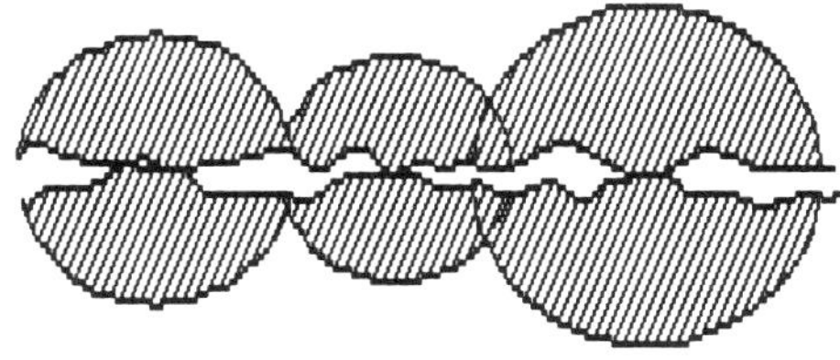

Fig. 18-10. Formation of plastic zones under the high pressure contact points between two surfaces.

Adhesive wear is also present in normal recorder operation, where continuing sliding causes the junctions to be sheared and new junctions formed. This leads to an oversimplified theory of friction where the coefficient of friction is equal to the shear strength divided by the yield pressure. With further rubbing some of the transferred material is detached to form loose wear particles.

The wear volume, per unit distance of sliding is given by (ref. Archard):

$$Q = kW/3p_o \qquad \textbf{(18.4)}$$

where

k = probability of an asperity contact producing a wear particle,
W = total load,
and p_o = yield pressure.

This law of wear is found to be true for a wide range of conditions, and expresses that the volume of wear material is:

- proportional to the distance of travel,
- proportional to the load,
- inverse proportional to the yield stress, or hardness, of the softer material.

At high loads we will find plastic zones below the contacting surfaces, as shown in Fig. 18-10. This leads to large scale welding and seizure, and the wear rate increases dramatically, as indicated in Fig. 18-11. This occurs when the load pressure increases beyond one third of the hardness of the softer material. A couple of tape abrasion tests operate in this region, where the pressure is in excess of normal tape transport operation (Shim test, Ampex test element, when new); results from these tests are therefore highly questionable.

B. Abrasive Wear. Basically involves various cutting actions in which a hard material, sliding under pressure across a weaker surface, tends to plough the weak surface. There are many variations of this basic action, depending on surface size, shape, velocity, pressure, number of passes and the properties of the materials.

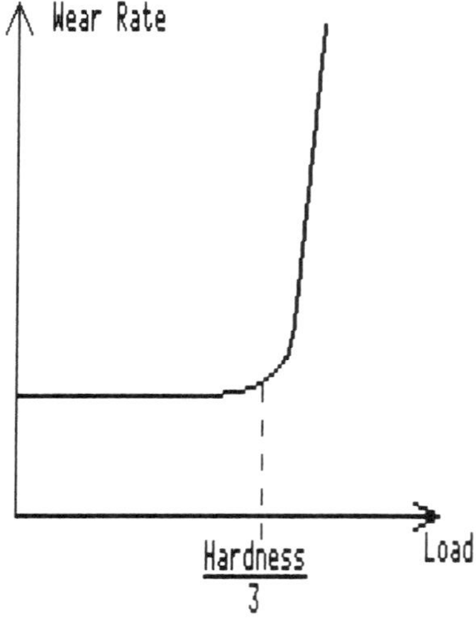

Fig. 18-11. Wear rate as function of load pressure.

Many abrasive situations are covered by an equation of the form:

$$Q = k_a W/H \qquad \textbf{(18.5)}$$

where

k_a = an abrasive constant,
W = total load,

and

H = hardness of the softer material.

This mode alone is not responsible for wear of magnetic heads. Figure 18-12 illustrates the results of a test carried out at the Jet Propulsion Laboratories in the late sixties, showing the wear pattern left after two different tapes were run over the head. The scalloped patterns show that there was no correlation between wear of the cores (wide laminations) and the shields (narrow laminations) and their hardnesses!

Further: The operating environment for many tape and disk drives are not exactly clean room conditions, and dust from the atmosphere may be a major cause of abrasive wear. This fact may contribute to the spread in wear data measured for a certain tape type, but under different conditions and by different methods.

C. Corrosive Wear. Involves corrosion and abrasive wear, either combined or alternate. Both involve removal of surface material; corrosion by chemical action (such as oxidation), wear by mechanical action.

D. Contact Stress Fatigue. Concerns the mechanism of fatigue in removal of comparatively large particles from surfaces under high contact pressure, combined with a degree of sliding and rolling.

Wear of Magnetic Heads

With so many ways to remove surface material by mechanical action, it is not surprising that the analysis of wear failures is difficult. Knowledge and study of the surface conditions

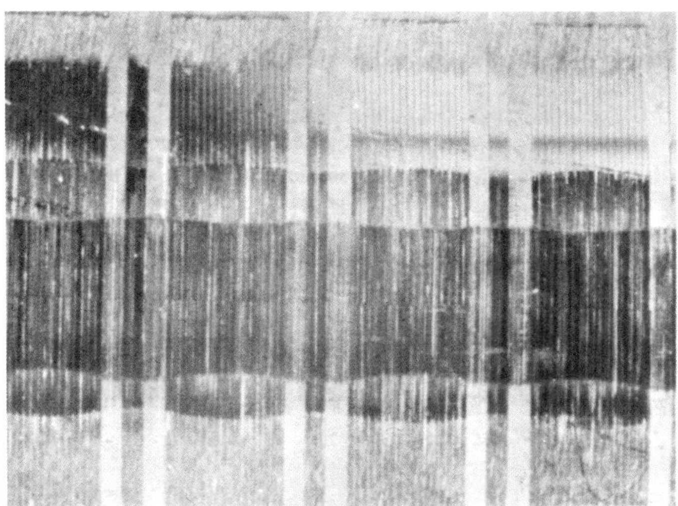

Fig. 18-12. Tape head surface showing reversal of wear intensity by changing from tape X to tape Y (courtesy JPL).

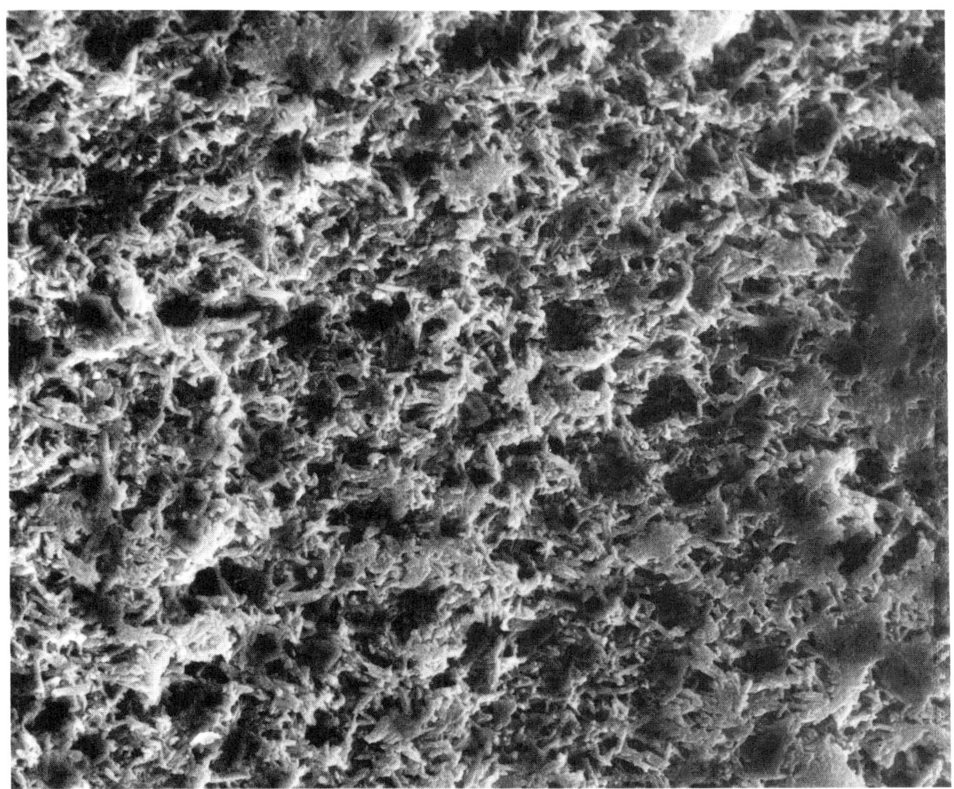

Fig. 18-13. Tape surface, untreated (courtesy Pfizer Pigments, Inc.).

Fig. 18-14. Tape surface, calendared and polished (courtesy Pfizer Pigments, Inc.).

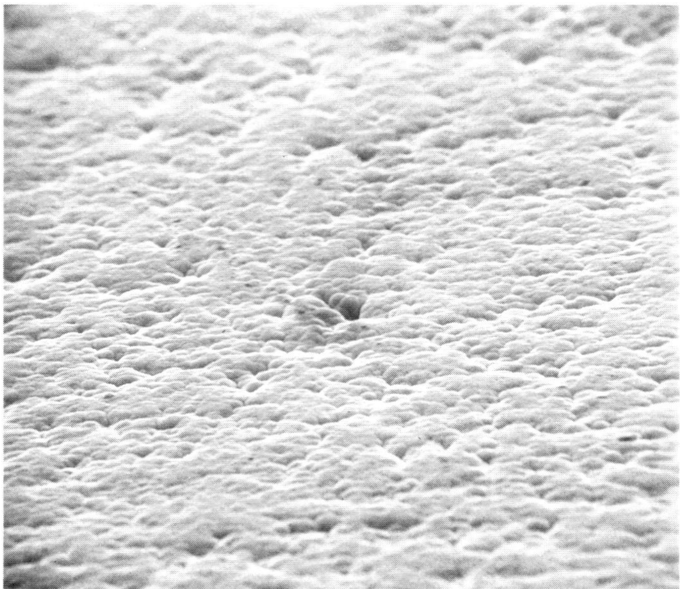

Fig. 18-15. Surface of metal coated tape (courtesy JPL).

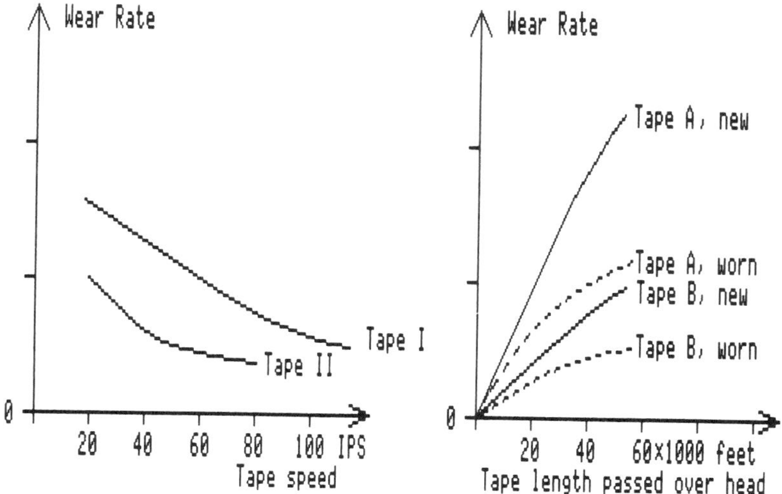

Fig. 18-16. Wear rate is a function of tape type, and reduces as the tape is worn.

causing wear are obviously necessary. A proper analysis will also depend upon careful visual examination, on both the macro and micro scales, as well as upon consideration of a number of factors such as hardness examination and possible chemical interactions. Figures 18-13 and 18-14 illustrate a tape surface before and after surface treatment. Figure 18-15 shows the surface of a metal coated tape (sputter process).

Abrasive wear and some amount of corrosive wear are largely responsible for head wear. The adhesive mode is observed in the *stick-slip* motion of a tape at slow tape speeds (½ IPS and slower), and may in part be the cause of changes in frictional forces (drag) which vary greatly for different coatings, different head materials and different environments. The coefficient of friction also increases with the number of passes whereby head and coating develop wear patterns that are replicate of one another with resulting larger contact area and reduced wear, see Fig. 18-16.

Abrasive wear is also observed as severe scratches on head and coating surfaces, caused by debris formation or foreign particles. This mode can lead to a destruction of the otherwise perfectly straight gap, and occasionally drags material across the gap (*gap-smear*), and causes a magnetic short of the head. This failure mode can only be remedied by re-surfacing the head by passing a highly abrasive tape across it.

EARLY TAPE ABRASION TEST METHODS

Abrasivity causes removal of material from the head surface, and test techniques have therefore aimed at measuring the quantity removed. One early method was reported by Rettinger in 1955 (see ref.). The head wear rate was determined by optical measurement of the reduction in the front gap depth. He found that the *wear rate was proportional to the pressure between the tape and the head*, which later techniques have confirmed.

In 1963, H.U. Ragle and E.D. Daniel (see ref.) described a method of determining the amount worn off the edge of a shim of suitable material held against a moving tape with a given force. The amount of wear was continuously measured by monitoring the output of a small reproducing head mounted to a pivoted arm assembly, which moves closer to the tape as the shim wears. The depth of the shim that was worn off, and hence the volume of material, was accurately calculated using the well established spacing law governing the reproduction of a sinusoidal signal, see formula (18.2). With a recorded wavelength of .55 mils and a spacing change of .1 mils, the corresponding output change is 10 dB.

Tapes were at that time rather highly abrasive and the method was helpful in screening tapes; the contact pressure was 200 psi, about 50 times the normal head-coating pressure—a limitation recognized by the authors.

In 1966 Carroll and Gotham (see ref.) published a paper dealing with an abrasion test utilizing a metal rod of diameter ½ inch. Different materials were used for the rod, with emphasis on the 80-20 Ni-Fe alloys (Permalloy, MuMetal etc.). A rod was carefully polished and cleaned, weighted, and installed on a tape transport. Then, after passage of several thousand feet of tape, the rod was removed and again weighted. The weight loss represented the material removed, proportional to the abrasivity of the tape sample as expressed in milligrams per 1000 feet.

A couple of their results are shown in Fig. 18-16, illustrating the progressive wear of two samples from which we can learn two things: the decreasing slope of the four curves shows a decrease in abrasivity with increasing number of passes (this is found in general usage of tapes and disks also), and the first pass is most abrasive. The explanation for this may be that high points on the tape/disk are worn off, and voids in the coating surface are filled up with wear products. This was particularly true for a short loop sample, and when the coating and head surfaces eventually conform to each other, then the pressure per unit area reduces and so does the wear rate.

The effect of tape speed is also shown in Fig. 18-16. It is unfortunate that these measurements were done at the high pressure of 35 psi, or we should have seen a reduction of wear to a very low value at 60-80 IPS (and above), where tapes actually start to fly over the head surface (which acts as an air bearing). (Their measurements of friction forces at the too low pressure of 0.4 psi did show this effect.)

Of further interest in Carroll and Gotham's work was the investigation of humidity effects. They found that the wear rate of heads increased with the relative humidity of the environment. However, the humidity conditions of the tape was found to have little effect.

They also reported that the 80-20 Ni-Fe alloys responded strongly to humidity conditions, while aluminum or copper showed little or no response. Brass showed approximately half the humidity response of MuMetal. These findings suggest that the presence of humidity causes corrosive wear.

In 1968 Kolb and Perry (see ref.) reported on wear of permalloy heads against striped motion picture film, and expressed the wear in a formula:

$$dA/dL = K_f K_h P^x M^y V^z \qquad \textbf{(18.6)}$$

where A = loss of metal per unit width of the head and L = length of stripe run against the head. K_f and K_h are abrasiveness factors characteristic of the film and head respectively. P, M and V are pressure, contact area, and film velocity, respectively.

Kolb and Perry found the following values for virgin film stripes against permalloy heads:

$$x = 1.1$$
$$y = 1.0$$
$$z = -0.3$$

A more sophisticated technique was reported by Buchanan and Tuttle (see ref.), also in 1968. Their method, called the Radicon test, employed a special transport and a radioactive dummy head. As the tape passes over the neutron activated Kovar head its surface is abraded, and the wear particles become imbedded in the tape surface. The tape sample (50-200 feet) was removed and placed in a special chamber for counting its radioactivity, which is proportional to the amount of material worn, indicating the tape abrasivity.

CURRENT TAPE ABRASION TEST METHODS

Thin Film Element Test

An abrasion test element is fabricated by deposition of a thin metal film resistor onto a curved substrate (ref. Jorgensen, Williams); it may be a straight pattern, or meander shaped to provide better insensitivity to deep scratches. The element's resistance value is inversely proportional to its thickness, or height. This dimension will be reduced during the passage of an abrasive medium over its top, see Fig. 18-17.

The element's resistance value can be monitored while a tape is passed over it, and can be calibrated to provide a dynamic reading of the wear rate. The method is repeatable and inexpensive, and it is easy to realize an abrasive ranking on a small sample of only 100 feet of tape.

The results of wear rates correspond closely to what is experienced with actual head wear measurements (see next section). Yet it can be argued that the metallurgical structure of the deposited films are different from the core materials found in actual heads. This argument has quite some weight since the abrasion of heads in high humidity appears to be a strong function of the head material, rather than depending on tape coating.

Ampex Test Element

A bar of square cross section with perfectly sharp edges has been evaluated by Ampex, see Fig. 18-18 (ref. Kelly). The material is an Al-Fe alloy, commonly used in hard-tipped heads. The tape will quickly wear a pad, with a width proportional to the tape wear. It is reported that this width stabilizes, after initial wear. The area must obviously increase in size such that the pressure falls below one third of the bar's hardness, ref. Fig. 18-11.

The element can be used on all edges, economizing its use. The method provides a quick measurement, but appears inaccurate; in one set of measurements it reversed the ranking of

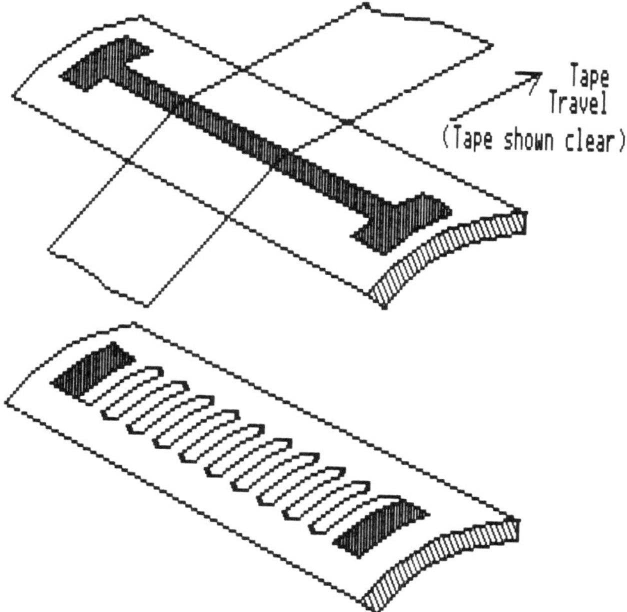

Fig. 18-17. Thin film wear test element.

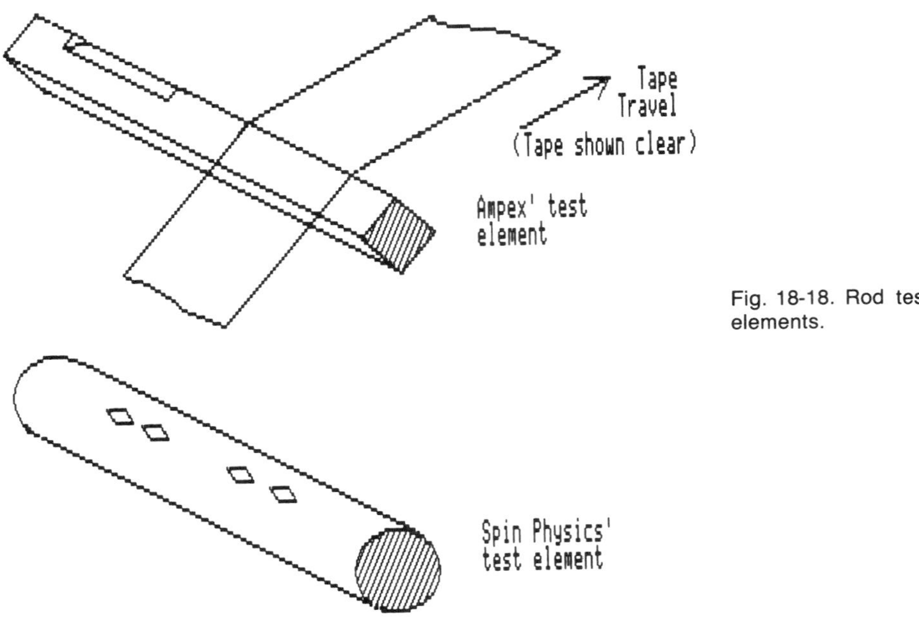

Fig. 18-18. Rod test elements.

two tapes (ref. Schultz). It operates, like the Daniel/Ragle test, with an unrealistically high tape pressure.

Spin Physics Test

A round test element is used and, like the Ampex element, fabricated from an Al-Fe alloy (Spin-Alloy). It is perfectly round, polished, and provided with diamond shaped indentations. Tape wear will reduce the visible area of the diamond shapes, and thus offer a measure of abrasion.

This element can also be used in several tests by merely turning it in its holder.

A similar element has been proposed by IITRI (ref. Schultz), and appears more accurate.

Philips-Ball Test Method

In this quick test a small ball is pressed against the backside of the tape, see Fig. 18-19 (ref. Groenou). When the tape moves between the ball and the head (real, or dummy) a round indentation will wear into the head surface. The method gives a rapid assessment of the tape sample's abrasive properties at high pressures, which occurs in video recorders where the head penetrates the tape by .5 to 2 mils.

RESULTS OF TAPE WEAR TESTING

Actual testing of head wear has been done by government agencies, tape and disk manufacturers, and reported by several equipment manufacturers (Ampex, Honeywell, Kodak Datatape) (ref. Levy). Oftentimes only first pass tests of tapes are valid, making the test procedure a very costly affair; hence the need for a test method with good correlation to actual head wear.

A summary of the results for actual head wear and data from some of the test methods is in place. This task was made difficult by the many different conditions and methods used, and units of measurement.

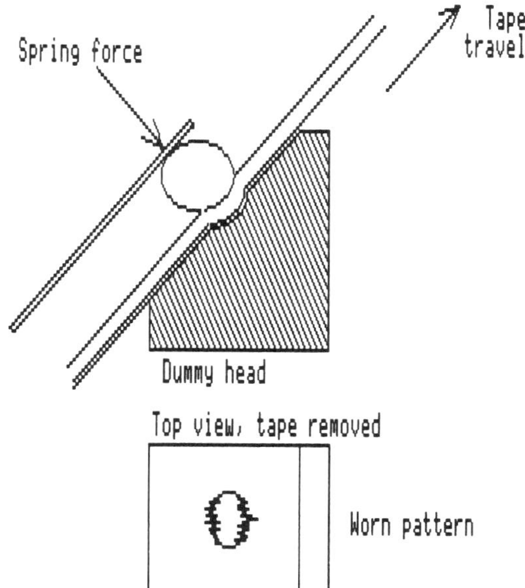

Fig. 18-19. Ball test method.

Effect of Hardness

Wear predictions based upon head core material hardness are unreliable.

This was evident from JPL's investigations (Fig. 18-12). Hahn also found that the degree of wear had no correlation to Knoop hardness (ref. 1975); There was no correlation for one tape, while some existed for another.

It does remain a fact that head cores fabricated from Al-Fe alloys wear about ten times slower than Ni-Fe cores (Mu-Metal, Permalloy etc).

This should hold also for the amorphous Met-Glass cores, that have a very high hardness? This has been the author's experience, but not others (ref. Ozawa).

The wear rate data for ferrites are scattered, and do not permit any summary. They vary from very good to very poor. Yamada reports that the rate is inversely proportional to the squareroot of the hardness (see ref.).

Effect of Load Pressure

The wear rate is proportional to the load pressure.

This was confirmed by Kolb and Perry in 1968. Clurman's analysis of head contours after the lapping effect (wear) of tapes correlates very well with actual observations; he assumed that the wear rate is proportional to the local pressure. He also found that the pressure contour is practically constant over the contact area.

Effect of Velocity

The wear rate is constant for velocities less than 10-15 IPS.
The wear rate decreases in inverse proportion to the speed above 15 IPS.

This observation has no theoretical counterpart. The reduced wear is due to the formation of an air bearing. Measured wear rates are shown in Fig. 18-20, using 3-4 degrees wrap angle and a tape tension of 8 oz per ½ inch wide tape.

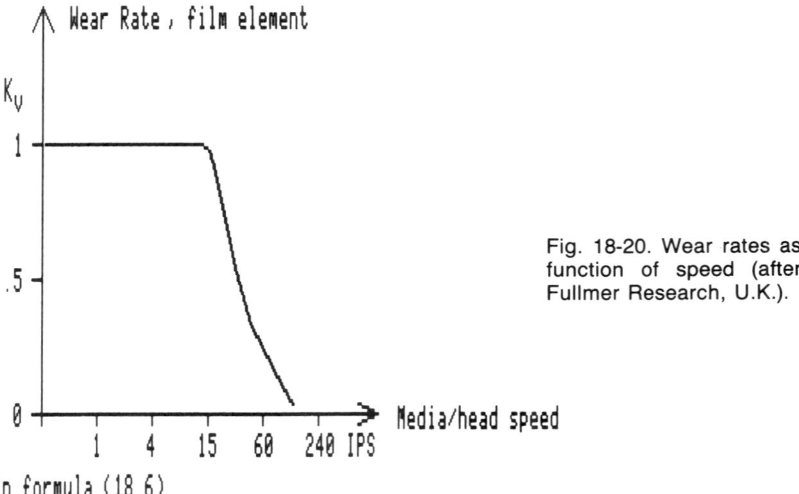

Fig. 18-20. Wear rates as function of speed (after Fullmer Research, U.K.).

Effect of Multiple Passes

The wear rate decreases after multiple passes, possibly by an order of magnitude.

The general data are shown in Fig. 18-21, and are the results of loop testing (ref. author's notes).

Effect of Humidity

The wear rate increases dramatically with relative humidity of the operating environment. The wear rate is affected only slightly by the humidity condition of the medium (tape or disk).

These findings are the results of extensive testing at JPL and Ampex (ref. Kelly, Levy). The results are shown in Fig. 18-22.

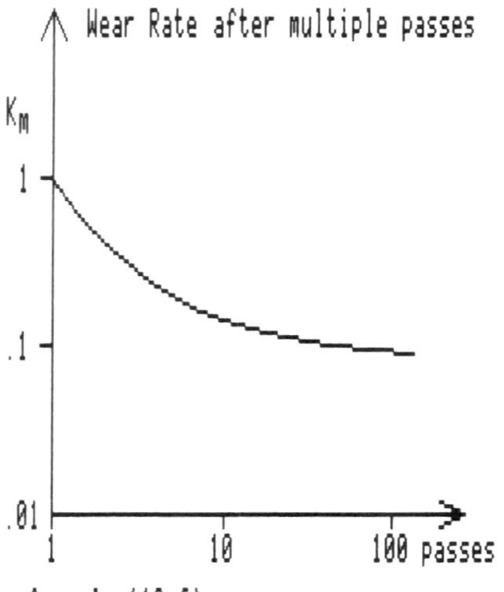

Fig. 18-21. Wear rates reduce after multiple passes.

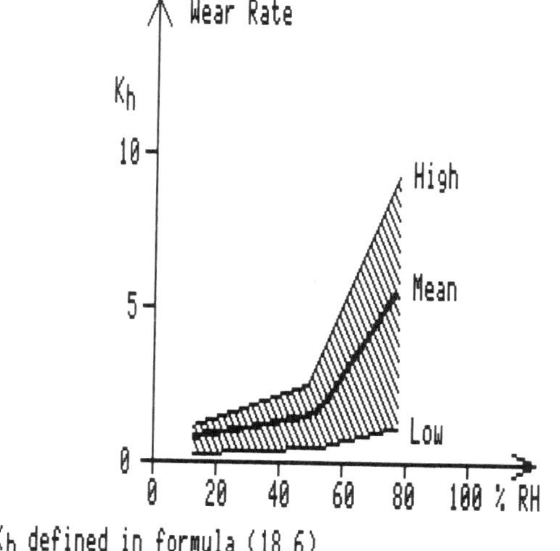

Fig. 18-22. Wear rate dependence on relative humidity. Curves for typical tapes fall within cross-hatched region (after Kelly, Levy).

CONCLUSIONS AND GUIDELINES

A wear rate unit of APF (*angstroms per foot*) has been selected to provide the reader with a convenient number for wear rates.

The chart in Fig. 18-23 summarizes the limited data found scattered in the literature, representing actual head wear measurements. Many sets of data from simulated tests were not applicable for this chart since they were in arbitrary units that either were non-convertible,

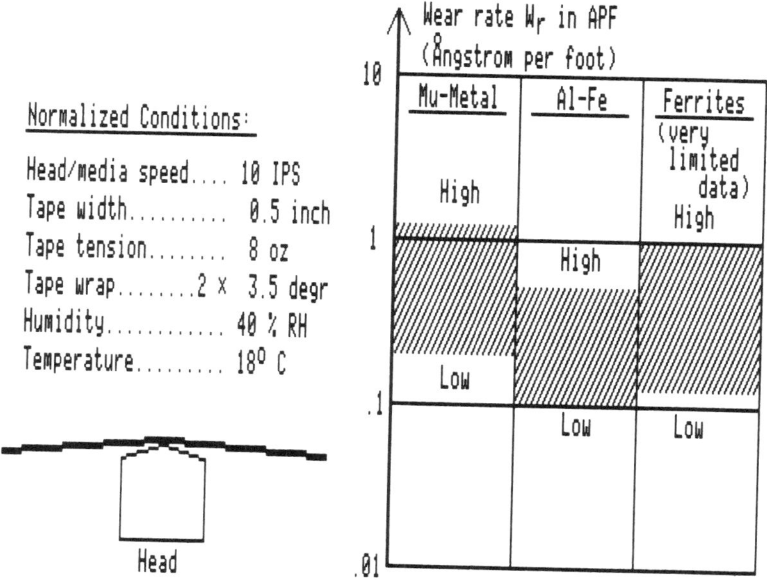

Fig. 18-23. Measured wear rates: normalized values and the range of values observed to date. Use formula (18.6) to calculate wear rate for conditions other than the normalized case.

or one or more test conditions were ill defined or lacking. No contradictory results were found, however.

Figure 18-23 will, in conjunction with the K-multipliers from Figs. 18-20, 18-21 and 18-22, allow for a *budgetary* estimate of head wear, from a formula:

$$\text{Wear rate} = W_r * K_t * K_v * K_m * K_h \qquad \textbf{(18.6)}$$

where

W_r = Wear rate from Fig. 18-23.
K_t = tension constant =
= (Tension in oz)$*$(tan α)$*$(0.5/tape width) or
= head-media pressure in oz.
K_v = velocity constant from Fig. 18-20.
K_m = repeated pass constant from Fig. 18-21.
K_h = humidity constant from Fig. 18-22.

Having decided that a model is better than no model the reader must use caution in applying formula (18.6). It does not include the severe wear problems that occur when a diskette drive starts to "sing," i.e., exhibit vibration problems that very likely are to be found at the head/media interface (remedy: use a cleaning diskette; repair drive; or replace). This vibration will aggravate wear just as a vibrating support for sandpaper accelerates the abrading action.

The author has found the formula quite good for estimating the wear of heads; and experience can only improve the model as expressed by the four graphs and formula (18.6).

Example 18.1: What is the estimated head wear produced by ten passes of a 30 minute audio cassette, of good quality (low abrasion)? A MuMetal head is used; tape tension is 2 oz, the width .15″, wrap angle 5°, and humidity conditions are 50 percent RH.

The tape tension constant is:

$$K_t = 2*(\tan 5°) * (0.5/0.15) = .58.$$

The tape speed is 1⅞ IPS, so $K_v = 1$.

After ten passes K_m has fallen to .2; the mean value for the *ten passes* is .7.

At 50 percent RH we read $K_h = 1.3$.

The wear rate from Fig. 18-23 is judged at approx. .1 APF, a low abrasivity tape; hence the estimated wear rate is .1$*$.58$*$1$*$.7$*$1.3 = .053 APF. The length of tape passed over the head for ten single passes is 10$*$2$*$30$*$60/12 = 3000 feet, and the total wear equals 3000$*$0.053 = 159 Angstrom = .016 μm = 0.64 μinches.

If the gap depth is 2 mils, then 10 cassette playings can be performed 2000/0.64. = 3144 times—or a total of 31,400 playings = 654 days of continuous playing. A poor grade tape (i.e., no brand name type) may, on the other hand, wear out the head 10 to 100 times as fast.

CHANGES IN RECORDER PERFORMANCE WITH HEAD WEAR

Changes in the sensitivities and impedances of magnetic heads will occur when the front gap depth is reduced due to wear from magnetic tapes. The changes are, in general, so small that they only develop after many passes of reels of tape.

The trend is always an increase in sensitivity and a decrease in impedance. They should therefore be considered in programs that assess tape interchangeability and certainly in any operation pertaining to standards work.

When the extent of changes versus wear have been established for a particular family of heads, then there may be additional steps to be included in a recorder maintenance schedule.

The 1980 edition of this handbook showed several graphs that illustrated what happens to a head's read efficiency, write current requirements and impedance levels as a function of the gap depth (representing wear). The improved models of magnetic heads that have been developed since will show different behaviors for different heads.

All heads will still exhibit improved read efficiency as the head wears. The write efficiency should include the stray fields around the gap and the coil, and each head must be analyzed separately before a prediction can be made of its changes in write efficiency and impedance during wear.

BIBLIOGRAPHY TO CHAPTER 18

Talke, F.E., and Bogy, D.B., "Head-Medium Interface," in *Magnetic Recording*, Ed. by C.D. Mee and E.D. Daniel, McGraw-Hill 1987, pp. 427-506.

Halling, J. (Editor), *Principles of Tribology*, The Macmillian Press Ltd., London 1978.

Rabinowicz, E., *Friction, Wear and Lubrication*, MIT, Cambridge, Mass. 1974.

REFERENCES TO CHAPTER 18

Losses

Yeh, N.H., "Review of Head-Media Coupling in Magnetic Recording," *IEEE Trans. Magn.*, Sept. 1985, Vol. MAG-21, No. 5, pp. 1338-1343.

Friction

Schnurmann, R., and Warlow-Davies, E., "The Electrostatic Component of the Force of Sliding Friction," *Physical Society,* Jan. 1942, Vol. 54, No. 1, pp. 14-27.

Ristow, J., "Die elektrische Nachbildung des stick-slip-Vorganges," I and II, *Hochfreq. und Elektroakustik,* Dec. 1965, Vol. 74, No. 12, pp. 191-198 and Vol. 75, No. 1, pp. 1-9.

Sakai, Y., and Hori, K., "Measurement of Video Tape Temperature," *Television* (Japan), Jan. 1966, Vol. 17, No. 11, pp. 688-700.

Kalfayan, S.H., Sil, R.H., and Hoffman, J.K., "A Study of the Frictional and Stick-Slip Behavior of Magnetic Recording Tapes," *JPL Tech. Report,* April 1972.

Mukhopadhyay, A., "Study of the Effect of Static/Dynamic Coulomb Friction Variation at Tape-head Interface of Spacecraft Tape Recorder by Nonlinear Time Response Simulation," *IEEE Trans. Magn.,* Sept. 1978, Vol. MAG-14, No. 5, pp. 333-335.

Clurman, S., "A Predictive Analysis for Optimum Contour-After-Wear for Magnetic Heads," *IEEE Trans. Magn.,* Sept. 1978, Vol. MAG-14, No. 5, pp. 339-341.

Hahn, F.W., "Design Considerations for a Helical-scan Recording Head Contour," *Proc. Intl. Conf. Video and Data,* April 1976, pp. 261-266.

Diskette Head Interface

Charbonnier, P.P., "Flight of Flexible Disk over Recording Heads," *IEEE Trans. Magn.,* Nov. 1976, Vol. MAG-12, No. 6, pp. 728-730.

Greenberg, H.J., "Flexible Disk - Read/Write Head Interface," *IEEE Trans. Magn.,* Sept. 1978, Vol. MAG-14, No. 5, pp. 336-338.

Adams, G.G., "Analysis of the Flexible Disk/Head Interface," *ASME Jour. of Lubr. Tech.,* Jan. 1980, Vol. 102, No. 1, pp. 86-90.

Klevesahl, B., and Stromsta, R., "The Double-sided Floppy is Reborn," *Mini-Micro Systems,* May 1981, Vol. 14, No. 5, pp. 159-163.

Licari, J., and King, F., "Elastohydrodynamic Analysis of Head to Flexible Disk Interface Phenomena," *Journal of Applied Mechanics,* Dec. 1981, Vol. 48, pp. 763-768.

Flying Heads

Stahl, K.J., White, J.W., and Deckert, K.L., "Dynamic Response of Self-Acting Foil Bearings," *IBM Jour. Res Devel.,* Nov. 1974, Vol. 18, pp. 513-520.

Vogel, S.M., and Groom, J.L., "White Light Interferometry of Elastohydrodynamic Lubrication of Foil Bearings," *IBM Jour. Res. Devel.,* Nov. 1974, No. 18, pp. 521-528.

Tseng, R.C., and Talke, F.E., "Transition from Boundary Lubrication to Hydrodynamic Lubrication of Slider Bearings," *IBM Jour. Res. Devel.,* Nov. 1974, Vol. 18, pp. 534-540.

Fleischer, J.M., and Lin, C., "Infrared Laser Interferometer for Measuring Air-bearing Separation," *IBM Jour. Res. Devel.,* Nov. 1974, Vol. 18, pp. 529-533.

Ono, K., "Dynamic Characteristics of Air-Lubricated Slider Bearing for Non-contact Magnetic Recording," *J. Lubrication Tech.,* April 1975, Vol. 97, pp. 250-260.

Ragle, H., and Smith, P., "High Speed Flexible Disk - Head Interface," *IEEE Trans. Magn.,* Nov. 1979, Vol. MAG-15, No. 6, pp. 1459-1461.

Kita, T., Kogure, K., Mitsuya, Y., and Nakamishi, T., "New Method of Detecting Contact Between Floating-Head and Disk," *IEEE Trans. Magn.,* Sept. 1980, Vol. MAG-16, No. 5, pp. 873-875.

Miu, D.K., Bouchard, G., Bogy, D.G., and Talke, F.E., "Dynamic Response of a Winchester-type Slider Measured by Laser Doppler Interferometry," *IEEE Trans. Magn.,* Sept. 1984, Vol. MAG-20, No. 5, pp. 927-929.

Kawakubo, Y., Ishihara, H., Seo, Y., and Hirano, Y., "Head Crash Process of Magnetic Coated Disk during Contact Start / Stop Operations," *IEEE Trans. Magn.,* Sept. 1984, Vol. MAG-20, No. 5, pp. 933-935.

Tagawa, N., and Mashimoto, M., "Submicron Spacing Dynamics for Flying Head Slider Mechanisms Using Building Block Approach," *IEEE Trans. Magn.,* Sept. 1985, Vol. MAG-21, No. 5, pp. 1506-1508.

Mizoshita, Y., Aruga, K., and Yamada, T., "Dynamic Characteristics of a Magnetic Head Slider," *IEEE Trans. Magn.,* September 1985, Vol. MAG-21, No. 5, pp. 1509-1511.

Knudsen, J.K., "Flying Head for Stretched Surface Recording Disk," *IEEE Trans. Magn.,* Sept. 1985, Vol. MAG-21, No. 5, pp. 1512-1514.

Kumaran, A.R., and Chou, Y.S., "Effect of Head/Disc Imperfections on Gas Lubricated Slider Performance," *IEEE Trans. Magn.,* Sept. 1985, Vol. MAG-21, No. 5, pp. 1515-1517.

Mechanisms of Wear

Archard, F.J., "Contact and Rubbing of Flat Surfaces," *Jour. Appl. Phys.,* Aug. 1953, Vol. 24, No. 8, pp. 981-988.

Bhushan, B., "Analysis of the Real Area of Contact Between a Polymeric Magnetic Medium and a Rigid Surface," *Jour. Tribology, Trans. of the ASME,* January 1984, Vol. 106, No. 1, pp. 26-34.

Wear Test Methods; Results

Rettinger, Michael, "Magnetic Head Wear Investigation," *Jour. SMPTE,* April 1955, Vol. 66, No. 4, pp. 179-183.

Daniel, E.D., and Ragle, H.U., "Head Wear," *Memorex Bulletin,* 1963.

Carroll, Jr., J.F., and Gotham, R.C., "The Measurement of Abrasiveness of Magnetic Tape," *IEEE Trans. Magn.,* March 1966, Vol. Mag-2, No. 1, pp. 6-13.

Buchanan, J.D., and Tuttle, J.D., "A Sensitive Radiotracer Technique for Measuring Abrasivity," *International Jour. of Appl. Radiation and Isotoped,* Jan. 1968, Vol. 19, pp. 101-102.

Kolb, F.J., and Perry, R.S., "Wear of Permalloy Magnetic Heads Against Striped Motion-Picture Film," *Jour. SMPTE,* Sept. 1968, Vol. 77, No. 9, pp. 912-919.

Jorgensen, F., "Investigation and Measurement of Magnetic Head/Tape Interfacer Problems I-II," *Telemetry Journal,* February 1972, Vol. 7, No. 2, pp. 13-16 and pp. 21-24.

Hahn, F., "An Evaluation of Wear Resistant Coatings for Recording Heads," *IEEE Trans. Magn.,* Sept. 1975, Vol. MAG-11, No. 5, pp. 1242-1244.

Yamada, T., Shimizu, Y., and Ito, T., "Effect of Non-Stoichiometry on Wear of Ferrite Recording Heads," *IEEE Trans Magn.,* Sept. 1975, Vol. MAG-11, No. 5, pp. 1227-1229.

Williams, G.I., and Lewis, H.M., "Review of Methods for Measuring Abrasiveness of Magnetic Tape," *Intl. Conf. Video and Data Rec.,* April 1982, IERE Publ. No. 54, pp. 11-21.

Levy, A., "Wear Testing," *Magnetic Tape Recording for the Eighties*. NASA Ref. Publ. 1075, April 1982, pp. 23-34.

Kelly, J.J., "Tape and Head Wear," *Magnetic Tape Recording for the Eighties*, NASA Ref. Publ. 1075, April 1982, pp. 7-22.

Groenou, A., and Uijterschout, M., "A Quick Test on Wear of Head Materials by Recording Tapes," *IEEE Trans. Magn.,* Sept. 1983, MAG-19, No. 5, pp. 1674-1676.

Bhushan, B., "Influence of test parameters on the measurement of the coefficient of friction of magnetic tapes," *WEAR,* Jan. 1984, Vol. 93, No. 1, pp. 81-100.

Ozawa, K., Wakasugi, H., and Tanaka, K., "Friction and Wear of Magnetic Heads and Amorphous Metal Sliding Against Magnetic Tapes," *IEEE Trans. Magn.,* March 1984, Vol. MAG-20, No. 2, pp. 425-430.

Schultz, R.A., "Equal Pressure Contour Abrasivity Test for Magnetic Tape," *IITRI Prof. No. E06571,* Jan. 1986, 21 pages.

Part 6

Write/Read Process

Chapter 19

Direct Recording — Digital

The quest for increased packing density of bits on a recorded track requires a detailed understanding of the magnetic process whereby a pulse is recorded and read back. Several pulses in sequence will also interfere with each other and cause peak shifts and reduced output.

The mechanism of pulse recording and the influence of head/media parameters have been studied extensively in the past. The recording or write process has not been open to a rigorous and accurate treatment due to the nonlinear behavior of the magnetics, interaction fields and simultaneous satisfaction of several conditions (changing demagnetization during the record process). A fair agreement between measurements and some of the theories have been established for pulse recording at low densities, but modern recordings with ten thousand or more bits per cm of track length have not been adequately explained.

The following section will first review earlier models for the record process, and then collect what we know today about models of the record process.

MAGNETIC RECORDING PROCESS

Direct recording was introduced in Chapter 5 and was initially explained through a series of graphical projections. This recording (or write) model is extraordinarily simple; the recording zone is merely a thin sheet, placed perpendicular to the front plane of the write head (Fig. 19-1A). The coating is magnetized in its full thickness in accord with the longitudinal field strength H_x = gap field strength H_g. The perpendicular field H_y is equal to zero everywhere, and the track width w is equal to infinity (as is generally the case).

The remanent magnetization in this simple *remanence model* is calculated as the graphical projection onto the B_{rs}-axis in the $B_{rs}H$ graph (Figs. 5-4 through 5-9, and corresponding text in Chapter 5). We will now review developments of models that represent the actual situation in better ways.

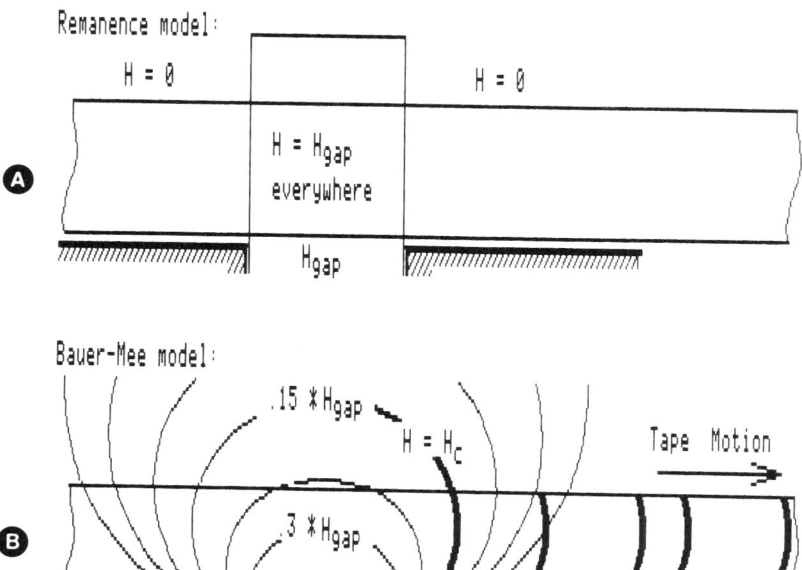

Fig. 19-1. Models of the recording process. (A) Remanence model; (B) Bauer-Mee model.

These models can be of great value for us in understanding the digital record process and explaining some of the things we observe, for instance *peak shifts*; these are phenomena where the timing between the reproduced voltage peaks differ from the timing between the write current pulse transitions.

We should also like to obtain a correlation between write/read performance and the magnetic properties of the magnetic coating. If we can first design a model by using our knowledge and past experience, and then verify the model by experiments, then we may use it both as a design guide and in calculations of performances.

No less than three recording models were proposed in publications in 1961: these were proposed by Stein, Bauer and Mee and Schwantke.

Stein modified the remanence model so the field strength across the gap varied as a cosine function. It was purely longitudinal and confined to the gap:

$$H_x = H_g \cos(\pi x/s) \text{for } |x| \leqq L/2$$
$$H_x = 0 \quad \text{for } |x| > L/2$$
$$H_y = 0 \text{ everywhere}$$

The initial magnetization curve was linear, and the B-H loops were simple parallellograms. The use of the model in calculations gave reasonable results when compared with experimental data for thin coatings and at long wavelengths.

Bauer and Mee recognized that the magnetization in a coating was not necessarily uniform in depth. Also, the remanence was obtained in a zone somewhere after the gap. They analyzed the recording field and postulated their bubble theory in 1961. Certain restrictions were necessary:

- All particles are identical, with the same coercive forces.
- There is no interaction between particles.
- The particles exhibit longitudinal anisotropy, and cannot support a perpendicular magnetization.
- The recorded wavelength $\lambda >> L_{gap}$.
- The coating thickness $c >> L_{gap}$.

It is then permissible to consider only the horizontal component $H_x(x,y)$ of the total recording field:

$$H_x(x,y) = H_{total} \cos \beta$$

where

$$H_{total} = (L_{gap}/\pi)*H_{gap}/R$$
$$R = \sqrt{x^2 + y^2}$$

or

$$H_x(x,y) = (L_{gap}/\pi)*H_{gap}*y/R^2 \qquad \textbf{(19.1)}$$

The field lines are circular and the total field is formed from cylinders (bubbles) of varying field intensity: the small cylinders near the gap have the largest field strength (Fig. 19-1B). The first three approximations listed above imply that the medium has a square hysteresis loop and remagnetization occurs for fields exceeding the coercive force H_c. The particles inside a cylinder whose periphery corresponds to H_c, are consequently magnetized to the polarity of the field. Outside the cylinder the particles are not affected. It is easy to see that the magnetization pattern on the medium ends up as shown in Fig. 19-1B.

The Bauer-Mee model gave some insight into the distortion mechanism at long wavelengths, when recording with AC-bias was used. It also explained the associated "self-erasure" effects at short wavelengths. Its drawback was limitation to longitudinal magnetization and an "almost" ideal media coating; the method did not correlate the BH-loop and its magnetic properties to recording performance.

Schwantke's proposal offered understanding of the internal magnetization process through the concept of particle interaction. This was explored as early as in 1935 by the German physicist Preisach, and has been invaluable in explaining the recording process with AC-bias (see Chapter 20).

Early studies of digital recording concentrated on investigations of the recording of a single transition, (i.e., the leading or trailing edge of a bit). This is the position where the maximum flux change occurs and the maximum read voltage is sensed. It is also the position where a digital signal carries its information or clock signals.

A short transition zone is desirable in order to achieve a maximum output voltage and a maximum packing density. The latter is, as we shall see, partly diluted by the playback process that involves a fairly broad sensing function as determined by the gap field.

The Bauer/Mee model did find use with magnetic coatings that had normal hysteresis loops. The magnetization on each side of a transition is longitudinal, and the transition zone is a magnetic pole, or a curved sheath of thickness Δx. We will now analyze the formation of this transition zone and its final shape after removal of the recording field and subsequent removal of the high permeable write head.

RECORDING OF A SINGLE TRANSITION

Let us see how the recording of a transition takes place. A presaturated coating of thickness c is moving past a write head at a distance of d (Fig. 19-2). The record field is turned on in Fig. 19-2A and we shall, for now, consider only the longitudinal field components $H_x(x,y)$ of the head field. The lines of constant field strength are therefore equal to the circles (bubbles) in the Bauer-Mee model.

The coating's hysteresis loop is idealized as shown on the right, and the cylinder of strength H_1 will remagnetize all of the coating inside it. The magnetic particles (or domains) lying on a line of strength H_2 are brought to a non-magnetic state, while those on a line of strength H_3 are left in the original state of magnetization.

When the record field is removed (turned off), as shown in c, or the coating moved away, then a transition zone of width Δx has been recorded. The magnetization changes by an amount of $2J_s$ over the distance Δx, and from this we find the pole strength (by definition):

$$\text{Pole strength } p = 2J_s/\Delta x. \qquad \textbf{(19.2)}$$

The read voltage from the read head will be proportional to the pole strength, which will require that Δx is as short as possible. This will simultaneously assure the highest packing density, and the recording will in the limit consist of a series of closely spaced transition zones.

It is therefore of great practical value to determine what the minimum transition length can be. Δx may be estimated as the distance between the two field lines of strength $H + \Delta H$ and $H - \Delta H$ as shown in Fig. 19-2D, for the appropriate scaled values of c and d. The example chosen has a transition zone length Δx of 0.2 to 0.4 μm, which corresponds to maximum packing densities of 50,000 to 25,000 flux reversals per cm. Such packing densities are made today with short gap heads. The flux reversals can, of course, be made at a rate so high that a new

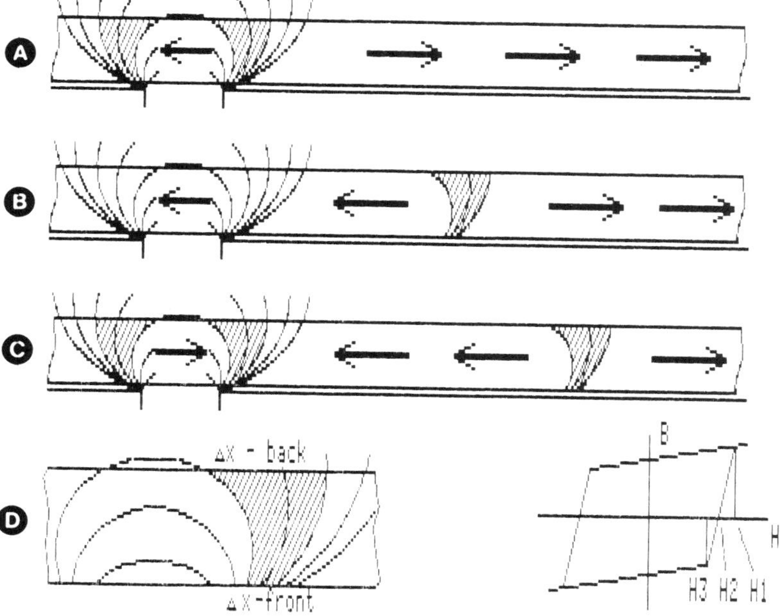

Fig. 19-2. Recording of transitions; (A) write field turned on; (B) write field turned off; (C) write field turned on with opposite polarity of that in (A); (D) graphical estimate of the length of the transition zone.

transition zone is started before the preceding one is ''completed,'' with loss in overall levels (see under Resolution, later this chapter).

Calculation of the transition zone length Δx has for years been the focal point of much research. In order to do this many simplifying assumptions were necessary. A detailed account is beyond the scope of this chapter and the interested reader is referred to the bibliography. Most of the reported work was done for the longitudinal case only and not until recently has the difficult task of calculations with vector magnetization begun. This is necessary for the high packing densities used today, and much work theoretical as well as experimental, is required.

The earliest approach was to assume a certain pattern for the distribution of the magnetization $2J_s$, and calculated the field from the pole (the transition). This is the demagnetizing field we treated earlier in Chapter 4, and its maximum value is always limited to the material's coercivity H_c. This permits a calculation of the minimum Δx; results were inconsistent with observations, and the magnetization pattern changed.

The best fit is achieved by assuming a gradual change in magnetization across the transition between two bits, following an arctangent function. Of the many models two appear to give the best fit to measured data, the model by Williams and Comstock, and by Middleton and Wright. The reader is referred to an update account by B.K. Middleton of several approaches to analysis of the record process in Chapter 2, *Magnetic Recording,* Vol. I, McGraw-Hill, 1987.

SELF-CONSISTENT, ITERATIVE COMPUTATIONS

The field strength in the immediate vicinity of the transition equals the demagnetizing field, which approaches the value of the material's coercivity H_c. This field is therefore not small compared to the recording field at that point, and it was in the late sixties suggested that it should be added to the record field and would modify it so the transitions end up being recorded at slightly different positions.

This addition to the head field is identical to the argument that the field acting upon a magnet includes its own demagnetizing field:

$$H_{core} = H_{applied} + H_{demag} \tag{4.3}$$

The net effect is a weakening of the head field on the inside of the transition (near the gap), and a strengthening on the outside. The field patterns without and with a transition are shown in Fig. 19-3 and Fig. 19-4. The corresponding field magnitudes along the medium surface are shown below, and are useful for estimating the length of the transition zone Δx at that location. Note that this illustration reflects an assumed value of one for the coating's relative permeability; this is customary in most theories for the recording process.

It is evident that the inclusion of the transitions' own field will broaden the transition, and shift it outward (away from the gap). At a certain point there will be a stable magnetization such that its demagnetizing field together with the head field results in that pattern. The computations are iterative and must be selfconsistent, i.e., the final magnetization pattern must correspond to one that produces the field that together with the head field produces that pattern.

The details of these computations are beyond the scope of this chapter. Two interesting results of computations are shown in Fig. 19-5, computed by Iwasaki and Suzuki in 1976. The short transition zone is the result of applying the head field alone to a saturated medium, while the longer zone results when both the head field and the demagnetization field H_d are present.

The *result is an outward shifted and longer transition zone*. When a following transition zone is recorded it is also shifted, *but lesser than the first zone was shifted, if the two are close together*. The reversed head field will assist the demagnetizing field from the prior recorded transition and will reduce its strength (partial erasure).

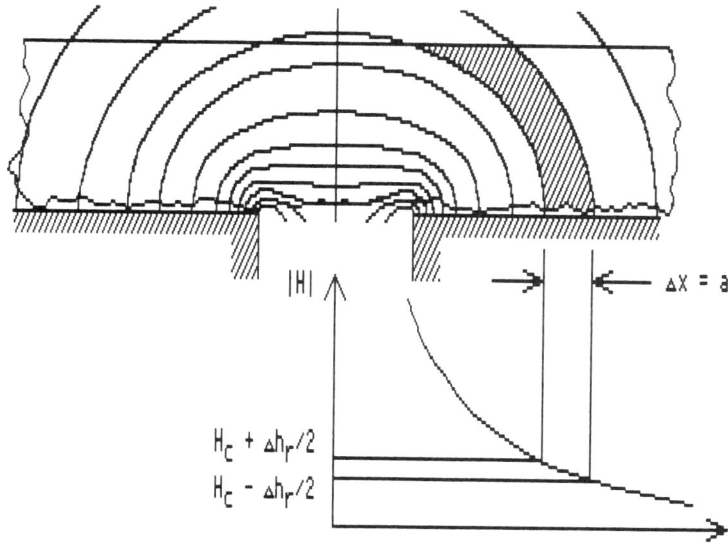

Fig. 19-3. Total head field alone produces a transition zone of length a.

All data patterns have varying timings between their transitions, and the timing relationship between the recorded transitions will be destroyed by the *peak shift* introduced by the recording process.

PERPENDICULAR RECORDING

The conventional scalar models of the magnetic recording process are insufficient in analyzing high density recording. It was also found that even for rather thin magnetic medium the magnetization vector still contains a surprisingly large perpendicular component (ref. Suzuki, 1976).

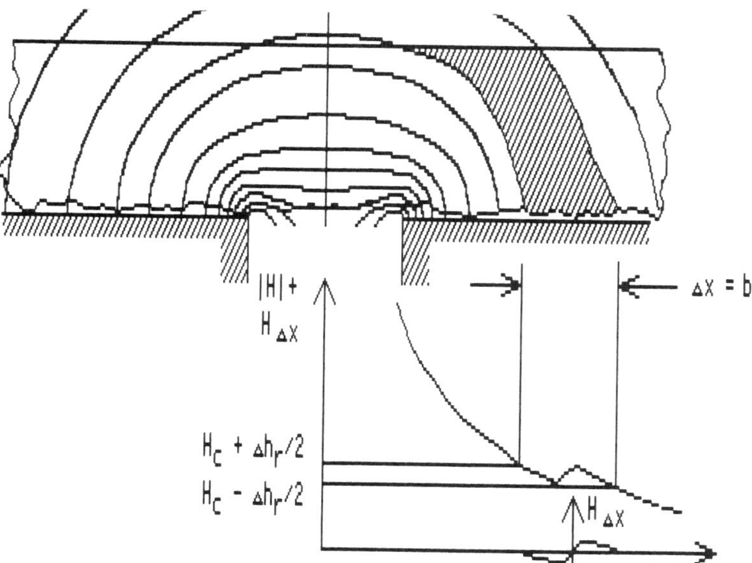

Fig. 19-4. Total head field plus field from transition produces a transition of length b > a.

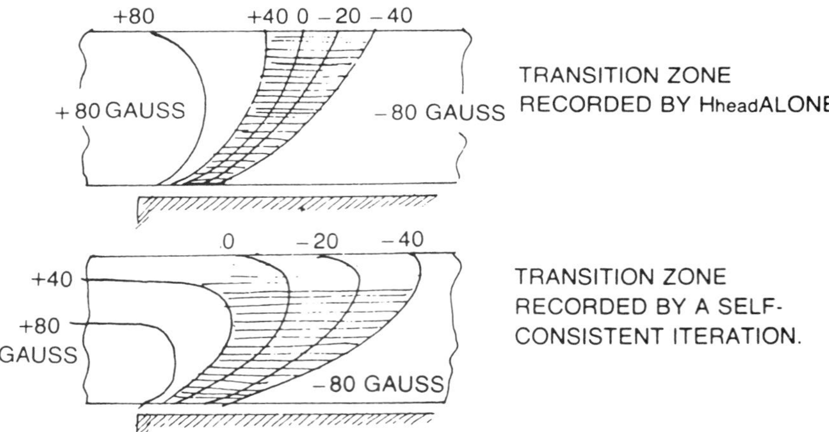

Fig. 19-5. Computed transition zones; longitudinal head field alone produces transition zone shown top, a self-consistent iterative computation produces bottom (after Iwasaki and Suzuki).

Perpendicular recording was therefore proposed in 1977 (ref. Iwasaki), and many theories and computations have been presented since (ref. Potter and Ortenburger, 1980, Middleton and Wright, 1982), while measured data have been scarce. The latter is probably due to the limited availability of good perpendicular media. This is in particular true for the tape format, which is exactly where perpendicular media should excel. The advantages in rigid disk systems appear non-existent when the head flying height over the disk exceeds 0.12-0.2 μm (5-8 μin).

The method of making perpendicular recordings has been discussed in the section on heads as well as on media. The interested reader is referred to the bibliography for further details.

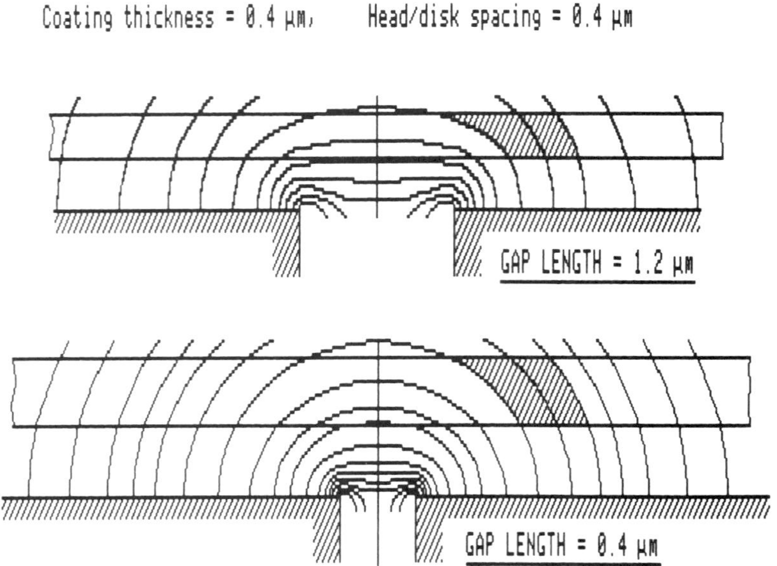

Fig. 19-6. Recording fields and freezing zones for a Winchester disk system.

THE TRANSITION ZONE AND RESOLUTION

We saw in Chapter 5 how the combination of the head field gradient and the medium's switching field distribution SFD controlled the length of the transition recorded between bits. Figure 19-6 shows the recording of transitions on a disk system with 0.4 μm head to disk spacing. The transition zone is fairly long, and the field lines are not perpendicular to the medium front. (For perpendicular recording a medium with high permeability material undercoat would be required.)

Figure 19-7 shows the situation for an in-contact recording. Clearly, the recording zone can be shortened by reducing the record field. That is, the length of the transition zone is proportional to the record field strength.

A very shallow recording is made with the field adjusted as shown in Fig. 19-7, middle. Now we do have two recording zones, which causes a very poor quality recording — one on top of another. This situation can only be remedied by reducing the gap length, as shown in Fig. 19-7 bottom. This will require an increase in the gap field strength, and such heads are therefore provided with pole shoes of high saturation material, such as an Al-Fe alloy (MIG, meaning Metal-In-the-Gap).

This brings about a ripple in the read voltage response versus frequency, possibly due to the formation of secondary gaps due to eddy currents in the pole shoes.

We learned earlier, in Chapter 5, that the length of the gap is of little influence on the length of the transition zone. That situation was different: When the recording depth remains fixed, *then* the gap length has little effect.

The length of the transition zone equals the length of the freezing zone, which is located by using field lines of constant strength in combination with the SFD to establish the borders of Δx.

This chapter concludes with some further computations on the record model from Chapter 5, used to generate pictures of the broadening of the transition zone (Figs. 5-17 and 5-19), and the reduction in resolution for high write levels and/or large SFD (Fig. 5-22).

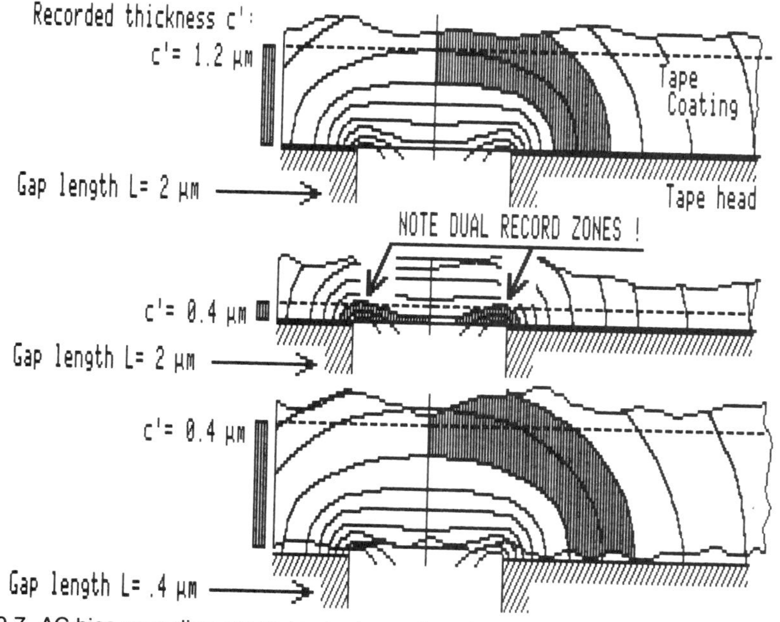

Fig. 19-7. AC-bias recording zones in short wavelength recordings.

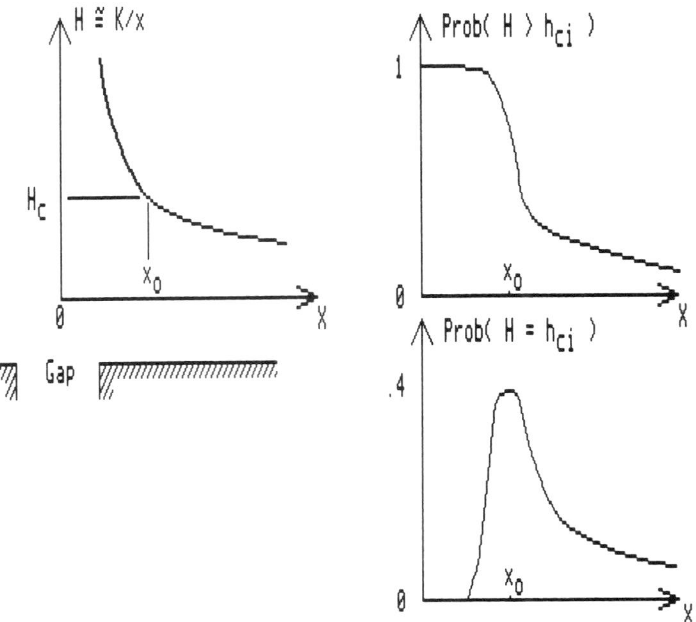

Fig. 19-8. Switching field probabilities in front of coating near head surface (see text).

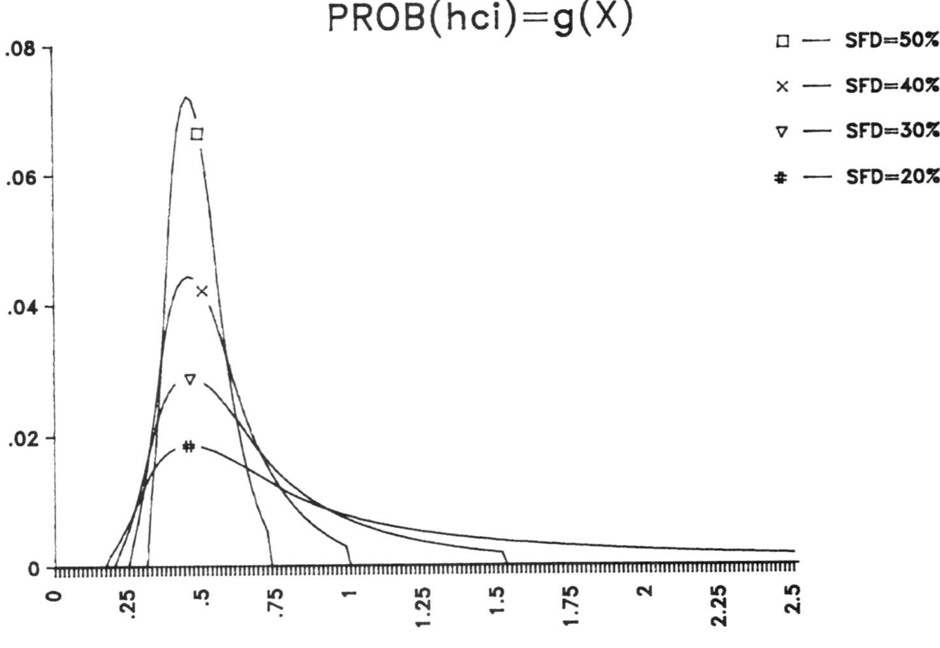

Fig. 19-9. Computed probabilities P(hci) = head field.

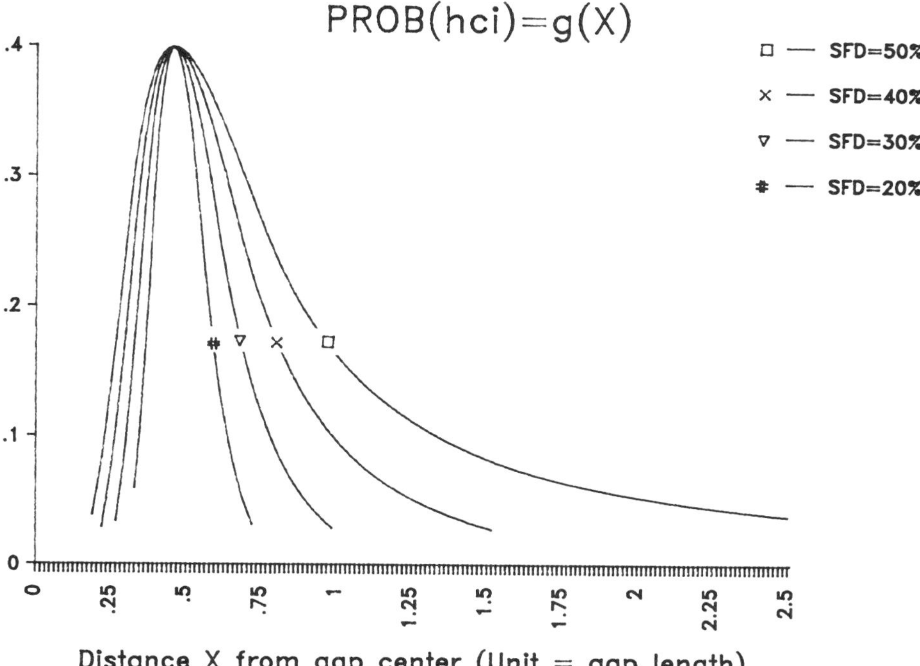

Fig. 19-10. Computed probabilities P(hci) = head field, normalized.

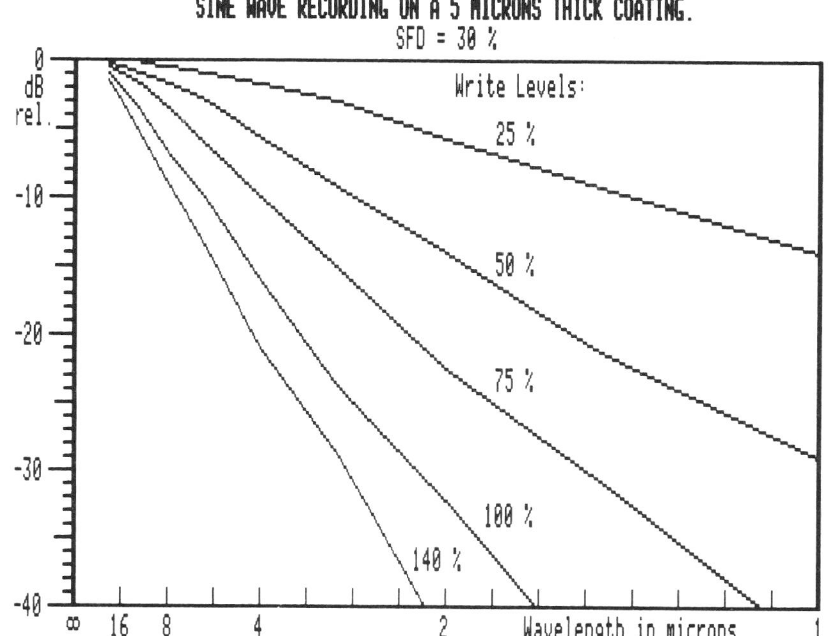

Fig. 19-11. Computed magnetization levels versus wavelength at various record level; SFD = 30 percent.

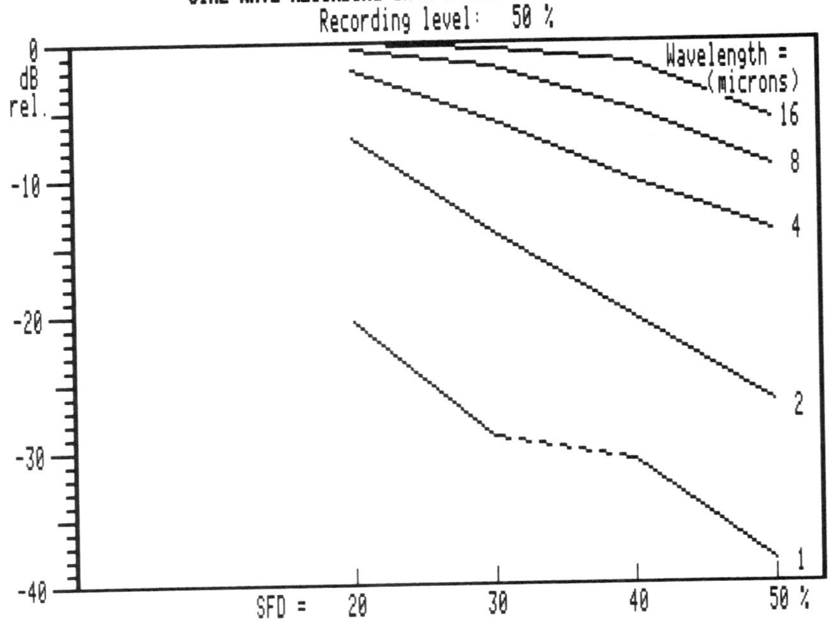

Fig. 19-12. Computed magnetization levels versus SFD, at five different wavelengths. Dotted lines are measured results (see text).

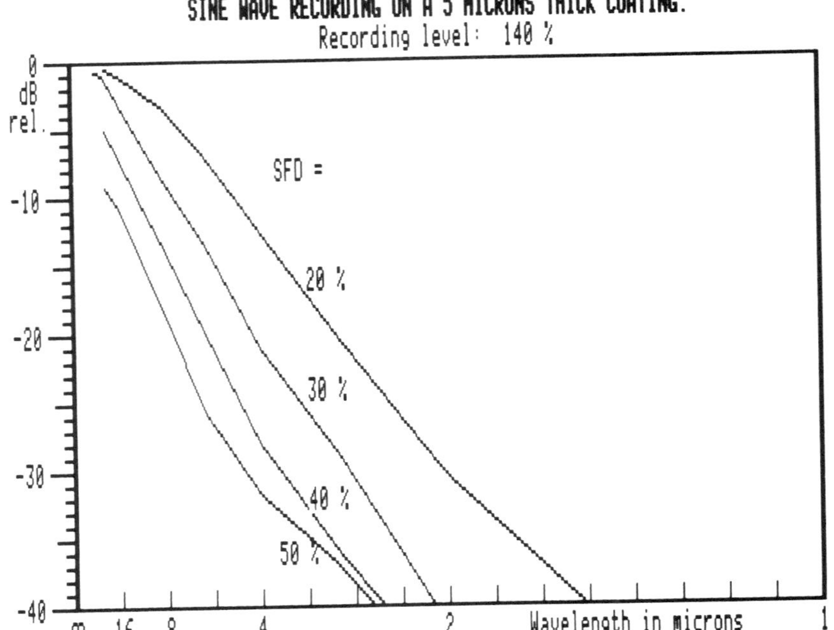

Fig. 19-13A. Magnetization levels versus wavelength at a write level of 140 percent.

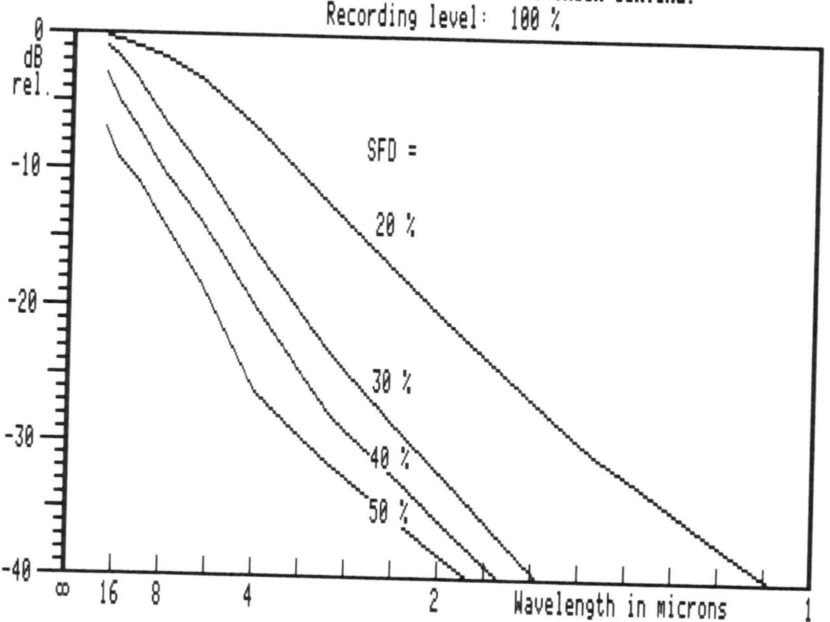

Fig. 19-13B. Magnetization levels versus wavelength at a write level of 100 percent.

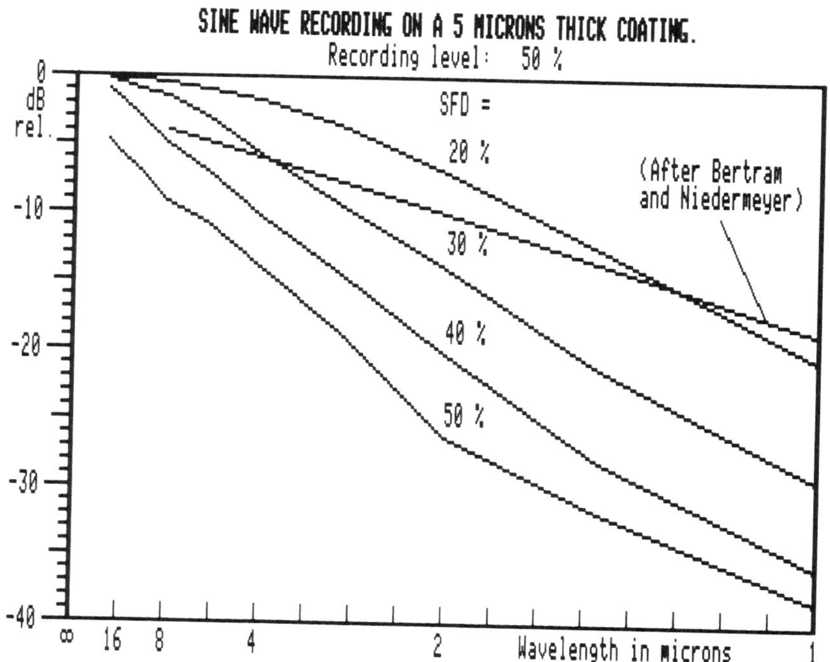

Fig. 19-13C. Magnetization levels versus wavelength at a write level of 50 percent. Dotted curve shows measured data (see text).

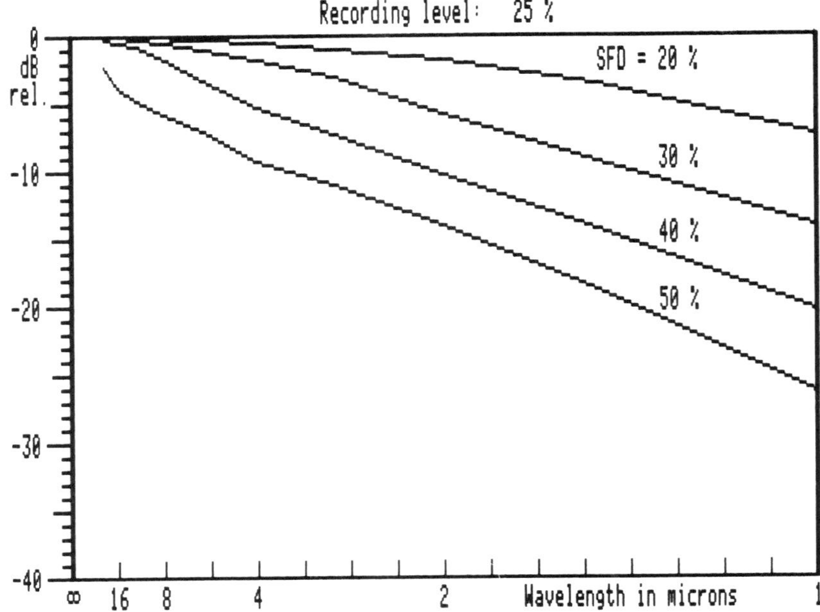

Fig. 19-13D. Magnetization levels versus wavelength at a write level of 25 percent.

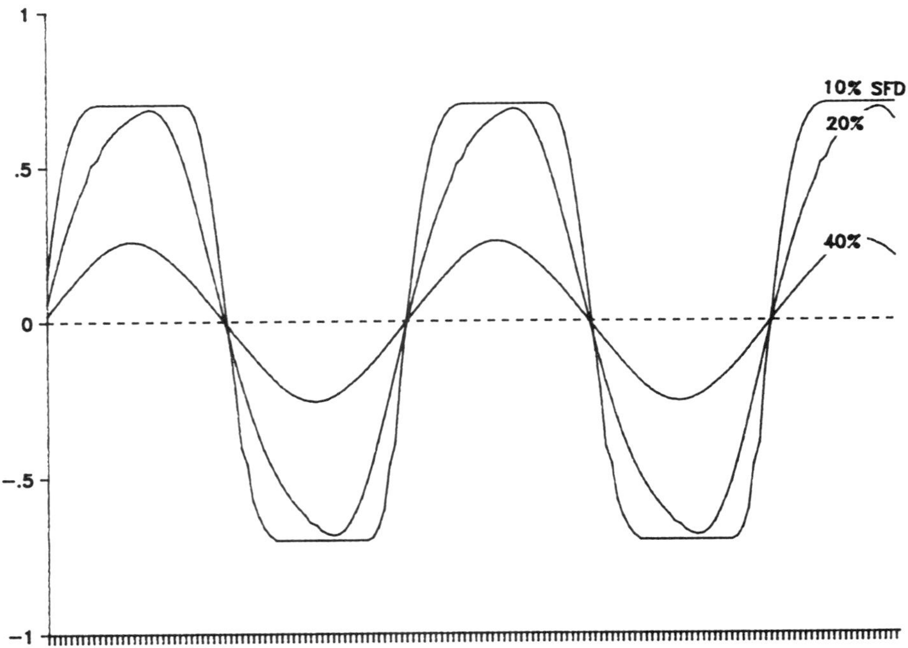

Fig. 19-14. Resolution in magnetic recording is strongly influenced by the particles switching field distribution. Recorded bit length is c/4 (c = coating thickness), and the write level is 70 percent. Compare with Fig. 5-22.

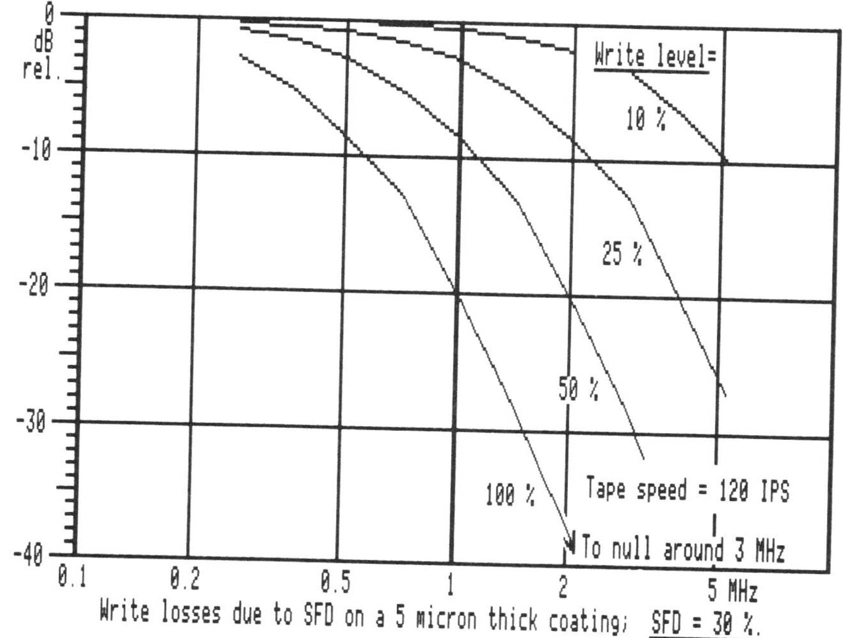

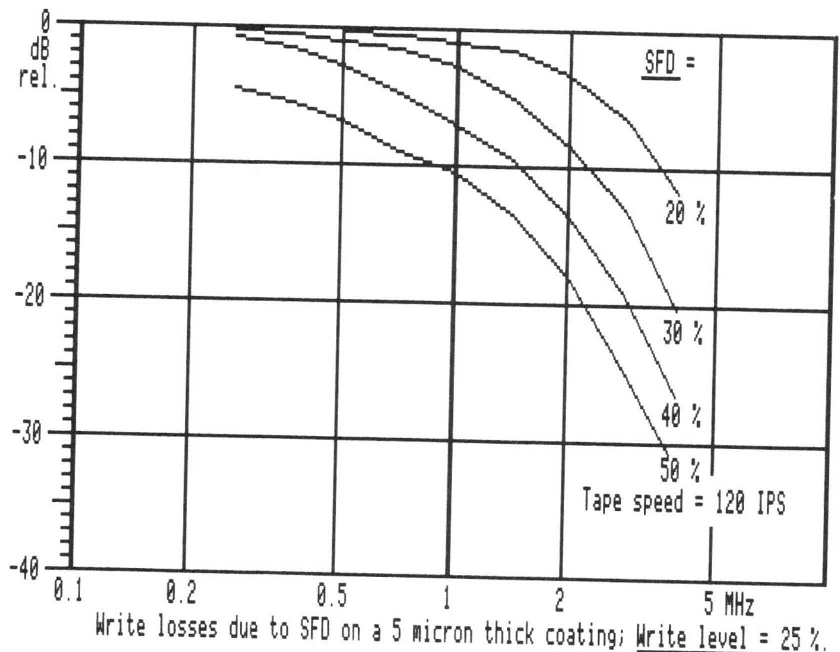

Fig. 19-15. Graphs of the record losses in a tape recorder.

That model does not include the effect of demagnetization, which appears justified in view of the lack thereof for in-contact recordings (ref. Bertram, and Bertram and Niedermeyer). Several measurements have in the past reached the same conclusion (ref. Daniel, Smaller).

In reviewing Fig. 5-22 there will appear to be a signal loss, and possibly a null, as shorter and shorter bit-lengths (wavelengths) are recorded. The resolution is controlled by the write field gradient and the SFD. The head field gradient is shown in Fig. 19-8, top left, which in combination with the SFD function, top right, results in a new probability density, namely that of finding particles of certain coercivities along the trailing pole piece of the record head, Fig. 19-8, bottom. This function is shown in Figs. 19-9 and 19-10 (normalized).

A program is listed in the appendix for computation of the resulting loss in the recorded magnetization. Results are plotted in Figs. 19-11, 19-12 and 19-13A through 19-13D. The frequency scale is linear.

Comparison with measured data are shown in Fig. 19-13C (ref. Bertram and Niedermeyer).

Computed signal waveforms are shown in Fig. 19-14, and correspond to the graphic plots in Fig. 5-22.

The data have been reduced to the graphs in Fig. 19-15, showing the record losses in a tape recorder. These losses can be summarized in a simple expression:

$$\text{SFD Losses} = K \bullet (1 - S^*) \bullet (I_{write}/I_{100\%}) \bullet (c + d)/BL \quad \text{dB} \tag{19.1}$$

The multiplier K is approximately equal to 20 for in-contact sine wave recordings, a few percent higher for a square wave (simulating digital recording). K increases with the spacing between head and media; its value can be determined from a short series of computations.

BIBLIOGRAPHY TO CHAPTER 19

Pulse Recording

Bauer, B.B., and Mee, C.D., "A New Model for Magnetic Recording," *IRE Trans. Audio*, 1961, Vol. AU-9, pp. 139-145.

Stein, I., "Analysis of the Recording of Sine Waves," *IEEE Trans. Audio*, Vol. AU-9, pp. 146-155.

Hoagland, A.S., *Digital Magnetic Recording*, Wiley, New York, January 1963, 154 pages; reprinted 1983 by Krieger Publ. Co., Malabar, Fla 32950.

Speliotis, D.E., and Morrison, J.R., "A Theoretical Analysis of Saturation Magnetic Recording," *IBM Journal*, May 1966, pp. 234-243.

Lee, J.E., and Truman, N.N., "The Effect of Vertical Head Field Component on the Switching of Thin Metallic Film," *IEEE Trans. Magn.*, Sept. 1968, Vol. MAG-4, No. 3, pp. 277-281.

Kostyshyn, B., "The Write Process in Magnetic Recording," *IEEE Trans. Magn.*, Dec. 1971, Vol. MAG-7, No. 4, pp. 880-885.

Williams, M.L., and Comstock, R.L., "An Analytical Model of the Write Process in Digital Magnetic Recording," *17th Annual AIP Conf.*, Part 1, No. 5, 1971, pp. 738-742.

Comstock, R.L., and Williams, M.L., "Frequency Response in Digital Magnetic Recording," *IEEE Trans. Magn.*, Sept. 1973, Vol. MAG-9, No. 3, pp. 342-346.

Fluitman, J., "Effect of Nonzero Write Field Rise Time in Digital Magnetic Recording," *IEEE Trans. Magn.*, May 1976, Vol. MAG-12, No. 3, pp. 218-224.

Iwasaki, S.I., and Nakamura, Y, "An Analysis for the Magnetization Mode for High Density

Magnetic Recording," *IEEE Trans. Magn.*, July 1977, Vol. MAG-13, No. 5, pp. 1272-1278.

Bertram, H.N., "The Effect of the Angular Dependence of the Particle Nucleation Field on the Magnetic Recording Process," *IEEE Trans. Magn.*, Nov. 1984, Vol. MAG-20, No. 6, pp. 2094-2104.

Lefranc, C., "Micromagnetic Models in Magnetic Recording Simulation," *IEEE Trans. Magn.*, Aug. 1985, Vol. MAG-21, No. 5, pp. 1417-1422.

Beardsley, I.A., "Modeling the Record Process," *IEEE Trans. Magn.*, Sept. 1986, Vol. MAG-22, No. 5, pp. 454-459.

Perpendicular Recording

Iwasaki, S.I., and Takemura, K., "An Analysis for the Circular Mode of Magnetization in Short Wavelength Recording," *IEEE Trans. Magn.*, Nov. 1975, Vol. MAG-11, No. 5, pp. 1173-1176.

Suzuki, K., "Theoretical Study of Vector Magnetization Distribution Using Rotational Magnetization Model," *IEEE Trans. Magn.*, May 1976, Vol. MAG-12, No. 3, pp. 224-230.

Iwasaki, S., "Perpendicular Magnetic Recording," *IEEE Trans. Magn.*, Jan. 1980, Vol. MAG-16, No. 1, pp. 71-76.

Lemke, J.U., "An Isotropic particulate medium with additive Hilbert and Fourier field components," *Jour. Appl. Phys.*, Mar. 1982, Vol. 53, No. 3, pp. 2561-2566.

Middleton, B.K., and Wright, C.D., "Perpendicular Recording," *Intl. Conf. Video and Data 82*, April 1982, IERE Publ. No. 54, pp. 181-192.

Wielinga, T., *Investigations on Perpendicular Magnetic Recording*, Twente Univ. of Technology, Holland, April 1983, Thesis, 181 pages.

Chi, C.S., Frey, K.A., Hohnson, R.A., and Maloney, W.T., "Experimental Studies of Longitudinal and Perpendicular High Density Recording," *IEEE Trans. Magn.*, Nov. 1983, Vol. MAG-19, No. 6, pp. 1608-1610.

Wright, C.D., and Middleton, B.K., "The Perpendicular Record and Replay Processes," *Intl. Conf. Video and Data 84*, April 1984, IERE Publ. No. 59, pp. 9-16.

Mallinson, J.C., and Bertram, H.N., "A Theoretical and Experimental Comparison of the Longitudinal and Vertical Modes of Magnetic Recording," *IEEE Trans. Magn.*, May 1984, Vol. MAG-20, No. 3, pp. 461-467.

Soohoo, R.F., "Micromagnetics of Domain Walls in Vertical Recording," *Jour. Appl. Phys.*, June 1984, Vol. 55, No. 6, pp. 2211-2213.

Lopez, O., "Analytic Calculation of Write Induced Separation Losses," *IEEE Trans. Magn.*, Sept. 1984, Vol. MAG-20, No. 5, pp. 715-717.

Sharrock, M. P., and Subbs, D.P., "Perpendicular Magnetic Recording Technology: A Review," *Jour. SMPTE*, Dec. 1984, Vol. 93, No. 12.

Bertram, H.N., "Interpretation of Spectral Response in Perpendicular Recording," *IEEE Trans. Magn.*, Sept. 1985, Vol. MAG-21, No. 5, pp. 1395-1397.

Wright, C.D., and Middleton, B.K., "Analytical Modelling of Perpendicular Recording," *IEEE Trans. Magn.*, Sept. 1985, Vol. MAG-21, No. 5, pp. 1398-1400.

Beusekamp, M.F., and Fluitman, J.H., "Simulation of the Perpendicular Recording Process including Image Charge Effects," *IEEE Trans. Magn.*, Sept. 1986, Vol. MAG-22, No. 5, pp. 364-366.

Zhu, J., and Bertram, H.N., "Computer Modeling of the Write Process in Perpendicular Recording," *IEEE Trans. Magn.*, Sept. 1986, Vol. MAG-22, No. 5, pp. 379-381.

The Write Process Using Self-Consistent Numerical Methods

Iwasaki, S. and Suto, Y., "Investigations on the Recording Demagnetization Phenomena in Magnetic Recording Process," *Rep. of Res. Inst. El. Comm.*, Tohoku Univ., March 1966,

Vol. 18, No. 3-4, pp. 223-236.

Iwasaki, S., and Suzuki, T., "Dynamical Interpretation of Magnetic Recording Process," *IEEE Trans. Magn.*, Sept. 1968, Vol. MAG-4, No. 3, pp. 269-277.

Suzuki, T., and Iwasaki, S., "Theoretical Analysis of Recording Process in Digital Magnetic Recording," *Rep. of Res. Inst. of El. Comm.* Tohoku Univ., Jan. 1971, Vol. 23, No. 2, pp. 51-67.

Curland, N., and Speliotis, D.E., "An Iterative Hysteresis Model for Digital Magnetic Recording," *IEEE Trans. Magn.*, Sept. 1971, Vol. MAG-7, No. 3, pp. 538-543.

Potter, R.I., and Schmulian, R.J., "Self-Consistently Computed Magnetization Patterns in Thin Magnetic Recording Media," *IEEE Trans. Magn.*, Dec. 1971, Vol. MAG-7, No. 4, pp. 873-880.

Ortenburger, I.B., Cole, R.W., and Potter, R.I., "Improvements to a Self-Consistent Model for the Magnetic Recording Properties of Non-Particulate Media," *IEEE Trans. Magn.*, July 1977, Vol. MAG-13, No. 5, pp. 1278-1284.

Hartman, K., Potter, R.I., and Ortenburger, I.B., "A Vector Model for Magnetic Hysteresis Based on Interacting Dipoles," *IEEE Trans. Magn.*, July 1978, Vol. MAG-14, No. 4, pp. 223-227.

Bertram, H.N., and Niedermeyer, R., "The Effect of Demagnetization Fields on Recording Spectra," *IEEE Trans. Magn.*, Sept. 1978, Vol. MAG-14, No. 5, pp. 743-746.

Middleton, B.K., and Wisely, P.L., "Pulse Superposition and High-Density Recording," *IEEE Trans. Magn.*, Sept. 1978, Vol. MAG-14, No. 5, pp. 1043-1050.

Potter, R.I., and Beardsley, I.A., "Self-Consistent Computer Calculations for Perpendicular Magnetic Recording," *IEEE Trans. Magn.*, Sept. 1980, Vol. MAG-16, No. 5, pp. 967-973.

Hughes, G.F., Bloomberg, D.S., Castelli, V., and Hoffman, R., "Not Just Another Self-Consistent Magnetic Recording Model," *IEEE Trans. Magn.*, March 1981, Vol. MAG-17, No. 2, pp. 1192-1199.

Bertram, H.N., and Niedermeyer, R., "The Effect of Spacing on Demagnetization in Magnetic Recording," *IEEE Trans. Magn.*, Nov. 1982, Vol. MAG-18, No. 6, pp. 1206-1209.

Beardsley, I.A., "Three Dimensional Write Model for Magnetic Recording," *IEEE Trans. Magn.*, Sept. 1986, Vol. MAG-22, No. 5, pp. 361-363.

The Transition

Bonyhard, P.I., Dvies, A.V., and Middleton, B.K., "A Theory of Digital Magnetic Recording on Metallic Films," *IEEE Trans. Magn.*, March 1966, Vol. MAG-2, No. 1, pp. 1-5.

Hsieh, E.J., Soohoo, R.F., and Kelly, M.F., "A Lorentz Microscopic Study of Head-On Domain Walls," *IEEE Trans. Magn.*, June 1974, Vol. MAG-10, No. 2, pp. 304-308.

Dressler, D.D., and Judy, J.H., "A Study of Digitally Recorded Transitions in Thin Magnetic Films," *IEEE Trans. Magn.*, Sept. 1974, Vol. MAG-10, No. 3, pp. 674-677.

Ramachandran, K., and Soohoo, R.F., "Magnetization Reversal Behavior in High Density Recording," *IEEE Trans. Magn.*, Sept. 1974, Vol. MAG-10, No. 3, pp. 780-783.

Ortenburger, I.B., and Potter, R.I., "A Self-consistent Calculation of the Transition Zone in Thick Particulate Recording Media," *J. Appl. Phys.*, March 1979, Vol. 50, No. 3, pp. 2393-2395.

Gulak, G., Shwedyk, E., and Card, D., "An Improved Approximation for the Isolated Transition in Saturated Magnetic Recording," *IEEE Trans. Magn.*, Sept. 1982, Vol. MAG-18, No. 5, pp. 989-993.

Lin, C.J., Best, J.S., Bullock, D.C., and Stukey, F.W., "Transitions on Perpendicular Rigid Disks in Quasi-Contact," *IEEE Trans. Magn.*, Sept. 1985, Vol. MAG-21, No. 5, pp. 1359-1364.

Wright, C.D., Loze, M.K., Hudson, V.N., and Edwards, J., "Spectral Determination of Transition Length in Digital Magnetic Recording," *Video, Audio and Data Recording Conf.*, 1986, IERE Publ. No. 63, pp. 23-30.

Peak Shift

Sierra, H.M., "Bit Shift and Crowding in Digital Magnetic Recording," *Electro-Technology*, Sept. 1966, pp. 56-59.

Morrison, J.R., and Speliotis, D.E., "Study of Peak Shift in Thin Recording Surfaces," *IEEE Trans. Magn.*, Sept. 1967, Vol. MAG-3, No. 3, pp. 208-211.

Behr, M.I., and Blessum, N.S., "Technique for Reducing Effects of Pulse Crowding in Magnetic Recording," *IEEE Trans. Magn.*, Sept. 1972, Vol. MAG-8, No. 3, pp. 55-58.

Helle, M., and Lazzari, J.P., "Experimental Study of the External Fringing Field on Integrated Head," *IEEE Trans. Magn.*, Sept. 1975, Vol. MAG-11, No. 5, pp. 1221-1223.

Jacoby, G.V., "A New Look-Ahead Code for Increased Data Density," *IEEE Trans. Magn.*, July 1977, Vol. MAG-13, No. 5, pp. 1202-1204.

Thornley, R.F.M., "Compensation of Peak Shift with Write Timing," *IEEE Trans. Magn.",* Nov. 1981, Vol. MAG-17, No. 6, pp. 3332-3335.

Overwrite; Erasure

Bloomberg, D.S., Hughes, G.F., and Hoffmann, R.J., "Analytic Determination of Overwrite Capability in Magnetic Recording Systems," *IEEE Trans. Magn.*, Nov. 1979, Vol. MAG-15, No. 6, pp. 1450-1453.

Fayling, R. E., Szczech, T.J., and Wollack, E.F., "A Model for Overwrite Modulation in Longitudinal," *IEEE Trans. Magn.*, Sept. 1984, Vol. MAG-20, No. 5, pp. 718-720.

Wachenschwanz, D., and Jeffers, F., "Overwrite as a Function of Record Gap Length - A New Effect," *IEEE Trans. Magn.*, Sept. 1985, Vol. MAG-21, No. 5, pp. 1380-1382.

Chapter 20

Recording with AC-Bias

The reader was introduced to the concept of AC-bias recording in Chapter 5 and it may be worthwhile to review it in these few pages. In this chapter we will develop a model for the AC-bias recording technique that will allow us to calculate the recording performance in terms of record current, head fields and tape coating parameters.

What we know today about AC-bias recording is the sum total of many researchers' efforts to achieve an in-depth understanding of the anhysteretic record process, where two signals of different frequencies and amplitudes generate a remanent magnetization in the tape coating.

The graphic approach, using projections on the hysteresis loop, was tried early and did not provide a meaningful insight into the record process and it did at one point or another disagree with experiments. It became necessary to go back to the mechanism that causes hysteresis loops, namely the irreversible rotation of magnetization in single particles (Figs. 3-23 and 3-24).

THE PREISACH MODEL

The behavior of an assembly of small magnetic particles is best described by a model that ties together the individual coercivities and their collective interaction fields. Such a model dates back to 1935, when it was first described by Preisach, in *Zeitschrift fuer Physik*. The application to anhysteretic recording was presented first by G. Schwantke. Independent and almost simultaneous, are presentations by Woodward and Della Torre, and by Daniel and Levine.

The effect of interaction between the particles in a tape coating can be illustrated by placing the particles in a coordinate system where particles with zero interaction field are placed on an H_c-axis in order of intrinsic coercivities. The ordinate axis is used to locate particles that are influenced by positive or negative interaction fields.

This is the Preisach diagram, and is shown in Fig. 20-1. It is a three-dimensional distribution $h(h_{ci}, \Delta H)$, as shown in Fig. 20-2. Illustration A is a conceptual drawing of such a distribution, while B is a histogram of a measured audio tape (ref. Brock).

The remanent magnetizations are in Fig. 20-1 indicated by the dots, assuming that this assembly of particles was completely demagnetized, and therefore has a sum total of $J_r = 0$. The demagnetization can be performed by a slowly decaying AC-field, and all particles with zero interaction field will end up magnetized at either $+J$ or $-J$; the chances for one or the other is fifty-fifty.

If the particle is under the influence of a positive interaction field (see upper right Fig. 20-1), then the decaying field will at first be strong enough to alternate the particle's magnetization, but can no longer magnetize the particle to $+J$ after the field's peak value has decayed to below $\Delta H + H_c$. All particles in the upper half will therefore end up magnetized at $-J$.

Particles in the lower half will similarly be magnetized at $+J$. The total magnetization of the particle assembly will be zero given that the distribution of the particles in the diagram is symmetrical about the H_c axis.

A field is now applied, *increasing* from zero to $+H_1$ (see Fig. 20-1, A). All particles below the H_c-axis are magnetized to $+J_s$, and those above will require that H_1 equals, and exceeds, $H_c + \Delta H$ before they switch to $+J_s$.This will first happen to those particles having small H_c and ΔH (i.e., near the origin).As H_1 increases, the particles will change magnetization to $+J_s$ as the line:

$$H = \Delta H + H_c \tag{20.1}$$

sweeps up in the first quadrant. The line has an angle of 135° with the H_c-axis, and it crosses the ΔH axis at $\Delta H = H_1$ ($H_c = 0$).

When H reaches H_1 all particles in the triangular area behind the line have been remagnetized and the net magnetization of the assembly is positive, as shown. If we now *decrease* H from H_1 to a negative value, another line will sweep into the fourth quadrant, at an angle of +45° with the H_c-axis. This line is determined by:

$$H = \Delta H - H_c \tag{20.2}$$

where the proper sign (−) applies for $\Delta H$when the line moves into the fourth quadrant due to a decreasing field strength.

When H reaches zero (b) we can readily see that a net magnetization remains, J_r; as H changes sign, the magnetization changes through zero to the negative value $-J_n$ (c).

We can finally reduce H back to zero, and $-J_r$ results (d). From this little exercise we obtain the following rules for the use of the Preisach diagram:

- An *increasing field strength* H moves a 135° line upward; this line intersects with the ΔH axis at the present value of H. All particles below this line become magnetized to $+J_s$, if not already at $+J_s$.
- A *decreasing field strength* H moves a 45° line downward; this line intersects with the ΔH axis at the present value of H. All particles above this line become magnetized to $-J_s$, if not already at $-J_s$.

These rules are summarized in Fig. 20-3 and applied in Fig. 20-4 and in several examples this chapter. NOTE: The distinction is between increasing or decreasing field strength, NOT positive or negative!

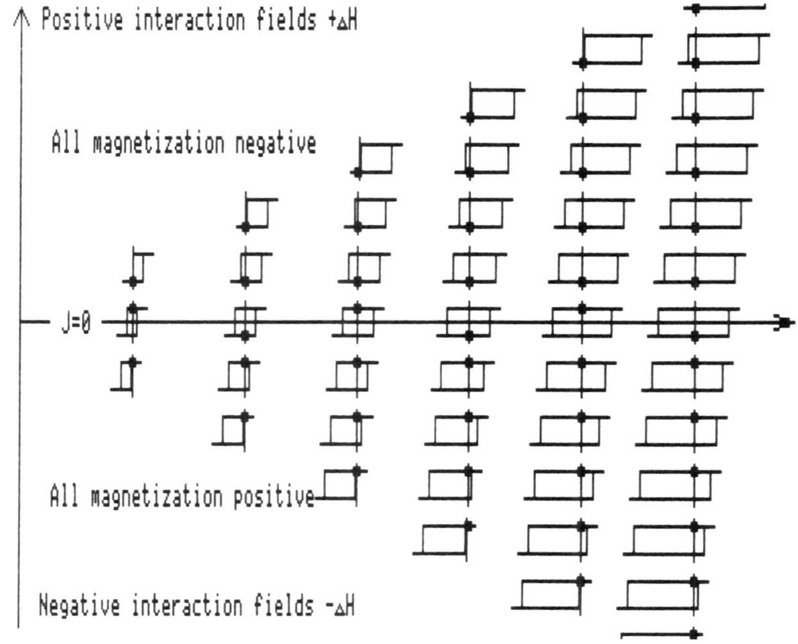

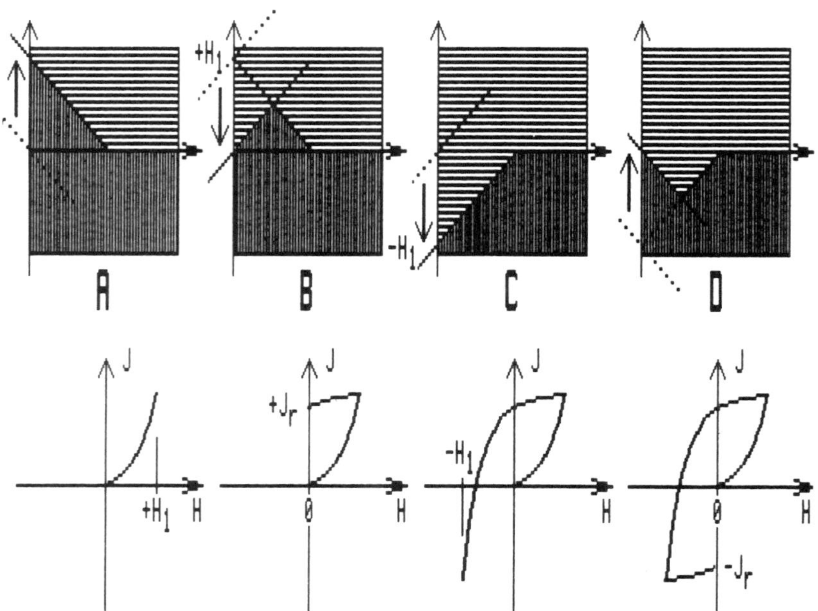

Fig. 20-1. Placement of magnetic particles in the Preisach model coordinate system (top); Changing of magnetization distribution by the use of two simple rules (bottom; see text).

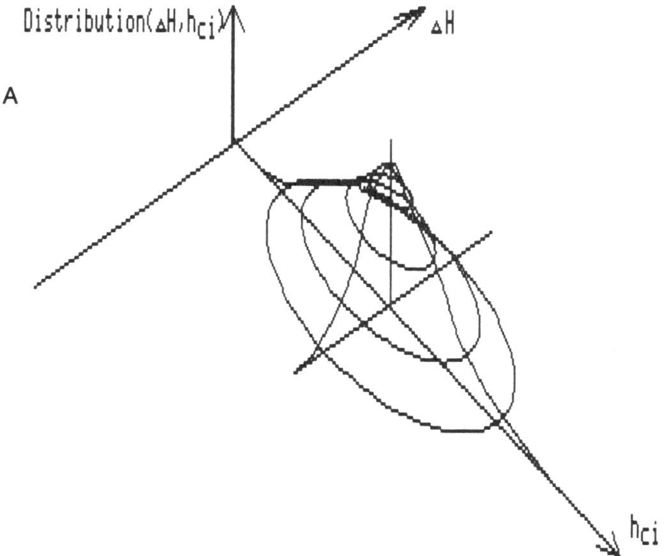

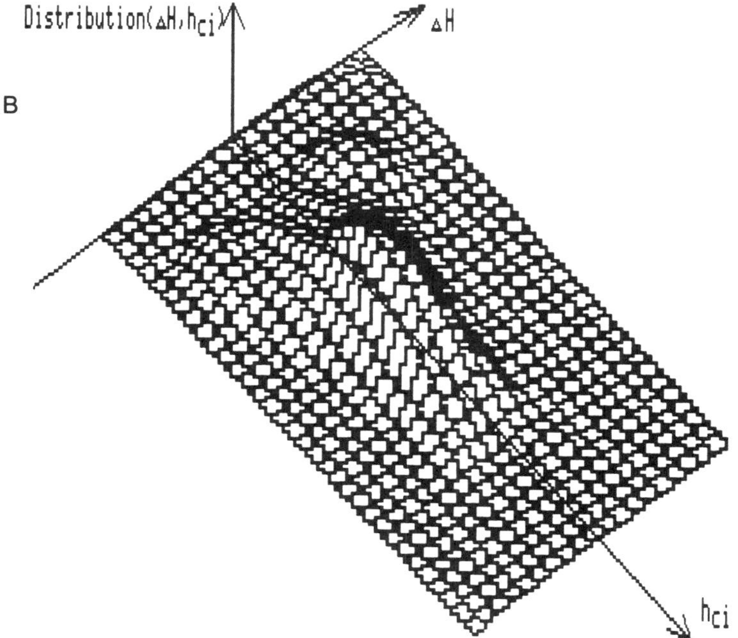

Fig. 20-2. (A) Statistics of the Preisach diagram; (B) measured density distribution (after Brock, 1979).

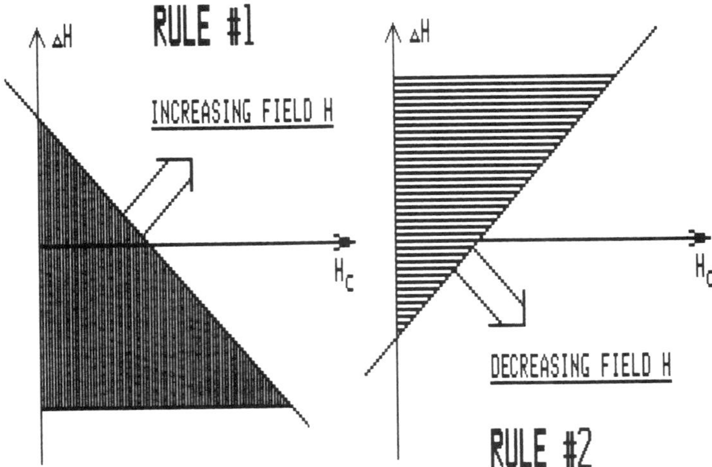

Fig. 20-3. The two rules for changing of magnetization in the Preisach diagram, following a change in field strength.

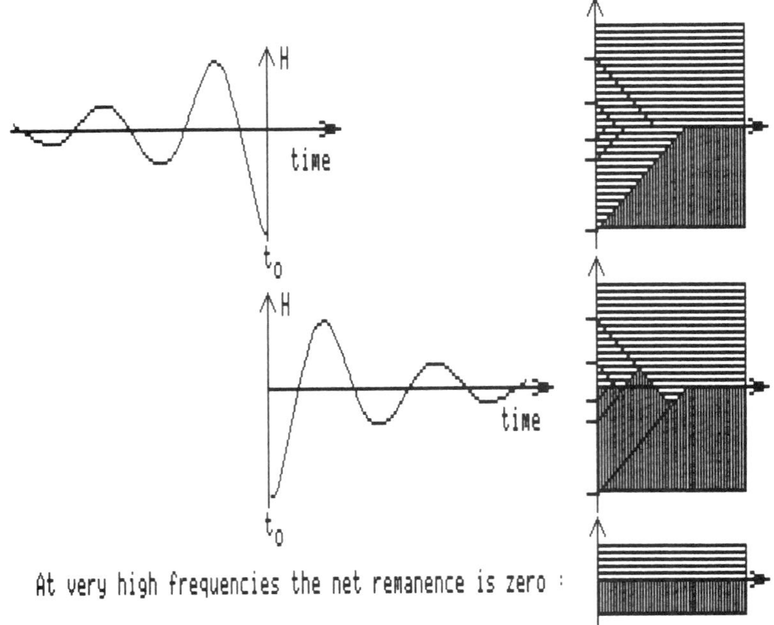

Fig. 20-4. Principle of AC-erasure.

We will next apply these two rules and observe the outcome; first we will treat the problem of erasure, and then work into AC-bias recording, also called anhysteretic recording.

AC ERASURE PROCESS

AC erasure of magnetic tapes, disks or heads is done by applying an alternating field that at some time has a value that exceeds the material's coercivity by a factor of at least five.

The field strength is next decreased slowly to zero, in a time span that corresponds to several hundred cycles duration, i.e., longer than 5 seconds for a demagnetizing field from a 60 Hz line.

The alternating field that increases in amplitude is initially applied as shown in Fig. 20-4. A maximum is reached at t_0 whereafter the field decays. This decay is in practice accomplished by removing the degausser from tape, disk or head, and in a recorder when a segment of a tape leaves the gap in an erase head.

The illustration shows that alternating, and decreasing, amounts of magnetization are left in the tape. Only a proper timing of the alternating cycles would leave the tape with a net magnetization zero. A higher frequency must be used for complete erasure (see bottom Fig. 20-4).

This first example of the use of the Preisach diagram is in complete agreement with our practical experience.

RECORDING WITH AC-BIAS

We will in this section examine the record process by following a small coating segment dx by dy (by track width) on its way across the record head. It will experience an increasing and then decreasing field strength H, with a maximum in front of the gap in the head. This is illustrated in Fig. 20-5 for both a DC and an AC current through the head winding.

In AC-bias recordings a high-frequency AC-bias signal is added to the data signal and both fed into the head winding (Fig. 20-6). Note the waveform of the added signals; the process of addition is linear.

The coating will experience a strong field in its front, in contact with the head while the strength will be less in the back. This will complicate an explanation of the AC-bias process a great deal, and we will therefore limit our initial discussion to the small element dxdy, as shown in Figs. 20-5 and 20-6; the third dimension of the element equals the track width.

The field variations experienced by the element as it moves across the head is shown in the amplitude varying signal above the head.

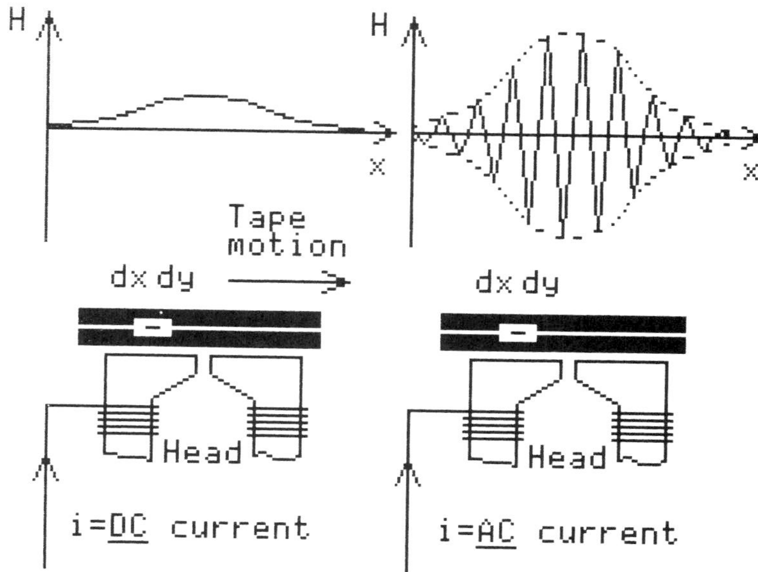

Fig. 20-5. Fields H experienced by dxdy as it moves past a head energized by a DC-current and an AC-current.

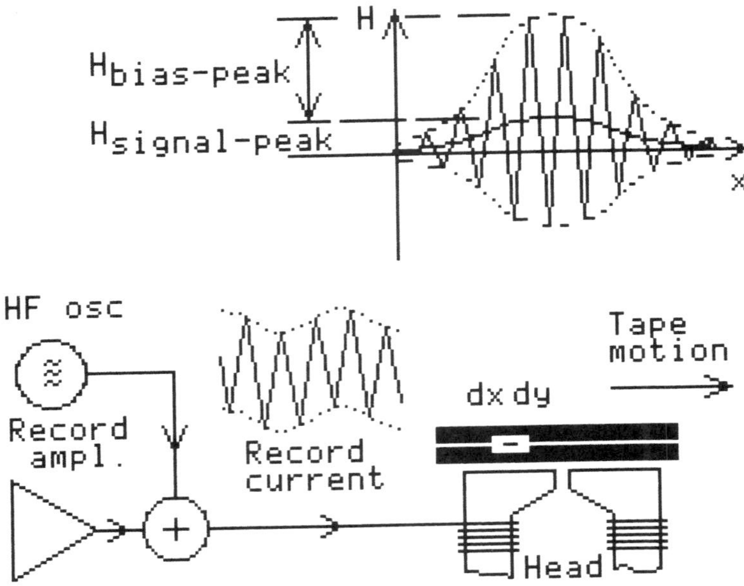

Fig. 20-6. Addition of AC-bias to the data signal current produces an alternating field for tape element dxdy as it moves across the gap. Note the summation of currents, NOT modulation.

Magnetization Level Versus Data Current Level

Figure 20-7 shows the remanences left after three levels of DC current only. They are nonlinearly related to the current levels, and illustrate the curvature of the initial magnetization curve. The magnetized regions of the Preisach diagram are shown by the clear triangles, and their areas increase as the square of the current value. This is in agreement with Fig. 20-1, and with the beginning of the initial magnetization curve from Fig. 3-12. It also illustrated the highly non-linear nature of direct recording of signals, ref. Fig. 5-4.

Now we add an AC current to the data current and the process will become linear. When at first a low frequency AC-bias current is added to the DC current then the field signal strength varies as shown in Fig. 20-8, leaving the element magnetized according to rules No. 1 and 2, from Fig. 20-3. It is clear that the remanence is "cleaner" the higher the AC-bias frequency is.

This finding is the cause for a rule-of-thumb: *The bias signal frequency should be as high as feasible, at least five times the highest data frequency, and no less than three times this frequency.*

A very high AC-bias frequency will result in an almost straight line, with very small "jaggies," and it will appear that the magnetized region in the Preisach diagram is a triangle that now is proportional to the data current since the height of the triangle equals the corresponding data field level.

Figure 20-9 illustrates this relationship, summarized in Fig. 20-10. The level's magnitudes relate as 1:2:3, and so do the heights of the triangles. This should lead to increases in magnetization as 1:2:3, but does not, as shown in the measured graph. At levels greater than 3 the curve levels off and we say that the material saturates.

It is more correct to say that there are no more particles to magnetize, as illustrated in the accompanying 3-D views of the number of particles magnetized (see Fig. 20-11). The height of the triangular region will at saturation reach into a region where only few particles are located, i.e., particles under the influence of very large interaction fields. (Note: These fields can theoretically not be larger than the particle coercivities.) The measured distribution in Fig.

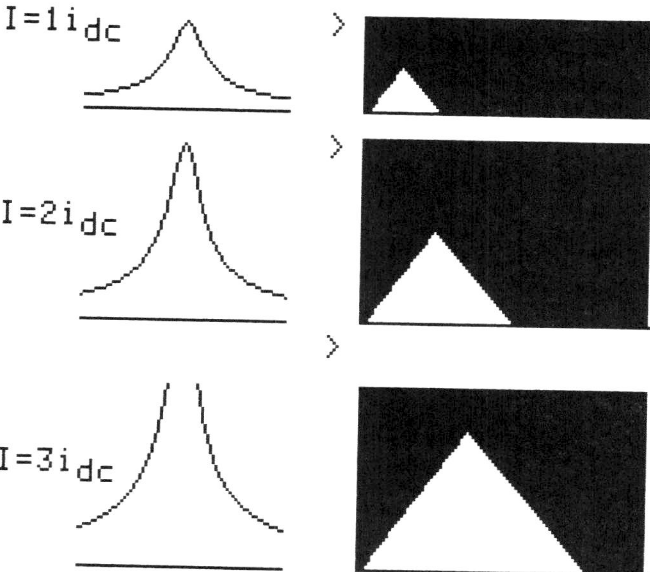

Fig. 20-7. The remanence will ideally quadruple when the signal current is doubled.

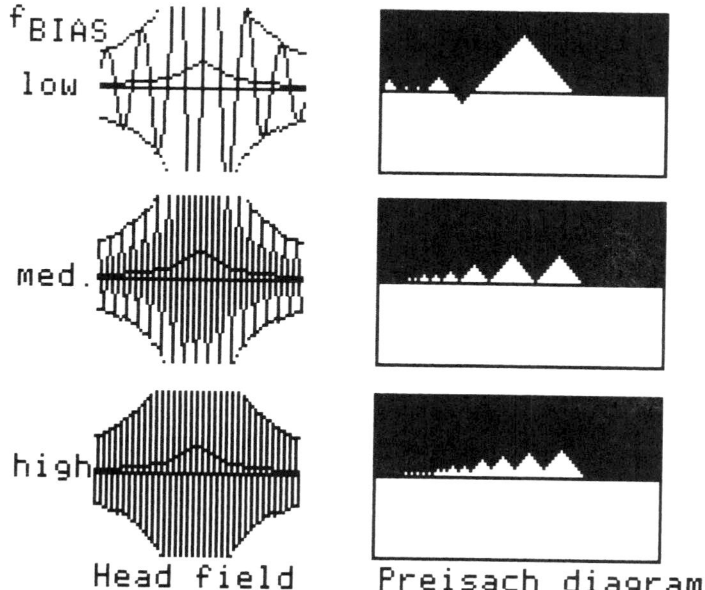

Fig. 20-8. AC-bias recording improves with higher bias frequency.

20-2B shows two humps at low field values at regions where ΔH is larger than H_c, and the blame placed on measurement errors. But the diagram merely indicates that some large particles (with low h_{ci}) are located in the vicinity of smaller particles with higher interaction fields.

The relationship between the magnetization M and the record current I_{data} is the recorder's transfer curve, and we observe its excellent linearity when using AC-bias. It clearly reflects the distribution curve for the particle interaction fields (Fig. 20-12).

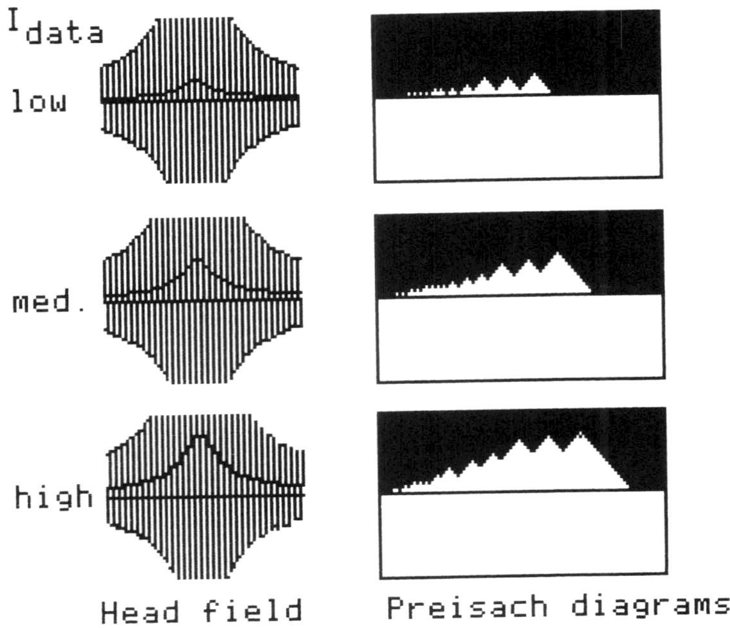

Fig. 20-9. Magnetization is proportional to data current level.

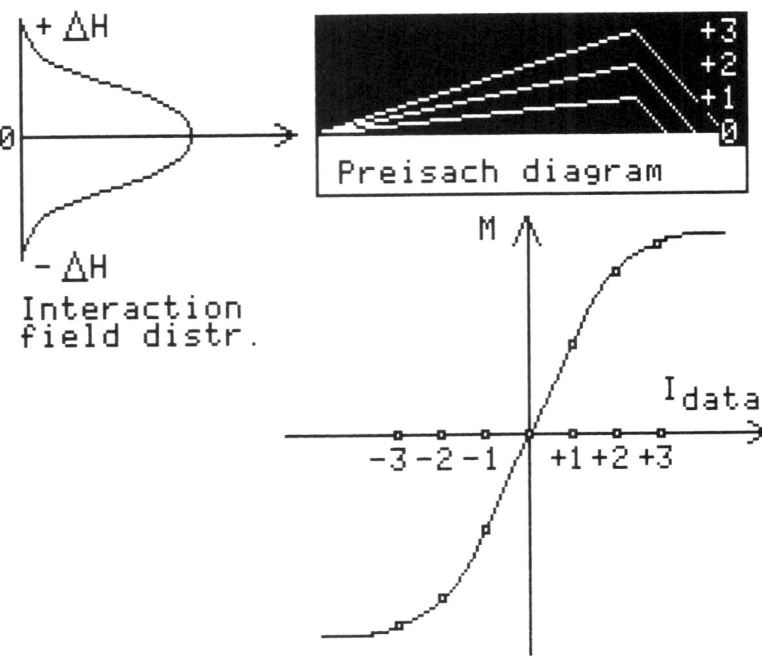

Fig. 20-10. Magnetization M is limited by interaction field distribution at high data signal currents.

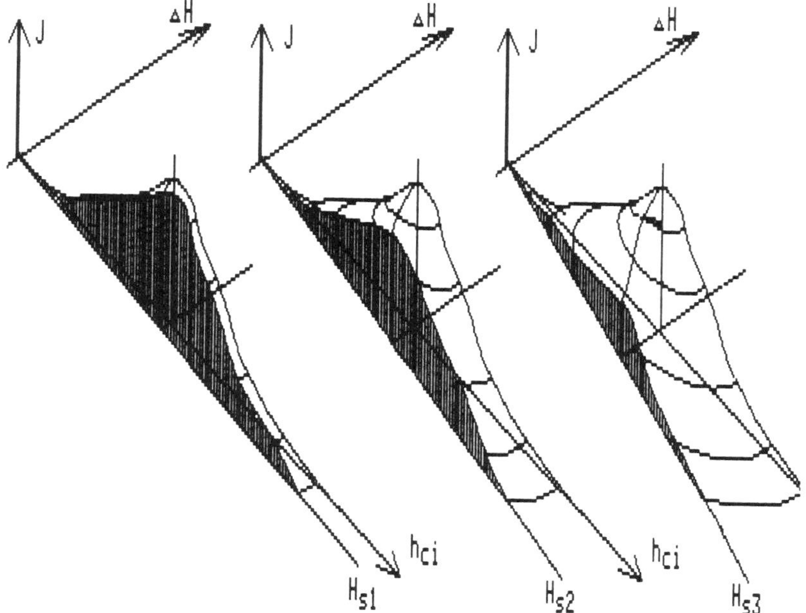

Fig. 20-11. Magnetization limitation illustrated by the 3-D Preisach diagram.

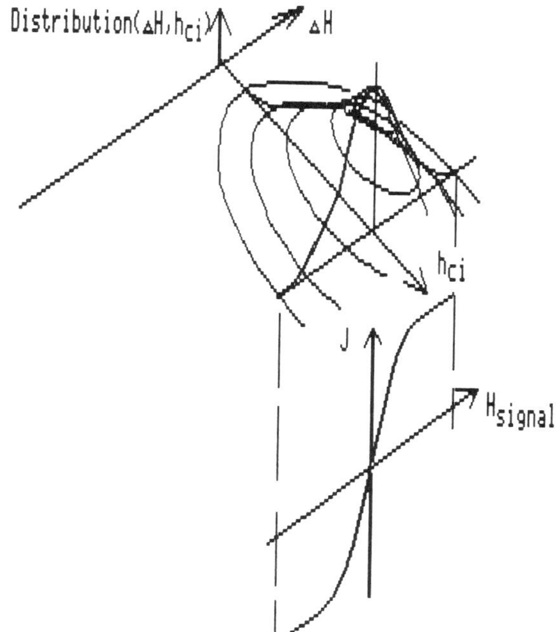

Fig. 20-12. Correlation between the interaction field distribution and the overload curve.

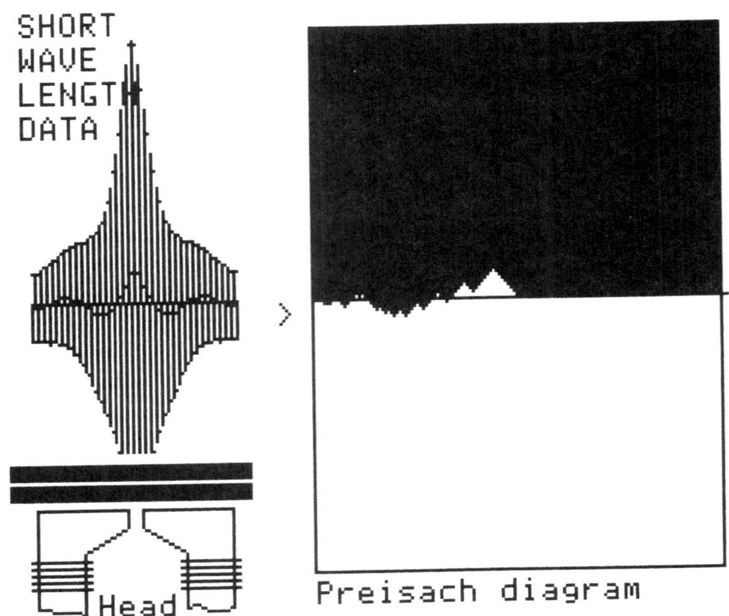

Fig. 20-13. Recording head data field alternates too fast during the recording of very short wavelengths.

We will later return to the transfer curve for a discussion of linearity and distortion in AC-bias recordings.

Magnetization Level Versus Data Frequency

When the frequency of the data signal is very high then the magnetization process changes as the element dxdy moves across the head. This has been discussed in connection with the length of the recording, or freezing zone at the trailing edge of the gap in the record head, and the actual loss occurring computed in Chapter 19.

The Preisach diagram offers an additional view in Fig. 20-13, where the magnetized region now alternates around the zero line. Figure 20-14 offers a 3-D view of the number of particles magnetized; the + portion subtracts from the − portion, and the net magnetization is zero when the two are equal.

Magnetization Versus AC-Bias Current Level

The AC-bias current level is now varied while keeping the data current level constant, at a small value (in the linear range). Successive applications of rules no. one and two result in the three graphs shown in Fig. 20-15, summarized in Fig. 20-16.

The magnetization level should ideally increase linearly, but the curve is S-shape as shown in the actually measured curve. By placing the switching field distribution curve below and in proper relation to the Preisach diagram we quickly see that there are no particles to magnetize at very low levels, and there are no further particles to magnetize at high levels; the result is the S-shape.

As we increase further the AC-bias level we will in a recorder measure a decline in magnetization (overbias, see Fig. 20-17). This phenomenon has elsewhere in this book been related

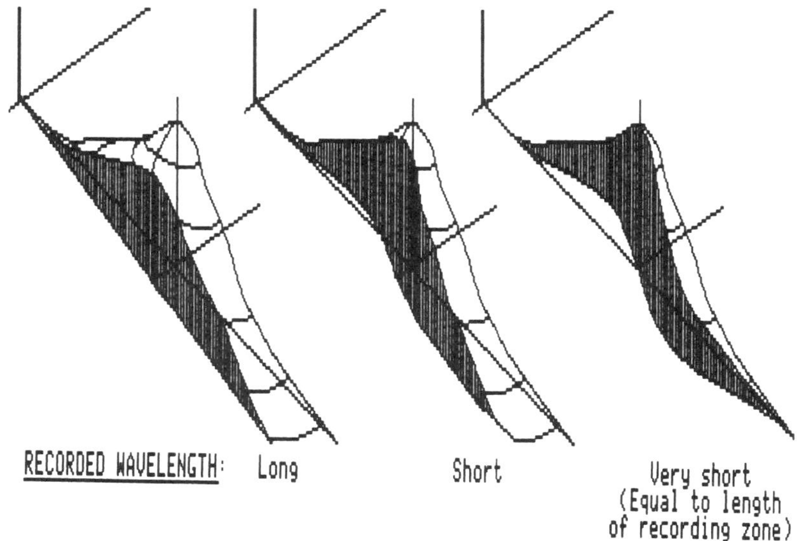

Fig. 20-14. Remanent magnetizations at short wavelengths.

to the steeper record field and subsequent perpendicular remanence in the otherwise longitudinally oriented coating.

An additional factor now enters the picture: Notice how the top side in the triangles in the Preisach diagram (Fig. 20-17) gently moved closer to the horizontal line at H_2 at very high AC-bias levels. The extension to the right of the triangles does not increase the magnetization since there are no particles there, and the lowering of the top side will on the other hand *reduce* the magnetized area where the particles are located.

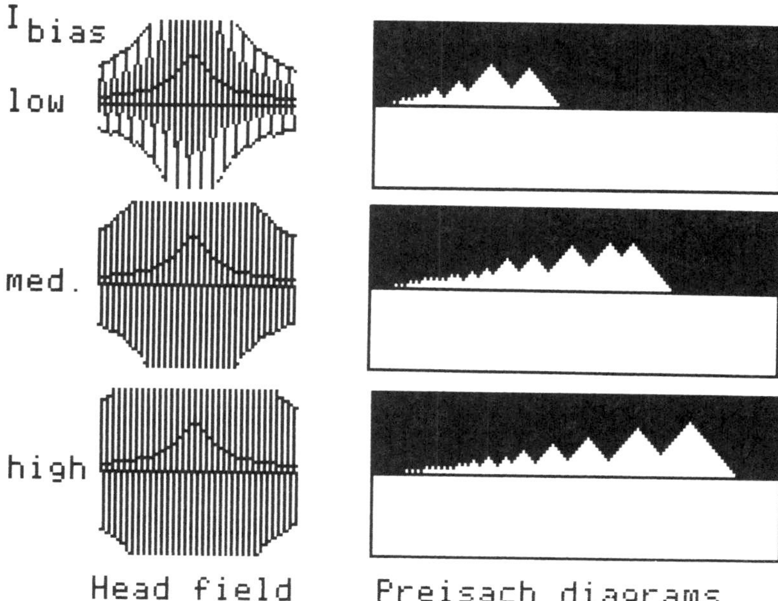

Fig. 20-15. Magnetization is ideally proportional to the bias current level.

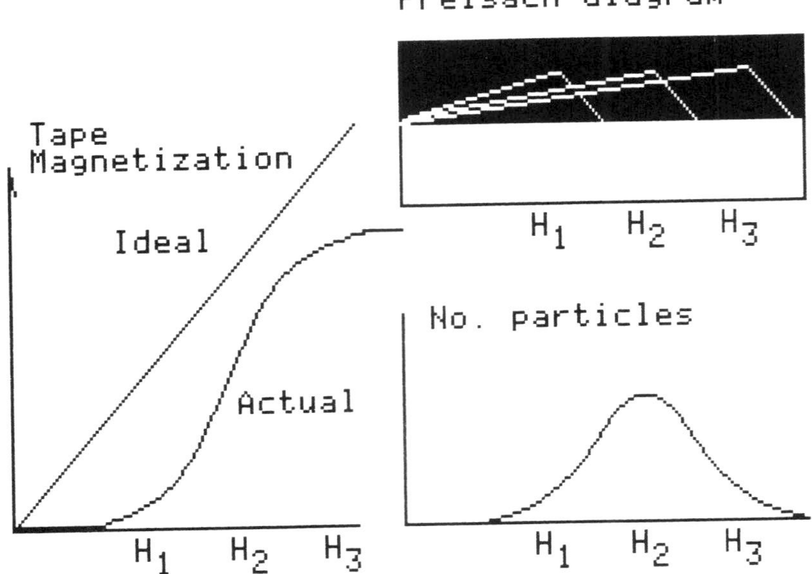

Fig. 20-16. Magnetization is limited by the switching field distribution.

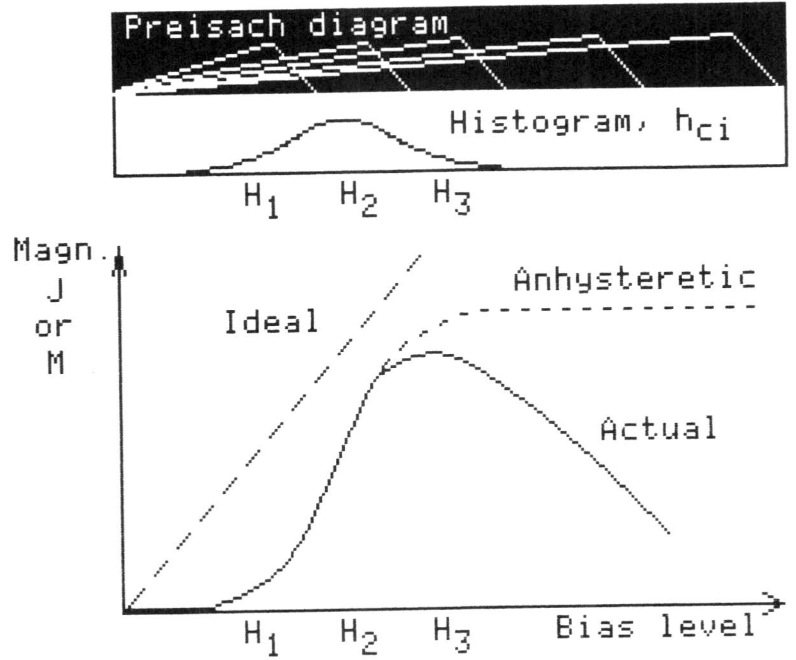

Fig. 20-17. Increasing the AC-bias field will at first increase the remanence, thereafter it will decrease it.

This topic was investigated by Daniel and Levine. They made measurements on tape samples placed in the center of a long solenoid, and their instrumentation would decrease the bias field while keeping the data field constant. Their output versus bias level is shown as the broken line curve, called the *ideal anhysteretic* remanence, where the signal level is the same everywhere during the decay of the bias field. The AC-bias process when both field decays simultaneously is named *modified anhysteretic.*

In early investigations a fairly large discrepancy was found between the measured and the calculated output results, in some cases as much as 8 dB. This mystery was resolved by Bertram who included the reduced coating magnetization due to the perpendicular magnetization which is pronounced at high bias levels; his analysis reduced the difference between measured and calculated remanences to about 1 dB.

The difference between the ideal anhysteretic curve (broken line) and the actual magnetization (solid line) in Fig. 20-17 is therefore made up of a lowering of the number particles to become magnetized (area under triangles, between H_1 and H_3), and the transition toward perpendicular magnetization (dead layer, velour effect).

An AC-bias tape specification can be expressed by its susceptibility, which is its low-level signal sensitivity:

$$\text{Magnetization J} = \text{Susceptibility } \chi * \text{Signal field } H_s$$

$$J_r = \chi * H_s \text{ (Unit: Gauss/Oersted)} \quad \textbf{(20.3)}$$

χ has its electronic counterparts in the amplification factor μ of a vacuum tube or β of a transistor.

Most modern tapes are manufactured to provide a coating with longitudinally oriented particles. This provides for a several dB higher output across the frequency spectrum. We may measure a value $\chi_x$for the longitudinal sensitivity of susceptibility; we will also find that the coating is sensitive to perpendicular magnetization. The *perpendicular susceptibility* χ_y is quite a bit less than χ_y for oriented tapes:

$$\chi_y \cong \chi_x/4 \quad \textbf{(20.4)}$$

It has also been found that the perpendicular remanence is significant in many short wavelength recordings. This is the area where all fields of application are looking for improvements.

REMANENT MAGNETIZATION

We will first examine the remanence after a recording is made with normal bias setting (Fig. 20-18). We will assume a coating permeability, μ_r,of one, so the field lines are semiarcs (the latter approximation is acceptable so long as $R = \sqrt{x^2 + y^2} > L_g$:

$$J_x = \chi_x * H_x \quad \textbf{(20.5)}$$
$$J_y = \chi_y * H_y \quad \textbf{(20.6)}$$

The fields H_x and H_y can be expressed as:

$$H_x = H_s * \sin \alpha$$
$$= H_s * y/R$$

and

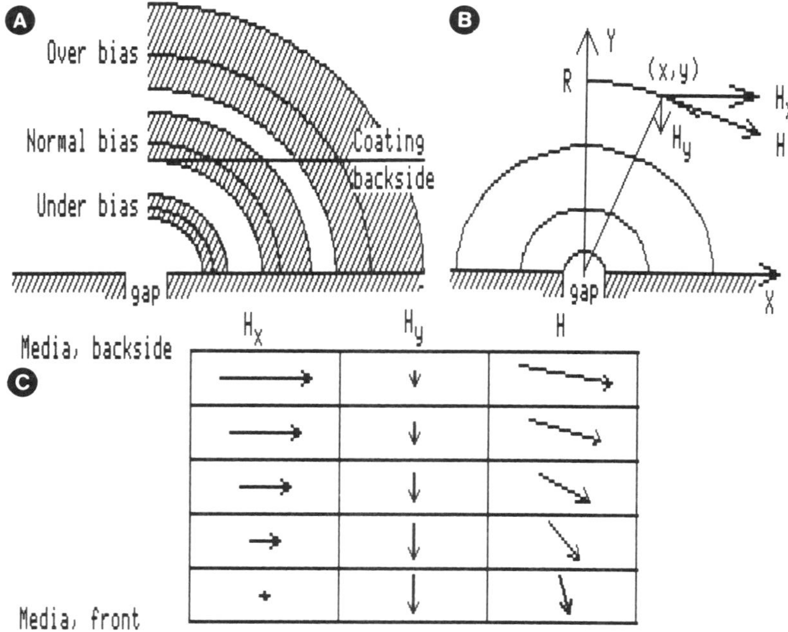

Fig. 20-18. Vector magnetization after recording with AC-bias. (A). recording zones shown cross hatched; (B). components of the recording field; (C). remanent magnetization in the coating.

$$\begin{aligned} H_y &= H_s * \cos\alpha \\ &= H_s * x/R \\ &= H_s * \sqrt{R^2 - y^2}/R \end{aligned}$$

The remnant magnetization components are now:

$$J_x = \chi_x * H_s * y/R \qquad \textbf{(20.7)}$$
$$J_y = \chi_y * H_s * x/R \qquad \textbf{(20.8)}$$

from which we can compute the composite magnetization vector.

The remanent magnetization J is therefore proportional to the total field strength H_s multiplied with susceptibility χ_x.

Our previous formula will also apply for the underbias condition, provided that gap length is smaller than the effective coating thickness c′ of the magnetized layer and that the expression in the formula is *multiplied by* (c′/c) for underbias (Fig. 20-18).

The overbias condition broadens the recording zone and the remanence becomes more perpendicular.

A vectorial picture of the calculated remanent state in the recorded tape is shown in Fig. 20-18. Comparisons with Fig. 5-2 and Fig. 21-7 show good agreement.

Demagnetization and interaction between the magnetized areas will modify the final remanent state.

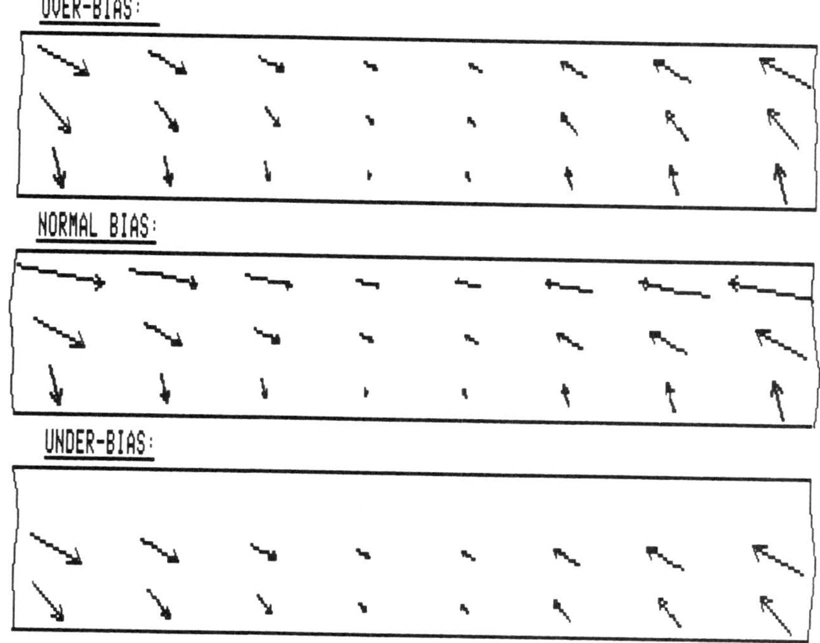

Fig. 20-19. Magnetization patterns in AC-bias recorded coating.

Two observations are important:

- *The remanent magnetization is stronger at the back side of the coating than in the surface.*
- The surface remanence is predominantly perpendicular.

The first statement has been verified by other workers in the field. Bertram found a linear increase in H_x from zero at the coating surface to a maximum at the backside and he also includes the perpendicular component. He proved, in the latter paper, that the y-component must be included into a total vector field in order to achieve agreement between theory and measurements of the effective anhysteretic susceptibility.

Hersener conducted experiments with a scale model and his results verify both statements. Bate and Dunn found a perpendicular magnetization that ranged from 2 to 15 percent of the in-plane component. Further proof of the perpendicular components is evident from recent years' work, which goes further to make media that will enhance the perpendicular magnetization.

Figure 20-19 shows three recorded patterns, using our previous formula for calculation of the magnetization in segments of the coating, or packets each consisting of a very large number of particles. The similarity of Fig. 20-19 with the experiments of Tjaden and Leyten, Hersener and Iwasaki's is evident.

LINEARITY AND DISTORTION

Figure 20-20 shows the transfer curve for AC-bias recording; the low-level cross-over distortion from direct recording without bias is non-existent. This transfer curve was produced by Koester, who made measurements on a large number of tapes. Some 6 to 10 specimens

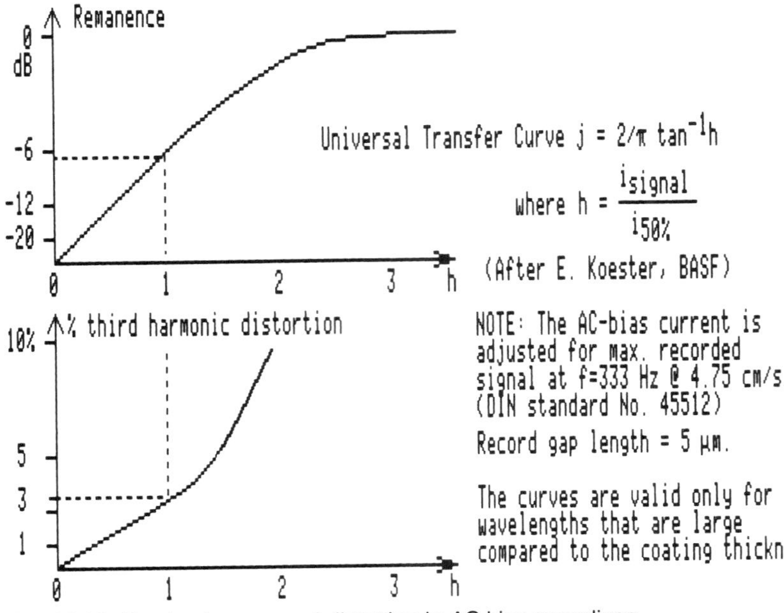

Fig. 20-20. Overload curve, and distortion in AC-bias recordings.

of different tape materials (γFe_2O_3, $Co-\gamma Fe_2O_3$, CrO_2, Fe), in thicknesses from 2 to 6 μm were measured and their transfer curves all normalized to a number j_{ar} for the anhysteretic remanence, $j_{ar} = J_r/J_{sat}$. All curves were found to be *equal to a universal transfer curve* to within + − 2.5 percent!

This is the curve shown in Fig. 20-18 and it can be expressed:

$$j_{ar} = (2/\pi)\ \text{atan}\ (h) \qquad \textbf{(20.9)}$$

where h is the normalized record current i_{data} divided by the record current $i_{50percent}$ required for a remanence that is 50 percent of the saturation value.

Its universal applicability makes it a quite useful curve, and it can be expressed as:

$$j_{ar} = a_1h + a_3h^3 + a_5h^5 \qquad \textbf{(20.10)}$$

where:

$a_1 = 0.57$
$a_3 = -0.076$
$a_5 = 0.0065$

By substituting $h = h_o \sin \omega t$, the *third harmonic distortion* can be calculated, and the result is shown below the universal transfer curve.

Note that the levels for one percent distortion are 10 dB below saturation and it is only 5 dB below for 5 percent distortion.

This margin, or *headroom*, is essentially the same for all tapes by nature of their close fit to the universal curve, which is valid for bias set to optimum output as noted in the figure.

It is common practice, during adjustment of bias, to record a low frequency signal and observe the playback signal level. The bias is often adjusted for peak signal (maximum susceptibility), or until the signal level has dropped one or two dB. The latter provides for better recording with less noise and distortion. If the operator also monitors the third harmonic distortion he will see a pronounced *dip in distortion* somewhere before the signal level reaches its peak. This is illustrated in Fig. 20-21.

The reason for this dip can be understood if we examine the transfer curves for three different bias levels (Fig. 20-21, right). Three bias levels are shown: normal (optimum bias) "b," underbias "a" and optimum "o." Curves "a" and "b" have opposite initial curvatures and the third harmonic distortion does therefore change phase in relation to the fundamental when the bias level changes from "a" through "b." The transfer curve for "o" is the straightest line and at that bias level there is minimum distortion.

One may object by referring back to Fig. 20-18, and pointing out that the recording conditions vary through the tape coating thickness due to anisotropy. And indeed the curves in Fig. 20-21 are valid only for a very *thin* coating. The *distortion analysis* was verified in the first published analytical treatment of distortion by Fujiwara in 1979. His treatment of the subject is a testimony to the Preisach statistical approach to recording theory, as described in this chapter. Fujiwara calculated the magnetization curves based on the distribution of particles switching fields, measured the material constants for these and next calculated the sum total of distortion in the magnetization in several layers of the coating. A computed playback signal, using reciprocity, gave results that correlated well with measurements.

The remanence from an AC-biased recording can be determined from the two-dimensional density function $h(H_c, \Delta H)$ in the Preisach diagram. The net magnetization can be expressed as the difference between the particle magnetizations in the upper $(+\Delta H)$ and lower $(-\Delta H)$ portions of the diagram.

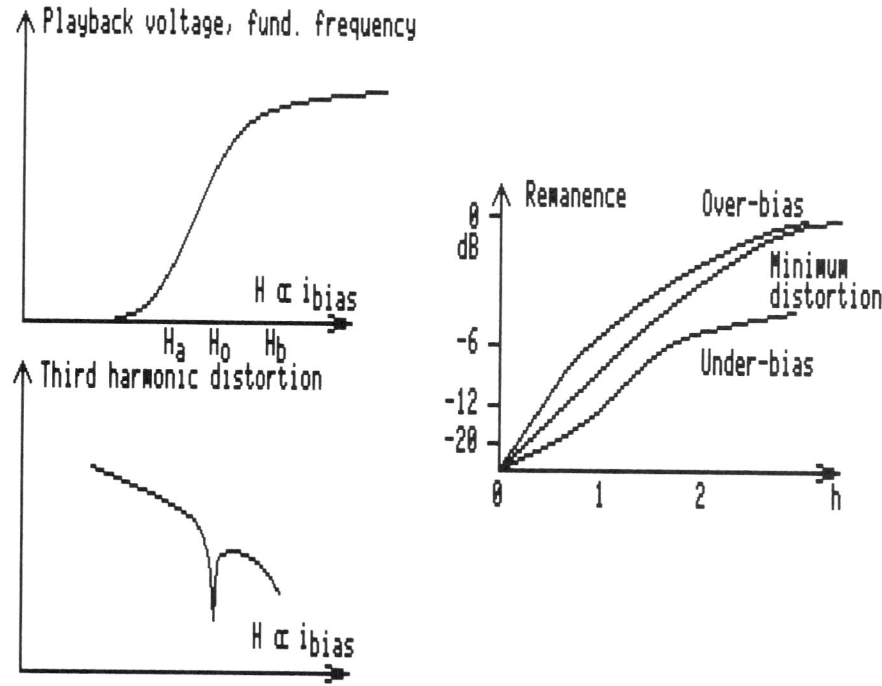

Fig. 20-21. Output and distortion as functions of bias current level (left); Remanence versus signal level at three bias levels.

coercivity distribution as well as to the interaction fields. The *net magnetization J* can then be expressed as:

$$J = 0.5 (\text{erf} (\alpha((H_s - H_c + H_b)/H_c) + \text{erf} (\alpha ((H_s + H_c - H_b)/H_c)) \quad \textbf{(20.11)}$$

where:

erf = error function (see tables in mathematical handbooks).
α = material constant = $H_c/\sqrt{2} * \sigma_c$.
H_c = mean value of coercivity.
σ_c = standard deviation of switching field distribution.
H_b = bias field.
H_s = data signal field.

When the bias field equals H_c, we find:

$$J = 0.5 (\text{erf} (\alpha H_s/H_c) + \text{erf} (\alpha H_s/H_c))$$
$$= \text{erf} (H_s/\sqrt{2} * \sigma_c$$

This expression can also be expanded into a series (Lagrange polynomial):

$$J = AH_s + BH_s^3 + CH_s^5$$

where:

A = 0.398 → 0.570
B = −0.062 → −0.089
C = 0.002 → 0.003

This transfer curve compares well with the curve found by Koester.

INTERMODULATION; EVEN HARMONIC DISTORTION

Two other distortion figures need to be mentioned: *intermodulation*, which occurs when two or more signals mix in a non-linear transfer curve, and *second harmonic distortion*. Intermodulation is annoying in music recording and a source of errors in instrumentation recorders. Its calculation and prediction is difficult to the point of not being worth the effort and its occurrence should be avoided by using fast, peak level indicators (oscilloscopes, L.E.D. devices or similar rather than VU-meters).

The second harmonic distortion does not really exist in tape recording since the transfer curve is symmetrical. So when it shows up, something is wrong in the form of a dc-component:

- Magnetized heads.
- Magnetized tape guides.
- Leakage of DC-current into record head (Faulty coupling capacitor).
- Amplifier distortion.
- Asymmetrical AC-bias waveform.

The reader is referred to Chapter 30 on maintenance, and of course the manual for the recording equipment, in order to correct any of these malfunctions.

This difference is shown graphically in Figs. 20-10 and 20-11 for three signal levels $H_{s1} < H_{s2} < H_{s3}$. Clearly, as H_s increases, the net remanence increases. The relationship is linear for small values of H_s. The magnetization J versus H_s is shown in the middle of Fig. 20-12 and can be calculated if we assume that the density function is gaussian with respect to the

BIAS OSCILLATOR

The high-frequency bias current added to the record-head current improves the linearity of the magnetic recording process. Bias is generated by an *oscillator* and in audio and home-type recorders the same oscillator is used to produce current for the erase head. The bias current, as mentioned earlier, may cause intermodulation in the output stage of the record amplifier (beats). The bias frequency in audio or instrumentation recorders is normally five times the highest frequency to be recorded. For home recorders this means that a 60- to 75-kHz bias oscillator frequency is required. This is also what the erase current for the erase head requires. In instrumentation recorders with extended bandwidth to 4 MHz or higher, the bias frequency has to be 14 to 20 MHz, which in turn imposes a design limitation upon the record head, extending its self-resonance beyond 25 MHz.

The bias current level is normally 8 to 10 times that of the data current. In order to *avoid* beats between bias oscillators in multi-channel recording, it is a common design practice to use one master oscillator with buffer amplifiers in each record-amplifier section. This applies in particular to instrumentation recorders where as many as 14 channels may be recorded simultaneously.

While high-frequency bias added to the record current improves the linearity of analog magnetic recordings, its waveform must be purely sinusoidal in order to avoid excessive noise and distortion.

PLAYBACK AMPLITUDE VERSUS FREQUENCY RESPONSE

We can now, finally, determine the recorded flux levels. We will consider three bias conditions: under bias, normal bias, and over bias.

Normal bias is here defined as that bias field which records the tape in its exact coating thickness (Fig. 20-19, B). We then have, at the shortest wave length, a demagnetization loss and a recording zone loss. To this we must add the normal coating thickness loss, and the net available flux is shown as the bottom curve in Fig. 20-22.

An *under-bias* condition (see C in Fig. 20-19) only records the surface portion of the coating. Hence both demagnetization loss and coating thickness losses are less than for a normal bias setting. And so are the recording losses due to the shallower recording zone.

This appears deceitfully promising for high frequencies, but the quality is poor. Distortion is higher, the long wave length output lower, and the sound in audio recording is unsteady and "gurgled."

Over-bias, on the other hand, does not increase the level at long wave length beyond the levels for normal bias; on the contrary, the recording zone now becomes so wide that the record losses extend into the long wavelength area. See Fig. 20-19A. Coating thickness and demagnetization losses are, on the other hand, unchanged from normal bias.

Concluding Comments

Over bias is used in top-quality audio recording because of its better sound fidelity; the excessive short wave length loss is traded for a higher tape speed.

It is beneficial to control the bias level by the record signal to achieve less distortion and better high frequency response (Ref. Selmer Jensen).

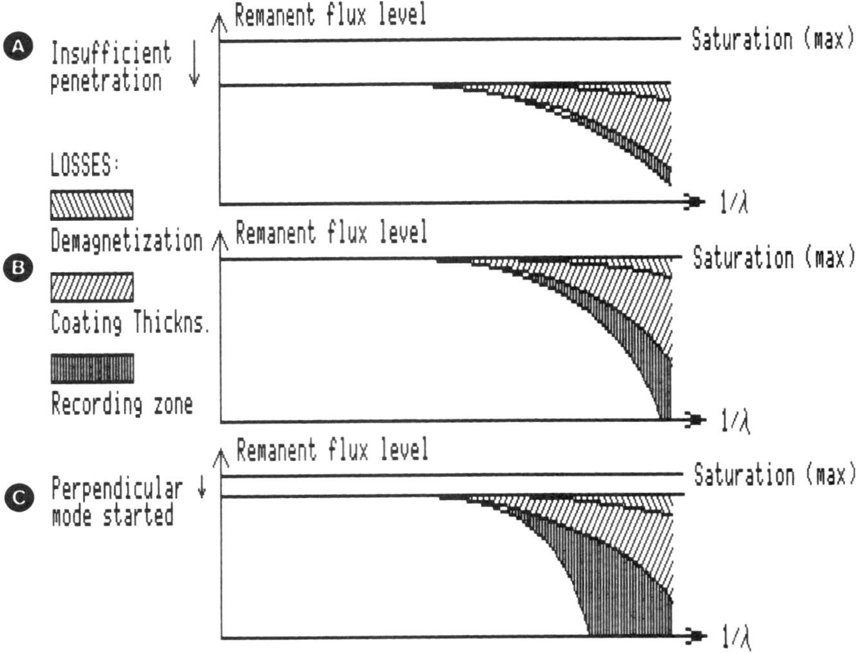

Fig. 20-22. Available flux levels in AC-bias recordings, as function of frequency (1/wavelength). (A) Under bias. (B) Normal bias. (C) Over bias.

The information in this chapter pertains to recording of analog as well as digital signals, although the latter rarely occurs (digital data are recorded directly). When recording digital data it often becomes necessary to synchronize the bias oscillator to the data clock rate, and assure that the bias frequency has an integer relationship to the clock frequency.

The ratio between data and bias signal amplitudes is much less in digital recording than in audio/instrumentation. The maximum ratio of bias-to-data levels is generally between 10:1 and 30:1 in audio (at high signal levels), while it can be as low as 6:4 in digital. The author has found very little documentation on the latter issue.

REFERENCES TO CHAPTER 20

Woodward, J.G., and Della Torre, E., "Magnetic Characteristics of Recording Tapes and the Mechanism of the Recording Process," *Jour. AES*, Oct. 1959, Vol. 7, No. 4, pp. 189-195.

Daniel, E.D., and Levine, I., "Experimental and Theoretical Investigation of the Magnetic Properties of Iron Oxide Recording Tape," *JASA*, Jan. 1960, Vol. 32, No. 1, pp. 1-15.

Daniel, E.D., and Levine, I., "Determination of the Recording Performance of a Tape from Its Magnetic Properties," *JASA*, Feb. 1960, Vol. 32, No. 2, pp. 258-267.

Schwantke, G., "The Magnetic Recording Process in Terms of the Preisach Representation," *Jour. AES*, Jan. 1961, Vol. AU-9, pp. 139-145.

Bate, G., "Statistical Stability of the Preisach Diagram for Particles of Gamma-Ferric-Oxide," *Jour. Appl. Phys.*, July 1962, Vol. 33, No. 7, pp. 263-269.

Koester, E., "The Reversible Susceptibility of Fine-Particle Assemblies and the Magnetic Anisotropy of Gamma-Fe2O3 and CrO2 Particles," *IEEE Trans. Magn.*, Sept. 1969, Vol. MAG-5, No. 3, p. 263.

Hersener, J., "Modelluntersuchungen zur Magnetisierungsverteilung im Magnetband bei sinusformigen Signalen," *Wiss. Ber. AEG-Telefunken*, 1973, Vol. 46, No. 1, pp. 15-24.

Bertram, H.N., "Long Wavelength AC Bias Recording Theory," *IEEE Trans. Magn.*, Dec. 1974, Vol. MAG-10, No. 4, pp. 1039-1048.

Bertram, H.N., "Wavelength Response in AC Biased Recording," *IEEE Trans. Magn.*, Nov. 1975, Vol. MAG-11, No. 5, pp. 1176-1179.

Bertram, H.N., and Niedermeyer, R., "The Effect of Demagnetization fields on Recording Spectra," *IEEE Trans. Magn.*, Sept. 1978, Vol. MAG-14, No. 5, pp. 743-745.

Brock, M., "Preisach Distribution in Magnetic Tapes," *Thesis for M.Sc.*, Tech. Univ. of Cph., 1979.

Fujiwara, T., "Nonlinear Distortion in Long Wavelength AC Bias Recording," *IEEE Trans. Magn.*, Jan. 1979, Vol. MAG-15, No. 1, pp. 894-899.

Fujiwara, T., "Even-Order Harmonic Distortion in AC-Bias Recording," *IEEE Trans. Magn.*, Sept. 1979, Vol. MAG-15, No. 5, pp. 1336-1339.

Koester, E., "A Contribution to Anhysteretic Remanence and AC Bias Recording," *IEEE Trans. Magn.*, Nov. 1975, Vol. MAG-11, No. 5, pp. 1185-1188.

Fujiwara, T., "Wavelength Response of Harmonic Distortion in AC-Bias Recording," *IEEE Trans. Magn.*, May 1980, Vol. MAG-16, No. 3, pp. 501-507.

Selmer Jensen, J., and Pramanik, S.K., "Dynamic Bias Control with HX Professional," *Audio*, Aug. 1984, Vol. 68, No. 8, pp. 34-41.

BIBLIOGRAPHY TO CHAPTER 20

von Braumuhl, H.J., and Weber, H., "An Improved Magnetron Process," *Z. Ver. deuts. Ing.*, July 1941, Vol. 85, p. 628.

Holmes, L.C., and Clark, D.L., "Supersonic Bias for Magnetic Recording," *Electronics*, July 1945, Vol. 18, pp. 126-136.

Daniel, E.D., Axon, Ph.D., P.E., and Frost, W.T., "A Survey of Factors Limiting the Performance of Magnetic Recording Systems," *Proceedings I.E.E.*, March 1957, Vol. 104, No. 14, part B, pp. 158-168.

Stewart, W.E., *Magnetic Recording Techniques*, McGraw-Hill, New York, January 1958, 272 pages.

Woodward, J.G., and Pradervand, M., "A Study of Interference Effects in Magnetic Recording," *Jour. AES*, April 1961, Vol. 9, No. 4, p. 254.

Straubel, R., "Der Aufzeichnungsvorgang der Magnetspeichertechnik mit Wechselfeldvormagnetisierung in phaenomenologischer Sicht," *Hochfrequenztechnik und Elektroakustik*, May 1966, Vol. 75, pp. 153-162 and 2160 -225.

Iwasaki, A., "An Analysis on the State of AC-Biased Recording," *Ann. N. Y. Acad. Sci.*, March 1972, Vol. 189, pp. 3-20.

Dolivo, F., and Closs, F., "A Simulator for the AC-Biased Magnetic-Recording Channel," *IEEE Trans. Magn.*, Sept. 1978, Vol. MAG-14, No. 5, pp. 737-740.

Thurlings, L. and Kipzen, W. "On the Mechanism of Particle interaction in Magnetic Recording Media," *IEEE Trans. Magn.*, Sept. 1980, Vol. MAG-16, No. 5, pp.1120-1122.

Thiele, H., "On the Origin of High-Frequency Biasing for Magnetic Audio Recording," *Jour. SMPTE*, July 1983, Vol. 92, No. 7, pp. 752-754.

Principle of AC-Bias

Camras, M., "Graphical Analysis of Linear Magnetic recording Using High-Frequency Excitation," *Proc. I.R.E.*, May 1949, Vol. 37, pp. 569-573.

Axon, P.E., "An Investigation into the Mechanism of Magnetic Tape Recording," *Proc. Inst. Elec. Engrs.* (London), May 1952, Vol. 99, pt. III, pp. 109-126.

Daniel, E.D., "The Influence of Some Head and Tape Constants on the Signal Recorded on Magnetic Tape," *Proc. IEE (London)*, May 1953, Vol. 100, Pt. III, pp. 168-175.

Albach, W., "The Action of High-Frequency Bias in Magnetic Tape Recording," *Funk und Ton*, Dec. 1953, Vol. 7, pp. 628-630.

Straubel, R., and Nietzsch, J., "Analoge Simulation des Aufzeichnungsvorganges der Dynamischen Magnetspeichertechnik," *Hochfreq. und Elektroakustik*, Jan. 1967, Vol. 76, pp. 182-187.

Sawamura, S., and Iwasaki, S.I., "Application of Internal Reaction Field on the Analysis of Anhysteretic Magnetization Process," *IEEE Trans. Magn.*, Sept. 1970, Vol. MAG-6, No. 3, pp. 646-649.

Mallinson, J.C., and Bertram, H.N., "Write Processes in High Density Recording," *IEEE Trans. Magn.*, Sept. 1973, Vol. MAG-9, No. 3, pp. 329-331.

McCown, D.P., Barbosa, L.C. and Howell, T.D., "Comparative Aspects of AC Bias Recording," *IEEE Trans. Magn.*, Nov. 1981, Vol. MAG-17, No. 6, pp. 3343-3346.

Wohlfarth, E.P., "Review of the Problems of Fine Particle Interactions with Special Reference to Magnetic Recording," *Jour. Appl. Phys.*, March 1964, Vol. 35, pp. 783-790.

Fritsch, K. and Scholz, Chr., "Zur Verwendung des Preisach-Modells in der magnetslpeichertechnik.," *Hochfreq. und Elektroakustik*, 1965, Vol. 74, pp. 25-30.

Kneller, E., Friedlander, F.J., and Pushert, W., "AC Field 'Freezing' and 'Melting' of Magnetization in Fine -Particle Assemblies," *Jour. of Appl. Phys.*, March 1966 Vol. 37, No. 3, pp. 1162-1163.

Kneller, E., Puschert, W., "Pair Interaction Models for Fine Particle Assemblies," *IEEE Trans. Magn.*, Sept. 1966, Vol. MAG-2, No. 3, p. 250.

Arndt, W., and Carraro, U., "Elektromekanische Simulation des Preisach-Modells," *Hochfreq. und Elektroakustik*, 1968, Vol. 77, pp. 7-11.

Volz, H., "Gerat zur halbautomatischen Messung von Preisachbelegungen bei Magnetbandern," *Hochfreq. und Elektroakustik*, Jan. 1969, Vol. 78, pp. 101-105.

Straubel, R., "A Method for Calculating Hysteresis Loops of Interacting Single-Domain Particle Systems in a Nonmagnetic Binder," *IEEE Trans. Magn.*, Sept. 1969, Vol. MAG-5, No. 3, pp. 263-264.

Minnaja, N., "Magnetization Reversal of a Pair of Interacting Linear Dipole Distributions," *IEEE Trans. Magn.*, Sept. 1970, Vol. MAG-6, No. 3, pp. 649-662.

Bertram, H.N., "Monte Carlo Calculation of Magnetic Anhysteresis," *J. Phys.* (Paris), March 1971, Vol. 32, pp. C1 684-685.

Dunlop, D.J., "Grain Size Dependence of Anhysteresis in Iron Oxide Micropowders," *IEEE Trans. Magn.*, June 1972, Vol. MAG-8, No. 2, pp. 211-213.

Spindler, S., "Eine Analyse des Preisachmodells," *Arch. elektr. Inform. und Energietechnik*, Jan. 1973, Vol. 3, No. 1, pp. 31-35.

Dolivo, F., and Closs, F., "A Simulator for the AC-Biased Magnetic-Recording Channel," *IEEE Trans. Magn.*, Sept. 1978, Vol. MAG-14, No. 5, pp. 737-739.

Korolev, "Towards Analysis of Magnetic Recording on a relatively Thin Carrier by Preisach's Method," *Elektrichestvo*, September 1980, No. 9, pp. 66-69.

Huisman, H. F., "Particle Interactions and Hc: Experimental Approach," *IEEE Trans. Magn.*, Nov. 1982, Vol. MAG-18, No. 6, pp. 1095-1098.

Kneller, E., "Magnetic-Interaction Effects in Fine-Particles Assemblies and in Thin Films," *Jour. of Appl. Phys.*, Feb. 1970, Vol. 39, No. 2.

Kneller, E., "Relation Between Anhysteretic and Static Magnetic Tape Parameters," *IEEE Trans. Magn.*, Sept. 1977, Vol. MAG-13, No. 5, pp. 1388-1390.

Kneller, E., "Static and Anhysteretic Magnetic Properties of Tapes," *IEEE Trans. Magn.*, Jan. 1980, Vol. MAG-16, No. 1, pp. 36-41.

Chapter 21

Playback Waveforms (Read)

Faraday's law provides the simple expression for the playback voltage from a magnetic recording:

$$e = -n\, d\psi/dt \tag{21.1}$$

where ψ is the flux from the recording. We have used this formula for a sinusoidal flux pattern, and found:

$$e = -n\, \psi_m\, \omega\, \cos\omega t \tag{21.2}$$

The amplitude of the flux ψ_m can be determined by the principle of reciprocity, where for each element $w\Delta x\Delta y$ in the coating we have:

$$\Delta(x,y) = H_s(x,y) \bullet j(x,y) \bullet \cos\beta \tag{21.3}$$

where $j(x,y)$ is the magnetization (in Wbm) from element $w\Delta x\Delta y$ in the coating and $H_s(x,y)$ is the reproduce head's sensing function (in m^{-1}, see Chapter 6). β is the angle between the magnetization and the sensing function vectors. The total flux is found by substituting $j(x,y)$ with the recorded magnetization pattern and then carrying out a double integration, with x going from plus to minus infinity, and letting the y component go from d to $c'+d$ (d = spacing, or distance between head and tape; c' = recorded thickness):

$$\psi(x) = w \int_{-\infty}^{\infty} \int_{d}^{c'+d} H(x-\tau,y)\, j(\tau,y)\, \cos\beta\, d\tau dy \tag{21.4}$$

This integration is manageable when j(x,y) is zero in the y-direction and varies sinusoidal in the x-direction. This is the classical longitudinal magnetization, and the result of the integration was presented in Fig. 6-9 in Chapter 6.

We can also integrate when the magnetization pattern is a step from negative to positive magnetization. It is common practice to use the arctangent model for the transition zone, and to calculate the resultant flux change and its derivative to find the voltage pulse.

PULSE WIDTH; SENSING FUNCTION

The reproduced pulse is characterized by its length, as measured between the two points where the amplitude is 50 percent of the pulse peak voltage. This 50 percent pulse width, also named PW50, can be estimated from the calculations, and an approximation for a thin coating is:

$$PW50 = 2\sqrt{d^2 + (L/2)^2} \qquad \textbf{(21.5)}$$

where d is the head to media spacing, and L is the gap length.

There have been considerable difficulties in finding a good agreement between the theories and measurements in pulse recording. The advances in high packing densities indicated several years ago that the perpendicular y-component of the magnetization should be included in the integration in formula (21.4).

The complexity of the playback process becomes obscured in the integration. A picture of the interplay between the head and the tape magnetization is biased by the awareness of a short gap, almost to the point where one envisions the flux being sensed in front of the gap only. This gives the playback head a sensitivity function much like the Stein model of the record process, a sheet in front of the gap, perpendicular to the head. The mathematical solutions, where differentiation of a sinusoidal flux gives a cosine voltage, emphasizes this simple picture, as shown in Fig. 21-1A.

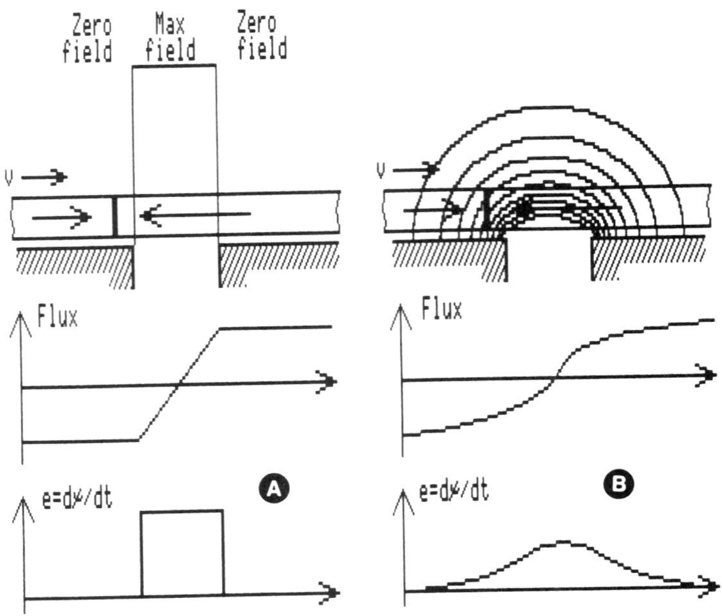

Fig. 21-1. The read pulse from a single transition is ideally a very short pulse (A), while a true read pulse broadens out (B).

The total flux is determined by the vector products of the magnetization vector and the head ''field'' vector. This results in a considerably broader sensitivity function (= the head field pattern). For a purely longitudinal magnetization the playback of a transition gives a fairly broad voltage pulse (Fig. 21-1B).

The frequently observed asymmetry in the read pulses from recorded transitions has its origin in the vector magnetization of the magnetic coating. This pattern is the result of the media magnetization anisotropy and the write field shape at the trailing edge of the write gap. The transition zone itself, lengthened by demagnetization, has less influence due to the long sensing field of the read head, extending many gap lengths on both sides of the gap: the read sensing function encompasses flux many gap lengths away from the gap.

It is relatively simple to compute the shape of a read pulse, when the magnetization pattern in the coating is known. Such information is useful in at least four areas:

A. The effects of various magnetization patterns upon the shape of the read pulse can quickly be evaluated; this relates to the media magnetization anisotropy.

B. A predicted magnetization pattern is verifiable by comparing the computed and the measured read pulses from isolated transitions.

C. A Fourier transform of the response to a single transition will reveal the write/read systems transfer function (see Chapters 22 and 23).

D. The isolated pulse response is useful in a convolution with a write data pattern (corrected for write-peak shift) to determine the read voltage pattern.

COMPUTATION OF READ VOLTAGE

The output voltage from a reproduce head reflects the magnetization pattern in the tape or disk coating. The purpose of this chapter is to outline a method of voltage computation, and to highlight some general relations between magnetization patterns and voltage waveforms.

The approach is elementary: We know from Chapter 6 that the flux contribution from a magnetization is:

$$\Delta\psi = \Delta\mathbf{j} \cdot \mathbf{H}_s \qquad \textbf{(21.6)}$$

where $\mathbf{H}_s$ is the reproduce head sensing function; this may be the Karlquist field for a long-pole head (Fig. 4-24), or as given by Potter for a thin-film head (see ref. in Chapter 8), or by Szczech and Fayling (see Bibliography of this chapter).

We can also express (21.6) as:

$$\Delta\psi = j_x H_{sx} + j_y H_{sy} \qquad \textbf{(21.7)}$$

where j_x and j_y are the magnetization components (see Fig. 21-2), and H_{sx} and H_{sy} are the sensing function components.

MAGNETIZATION PATTERNS

We will use a simple method for determination of the magnetization in the coating. It appears equally well suited for biased or non-biased, direct recordings (Fig. 20-18).

Most coatings are anisotropic due to the use of oriented, elongated particles. A modern tape will typically have a longitudinal squareness S_x equal to 0.8 to 0.85, while the perpendicular squareness S_y is only 0.2. The magnetizations' x – and y – components are determined from equations (20.5) and (20.6), or from (20.7) and (20.8). For saturation recordings use (Fig. 21-2):

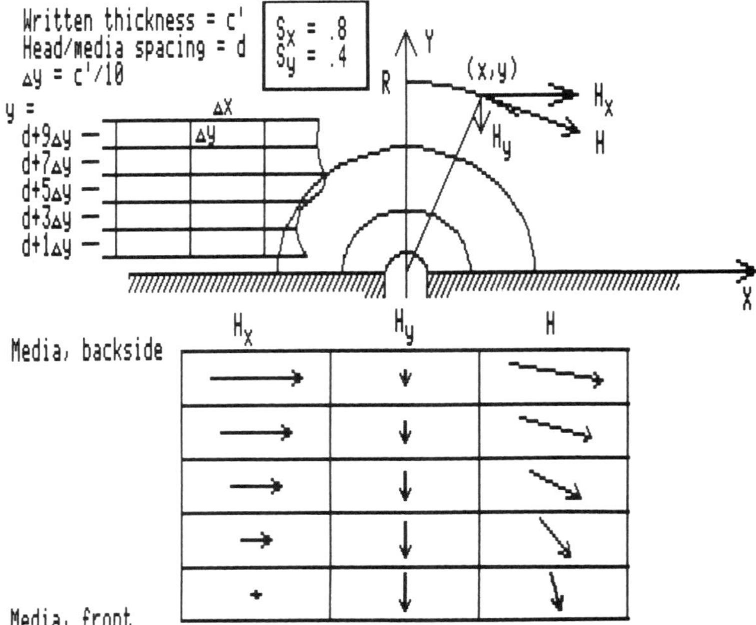

Fig. 21-2. Computation of the magnetization vectors in a magnetic coating (see text).

$$j_x = S_x J_{sat} y/R \tag{21.8a}$$
$$j_y = S_y J_{sat} x/R \tag{21.8b}$$

The resulting variation in magnetization through the coating is shown in Fig. 21-2.

RESPONSE TO A TRANSITION

Very Short Transitions

Let us, for a moment, consider how we could determine the read system's response to magnetization changes. If we use a single impulse signal then its resulting output is an impulse response. Such an impulse is obtained by differentiating a step function. The system is linear, and the differentiation can take place anywhere.

Let this be the action of the inductive read head—it differentiates a step change in magnetization into a voltage that ideally is an impulse. Thus a single magnetization transition at one ΔxΔy element produces a contribution to the impulse response. If done for all elements we will obtain the read system's "impulse response," placed in quotation marks since it really is the response to a step change in magnetization.

We will therefore, with this short method, obtain the impulse response, without moving a transition past the head. Merely sum up all for each column at a given x value, and use it for a point on the response curve, at that x value. The resulting total curve is the response that an ideal, vertical transition of width Δx will produce (Fig. 21-3).

If the transition zone is curved then the vertical summation is changed to using the flux values that are displaced in accord with the curvature. One step in this summation is illustrated to the right of the gap in Fig. 21-4. The flux contribution is—at that point—higher than for a straight transition; and it is lower for summations to the left of the gap.

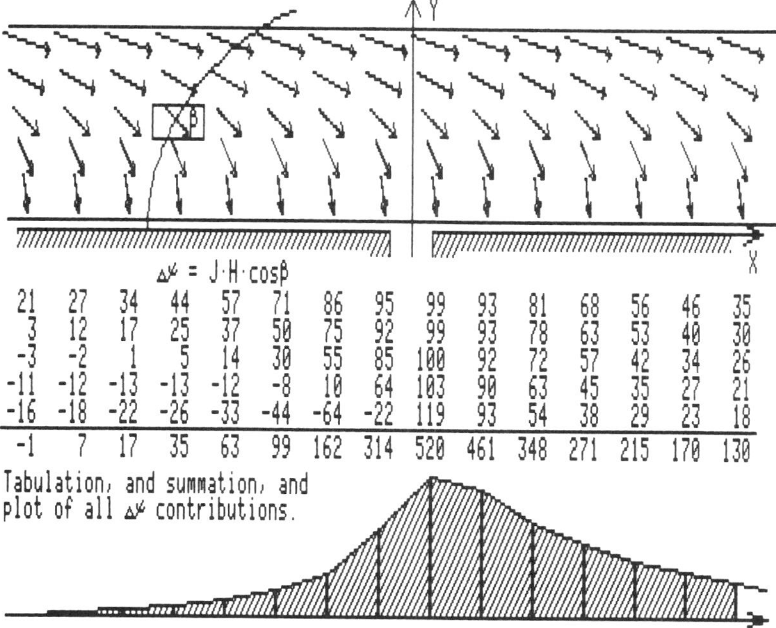

Fig. 21-3. Computation of the read flux patterns, and voltage, from a magnetized coating (gives ideal, short transition response).

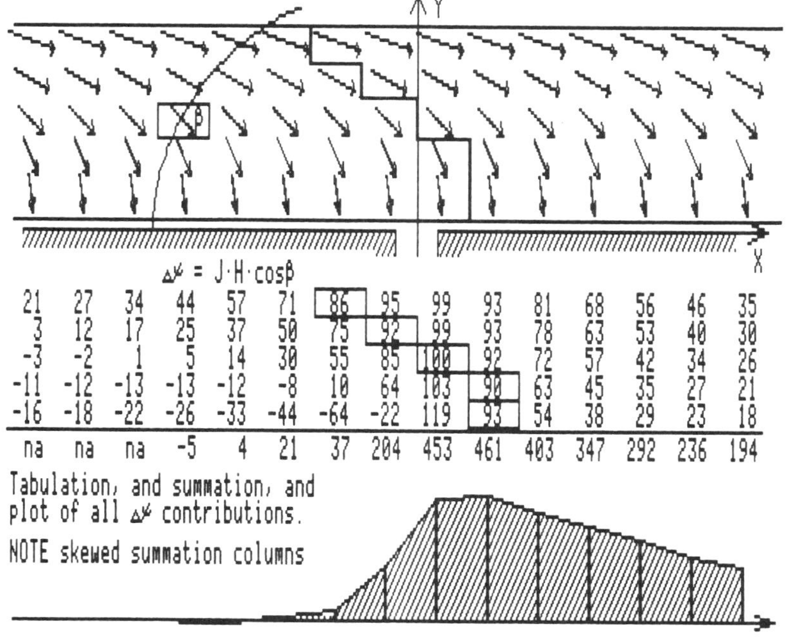

Fig. 21-4. Computation of the read flux, including the effect of a curve recording zone, and hence transition zone.

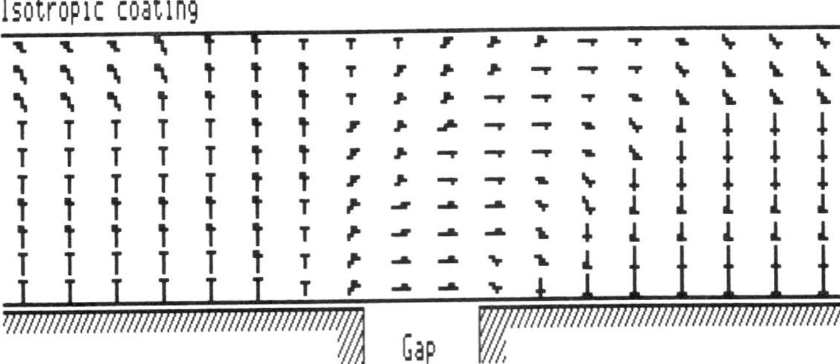

Fig. 21-5. Magnetization patterns in a long transition.

Long Transitions—the Complete Computation

We shall now determine the voltage response for a transition recorded on a coating, with an arbitrary (but known) magnetization pattern, such as, for instance, shown in Fig. 21-5.

The long way to compute the voltage is to sum all flux contributions, with the transition first at the leftmost value of x (say x = > 100 times L (gap length)), the last at the rightmost value; and for all layers, from y_{front} to y_{back}. This represents the flux value at the first position on the flux curve.

Next shift the transition one Δx to the right and re-compute the flux sum, which is the value at the next point; then make another shift and re-compute, etc. The waveform:

$$\Phi_n = \psi_n(n\Delta x) \tag{21.8}$$

is the flux waveform, and a differentiation will produce the induced voltage waveform.

Long Transitions—the Quick Computation

For a broadened transition zone it becomes necessary to use a mixture of two computational methods. First compute the response (ψ = f(x)) for the uniform magnetization pattern that exists away from the transition zone, say in the middle of a bit (or in the coating after DC-magnetization by the write head). Store the values for each x in array elements. This corresponds to the response for a very short transition (Fig. 21-3).

Now start stepping the transition zone along the x axis, starting with the transition just outside the leftmost value of x. Compute the total flux from all elements in the transition zone; add to this number the summation of all stored flux values to the left and to the right of the present position of the zone-block, see Fig. 21-6. This gives one point on the flux curve. Now step the zone one Δx to the right, and repeat the computation; repeat as many times as needed to make the transition zone move past the x-value x_{END}.

This will give a curve that represents the flux level as a function of the position of the zone center line, along the x-axis. Now differentiate this curve to get $d\psi/dx$, substitute x = vt, and we have now determined $d\psi/dt$.

Neither the first complete summation procedure nor the last shorter way are necessary for a 5 to 10 percent accurate determination of the output pulse shape. The short "once-through" computation will frequently suffice since the vector magnetization pattern has a larger influence upon the pulse shape than the zone broadening and/or curvature.

The saving in computation time is formidable: A reasonable resolution and accuracy can be obtained by using 10 Δy layers for the coating, and a couple of hundred Δx elements for

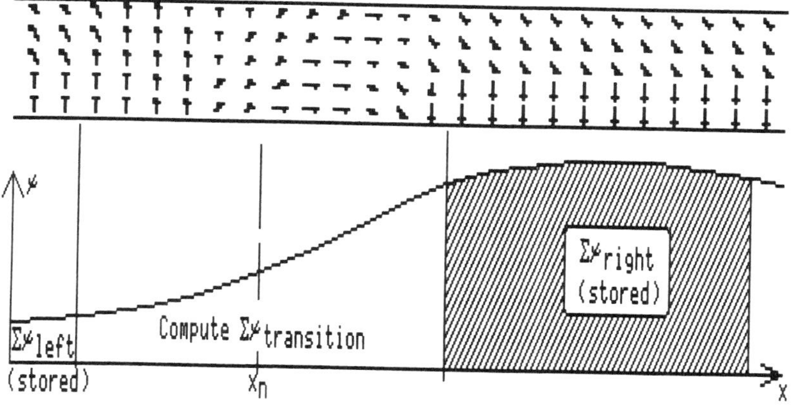

Fig. 21-6. Method of fast computation of the response to a long transition (see text).

the x scan (equal to a minimum of 20 gap lengths; verify that additional contributions from elements further out are not needed).

The number of computations required are:

The long method	150 * 150 * 10	=	225,000 computations
The short method	1 * 150 * 10	=	1,500 computations
The medium method	1 * 150 * 10		
	+ 20 * 190 * 10	=	39,500 computations

The computation time ratios are 150:1:25. On an Apple II the run times are 38:1/4:5 hours. On an IBM PC with the 8087 the time is much shorter. See Fig. 21-6 for the quick computation.

INVESTIGATE MAGNETIZATION PATTERNS

The read waveform from a variety of magnetization patterns can now be evaluated, as a few examples will illustrate. The pattern from Fig. 21-5 results in curve shown in Fig. 21-7 (broken line). It compares well with the read pulse from Tjaden and Leyten's experiment (from Fig. 5-2). Both show a short rise and a long decay time.

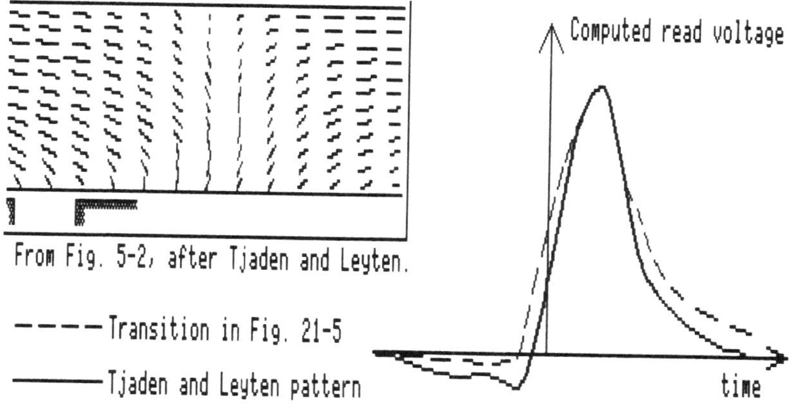

Fig. 21-7. Computed read pulses from the transition shown in Fig. 21-5 and the Tjaden-Leyten transition.

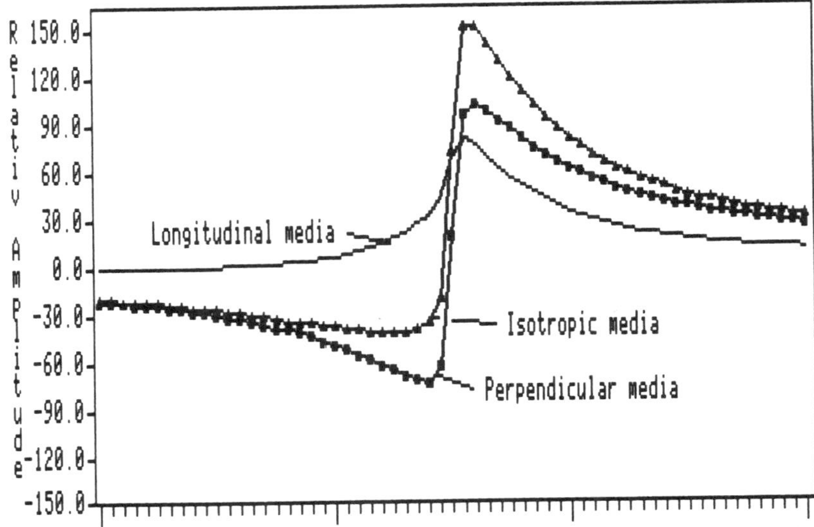

Fig. 21-8. Read pulses from three different media: longitudinal, isotropic and perpendicular.

This is characteristic for vector magnetization patterns. A purely longitudinal magnetization results in a perfectly symmetrical pulse, while a purely perpendicular magnetization results in a dipulse. Both are shown in Fig. 21-8, that further shows another response to a short transition with vector magnetization.

The usefulness of separate arrays for the computational result, using formula (21.7) is evident from Fig. 21-9. Here the response to the transition in Fig. 21-5 is shown, made up from fluxes from the longitudinal and perpendicular components.

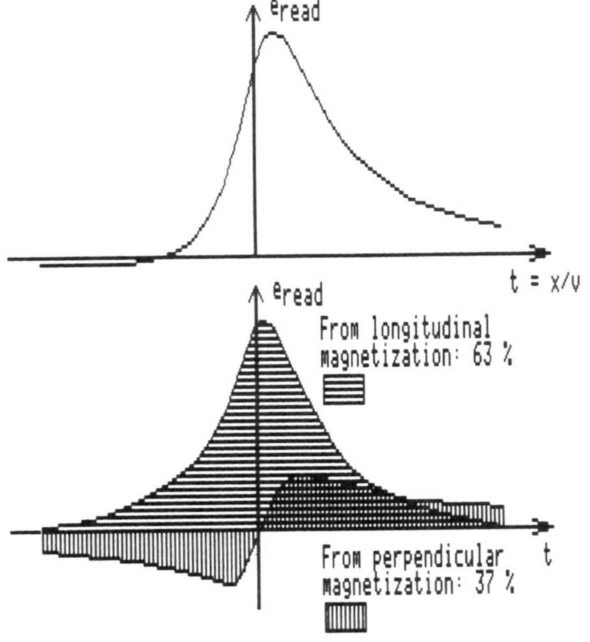

Fig. 21-9. Read pulse for the transition shown in Fig. 21-5 is composed of contribution from longitudinal and perpendicular magnetizations.

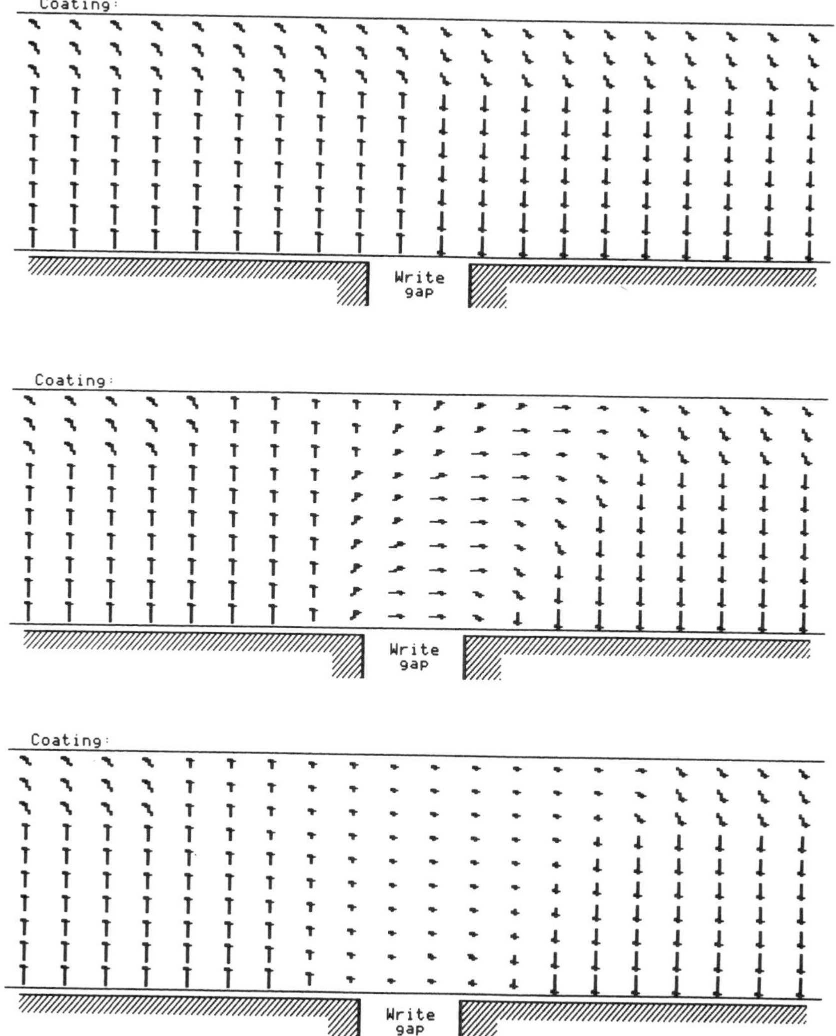

Fig. 21-10. Three possible transition patterns from recording of a single transition on an isotropic medium: Short and strong; broad and strong; broad and weak.

Figure 21-10 shows three transitions: A short, strongly magnetized transition (top), a broad and strong transition (middle), and a broad and weak transition (bottom). The last should give a weak and broad read pulse, as compared to the top transition.

The computed responses are shown in Fig. 21-11 which verifies our predictions for the output levels. When the levels are normalized we can also verify that the shortest response is obtained for the short transition. Notice the dipulse response; they are all predominantly perpendicular magnetized. See Fig. 21-12.

The effect of head-media spacing can easily be computed, and the result shown in Fig. 21-13 for a 1.25 μm coating with a 0.7 μm gap head. Increased spacing causes reduced output and broadened pulses.

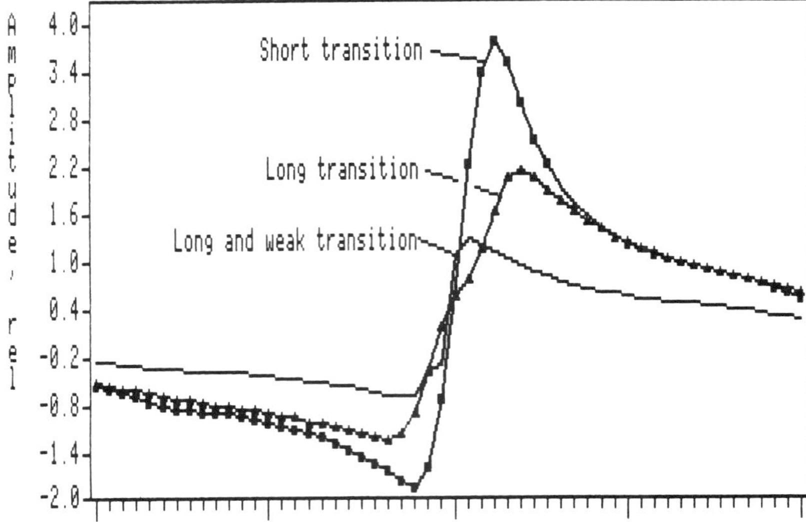

Fig. 21-11. Computed response from the three transitions in Fig. 21-10.

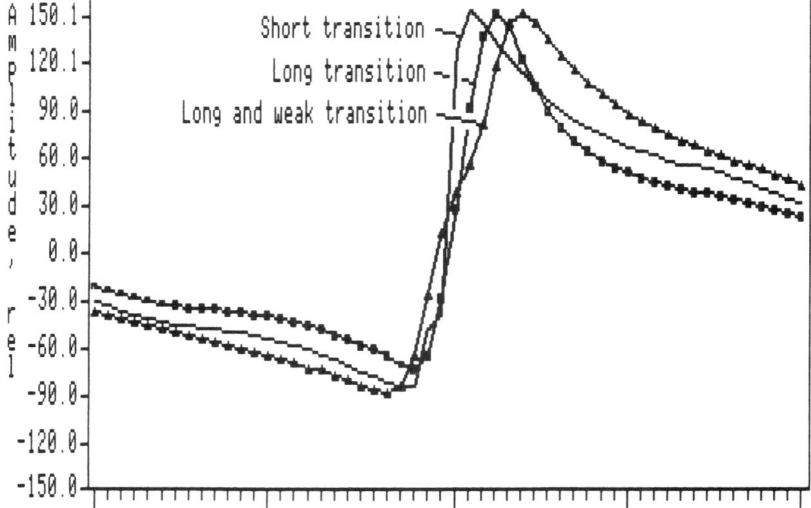

Fig. 21-12. Normalized levels for responses in Fig. 21-11.

The reader will find further discussions on the response to a single transition in Chapters 22 and 23, and will hopefully benefit from the knowledge of the various responses from various magnetization patterns in and around written transition. This chapter's modelling is a valuable tool for all sorts of trend analysis.

BIBLIOGRAPHY TO CHAPTER 21

Read Sensitivity Function

Daniel, E.D., and Axon, P.E., "The Reproduction of Signal Recorded on Magnetic Tape," *Proc. IEE* (London), Paper 1499R, May 1953, Vol. 100, Pt. III, p. 157.

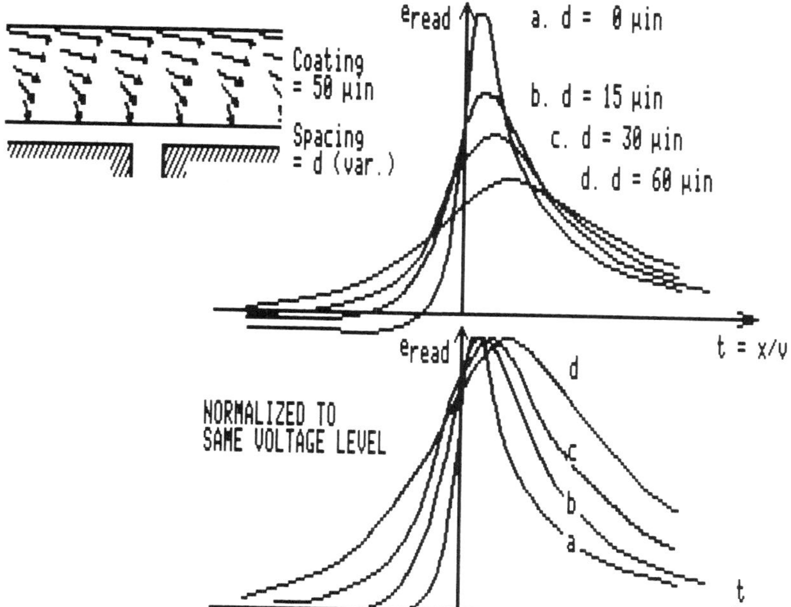

Fig. 21-13. Effect of distance between a read head and the recording surface: Increased distance causes reduced output level, and a broadening of the pulse.

Straubel, R., and Willascheu, K., "Ein spaltloser Wiedergabekopf zur Fluempfindlichen Abtastung von Magnetbandaufzeichnungen," *Int. Elektr. Rundschau*, March 1965, pp. 131-132.

Herbert, J.R., "The Readback Process in Digital Magnetic Recording," *IEEE Trans. Magn.*, Sept. 1966, Vol. MAG-2, No. 3, pp. 247-250.

Greiner, J., "Der magnetischen Kraftfluss im Lesekopf bei der Wiedergabe digitaler Informationen vom Magnetband," *Frequenz*, March 1967, Vol. 21, No. 3, pp. 81-94.

Nishikawa, M., "Characteristics of the Readback Signal in Digital Magnetic Recording," *IEEE Trans. Magn.*, Dec. 1970, Vol. MAG-6, No. 4, pp. 811-818.

Szczech, T., and Fayling, R., "The Use of Perpendicular Head Field Equations for Calculating Isolated Pulse Output," *IEEE Trans. Magn.*, Nov. 1982, Vol. MAG-18, No. 6, pp. 1176-1177.

Heim, D., "The Sensitivity Function for Shielded Magnetoresistive Heads by Conformal Mapping," *IEEE Trans. Magn.*, Sept. 1983, Vol. MAG-19, No. 5, pp. 1620-1622.

Computations of Read Voltage

Camras, M., "Theoretical Response from a Magnetic Wire Record," *Proc. I.R.E.*, Aug. 1946, Vol. 34, No. 8, pp. 593-602.

Laubsch, H., Kisch, S., and Sladek, J., "Modelluntersuchungen zum Wiedergabevorgang," *Hochfrequenztechnik und Elektroakustik*, Jan. 1970, Vol. 79, pp. 144-149.

Middleton, B. K., "The Replay Signal From a Tape with Magnetization Components Parallel and Normal to Its Plane," *IEEE Trans. Magn.*, Sept. 1975, Vol. MAG-11, No. 5, pp. 1170-1173.

Baker, B. R., "Response of Asymmetric Thin Film Heads to Vertical and Horizontal Magnetization," *IEEE Trans. Magn.*, Nov. 1981, Vol. MAG-17, No. 6, pp. 3123-3125.

Yen, N., "Anomalous Wavelength Response of a Thin Film Head from Double-Layer Perpendicular Media," *IEEE Trans. Magn.*, Sept. 1983, Vol. MAG-19, No. 5, pp. 1611-1613.

Lopez, O., "Reproducing Vertically Recorded Information-Double Layer Media," *IEEE Trans. Magn.*, Sept. 1983, Vol. MAG-19, No. 5, pp. 1614-1616.

Baker, B., Herte, L., and Labmeier, C., "Analysis of Readback from Duplex Vertical Recording Films," *IEEE Trans. Magn.*, Sept. 1984, Vol. MAG-20, No. 5, pp. 736-738.

Minuhin, V.B., "Theory of Playback Process with Soft Magnetic Underlayer," *IEEE Trans. Magn.*, Jan. 1985, Vol. MAG-21, No. 1, pp. 28-35.

Minuhin, V.B. (pt. 2 with Steinback, M.), "Dependence of Readback Output on Medium Thickness in the Presence of a Permeable Underlayer - Part 1 and 2," *IEEE Trans. Magn.*, Nov. 1985, Vol. MAG-21, No. 6, pp. 2595-2619.

Stubbs, D.P., and Alexander, J.L., "Transition Lengths in Magnetic Recording on Particulate Disks," *IEEE Trans. Magn.*, Sept. 1986, Vol. MAG-22, No. 5, pp. 382-384.

Vos, M.J., Luitjens, S.B., de Bie, R.W., and Lodder, J.C., "Magnetization Transitions Obtained by Deconvolution of Measured Replay Pulses in Perpendicular Recording," *IEEE Trans. Magn.*, Sept. 1986, Vol. MAG-22, No. 5, pp. 373-375.

Magnetization in Media

Heidenwolf, H., "Method for Microscopic Viewing of Magnetic Sound Recordings," *Lorenz Ber.*, December 1941, pp. 119-122.

Lemke, James U., "Remanence of Tape; Effect of Vector History," *IEEE Trans. Magn.*, Sept. 1966, Vol. MAG-2, No. 3, pp. 230-232.

Hersener, Jurgen, "Modelluntersuchungen zur Magnetisierungsverteilung im Magnetbandbei sinusformigen Signalen," *Wiss. Ber. AEG-Telefunken*, 1973, Vol. 46, No. 1, pp. 15-24.

Bate, G. et al., "Magnetization in Recorded Tape," *IEEE Trans. Magn.*, Sept. 1974, Vol. MAG-10, No. 3, pp. 667-669.

Bate, G. et al., "The Remanent State of Recorded Tapes," *IBM Jour. Res. Develop.*, Nov. 1974, pp. 563-569.

Speliotis, D.E., and Chi, C.S., "The Remanent Magnetization in Particulate Disk Coatings," *IEEE Trans. Magn.*, Nov. 1976, Vol. MAG-12, No. 6, pp. 734-737.

Valstyn, E.P., and Monson, J.E., "Magnetization Distribution in an Isolated Transition," *IEEE Trans. Magn.*, Nov. 1979, Vol. MAG-15, No. 6, pp. 1453-1456.

Knowles, J.E. et al., "The Angular Distribution of the Magnetization in Recording Tapes," *IEEE Trans. Magn.*, Jan. 1980, Vol. MAG-16, No. 1, pp. 42-44.

Williams, E.M., "The Dorf Effect: Magnetization Ripple In Particulate Media," *IEEE Trans. Magn.*, Nov. 1982, Vol. MAG-18, No. 6, pp. 1086-1088.

Chung, S., "A One Dimensional Analysis of Magnetization Ripple in Uniaxial Thin Films with Perpendicular Anisotropy," *IEEE Trans. Magn.*, Nov. 1983, Vol. MAG-19, No. 6, pp. 2694-2697.

Yoshida, K., Okuwaki, T., Osakabe, N., Tanabe, H., Horiuchi, Y., Matsuda, T., Shinagawa, K., Tonomura, A., and Fujiwara, dH., "Observation of Recorded Magnetization Patterns by Electron Holography," *IEEE Trans. Magn.*, Nov. 1983, Vol. MAG-19, No. 6, pp. 1600-1604.

Minuhin, V.B., "Theoretical Comparison of Readback Harmonic Responses for Longitudinal Recording and Perpendicular Recording with Probe Head over a Medium with Permeable Underlayer," *IEEE Trans. Magn.*, Sept. 1986, Vol. MAG-22, No. 5, pp. 388-390.

Fields from Media

Schmidbauer, O., "The Field of the Harmonically Magnetized Tape," *Frequenz*, Oct. 1952, Vol. 6, pp. 319-334.

Westmijze, W.K., "Studies on Magnetic Recording, IV: Calculation of the Field In and Around the Tape," *Philips Res. Rep.*, August 1953, Vol. 8, No. 4, pp. 255-269.

Iwasaki, S., and Suzuki, T., "The Effect of Demagnetizing Field on 2-Bit Pattern in Digital Recording Process," *Rep. of Res. Inst. of El. Comm. Tohoku Univ.*, January 1967, Vol. 19, No. 3, pp. 171-186.

McKnight, J.G., "Tape Flux Measurements Theory and Verification," *Jour. Audio Engr. Soc.*, June 1970, Vol. 18, No. 3, pp. 250-259.

Speliotis, D.E., and Judy, J.H., "Calculations of External Bit Fields," *IEEE Trans. Magn.*, March 1971, Vol. MAG-7, No. 1, pp. 158-163.

Druyvesteyn, W.F., Tjaden, D.L.A., and Dorleijn, J.W.F., "Calculation of the Stray Field of a Magnetic Bubble, with Application to Some Bubble Problems," *Philips Res. Repts.*, 1972, pp. 7-27.

Mallinson, J.C., "One-Sided Fluxes—A Magnetic Curiosity?," *IEEE Trans. Magn.*, Dec. 1973, Vol. MAG-9, No. 4, pp. 678-683.

Tagami, K., Suganuma, Y., and Nagao, M., "A New External Bit Field Observation Method," *IEEE Trans. Magn.*, July 1977, Vol. MAG-13, No. 5, pp. 1689-1692.

Tagami, K., Suganuma, Y., and Nagao, M., "External Bit Field Analyses," *IEEE Trans. Magn.*, May 1979, Vol. MAG-15, No. 3, pp. 1054-1060.

Fayling, R.E., "Edge Profile Studies of Recorded Flux Transitions," *IEEE Trans. Magn.*, Nov. 1979, Vol. MAG-15, No. 6, pp. 1469-1470.

Yeh, N.H., "Ferrofluid Bitter Patterns on Tape," *IEEE Trans. Magn.*, Sept. 1980, Vol. MAG-16, No. 5, pp. 797-982.

Mallinson, J.C., "On the Properties of Two-Dimensional Dipoles and Magnetized Bodies," *IEEE Trans. Magn.*, Sept. 1981, Vol. MAG-17, No. 5, pp. 2453-2460.

Baird, A.W., Chaurette, W.F., and Lustig, C.D., "Field Measurements Near High-Density Statically Recorded Transition in a Thin-Film Medium," *IEEE Trans. Magn.*, Nov. 1981, Vol. MAG-17, No. 6, pp. 2553-2520.

Baird, A.W., Johnson, R.A., Chaurette, W.F., and Maloney, W.T., "Measurements and Self-Consistent Calculations of the Magnetic Fields Near a Perpendicular Medium," *IEEE Trans. Magn.*, May 1984, Vol. MAG-20, No. 3, pp. 479-485.

Indeck, R.S., and Judy, J.H., "Measurements of Surface Magnetic Fields of Perpendicular and Longitudinal Magnetic Recorded Transitions Using a Magnetoresistive Transducer," *IEEE Trans. Magn.*, Sept. 1984, Vol. MAG-20, No. 5, pp. 730-732.

Time and Frequency Response

Woodward, J.G., "Approaches to Wideband, High-Resolution Magnetic Recording," *Jour. AES*, Feb. 1962, Vol. 10, No. 2, p. 53.

Bloomberg, D.S., "Spectral Response From Perpendicular Media with Gapped Head and Underlayer," *IEEE Trans. Magn.*, July 1983, Vol. MAG-19, No. 4, pp. 1493-1502.

Peak Shifts

Roscamp, Thomas, A., "Origins of Playback Pulse Asymmetry in Recording with Thin Film Disk Heads," *IEEE Trans. Magn.*, Nov. 1981, Vol. MAG-17, No. 6, pp. 2902-2904.

Jorgensen, F., "Asymmetry Peak Shift on the Read Pulse for a Single Recorded Transition," *IERE Conf. Video and Data*, April 1982, IERE Publ. No. 54, pp. 165-179.

Bloomberg, D., "Readback Bit Shift with Finite Pole-Length Heads on Perpendicular Media," *IEEE Trans. Magn.*, Sept. 1983, Vol. MAG-19, No. 5, pp. 1617-1619.

Saito, I., Satake, S., Kawazoe, K., and Hokkyo, J., "Peakshift Characteristics of Perpendicular Magnetic Recording," *IEEE Trans. Magn.*, Sept. 1984, Vol. MAG-20, No. 5, pp. 903-905.

Murayama, H., "Read/Write Evaluation Considering Peakshift Characteristics for Perpendicular Recording," *NEC Res. Devel.*, October 1985, No. 79, pp. 35-42.

Part 7

Equalization and Coding

Chapter 22

Signal Response and Equalization

Faithful reproduction of magnetically recorded information requires compensation for a number of losses. Some of these occur during the recording (write) cycle, such as the loss in resolution due to the switching field distribution, the gradual decrease in head field strength, and are due to head-medium spacing. There may also be signal attenuation at high frequencies due to head core losses. Further attenuation occurs in the media magnetization due to demagnetization, and during storage a small decay with time may occur. During the read process there are additional losses due to geometrical factors (recorded thickness, head-medium spacing, read gap length) and high frequency core losses.

When the losses are subtracted from the signal amplitude versus frequency curve, that has a slope of 6 dB/octave, then a response as shown in Fig. 6-2 results. Compensation is made by an amplifier circuit that boosts the low and the high frequencies, a process named *amplitude equalization*. The choice of equalizer circuits should not be made arbitrarily since the majority of the losses have little or no associated phase shifts; ordinary RLC-circuits for compensation can equalize the amplitude-versus-frequency response very well but may at the same time add considerable phase error into the signal.

This chapter will discuss the means of restoring the signal to become as close a representation as possible of the original input signal to the record circuit. This implies restoration of the amplitude response to a flat response, and the phase to become a linear function of the frequency, and that intercepts the origin (frequency equal to zero) at zero phase (or $+-n\pi$). The amplitude equalization is not necessarily carried out for low frequencies in digital recording, i.e., the read preamplifier is not followed by an integrator, but possibly by a differentiator to facilitate zero crossing detections, which corresponds to the induced voltage peaks occurring at transitions.

It is a mistake to believe that digital recording is a matter of digital processing only—the signal passes through the write amplifier, media magnetization, and read amplifier and filter/equalizer circuits, all members of the analog world. These circuits are discussed in this chapter.

The phase response of a recorder must be correct in pulse (digital) recording equipment, which is not necessarily assured by a flat amplitude versus frequency response. An evaluation of the response to a single transition is important in optimizing a disk or tape recorder for digital data storage. The record/reproduce (write/read) channel can be modelled, optimized and verified by measurements. And powerful instruments are available, using the most up-to-date signal analysis methods (see Chapter 23).

RESPONSES OF AMPLITUDE AND PHASE VERSUS FREQUENCY

The amplitude response of a record/playback channel is well known with fairly precise characterizations of the various losses (see Fig. 6-2); a signal phase shift is not evident from the expression for the read voltage in Fig. 6-9 since it was derived for purely longitudinal magnetization in the medium. We would have found a similar formula for purely perpendicular magnetizations, but with a 90 degree phase shift.

Signal attenuation in ordinary electric networks made up of resistors, capacitors and inductors have an associated change in the phase characteristic. (Certain electrical networks are exceptions to this rule, such as the class of all-pass networks, and certain lossy filters, ref. Kallmann). A phase shift will also occur in magnetic recording, as we will now discuss.

The direction of the remanent magnetization changes along and through the coating thickness. Its pattern may look like Fig. 22-1. A simple experiment verifies such phase shifts: a row of dipole magnets are placed in groups, each corresponding to one bit length (or half a wavelength). This is shown in Fig. 22-2 where four bits are made up of three, three, two, one and one dipole, respectively.

The external flux lines are made visible with iron powder and they clearly show the 90 degree phase shift between the longitudinal (top) and perpendicular (bottom) magnetization patterns. The in-between pattern for the 45 degree position of the dipoles (polarization) results in a phase shift that is somewhere between zero and 90 degrees. The bottom flux pattern also verifies the observed change in read waveforms that occur when a tape is played back in the forward and then in the reverse directions; playback equalizers require a polarity change in the phase equalizer adjustment.

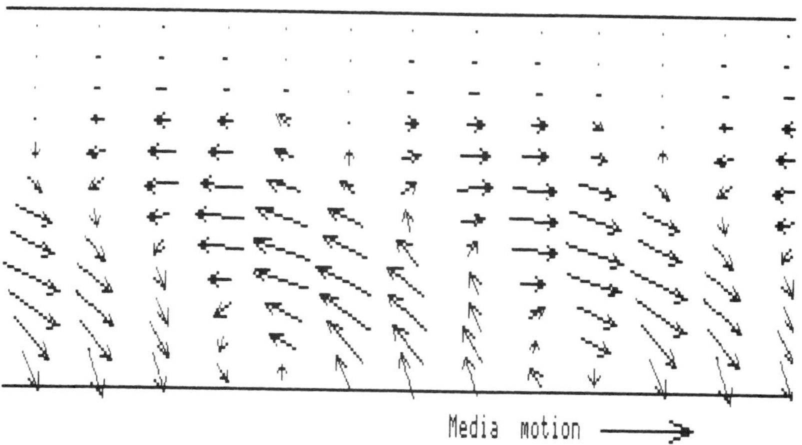

Fig. 22-1. Magnetization inside a magnetic coating recorded at short wavelengths (after Tjaden and Leyten).

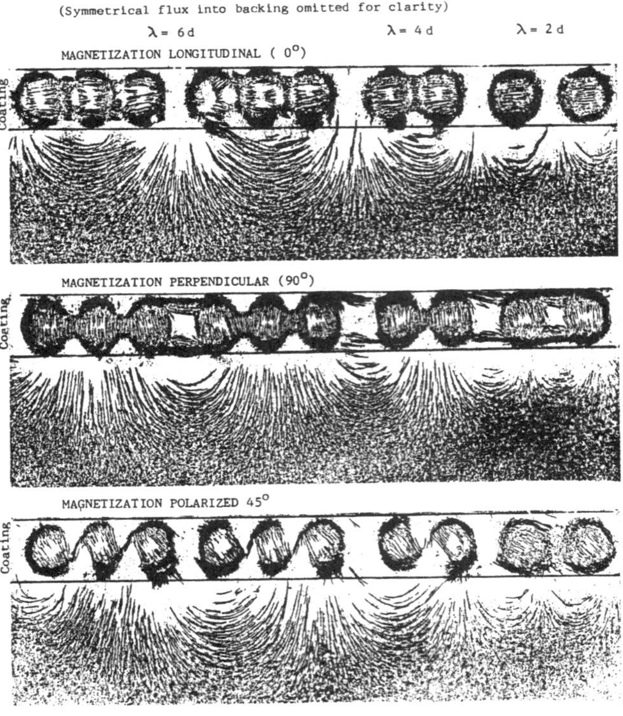

Fig. 22-2. Dipole model of magnetized coating. Flux lines are made visible by application of iron powder. Top: Longitudinal magnetization; Middle: Perpendicular magnetization; Bottom: 45 degrees magnetization (polarization).

The magnetization in the coating is only purely longitudinal if the medium's susceptibility is zero in the perpendicular direction. The read voltage from a medium with both longitudinal and perpendicular magnetizations will be an asymmetrical pulse, as shown in Fig. 21-9, and is composed of signals from the longitudinal as well as the perpendicular magnetizations (see also Fig. 6-8).

The write/read process has thus split the input signal into two components. The first represents the original signal, while the other is a transform thereof, called the Hilbert Transform. The latter is always in quadrature with the first (ref. Bracewell).

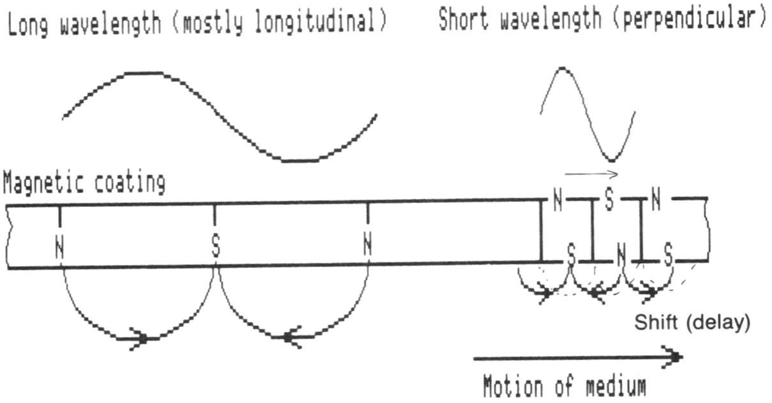

Long wavelength (mostly longitudinal) Short wavelength (perpendicular)

Magnetic coating

N S N N S N S N S

Shift (delay)

Motion of medium

Fig. 22-3. Short wavelength fluxes arrive "late" at the read head relative to the long wavelengths, i.e., the high frequencies are delayed.

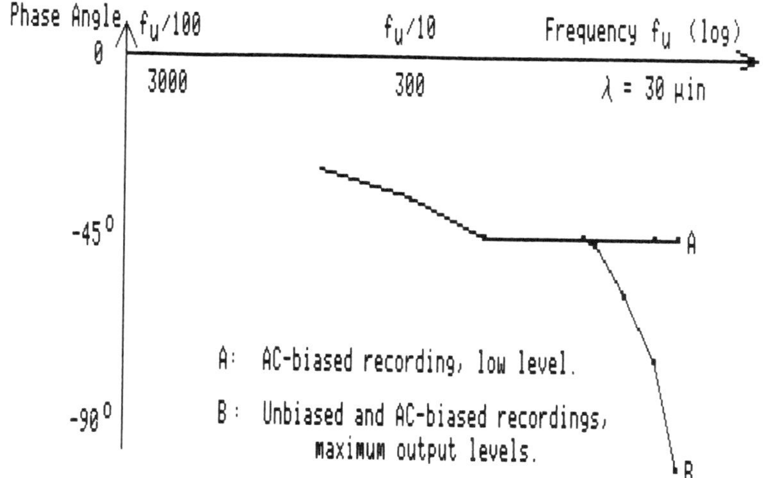

Fig. 22-4. Phase shift curves for the flux into the read head (after Haynes).

The magnetization versus frequency (or correctly: versus 1/wavelength) is lagging at short wavelengths, as shown in Fig. 22-3. The magnetization poles are shifted and turned from the bit cell ends (transitions) toward the middle of the bit cells in such a way that they arrive later at the read head, and present a lag.

This results in a phase versus wavelength response for the coating flux, as shown in Fig. 22-4, that agrees with the limited information available in the literature (ref. Jacoby, Minukhin, Haynes). A different presentation is seen in Fig. 22-5, employing a linear frequency scale. The curves are quite linear over the recorder passband, but an extrapolation intersects the phase angle scale at a point different from zero degrees. This shift is carried along when the flux is differentiated in the read head winding, resulting in a +90 degree phase shift, and we obtain Fig. 22-6. The intersect different from zero (or $+-n\pi$) is the source of the problem that requires phase equalization.

Phase distortion is illustrated in Fig. 22-7: A square wave is made up of a fundamental cosine wave form plus odd harmonics. The illustration uses only the third harmonic to recompose the waveform, and shows asymmetrical distortion of the waveform when the high frequencies arrive late.

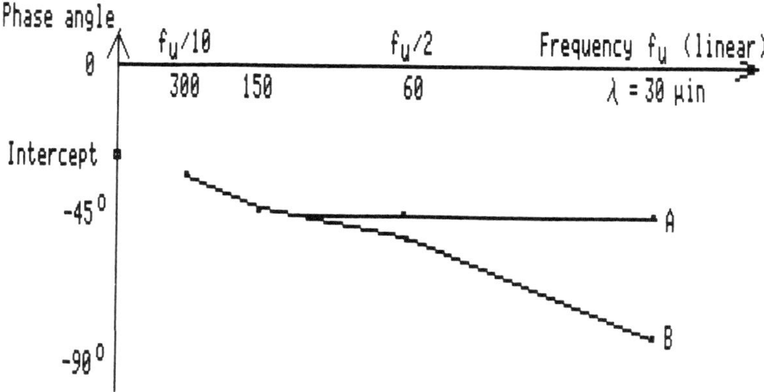

Fig. 22-5. Phase shift curves for the flux, shown with a linear scale for the frequencies.

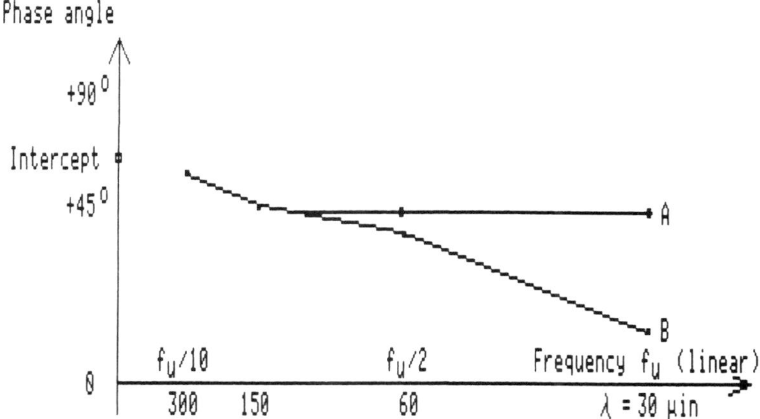

Fig. 22-6. Phase shift curves for the read voltage (i.e., differentiated flux, at +90 degrees phase shift).

A correct waveform can be obtained by shifting the phase curve so it intersects the phase axis at zero degrees, see Fig. 22-8. This is simplest done with an all-pass network in the playback amplifier.

The photos in Fig. 22-9 illustrate how a square wave input is distorted in an ordinary tape recorder. It can be restored to a symmetrical waveform by adding phase equalization to the electronics, and the result is the waveform shown in the bottom.

An estimate of the phase correction needed is possible from a knowledge of the ratio between the perpendicular and longitudinal magnetizations, J_{perp} and J_{long} (after Minukhin):

$$\text{Phase intersect at } (\omega = 0) = \arctan (J_{perp}/J_{long}) \qquad \textbf{(22.1)}$$

For purely longitudinal magnetization the angle is zero degrees, while it is 90 degrees for purely perpendicular magnetization.

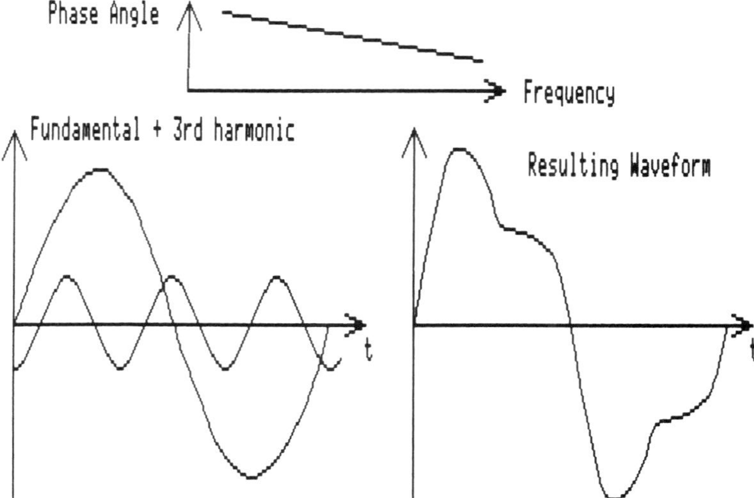

Fig. 22-7. A linear phase versus frequency relation can result in phase distortion if the line intersects the phase axis at points different from 0 or + − n * 180 degrees.

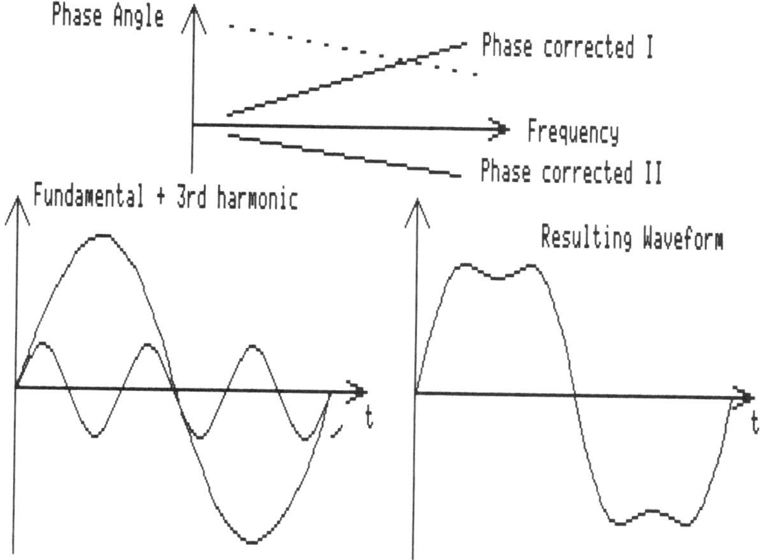

Fig. 22-8. Phase correction of the phase versus frequency curve results in correct equalization.

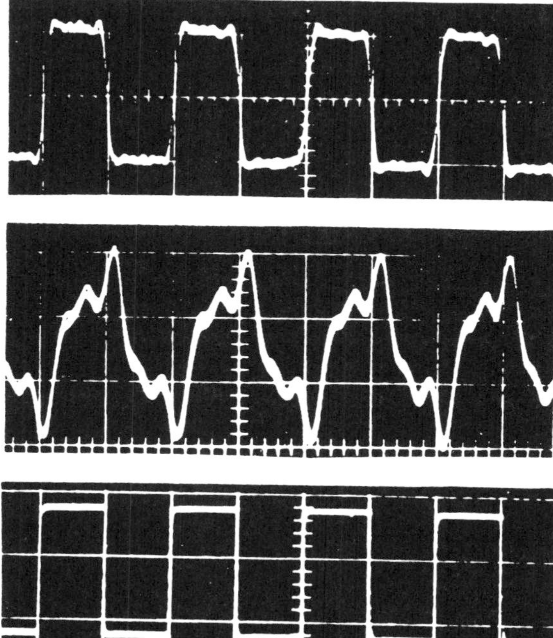

Fig. 22-9. (Top) Square wave input. (Center) Reproduction from a recorder with no phase equalization. (Bottom) Properly equalized playback waveform.

EQUALIZATION IN LINEAR (AC-BIASED) RECORDERS

A block diagram of an AC-biased recorder, audio or instrumentation, is shown in Fig. 22-10. The input voltage V_{in} is converted into a current and passed through a pre-equalization network into the head winding. The high-frequency bias signal is added through capacitor C_{bias}.

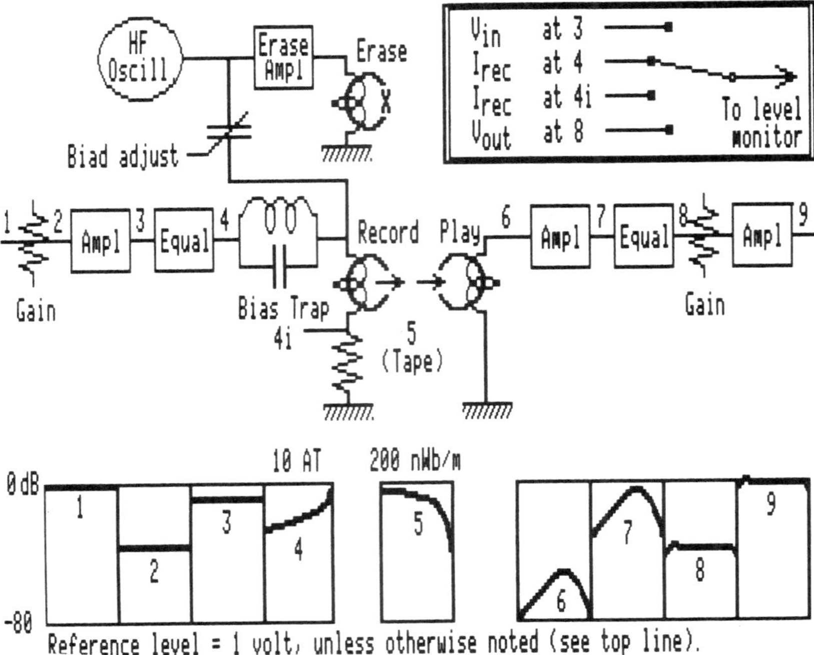

Fig. 22-10. Block diagram of electronics in a typical AC-biased recorder, and signal level diagram.

The bias signal may cause intermodulation distortion at the output stage of the record amplifier; consequently, a tuned bias trap is inserted between the amplifier and the recording head.

The record amplifier, bias oscillator and playback amplifier do not differ very much from conventional circuit designs, except that particular attention must be paid to such factors as low distortion (amplitude and phase) and low noise.

During playback, the flux from the recorded tape is differentiated in the reproduce head, amplified and then equalized. The quality of a recording can be monitored by a metering circuit, which can be switched to the playback position for an A-B comparison between recorder input and output. The level indicator can also be used for monitoring the record and bias current through a resistor (typically 10 ohms) in the ground leg of the record head—a useful feature during checkout or service of a recorder. The lower portion of Fig. 22-10 is a level diagram for the signal as it passes through the record amplifier, is recorded, played back, equalized, and amplified.

Record Circuitry

The *record amplifier* is an amplifier with a flat amplitude versus frequency response that converts the input voltage into a record current. Its input impedance varies from 91 ohms for a wideband recorder (several MHz bandwidth) to 100 kilohms for a microphone audio amplifier. The corresponding sensitivities range from 1 volt rms to 1 millivolt rms, which in either case gives an input sensitivity on the order of 10 milliwatts. The record amplifier must be designed to provide a current that will saturate the tape prior to any amplifier distortion. If the record current is adjusted to produce one percent 3rd harmonic distortion on a tape (any type), then saturation will occur at a record current that is approximately 4 times higher (12 dB). The amplifier should have less than one percent distortion at that level.

The record head is an inductive load for the record amplifier, and its impedance, consequently, increases with frequency. This in turn requires that the record head be driven by a constant-current source which again dictates that the record amplifier has a high output impedance. A constant record current should, in reality, mean a constant record flux but eddy-current and hysteresis losses make it necessary to increase the record current toward the higher frequencies in order to obtain a constant recording field strength versus frequency.

Record Equalization

Pre-equalization in the record amplifier compensates for eddy current and hysteresis losses in the record head, and should also compensate for record zone losses (resolution) and demagnetization losses. The compensations for the latter must be done with a circuit that does not introduce phase shifts (see later: the delay line equalizer).

It is also desirable to compensate for the magnetization phaseshift toward short wavelengths; this should be done with an all-pass network that preserves a flat amplitude response.

This nominal equalization will assure that the recorded magnetization level is uniform across the frequency, or better, wavelength range. Subsequent playback will require an additional boost of both low and high frequencies in order to compensate for thickness, spacing, and gap losses. This playback, or post-, equalization causes an increased amplification of any low frequency noise (power supply hum, 1/f-noise) and high frequency (tape, head and amplifier noise), and can be minimized by applying the maximum permissible record current boost.

Most program materials do not have a flat energy spectrum, and pre-equalization may therefore be used to fully utilize the media magnetization; this will lessen the requirements for post-equalization, leading to an improvement in the overall signal-to-noise ratio.

Pre-equalization in audio recording was based originally upon the energy spectrum of music (i.e., level versus frequency). At first the classical work of Sivian, Dunn and White was used in establishing pre-equalization standards. However, later findings and experience have shown that at times it is tolerable to record with a higher pre-emphasis than dictated by the energy spectra.

Here is an area where audio recordings differ from instrumentation recordings. In audio recordings, harmonic distortion is less noticeable and intermodulation distortion is often confused and intermixed with that of scrape flutter. In instrumentation recording, on the other hand, harmonic distortion and intermodulation in a multiplexed channel is readily observed as a foreign signal.

Record equalization in instrumentation recorders is therefore limited to that amount required to achieve a *constant flux* on the recorded tape. The pre-equalization is designed to correct for:

- Head core losses (incl. phase shifts).
- Write resolution losses (recording zone, SFD - no phase shift).
- Demagnetization losses (some phase shift).

Adherence to this method of pre-equalization will produce tapes that are interchangeable from machine to machine since they all will have a constant flux level independent of wavelength. The write resolution losses listed above correspond to what generally is called bias self-erasure losses.

The head response, i.e., current in the winding versus frequency for a constant input current, must enter into the design considerations. A square wave write current can produce undesirable ringing, as shown in Fig. 22-11.

Pre-equalization should also include a phase correcting network that will produce tapes that have identical phase response whether reproduced in forward or reverse motion (Ref. Johnson and Gregg).

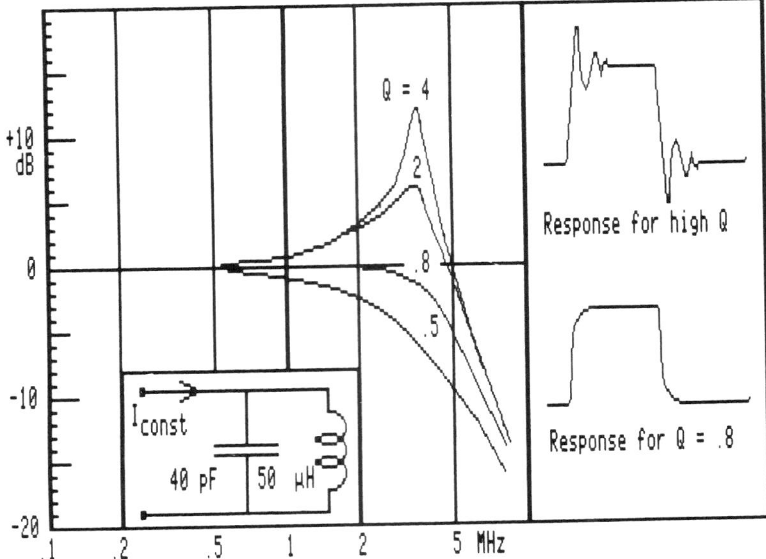

Fig. 22-11. Response of current in a magnetic head for four values of Q.

Pre-equalization in audio recorders is tied in with the standards for playback equalization, and will be covered later in the chapter.

Record Level Indicator

A recording level indicator is required to warn the operator of excessive recording levels. In its infancy, the magnetic recorder was provided with a VU meter (which stands for volume meter), a leftover from the early days in broadcasting, where it was used to monitor the signal level sent to the transmitter.

The VU meter is an averaging type instrument, and quite inadequate for informing the recording engineer about the proper recording level. Distortion takes place at instantaneous peaks of the signal to be recorded, causing both harmonic and intermodulation distortion. Peaks are much better detected by peak-reading indicators, which may be moving-coil instrument with a suitable peaking amplifier or a row of light emitting diodes (LED).

A record-level indicator should be connected (as shown in Fig. 22-10) to the electronic circuitry after equalization, since constant-current recording (which produces a constant flux) provides a nominally constant distortion level. The disadvantage of connecting the level indicator at this point is that it will not show a true comparison between V_{in} and V_{out} (A-B test). A fourth position, as shown by the broken line in Fig. 22-10 provides this function.

Playback Amplifier

The playback amplifier's function is to amplify the weak reproduce signal to suitable high output level. The stages in the reproduce amplifier are normally of conventional design, with the exception that extreme care must be taken in the selection of the input circuit, immediately following the reproduce head. The available voltage from the reproduce head at very low frequencies (20 Hz) is in the microvolt region. This means that the input stage in the playback pre-amplifier must have an internal noise voltage that is only a fraction of a microvolt. Since the reproduce head is an inductive generator, it is further important that the input impedance be high to avoid signal roll-off at high frequencies.

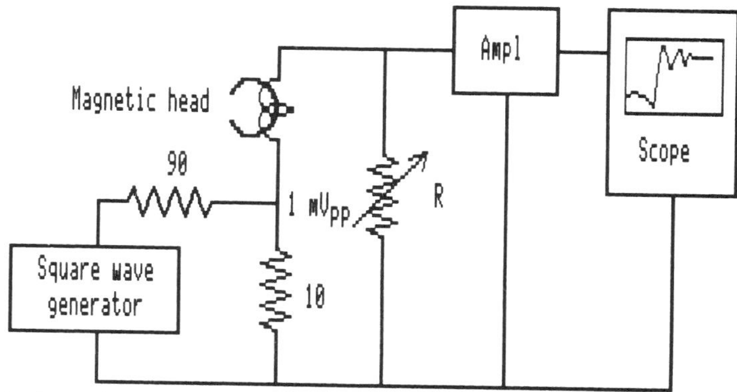

Fig. 22-12. Test set-up to determine proper matching of a magnetic head.

The output voltage from the reproduce head can be increased by increasing the number of turns on the reproduce head core. But this lowers the self-resonance of the head and limits the high-frequency response. Therefore, the head design criteria becomes a compromise between bandwidth and output level. The self-resonance of a magnetic head can readily be measured, see Figs. 9-24 and 9-25; care must be taken to include the head cable capacitance and the pre-amplifier input impedance.

The reproduce head is an inductive component with a certain quality $Q = \omega L/R_s$. Values of Q are quite low for magnetic heads, see Fig. 10-11; it may be around 5 to 15 at low frequencies, decreasing towards higher frequencies due to eddy-currents losses, and approach a value of one at very high frequencies. A Q of value greater than two at high frequencies may appear advantageous if the head resonance is "tuned" to the upper limit of the desired frequency response, and thus peaks the output voltage at those frequencies.

However, in reading pulses this may be a disadvantage since the resonant frequency introduces an additional phase shift in the playback signal. It may be necessary to design the read amplifier input impedance to such a value that the head inductance plus input capacitance tuned circuit is critically dampened. The proper value of the pre-amplifier input impedance can easily be determined with the test setup shown in Fig. 22-12. A square-wave generator is connected to a 10-ohms resistor in the ground leg of the reproduce head. (Assuming that the normal load for the square-wave generator is 600 ohms, a 590-ohm resistor must be used in series with the connection.) With a one millivolt peak-to-peak signal across the 10-ohm resistor, the variable input resistor (R) is adjusted until the amplified square wave shows little or no ringing; read the value of R to be used for the preamplifier input impedance, or head termination.

Another factor to be considered is the large dynamic range the first stage in the playback amplifier must handle: At low frequencies the induced voltage in the head is very low compared to the mid-frequency range signals. At the same time, the first amplifier stage must be able to handle a signal range of 50 to 60 dB. A high signal-to-noise ratio is therefore particularly difficult to achieve at the low frequencies, i.e., the long wavelengths.

Playback Equalization; Amplitude and Phase Response

The flux level (in nWb/m) from a recorded tape is generally specified at a certain median wavelength (for instance at 333 Hz at v = 4.75 cm/sec (1⅞ IPS), or at 1 kHz at higher tape speeds). That topic was discussed in Chapter 14 under Normal Record Level.

The coating thickness loss causes a reduction in flux level toward short wavelengths. This led to early definitions of the crossover point in equalization, i.e., where the low frequency read signal gain of 6 dB/octave ceases. The associated time constants were listed in Table

14-2, and the standard flux level versus frequency curves are shown for four audio tape speeds in Figs. 28-1 and 28-2.

An amplitude-versus-frequency correcting network, with an inverse amplitude-versus-frequency response compared to the head output, may be inserted as a passive loss element between two amplifier stages or it may be incorporated in a feed-back loop. It is important to notice that several of the losses in the playback process have no phase shift, while most passive networks made up of R, L, and C components will have a phase shift.

For example, from electrical network theory it is well known that an RC network has a 6 dB per octave slope associated with a 90-degree phase shift. Therefore, such an equalization network will introduce a phase shift which is detrimental to a true reproduction of the recorded signal.

There are *post-equalization* networks that do not introduce phase shift. Let us first consider a simple RC network to produce an amplitude versus frequency response that is complementary to the playback head voltage response.

The circuit is shown in Fig. 22-13, and an analysis of this circuit, and any others considered for equalization, can readily be made by using Kirchhoff's laws and write circuit equations for currents and voltages, and solve for the desired voltage or current, and possibly impedances. The equations are invariably first or second order differential equations, that must be solved together.

A better, faster, and more informative way is to transform the equations by the *Laplace transformation*. Differentiation and integration transform, respectively, into multiplication and division; and signal waveforms transform into easy-to-handle algebraic functions. The Laplace transform is somewhat analogous to logarithmic transformation, that enables us to add numbers rather than multiplying them. The interested reader can study an elementary introduction (ref. Bogart), or read in books by Cheng, Thomas and Rosa, or Maddock.

The bonus in using the Laplace transform to find the network response lies in its result: the *transfer function* $H(j\omega)$, which is easily applied to find either the amplitude-versus-frequency response, or the time response. (A couple of programs for these tasks are available at a reasonable cost, ref. Dynacomp, Inc. and BV Engineering.) Further, if $g_i(\omega)$ is the *Fourier transform*

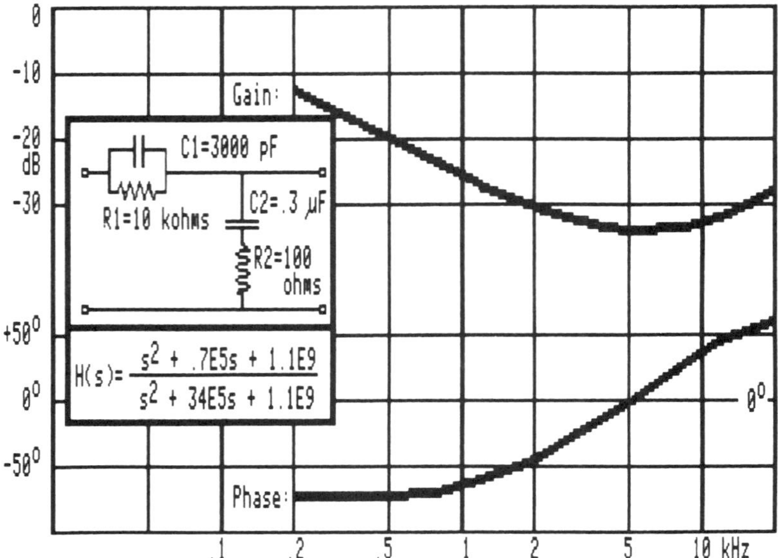

Fig. 22-13. Amplitude and phase response of lead-lag network used for equalization.

of the input, then we find the output spectrum function $g_o(\omega) = g_i(\omega)H(j\omega)$; and the response function $f_o(t)$ is found by taking the inverse Fourier transform of $g_o(\omega)$.

The analysis of the circuit in Fig. 22-13 shows that it has the desired amplitude-versus-frequency response, but is accompanied by a full 180 degree phase shift. This can be corrected by adding a single amplifier stage, shown in Fig. 22-14, where we take the output from the emitter at low frequencies, and from the collector at high frequencies; this provides a 180 degree phase correction, with a net overall response of −90 degrees (complementing the head's differentiation at +90 degrees). The transfer functions for the two circuits are listed in Fig. 22-15.

A similar effect is achieved by the network shown in Fig. 22-16, where the essential element is a 1:1 transformer; the network operates as follows:

Low frequency input voltages are integrated across capacitor C_1; this action is shelved by R1, and a subsequent increase in amplitude provided for by L_1.

The signal phase is shifted 180 degrees, very much like in the circuit in Fig. 22-13. Here the phase is reversed, and hence corrected, by the opposite coupling of L_1 and L_2.

The additional boost at band edge is accomplished by tuning L_2 with C_2, damped by R_2. The variable value of R_1 makes it possible to shift midrange equalization; C_2 determines the band-edge frequency, and R_2 the amount of boost.

In addition to the equalizer phase shift, there is also the phase shift in the tape flux itself. This was discussed earlier in this chapter and can be proved by recording a square wave on a tape, controlling the amplitude and then equalizing the phase of the signal. When the tape is played in the reverse direction the waveform becomes highly distorted. This explains the degradation occurring in tape copies made in sequences, where the transient response (like in piano music) of copy number six, for example, has been severely distorted.

The phase shift from an LCR equalizer used to peak the response at the upper frequency band edge can be eliminated by replacing the LCR equalizer with the *Delay line* or *Aperture Equalizer*. Its circuits and response curves are shown in Fig. 22-17 (Ref. Dennison, Jorgensen). The phase shift of the circuit is constant across the pass band (in the example shown in Fig. 22-17 the phase is +90 degrees up to 300 kHz), and does therefore not contribute to a phase shift, as an ordinary tuned LC-circuit would.

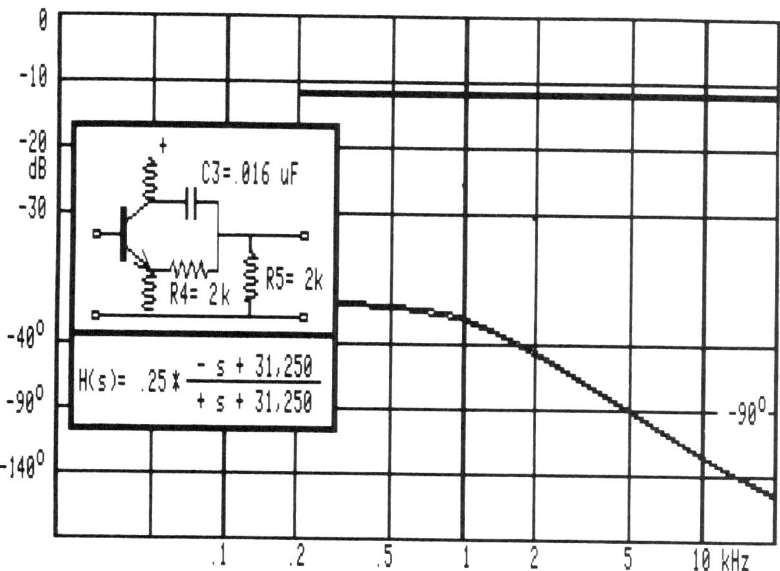

Fig. 22-14. Phase correction network for network in Fig. 22-13.

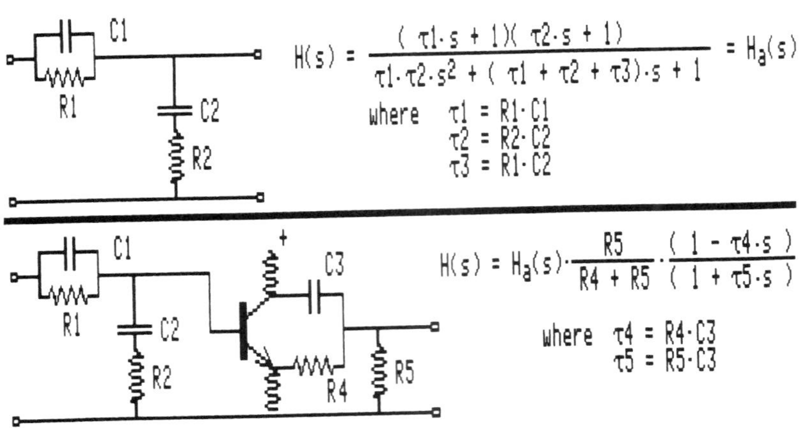

Fig. 22-15. Transfer functions for networks in Figs. 22-13 and 22-14.

The LC-network represents a delay line, that in practice can be constructed from lumped elements, as shown. The 91 ohms resistor equals the line characteristics impedance Z_o. The input voltage is transmitted along the delay line and received without attenuation at the amplifier. Since the termination here represents an open circuit, the signal is reflected and combines with the input voltage at the input. The termination at that point equals the characteristic impedance, and there is no further reflection. The voltage amplitude at the input will therefore vary cosinusoidally versus frequency because of the reflected signal.

The circuit gain can be expressed:

$$\text{Gain} = (\mu 741 \text{ Gain}) * (1 - T * \cos \omega\tau) \qquad \textbf{(22.2)}$$

where $\omega = 2\pi f$, and T represents how far toward the high end the slider is positioned on the 91 ohms potentiometer. The gain rises cosinusoidally with frequency attaining a maximum value at a frequency f_p at which the delay line is one-half wavelength long ($\omega\tau = \pi$). Thus the required line delay is $\tau = 1/f_p$.

The curves A, B and C correspond to T = 90, 99 and 99.7 percent of the potentiometer, i.e., always close to the top. This adjustment makes the delay line equalizer well suited for

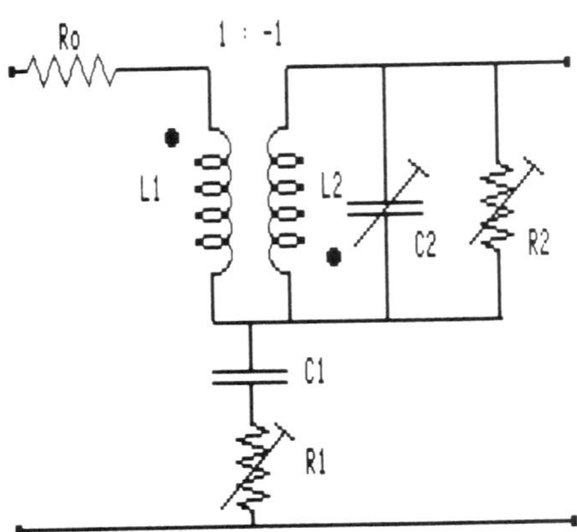

Fig. 22-16. Transformer coupled network for proper amplitude and phase equalization. Phase is constant = + 90 degrees across the frequency range of the equalizer.

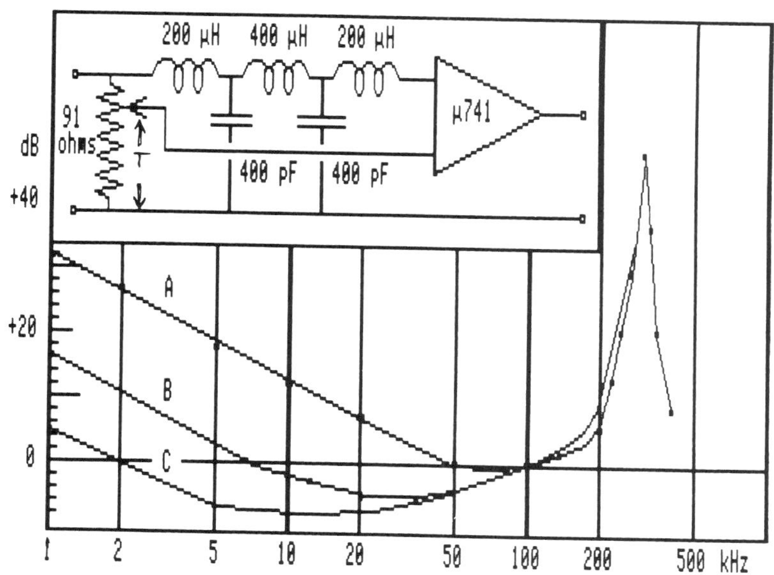

Fig. 22-17. Gain of Delay line, or Cosine, Equalizer. Phase remains constant across the frequency range of the circuit. See text for explanation of A, B and C.

an all-round equalizer with flat response obtained for curve A, and boost of the high frequencies only for curve C; the latter is advantageous in equalization of digital signals where the encoded signal has none or few low frequency components.

It is assumed that all amplifier sections and couplings are free from phase errors, which can be assured by proper design (ref. Horwitz, Clark and Hess). The residual phase error from the magnetic write/read process, as indicated by the intersection of the phase versus frequency curve in Fig. 22-5 needs correction, and this is generally done by inserting an all-pass network. The two generic types are shown in Fig. 22-18, the first (named Type I) changes the phase from zero degrees at low frequencies to −180 degrees at high frequencies, while the second (Type II) starts at −180 degrees at low frequencies and changes to zero degrees (Ref. van Valkenburgh). The Type II network is suitable.

Bridged-T networks are also candidates for phase equalization (ref. Avins et al., Cunningham), and can be designed to be adjustable (ref. Pfitzenmaier). Band-pass filters may also serve to correct phase (ref. Herzog).

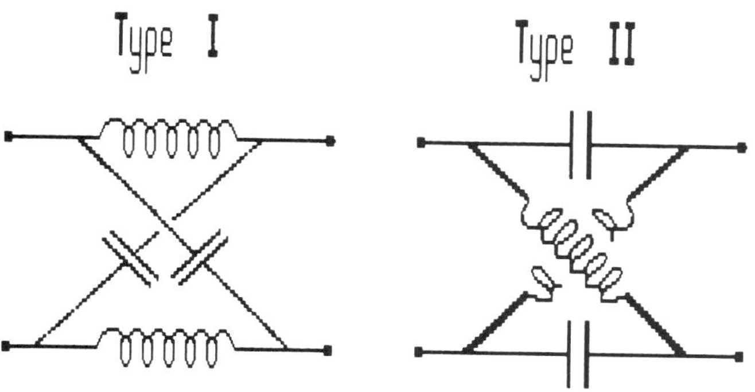

Fig. 22-18. All-Pass filters can correct phase while preserving flat amplitude response.

A final topic that belongs under signal restoration is the recovery of the original recording signal that has been nonlinearly distorted. This signal can be computed by an iterative process, given the distorted recording and the actual recording device or a model of the record-reproduce process (ref. Preis and Polchlopek).

EQUALIZATION, DIGITAL SIGNALS

In Chapter 21 we learned that the vector magnetization pattern and the curved transition zone cause a peak shift, and a broadening of the pulse width PW50. The asymmetry of the pulse is also very evident. An equalization of this signal is made in anticipation of optimum detection of the earlier written bit pattern. The detection can be carried out by either detecting the level of the bit (voltage greater than zero for a "1," and voltage less than zero for a "0"), or by detecting the transitions; either method will work.

Level detection requires a broadband equalization of the signal, very much like discussed in the earlier sections. This method is often used when wideband instrumentation recorders are used to record digital data (HDDR—High Density Digital Recorders).

Transition detection is more appealing to the digital engineers since each transition produces a pulse, and its position determined by detecting the peak of the signal. This is accomplished by differentiating and sampling for the signals zero-crossing, that corresponds to the peak of the original pulse.

It is still desirable to equalize the signal, prior to any differentiation or other processing. Equalization means restoring the signal, which for the read pulse means to make it symmetrical, and of shortest possible duration. Otherwise adjacent pulses will overlap, or *interfere*, and result in a pattern that becomes more and more difficult to detect, as the packing density increases.

Most work on equalizer designs has therefore concentrated on *pulse slimming* (ref. Schneider). An early method used the aperture equalizer from Fig. 22-17, and offered a circuits description as shown in Fig. 22-19 (ref. Kameyama et al.). The delay line is matched at the input, while open at the other end. An incoming pulse will travel down the line and reflected at the open termination, and its voltage double. Hence, for an input waveform of $f_i(t + \tau)$ the output of the equalizer can be expressed as:

$$f_o(t) = f_i(t) - (K)*(f_i(t - \tau) + f_i(t + \tau)) \tag{22.3}$$

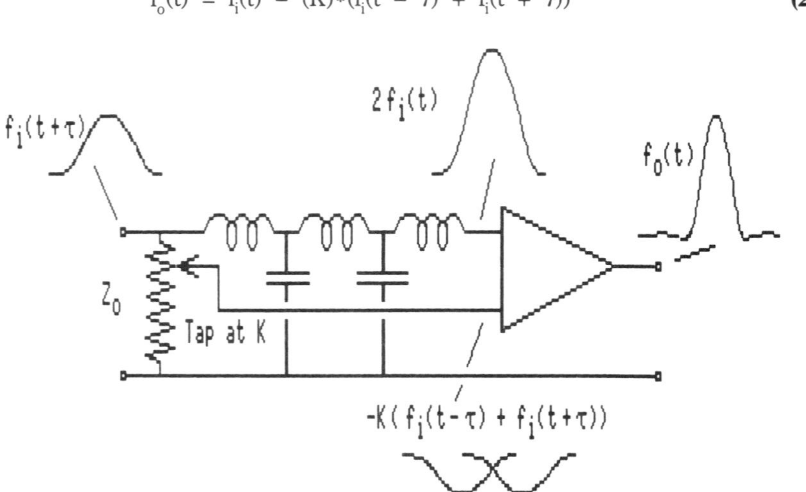

Fig. 22-19. Time domain analysis of delay line equalizer (after Kameyama et al.); see text for details.

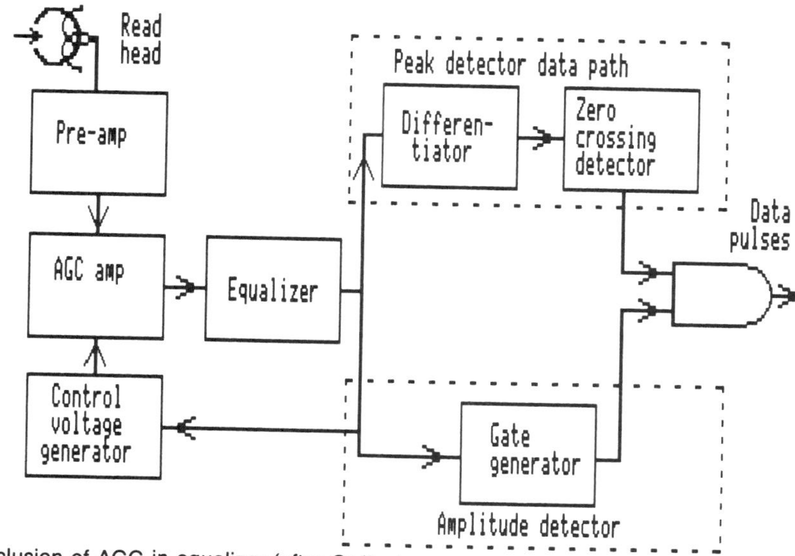

Fig. 22-20. Inclusion of AGC in equalizer (after Geffon).

where K is the fraction of the signal across Z_o. The half width of an isolated pulse is reduced to about 60 percent of the original pulse. It was found, however, that a higher SNR is required, so as to obtain the improvement effectively. As an example, the recording density was increased by about 30 percent at $K = 0.7$ and $\tau = PW50/2$, when the SNR at the input is 35 dB (ref. Kameyama et al.).

Refinements to the equalization, and subsequent detection, have been made by adapting results from communication theory, in particular from radar where distinction between return echos (pulses) is important for high resolution. The analytical results (ref. Huber, 1977) were employed in the design of a preamplifier, a $\cos^4$ equalizer, and AGC, and shown in Fig. 22-20 (ref. Geffon). The system was used in the first applications of 3PM coding (ref. Jacoby. See also Table 24-1.) The topic of raised cosine equalizers was subsequently examined (ref. Chi), and the models for the read pulses examined (ref. Mackintosh). But the generic block diagram of the read electronics remains pretty much as shown in Fig. 22-21.

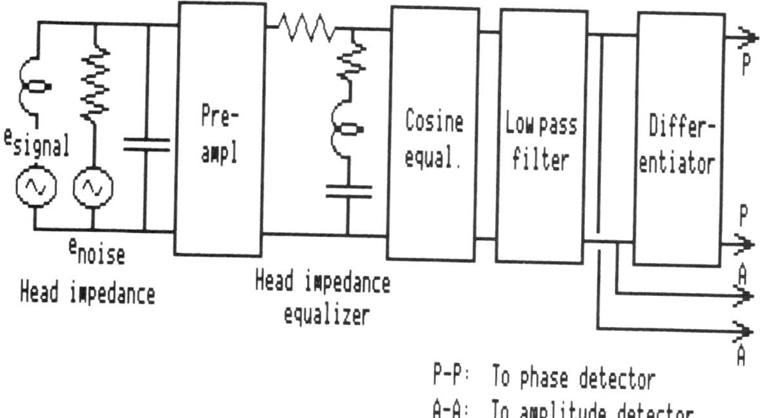

Fig. 22-21. General block diagram of read circuitry in digital recorders.

The delay line equalizer is subject to improvements, such as providing it with taps (ref. Nishimura and Ishii), which of course is the original Kallman filter.

Equalization of NRZ signals can also be accomplished, in spite of their heavy DC-content (ref. Wood and Donaldson, Huber (1981)).

Latest member in the equalization techniques is the employment of decision feedback equalization (ref. Bergman). An explanation of these equalizers is beyond the scope of this book, but the interested reader is referred to a recent book by A.P. Clark, on the topic of equalizers for digital signals (see ref.).

WRITE EQUALIZATION

The read waveform should, from the viewpoint of the communications engineer, look exactly like the input waveform. Equalization is a means to achieve this. If the record (write) current waveform is a square wave, we would also like to read a square wave, after signal integration. Figure 22-22 shows the actual read voltage e′ (integrated), and we note the asymmetry and lack of sharp corners.

The droop in the signal is caused in part by phase shift (the low frequencies arriving early, see also Fig. 22-7), and in part by the lack of read-out response down to DC. The lack of well defined corners is caused by the filtering action of the write-read channel.

The waveform can be analyzed by a mathematical method called a *Fourier Transformation* that converts the amplitude versus time relationship for a periodic waveform into an amplitude versus frequency spectrum together with a phase versus frequency spectrum. The process of doing this used to be very time consuming, but today it can be done in a very short time on a computer. Programmable engineering calculators can be provided with an electrical engineering module where a program carries out a Fourier transform in a short time.

The result of such an analysis is shown in Fig. 22-22, and we observe the following two important results:

The amplitude spectrum decreases faster with frequency than the sequence 1 - ⅓ - ⅕ - ⅐, etc., that is the normal amplitude distribution for the components in a squarewave. The relationship can be partly restored by amplitude equalization. Frequency components beyond the upper frequency limit of the recorder can obviously not be restored.

The phase relationship between the fundamental and the harmonics is not linear. This means that the components will make up a new waveform that can be expressed as (for the example shown):

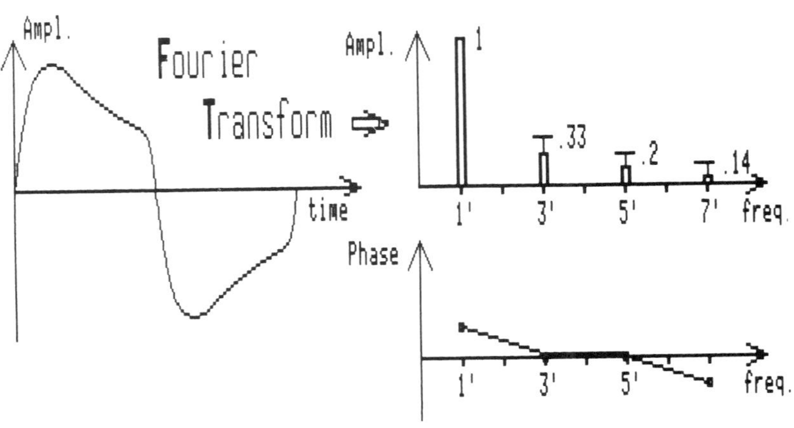

Fig. 22-22. Integrated read waveform has spectral amplitude response different from a squarewave.

$$
\begin{aligned}
V &= E \sin\omega t \\
&+ (E/\ 6.9) * \sin(3\omega t + \alpha_3) \\
&+ (E/20.5) * \sin(5\omega t + \alpha_5) \\
&+ (E/51.0) * \sin(7\omega t + \alpha_7) \\
&+ \ldots
\end{aligned}
$$

The angles α_n represent the phase distortion of the read signal. If they produce a phaseshift versus frequency curve that deviates from a straight line that goes through zero at $\omega = 0$, then the phase distortion must be corrected. This is done by the earlier mentioned techniques.

The absolute amount of phase equalization is unimportant. The goal is to restore a linear relationship: if we include a phase angle β in the fundamental signal, then the requirement is:

$$\alpha_n = n * \alpha_1 \qquad \textbf{(22.3)}$$

For example: If $\alpha_1 = 20°$, then $\alpha_3 = 60°$ and $\alpha_5 = 100°$ etc. Phase Distortion can also be evaluated by the *envelope delay*, defined as:

$$\text{Envelope Delay } \tau = d\alpha/d\omega = d\alpha/2\pi df \qquad \textbf{(22.4)}$$

where α = function(ω) is the phase versus frequency characteristics. If the slope is constant, then the envelope delay τ is also constant, and all harmonics arrive simultaneously at the network's output and a square wave is restored to its original shape, with the exception of missing high frequencies.

Figure 22-23 illustrates one method of pre-compensation by using a suitable write current. Its waveform is determined by correcting for the deficiencies shown in Fig. 22-22, i.e., we wish to emphasize the amplitudes of the harmonics, in connection with a phase shift as shown in Fig. 22-23. After a Fourier transform from the frequency to the time domain we arrive at a double-step waveform.

A time domain analysis of the double-step write waveform is shown in Fig. 22-24 (ref. Jacoby). An isolated write transition (a) produces a read pulse as shown in (b). A reduction of its PW50 can be obtained by superimposing a doublet on the transition, as shown in (c). Every transition in the write current is now preceded and followed by a minor transition of opposite polarity (d). The transition will, itself, still produce the read pulse from (b), but addi-

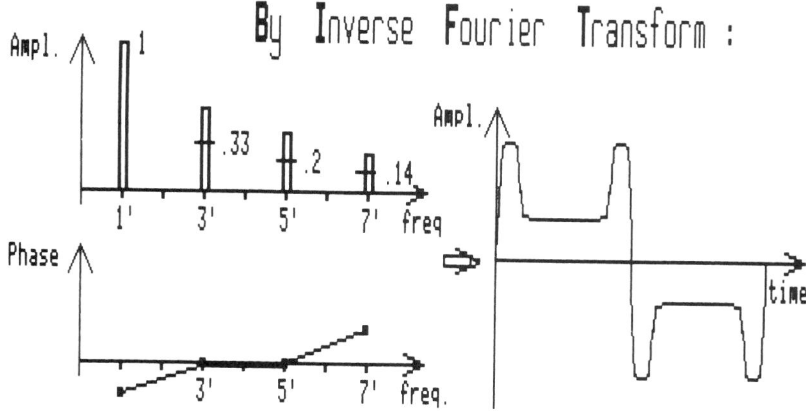

Fig. 22-23. Write waveform required to compensate for amplitude and phase responses shown in Fig. 22-22.

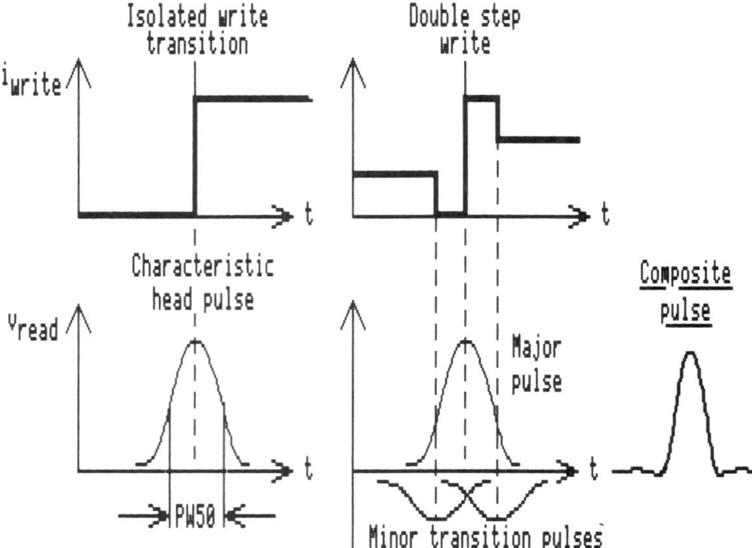

Fig. 22-24. Double-Step write waveform derived from time domain (after Jacoby).

tional two responses are now added from the minor transition's pulses. This method affords a reduction of PW50 to 60 percent of its original value (ref. Jacoby).

This form of recording requires a linear recorder, which means that AC-bias has to be applied. Its frequency must be synchronized to the clock frequency, and its level is adjusted so the bias current is approximately equal to the data current. This results in minimum phase distortion of the recording. AC-bias further results in a quieter recording, with an overall improvement in the SNR.

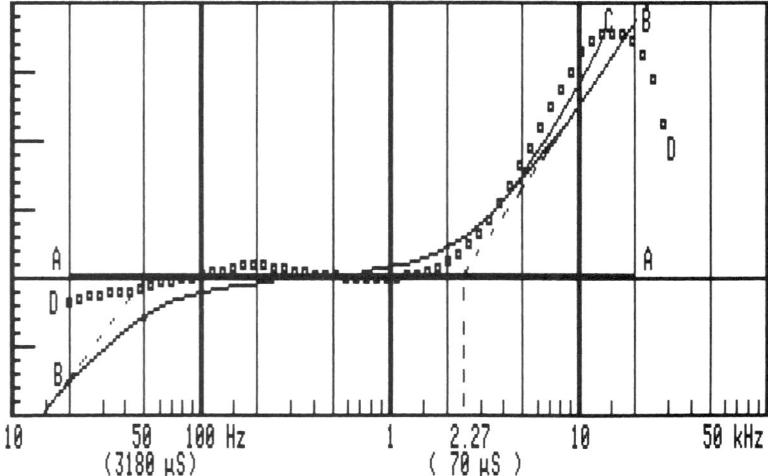

Fig. 22-25. Curves to establish the proper post-equalization in a linear (AC-biased) recorder.

Pre-compensation, including the transition time shifts discussed in Chapter 19, has received little attention in the literature (ref. Jacoby, Schneider, Kato et al.). Kato illustrates how the fast rise time in a high-Q (8-10) head can be utilized by superimposing a compensating current with ringing complementary to the head.

The fast rise time of the write current is beneficial to high packing density. It seems like another possibility exists in AC-bias recorders where a control of the bias current could be used to shift the write zone back or forth to compensate for transition time equalization.

A final consideration must be given to the fact that write current adjustment in disk drives is a compromise, since lesser current is required when the head is located at an inner track, where it is flying close to the disk surface, as compared to writing on an outer track.

Adjusting the Equalization Circuits (AC-Bias Recorders)

The simplest and easiest way to adjust a recorder's playback equalization is to use pre-recorded standard tapes. Such tapes are available from several manufacturers for the tape speeds used in audio work, and for instrumentation recorders as well. A typical alignment tape starts with a section that contains a short wavelength recording for proper alignment of the reproduce head. A following section has a normal record level (one percent distortion) for VU meter zero setting.

After these two tones there follows a series of signals of varying frequency and upon playback of these tones the controls in the equalization circuitry are adjusted for optimum flat frequency response. This procedure will assume that all pre-recorded tapes for a given standard will be reproduced correctly. Prior to using a test tape, it is advisable to clean the magnetic heads and demagnetize them with a head demagnetizer. It is also advisable not to start and stop the recorder in the middle of a test tone since this may produce weak transients on the tape and, consequently, shorten its useful life.

A word of caution: When a full-track alignment-tape is used for narrow tracks there will be excessive output at long wavelengths due to fringing response.

If a test tape is not available a different approach must be taken. This technique utilizes the figure-eight loop (see Fig. 9-24), or a straight wire in front of the reproduce gap, which is fed with current that has an amplitude versus frequency response as established below. (The input signal to the reproduce amplifier can also be provided by a voltage generated across a small resistor placed in the ground side of the magnetic head.) It is important that the signal be introduced to the playback amplifier through the reproduce head. If the signal is connected directly to the input stage, the head core losses and the self-resonance will not be included in the measurement and the reading will be incorrect.

Assuming, at first, that the current through the wire in front of the head is constant and independent of frequency, the flux through the head will be constant but limited by core losses. This corresponds, at low frequencies, entirely to the conditions existing during playback of a tape. But the external flux from a medium decreases due to coating thickness losses toward the shorter wavelengths. The flux through the head core will be further reduced toward the shorter wavelengths because of read gap losses, which can be determined as shown in Fig. 22-25.

Assuming that the equalization has already been adjusted, the output voltage from the playback amplifier will increase toward the high frequencies when the wire in front of the head is fed with a constant current. It is relatively easy to pre-determine how much this rise in output voltage should be for proper equalization. Referring to Fig. 22-25, the following steps show how to determine the rise in the output voltage:

- On a sheet of lin-log graph paper, with the Y-axis for amplitude and the X-axis for frequency, draw a straight line representing the desired final frequency characteristic (curve A in Fig. 22-25).

• On a sheet of lin-log graph paper, with the Y-axis for amplitude and the X-axis for frequency, draw a straight line representing the desired final frequency characteristic (curve A in Fig. 22-25).

• Add to this curve the difference between a constant flux recording and a standardized tape flux. Figures 28-1 and 28-2 show the currently used standards.

• Measure the effective gap length of the reproduce head as shown in Fig. 9-23. Determine the curve for the gap loss from Fig. 6-14. Add this loss to curve B (Fig. 22-25), obtaining curve C.

• Place the straight wire or the figure-eight loop in front of the reproduce head gap in such a position that the output voltage from the amplifier is maximum. If a constant current (versus frequency) is now fed through the wire, then the output voltage should closely match curve C in Fig. 22-25.

An actual measured curve may have a response as shown by curve D, where the low end will roll off due to amplifier limitations. It may have a peak at high frequencies due to self-resonance in the reproduce head. If the deviation between curve D and curve C is within 1 dB, the equalization is as good as can be, but deviations in the order of several decibels require readjustment (and possible redesign) of the playback equalizer.

Prior to measuring curve D, it is advisable to connect an oscilloscope to the playback amplifier and monitor it so that the induced voltage in the playback head does not cause overload and distortion, and therefore, an erroneous reading of the output signal.

Adjustment of Record Pre-Equalizer

A prerequisite for proper adjustment of the record amplifier's equalization network, and record level setting, is a properly functioning bias oscillator/amplifier. Even small amounts of distortion in the bias supply will generate both noise and distortion on the tape. The latter will affect the proper setting for normal record level (1 percent third-harmonic distortion on the tape).

The best way to check for a *proper bias waveform* is to connect the figure-eight loop to an oscilloscope and place it in front of the record head and next energize the record and bias circuitry. The waveform of the bias signal (with data input short circuited) must be a perfect symmetrical sine wave.

Most recorders have a 10-ohm resistor in the ground leg of the record head winding which provides a suitable test point to check the bias signal. Many bias oscillators, which also function as erase oscillators, have a balancing potentiometer. An easy way to check for a proper bias waveform is to listen to a tape via the playback amplifier and speaker and to adjust the balancing control for minimum noise. The nature of poor bias or erase noise is a gurgling sound rather than a hiss.

Now connect a signal generator to the record amplifier input and observe the waveform across the test resistor (10 ohms) in the ground leg of the record head circuit (or from the figure-eight loop in front of the record head). As a first approximation in setting the correct record level, adjust the ratio between the bias amplitude and the data signal amplitude to approximately 20:1. The tape selected for use on the recorder is now threaded on the tape deck and tape motion started. In *adjusting the proper bias level*, there are, unfortunately, no set standards, but the two following rules of thumb apply fairly well:

• For high-quality audio recordings, the bias current is adjusted for 2 dB over bias at 1 kHz. That is, the bias current is increased from a low value until the recorded signal level reaches a maximum; the bias current is now increased until the signal level has decreased 2 dB. This setting will assure the largest dynamic range with a minimum of distortion.

- In instrumentation recorders with wide signal bandwidths the bias setting is adjusted to achieve a high output at the shortest wavelength. For example, it is common practice in an instrumentation recorder operating at 120 IPS with a bandwidth of 4 MHz to adjust the bias while recording a 4 MHz tone. The bias level is increased from a low value until the signal on the tape reaches a maximum and then it is increased slightly until the signal level drops 2 dB.

The recorder designer has a choice between these two extreme bias settings and it is up to him/her to establish the compromise between bandwidth and dynamic range. The recorder user is referred to the manual for the adjustments on the particular recorder. It should be emphasized again that the bias level is dependent upon the type of tape used and the thickness of its oxide coating, particularly with audio recordings where the higher bias levels are used.

SETTING OF THE NORMAL RECORDING LEVEL; MEASUREMENT OF DISTORTION

The "normal recording level" is that level which produces one percent harmonic distortion off the tape. With freedom from DC magnetization and a clean bias waveform, this distortion will be odd, from the third-harmonic component. The simplified formula for this distortion is:

$$\text{Distortion} = (V_3/V_1) * 100 \text{ percent} \qquad \textbf{(22.5)}$$

where V_1 is the level of the fundamental signal and V_3 is the level of the third-harmonic component.

Until recently, a frequency of 1 kHz was used in setting this level for instrumentation as well as for audio recorders. But it will be found that the distortion in wideband instrumentation recorders often increases with frequency. Therefore, it has become standard to set the record level in a wideband instrumentation recorder at a frequency that is one-tenth of the upper-frequency limit (for example, 150 kHz for a 1.5 MHz recorder).

Measurement of the third harmonic signal in a tape system differs from the measurement of distortion in an amplifier. A sharply-tuned selective wave analyzer will not accurately measure the harmonic components in the presence of wow and flutter from the tape (or disk) transport. The bandwidth of the signal analyzer must be quite wide in order to capture the third harmonic. The simplest way to do this is to measure V_1 with a voltmeter, then insert a high-pass filter which attenuates V_1 by at least 50 dB and then measure V_3 (plus any higher harmonics). The proper record level is now adjusted by trial and error to achieve a one percent distortion (= −40 dB); the VU meter or other record level indicating device is then adjusted for a reading of 0 dB.

The recorder is now ready for *adjustment of pre-equalization* in the record amplifier. Since pre-equalization in audio recordings requires a boost of the high frequencies, which cause overload and distortion at the normal record level, the following adjustments should be made at a level that is as many dB below the normal record level as the amount of maximum equalization (boost) in dB. It is generally safe to carry out the measurements at a level of −20 dB re 0 dB VU.

A series of frequencies (in audio, for instance: 50, 100, 200, 500, 1000 Hz and 2, 5, 10, 12, 15 kHz) are recorded and played back and the frequency response plotted on an amplitude versus frequency graph. Without pre-equalization it will be found that the frequency response curve will roll off toward higher frequencies and that a corrective network must be installed in a record amplifier or adjustments made to an existing network in order to achieve a flat frequency response. The amount of pre-equalization can be evaluated by measuring the record current across the 10 ohm test resistor in the ground leg of the record head, with the bias oscillator disabled.

REFERENCES TO CHAPTER 22

Amplitude and Phase Response in Magnetic Recording

Jorgensen, F., "Phase Equalization is Important," *Electronic Industries*, Oct. 1961, Vol. 20, No. 10.

Star, J., "Envelope Delay in a Tape Recorder System," *Solid State Design*, June 1965, pp. 27-33.

Jacoby, G.V., "Signal Equalization in Digital Magnetic Recording," *IEEE Trans. Magn.*, Sept. 1968, Vol. MAG-4, No. 3, pp. 302-305.

Hedeman, W.R., "Phase Distortion in Magnetic Tape Recorders," *Intl. Telem. Conf. Proc.*, Jan. 1972, Vol. 8, pp. 64-71.

Minukhin, V.B., "Hilbert Transform and Phase Distortions of Signals," *Radio Engineering and El. Phys.*, Aug. 1973, Vol. 18, No. 8, pp. 1189-1193.

Minukhin, V.B., "Phase Distortion of Signals in Magnetic Recording Equipment," *Telecom. and Radio Engr.*, 1975, Vol. 29/30, pp. 114-120.

Haynes, M.K., "Experimental Determination of the Loss and Phase Transfer Functions of a Magnetic Recording Channel," *IEEE Trans. Magn.*, Sept. 1977, Vol. MAG-13, No. 5, pp. 1284-1286.

Amplitude and Phase Equalization

Bracewell, R.N., *The Fourier Transform and Its Applications*, McGraw-Hill, 1978, 444 pages.

Cheng, D.K., *Analysis of Linear Systems*, Addison-Wesley, 1959, 431 pages.

Clarke, K.K., and Hess, D.T., *Communication Circuits: Analysis and Design*, Addison-Wesley, Jan. 1971, 658 pages.

Van Valkenburg, M.E., *Analog Filter Design*, Holt, Rinehart and Wilson, 1982, 608 pages.

Maddock, R.J., *Poles and Zeros in Electrical and Control Engineering*, Holt, Rinehart and Wilson, 1982, 216 pages.

Bogart, F., *Laplace Transforms, Theory and Experiments*, John Wiley and Sons, 1983, 148 pages.

"s-Plane Tutorial," "Transfer Function Analyzer," and "Active Circuit Analysis," *Dynacomp programs*, 1427 Monroe Avenue, Rochester, NY 14618.

Thomas, R.E., and Rosa, A.J., *Circuits And Signals: An Introduction To Linear and Interface Circuits*, John Wiley and Sons, 1984, 758 pages.

Kallmann, H.E., "Transversal Filters," *Proc. IRE*, July 1940, Vol. 28, pp. 302-310.

Dennison, R.C., "Aperture Compensation for Television Cameras," *RCA Review*, Dec. 1953, Vol. 14, pp. 569-585.

Avins, J., Harris, B., and Horvath, J.S., "Improving the Transient Response of Television Receivers," *Proc. IRE*, Jan. 1954, pp. 274-283.

Sivian, L.J., Dunn, H.K., and White, S.D., "Absolute Amplitudes and Spectra of Certain Musical Instruments and Orchestras," *IRE Trans. Audio*, May-June 1959, Vol. AU-7, No. 3, pp. 1-29.

Johnson, K.O., and Gregg, D.P., "Transient Response and Phase Equalization in Magnetic Recorders," *Jour. Audio Engr. Soc.*, Oct. 1965, Vol. 13, No. 4, pp. 323-330.

Cunningham, V.R., "Pick a Delay Equalizer, and Stop Worrying About Math," *Electronic Design*, May 1966, pp. 62-66.

Horwitz, J.H., "Reduce Delay Distortion at the Source," *Electronic Design*, Aug. 1967, pp. 116-120.

Herzog, W., "Linearizing of the Phase at Band-passes" (in German), *Archiv fuer Elektrotechnik*, April 1983, Vol. 66, No. 4, pp. 187-194.

Preis, D., and Polchloek, H., "Restoration of Nonlinearly Distorted Magnetic Recordings," *Jour. AES*, Jan/Feb 1984, Vol. 32, No. 1/2, pp. 26-30.

Pfitzenmaier, G., "A Contribution to the Realization of Adjustable Delay Equalizer with Fixed Points" (in German), *FREQUENZ*, Ma. 1984, Vol. 38, No. 3, pp. 54-62.

Signal Equalization, Digital

Barna, A., *High Speed Pulse and Digital Techniques*, John Wiley and Sons, 1980, 185 pages.

Graham, I.H., *Magnetic Recording Circuits*, Lecture Notes, Bookstore, Santa Clara University, CA 95052, 1983, 443 pages.

Baumgarten, W.H., *Pulse Fundamentals in Small Scale Digital Circuits*, Reston (Prentice Hall), 1985, 574 pages.

Clark, A.P., *Equalizers for Digital Modems*, Halsted Press (John Wiley), 1985, 468 pages.

Schneider, R.C., "An Improved Pulse-Slimming Method for Magnetic Recording," *IEEE Trans. Magn.*, Nov. 1975, Vol. MAG-11, No. 5, pp. 1240-1242.

Kameyama, T., Takanami, S., and Arai, R., "Improvement of Recording Density by Means of Cosine Equalizers," *IEEE Trans. Magn.*, Nov. 1976, Vol. MAG-12, No. 6, pp. 746-748.

Geffon, A.P., "A 6 KBPI Disk Storage System using Mod-11 Interface," *IEEE Trans. Magn.*, Sept. 1977, Vol. MAG-13, No. 5, pp. 1205-1207.

Jacoby, G.V., "A New Look-Ahead Code for Increased Data Density," *IEEE Trans. Magn.*, Sept. 1977, Vol. 13, No. 5, pp. 1202-1204.

Huber, W.D., "Maximization of Lineal Recording Density," *IEEE Trans. Magn.*, Sept. 1977, Vol. MAG-13, No. 5, pp. 1208-1210.

Wood, R.W., and Donaldson, R.W., "Decision Feedback Equalization of the DC Null in High Density Digital Magnetic Recording," *IEEE Trans. Magn.*, July 1978, Vol. MAG-14, No. 4, pp. 218-222.

Mackintosh, N.D., "A Superposition-Based Analysis of Pulse-Slimming Techniques for Digital Recording," *Intl. Conf. on Video and Data 79*, July 1979, IERE Publ. No. 43, pp. 121-147. Also in *The Radio and Elec. Engr.*, June 1980, Vol. 50, No. 6, pp. 307-314.

Chi, C.S., "Characterization and Spectra Equalization for High Density Disk Recording," *IEEE Trans. Magn.*, Sept. 1979, Vol. MAG-15, No. 5, pp. 1447-1449.

Langland, B.J., "Phase Equalization for Perpendicular Recording," *IEEE Trans. Magn.*, Nov. 1982, Vol. MAG-18, No. 6, pp. 1247-1250.

Huber, W.D., "Equalization of the D.C. Null in High Density Digital Magnetic Recording," *IEEE Trans. Magn.*, Nov. 1981, Vol. MAG-17, No. 6, pp. 3352-3354.

Nishimura, K., and Ishii, K., "A Design Method for Optimum Equalization in Magnetic Recording With Partial Response Channel Coding," *IEEE Trans. Magn.*, Sept. 1983, Vol. MAG-19, No. 5, pp. 1719-1721.

Bergmans, J.W.M., "Decision Feedback Equalization for Digital Magnetic Recording Systems," *Intl. Video and Data Conf.*, 1986, IERE Publ. No. 67, pp. 141-145.

Write Equalization

Jacoby, G.V., "High Density Recording with Write Current Shaping," *IEEE Trans. Magn.*, July 1979, Vol. MAG-15, No. 3, pp. 1124-1130.

Schneider, R.C., "Write Equalization in High-linear-density Magnetic Recording," *IBM Jour. Res, and Devel.*, Nov. 1985, Vol. 29, No. 6, pp. 563-568.

Kato, T., Arai, R., and Takanami, S., "Write-Current Equalization for High-Speed Digital Magnetic Recording," *IEEE Trans. Magn.*, Sept. 1986, Vol. MAG-22, No. 5, pp. 1212-1214.

Chapter 23

Signal Recovery and Detection; Measurements

The traditional illustration for the transmission of a signal from source to destination is shown in Fig. 23-1. It shows the application of analog-to-digital conversion, which today is far more used than direct transmission of the signal. The transmission channel can be many things: A telephone wire, antenna-to-antenna transmission of radio, TV or other signals, and storage of data on magnetic or optical media for later playback.

The reasons for digitizing the signal are many. Small perturbations of digital signals caused by the channel can be recognized and corrected, whereas analog signals are left at their perturbed values, with no possibility of correction. However, the channel does not make any distinction whether the signal is analog or digital, they are treated alike. The superiority of digital signals is that they can be regenerated.

The noise that occurs in the write/read channel has many possible sources: It may be random noise from the magnetic medium, from preamplifier or the power supplies. It may also be impulse noise of very short duration, or RFI (Radio Frequency Interference). The signal may also disappear for a very short duration due to drop-outs.

SIGNAL WAVEFORMS THROUGH THE RECORDER CHANNEL

We will briefly review the origin of the train of bits, or ''On-Off'' signals. An example is shown in Fig. 23-2, illustrating the pulse signals that represent the letters TAPE.

Each letter in the alphabet is represented by a byte that consists of 8 bits, where the last bit is an added ''0,'' and the remaining seven are the ASCII code. (This 8-bit code is also called ASCII-8, and is part of the commonly used code called EBCDIC, which has $2 \times 128 = 256$ different bytes.)

The sequence of the four bytes that represent TAPE forms a 32-bit signal (B), that in many systems would be recorded in NRZI (Non-Return-to-Zero, inverted—see Chapter 24).

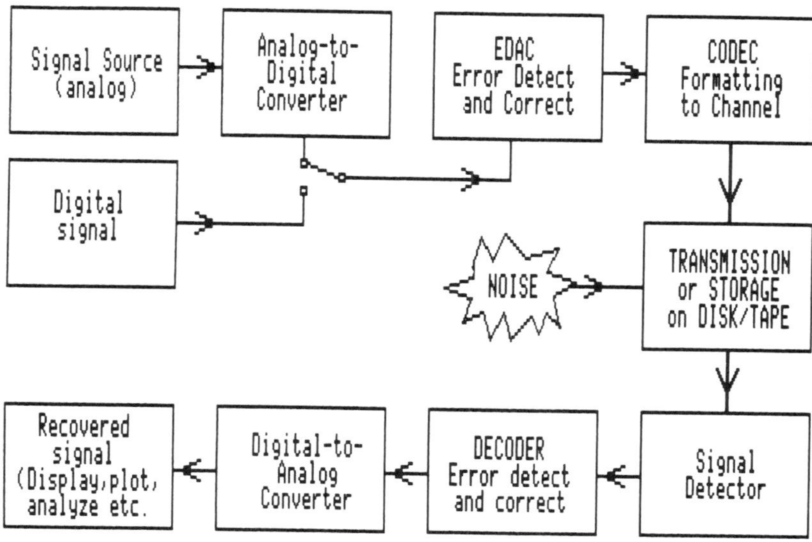

Fig. 23-1. Block diagram of communication channel.

The NRZI pattern is shown in line (C) and the resulting media magnetization in line (D).

The read head output is obtained by differentiating the flux from the magnetization (Faraday's law), line (E), and may further be processed by integration (line (F). The waveforms shown are ideal; the real world signal is shown in line (G).

The high-frequency limitation decreases the rise-time of the leading and trailing edges of the individual pulses, while the lack of low-frequency response in the magnetic read process

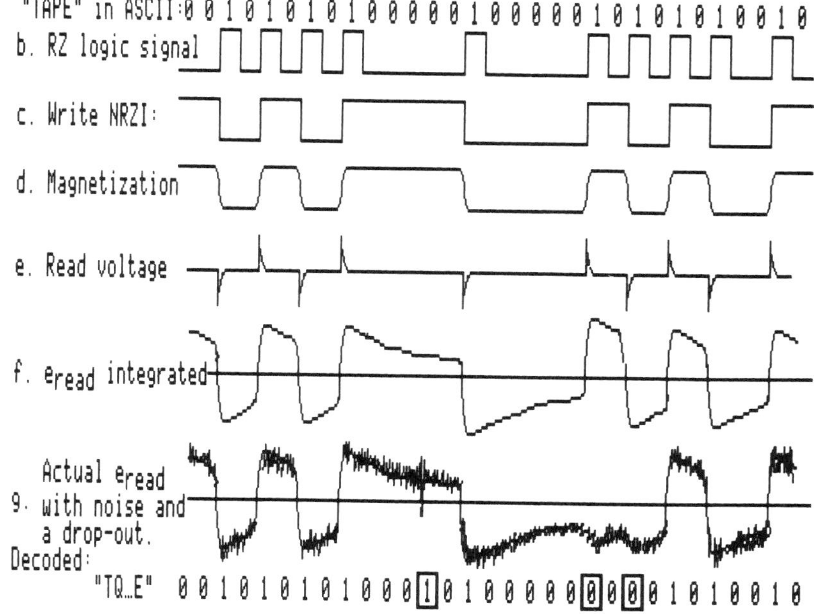

Fig. 23-2. Signal degradation by passage through a tape or disk drive signal channel.

causes a droop in the flat portions of the pulse train. This lack of a DC component in the read signal may cause serious signal deterioration, and is avoided by use of modulation codes (see next chapter).

An examination of the RZ code versus the NRZI code for just the letters A and P shows a smaller DC-component in the NRZI version. The excessive droops are caused by too many consecutive "0"s (5 in the letter A).

The addition of noise can cause mistriggering of the detection circuits, in particular if noise spikes are present on a droop. One example is shown, causing a false reading of a zero-crossing, and hence a change of the word A's code 1000 0010 to 1000 1010, which is the letter Q!

An "open circuit" in the transmission can be caused by a drop-out in the head-media interface area. It may be a fault in the coating, but is much more likely to be caused by a foreign dust particle (smoke-particles, floor dust) that lifts the coating away from the read gap (like a pole in a tent). A recorded change in flux polarity may not be sensed, as shown at the second bit in the letter P. It may further affect the correct detection of following flux-changes, as for the second "1" in the example shown. The original letter "E" comes out as 0000 0000, which is "NUL" = Null.

This example has illustrated how the write/read channel's imperfections will degrade the digital signal. We shall now take a closer look at the write/read process for a single transition and thereafter a larger view of the entire process for a sequence of transitions.

The Read Signal; Response to a Transition

The read (playback) process was analyzed in Chapter 21 and the result is shown in Fig. 23-3. The read voltage waveform from a single, isolated transition has a shape that depends upon, among other things, the pattern of the vector magnetization. A prediction of this magnetization presents somewhat of a difficulty since the saturation recording process is non-linear, and our knowledge of the recorded pattern is at best a rough estimate.

We may, however, modify the estimated pattern within a certain framework so the calculated response will closely match the measured. We can, in other words, learn about the mag-

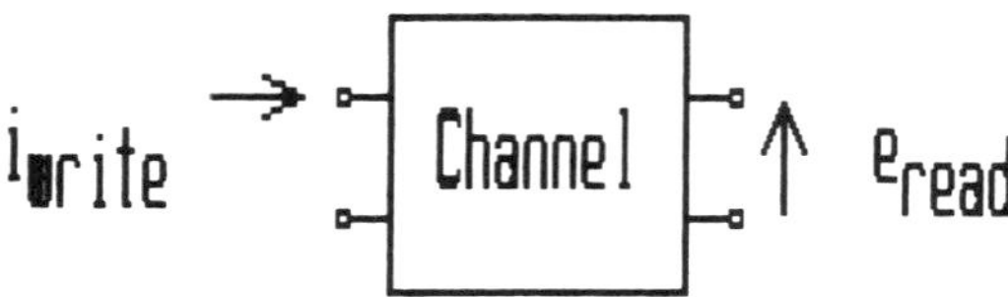

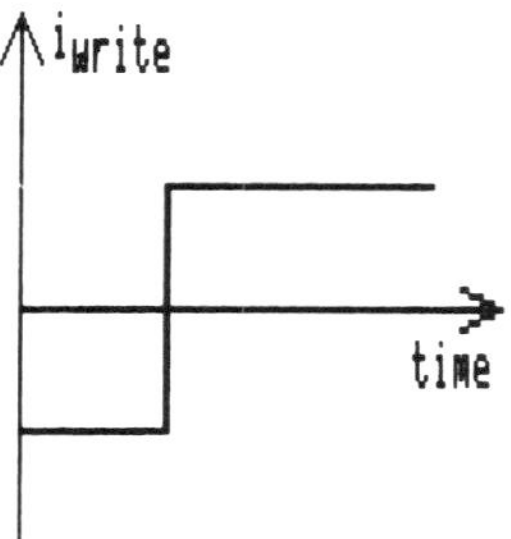

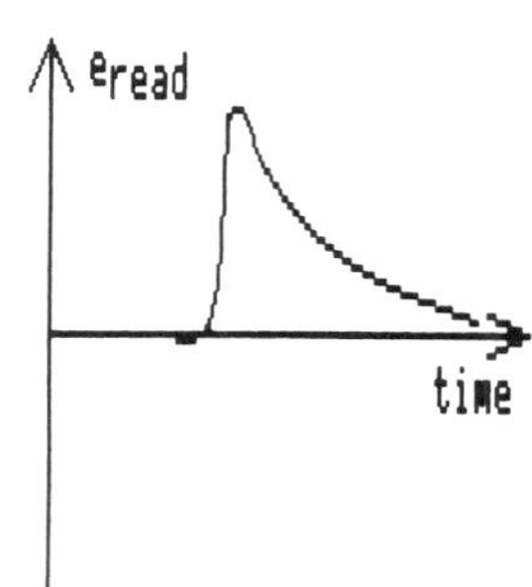

Fig. 23-3. Tape or disk channel response to a single input step current.

Estimating system response from observation of square wave response:

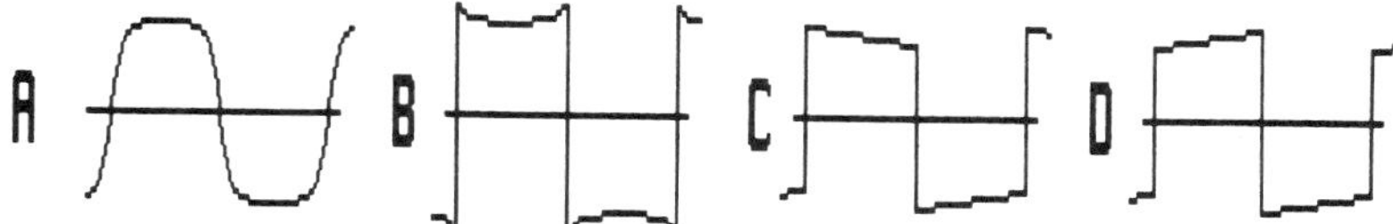

A. High frequencies attenuated; No phase error.
B. Fundamental attenuated; No phase error.
C. Phase distortion: Harmonics lagging (less than 5%).
D. Phase distortion: Harmonics leading (less than 5%).

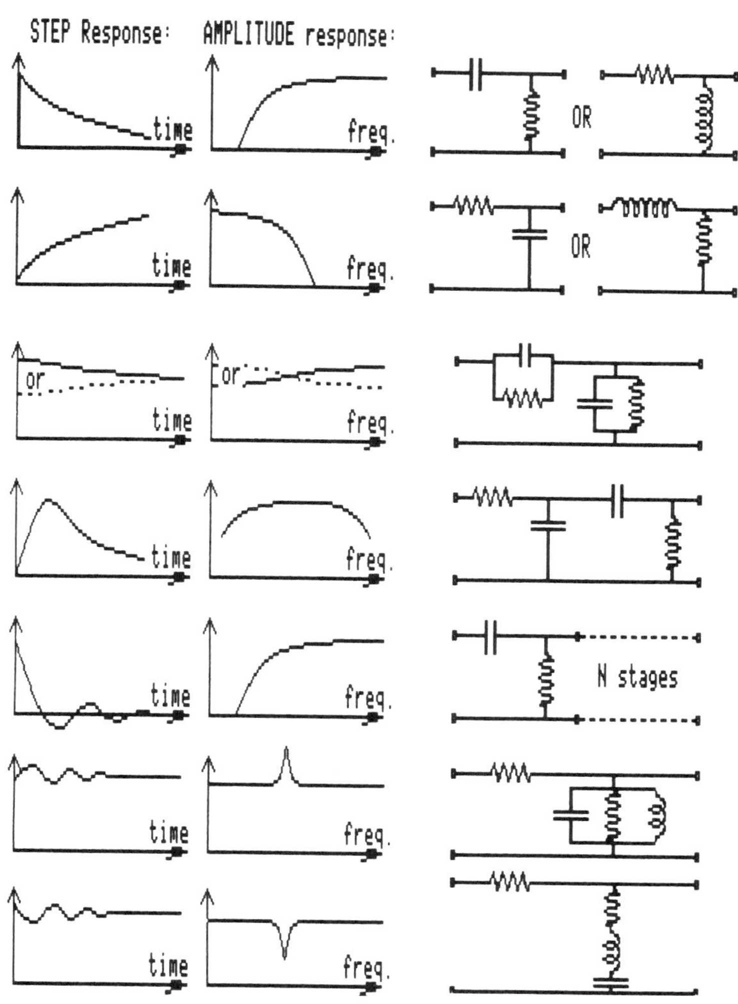

Fig. 23-4. General system responses to a step input function.

netization pattern through a reverse or backward analysis. This method must, however, be applied cautiously since different magnetization patterns can produce the same response.

The response to an input step can now be applied to the write-read channel, because we have defined and "frozen" the non-linear recording process. It is safe to assume that the read process itself is linear, and that we therefore can superimpose the response from several transitions to form a resulting read voltage waveform.

Comparisons with other system tests are in place. An often used test signal is the square wave signal, which in a tape or disk system with inductive read heads will result in a response as shown in Fig. 23-3. Rounding of the corners is caused by the lack of high frequency response, and the droop in signal level by lack of response down to DC. Such a channel is said to have a *partial response*.

The response to just the leading (or trailing) edge of the square wave is called the *step response*. An examination of this response can reveal the network function of the channel; examples are shown in Fig. 23-4.

Another popular test signal is the impulse, which gives a response called the *impulse response*. The impulse is a differentiation of the step, and the response is likewise the differentiated step response. Further, since the system (channel) is assumed linear, then a differentiation somewhere between input and output will result in the output being the impulse response, when the input is a step.

In other words: a unit step write current results in a unit step response in the flux, which differentiated results in the impulse response.

This is very useful since the read voltage now is determined by straightforward convolution of the write current with the impulse response, which is the recorder channel's response to a step write current. Any peak-shifts that occur during the record process must be determined and included in the write current. Also, the method assumed inductive read heads.

We can even go so far as naming the response an *impulse response* $h(\tau)$, similar to the notation used in systems theory.

Measurements of h(τ)

$h(\tau)$ is easily measured for a read-write channel by first recording a square wave with a low enough frequency so the recorded transitions do not interfere. The rise time of the input square wave form should be less than one fifth of $1/f_u$, where f_u corresponds to the system's upper frequency limit. The voltage from the read head is observed, and its shape is the shape of the impulse response. Comparison of calculations will provide information about both the correlation between assumed and actual vector magnetization and the shape of the transition zones.

The Transfer Function

The plot of a read-write channel's impulse response is truly a *signature of that particular channel*. It contains information about:

Write field: The penetration of and curvature of the field lines corresponding to a strength equal to the coating's coercive force H_c. This will include effects of head-medium spacing.

Coating: The magnetization of each segment is determined by the write field lines in combination with the longitudinal and the perpendicular susceptibilities. The coating permeability is assumed isotropic and equal to one for both the write and read field patterns.

Read Process: The read signal is determined by reciprocity, and the Karlqvist equations are used for the read field (or similar expressions for thin film heads). Hence the following dimensions are included:

Gap length
Head-medium spacing
Recorded coating thickness.

We may also use a tabulated set of values for h(τ) in a Fourier transform to find the channel's amplitude response H(ω) and phase response ψ(ω), and from the latter the envelope delay.

Measurements have recently been made on tape recorders using the impulse as input function.

Superposition (Convolution); Symbol Interference

The single pulse voltage from a transition has a pulse duration PW50, measured at half pulse height, that is considerably longer than the time for passage of the transition zone length Δx past the read gap; it has been said that the read process has a resolution that is several times broader than the write process. (It is, however, still important to aim at short transition zones in the write process since shorter transitions provide faster changes of flux and therefore higher read voltages.)

When the coating is recorded with some information from a write current i = I(τ), then it is a simple matter to determine the resulting read voltage waveform by convolution:

$$e(\tau) = \int I(th)(\tau - t)\,dt \qquad \textbf{(23.1)}$$

or

$$e(\tau) = I(\tau) * h(\tau) \qquad \textbf{(23.2)}$$

This is equivalent to sampling the waveform with the impulse response. Examples are shown in Figs. 23-5 and 23-6. The latter illustrates the serial product method that is readily carried

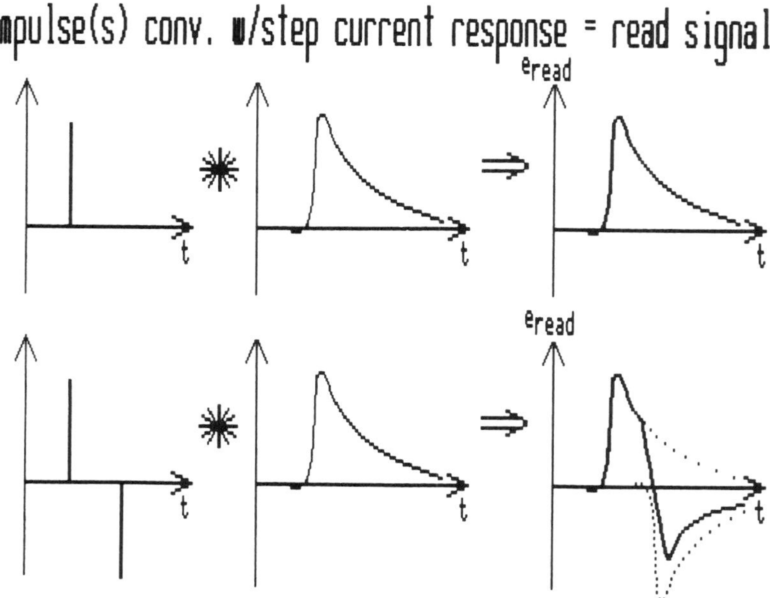

Fig. 23-5. Tape or disk channel response to one and two steps are obtained by taking the convolution of the corresponding impulse inputs with the transfer function.

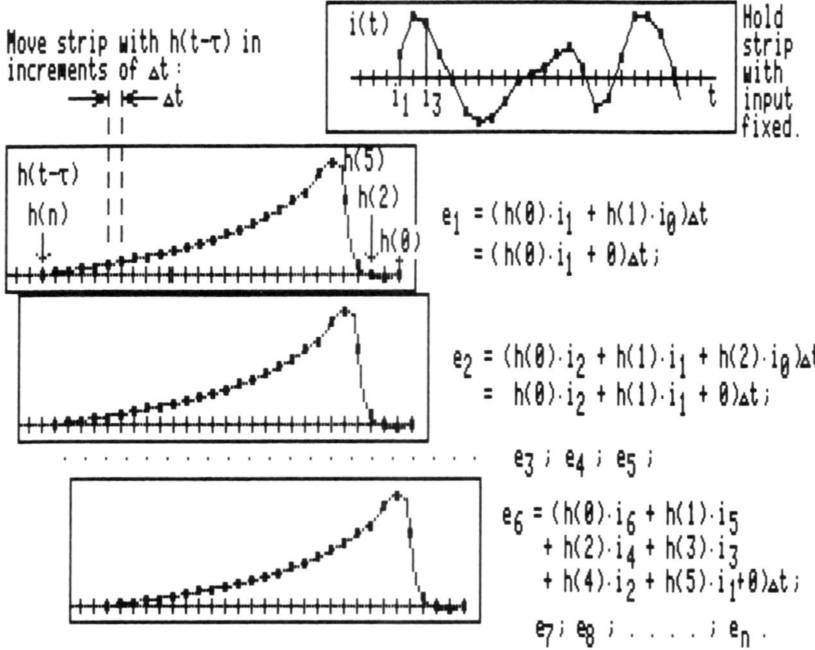

Fig. 23-6. Method for numerical/graphical convolution (after Bracewell).

out by using both a listing of the values of I(τ) on a sheet of paper and a reversed listing of the h(τ) values on a separate strip of paper. This strip is moved from left to right in increments of $\Delta\tau$ and a serial product summation carried out for each value of $\Delta\tau$. These results will form e(τ).

This method is adapted from Bracewell's *The Fourier Transformer and Its Applications*, Chapter 3.

NOTE: Peak shifts due to the interaction (during recording) between closely spaced transitions *are not* included in the convolution and must be added as corrections in the timing of the transition sequence, *prior* to the convolution computation.

The result in Fig. 23-7, shows not only a reduction in amplitudes, but also a broadening of the timing between the voltage peaks. The result is similar for three or more impulses. This phenomenon is called *peak shift.*

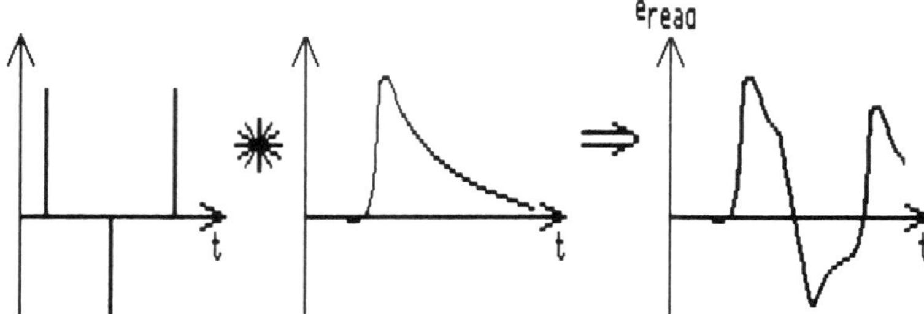

Fig. 23-7. Response to three transitions results in pulse crowding: peak shifts and reduced amplitudes (asymmetrical).

DETECTION OF DIGITAL SIGNALS

A digital read circuit extracts the digital data by detecting either amplitudes (higher or lower than a certain threshold), or by detecting the zero-voltage line crossovers. The first method is generally selected by the engineers with background in signal transmission, while the other is selected by engineers with digital circuit experience.

The detection circuit needs timing information in order to interpret the pulse stream. The timing information is extracted from the coding of the bits and often from a "Ready-Phasing" signal just prior to the data bits. The timing is used to control a *clock* that will produce pulses synchronous with the bit rate.

There does not appear to be a thorough theoretical treatment that compares signal amplitude detection at MBT (Mid-Bit-Time) with the zero-crossing detection of the signal at EBT (End-of-Bit-Time). The amplitude detection assumes an amplitude equalized signal channel, while the zero-crossing detection assumes a differentiated signal (after the read head preamplifier).

Amplitude Detection

Amplitude detection circuits rely upon the sampling of the pulse stream. The signal waveform is strobed at the center of each bit and the sample value is compared with a preset discrimination level. If the level is above the discrimination level, a "1" will result and if it is below, a "0" will result.

The strobing time should be much shorter than the bit length. The timing tolerances are reduced by peak-shift, and the amplitude sensing mechanism is affected by bit amplitude changes caused by pulse crowding, and by the general amplitude variations in a magnetic recording channel.

The effect of interference and noise can readily be observed in the laboratory by using an oscilloscope to sweep the bit stream, with a horizontal sweep rate of 1/BR, where BR is the bit rate. The resulting display is widely known as an *eye pattern* from its resemblance to the human eye.

Two binary waveforms are shown in Fig. 23-8, one distorted, the other undistorted.

When the waveform is undistorted and all sampled values are + or -1, the eye pattern is wide open.

Intersymbol interference and noise will close the eye pattern due to both amplitude and timing shifts. The pattern will provide a great deal of information about the performance characteristics of a recording system. If the noise is minimal, we can evaluate signal amplitude distortion (A), and timing errors by the distance B′, leaving only B for detection. The best sampling time is obviously in the middle of the bit period.

Zero Crossing Detection

The differentiated voltage e' of the inductive read head exhibits very sharp and well defined crossings of the 0-level axis, making it well suited for the detection of bits.

The optimum method of bit detection depends upon the selected modulation code, packing density, pulse crowding effect, and noise. Each system can be described in terms of an *error budget*, which sums up errors from all conceivable sources, including the mechanics of the media transport. Several alternatives should be analyzed and estimates compared with data measured on experimental recording systems. Such analysis will lead to decisions regarding coding schemes and detection methods.

BIT ERROR RATES

The performance of any digital data channel is measured as its freedom from errors (missing or false bits). Ratings are typically 10^{-6} to 10^{-9} errors (raw).

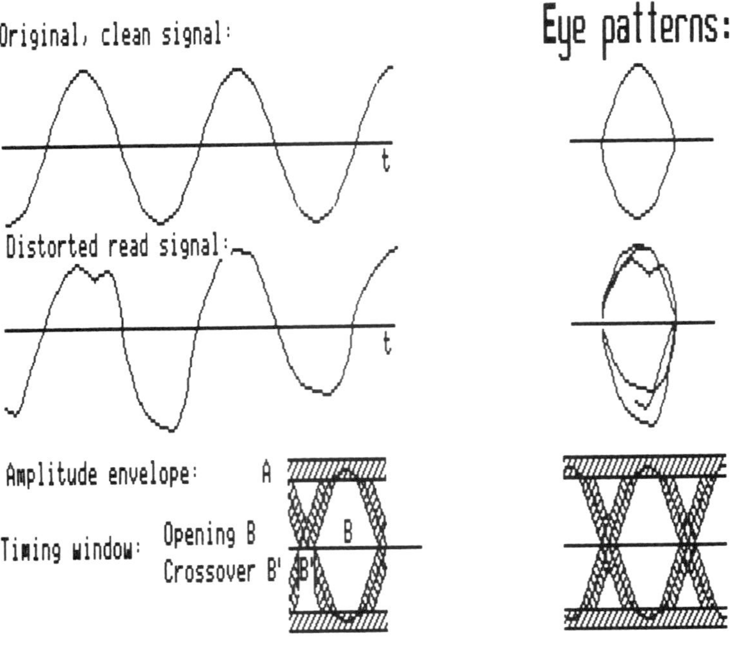

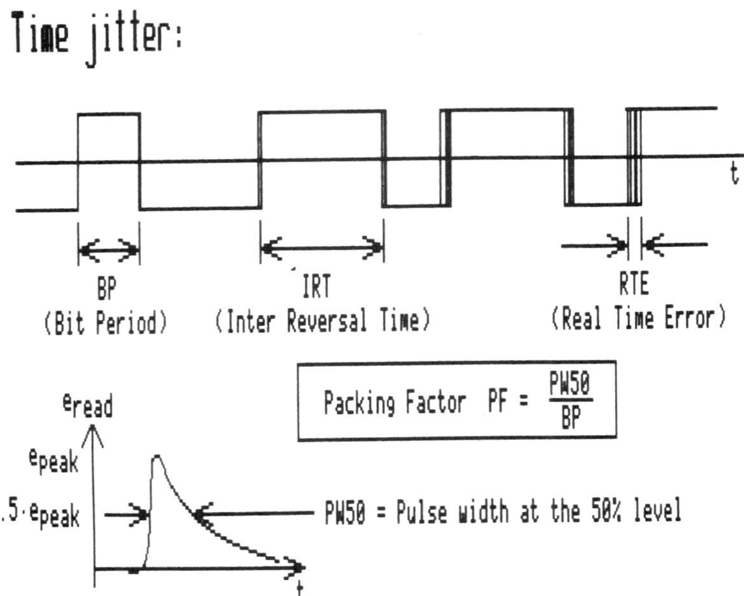

Fig. 23-8. Signals are displayed in Eye-Patterns on an oscilloscope when the horizontal sweep starts every bit. Jitter can be estimated by displaying many bit periods.

A bit-error-rate, or BER, of 10^{-6} means that, on the average, there is no more than one error per million bits.

A poor BER is in general caused by mechanical errors such as drop-outs on a recording surface, or foreign dust particles. The latter can often accumulate at an alarmingly fast rate unless cleanliness is strictly adhered to. Other causes may be RFI.

If a system is free from drop-outs, dust, and RFI, then BER is caused by the noise from the media and/or heads plus the electronics. Both have approximate Gaussian distributions and the achievable BER can be calculated with the aid of probability mathematics. This results in a curve as shown in Fig. 23-12, correlating the system SNR and the anticipated BER.

There are error detecting and correcting codes that can significantly improve a system's BER. This is invariably paid for by added electronics, which are often highly sophisticated circuits. Today LSI circuits are making fast advances where ready made chip for EDAC (Error Detection And Correction) can be installed in the write and read circuits.

PROBABILITY OF ERRORS IN A MESSAGE

The signal elements in digital recording are one's and zero's, which during the write process are changed into magnetizations of opposite signs. This is then a *polar PCM* signal (PCM = Pulse Code Modulation). It will ideally look like the waveform shown in Fig. 23-9(A).

If noise sneaks into an otherwise ideal system the waveform appears like shown in Fig. 23-9(B). A histogram can be plotted, showing the likelihood of signal levels resulting from this noisy signal.

We will assume that the signal detection is done by distinguishing between signals above or below zero volts, and will further assume a noise voltage with a normal (Gaussian) distribution and a mean noise power σ^2. This distribution or density is shown in Fig. 23-10. The cross hatched areas under the density curve represent the probability of a "0" sent and a "1" received, or a "1" received when a "0" was sent.

An error may occur whenever the noise voltage exceeds a level equal to the peak signal voltage V. We wish to determine the probability thereof, based upon the signal-to-noise ratio (SNR) of the system. This SNR is specifically the peak-signal (V) to rms-noise (σ) ratio, which we will call K:

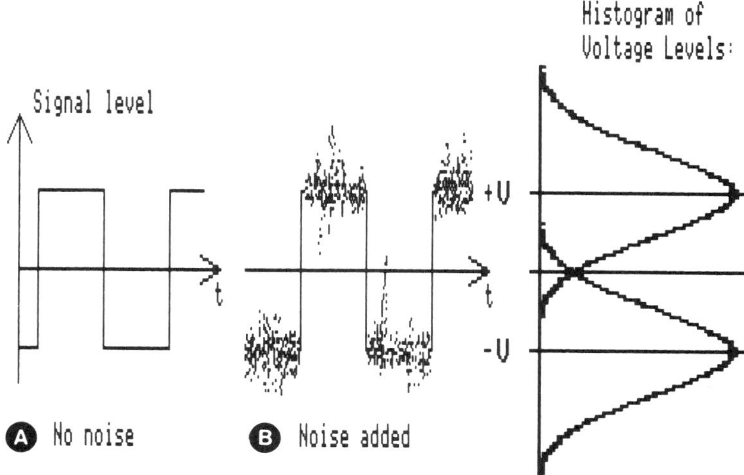

Fig. 23-9. Binary signal levels, shown without (A) and with (B) added noise. The variations in signal levels can be displayed in the histogram shown on the right.

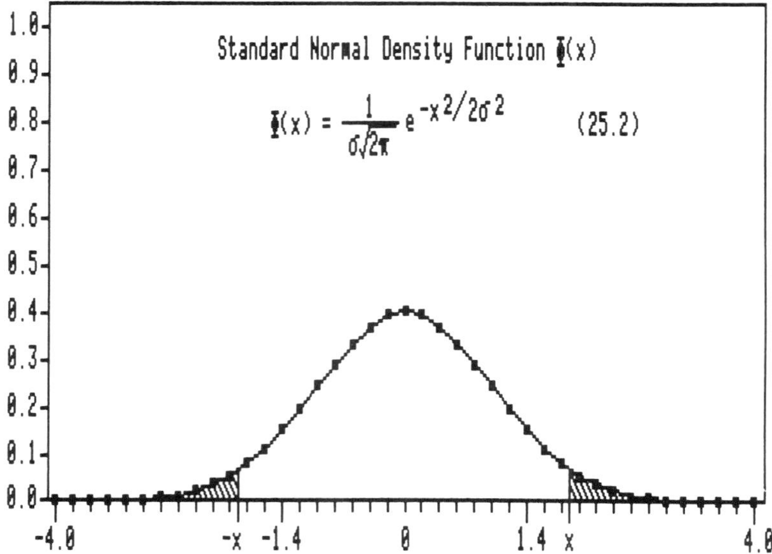

Fig. 23-10. The Standard Normal Density Function.

$$K = V_{peak}/\sigma \tag{23.3}$$

The standard normal density function is shown in Fig. 23-10, and the probability of a noise voltage exceeding the numerical value of V is found as the sum of the two cross-hatched areas outside of $-x$ and $+x$, or twice the integral of $\Phi(x)$, taken from minus infinity to $x = -k\sigma$.

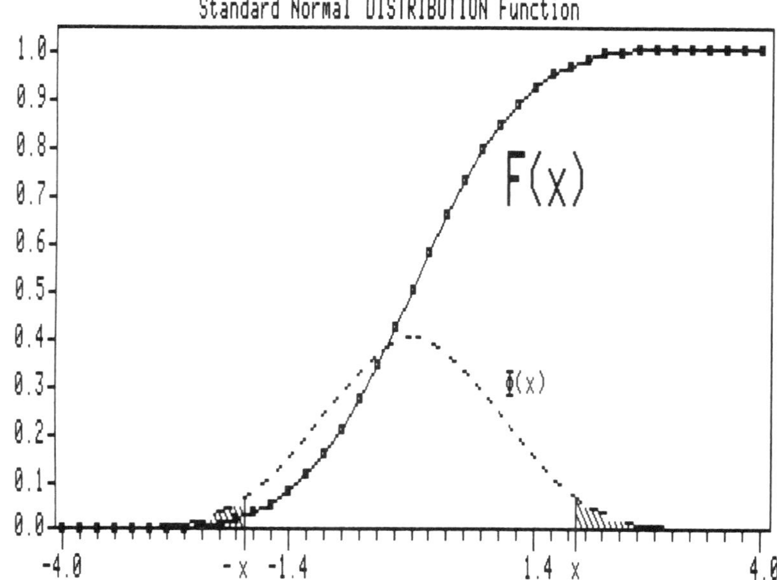

Fig. 23-11. The Standard Normal Distribution Function.

The integral of $\Phi(x)$ is called the normal distribution function F(x), and it changes in value from zero at x = −infinity to one at x = +infinity, as shown in Fig. 23-11. The values of the integral F(x) are listed in several textbooks as the function erf(x), which has integration limits from zero to some positive value of x. This function is called the error function; we have erf(0) = 0.0 and erf(+infinity) = 0.5000.

For values of x beyond $+K\sigma$ and $-K\sigma$ we can therefore write the probability of measuring a noise voltage greater than V:

$$P(x > K\sigma \text{ or } x < -K\sigma) = 1 - \text{erf}(K/\sqrt{2}) = \text{erfc}(K/\sqrt{2}) \quad \textbf{(23.4)}$$

where erfc(x) is named the complementary error function.

The end result for the error probability for $x > K\sigma$ or $x < -K\sigma$ is:

$$\boxed{P(x < -K\sigma) = P(x > K\sigma) = 0.5\ \text{erfc}(K/\sqrt{2})} \quad \textbf{(23.5)}$$

Notice that the error probability depends strictly upon the SNR K. The error probability given in formula (23.5) is a reasonable first approximation for the BER. This BER versus SNR is plotted in Fig. 23-12. The BER versus SNR relationship for communication channels is determined by the method used to detect whether a "1" or a "0" was sent. A three dB improvement is possible for a well designed data channel with matched filters (shown by the broken line).

The BER analysis presented is ideal, and never found in a magnetic recording channel: the noise is not truly random, nor is the signal voltage constant. The amplitude can be expected to vary ± 1 dB, with occasional large drops in level due to drop-outs. These amplitude variations are named multiplicative noise, and must be included in the SNR (see Chapter 27, under Amplitude Fluctuations). The BER evaluation must also include the signal deterioration that occurs due to pulse interference and timing errors.

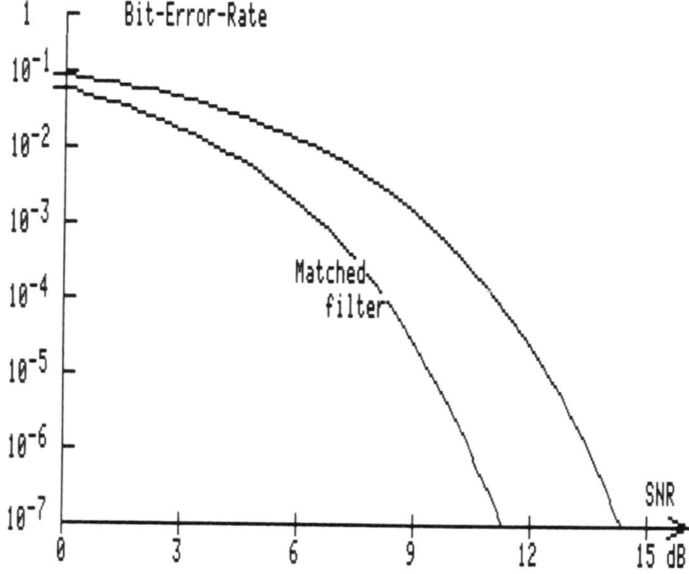

Fig. 23-12. Bit-Error-Rate as a function of the peak-to-rms signal-to-noise ratio for a bipolar signal (magnetic recording). A 3 dB improvement is possible by using matched filters prior to detection.

Multi-Level signalling for recorders has been evaluated as a means of increasing the information stored. It can only be implemented successfully if automatic gain circuits are employed to stabilize the signal amplitude variations in the read signal, in combination with sophisticated error detecting and correcting codes.

SOURCES OF NOISE

We will now discuss the various sources of noise as it relates to magnetic recording.

Media Noise

The number of magnetic particles per unit volume of the coating and their dispersion will determine the noise voltage that is added to the recorded signal. Metallic films have noise spectra of their own.

The reader is referred to Chapter 12 for details (see index under noise, media).

Crosstalk

Crosstalk is the appearance of unwanted signals from adjacent tracks or from incompletely overwritten data. Crosstalk may also occur through strayfields from the recording head in the write mode in a digital tape drive.

The reader is referred to Chapters 8 and 10 (see index under crosstalk).

Head Core Noise

The high frequency losses in a magnetic core can be represented as a resistor that generates a noise voltage. The reader is referred to Chapter 7 (see index under noise, resistance).

Electronics Noise

Solid state devices generate noise that may impair the overall signal-to-noise ratio. All preamplifiers should be constructed from low noise devices and the circuit's noise performance analyzed and measured.

Radio Frequency Interference (RFI)

Man-made noise often limits the performance of a system. The sources of this noise range from noise spikes from DC-motor commutators, car engine's switching circuits and so on, to the pick-up of broadcast stations by the read head.

The rules are that such noises shall be contained at their source. It is still necessary for the manufacturer of recording equipment to guard against RFI by using shielded enclosures and proper grounding techniques. Filtering of the AC-power line is often necessary as are low-pass filters on all ingoing and outgoing data lines.

System Noise

We will name the sum total of media noise, crosstalk, head noise, electronics noise and RFI as *system noise*. Some of the noise components are *randomly distributed* (coating noise, head noise, electronics noise), others are occasionally occurring noise spikes (drop-outs, some RFI) and have a *Poisson distribution*. Finally there are *cyclic components* (RFI).

Each system has its individual system noise and it is best represented by a spectrum showing band pass filtered noise versus frequency, *plus* statistical data for levels. The statistical information is needed to determine the number of times the noise level will be high enough to offset the detection of a bit, or actually appear itself as a bit. These occurrences will determine the number for the system's Bit-Error-Rate (BER).

SYSTEM MEASUREMENTS

Frequency Response

The measurement of a tape recorder's frequency response can be accomplished by the use of a standard alignment tape, which checks the frequency response of the playback amplifier at the same time. To check the overall frequency response, a signal generator is connected to the record amplifier input terminal and the output is recorded at a number of frequencies and plotted on graph paper. When measuring the response of an analog recorder, it is necessary that the record level be 20 dB lower than the normal record level, since distortion and overload will occur at the higher frequencies, resulting in a frequency response curve showing a roll-off toward high frequencies. This roll-off is unrealistic and is caused by tape saturation due to pre-equalization.

A sweep generator is a very useful tool in setting up and aligning a tape recorder. The sweep generator output frequency is swept from low frequencies up to the highest frequency interest, many times a second. This may be 50 or 60 times per second, providing an excellent display of the amplitude versus frequency response on an oscilloscope. This greatly facilitates head alignment and bias and equalizer adjustment. The single-frequency generator is more useful in the final plot of a recorder's frequency response.

Signal-to-Noise Ratio, SNR

A tape recorder's noise is always expressed in terms of a number that gives the ratio of the normal signal voltage to the noise voltage. This is referred to as the *Signal-to-Noise Ratio (SNR)*.

The SNR will depend strongly upon the method used for its measurement. The *normal level* will in general correspond to a level associated with a one percent total harmonic distortion (although the 3rd harmonic is predominant). It may equally well correspond to three percent harmonic distortion, which gives a 4-5 dB higher SNR. The three percent distortion level is more reliable as a reference.

The *noise level* is more ambiguous: It can be measured with the tape transport stopped, or while moving a bulk erased tape over the playback head.

It is important that the tape is in motion so any electric noise from the tape transport is included. A realistic measurement will further measure the noise from a tape recorded on the unit with input terminal shorted. This will include bias noise and thus be a true SNR value for the recorder.

It may not be a realistic number for an audio recorder, where the ultimate judgment of an acceptable SNR is the human ear. It is fortunate that the ear's sensitivity versus frequency varies with the sound level. The response is fairly flat at normal listening levels, but less sensitive at low and high frequencies at low sound levels. The latter means that the ear is less sensitive to the amplifier hum and the tape bias, which were both boosted in amplification along with the playback equalization. This results in a subjectively better judgment of the SNR than the bare number gives.

It is therefore practice to measure an audio recorder's SNR as the ratio (in decibels - dB) between the normal level and the *weighted noise level*, resulting in for example 70 dB instead of 50 dB.

This method is not applicable to instrumentation recorders, where, at most, a band-pass filter with cut-off frequencies set at lower and upper frequency limits are used.

Instrumentation recorders are often used to record several frequency multiplexed signals onto one channel. It is important that none of the individual signals cause harmonic distortion and the harmonics then interfere with the reception and detection of the other channels.

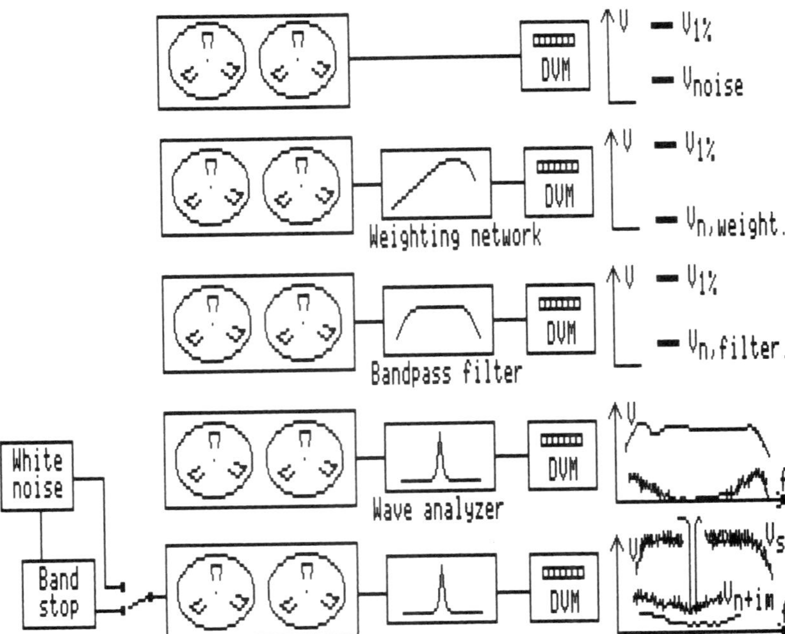

Fig. 23-13. Test methods for measurements of noise and SNR.

It is even more important that the several signals do not interfere with each other via the tape's overload curve and thus cause *intermodulation*. A good test can be performed to assess a recorder's suitability for multiplex recording of several signals simultaneously on one channel. The record signal is simply *white noise*, which contains signals of uniform amplitude at all frequencies.

The components in the white noise will all cause intermodulation among each other, and the amount thereof is measured in the following way (Fig. 23-13, bottom). The white noise is recorded in full bandwidth or with part of its spectrum stopped by a band-stop filter, set at a frequency f_o. The filter bandwidth may be 10 kHz. The playback signal is measured through a narrower band-pass filter, also set at f_o.

When the band-stop filter on the record side is bypassed, a signal voltage V_s is measured. Next the filter is inserted and a voltage V_n measured. The latter voltage is the sum of the reproducer noise plus any intermodulation products from the recording of the white noise outside the band-stop filter.

At low record levels the intermodulation is insignificant, only reproducer plus tape noise is measured. The record level is below the normal level and the SNR $= V_s/V_n$ is quite low (its value will differ from the *slot-noise* SNR, which is displayed as a noise-voltage-versus-frequency curve, measured with a wave analyzer. It is expressed in dB relative to a normal record level of a sine wave.)

The signal-to-noise ratio in the white noise loading test is called *Noise-Power-Ratio* (NPR) since two noise voltages are compared rather than a sine voltage and a noise voltage. The NPR range is therefore considerably smaller in dB than the normal SNR or slot-noise SNR. The first measured NPR can be improved by raising the record level. At some level the intermodulation will increase drastically, and the intermodulation products will spill into the slot around f_o and add to V_n. One may optimize the NPR by trial-and-error setting of the record level, or settle for a NPR number at the level where the noise voltage has increased 3 dB over the pure noise voltage (where the record input signal is equal to zero).

Phase Distortion

Measurements of Phase Distortion or its companion, *envelope* or *group delay* requires a specialized and difficult technique and is beyond the scope of this book.

Envelope delay is related to the delay that individual signals at different frequencies incur when passing through a system such as an amplifier, a cable, or a tape recorder. The definition of envelope delay is:

$$\text{Delay } \tau = d\psi/d\omega \tag{23.6}$$

which is the slope of the phase versus frequency response curve. When the relationship between phase and frequency is linear, then τ is constant. All signals of various frequencies arrive at the same time and they will add up to a faithful reproduction of the original signal.

IMPROVEMENTS TO THE SNR

There are several ways in which the SNR of a recording system can be improved. They all involve some form of pre- and post-treatment of the signal and they are not part of the magnetic record and/or playback processes.

Level Stretching

The transfer curve for tape magnetization versus record current (field) departs from linearity at a level corresponding to 30 percent below saturation. Recordings at higher levels will produce harmonic and intermodulation signals that upon playback show up as distortion.

It is possible to stretch the peak level of the record current to counterbalance the curvature of the overload curve and thus record a higher level on the tap at a lower distortion.

A simple form of a *record current stretcher* is shown in Fig. 23-14. It is simply a resistive voltage divider with one resistor shunted by two identical diodes, back-to-back. Its implemen-

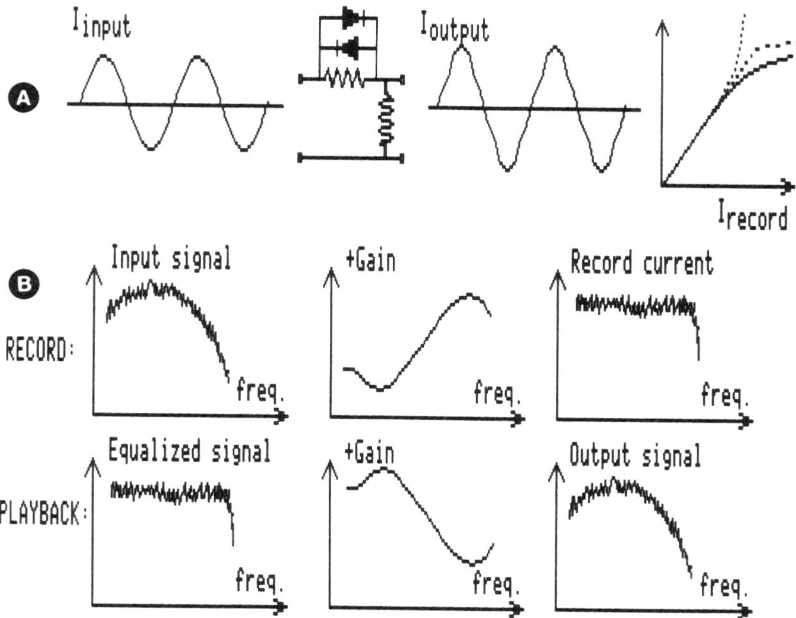

Fig. 23-14. Techniques for improvement of the SNR: (A) Stretching of the record current level; (B) Additional pre- and post-equalization.

tation is difficult, since the network must produce the exact inverse of the overload curve of the tape and that overload changes with the recorded wavelength and it differs among the various tape types.

Add to these difficulties the fact that accidental overloads cause hard clipping and in audio therefore a very harsh sound. Only a carefully designed circuit of fair complexity will provide a significant improvement in the apparent SNR (10 dB).

The stretcher circuit can be equally well employed on the playback side. It now restores occasional peaks that were "squashed" during recordings. Needless to say, these circuits must also be matched to the tape's overload curve.

Additional Equalization

It follows from the discussions in Chapter 22 of equalization that it is desirable to record all signals in the record signal at the highest level possible. This reduces the requirements for post-equalization, with a resulting reduction in the equalized noise level.

Equalization standards need not be a common law in a closed organization such as a research laboratory or telemetry group.

It may be very advantageous to examine the level versus frequency spectrum of the signal to be recorded then design a simple pre-equalizer to provide a flat spectrum.

A matching post-equalizer will restore the data to its original spectrum. Neither amplitude nor phase distortion will be introduced if the pre- and post-equalizers are complementary.

Bandpass Filtering

SNR improvements are readily obtained by band-pass filtering of the reproduce signal. The desired fidelity determines the extent of filtering required (amplifier's treble control is an example of a band-pass filter). Filtering of extraneous noise may also be advantageous during recording.

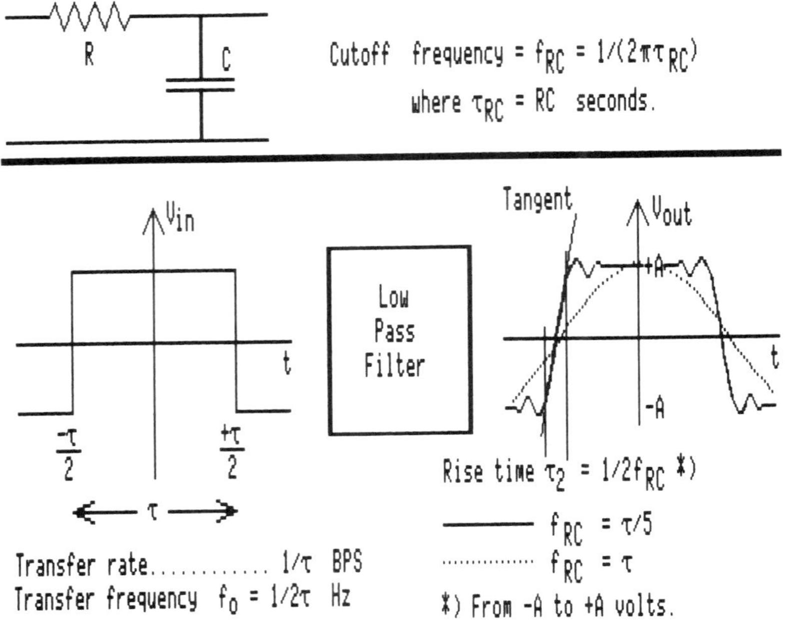

Fig. 23-15. Rounding of pulse waveform by low-pass filtering.

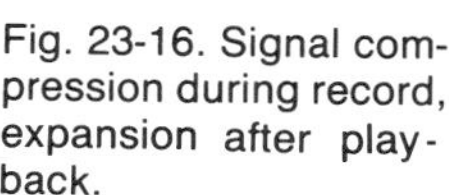

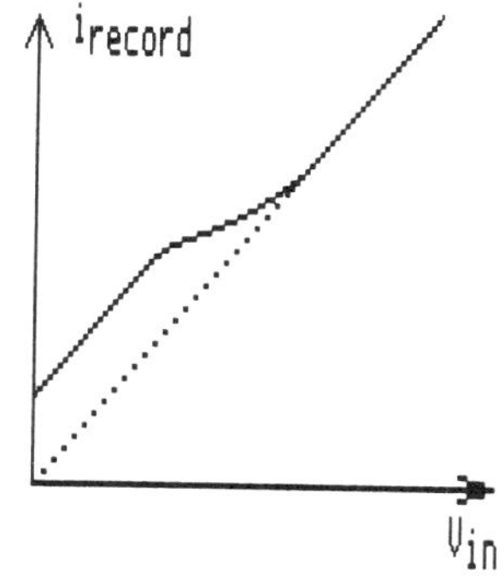

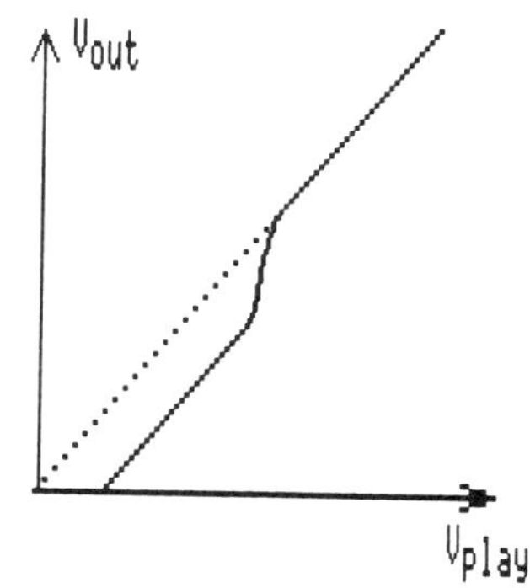

Fig. 23-16. Signal compression during record, expansion after playback.

Simple RC-filtering will work, but is inferior to the use of low-Q filters. The rounding of pulses is shown in Fig. 23-15.

Compression/Expansion

Finally, there exists the methods of compression of the signal prior to recording and subsequent expansion after playback. This technique is useful when a signal with a large dynamic range is recorded and it offers a subjective increase in the SNR.

The transfer curves for the Expander and the Compressor are shown in Fig. 23-16. The scheme is simple in theory, but complex in reality.

It is not suitable for any instrumentation applications and there are several requirements of the activation of the compression and expansion circuits with regard to response time, hold time, and decay time.

BIBLIOGRAPHY TO CHAPTER 23

Signal and Communication Theory

Frans, S., *SPIA, an Interactive Guide to Signal Processing,* Moonshadow Software, 5016 Castlewood Drive, San Jose, CA 95129. FFT, Convolution etc. program for the IBM PC.

Shannon, C.E., and Weaver, W., *The Mathematical Theory of Communication,* 1948, reprinted by Dover, New York, Jan. 1975, 125 pages.

Cheng, D.K., *Analysis of Linear Systems,* Addison-Wesley, 1961, 431 pages.

Wozencraft, J.M., and Jacobs, I.M., *Principles of Communication Engineering,* John Wiley and Sons, 1965, 720 pages.

Singh, J., *Great Ideas in Information Theory, Language and Cybernetics,* Dover, Jan. 1966, 338 pages.

Cooper, G.R., and McGillem, C.D., *Probabilistic Methods of Signal and System Analysis,* Holt, Rinehart and Winston, Jan. 1971, 258 pages.

Brigham, E.O., *The Fast Fourier Transform,* Prentice-Hall, New Jersey, Jan. 1974, 252 pages.

Liu Liu, *Linear Systems Analysis,* McGraw-Hill, Jan. 1975.

Bracewell, R.N., *The Fourier Transform and Its Application,* McGraw-Hill, Jan. 1978, 444 pages.

Schwartz, M., *Information Transmission, Modulation and Noise,* McGraw-Hill, January 1980, 646 pages.

Lathi, B.P., *Modern Digital and Analog Communication Systems,* Holt, Rinehart and Winston, 1983, 708 pages.

Thrane, N., "The Hilbert Transform," *Bruel and Kjaer Technical Review*, July (No. 3) 1984, 13 pages.

Chambers, W.G., *Basics of communications and coding*, Clarendon Press, Oxford, 1985, 240 pages.

Bendat, J.S., "The Hilbert Transform", *Bruel and Kjaer*, 42 pages, 1986.

Senior, Th. B.A., *Mathematical Methods in Electrical Engineering*, Cambridge Univ. Press, 1986, 272 pages.

Bylanski, P., and Ingram, D.G.W., *Digital Transmission Systems*, Peter Peregrinus Ltd. on behalf of the Inst. of El. Engrs., London, 1980, 431 pages.

Biering, H., and Pedersen, O.Z., "System Analysis and Time Delay Spectrometry," *Bruel and Kjaer Technical Review*, Jan. (No. 1) and Apr. (No. 2) 1983, 52 and 50 pages.

Quarmby, D. (Editor), *Signal Processor Chips*, Prentice-Hall, 1985, 179 pages.

Modelling of the Signal Channel

Voelz, H., "Spurhoehe, Spurzahl und Kanalkapazitaet bei der magnetischen Speicherung," *Hochfrequenztechnik und Elektroakustik*, Jan. 1967, pp. 172-175.

Hodder, W.K., and Monson, J.E., "FM Distortion Caused by Head-to-Tape Spacing," *ITC Conf. Proc.*, 1970, Vol. 6, pp. 178-187.

Lindholm, D.A., "Fourier Synthesis of Digital Recording Waveforms," *IEEE Trans. Magn.*, Dec. 1973, Vol. MAG-9, No. 4, pp. 689-698.

Mallinson, J.C., "Applications of Fourier Transforms in Digital Magnetic Recording Theory," *IEEE Trans. Magn.*, Mar. 1974, Vol. MAG-10, No. 1, pp. 69-78.

Mallinson, J.C., "Correction of 'Applications of Fourier Transforms in Digital Magnetic Recording Theory'," *IEEE Trans. Magn.*, Dec. 1974, Vol. MAG-10, No. 4, p. 1137.

Price, R., "An Experimental, Multilevel, High Density Disk Recording System," *IEEE Trans. Magn.*, Sept. 1978, Vol. MAG-14, No. 5, pp. 315-318.

Melbye, H.E., and Chi, C.S., "Nonlinearities in High Density Digital Recording," *IEEE Trans. Magn.*, Sept. 1978, Vol. MAG-14, No. 5, pp. 746-748.

Wood, R., and Donaldson, R., "The Helical-Scan Magnetic Tape Recorder as a Digital Communication Channel," *IEEE Trans. Magn.*, Mar. 1979, Vol. MAG-15, No. 2, pp. 935-943.

Middleton, B.K., "Performance of a Recording Channel," *Intl. Conf. Video and Data 82*, Apr. 1982, IERE Publ. No. 54., pp. 137-146.

Wood, R.W., and Donaldson, R.W., "Signal Processing and Error Control on an Experimental High-Density Digital Magnetic Tape Recording System," *IEEE Trans. Magn.*, Sept. 1980, Vol. MAG-16, No. 5, pp. 1255-1265.

Loze, M.K., Middleton, B.K., and Riley, A., "Simulation of Digital Magnetic Recording Systems," *Intl. Conf. Video and Data Recording*, March 1986, IERE Publ. No. 67, pp. 185-193.

Noise in a Magnetic Recording Channel

Ott, H.W., *Noise Reduction Techniques in Electronic Systems,* John Wiley and Sons, 1976, 294 pages.

Wax, Nelson, Selected Papers on Noise and Stochastic Processes, *Dover Publications*, Jan. 1954, 337 pages.

Bennett, W.R., "Methods of Solving Noise Problems," *Proc. IRE*, May 1956, Vol. 24, pp. 609-638.

Weinberg, L., "Exact Ladder Network Design Using Low-Q Coils," *Proc. IRE*, April 1958, Vol. 46, No. 4, pp. 739-750.

Skov, E.P., "Noise Limitations in Tape Recorders," *Jour. Audio Engr. Soc.*, Jan. 1964, Vol. 12, No. 4, pp. 147-160.

Tavares, S.E., "A Comparison of Integration and Low-Pass Filtering," *IEEE Trans. Instr. and Meas.*, June 1966, Vol. IM-15, No. 1 and 2.

Mallinson, J.C., "Maximum Signal-to-Noise Ratio of a Tape Recorder," *IEEE Trans. Magn.*, Sept. 1969, IEEE Trans. Magn., Vol. MAG-5, No. 3, pp. 182-186.

Sadsshige, K., "Study of Noise in Television Broadcast Equipment," *Jour. SMPTE*, Dec. 1969, Vol. 78, pp. 1069-1076.

Tahara, Y., Takagi, H., and Ikeda, Y., "Optimum Design of Channel Filters for Digital Recording," *IEEE Trans. Magn.*, Nov. 1976, Vol. MAG-12, No. 6, pp. 749-750.

Hughes, B.F., and Schmidt, R.K., "On Noise in Digital Recording," *IEEE Trans. Magn.*, Nov. 1976, Vol. MAG-12, No. 6, pp. 752-754.

Thurlings, L., "Statistical Analysis of Signal and Noise in Magnetic Recording," *IEEE Trans. Magn.*, May 1980, Vol. MAG-16, No. 3, pp. 507-514.

Netzer, Y., "The Design of Low-Noise Amplifiers," *Proc. IEEE*, June 1981, Vol. 69, No. 6, pp. 728-741.

Okuwaki, T., Koizumi, M., Uesaka, Y., and Fujiwara, H., "Noise and Overwrite Characteristics for Perpendicular Magnetic Recording on a Sputtered Co-Cr Medium," *J. Appl. Phys.*, March 1982, Vol. 53, No. 3, pp. 2588-2592.

Keshner, M.S., "1/f Noise," *Proc. IEEE*, March 1982, Vol. 70, No. 3, pp. 212-218.

Finkelstein, B.I., and Christensen, E.R., "Signal-to-Noise Ratio Models for High-Density Recording," *IEEE Trans. Magn.*, Sept. 1986, Vol. MAG-22, No. 5, pp. 898-900.

Madrid, M., and Wood, R., "Transition Noise in Thin-Film Media," *IEEE Trans. Magn.*, Sept. 1986, Vol. MAG-22, No. 5, pp. 892-894.

Arnoldussen, T.C., and Tong, H.C., "Zigzag Transition Profiles, Noise, and Correlation Statistics," *IEEE Trans. Magn.*, Sept. 1986, Vol. MAG-22, No. 5, pp. 889-891.

Roe, K., and Soohoo, R.F., "Longitudinal and Perpendicular Magnetic Thin Film Media Noise," *IEEE Trans. Magn.*, Sept. 1986, Vol. MAG-22, No. 5, pp. 886-888.

Howell, T.D., Kasiraj, P., Best, J.S., Chu, F., and Yerry, M. M., "A Study of Disk Noise Statistics," *IEEE Trans. Magn.*, Sept. 1986, Vol. MAG-22, No. 5, pp. 901-903.

Darling, T.F., "Mathematical Noise Modelling and Analysis of Some Popular Preamplifier Circuit Topologies," *Jour. AES*, Jan-Feb. 1987, Vol. 35, No. 1/2, pp. 15-23.

Bit-Error Rates

Schwarz, T.A., "A Statistical Model for Determining the Error Rate of the Recording Channel," *IEEE Trans. Magn.*, Sept. 1980, Vol. MAG-16, No. 5, pp. 634-637.

Howe, D.G., "The Nature of Intrinsic Error Rates in High-Density Digital Optical Recording," *SPIE* Vol. 421-Optical Disks Systems and Application, 1983, pp. 31-42.

Schluessler, H.W., "On Influence of Noiselike Errors in Digital Systems," *Signal Processing*, July 1983, Vol. 5, No. 4, pp. 319- .

Skritek, P., "Dynamic Distortion Measurements of Tape Recorders and Electroacoustic Transducers," *Jour. AES*, Jul-Aug. 1983, Vol. 31, No. 7, pp. 512-516.

Middleton, B.K., and Jack-Kee, T., "Performance of Digital Magnetic Recording Channel Subject to Noise and Drop-Outs," *The Radio and Electr. Engr.*, Nov-Dec. 1983, Vol. 53, No. 11/12, pp. 393-402.

Pasian, F., and Crise, A., "Restoration of Signals Degraded by Impulse Noise by means of Low Distortion Non-Linear Filter," *Signal Processing*, Jan. 1984, Vol. 6, No. 1, pp. 67-76.

Howe, D.G., "Signal-to-Noise Ratio (SNR) for Reliable Data Recording," *SPIE* Vol. 695 Optical Mass Data Storage II (1986), pp. 255-261.

Detection

Chick, D.M., and Walker, K.F., "A Survey of Retiming and Decoding Methods in Digital Magnetic Recording," *IERE Proc. on Video and Data Recording Conf.*, Jan. 1973, pp. 105-112.

Graham, I.H., "Data Detection Methods vs. Head Resolution In Digital Magnetic Recording," *IEEE Trans. Magn.*, July 1978, Vol. MAG-14, No. 4, pp. 191-193.

Price, R., Craig, J.W., Melbye, H.E., and Perahia, A., "An Experimental, Multilevel, High Density Disk Recording System," *IEEE Trans. Magn.*, Sept. 1978, Vol. MAG-14, No. 5, pp. 315-318.

Mackintosh, N.D., "A Margin Analyser for Disk and Tape Drives," *IEEE Trans. Magn.*, Nov. 1981, Vol. MAG-17, No. 6, pp. 3349-3352.

Huber, W.D., Newman, J., and Fisher, R.D., "Detection Window and Lineal Density in Digital Recording Systems," *IEEE Trans. Magn.*, Nov. 1981, Vol. MAG-17, No. 6, pp. 3355-3357.

Burkhardt, H., "An Event-Driven Maximum-Likelihood Peak Position Detector for Run-Length-Limited Codes in Magnetic Recording," *IEEE Trans. Magn.*, Nov. 1981, Vol. MAG-17, No. 6, pp. 3337-3340.

Montgomery, D., "Borrowing RF Techniques for Digital Design," *Computer Design*, May 1982, pp. 207-217.

Siegel, P.H., "Applications of a Peak Detection Channel Model," *IEEE Trans. Magn.*, Nov. 1982, Vol. MAG-18, No. 6, pp. 1253-1255.

Huber, D.W., "Simultaneously and Orthogonally Interactive Clock Recovery and D.C. Null Equalization in High Density Digital Magnetic Recording," *IEEE Trans. Magn.*, Sept. 1983, Vol. MAG-19, No. 5, pp. 1716-1718.

Batey, R.M., and Becker, J.D., "Second-Generation Disc Read/Write Electronics," *Hewlett Packard Journal*, Jan. 1984, Vol. 35, No. 1, pp. 7-12.

Deeley, E.M., and Mitchell, A., "Optimization of Nonlinear Clock Recovery Circuits for Digital Recording," *Intl. Conf. Video and Data 84*, Apr. 1984, No. 59, pp. 137-140.

Vinding, J.P., "Detector Circuits for Ternary Coded Magnetic Recording," *IEEE Trans. Magn.*, Sept. 1984, Vol. MAG-20, No. 5, pp. 894-896.

French, C.A., and Wolf, J.K., "Application of the Viterbi Algorithm to the Detection of Compact Spectrum," *IEEE Trans. Magn.*, Sept. 1986, Vol. MAG-22, No. 5, pp. 1200-1202.

System Measurements

Feher, K., and Engineers at HP, *Telecommunications Measurements, Analysis, and Instrumentation,* Prentice-Hall, 1987, 412 pages.

Randall, R.B., *Frequency Analysis Bruel and Kjaer,* Sept. 1977, 239 pages.

Schluter, K., "Audio Envelope-Delay Sweep-Frequency Measuring Setup for Data Transmission," *NTZ-CJ,* Jan. 1965, pp. 241-245.

Melbye, H.E., "Determination of Frequency-Dependent Magnetic Recording Losses," *IEEE Trans. Magn.*, Nov. 1976, Vol. 12, No. 6, p. 737.

Bates, C.M., "Total Harmonic Distortion Measurements in Magnetic Tape Equipment," Preprint No. 1847 (J-3), *AES Conv.* 1981, New York, 4 pages.

Bertram, H.N., and Fielder, L.D., "Amplitude and Bit Shift Spectra Comparisons in Thin Metallic Media," *IEEE Trans. Magn.*, Nov. 1983, Vol. MAG-19, No. 6, pp. 1605-1607.

Perman, A., "Dual Channel FFT Analysis for the Development and Evaluation of Tape Recorders," *Bruel and Kjaer App. Notes*, Jan. 1984, 16 pages.

Bruel and Kjaer, "Dual Channel FFT Analysis, I-II," *Bruel and Kjaer Technical Review,* Jan. 1984.

Measurements of Noise; NPR

Tant, M.J., *Multichannel Communication Systems and White Noise Testing*, Marconi Instruments, 1974, 104 pages.

Blackman, R.B., and Tukey, J.W., *The Measurement of Power Spectra*, Dover Publ., Jan. 1958, 190 pages.

Heidenreich, K. Heinz, ''DerRauschklirr-Belastungsversuch; Darstellung und Auswertung der Ergebnisse,'' *NTZ*, Jan. 1974, No. 12, pp. 457-463.

Newcombe, E.A., and Pasupathy, S., ''Error Rate Monitoring for Digital Communications,'' *Proc. IEEE,* Aug. 1982, Vol. 70, No. 8, pp. 805-828.

Pasian, F., and Crise, A., ''Restoration of Signals Degraded by Impulse Noise by means of Low-distortion Non-linear Filter,'' *Signal Processing*, Jan. 1984, Vol. 6, No. 1, pp. 67-76.

Lowe, K.A., ''Noise-Margin Analysis Automatically Lays Bare Hidden Logic Problems,'' *Electr. Des.*, Oct. 1984, Vol. 32, No. 21, pp. 229- .

Haynes, M.K., ''Density-Response and Modulation-Noise Testing of Digital Magnetic Recording Tapes,'' *IEEE Trans. Magn.*, Sept. 1984, Vol. MAG-20, No. 5, pp. 897-899.

Tang, Y.S., ''Noise Autocorrelation in Magnetic Recording Systems,'' *IEEE Trans. Magn.*, Sept. 1985, Vol. MAG-21, No. 5, pp. 1389-1391.

Chapter 24

Modulation Codes

The Institute of Electric and Electronic Engineers defines a code as "a plan for representing each of a finite number of values or symbols as a particular arrangement or sequence of discrete conditions or events." Thus, coding is the process of transforming messages or signals in accordance with a definite set of rules.

The stream of "0"s and "1"s from a digital signal source represents data that has been coded into a binary format. The original data can represent digitized measurements, sound, video, data files, etc. A bit is a binary digit.

The "raw" signal is not well suited for recording and subsequent playback from a magnetic tape or disk; there is no bound on the DC-content in a binary data stream and the recording channel will not transmit (store and playback) DC levels.

The solution is modulation of the data into a more suitable code format that is DC-free. At the same time, modulation offers a higher packing density so more data can be recorded per cm of track length. This form of coding is a modulation process where one pattern of transitions is changed into another pattern more suitable for magnetic recording. It is assumed that any error correction coding is done prior to modulation coding (see next chapter for error correction and detection coding.)

Finding a digital code that maximizes the net linear density while maintaining satisfactory data reliability is a challenging objective. There exists a host of codes, and each is touted as being superior to the others. These claims may be true for certain applications, and for a given recorder and its set-up.

Codes commonly used in data communication are RZ (return to zero), NRZ (no return to zero), and Bi-Phase. Further definitions are made by suffixes:

- L (level) means that logical values are represented by different levels or transitions.
- M (mark) means that logic "1" is defined as "mark" and is represented by a level transition, with logic "0" defined by no transition.

• S (space) means that logic "0" is defined as "space," and is represented by a transition, while logic "1" is represented by no transition.

We can appreciate the various codes if we follow the rationale for their developments.

DEVELOPMENT OF DC-FREE CODES

We have, in a couple of decades, come from a very simple RZ (Return-to-Zero) code to a sophisticated family of GCR (Group-Code Recording) and RLL (Run-Length-Limited) codes.

Foremost in our choice of a code lies the demand that it be essentially DC free, since the magnetic recording channel does not reproduce DC. Next is the desire to stuff as much data as possible into a code for maximum utilization of the recorder's signal bandwidth. And in our attempts to satisfy these two demands we must still have a code that is easy to manage in regard to clock extraction, synchronization, and signal detection.

RZ Code

The earliest code is the RZ code (Return-to-Zero), which uses a positive pulse for a "1" and a negative pulse for a "0". Each pulse produces two output voltages upon playback, where the first can be used to derive the clock for strobing to see if the pulse is positive (a "1") or negative (a "0"). The code is, in other words, self-clocking. The RZ code is shown in the top line of Fig. 24-1.

The flux changes are from the zero level to saturation and back, and a low data density results since each bit requires two transitions. Another drawback is that you cannot record over old RZ data with new data; the tape must be erased.

Digital logic circuits, including A/D converters, do not produce separate pulse for "1"s and "0"s. The signals are continuous trains of positive voltages for "1"s and zero voltage for "0"s (or vice versa in some circuits), as suggested by the NRZ-L code.

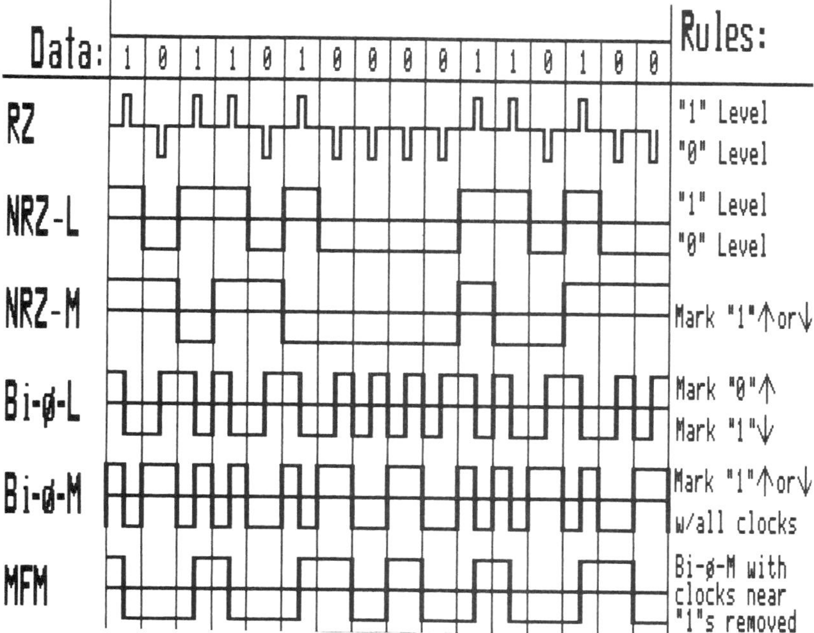

Fig. 24-1. Initial development of codes.

NRZ-L Code

We can record the digital signal as positive magnetization for all "1"s, and negative magnetization for all "0"s. This method is known as No-Return-to-Zero, Low, or NRZ-L, and is shown in the second line in Fig. 24-1.

The transition density of NRZ-L is less than half the density of RZ, but it is not self-clocking. The window for signal detection must be operated by an external oscillator that essentially is free running and then re-phased (or synchronized) as often as possible; the NRZ-L code has no built-in assurance that synchronization takes place often enough. The maximum time between transitions could be infinity.

The NRZ-L code is also prone to errors. If no flux change is detected at a clock timing then the read logic assumes that the next bit is the same as the preceding bit. Now, if a string of "1"s (or "0"s) is recorded, then a spurious noise spike can make the logic think that a "0" ("1") was present. The detection logic will continue calling every "1" a "0" and every "0" a "1" until another noise spike bumps it back on track. This is called error propagation and must be prevented so that we can distinguish between "1"s and "0"s.

NRZ-M Code (Also named NRZ-I)

Error propagation is prevented in the NRZ-M code by generating a transition for every "1," and no transition for any "0" (vice versa for NRZ-S (Space)). NRZ-M was used in many early tape systems. The "1" transition is a flux reversal, without regard for direction, Fig. 24-1, third line. When the read circuit detects a voltage from a flux change, it "knows" that a "1" is present. No flux change means that the bit is a "0."

This method will only be sensitive to single bit errors, since subsequent bit detection is independent of previous polarity. But the NRZ-M code cannot, like NRZ-L, distinguish between a drop-out and a "0," nor is it self-clocking. The latter problem is solved by having a transition at mid-bit time for each bit, a logic "1" starts out at a "1" level, but transitions to the "0" level at mid-bit time (vice versa for a logic "0"). Additional transitions must therefore be inserted between like bits, see line 4 in Fig. 24-1 for this next code, Bi-Phase-L.

Bi-Phase-L Code (Also named Phase-Encoding (PE), Split-Phase, or Manchester II)

The inserted transitions change the NRZ-M code into a baseband form of phase shift keying. It has essentially zero DC content, and is self-clocking. The latter is achieved by having the clock running at twice the bit rate so it can synchronize to all transitions, whether they represent data or are simply transitions inserted between like bits. This code is one of the recommended digital HDDR codes for the telemetry groups working in the aerospace industry(IRIG Doc. 106-86, section 6.11).

Bi-Phase-M Code (Also named binary FM, Manchester, PE-M (Phase-Encoding Mark), or DF (Double Frequency))

A code similar to Bi-Phase-L was developed by merely adding transitions at the end of each bit cell to the NRZ-M code. This is the most commonly used code in magnetic tape recording, and in a large number of communication systems, including fiber optics.

Its popular name FM (Frequency Modulation) came about by observing that two frequencies are present in the data pattern; the code is a form of frequency shift keying. The FM code writes two transitions whenever a "1" is recorded. This is, of course, the double frequency and we can consider each "1" to be twice the clock frequency, while a "0" is equal to this frequency.

Appraisal of an FM-system is made on such items as resolution, which is the ratio between the signals at frequencies 2f and 1f. Also when considering the erasure of old data while new data is written (overwrite), reference is made to the two frequencies 1f and 2f.

Before we go on, let us pause and review the codes introduced so far. We paid a price for including a clock with a code: double bandwidth of the coded signal. It also affected the frequency spectrum of Bi-Phase-L and Bi-Phase-M. We see that Bi-Phase-M has little or no requirement for channel response to DC. The computer industry uses NRZ-M for tapes at low densities. Higher densities use phase encoding (Bi-Phase-L), while FM (Bi-Phase-M) is common for disks.

If we reexamine the Bi-Phase-M pattern in Fig. 24-1 we notice a couple of extra transitions we really do not need: At bit cell beginning and ending when writing a "1"; the mid-bit "1" transition can serve as a clock synchronization signal. We will still want the clock transitions for a string of "0"s.

MFM Code (Also named Double Density, Miller, DM (Delay Modulation)

Let us now delete the excess clock transitions, leaving only those between successive "0"s (again, a "1" is a positive or negative going mid-bit transition). We have now arrived at the modified Frequency Modulation, MFM, code from the FM code, see Table 24-1, bottom.

This particular code was patented by Miller of Ampex in 1963 and is therefore often called the Miller code.

The data information from FM has been preserved and the potential problem of long strings of "0"s has been eliminated, but the self-clocking feature for each bit has been lost. The major gain, though, is a reduction of about 2:1 in the required bandwidth and a large reduction in the requirement for DC. The signal spectrum is shown in Fig. 24-2. The signal's stability, in comparison with NRZ, is evident from the oscilloscope pictures in Fig. 24-3. Note the reduction in zero level drift and baseline galloping when no DC is present.

The upper frequencies required to faithfully record and reproduce the three basic codes (Miller, NRZ-M and Bi-Phase-M) are shown on the graphs in Fig. 24-2 as 1/2T, 1/T and 3/2T.

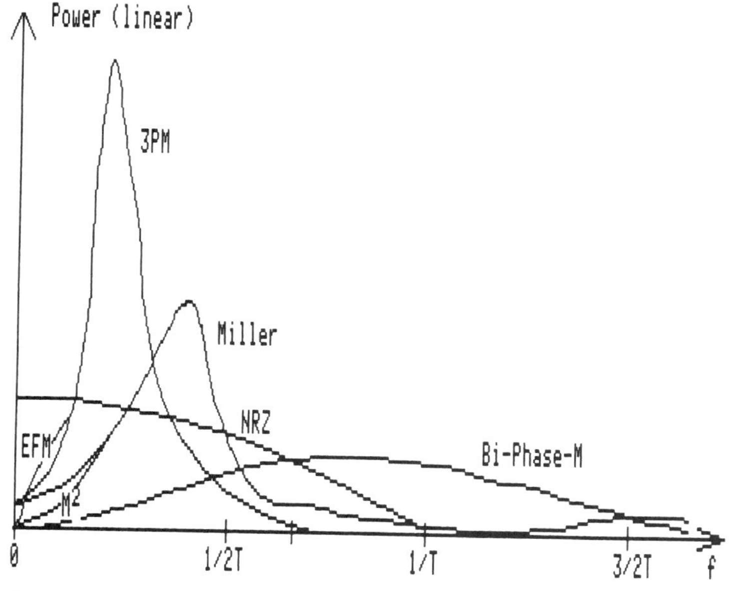

Fig. 24-2. Signal power spectra for various codes.

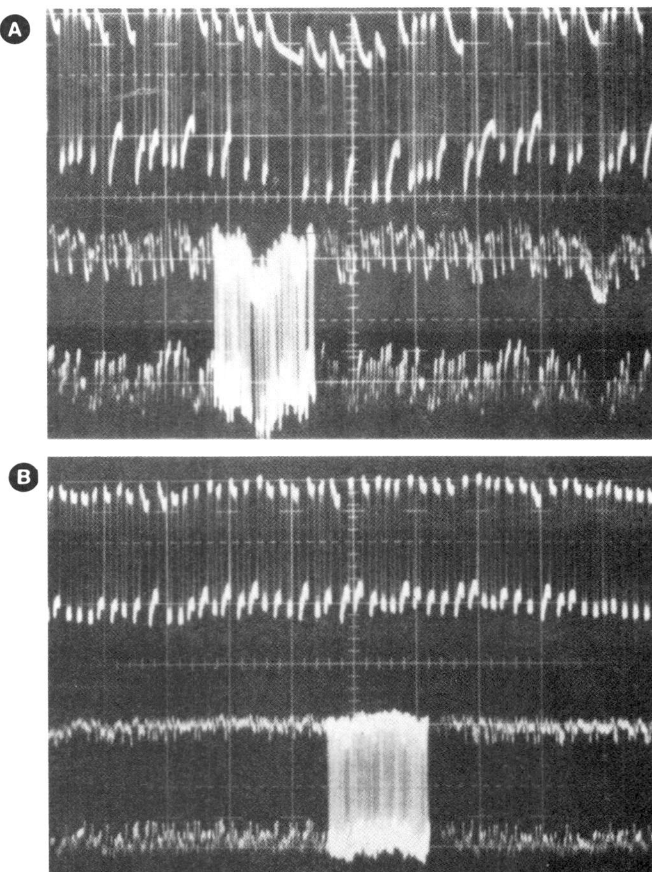

Fig. 24-3. (A). Codes with a large DC-content causes "galloping" baselines at amplifier coupling stages. (B). Codes with small or no DC content (courtesy Ampex Corp.).

We should note that the transitions in MFM signals have become critical; they are no longer spaced equally, but rather with 1, 1.5, or 2 times the bit cell length. We have deleted some clocking information, which previously reduced the transition spacing to ½ times the bit cell length. That is why we can increase the packing density to almost twice that of Bi-Phase-M, for the same amount of intersymbol interference and bit crowding.

In the code developments we have so far achieved a complete removal of the DC-content in the FM-code, with a small residual amount in the MFM-code. The effect of DC in a digital signal is shown in Fig. 24-3. It causes charge built-up on any capacitive element in the read amplifier chain; DC-restorer circuits are required.

Placing DC Constraint on MFM

The Bi-Phase-L and -M codes are DC-free; they effectively remove any DC component that may be in the NRZ-L data stream supplied by digital electronics.

The price is a doubling of the number of transitions in certain portions of the data stream. We can express this as a halving of the minimum time T_{min} between transitions in the code:

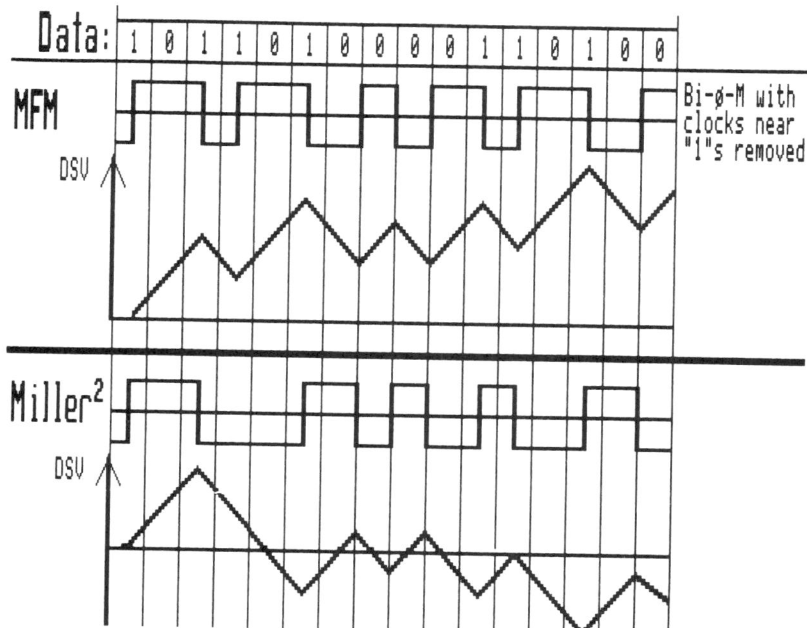

Fig. 24-4. Average DC-level in a code can be controlled, such as in the Miller-squared code. The running integral of the DC voltage is measured as DSV (see text).

$$T_{min} = (1/2)*T$$

where T is the original time between clock signals in the data.

The net effect is a necessary reduction in the length of the detection window T_d to half its original size:

$$T_d = (1/2)*T$$

The MFM code was derived by deleting all the extra transitions around "1"s. This brought T_{min} back to T, but remained the same data. T_d remains at T/2 in order to separate the "1" transitions from clock transitions between "0"s.

The DC content can be evaluated by the *Digital Sum Variation*, abbreviated DSV. This is the cumulative DC-level, measured as the running integral of the area beneath the coded waveform. In computing the DSV the binary levels are assumed to be ± 1. If the DSV of the code is bounded, then the code is DC-free. Figure 24-4 illustrates the DSV for MFM and Miller Squared.

The MFM code can be made DC free by occasionally leaving out a transition: Whenever an isolated "0" is followed by an even number of "1"s, then the final "1" will not be marked (i.e. no transition); this modified code is named **Miller Squared**, and assures freedom from DC (ref. Mallinson and Miller).

GROUP CODES

When coding is done by translation of blocks of m data bits each into another set of blocks with n bits each, where $m < n$, then the coding process is called group coding (recording), GCR.

This technique actually belongs in the next chapter on error detection and correction. However, the magnetic recording industry has historically favored the word "code" when developing techniques which borrowed from the sciences of both modulation and coding.

The rules for the re-coding can be simple table look-up, or may be rather complex. The coding is in essence done by breaking the data code up in groups of m bits and mapping them into code words of n bits. The 2^m patterns of m-bit data words are stored in a library, with corresponding patterns for the n bit code words. Such an arrangement is shown for the popular **4/5 GCR** code in Table 24-1, left. Also, see Fig. 24-5.

GCR basically uses the MFM format for "1"s and "0"s, but a restriction is added: there can be no more than two "0"s in sequence (k=2 - see next section). This guarantees that flux changes occur at least once every three bit cells, and the variable-frequency clock need only be able to lock onto three pulses, corresponding to a succession of "1"s, alternate "1"s and "0"s, or a "1" followed by two "0"s.

The GCR-code is also advantageous when error detection and correction is considered. 16 of the available 32 code words are not used, see Table 24-1. If a detected 5-bit word does not match any of the 16 allocated words then an error has occurred. This is important information although we do not have sufficient information to correct the error (see next chapter).

The reader is referred to papers by Ringkjoeb and by Newton for details about the implementation of the GCR code in a 6,250 BPI 9-track tape system.

A substitution of bits plus the addition of a parity check bit is found in a 7/8 code named **E-NRZ**, Enhanced NRZ. The encoding entails separating the NRZ-L data stream into seven-bit words. Bits 2,3,6 and 7 of each word are then inverted, and a single bit added as a parity (enhancement) bit to each of the words. This parity bit is added at the end of each word to

Table 24-1. Look-Up Libraries for the 4/5 and the 3PM Codes.

4/5 GCR			3 PM	
Number	Data Bits	Code Bits	Data Bits	Code Bits
0	0 0 0 0	1 1 0 0 1	0 0 0	0 0 0 0 1 0
1	0 0 0 1	1 1 0 1 1	0 0 1	0 0 0 1 0 0
2	0 0 1 0	1 0 0 1 0	0 1 0	0 1 0 0 0 0
3	0 0 1 1	1 0 0 1 1	0 1 1	0 1 0 0 1 0
4	0 1 0 0	1 1 1 0 1	1 0 0	0 0 1 0 0 0
5	0 1 0 1	1 0 1 0 1	1 0 1	1 0 0 0 0 0
6	0 1 1 0	1 0 1 1 0	1 1 0	1 0 0 0 1 0
7	0 1 1 1	1 0 1 1 1	1 1 1	1 0 0 1 0 0
8	1 0 0 0	1 1 0 1 0		
9	1 0 0 1	0 1 0 0 1		
10	1 0 1 0	0 1 0 1 0		
11	1 0 1 1	0 1 0 1 1		
12	1 1 0 0	1 1 1 1 0		
13	1 1 0 1	0 1 1 0 1		
14	1 1 1 0	0 1 1 1 0		
15	1 1 1 1	0 1 1 1 1		

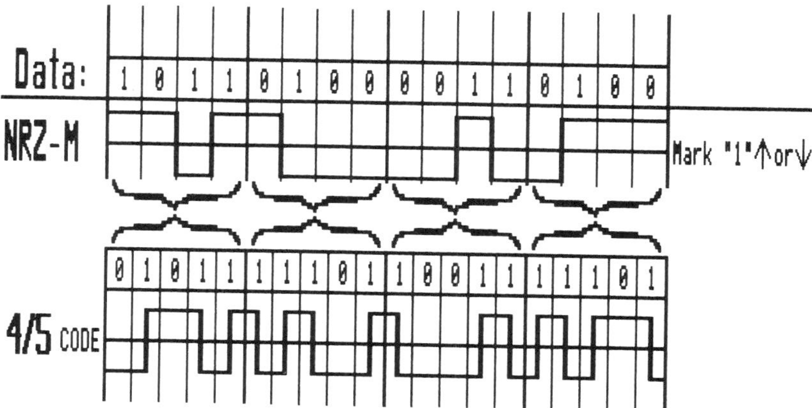

Fig. 24-5. Generation of the ⅘ GCR code (MFM with GCR error detecting code).

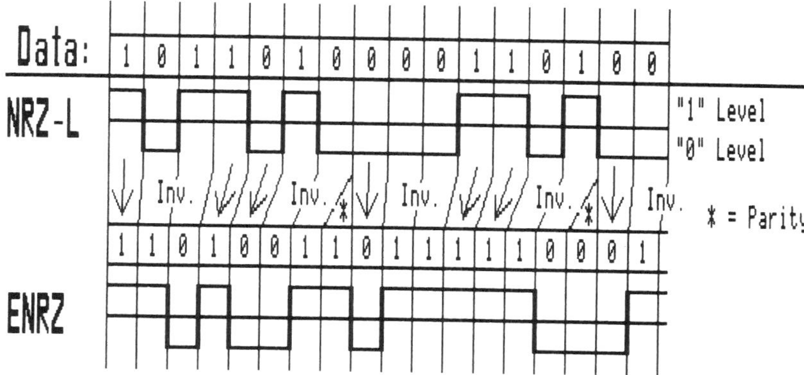

Fig. 24-6. Generation of the E-NRZ code (NRZ-L with error detection) (courtesy Datatape).

make the total number of "1"s in an eight bit word an odd count (ref. Severt). Figure 24-6 shows the implementation.

Zero Modulation (**ZM-code**) is a group code that modifies a Miller-code to provide DC-free sequences. The preceding bits in a coded block are entered into the evaluation of a block's DC content, and a single transition may be added or deleted to make the net DC in the bit stream equal to zero. For details see Patel's paper.

A further advanced group code is the **3PM code**, where three data bits are coded into six code bits. The encoding is made in accordance with the look-up library shown in Table 24-1, right. At the point of joining code word bits a pattern 101 may appear. This is a violation of the code, and it is replaced by 010. In order to make this possible at any time, the last code bit is always a 0.

The coding logic must therefore *look ahead* two data words (2×3 bits) so any merging resulting in 101 can be corrected prior to writing of the final code words. For details see Jacoby's papers.

RUN-LENGTH-LIMITED CODES; FORMAL CODE DEVELOPMENT

It is constructive to examine the MFM code as a string of transitions ("1"s) and spaces ("0"s), at the bit rate of $(\frac{1}{2}) * T$. Note that for the MFM code there is a minimum of one "0" and a maximum of three "0"s between two successive "1"s.

There is no distinction between data and clock transitions, and the decoder must know what the encoder did. Formalizing this approach to describing codes uses the parameters "d" and "k" as the respective minimum and maximum number of "0"s between "1"s.

$$d = < z = < k \qquad \textbf{(24.1)}$$

For the MFM code we have: d = 1 and k = 3. Codes with such (d,k) properties are called **RLL (Run-Length-Limited)**. The constraint d is used to control pulse crowding effects, while k is used to assure self-clocking ability (ref. Franaszek, Davidson et al, Norris and Bloomberg).

To meet the (d,k) constraint, m bits of data are mapped into n bits of code, on the average, where n>m. We can use the numbers m and n, plus the (d,k) to assist in evaluating codes, and to develop new coding schemes. Some of the descriptive numbers are defined in the following paragraphs. Good tutorial papers on the topic are by Watkinson, and by Siegel. Early coding is described by Kiwimagi et al., and coding in general communication by Sanders.

Mapping determines a code rate R, which is:

$$R = m/n \qquad \textbf{(24.2)}$$

Codes with a high R are less sensitive to timing jitter caused by noise and peak shift, than those with small R.

The minimum and maximum intervals between transitions are:

$$T_{min} = R*(d + 1)*T \text{ sec.} \qquad \textbf{(24.3)}$$
$$T_{max} = R*(k + 1)*T \text{ sec.} \qquad \textbf{(24.4)}$$

where:

T = Original bit cell time in seconds.

The detection window T_d for the code is:

$$T_d = R*T \text{ sec.} \qquad \textbf{(24.5)}$$

and the clock rate is (1/R)*(1/T). *It is desirable to have a long detection window.*

The density ratio (or minimum transition interval) is:

$$DR = (m/n)*(d + 1) = T_{min}/T \qquad \textbf{(24.6)}$$

Codes with high DR may provide read signals with a high signal-to-noise ratio.

Another ratio is:

$$P = T_{max}/T_{min} = (k + 1)/(d + 1) \qquad \textbf{(24.7)}$$

Codes with a high P tend to have large peak-shifts, poor self-clocking ability and a large spectral DC component.

RANKING OF CODES, ERROR BUDGET

The matter of selecting a code for recording and playback of digital data appears to involve more trade-offs and weighing of decisions than most engineering jobs require. The reader has been presented with a variety of codes and a fair amount of definitions to describe them (m,

Table 24-2. Table of Various Codes with Listing of Code Parameters.

Data:	Characteristics: 1 0 1 1 0 1 0 0 0 0 1 1 0 1 0 0	d	k	m	n	$\frac{m}{n}$	T_{mi}	T_{ma}	Wi	Cl. R.	Max DSV	Max DC
RZ												
NRZ-L		0	∞	1	1	1	T	∞	T	$\frac{1}{T}$	±∞	±1
NRZ-M		0	∞	1	1	1	T	∞	T	$\frac{1}{T}$	±∞	±1
Bi-φ-L		0	1	1	2	$\frac{1}{2}$	$\frac{T}{2}$	T	$\frac{T}{2}$	$\frac{2}{T}$	±T	(0)
Bi-φ-M		0	1	1	2	$\frac{1}{2}$	$\frac{T}{2}$	T	$\frac{T}{2}$	$\frac{2}{T}$	±T	(0)
MFM		1	3	1	2	$\frac{1}{2}$	T	2T	$\frac{T}{2}$	$\frac{2}{T}$	±∞	$\pm\frac{1}{3}$
E-NRZ		0	7	7	8	$\frac{7}{8}$	$\frac{7T}{8}$	7T	$\frac{7T}{8}$	$\frac{8}{7T}$	±∞	$\pm\frac{3}{4}$
4/5 CODE		0	2	4	5	$\frac{4}{5}$	$\frac{4T}{5}$	$\frac{12T}{5}$	$\frac{4T}{5}$	$\frac{5}{4T}$	±∞	$\pm\frac{2}{5}$
Miller^2		1	5	1	2	$\frac{1}{2}$	T	3T	$\frac{T}{2}$	$\frac{2}{T}$	$\frac{\pm 5}{2T}$	0
ZM		1	3	1	2	$\frac{1}{2}$	T	2T	$\frac{T}{2}$	$\frac{2}{T}$	$\frac{\pm 3}{2T}$	0
3PM	?	2	11	3	6	$\frac{1}{2}$	$\frac{3T}{2}$	6T	$\frac{T}{2}$	$\frac{2}{T}$		

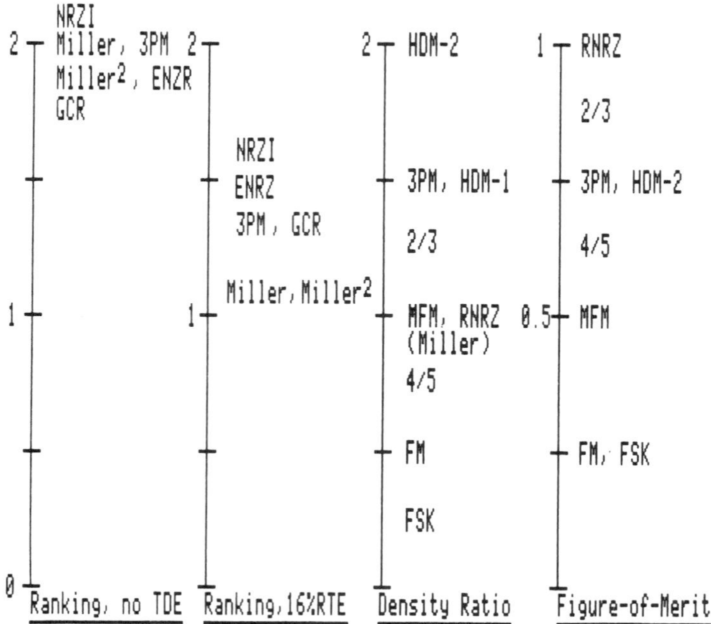

Fig. 24-7. Ranking of codes: Left scales are for no timing errors (TDE) and with a 16% error (after Mackintosh); Right scales showing ranking in accord with DR and FoM (after Watkinson).

n, k, d) and other definitions for comparing codes (R, T_{min}, T_{max}, T_d,DR, P, DSV). See Table 24-2 for a ranking of codes.

Selection of a code must be based upon the characteristics of the hardware (heads, media, resolution, time error budget) such as RTE (real time errors) and IRT (interreversal time). The resulting maximum possible packing factor:

$$PF = PW50/(\text{bit period}) \qquad \textbf{(24.8)}$$

has been analyzed for several codes by Mackintosh. All codes degrade under increasing time errors. Their maximum PF are within $\pm$ 10 percent of each other for a RTE of approximately 16 percent PW50, so they perform quite similar. The moral is: Select a code, make the best out of the system—and you will be near optimum performance (other papers on ranking of codes are by Huber, by Osawa et al, and by Gambe et al.).

The RTE is typically composed of many factors, such as skew, time displacement errors, incomplete erasure, crosstalk, print-through, medium noise, component tolerances and phase-locked loop errors.

A *Figure of Merit* has been suggested (ref. Watkinson) as the product of the density ratio DR (from (24.6)) and the detection window T_d (from (24.5)):

$$\text{Figure of Merit} = \text{FoM} = \text{DR} * T_d \qquad \textbf{(24.9)}$$

Figure 24-7 displays the calculated FoM, compared with Mackintosh's findings.

APPLICATIONS

Computer Tape Drives and Streamers use 4/5 GCR in the 6,250 BPI drives, while Computer Disk Drives use binary FM in the single density floppies, and MFM in double density floppies, and in rigid disk drives. Many papers pertaining to these encoding methods have already been mentioned.

Multilevel encoding can increase the stored information by providing 2^n signal levels within each data bit, where n is the number of levels. The application to a disk system has been examined and judged unsuitable in most practical systems due to SNR limitations (ref. Mackintosh and Jorgensen), while it nevertheless has been implemented with complex channel modulation and error correction and detection coding (ref. Price et al.). The topic of maximum packing density on a disk system is discussed in a paper by Wood et al.

High Bit Rate Data—HDDR

Several manufacturers of precision instrumentation recorders have developed codes for modulation of the data stream, and are using different methods:

Ampex	(Ref. Kelly)	—Miller Squared
Datatape	(Ref. Staff)	—ENRZ-L
EMI	(Ref. Howard and Nottley)	—3PM
Honeywell	(Ref. Meeks)	—NRZ-L
Sangamo	(Ref. Stein and Kessler)	—RNRZ-L

This makes interchange of recorded tapes impossible. The ANSI X3B6 and the ISO TC97-SC12 committees are still trying to make decisions on a standard, but the differences between the different systems are not all that subtle (ref. Reynolds).

The Range Commanders Council has recently issued IRIG Standard 106-86 with a choice between Bi-Phase-L for standard data gathering, and RNRZ-L for extended bandwidth data.

RNR-L is a randomized code, related to convolutional codes. The randomizing is a form of polynomial division, modulo 2, of the data stream by a special number defined by a shift register circuit. It produces an output resembling, to a degree, a random sequence. This sequence, however, is an encoded form of the original sequence and can be decoded by the cor-

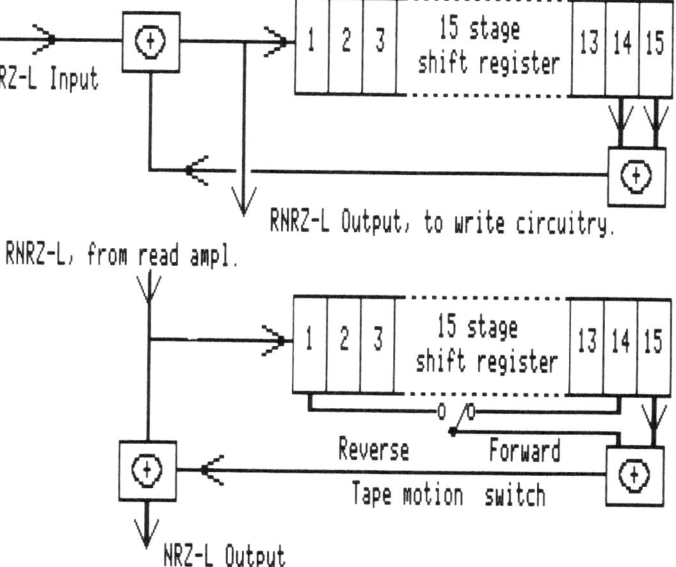

Fig. 24-8. Randomizers for generation and decoding of RNRZ-L codes (courtesy Sangamo/Fairchild Weston).

responding polynomial multiplication. The concept originated in scramblers for communication (ref. Savage).

Figure 24-8 shows the standard 15-cell randomizer and derandomizer. The reader may easily test, on paper, how these work by applying a data stream containing all "0" values except for a single "1" (ref. Stein and Kessler).

The choice of Bi-Phase-L was made for the following reasons (ref. IRIG 106-86, Appendix D, paragraph 3.6):

- Only a small portion of the total signal energy occurs near DC.
- The maximum time T_{max} is T, one bit period.
- The symbols for a "1" and a "0" are antipodal; that is the symbols are exact opposite of each other (i.e., which is better than FM, where there is no symbol for "0"s). Therefore, the bit error probability versus SNR performance is optimum.

Experiments further supported Bi-Phase-L since it is less sensitive to misadjustments of AC-bias and read equalizers than most other codes. Bi-Phase-L also performs well at low tape speeds and low bit rates.

RNRZ-L was chosen over the other codes for several reasons:

- RNRZ-L requires approximately one-half the bandwidth of Bi-Phase-L.
- The symbols for a "1" and a "0" are antipodal.
- The RNRZ-L decoder is self-synchronizing.
- The RNRZ-L data bits can easily be decoded in the reverse mode of tape playback direction.
- The RNRZ-L data are bit detected and decoded using a clock at the bit rate. Therefore, the phase margin is much larger than that of codes that require a clock at twice the bit rate for bit detection.
- The RNRZ-L code does not require overhead bits.

It is recognized that the RNRZ-L can have a large DC-content. Therefore, reproducing data at tape speeds which produce PCM bit rates less than 200 kB/s is not recommended unless a bit synchronizer with a specially designed DC and low frequency restoration circuitry is used.

The codes listed above are used on multichannel high speed recorders. High data rates can also be stored on rotating head machines, and suitable channel codes are discussed by Newby and Yen, and by Coleman et al.

The field of audio recording employed MFM in the first digital recordings. The compact disks use an 8/14 group code, while the masters for these disks are tapes made on rotary head machines with transformer coupling to the heads, using FSK (Frequency Shift Keying). Other codes are constantly being conceived and evaluated, such as HDM-1 (ref. Doi) and 8/10 coding (Ref. Fukuda et al., and Tazaki et al.).

Papers from an AES convention in 1982 (see ref.) plus papers by Moriyama et al. and by Bellis provide further insight into channel coding for digital audio.

In video recording binary FM is used for the SMPTE/BEU time code, and RNRZ-L in other instances; the latter code appears quite resistant to jitter. Four papers listed in the references section provide a good insight into the particular problems of digital video recording.

REFERENCES TO CHAPTER 24

Introduction

Kiwimagi, R.G., McDowell, J.A., and Ottesen, H.H., "Channel Coding for Digital Recording," *IEEE Trans. Magn.*, Sept. 1974, Vol. MAG-10, No. 3, pp. 515-518.

Sanders, L.S., "Pulse Codes in Serial Data Communications," *Computer Design*, Jan. 1982, Vol. 21, No. 1, pp. 203-210.

Siegel, P.H., "Recording Codes for Digital Magnetic Storage," *IEEE Trans. Magn.*, Sept. 1985, Vol. MAG-21, No. 5, pp. 1344-1349.

Watkinson, J.R., "Channel Codes for Digital Recording," *Intl. Conf. on Video, Audio and Data*, 1986, IERE Publ., No. 67, pp. 129-140.

Development of DC-Free Codes

Patel, A.M., "Zero-Modulation Encoding in Magnetic Recording," *IBM Jour. Res. and Dev.*, July 1975, Vol. 19, pp. 366-378.

Mallinson, J.C., and Miller, J.W., "Optimum Codes for Digital Magnetic Recording," *Radio and Electronic Engr.*, Apr. 1977, Vol. 47, No. 4, 4 pages.

Lindholm, D.A., "Power Spectra of Channel Codes for Digital Magnetic Recording," *IEEE Trans. Magn.*, Sept. 1978, Vol. MAG-14, No. 5, pp. 321-324.

Schouhamer Immink, K.A., and Gross, U., "Optimization of Low Frequency Properties of EFM Modulation," *Intl. Conf. Video, Audio and Data 1982*, IERE Publ., No. 54, pp. 375-383.

Run-Length-Limited Codes; Formal Code Development

Franaszek, P.A., "Sequence-state Methods for Run-Length-Limited Coding," *IBM Jour. Res. and Dev.*, July 1970, Vol. 14, pp. 376-383.

Ringkjoeb, E.T., "Achieving a Fast Data Transfer Rate By Optimizing Existing Technology," *Electronics*, May 1975, pp. 86-91.

Davidson, M., Haase, S.F., Machamer, J.L., and Wallman, L.H., "High Density Magnetic Recording Using Digital Block Codes of Low Disparity," *IEEE Trans. Magn.*, Sept. 1976, Vol. MAG-12, pp. 584-586.

Jacoby, G.V., "A New Look-Ahead Code for Increased Data Density," *IEEE Trans. Magn.*, Sept. 1977, Vol. MAG-13, No. 3, pp. 1202-1204.

Severt, R.H., "Encoding Schemes Support High Density Digital Data Recording," *Computer Design*, May 1980, Vol. 19, No. 5, pp. 181-190.

Newton, M., "GCR Increases Data Recording Rates and Reliability," *Digital Design*, July 1981, pp. 36-39.

Norris, K., and Bloomberg, D.S., "Channel Capacity of Charge-Constrained Run-Length Limited Codes," *IEEE Trans. Magn.*, Nov. 1981, Vol. MAG-17, No. 6, pp. 3452-3455.

Cohn, M., and Jacoby, G.V., "Run-Length Reduction of 3PM Code via Look-Ahead Technique," *IEEE Trans. Magn.*, Nov. 1982, Vol. MAG-18, No. 6, pp. 1253-1255.

Jacoby, G.V., and Kost, R., "Binary Two-Third Rate Code with Full Word Look-Ahead," *IEEE Trans. Magn.*, Sept. 1984, Vol. MAG-20, No. 5, pp. 709-714.

Applications; The Choice of a Recording Code

Mackintosh, N.D., "The Choice of a Recording Code," *The Radio and Electronic Engineer*, Apr. 1980, Vol. 50, No. 4, pp.177-193; Also: IERE Conf. Proc. No. 43, Jan. 1979, pp. 77-119.

Huber, W.D., "Selection of Modulation Code Parameters for Maximum Lineal Density," *IEEE Trans. Magn.*, Sept. 1980, Vol. MAG-16, No. 5, pp. 637-640.

Osawa, H., Tazaki, S., and Andoh, S., "Performance Comparison of Partial Response Systems for N.R.Z. Recording," *Intl. Conf. Video and Data 82*, Apr. 1982, No. 54.

Gambe, HY., Matsumura, T., and Matsuda, T., "Codec Evaluation Method Based on Measured DC Characteristics," *FUJITSU Scientific and Technical Journ.*, Aug. 1984, Vol. 20, No. 3, pp. 259-281.

Computer Disk Drives

Price, R., Craig, J.W., Melbye, H.E., and Perahia, A., "An Experimental, Multilevel, High Density Disk Recording System," *IEEE Trans. Magn.*, Sept. 1977, Vol. MAG-14, No. 5, pp. 315-317.

Mackintosh, N.D., and Jorgensen, F., "An Analysis of Multi-Level Encoding," *IEEE Trans. Magn.*, Nov. 1981, Vol. MAG-17, No. 6, pp. 3329-3332.

Wood, R., Ahlgrim, S., Hallamasek, K., and Stevenson, R., "An Experimental Eight-Inch Disc Drive with One-Hundred Megabytes per Surface," *IEEE Trans. Magn.*, Sept. 1984, Vol. MAG-20, No. 5, pp. 698-702.

HDDR

Savage, J.E., "Some Simple Self-Synchronizing Digital Data Scramblers," *Bell Systems Tech. Jour.*, Feb. 1967, Vol. 46, pp. 449-487.

Newby, P.S., and Yen, J.L., "High Density Digital Recording Using Videocassette Recorders," *IEEE Trans. Magn.*, Sept. 1983, Vol. MAG-19, No. 5, pp. 2245-2252.

Coleman, C., Lindholm, D., Peterson, D., and Wood, R., "High Data Rate Magnetic Recording In a Single Channel," *Intl. Conf. Video, Audio and Data 1984*, IERE Publ., No. 59, pp. 151-157.

Kelly, J., "Miller Squared Coding," *High-Density Digital Recording*, NASA Ref. Publ. No. 1111, Sept. 1985, pp. 127-142.

Staff at Datatape, "Parallel Mode High-Density Digital Recording: Technical Fundamentals," *High-Density Digital Recording*, NASA Ref. Publ. No. 1111, Sept. 1985, pp. 143-194. Also available in book of same title from Datatape in Pasadena, Calif.

Howard, J.M., and Nottley, G.C., "The Application of 3-Position Modulation Coding to Longitudinal Instrumentation Recording," *High-Density Digital Recording*, NASA Ref. Publ. No. 1111, Sept. 1985, pp. 195-215.

Meeks, L., "The Honeywell HD-96 High-Density Digital Tape Record/Reproduce System," *High-Density Digital Recording*, NASA Ref. Publ. No. 1111, Sept. 1985, pp. 215-230.

Stein, J.A., and Kessler, W.D., "The Development of a High-Performance Digital Recording Error Correction System," *High-Density Digital Recording*, NASA Ref. Publ. No. 1111, Sept. 1985, pp. 231-246.

Reynolds, S., "High-Density Digital Recording (HDDR) Users Subcommittee Evaluation of Parallel HDDR Systems," *High-Density Digital Recording*, NASA Ref. Publ. No. 1111, Sept. 1985, pp. 281-292.

Audio

Blesser, B., Locanthi, B., and Stockman, T.G. (Editors), *Digital Audio*, Collected Papers from the *AES Premiere Conference*, Rye, New York 1982, June 3-6, 262 pages.

Moriyama, Y., Yamagata, K., Suzuki, T., and Iwasawa, T., "New Modulation Technique for High Density Recording on Digital Audio Discs," *An AES Preprint No. 1827(I-3)*, AES Conv. Nov. 1981, 7 pages.

Doi, T.T., "Channel Coding for Digital Audio Recordings," *Jour. AES*, April 1983, Vol. 31, No. 4, pp. 224-238.

Bellis, F.A., "Introduction to Digital Audio Recording," *The Radio and Electronic Engr.*, Oct. 1983, Vol. 53, No. 10, pp. 361-368.

Fukuda, S., Kojima, Y., Shimpuku, Y., and Odaka, K., "8/10 Modulation Codes for Digital Magnetic Recording," *IEEE Trans. Magn.*, Sept. 1986, Vol. MAG-22, No. 5, pp. 1194-1196.

Tazaki, S., Kaji, T., and Osawa, H., "An Analysis of DC-Free Property on Run-Length Limited Code," *Intl. Conf. on Video, Audio and Data*, 1986, IERE Publ., No. 67, pp. 151-156.

Video

Baldwin, J.L.E., "Codes for Digital Video Tape Recording at 10 M.bit/Sq. Inch," *Intl. Conf. Video and Data 79*, July 1979, No. 43, pp. 147-163.

Baldwin, J.L.E., "Channel Codes for Digital Video Recording," *Intl. Conf. Video and Data 84*, Apr. 1984, No. 59.

Furukawa, T., Ozaki, M., and Tanaka, K., "On a DC-Free Block Modulation Code," *IEEE Trans. Magn.*, Sept. 1984, Vol. MAG-20, No. 5, pp. 878-880.

Heitman, J.K.R., "Digital Video Recording: New Results in Channel Coding and Error Protection," *Jour. SMPTE*, Feb. 1984, Vol. 93, No. 2, pp. 140-144.

Chapter 25

Error Detection and Correction Coding

Communication systems are subject to the affects of noise added to signals, or disruptions of a signal. The latter are in magnetic recording known as drop-outs. A common measure of the system's freedom from these ill effects is a rating called bit-error-rate, abbreviated BER. The BER number states what the probability of errors in the data stream are, for instance if BER $= 10^{-6}$, then an average of one bit out of one million may be in error, either missing or changed by a noise spike.

We will look at ways of adding extra bits to the signal and thereby making it possible to detect if an error occurred, and possibly correct it. This will lead us to the important concept of Hamming-distance, and the design of block-codes.

The technique of error correction will not eliminate errors nor may it provide a drastic change of a BER rating of say 10^{-6} to 10^{-12} (from one error in a million bits to one error in one trillion bits). We will also examine bit error patterns, and find methods of predicting the BER before and after error- detection- and- correction (EDAC) has been implemented into the system.

The chapter will provide the reader with an insight into the concept of error detecting and correcting codes, using the basic block codes as examples. Longer codes can only be studied and appreciated if the reader is well versed in abstract algebra (rings, fields, convolution etc.), and is clearly beyond the scope of this book.

ERROR DETECTION AND CORRECTION

Error detection and correction is accomplished by adding extra information to the transmitted data. Take for instance a message you receive over the phone: If you are uncertain about its meaning you ask to have it repeated. And to make absolutely clear what was meant a third repeat may be requested.

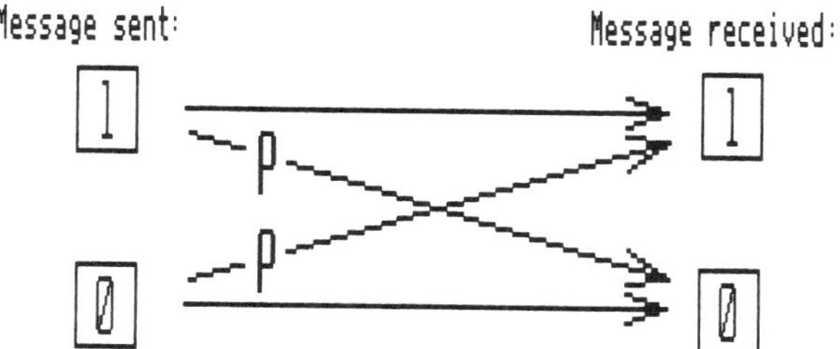

Fig. 25-1. Signal transmission of unprotected message.

This very concept forms the basis for EDAC. Figure 25-1 shows a communication channel that transmits the status of a two-state condition; this can be the range of a temperature, high or low with regard to a reference, or an ON-OFF condition of a relay controlled by a sensor circuit. We decide to transmit a "1" if the relay is closed, a "0" if it is open.

The possibility now exists that a "1" is received as a "0"; and vice versa. The *transition probability p* for this happening is equal to the transmission system's Bit-Error Rate, BER. There is no way of telling whether the received signal is correct or not.

We can improve this by always repeating the message, and will therefore transmit an "11" for the ON-condition, and "00" for the OFF-condition, see Fig. 25-2. We have, in terms of coding language, added a *parity check bit*. Also, we have changed the *message words* "1" and "0" into *code words* "11" and "00," respectively.

A single error may occur with a probability of p. We will then receive a "10" or a "01," and will know that an error occurred since neither belongs to the selected codewords of "11" and "00". But we cannot tell what the original message was, i.e., we cannot perform error correction. This is why we in a telephone conversation may ask what was said.

The probability of two errors, which will invert the transmitted messages without detection ("00" → "11," or "11" → "00") is equal to p^2 since the errors in two bits are independent of each other. The possibility of any double error is therefore extremely remote. (We will find, however, that the BER for a 2-bit word is different from the BER for a 1-bit word, see later).

Let us continue with our "intuitive" coding scheme—we will add another bit, so now three bits are transmitted for each single bit message. The codewords that belong to the "1" or the "0" will be "111" or "000". One, two or three errors may occur, and the received code words may have any of the eight (2^3) patterns shown in Fig. 25-3.

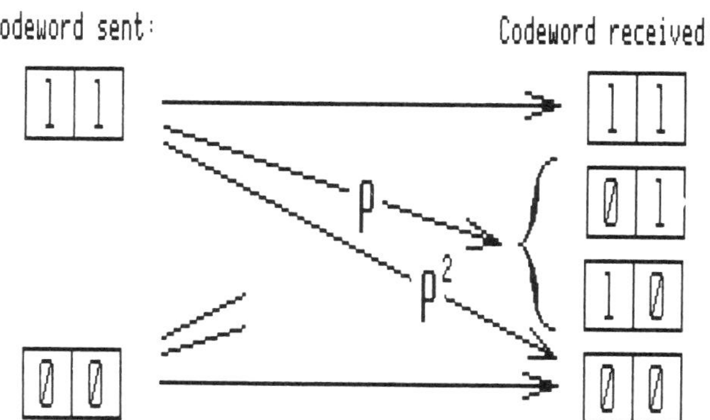

Fig. 25-2. Signal transmission of single repeat code.

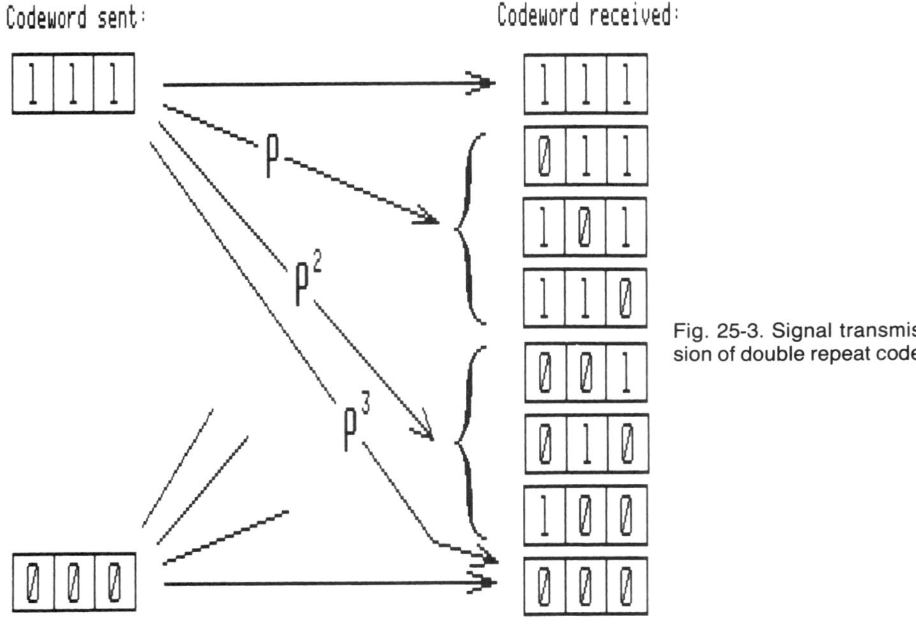

Fig. 25-3. Signal transmission of double repeat code.

A single error in the "111" code word will result in the received words no. 2 through 4, and the decoder will assign "111" to any of these. It will do so because that is the *most likely* choice. The similar argument holds for the "000" code word. We have thus detected and corrected for one error.

Two errors will result in wrong decisions by the decoder, and an error in the decoded message results ("0" → "1," or "1" → "0").

More bits could be added to each of the example's two codewords, but we are already dealing with a rather inefficient way of providing error protection: Each bit is replaced with three bits, so the overhead is 200%, or the code is 33% efficient. It would be better to take a few bits at a time, and replace them with a longer set of bits, where the extra bits will be used to detect for errors, and possibly correct them.

This changes the coding from the *repeat coding method* to the commonly used *block coding*, where k message or data bits are replaced by n code bits, by adding r *check bits*. This method is similar to the 4/5 GCR and the 3/6 3PM coding schemes, except this time we must select additional bits so that they will assist in error detection and correction, instead of reducing the code's DC content to zero (This is to say that the modulation encoding at times can merge with EDAC).

A POINT OF VANISHING RETURN

The incoming stream of data bits are divided up in message words which in turn are altered by the addition of check-bits. The codewords would quickly jam the encoder unless the data words are compressed in time so they, with the added check-bits, occupy the same time frame as the original dataword.

This shortened time per bit corresponds to a widening of the signal bandwidth for the PCM signal, which results in a lowering of the SNR. Let us assume that SNR for the channel is 13.5 dB, which results in a BER of 10^{-6}. We now add one check bit to each message bit, and the time allotted for each bit must be cut in half. This means a reduction of 3 dB in the SNR to

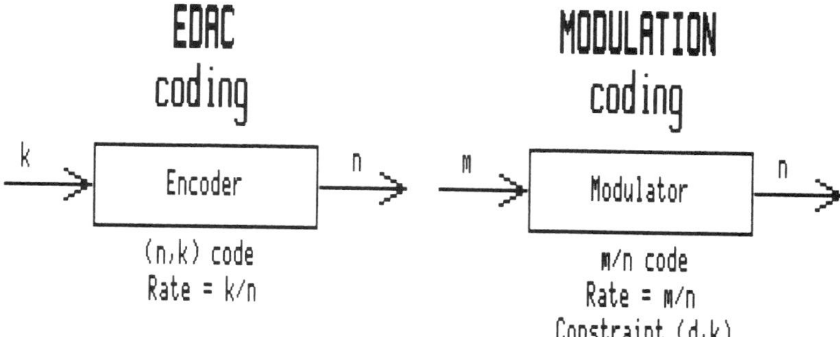

Fig. 25-4. Conflicting use of symbols in two related disciplines.

5 dB, and the new BER (or p) is $0.5 \bullet 10^{-3}$ (See Fig. 23-12). This is the new probability of a single error (which we can detect, though).

The probability of 2 errors, which we can not detect, is $p \bullet P = 0.25 \bullet 10^{-6}$, so we have only improved this system by a factor of four! Much longer codewords will be needed, as we shall see later.

First *a word of caution* before we proceed: There exists an unfortunate overlap in the use of symbols in coding theory, as shown in Fig. 25-4. The reader who plans to read further must be aware that the symbol n is always used for the number of bits coming out of an encoder. Then look for the use of n or d in either an (n,k) code (EDAC), or the (d,k) constraint in RLL codes, with corresponding rates k/n and m/n.

Let us now evaluate the (3,1) code (from the terminology: (n,k) code, where k incoming bits are changed into n transmitted bits). The time allotted per bit in the codeword of the (1,3) code is reduced by a factor of 3, resulting in a 4.8 dB lesser SNR. Continuing our previous example, we find the error probability corresponding to SNR = 8.8 dB equals $p = 2 \bullet 10^{-3}$. No final error results for one bit error in any of the three transmitted bits.

The probability of 2 bit errors equals $3 \bullet p^2 = 12 \bullet 10^{-6}$; the factor 3 accounts for the fact that there are 3 different patterns for two errors in three bits (this is a coincidence; b errors in a pattern of n bits results in a multiplier of n!/(b!(n-b)!). This result is 12/(0.25) = 48 times worse than the case where only one parity bit was added. The (3,1) code is therefore, at the SNR levels used, worse than no coding. We will later return to examine the value of longer codes.

The simple example with only two signal conditions helped us in getting a feel for the concept of error detection and correction. We will now expand the technique of adding check bits, and this time do it for a multiple bit data word.

BLOCK CODES

The incoming data stream is chopped up into *message words* **u** each of length k bits (Following the notation used in Lin and Costello's book, see ref.) A r = (n-k) bits long *check word* **r** is generated, and added to the message word, forming a *code word* **v** of length n bits; Fig.25-5 illustrates the process. (Lin and Costello place the check word ahead of the message, while earlier textbooks did the opposite. It doesn't really matter which way it is done; Lin and Costello's method does make the next step into cyclic codes smoother.)

We can write the code generation as:

$$\{ \text{ n-k check bits } \} + \{ \text{ k message bits } \} \rightarrow \{ \text{ n code bits } \}$$

The task at hand is to find the n-k = r check bits that will allow a decoder to determine and possibly correct errors. This is a formidable mathematical task for longer codes, and cod-

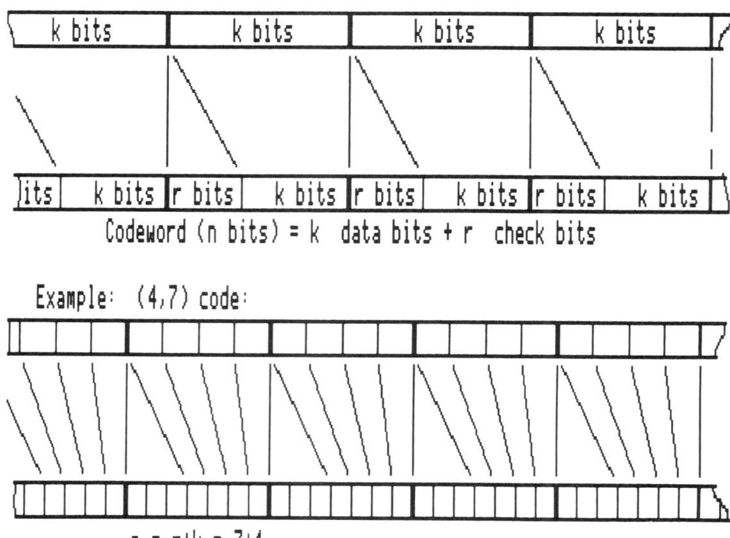

Fig. 25-5. Block coding, simplified. Example shows the (4,7) code.

ing specialists must be apt mathematicians in advanced abstract algebra to delve into this challenge.

The term "abstract" reflects that the mathematicians divorced themselves from the physical world in order to develop logical, exact mathematical systems. Here the symbols like a, b, A, B, C etc. do not stand for anything physical, i.e. they do not represent current, voltage, or time, for example.

The idea is then to see whether or not a physical system, and in the present case: a coding system, falls within the definitions of one of the abstractly developed mathematical systems. If it does, then we can apply the rules of operations (addition, subtraction, multiplication etc.) for that system, and remain assured that the outcome of the operation stays within the system, and therefore is correct.

The introduction to Error Detection leads us into the world of multi-dimensional vectors, and we will be required to carry out certain operations on these vectors, such as addition and multiplication. The result of these operations must fall within the set of vectors that are allocated to the system.

A couple of examples will allow us to appreciate the processes, without getting too deeply involved (there are plenty of references for those who wish to do so).

We saw earlier how addition of at first one, and then another bit, took us from a one-dimensional to a three-dimensional solution set. This points toward treating the collection of bits as vectors, i.e. each n bits word is a n-dimensional vector. This sort of presentation is clearly beyond physical comprehension for humans in a 3-D world, but rather a delight for mathematicians. How can he/she add or multiply or otherwise operate with these vectors and be sure that the results are correct?

Consider first our familiar decimal number system. We can add, subtract and multiply without much hesitation—the result is always right. But only a few hundred years ago subtraction did not always exist because negative numbers were not heard of. Numbers were used to keep track of quantities of items, dimensions, etc. - and negative numbers were invented by Italian bankers in order to provide loans to their customers.

We all do apply certain restrictions though, subconsciously. If we add 4 dogs to 6 cats we know the result is 10 animals, and we would never dream of multiplying the two! Then how can we be sure of the operations we do on vectors that represent words?

Our present concern is the application of algebra to codewords, which we block out in sets of n bits—how do we represent them in mathematical terms?

We are, first of all, only concerned about a number set consisting of a 0 or a 1. (There are multi-level codes but they are rarely employed in magnetic recording due to the poor signal amplitude stability of typically ± 10%).

This small mathematical system works in accordance with the so-called *Modulo-2 arithmetic*, where the outcome of any operation is determined by dividing the result with 2—and the remainder used for the result. The symbol for this operation is $\oplus$.

For example:

$$0 \oplus 1 = 1 / 2 = 0 + \text{remainder } 1 = 1 .$$
$$3 \oplus 4 = 7 / 2 = 3 + \text{remainder } 1 = 1 .$$
$$3 \oplus 1 = 4 / 2 = 2 + \text{remainder } 0 = 0 .$$

The addition rules for the *finite set* (0 and 1) are:

$$0 \oplus 0 = 0$$
$$0 \oplus 1 = 1$$
$$1 \oplus 0 = 1$$
$$1 \oplus 1 = 0$$

For multiplication we find, similarly:

$$0 \bullet 0 = 0$$
$$0 \bullet 1 = 0$$
$$1 \bullet 0 = 0$$
$$1 \bullet 1 = 1$$

The results are easy to remember if we just keep in mind a clock with only two hours: 0 where 12 normally is located, and 1 where 6 is.

We then immediately understand why in modulo-2: $1 = -1$ - add or subtract to your hearts delight, don't worry about signs, the remainder is always a 0 or 1.

CODEWORDS AND VECTOR SPACE

Let us now define a mathematical description for the block of 0 and 1 bits that make up a code word. We will define a *vector* in the sense it is used in algebra and coding theory, and will refer to it as a "vector **v** having n components over a field F."

We define a vector (in two dimensions) to be a pair of numbers { x, y }. The numbers, or scalars, x and y may be real or complex or indeed elements of any field, and the set of vectors over this field consists of all possible ordered pairs (x, y). Two vectors are equal if and only if both components x and y are equal.

A three-dimensional vector is defined as an ordered triad of the numbers (x, y, z), where the components may be real, complex or in any field.

The n components that make up the n-dimensional vector **v** are of course the n bits in the code word, and these components of the field F, the Galois field (0, 1), form what we can call a n-tuple:

$$\mathbf{v} = \{ v_1 , v_2 , v_3 , \ldots , v_n \}$$

The vector given by all zero components except the i'th, which is 1, is denoted by **w** and any vector may be expressed in the form:

$$\mathbf{v} = a_1\mathbf{w}_1 + a_2\mathbf{w}_2 + \ldots\ldots a_n\mathbf{w}_n$$

Set $\mathbf{w}_1$, $\mathbf{w}_2$, . . $\mathbf{w}_n$ is called a *base* for the vectors. Normally we shall use binary values for the a_1 terms.

We are now defining the *rules of operation on vectors*:

Addition:

$$\mathbf{v}_1 + \mathbf{v}_2 = \{ a_1 + b_1 , a_2 + b_2 , \ldots , a_n + b_n \}$$

Scalar multiplication:

$$c\bullet\mathbf{v}_1 = \{ c \bullet a_1 , c \bullet a_2 , \ldots , c \bullet a_n \}$$

Product of two vectors:

$$\mathbf{v}_1\bullet\mathbf{v}_2 = a_1\bullet b_1 + a_2\bullet b_2 + \text{- - -} + a_n\bullet b_n.$$

Our final definitions pertain to a *vector space over a field F*: It is a commutative group under addition (with components from (0, 1) = GF(2), the Galois Field over 2 , with modulo-2 arithmetic). It is closed under the operation of scalar multiplication, and the distributive and associative laws hold.

If we add a couple of 7-dimensional vectors we could write something like:

$$\{1,1,0,0,1,0,1\} + \{0,1,1,0,1,0,1\} = \{1,2,1,1,2,0,1\}$$

The two bi-level vectors have now become a tri-level vector, something our binary system is unable to handle. From abstract algebra we learn that we must apply *modulo-2 algebra* to force the solution into the binary system. A modulo-2 operation takes the outcome and divides it by two, and uses the remainder as its result. We did correctly operate on corresponding vector components in the example above, but should have used the modulo-2 operation on the end result:

$$\{ 1,1,0,0,1,0,1 \} + \{ 0,1,1,0,1,0,1 \} = \{ 1 \oplus 0,1 \oplus 1,0 \oplus 1,0 \oplus 0,1 \oplus 1,0 \oplus 0,1 \oplus 1 \}$$
$$= \{ 1,0,1,0,0,0,0 \}$$

HAMMING DISTANCE: WEIGHT

Let us review the ON-OFF example, before we go on to consideration of longer codes. The choice between a "1" or a "0" can be considered a one-dimensional matter—the selection of a point. With the addition of one check bit the choice is two-dimensional—the selection of a vector: the zero vector {0,0} or the vector to the point {1,1}. This is illustrated in Fig. 25-6, where also the erroneous vectors {1,0} and {1,1} are shown.

Going on to the three bit code words the selection is a three-dimensional choice. Vectors {0,0,0} and {1,1,1} correspond to the allowed code words, the other six vectors do not. Or better expressed: {0,0,0} and {1,1,1} is the *set* of transmitted code words, while all eight words belong to the set of possible received code words. We can show this by a three-dimensional illustration, or by a Venn diagram (Fig. 25-6).

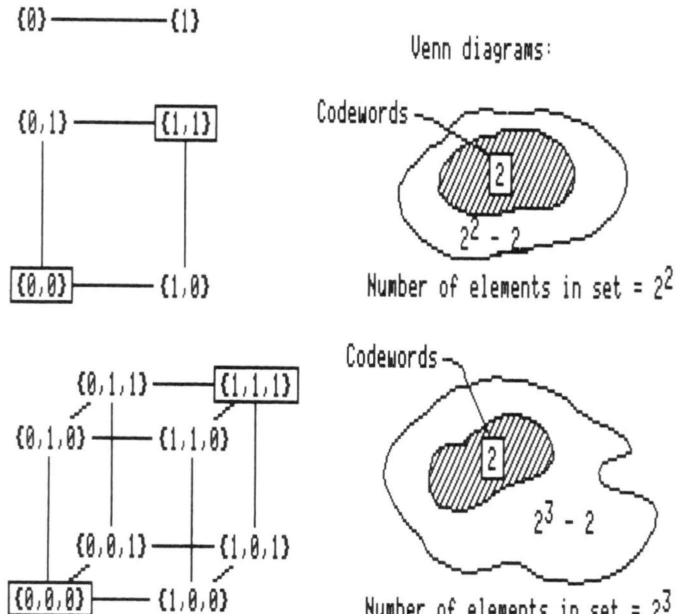

Fig. 25-6. Message (0 or 1), single repeat and double repeat coding illustrated by one, two and three-dimensional drawings, or by Venn diagrams.

Notice that the vectors {0,0} and {1,1} are different in two positions, and {0,0,0} and {1,1,1} in three positions. We refer to this difference as the *Hamming distance d*, introduced by professor Richard W. Hamming.

Another characterization of a code word is its weight, which is defined as the number of positions where it differs from zero. This number will be related to its distance from the zero vector.

The error correcting capabilities of a code can be evaluated (after Hamming):

The number t of errors that may be detected and corrected is:

$$\boxed{t = (d-1)/2} \tag{25.1}$$

where d is the minimum distance between codewords. It can be shown that d is equal to the smallest weight of any of the codewords.

The Hamming distance can be observed directly in the three-dimensional drawing in Fig. 25-6: Moving from {0,0,0} to {1,1,1}, from point to point, takes three steps. The two-dimensional drawing shows two steps in moving from {0,0} to {1,1}; hence t = (2-1)/2 = .5 → 0 errors corrected, which is what we learned.

CODE GENERATION

The message, the check bits and the code word are considered vectors of dimensions k, r and n respectively. We will perform several mathematical operations on these 'numbers', such as addition, subtraction, multiplication and division.

Magnetic recording codes belong to the binary system, with symbols 0 and 1. This number system is also called a field, Galois field GF(2), and is simpler than fields corresponding to multilevel, or multi-symbol, codes. The rules for computations are simple, as shown.

Addition: Multiplication:

	0 1
0	0 1
1	1 0

	0 1
0	0 0
1	0 1

The results are always Mod(2) ($\oplus$), i.e., equal to the remainder after division with 2. Note also, that $1 \oplus 1 = 0$, or $+1 = -1$!

Example no. 1. Addition of two words:

$$\begin{aligned} \mathbf{a} &= \{ 0, 1, 1, 0, 1, 0, 1, 1, 1 \} \\ + \mathbf{b} &= \{ 1, 1, 1, 1, 0, 0, 0, 1, 0 \} \\ &= \{ 0 \oplus 1, 1 \oplus 1, 1 \oplus 1, 0 \oplus 1, 1 \oplus 0, 0 \oplus 0, 1 \oplus 0, 1 \oplus 1, 1 \oplus 0 \} \\ &= \{ 1, 0, 0, 1, 1, 0, 1, 0, 1 \} \end{aligned}$$

Multiplication of two words:

$$\begin{aligned} \mathbf{a} &= \{ 0, 1, 1, 0, 1, 0, 1, 1, 1 \} \\ \pm \mathbf{b} &= \{1, 1, 1, 1, 0, 0, 0, 1, 0 \} \\ &= \{0 \bullet 1 \oplus 1 \bullet 1 \oplus 1 \bullet 1 \oplus 0 \bullet 1 \oplus 1 \bullet 0 \oplus 0 \bullet 0 \oplus 1 \bullet 0 \oplus 1 \bullet 1 \oplus 1 \bullet 0\} \\ &= 1 . \end{aligned}$$

Note that this vector multiplication is an inner product, or dot product, and the result is not a vector, but a scalar.

The incoming message bits are divided up into blocks of each k bits, and r parity check bits are appended to each block, making new code words of length n = r + k bits each:

$$\begin{aligned} \mathbf{u} &= u_1, u_2, \ldots\ldots, u_k \\ &\rightarrow v_1, v_2, \ldots\ldots v_r, v_{r+1}, v_{r+2}, \ldots\ldots v_{r+k}. \\ &\quad \text{<— r parity bits —> <—— k message bits ——>} \\ &= \mathbf{v} \end{aligned}$$

The number of transmitted bits has increased from k to r + k. This reduction in efficiency is expressed by the *code rate*:

$$R = k/n = \text{information bits / transmitted bits}$$

The parity checks must be generated as linear combinations of the k-tuple message vectors, by mod-2 summations ($\oplus$). Hence these codes are named *linear codes*, and *systematic* since the parity check bits are appended in a block of r bits at the beginning (or end) of the message bits.

The parity bits are calculated from:

$$\begin{aligned} v_1 &= p_{11}u_1 \oplus p_{12}u_2 \oplus \ldots\ldots \oplus p_{1k}u_k \\ v_2 &= p_{21}u_1 \oplus p_{22}u_2 \oplus \ldots\ldots \oplus p_{2k}u_k \\ &\ldots\ldots\ldots\ldots \\ &\ldots\ldots\ldots\ldots \\ v_r &= p_{r1}u_1 \oplus p_{r2}u_2 \oplus \ldots\ldots \oplus p_{rk}u_k \end{aligned}$$

which we can write in matrix form:

$$\{v_1, \ldots, v_r\} = \{u_1, u_2, \ldots, u_k\} \cdot \begin{Bmatrix} p_{11} & p_{21} & & p_{r1} \\ p_{12} & p_{22} & & p_{r2} \\ \cdot & \cdot & & \cdot \\ \cdot & \cdot & & \cdot \\ p_{1k} & p_{2k} & & p_{rk} \end{Bmatrix}$$

$$= \mathbf{u} \bullet \mathbf{P}_{kr}.$$

P cannot be constructed in an arbitrary fashion. It can be shown that the rows must be uniquely defined, i.e. no rows alike, not all zero's and no single 1s in a row.

We can write the message bits in the code word as:

$$\{v_{r+1}, v_{r+2}, \ldots, v_{r+k}\} = \mathbf{u} \bullet \begin{Bmatrix} 1\,0\,0\,.\,.\,0 \\ 0\,1\,0\,.\,.\,0 \\ 0\,0\,1\,.\,.\,0 \\ \cdots\cdots \\ 0\,0\,0\,.\,.\,1 \end{Bmatrix}$$

$$= \mathbf{u} \bullet \mathbf{I}_{kk}$$

The total code word is therefore:

$$\mathbf{v} = \mathbf{u} \bullet \mathbf{P}_{kr} + \mathbf{u} \bullet \mathbf{I}_{kk}$$

or

$$\mathbf{v} = \mathbf{u}\ \{\mathbf{P}_{kr}\ \mathbf{I}_{kk}\}_{kn}$$
$$\mathbf{v} = \mathbf{u} \bullet \mathbf{G}$$

where $\mathbf{G} = \{\mathbf{PI}\}$ is named the *code generator matrix*.

Example no. 2.

Find the code word for messages $\mathbf{u}_1 = \{1\,1\,0\,1\}$ and $\mathbf{u_2} = \{0\,1\,0\,1\,\}$, using the **P** matrix:

$$\mathbf{P} = \begin{Bmatrix} 1\,0\,1 \\ 1\,1\,1 \\ 1\,1\,0 \\ 0\,1\,1 \end{Bmatrix}.$$

Answer:

$$\mathbf{u}_1 \bullet \mathbf{G} = \{1\,1\,0\,1\,\} \bullet \begin{Bmatrix} 1\,0\,1\,1\,0\,0\,0 \\ 1\,1\,1\,0\,1\,0\,0 \\ 1\,1\,0\,0\,0\,1\,0 \\ 0\,1\,1\,0\,0\,0\,1 \end{Bmatrix} = \{0\,0\,1\,1\,1\,0\,1\,\}$$

$$\mathbf{u}_2 \bullet \mathbf{G} = \{0\,1\,0\,1\,\} \bullet \begin{Bmatrix} 1\,0\,1\,1\,0\,0\,0 \\ 1\,1\,1\,0\,1\,0\,0 \\ 1\,1\,0\,0\,0\,1\,0 \\ 0\,1\,1\,0\,0\,0\,1 \end{Bmatrix} = \{1\,0\,0\,0\,1\,0\,1\,\}$$

Exercise no. 1: Find the remaining 14 code words.

ERROR PROBABILITY IN CODE WORDS

We evaluated, earlier in this chapter, the BER for a (3,1) code; we will now extend this to other codes.

Whenever we transmit word **x** and receive word **y**, then a test for error-free transmission is that:

$$\mathbf{x} - \mathbf{y} = \mathbf{x} + \mathbf{y} = \{0\},$$

where {0} is the zero vector:

$$\begin{aligned} \mathbf{x} &= \{ x_1, x_2, x_3, \ldots\ldots, x_n \} \\ -\mathbf{y} &= \{ y_1, y_2, y_3, \ldots\ldots, y_n \} \\ \hline &= \{0, 0, 0, \ldots\ldots, 0\} \end{aligned}$$

A reasonable assumption is that any error will affect a bit, or symbol (or component), independent of errors in the others. An error vector adds to a signal vector, and it can be expressed as:

$$\mathbf{e} = \{ e_1, e_2, e_3, \ldots\ldots, e_n \}$$

and the total error probability:

$$P(\mathbf{e}) = P(e_1) \bullet P(e_2) \bullet \ldots\ldots \bullet P(e_n) .$$

The error probability, or BER, is also called *the transition probability*, where transition refers to a "1" changing into a "0," and vice versa. The probability of no errors is 1-p, and n independent symbols are therefore received error free with a probability of $(1\text{-}p)^n$. We can summarize:

$$\begin{aligned} P(\text{no errors}) &= (1\text{-}p)^n \\ P(\text{ 1 error }) &= p(1\text{-}p)^{n-1} \\ P(\text{ 2 errors}) &= p^2(1\text{-}p)^{n-2} \end{aligned}$$

$$\boxed{P(\text{b errors}) = p^b(1\text{-}p)^{n-b}} \qquad \textbf{(25.2)}$$

The actual probability will be greater, since errors can produce different *error patterns* (E = error), shown for a five bit word:

ONE error: E X X X X
– or X E X X X
– or X X E X X
– or X X X E X
– or X X X X E .

For a n-dimensional word there are n different patterns for a single error, and the true error probability is therefore:

$$P(1 \text{ error}) = n \bullet p(1\text{-}p)^{n-1}$$

Two errors will result in n•(n-1)/2 patterns:

E E X X X	X E X E X
E X E X X	X E X X E
E X X E X	X X E E X
E X X X E	X X E X E
X E E X X	X X X E E

A total of b errors will result in:

$$(n/1)\bullet((n-1)/2)\bullet((n-2)/3)\bullet((n-3)/4)\bullet\ .\ .\ \bullet((n-b-1)/b)$$

$$= n!/(b!(n-b)!) \text{ patterns,}$$

where

$n! = n\bullet(n-1)\bullet(n-2)\bullet\ .\ .\ .\ .\ \bullet3\bullet2\bullet1$

The total error probability is therefore:

$$\boxed{P(b \text{ errors}) = [\ n!/(b!(n-b)!)\ ]\bullet p^b(1-p)^{n-b}} \qquad \textbf{(25.3)}$$

Example no. 3: A (7,4) code. **A**. *Error probability, uncoded 4 bits message*. The probability of one error is P(1 error) = $4\bullet p(1-p)^3 \approx 4p$ (for $p << 1$). We arrive at the following table, showing P(1 error) as a function of the system SNR:

SNR	p	P(1 error)
3.0 dB	10^{-1}	$4\bullet10^{-1}$
7.2 dB	10^{-2}	$4\bullet10^{-2}$
9.8 dB	10^{-3}	$4\bullet10^{-3}$
12.5 dB	10^{-4}	$4\bullet10^{-4}$
14.2 dB	10^{-7}	$4\bullet10^{-7}$
16.0 dB	10^{-10}	$4\bullet10^{-10}$

This table results from the use of formula (23.1).

B. *Error probability, (7,4) code word*. We have now added 3 check bits, and can for a reasonable selection of message words get a Hamming distance of d=3. This will correct:

$$t = (d-1)/2 = (3-1)/2 = 1 \text{ error.}$$

One error will therefore result when the number of errors in the code word is 2. Hence, using formula (25.3):

$$\begin{aligned} &P(1 \text{ error}) = \\ &P(2 \text{ code word errors}) = [\ 7!/(2!(7-2)!)\ ]p^2(1-p)^{7-2} \\ &\qquad = 21\bullet p^2 \end{aligned}$$

We must reestimate the values of p, in light of the reduced SNR. The message bit rate remains unaltered, and the code word rate must therefore be increased by a factor of n/k = 7/4; this corresponds to a bandwidth increase by 7/4, and will result in a 2.5 dB reduction of the SNR. A new table is:

$SNR_{7\ BITS}$	p	P(1 error)	(uncoded P(1))
0.5 dB	$1.2 \bullet 10^{-1}$	$3.0 \bullet 10^{-1}$	$4 \bullet 10^{-1}$
4.7 dB	$3 \bullet 10^{-2}$	$1.9 \bullet 10^{-2}$	$4 \bullet 10^{-2}$
7.3 dB	$1 \bullet 10^{-2}$	$2.1 \bullet 10^{-3}$	$4 \bullet 10^{-3}$
8.9 dB	$2 \bullet 10^{-3}$	$8.4 \bullet 10^{-5}$	$4 \bullet 10^{-4}$
11.7 dB	$2 \bullet 10^{-5}$	$8.4 \bullet 10^{-9}$	$4 \bullet 10^{-7}$
13.5 dB	$5 \bullet 10^{-7}$	$5.3 \bullet 10^{-12}$	$4 \bullet 10^{-10}$

The coding does provide improvement, about two orders of magnitude. This is characteristic for short codes—much longer code words are needed to demonstrate appreciable improvement.

PARITY-CHECK MATRIX

The received word **r** should now be checked for errors. We can take the last k bits of **r** (the data bits), multiply with {**PI** } and compare this generated word with the total **r** vector. If they are identical we can conclude that there are no single errors; the message is then found by stripping its n-k = r parity check bits. But if the comparison shows differences, then we know that there is at least one error, and the next question is to find it (them ?).

Note, during encoding:

$$\begin{aligned} \mathbf{v} &= \mathbf{u}\ \{\mathbf{PI}\} \\ &= \{\mathbf{uP}\ \mathbf{uI}\} \\ &= \mathbf{v}_r + \mathbf{u} \end{aligned}$$

where $\mathbf{v}_r$ is the parity bit sequence:

$$\mathbf{v}_r = \mathbf{u}_k \bullet \mathbf{P}$$

The received code word **r** has the R check bits $\mathbf{v}_R$. An error free transmission would result in:

$$\begin{aligned} \mathbf{v}_R &= \mathbf{v}_r \\ \mathbf{v}_R - \mathbf{v}_r &= \mathbf{0} \end{aligned}$$

where **0** is the null vector with all components equal to zero.

or, (mod-2):

$$\begin{aligned} \mathbf{v}_r + \mathbf{v}_R &= \mathbf{0} \\ \mathbf{u}_k \bullet \mathbf{P} + \mathbf{v}_R &= \mathbf{0}\,. \end{aligned}$$

Re-writing:

$$\mathbf{u}_k \bullet \mathbf{P} + \mathbf{v}_R = \begin{Bmatrix} \mathbf{u} \bullet \mathbf{P} \\ \mathbf{v}_R \bullet \mathbf{I} \end{Bmatrix} = \begin{Bmatrix} \mathbf{uv}_r \\ \mathbf{I}_r \end{Bmatrix} \ \mathbf{P} = \mathbf{0}$$

In other words, if the last vector-matrix product is zero, then the received parity checks are equal to the transmitted ones, and there are no detectable errors!

The last expression can be written:

$$\mathbf{vH}^{\mathrm{T}} = \mathbf{v}\begin{Bmatrix} \mathbf{P} \\ \mathbf{I}_r \end{Bmatrix} = \mathbf{0}$$

where:

$$\mathbf{H} = \{\mathbf{P}^{\mathrm{T}}\mathbf{I}_r\}$$

is called the parity check matrix.

A code word must therefore satisfy the condition:

$$\boxed{\mathbf{vH}^{\mathrm{T}} = \mathbf{0}} \tag{25.4}$$

Example no. 4. Is { 1 1 0 1 0 1 0 } a code word from the **G**-matrix used in example 2? The message word is { 1 0 1 0 }, assuming no errors. Its associated code word is:

$$\mathbf{v} = \{\,1\,0\,1\,0\,\}\bullet \begin{Bmatrix} 1\,0\,1\,1\,0\,0\,0 \\ 1\,1\,1\,0\,1\,0\,0 \\ 1\,1\,0\,0\,0\,1\,0 \\ 0\,1\,1\,0\,0\,0\,1 \end{Bmatrix} = \{\,0\,1\,1\,1\,0\,1\,0\,\}$$

This generated word has check bits { 0 1 1 } ≠ { 1 1 0 }, and is thus *not* a codeword. Let us use the parity-check matrix:

$$\mathbf{H} = \begin{Bmatrix} 1\,1\,1\,0\,1\,0\,0 \\ 0\,1\,1\,1\,0\,1\,0 \\ 1\,1\,0\,1\,0\,0\,1 \end{Bmatrix}$$

The word is only a proper code word if its product with the transposed H-matrix, $\mathbf{H}^{\mathrm{T}}$, equals a zero word:

$$\{\,1\,1\,0\,1\,0\,1\,0\,\}\bullet \begin{Bmatrix} 1\,0\,1 \\ 1\,1\,1 \\ 1\,1\,0 \\ 0\,1\,1 \\ 1\,0\,0 \\ 0\,1\,0 \\ 0\,0\,1 \end{Bmatrix} = \{\,0\,1\,0\,\} \neq \mathbf{0}\,.$$

The condition is not satisfied—the word does not belong to the code words generated by the generator matrix **G**.

THE SYNDROME

It may seem trivial that the product of the received code word vector and the transposed parity check matrix should equal zero when the word is a proper code word.

But the interesting fact is that the result will tell us something about the possible error, if it is different from **0**. Let us name the resulting vector the syndrome **S**:

$$\boxed{\mathbf{S} = \mathbf{r}\bullet\mathbf{H}^{\mathrm{T}}} \tag{25.5}$$

where **r** is the received word.

If the channel is noisy then **r** equals the sum of the transmitted code word **u** and a noise vector **e**:

$$\mathbf{r} = \mathbf{u} + \mathbf{e}$$

and **S** is therefore:

$$\begin{aligned} \mathbf{S} &= \mathbf{r} \bullet \mathbf{H}^T \\ &= \mathbf{u} \bullet \mathbf{H}^T + \mathbf{e} \bullet \mathbf{H}^T \end{aligned}$$

or:

$$\mathbf{S} = \mathbf{e} \bullet \mathbf{H}^T$$

It turns out that we should compare **S** with the rows of $\mathbf{H}^T$; *if it equals row no. i*, then we can correct the received word by changing bit number *i*!

Exercise no. 1. An encoder uses a P-matrix given by:

$$\mathbf{P} = \begin{matrix} 1 & 1 & 1 \\ 1 & 1 & 0 \\ 1 & 0 & 1 \\ 0 & 1 & 1 \end{matrix}$$

One received word is (1 0 1 0 0 0 1) ; it is not a code word (check). Correct the error.

EXAMPLES OF BLOCK CODES

Linear block codes are {n,k}-codes, where k message bits are changed into n transmitted {or stored} code words. There are several block codes in use in tape drives, such as the Hsiao or Patel and Hong coding schemes. Disk systems use Fire codes or Reed-Solomon codes.

A few simple codes are listed below:

Single-error correcting, t=1 {d=3}

n	k	Code	Eff=k/n
5	2	{5,2}	.4
6	3	{6,3}	.5
7	4	{7,4}	.57
15	11	{15,11}	.73

Double-error correcting, t=2 {d=5}

n	k	Code	Eff=k/n
10	4	{10,4}	.4
11	4	{11,4}	.36
15	5	{15,8}	.53

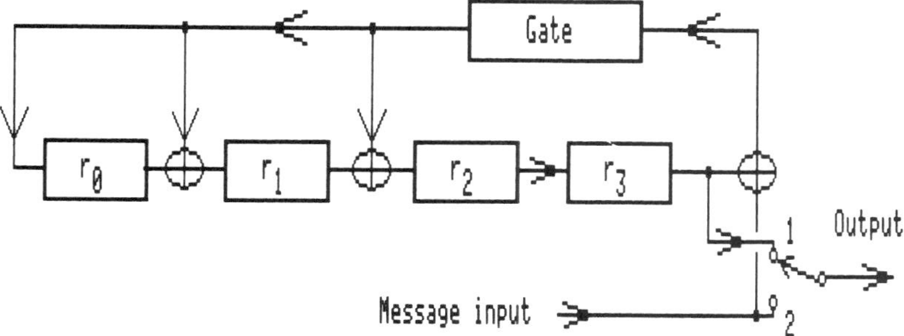

Fig. 25-7. Generation of cyclic (7,4) block code.

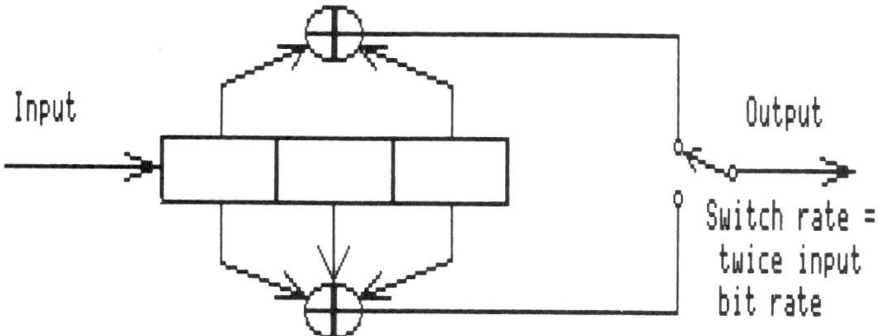

Fig. 25-8. Generation of a 1/2 convolution code.

Cyclic block codes are extensions of the codes covered in this chapter. An example of an encoder of a (7,4) code is shown in Fig. 25-7. The theory behind these codes are founded upon the algebraic concepts of rings, fields and polynomials, and would take us beyond the scope of this book.

Much longer codes are also needed to reduce the overhead, i.e. increase the efficiency. Members of these codes are the longer Fire codes, BCH (Bose- Ray- Chaudhuri- Hocquenghem) and RS (Reed- Solomon) codes. These codes use blocks consisting of many thousand bits. The interested reader is referred to books by Lin and Costello, and by Chambers.

Convolutional codes are mostly used in communication, and a simple encoder is shown in Fig. 25-8. It consists of a 3-stage shift register and two adders (mod 2). At the start of any stage the next bits from the source is fed into the right of the shift-register, pushing the left-hand bit out. Then the sum mod 2 of the first and the third bit is sent out, followed by the sum mod 2 of all three bits, providing an output consisting of bit-pairs.

A variation on the convolutional encoder was shown in Fig. 24-8 for encoding of NRZ-L data into randomized NRZ-L data.

Applications of codes for magnetic tapes and disks are covered in Chapter 16 of Lin and Costello's book.

INTERLEAVING

A very powerful technique will ensure protection against errors, using even a simple error correcting scheme. It consists of scrambling the data stream of encoded bits into a new sequence, that later is de-scrambled into the original sequence. This is illustrated in Fig. 25-9. Seven bit blocks of (7,4) codewords are stored in memory, and read out into a new memory as shown; at this point added protection can be made by generating check bits by taking mod 2 sums of the rows and columns.

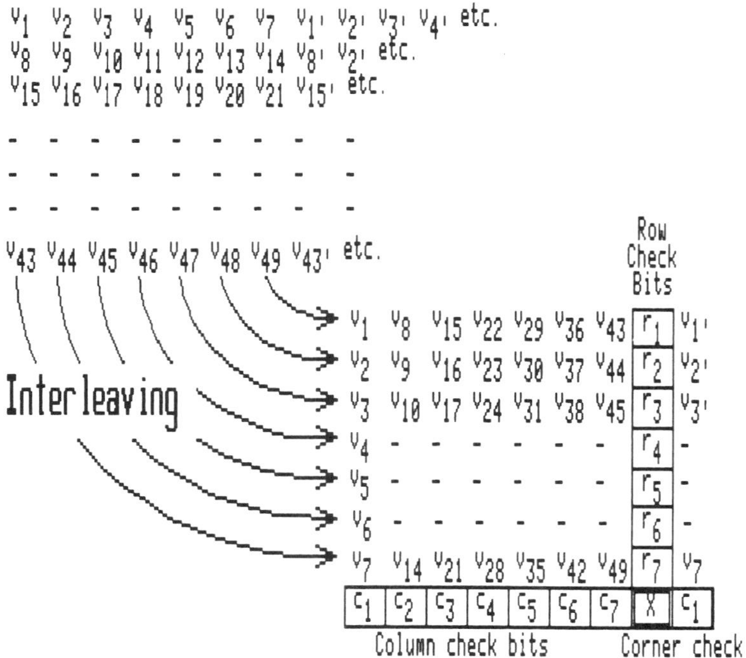

Fig. 25-9. Simplified diagram showing interleaving, column and row check bits, and corner check bit (X).

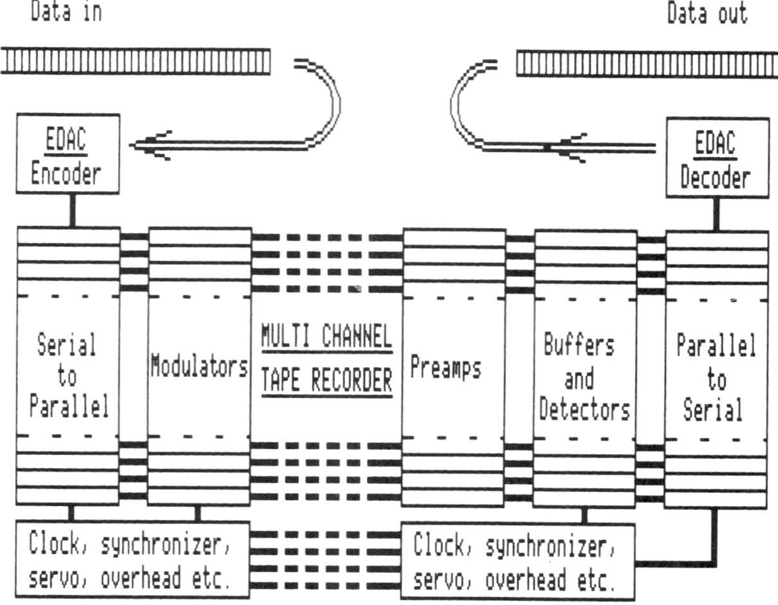

Fig. 25-10. Interleaving in HDDR recordings.

The data are now transmitted (stored/read) row by row, and the sequences have been scrambled. Now suppose that one entire row has been wiped out. After de-scrambling the upper-left pattern will have one column in error, but each word (1-7, 8-14, etc.) will have only one bit in error, which can be detected and corrected.

This method offers protection against burst errors, in single channel as well as multi-channel applications. The latter is illustrated in Fig. 25-10, illustrating the signal path through a HDDR recorder. These recorders accept unusually high data rates, and the data stream is first encoded to protect against errors, and then divided into a large number of channels. Each of these are now modulated into one of the codes from Chapter 24, and recorded on the tape.

The playback signals are amplified and temporarily stored in a buffer. They are clocked out in step with a reference clock (crystal controlled), thereby removing any time base errors. The buffers are monitored by a circuit that will signal the capstan servo to speed up or slow down. After detection (demodulation) a parallel- to- serial conversion follows. The resulting data stream is finally decoded and errors detected and corrected.

This scheme of EDAC plus interleaving is very powerful. Part of the overhead caused by the EDAC can even be recovered by not recording some of the tracks; the bit error rate will still be better than one in a trillion.

REFERENCES TO CHAPTER 25

Lin, S., and Costello, D.J., *Error Control Coding: Fundamentals and Applications*, Prentice-Hall, 1983, 603 pages.

Chambers, W.G., *Basics of Communications and Coding*, Clarendon Press, Oxford, 1985, 240 pages.

BIBLIOGRAPHY ABSTRACT ALGEBRA

Singh, J., *Great Ideas of Modern Mathematics*, Dover, New York, 1959, 312 pages.

Fraleigh, J.B., *A First Course in Abstract Algebra*, Addison-Wesley, Reading, Mass., 1976, 478 pages.

Sawyer, W.W., *Concrete Approach to Abstract Algebra*, Dover, New York, 1978 (Original 1959), 234 pages.

McClellan, J.H., and Rader, C.M., *Number Theory in Digital Signal Processing*, Prentice Hall, Inc, Englewood Cliffs, N.J., 1979, 276 pages.

Hall, F.M., *An Introduction to Abstract Algebra, Vol. 1 and 2.*, Cambridge University Press, Cambridge 1980 and 1981, 388 and 300 pages.

Cohn, P.M., *Algebra*, Vol. 1, John Wiley & Sons, 1982, 410 pages.

Childs, L., *A Concrete Introduction to Higher Algebra*, Springer-Verlag, 1983, 340 pages.

Schroeder, M.R., *Number Theory in Science and Communication*, Springer-Verlag, 1984, 324 pages.

BIBLIOGRAPHY, ERROR CORRECTION AND DETECTION CODING

Peterson, W.W., and Welden, E.J., *Error Correcting Codes*, M.I.T. Press, Jan. 1972.

Patel, A.M., "Signal and Error-Control Coding, Chapter 5, *Magnetic Recording*, Vol. 2 and/or Vol. 3, McGraw Hill, 1988.

Glover, N., *Practical Error Correction Design for Engineers*, Data Systems Technology Corp., 1801 Aspen Street, Broomfield, CO 80020, 1982, 379 pages.

Berlekamp, E.R., "The Technology of Error-Correcting Codes," *Proc. of IEEE*, May 1980, Vol. 68, No. 5, pp. 564-593.

Bhargava, V.K., "Forward Error Correction Schemes for Digital Communications," *IEEE Communic. Mag.*, Jan 1983, Vol. 23, No. 1, pp. 11-19.

Blahut, R.E.B., "Algebraic Fields, Signal Processing, and Error Control," *Proc. of IEEE*, May 1985, Vol. 73, No. 5, pp. 874-893.

Glover, N., "Error Correction and Detection (EDAC) Schemes," *Symposium on Mem. and Adv. Rec. Tech.*, Paper No. WS 4-C, May 1986, 17 pages.

Melas, C.M., "A New Group of Codes for Correction of Dependent Errors in Data Transmission," *IBM Jour. Res. Devel.*, Jan. 1966, Vol. 10, No. 1, pp. 58-65.

Brown, D.T., and Sellers, F.F., Jr., "Error Correction for IBM 800-bit-per-inch Magnetic Tape," *IBM Jour. Res. Devel.*, July 1970, Vol. 14, pp. 384-389.

Franaszek, P.A., "Sequence-State Methods for Run-Length-Limited Coding," *IBM Jour. Res. and Dev.*, July 1970, Vol. 14, pp. 376-383.

Patel, A.M., and Hong, S.J., "Optimal Rectangular Code for High Density Magnetic Tapes," *IBM Jour. Res. Devel.*, Nov. 1974, Vol. 18, pp. 579-588.

Varaiya, R., "On Ensuring Data Recoverability by the Use of Error-Correcting Codes," *IEEE Trans. Magn.*, July 1978, Vol. MAG-14, No. 4, pp. 207-210.

Hodgart, M.S., "High Performance Low Redundancy Serial Error Correcting Codes for Audio Digital Recorders," *Intl. Conf. Video and Data 79*, July 1979, IERE Publ. No. 43, pp. 163-177.

Swanson, R., "Matrix Technique Leads to Direct Error Code Implementation," *Computer Design*, Aug. 1980, Vol. 19, No. 8, pp. 101-108.

Lee, R., "Cyclic Code Redundancy," *Digital Design*, July 1981, Vol. 11, No. 7, pp. 77-85.

Meeks, L., "A Dropout-Based Error Correction Method for Use on High Density Digital Recording Systems," *Intl. Conf. on Video and Data 1982*, 22 pages.

Nelson, B., "Effortless Error Management," *Computer Design*, Feb. 1982, Vol. 21, No. 2, pp. 163-168.

Rennie, L.J., "Forward-Looking Error Correction via Extended Golay," *Computer Design*, June 1982, Vol. 21, No. 6, pp. 121-130.

Shenton, D., DeBenedictis, E., and Locanthi, B., "Improved Reed Solomon Decoding Using Multiple Pass Decoding," *Preprint No. 2035 (A-10), AES Convention*, Oct. 1982, 9 pages.

Galen, P.M., "Disc Drive Error Detection and Correction Using VLSI," *Hewlett Packard Journal*, Jan. 1984, Vol. 35, No. 1, pp. 12-13.

Parker, M.A., "A Range of Combined Error Correction and Recording Channel Code Schemes," *Intl. Conf. Video & Data 84*, Apr. 1984, No. 59.

Howell, T.D., "Analysis or Correctable Errors in the IBM 3380 Disk File," *IBM Jour. Res. and Devel.*, March 1984, Vol. 28, No. 2, pp. 206-211.

Watkinson, J.R., "Error Correction Techniques in Digital Audio," *Intl. Conf. Video & Data 84*, Apr. 1984, No. 59.

Couvreur, C., and Piret, P., "Codes Between BCH and RS Codes," *Philips Journ. of Research*, May 1984, Vol. 39, No. 4-5.

Adi, W., "Fast Burst Error-Correction Scheme with Fire Code," *IEEE Trans. COMP.*, July 1984, Vol. 33, No. 7, pp. 619-625.

Chi, C.S., "Triplex Code for Quartenary High Density Recording," *IEEE Trans. Magn.*, Sept. 1984, Vol. MAG-20, No. 5, pp. 888-890.

Hsu, I.S., Reed, Il S., Truong, T.K., Wand, K., Yeh, C.S., and Deutsch, L.J., "The VLSI Implementation of a Reed-Solomon Encoder Using Berlekamp Bit-Serial Multiplier Algorithm," *IEEE Trans. on Computers*, Oct. 1984, Vol. 33, No. 10, pp. 906-912.

Isailovic, J., ''Codes for Optical Recording,'' *Proc. of SPIE*, Jan. 1985, Vol. 529, pp. 161-168.

Erdel, K., ''Burst Error Correction for Digital 34 Mbit/s TV Signal Transmission with a Fire Code,'' *Frequenz*, June 1985, Vol. 39, No. 6, pp. 165-169.

Ferreira, H.C., ''The Synthesis of Finite State Magnetic Recording Codes with Good Hamming Distance Properties,'' *IEEE Trans. Magn.*, Sept. 1985, Vol. MAG-21, No. 5, pp. 1356-1358.

Ohr, S., ''Error Checking and Correcting IC Slashes Optical Disk Defects,'' *El. Design*, Dec. 1985, Vol. 33, No. 29, p. 37.

Gillard, C.H., ''Error Correction Strategy for the New Generation of 4:2:2 Component Digital Video Tape Recorders,'' *Intl. Conf. on Video, Data and Audio 1986*, IERE Publ. No. 67, pp. 165-175.

Part 8

Applications

Chapter 26

Computer Data Storage and HDDR

Direct recording is the elementary recording process whereby the magnetic remanence on the media is determined by the data current flowing through the write (record) head winding. Direct recording is applied in nearly all computer tape or disk storage units.

Tape storage offers the largest capacity for data, but are slower in finding data since they are stored in the sequence they arrived in at the recorder. Disk systems offer almost instantaneous access. The method of storage on tapes is known as *Sequential Access Storage* (SAS) while storage on disks is categorized as *Direct Access Storage* (DAS). We will discuss both SAS and DAS in this chapter, along with a detail description of storage on a 5¼ inch microcomputer floppy disk.

Direct recording has its proper home in digital recording, used for storage of sequences by bytes, each byte being equal to eight bits. It is an indispensable technology within computers, and it has entered the fields of audio, instrumentation and television recording (see following chapters). This adoption of digital recording provides the ultimate in signal-to-noise ratio, accuracy of signal waveforms and freedom from tape transport flutter.

Digital storage on magnetic tapes and disks has since 1940 been challenged by attempts to apply all conceivable physical effects to store On-Off signals. A few of these have found limited areas of practical use, leaving magnetic recording in clear superiority. Only optical recording will add a dimension of very large direct access files; the merit of optical recording lies in a very high track density, about 15,000 TPI which is an order of magnitude greater than the current state-of-art 1,400 TPI in a rigid disk drive system.

Recent years have seen a tremendous upsurge in applications and development efforts in digital magnetic recording in conjunction with the advances in the semiconductor industry.

The result of these efforts becomes clear when we examine both status and trends of some figures-of-merit, used to characterize digital storage devices. The most important of these are:

Storage Capacity in mega-bytes (one byte = 8 bits, one mega-byte = $8 \times 10^{+6}$ bits).

Access Time in milliseconds is the time the device spends in locating requested information. It may have to search through a portion of a reel of tape, or move the magnetic head assembly in a disk file.

Average Access Time is the average of a large number of times it takes to find the information. The average would seem to be equal to the time it would take to seek across half the number of tracks on a disk drive. But it fails to take into consideration that the DOS (Disk Operating System) tries to organize the data in such a way that data to be read sequentially will be stored on sequential tracks; much of the time we therefore only have to move one track distance.

The average access time is by the industry defined as the length of time required to seek across one-third of the tracks on a disk. It typically has values of 20 to 30 ms for high performance disks, 100 ms for the average microcomputer disk, and 500 ms for the CD-ROM.

Transfer Rate in bytes per second is the speed with which bytes (equal to eight data bits each) are stored (recorded, written) or retrieved (played back, read). Modern storage devices have rates from 100 kBytes/sec to 10 Mbytes/sec; the helical scan recorders go as high as 40 Mbytes/sec.

Error Rate is a measure for the accuracy of the data. If a device is rated with a BER (Bit Error Rate) of 10^{-7} it simply means that one out of every 10 million bits may be in error. This is, of course, an average count since errors are found to be of random nature. It is further assumed that the quoted rate is for the raw data, i.e., without error correction. The BER may be 10^{-12} with EDAC (Error Detection and Correction).

Table 26-1. Summary of State-of-Art Storage Devices, Spring 1988.

TAPE (SAS)	TPI	Capacity MBytes	BPI (bits)	Transfer rate MBytes/sec
1/2" Reel-to-reel	18	165	6,250	1
1/2" Cartridge (IBM 3480)	36	200	38,000	3
1/4" Cartridge (DC 600A)	96	150	10,000	.1-.5
1/4" Cassette	36	20	10,000	.1-.3
1" HDDR (15" reel)	100	100,000	66,000	.5 per track
1/2" VHS format	700	5,500	50,000	1
8 mm video format	1,000	2,300	60,000	1
R-DAT	3,000	1,000	70,000	.3
DISK (DAS)				
14" Hard disk ([illegible]de)	1,400			1.5
8" Hard disk "	1,400			1.5
5-1/4" Hard disk ""	1,400	50	30,000	1.5
3-1/2" Hard disk ""	1,400	25	30,000	1.5
5-1/4" Floppy w/servo	400	20	18,000	.2
5-1/4" Floppy	80	.4	6,000	.1
3-1/2" Floppy	135	1.2	10,000	.1
2" Floppy	200	1	20,000	.2
WORM (DAS) 12"disk	15,000	6,800	18,000	1

Two measures are used to increase the storage capacity of magnetic tapes or disks: An increase in the number of bits recorded per length along a track (*bits per inch*) and a decrease in track width to allow for more tracks per width of tape, or per disk (*tracks per inch*). When multiplied together they result in a number with unit of *bits per square inch*, which may be used to compare magnetic recording with semiconductor LSI, magnetic bubbles and optical storage disks.

The media cost is in the order of one dollar per megabyte, falling to about two tenth of a cent per megabyte in the new VHS format digital recorders.

There exists today a large number of different tape and disk drives on the market. A comprehensive survey is beyond the scope of this book and the following discussion will therefore be of a general nature. Up-to-date information about equipment specifications can be obtained from technical magazines, exhibits and manufacturer's literature. Table 26-1 summarizes the state-of-art storage devices early 1988. While the author has made every effort to write the book to have lasting accuracy and value, this table may be looked upon with a hind-sight smile by readers in the 90's.

The matter of data organization and flow is described in several books and papers on computers. Matick's book on Computer Storage Systems and Technology is recommended.

SAS AND DATA ORGANIZATION

In SAS, data is written and read in a sequential fashion on a moving tape, starting at the beginning of a reel of tape, cassette or cartridge. The operation may be continuous or intermittent, and the data is written in *blocks*. These blocks of data may consist of any reasonable number of *records*, where each contains a number of *bytes*.The blocks are separated by sections that contain addressing and synchronization information.

The *Stored Addressing Information* (SAI) contains location and serial number of the record (data), synchronization and deskew information. Allowance must also be made for the starting and stopping of the tape which results in an unrecorded portion of the tape called an *Inter Record Gap* (IRG). These gaps contain no data and are therefore made as short as possible and used sparingly. The physical length of the gap, or unrecorded portion, is 0.15 inches in some modern computer tape drives, as compared with 0.75 inches in earlier drives.

When the tape is started (from the center of an IRG) the first information read is a *preamble*. The preamble synchronizes the clock and detection circuits. Next follows a *deskew-byte*, and then the record. The end of a record is signalled by a *postamble*, which typically consists of an empty space followed by a *longitudinal parity* check and synchronizing signals to allow for reverse operation.

Figure 26-1 shows how blocks of records are separated by SAI's, which include postambles, IRG's and preambles.

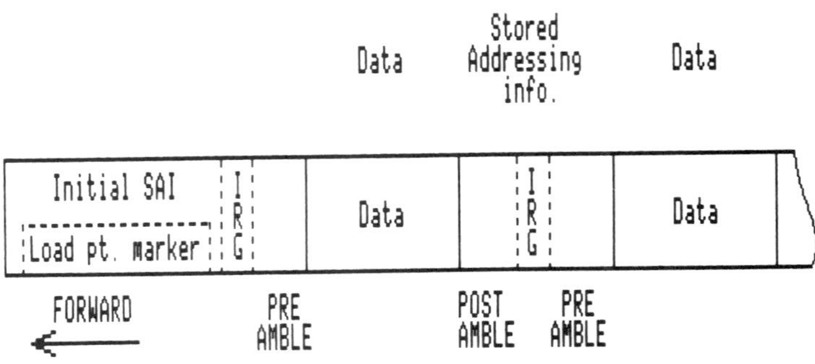

Fig. 26-1. General format for data and SAI on tape (after Matick).

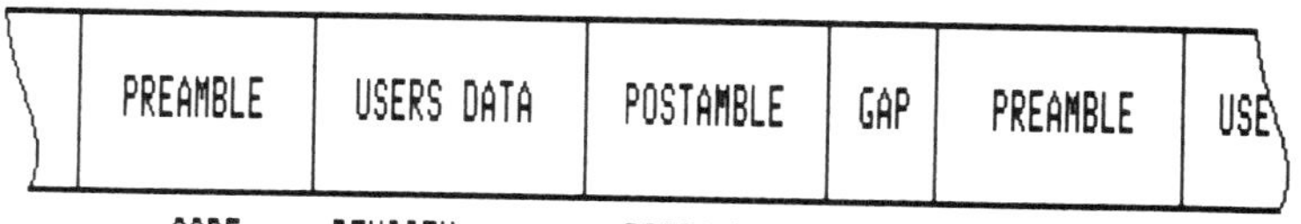

PREAMBLE	USERS DATA	POSTAMBLE	GAP	PREAMBLE	USE

CODE	DENSITY	POSTAMBLE	GAP	PREAMBLE
NRZI	800 BPI	CRC & LRC	0.6 in.	--
PE	1,600 BPI	41 Bytes	0.6 in.	41 Bytes
GCR	6,250 BPI	CRC, ECC, 80 Bytes	0.3 in.	80 Bytes

Fig. 26-2. Standard ½-inch tape formats.

The data is a serial stream of any of the number and symbol codes (6 bits BCDIC, 7 bits ASCII or the ASCII-8 (Extended BCDIC). They are also written as a serial stream in single track tape units, such as the cassette or cartridge drives. An odd parity check may be included after one or more bytes.

Larger computer drives use 7, 9 or 18 track parallel recording of the bytes. This allows for an odd parity bit for each byte and this will, in conjunction with an even, serial parity bit for each track, allow for error detection and correction. Standard tape formats are shown in Fig. 26-2.

It is common practice, during writing, to re-write if any errors occur.

The sequences of "1" 's and "0" 's are written in one of the earlier discussed codes. Most common is NRZI for low packing densities and *phase encoding*, or PE for higher densities. PE is defined as a code where a "1" is a transition to positive magnetization and an "0" is a transition to negative magnetization. A string of "0" 's or "1" 's will produce extra transition at each bit cell end and a higher frequency response required. The PE code is self-clocking. Both NRZI and PE codes are self-erasing. Other codes are the MFM, RLL, ⅘ GCR and 2,7 codes.

Tape drives have recently found application in the form of high capacity *streamers*, i.e., constant speed transports for storing the information from hard disk drives to provide backup. Tape formats are shown in Figs. 26-3 and -4.

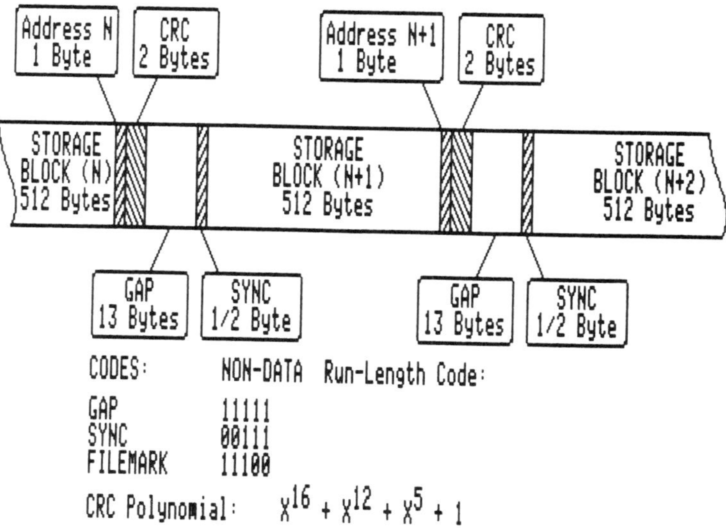

Fig. 26-3. Streaming-cartridge tape format.

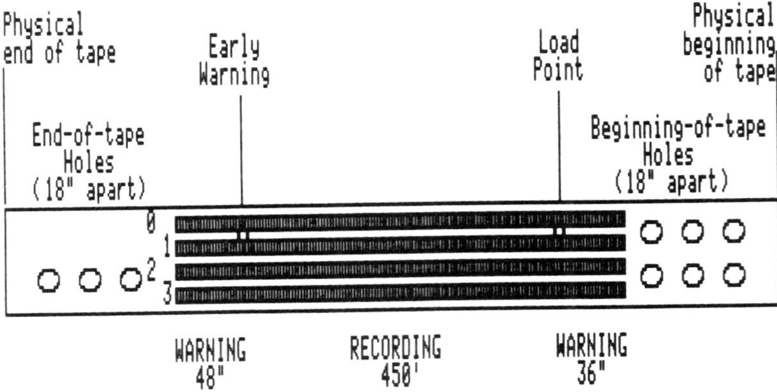

Fig. 26-4. Tape position holes, ¼-inch cartridge tape. Holes are sensed with light sources.

DAS AND DATA ORGANIZATION

DAS utilizes storage devices with one or more parallel disks for storing data. It is possible to get access to any data in a time that is equal to the travel time of a radially moving head plus half the time of a revolution of the disk (on the average). The first disk drive used a 14″ aluminum disk with a magnetic coating on each side. Several of these may be stacked together on a common spindle. They were introduced in the 60's by IBM as the *Disk-Pack*. The disks are rotated at a speed of 3-3600 RPM and can store anywhere from 1 to 40 mega bytes per disk surface.

There are many other disk sizes in use. Their operation and characteristics are generally identical to each other.

In 1970 IBM introduced the *Diskette*, which is a square envelope containing an 8″ diameter circular disk. An opening in the center of the envelope allows for a spindle to engage with the disk and rotate it at a speed of 300 or 360 RPM. Another opening, placed radially, allows for a movable head assembly to contact the disk surface(s) for the write or read operation.

The Diskette was introduced as a read-only device for loading diagnostic routines into IBM's 3300 Disk Controller. It became an industry standard in 1973 and was nicknamed the *Floppy*.

Shugart Associates introduced a 5¼″ floppy in 1976 and it quickly became popular for the micro computer market. In the eighties a smaller version, the 3½ inch diskette was introduced and standardized. It is shown in Fig. 13-20.

The 14 inch disks have decreased in size while simultaneously increasing their storage capacity. Today the 5¼ and 3½ inch hard disk drives are becoming very popular storage devices.

Supplementary mass storage in the form of optical disks called WORMs (Write Once Read Many) are finally making inroads on the market after many difficult years of development. Their potential is the storage of multiple computer programs, permanent data (catalogs, dictionaries), and many other areas.

Data Organization

In DAS, data is written and read sequentially on circular tracks on a *rotating disk*. It is practice to use only one or two heads per side of a disk, although systems with a *head-per-track* are available. The latter offers the fastest access time (no head movement, the time is, on the average, equal to the time for a half rotation of the disk). There are also disk systems with nine parallel tracks operation, which allows for faster transfer rates by recording bytes in parallel on interleaved tracks.

Some form of indexing is required as a starting point for finding the written data. The simplest is to apply a small notch in the disk. Since 2 or more disks of the standard 14″ diameter type are always used, the lower disk is utilized for an index marker plus sector markers. The smaller 8″ and 5¼ ″ *diskette* or *floppy* has a small hole notched near the center of the disk.

This once-per-revolution marker gives a starting point for data to be written (or read) at each track. All following data blocks on a track must then have assigned data bytes for unique location and identification. This method is called *soft-sectoring*, which is software controlled (i.e., written into a program that initially formats a disk). It offers a maximum utilization of the recording surface for data and it is a de facto standard, although the number of sectors per disk revolution varies among systems.

DATA STORAGE ON DISK SYSTEMS, MICRO COMPUTERS

A disk controller, part of the Disk Operating System (DOS), assists the disk drive in storing data and accounting for their whereabouts on the disk surface(s). The bit stream is divided up into chunks of typically 512 bytes each that are stored in individual sectors on the disk surface. The sector size used in the IBM DOS 1.0 - 3.2 is 512 bytes while Apple Computer's ProDOS uses 256 bytes sectors. The reason is that the 6502 microprocessor in the Apple computers has addressing modes that are best suited for manipulation of data blocks up to 256 bytes in size.

Several sectors are located around each track on a disk. Their number is only limited by the packing density of the particular disk drive: IBM uses 9 sectors per track on the double-sized, double-density 5¼ inch diskette (9 * 512 = 4,608 bytes per track), while Apple uses 16 sectors of 256 bytes each (16 * 256 = 4,096 bytes per track). A typical hard disk uses 17 sectors of 512 bytes each per track.

The maximum number of tracks possible on a flexible diskette was discussed in Chapter 10 (Select Trackwidth), and IBM uses 40 tracks per side and Apple uses 35. That brings the total storage capacities to 2 * 40 * 4,608 = 368,640 and 1 * 35 * 4,096 = 143,360 bytes, respectively. Hard disks have several hundred tracks per recording surface.

The following sections will further discuss the disk system as applicable for the IBM PC, XT and AT computers. Readers interested in details are referred to the books/disks by Norton, and for those who wish similar coverage for the Apple II computers will find it in the books by Worth and Lechner, Doms and Weishaar and Sather.

Interfaces to computers such as SCSI and ESDI will not be covered (SCSI—Small Computer Systems Interface operating at 1.5 Mbytes/sec; ESDI—Enhanced Small Disk Interface operating at 10 Mbytes/sec).

Formatting Disks; Sector Addressing

A good deal of housekeeping is required to write and read data to and from the sectors on a disk. It is necessary to know where the data start (what sector), and how long the stream of bits is, and where it ends. Some form of numbering of the sectors must be done, a directory (DIR) for the files on a disk generated, and a look-up table for the directory to find exactly where there are empty sectors that data can be written to, and later read from; the look-up table is called a FAT—File Allocation Table.

A blank disk must therefore be prepared for data storage by a process called formatting. Each sector will have space for the 512 bytes, and will in addition have an address and a synchronization pattern up front, separated by gaps from the prior data, and from the sector's own data.

The data are not necessarily laid down in chunks of 512 bytes since this may lead to very large DIR and FAT files on large disk systems (10 MB and up). The data are therefore written and read in clusters of sectors, typically one or two for the flexible disks, and as many as 16 for a 30 MB hard disk. The new high density 1.2 MB disks use one cluster for each sector in order to reduce wasted space at the end of each cluster (see later: slack).

The FAT has entries for clusters, not sectors. The entries are two-byte words that will advise of:

- Availability.
- In use (with number for the next cluster in the file).
- Bad cluster—do not use.
- Last cluster in file.

After this brief introduction to data organization let us see how we keep track of data by numbering things. First of all, the disk surfaces are assigned numbers: Zero for the top surface and one for the bottom side; if several disks on the same spindle are used the numbers continue with two for the next top surface, three for its bottom etc. (see Fig. 26-5).

Numbers are next assigned to the tracks, with the outermost track being number zero, the next inside number one etc. These numbers apply to similar tracks on the other disk surfaces, and one can therefore say that all tracks number X belong to *cylinder* X, that consists of all tracks No. X on the disk surfaces.

Now we are at the sectors: They start with number 1, somewhere on the periphery of track zero. Early disk drives had a mechanical marker (notch with magnet for induction of voltage, or a hole for light to shine through), and it would indicate the beginning of sector number one, with the others to follow. That method was called hard sectoring, and not used in many drives today since the controller does examine the read signal for the proper sector address. The sector address is therefore one of the items that must be generated during the preparation of the disk, the formatting process.

We finally arrive at the numbering of clusters, consisting of one or more sectors. Here we start with number two.

During formatting the disk spins with a nominal RPM (300 for flexible disks, 3,600 for hard disks), and the sectors are laid down with gaps between them. This will allow for later speed

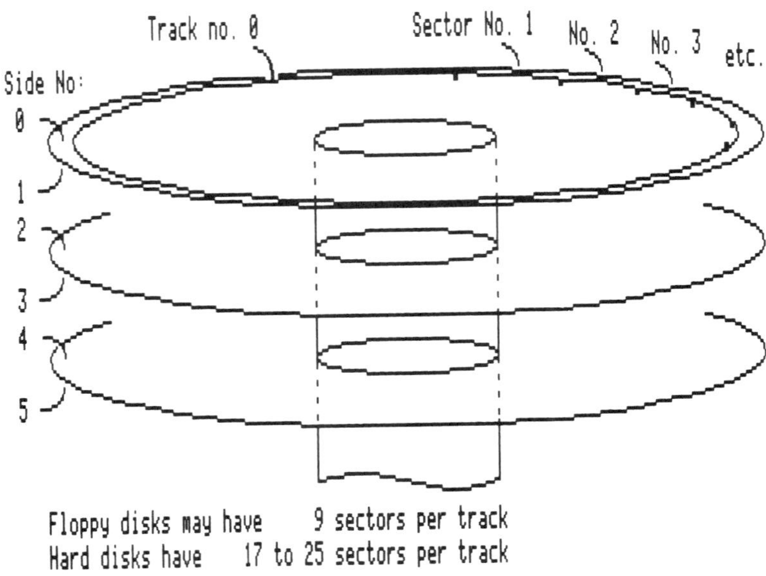

Fig. 26-5. Numbering system for disk surfaces, tracks, and sectors.

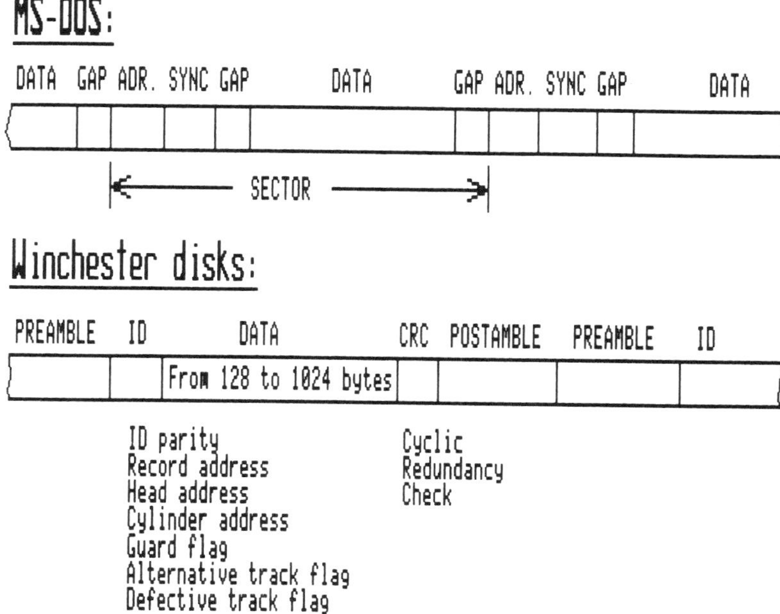

Fig. 26-6. Sector formats for the IBM MS-DOS diskettes, and for general hard disks.

variations, and up to three percent must be acceptable. Re-writing data on a disk that spins too fast will result in longer sectors, with the danger of overlapping (disastrous!). Hence gaps between the sectors are employed.

Each sector is formatted by first writing a gap, then a two byte address, a synchronization pattern (for use by the read circuit), another gap, and then a string of artificial data, using hexadecimal F6, which binary is 11110110 (see Fig. 26-6). It is then possible to verify the integrity of each sector during a following read cycle to assure that the read pattern is F6 also. If the sector fails that test it is then marked in the FAT as a bad sector, and will not be used in later data storage.

Sector Usage

The first sectors on all disks are used for housekeeping information. When a disk is first read by a disk operating system it provides some basic data about its own organization, stored in a BOOT record in—always—sector 0. The next sectors are used for the FAT area (sectors 1-4 on flexible disks, up to 1-16 on a 30 MB hard disk).

The DIR file is largest, using sectors 5-11 on flexible disks, allowing for 112 entries, each up to 32 bytes long. The DIR occupies sectors 17-80 on one 30 MB disk, allowing for 1024 entries.

The BOOT, FAT and DIR files do therefore not consume a large amount of disk space: 1.5 to 0.2 percent in the above cases. There is however a fair amount of space wasted by writing data in clusters of from one to many sectors each. This extra, wasted space is called *slack*, and can be a high percentage of the total disk space.

It arises from the fact that each file on a two-sector cluster system occupies integer multiples of 1024 bytes. Each file will therefore always have a slack. If the first file's length is 1040 bytes then there will be data in 1024 bytes of cluster No. 2, and only 16 bytes of cluster No. 3, leaving 1008 bytes filled with F6. The slack is 1008/20.48 = 49.2 percent.

Note that the file in a one sector cluster scheme will occupy 3 sectors, while it occupies 4 sectors in the two-sector cluster method. Hard disks use several sectors for each cluster, and a file will therefore appear to grow in size when copied onto a hard disk.

Problems can arise in the sector allocation when for instance a data base changes, or a program is revised and possibly expanded. The program was originally written as cluster No. 5 (clusters are assigned in sequence with new data blocks, starting with number 2). By resaving it there is not enough room in cluster No. 5, and the space following is occupied by cluster No. 6 etc. The FAT looks for empty sectors, and places the end of the program there.

This does not matter for the later readout, but the *fragmentation* of the program means that it will take longer time to read since it must search twice for sectors.

After prolonged use of particular hard disks during program development and/or changing data bases, the files may become severely fragmented, and the access times have increased to annoyingly long periods. The remedy is a restoration of the files, in which they are first backed up, and then re-written (restored) file by file, NOT sector by sector. There are several programs on the market for that sort of thing.

The data types are several, from program files to text files (ASCII), record of variable or fixed lengths, and binary picture files etc.

CRC; Interleaving

For each sector written a Cyclic Redundancy Check is made by taking the sum of the bytes and performing a division with a polynomial. The result is written at the end of each sector, and when it is read later the same CRC is generated and compared with the one written, and an error flag set if they are different. If another pass fails to read the data correctly an error has been encountered, and the process will stop with the dreaded message about I/O error printed on the screen.

Sectors are read into the controller's memory for the CRC check, and the memory is not immediately available for the next sector's data. They will nevertheless be there due to the

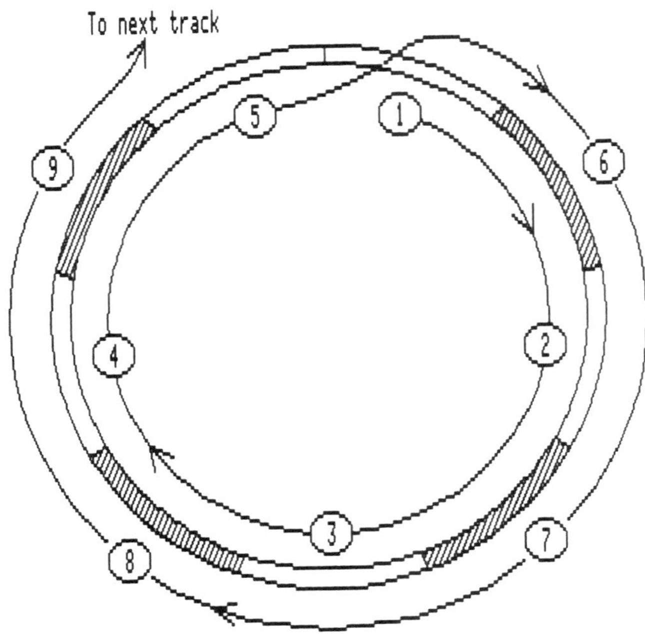

Fig. 26-7. Interleaving reduces access time.

disk's continuous rotation. If that sector had the next data then it will be necessary to wait for a full revolution of the disk to read it, perform the CRC check etc.

By sequencing the sectors within a given cluster, or group of clusters, to allow for skipping every other sector then writing and reading can be speeded up considerably. Figure 26-7 illustrates the method, where single sector clusters in sequence 2-3-4-5-6-7- etc. utilize sectors 1-3-5-7-9-2 and thereafter -4-6-8 → next side 10-12- etc.

File Entries in Directory

The directory, also called root or main directory, has entries for each file consisting of 32-byte words. Each word contains information about:

Description	Bytes	Format
Filename	8	ASCII characters
Extension	3	ASCII characters
Attributes	1	Each bit represents an attribute
Reserved	10	Unused
Time	2	Word,coded
Date	2	Word,coded
Starting FAT entry	2	word
File size	4	Long integer

The attributes represent one of several special properties of a file (read-only, hidden, subdirectory etc.).

HDDR—HIGH DENSITY DIGITAL RECORDING

Recording of very high data rates exceed the transfer rates of existing instrumentation recorders and the problem of recording them was solved in a straightforward manner by converting the *serial stream into parallel streams* that all would be within the recorder's transfer rate (Fig. 27-14).

The number of tracks required is determined in a straightforward fashion once the bit rate per channel has been determined. The data stream is normally in the NRZ-L digital logic format, which is not well suited for recording onto an analog recorder channel. The most severe limitation is the absence of response to DC, and it becomes necessary to encode the data stream into a format that is DC free, while keeping the new bit rate as low as possible (see Chapter 24).

The current packing densities are 67 kilobits-per-inch (kbpi), which equals 33.5 kHz/inch (4 MHz at 120 IPS). Higher densities (up to 200 kbpi) are at the laboratory stage, using refined heads (very short gaps) and modern tapes with high coercivity.

Two methods of recording are used, direct pulse recording (saturation) and AC bias recording. At the band edge there is very little difference between the two methods and the choice is often for no-bias direct recording because of simplicity (and lower cost). The recording level of the pulse signals is then much higher (approx. 20 dB) than with AC bias and it becomes a source of excessive channel-to-channel crosstalk during recording.

When crosstalk must be minimal, then the AC bias is used. This is for instance the case the HDDR tracks are interleaved with other analog tracks (direct or FM). Proponents of AC bias also claim to have better control over the phase response, which is a more important parameter than amplitude response in pulse recording. It is necessary to synchronize the data stream and the bias source, and that the bias frequency is an integer multiple of the clock frequency.

Recent HDDR recorders utilize rotating head recorders, as indicated in Table 26-1.

STANDARDS

The exchange of recorded information between data centers and users require that the media formats are interchangeable. They must be mechanically alike and written in the same format and these factors are controlled by national and international standards.

The standards are often set by the configurations used by IBM. This pertains to media size (tape width, reel sizes, cartridges and disk pack dimensions) and data format (track numbers, code, packing densities). IBM does not generate and issue the standards. They are generated and controlled by working committees with members from major manufacturers and users. Such committees are organized by several organizations here and abroad. The interested reader can write for free catalogs on recording standards from:

ANSI-American National Standards Institute
1430 Broadway
New York, NY 10018
Telephone (212)-354-3300

Their catalog covers the standards from ANSI and ISO (International Standards Organization).

ECMA-European Computer Manufacturers Association
144 Rye du Rhone
1204 Geneva, Switzerland

Testing and measurement procedures are spelled out in the pertinent specifications. These tests are generally related to the overall write/read unit. Separate testing of the tape or disk for amplitude uniformity and freedom from drop-out should be performed with additional testing of the timing errors in the media drive system.

HDDR systems have been standardized for the Telemetry Ranges in IRIG Standard 106-86, specifying NRZ-L for standard bandwidth systems, and RNRZ-1 (Randomized NRZ-L) for extended bandwidth systems.

The X3B6 Committee of the American National Standards Institute and its international counterpart, the ISO TC97-SC12 Committee, have prepared several standards for instrumentation magnetic tape recording, including HDDR. Status of the documents may be obtained by contacting the X3B6 Committee Chairman by writing ANSI.

REFERENCES TO CHAPTER 26

Matick, R.E., *Computer Storage Systems and Technology*, John Wiley Sons, June 1977, 667 pages.

Norton, P., *Inside the IBM PC*, 1983, Brady Co. (Prentice-Hall), 302 pages.

Norton, P., *The Norton Disk Companion*, 1987, Peter Norton Computing, Inc., 36 pages.

Norton, P., *The Norton Utilities Manual*, Vers. 4.0 and Advanced Edition, 1987, Peter Norton Computing, Inc., 174 pages plus two disks.

Doms, S., and Weishaar, T., *ProDOS—Inside and Out*, TAB Books, 1986, 260 pages.

Worth, D., and Lechner, P., *Beneath Apple ProDOS*, 1985, Quality Software, 21601 Marilla St., Chatworth, CA 91311, 368 pages plus one disk.

Sather, J., *Understanding the Apple IIe*, 1984, Quality Software, 21601 Marilla St., Chatworth, CA 91311, 280 pages.

BIBLIOGRAPHY TO CHAPTER 26

Information about the current state of the memory industry and their products can be monitored in the journals *COMPUTER DESIGN* (Advanced Tech. Group, PO Box 457, 119 Russell St., Littleton, MA 01460) and *MINI-MICRO SYSTEMS* (Cahners Publ. Co., 275 Washington Street, Newton, MA 02158).

Isailovic, J., *Videodisc and Optical Memory Systems*, Prentice-Hall, 1985, 350 pages.

Majumder, D.D., and Das, J., *Digital Computers' Memory Storage*, Halsted Press (Wiley), June 1984.

Streaming, 1982, Archive Corp., 174 pages.

Middlehoek, S., George, P.K., and Dekker, P., *Physics of Computer Memory Devices*, Academic Press, November 1976, 397 pages.

Madnick, S.E., and Donovan, J.J., *Operating Systems*, McGraw-Hill, 1974, 640 pages.

Principle of Digital Recording

Kagan, B.M., Adasko, W.I., and Pure, R.R., *Magnetomotorische Speicher fuer elektronische Datenverarbeitungsanlagen*, Akademische Verlagsgesellschaft, Leipzig, June 1973, 254 pages.

Wiehler, G., *Magnetic Peripheral Data Storage*, Siemens (Heyden), June 1978, 297 pages.

Hoagland, A.S., *Digital Magnetic Recording*, R.E. Krieger Publ. Co., Malabar, Florida, June 1983, 154 pages.

Huber, W.D., "Maximization of Linear Recording Density," *IEEE Trans Magn.*, September 1977, Vol. MAG-13, No. 5, pp. 1208-1210.

Gregory, W.T., "Write and Read Recovery Systems for a Half-Inch Tape Drive," *Hewlett Packard Journal*, March 1985, Vol. 36, No. 3, p. 16.

Mee, C.D., and Daniel, E.D., *Magnetic Recording*, Vol. 2, "Computer Data Storage" McGraw-Hill, 1988, 408 pages.

Tape Systems

Cory, C., and Jackson, G., "Efficient I/O Unleashes Benefits of Open Bus Concept," *Computer Design*, Oct. 1984, Vol. 23, pp. 157-165.

Ewins, A.J., "Micro-controlled Cassette Recorder," *Electronics and Wireless World*, Sept. and Nov. 1984, Vol. 90, No 1583, pp. 61-66, and No. 1585, pp. 17-25.

Thomas, R., "Interface Shifts Management Chores to Peripherals," *Computer Design*, May 1985, Vol. 24, pp. 107-113.

Bray, J.M., Nelson, V.P., deMaine, P.A.D., and Irwin, J.D., "Data-Compression Techniques Ease Storage Problems," *Computer Design*, Oct. 1985, Vol. 24, pp. 102-106.

Thomas, R., "Cache Memory splits Computer and Tape Operations," *Computer Design*, Oct. 1985, Vol. 24, pp. 89-93.

Williams, T., "Smart, Powerful Peripherals Build on Interface Standards," *Computer Design*, Dec. 1985, Vol. 24, pp. 69-74.

Thornley, R.F.M., "The Future of Digital Magnetic Tape," *Proc. of SPIE*, January 1985, Vol. 529, pp. 198-202.

Wright, M., "Cartridge tape drives," *EDN*, Oct. 31, 1985, pp. 109-118.

Gregory, W.T., "Write and Read Recovery Systems for a Half-Inch Tape Drive," *Hewlett Packard Journal*, March 1985, Vol. 36, No. 3, pp. 16-18.

Killmon, P., "For Computer Systems and Peripherals, Smarter is Better," *Computer Design*, Jan. 1986, Vol. 25, No. 2, pp. 57-70.

Staff, "Tape Backup, Measuring Speed and Cost per Megabyte," *PC Magazine*, Feb. 11, 1986, Vol. 5, No. 3, pp. 107-144.

Cutler, D.S., "Quarter and Half Inch Streaming Tape Drives," *Intl. Conf. Video and Data, 1986*, IERE Publ. No. 67, pp. 95-104.

Disk Systems

Disk/Trend Report is an annual market study of rigid, flexible, and optical disk drives. Publisher: Disk/Trend, Inc., 5150 El Camino Real Suite B-20, Los Altos, CA 94022.

Stevenson, T.J., "Disk File Optimization," *IEEE Trans. Magn.*, Sept. 1975, Vol. MAG-11, No. 5, pp. 1237-1239.

Geffon, A.P., "A 6k BPI Disk Storage System Using MOD-11 Interface," *IEEE Trans. Magn.*, July 1977, Vol. MAG-13, No. 5 pp. 1205-1207.

Katoh, Y., Nakayama, M., Chubachi, R., and Okamoto, N., "High Density Magnetic Recording on a 3.5 Inch Micro Floppy Disk Drive System," *IEEE Trans. Magn.*, Sept. 1983, Vol. MAG-19, No. 5, pp. 1707-1709.

Lerma, J., "Analyzer Integrates Disk Drives With Controllers, Systems," *Digital Design*, April 1984, pp. 132-134.

Thomasian, A. and Kiamanesh, K., "Queuing Analysis Compares Disk Organization," *Computer Design*, May 1984, pp. 145-150.

Kaneko, R., "Future of Hard Disk Technology," *Proc. of SPIE*, Jan, 1985, Vol. 529, pp. 190-197.

Bate, G., "The Future of Flexible Disk Technology," *Proc. of SPIE*, January 1985, Vol. 529, No. 1, pp. 182-189.

Schneider, H., "Disk System IC's Combat Data Errors," *Computer Design*, March 1985, pp. 147-152.

Young, M.S., "Controllers Wring Peak Performance Out of Disk Drives," *Computer Design*, May 1985, pp. 119-127.

Warren, C., "Disk Formats for Your Files," *Mini-Micro Systems*, April 1986, Vol. 19, No. 4, pp. 31-47.

Chen, P.P., "The compact disk ROM: how it works," *IEEE Spectrum*, April 1986, Vol. 23, No. 4, pp. 44-54.

Voelcker, J., "Winchester disks reach for a gigabyte," *IEEE Spectrum*, Feb. 1987, Vol. 24, No. 2, pp. 64-67.

Rosch, W.L., "Technical Knockouts: Why Everyone Needs a Hard Disk," *PC Magazine*, June 9, 1987, Vol. 6, No. 11, pp. 109-116. Same issue has several papers on hard disk drives, incl. evaluations.

Computer Applications

McLeod, J., *Winchester Disks in Microcomputers*, Elsevier Intl. Bull., 256 Banbury Road, Oxford, OX2 7DH, England, 1985, approx. 250 pages.

West, C.F., and deTurk, J.E., "A Digital Computer for Scientific Applications," *Proc. IRE*, Dec. 1948, Vol. 36, No. 12, pp. 1452-1460.

Miller, J., "The TBM Memory System and Future Projections in Transverse Digital Recording," *IEEE Trans. Magn.*, September 1974, Vol. MAG-10, pp. 498-501.

Speliotis, D.E., and Chi, C.S., "Design of Advanced Digital Magnetic Recording Systems," *IEEE Trans. Magn.*, November 1975, Vol. MAG-11, No. 5, pp. 1234-1237.

Mckintosh, N. and Miyata, J., "A Standard Disk for Calibrating Head-Disk Interference Measuring Equipment," *IEEE Trans. Magn.*, Nov. 1982, Vol. MAG-18, No. 6, pp. 1230-1232.

Chi, C.S., and Frey, K.A., "Directional 'CRA' for Ternary Digital Recording," *IEEE Trans. Magn.*, Nov. 1982, Vol. MAG-18, No. 6, pp. 1259-1262.

Hoagland, A.S., "Status and Future of High Density Magnetic Data Storage," *Proc. of SPIE*, January 1983, Vol. 382, pp. 2-16.

Scranton, R.A., Thompson, D.A., and Hunter, D.W., "The Access Time Myth," *Intl. Conf. Video and Data 84*, April 1984, IERE Publ. No. 59.

Shafer, J.L., "Digital Formatting and Control Electronics for Half-Inch Tape Data Storage," *Hewlett Packard Journal*, March 1985, Vol. 36, No. 3, p. 19.

Laub, L., "The Evolution of Mass Storage," *BYTE*, May 1986, Vol. 11, No. 5, pp. 161-172. (Note: There are several other papers of interest in that issue of *BYTE*.)

Allan, I.D., "ESDI Joins Interface Ranks," *Mini-Micro Systems*, Feb. 1986, Vol. 19, No. 2, pp. 83-91.

Millican, D.L., "SCSI Aids PC/AT Tape Drive," *Mini-Micro Systems*, Nov. 1986, Vol. 19, No. 11, pp. 11-20.

Freeman, R.C., "QIC standards stimulate tape drive markets," *Mini-Micro Systems*, April 1987, Vol. 20, No. 4, pp. 58-62.

Chapter 27
Instrumentation Recording

The applications of magnetic recording are widespread in the discipline of measuring technique. The user must be aware of several sources of error that may contaminate his measured data and subsequent analysis:

- Signal amplitude fluctuations.
- Tape speed variations.
- Errors in tape duplication.
- Noise.
- Distortion.
- Drop-Outs.
- Cross-talk.

All of these errors are predominant in direct analog recordings. They are of lesser importance in FM recordings, and in PCM recordings most of them can be reduced to a negligible level (see later this chapter).

This chapter will describe the first five errors. Noise, distortion, drop-outs, and cross-talk are described elsewhere in the book.

The reader may question the reasons for including this chapter's topics since digital recording appears to be the cure for all those ills. The SNR can be made arbitrarily high by selecting enough sampling levels, and signal accuracy made high by frequently sampling the signal waveforms (at least twice per signal cycle). Flutter and timing errors are easily removed by the use of sufficiently large data buffers from where the data are clocked out in perfect step with an oscillator. So why spend time on a dissertation on the ills of analog data handling?

One must remember that the digital signal is always recorded, stored, and read in an analog format. It is a nice, perfectly shaped rectangular waveform when it comes out of the TTL circuits; but after read-out it has been afflicted with all the problems of the analog signal process. The conscientious designer must therefore have in-depth knowledge about all facets of the analog signal channel. Hence an entire chapter is devoted to the specific weaknesses of tape recording and how to overcome some of them.

Amplitude Fluctuations and Signal Accuracy

Direct analog magnetic recording is limited in performance by modulation noise and tape coating irregularities (non-uniform dispersion, limited number of particles "under" the playback head gap, coating thickness variations and surface irregularities). The modulation noise is proportional to the level of the recorded signal and is also related to coating irregularities. The amplitude of the playback signal will fluctuate as if it was *amplitude modulated* with a modulation index m that typically has a magnitude of • .5 to + 1 dB.

A figure of merit for a transmission system is the product of signal bandwidth B and the logarithm of the number of discrete levels n. This product is defined as the maximum rate of transmitting information and called the *system capacity C*:

$$C = B * \log_2 n \tag{27.1}$$

The number of signal levels that can be distinguished from each other is proportional to the system's signal-to-noise ratio:

$$\begin{aligned} n &= (V_{signal} + V_{noise})/V_{noise} \\ &= 1 + (V_s/V_n) \\ &= SNR \end{aligned}$$

and C becomes:

$$C = B * \log_2(1 + SNR)$$

or

$$\boxed{C = 2B * \log_2 \sqrt{1 + SNR} \text{ bits}} \tag{27.2}$$

This classic law assumes a Gaussian distribution of the noise and freedom from amplitude variations; the data transmission channel associated therewith is generally referred to as the AWGN channel, i.e., average white Gaussian noise. The magnetic tape or disk channel is not AWGN, and we must modify (27.2) to include amplitude variations.

The fluctuations of the read level present a limitation on the number of amplitude levels that can be distinguished from one another. If the amplitude modulation index is: • m (m = the ratio $(m * V_{signal})/V_{signal}$) then the difference between two separate amplitude levels must be (see Fig. 27-1):

$$V_q(1-m) - V_{noise\ peak} = V_{q-1}(1+m) + V_{noise\ peak}$$

which results in a minimum amplitude difference of:

$$\begin{aligned} \Delta V &= V_q - V_{q-1} \\ &= m(V_q + V_{q-1}) + 2\ V_{noise\ peak} \\ &= 2mV_q + 2\ V_{noise\ peak} \end{aligned}$$

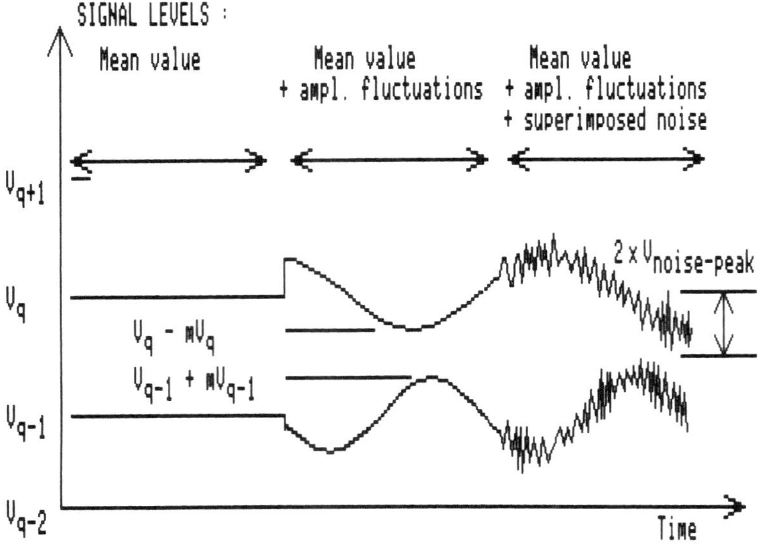

Fig. 27-1. Two adjacent signal levels must be separated by MVg plus 2 $V_{noise\text{-}peak}$.

for large signal-to-noise ratios. We may then also write:

$$\Delta V \cong 2m\ V_q$$

since $2mV_q >> 2V$ noise.

The maximum number of distinguishable amplitude levels is then:

$$n' = (1 + SNR)/2m$$
$$\cong SNR/2m \qquad (\text{for } SNR >> 1)$$

which equals a number of bits:

$$N' = \log_2 (SNR/2m)$$

Volz has determined n' without approximations (called n''):

$$\boxed{n'' = 1 + (\log_e(1 + 2m*SNR) - \log_e(1 + 2m))/2m} \qquad \textbf{(27.3)}$$

and the number of bits required:

$$N'' = \log_2(n'')$$

is plotted in Fig. 27-2. Shannon's law (formula (27.2)) is also plotted, but applies only for low signal-to-noise ratios. The scale shows the number of levels and the equivalent instrument accuracy. A typical performance specification for a 2 MHz recorder/reproducer is 30 dB SNR, with m = ± 1 dB. This corresponds to an equivalent instrument accuracy of about ± 4 percent. Improvements to the SNR alone will give diminishing returns unless means are found to also reduce the amplitude variations.

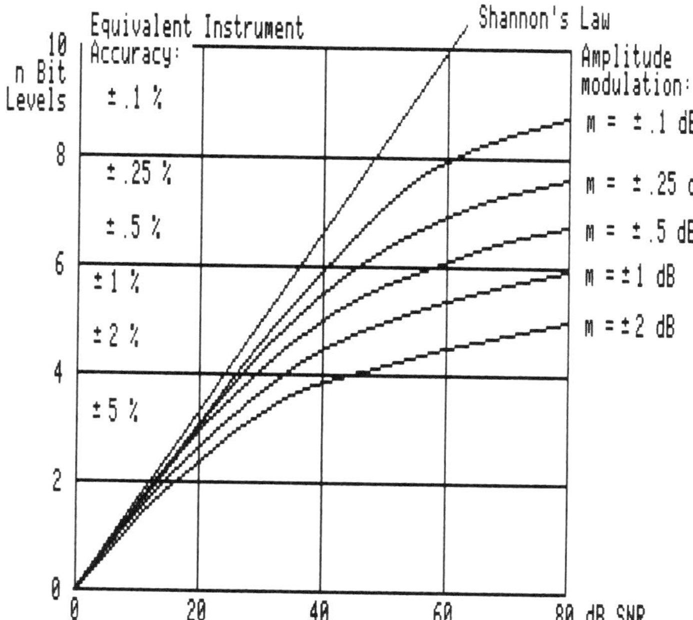

Fig. 27-2. Maximum number n″ of distinguishable levels, and the equivalent instrument accuracy, in analog magnetic recording (ref. Volz).

The amplitude accuracy will suffer during repeated replays of a tape due to mechanical *instability*. The gradual decrease in signal amplitudes, which are wavelength dependent, may quickly amount to errors greater than we have previously outlined. It is important to examine this effect for a given tape and recorder combination and to apply corrections to the analyzed data (see Chapter 15 under Tape Guidance).

The signal amplitude variations and the signal-to-noise ratio set a limit for the channel capacity of a magnetic tape recorder. It is important to recognize that the number of bit levels can be fairly large only if the *mid-frequency range* of a recorder is used (by band-pass filtering). Most recorders will in this range have a significantly higher SNR than the figure measured for the entire recorder bandwidth.

TAPE SPEED VARIATIONS

Magnetic tape transports for analog recording and playback are designed to provide a linear motion of the tape with as little speed variation as possible. This goal is limited by mechanical tolerances and all tape transports are therefore introducing errors into the data.

The speed variations are generally referred to as flutter. Their effect upon data in the form of noise in FM systems is well known and discussed elsewhere in the literature; methods have been developed for electronic compensation thereof.

Factors other that mechanical tolerances contribute to variations in the tape speed, since the data quite often are recorded on one transport and played back in the laboratory on another transport, or it may be a duplicate adding a third transport into the chain. Since it is most unlikely that they all run at exactly the same speed, the final speed may be .5-1 percent off of the speed at which the data were recorded. This results in an error if it is desired to reproduce "real time." Additional speed variations are often encountered since the recorder used in acquiring the data may have been subjected to heavy vibrations and gyrations, causing large variations and jumps in the tape speed (the latter from jerks from the reels) or the recorder may have been a lightweight unit with little control over moderate speed variations.

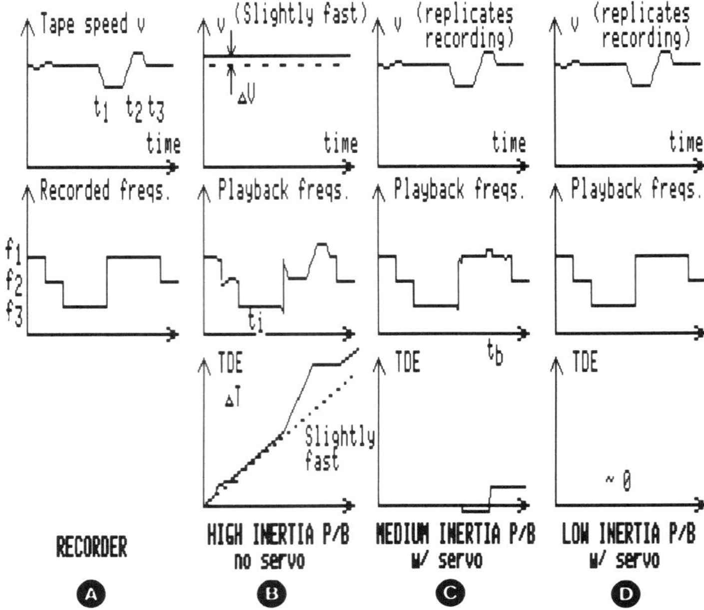

Fig. 27-3. Speed variations, signal frequency errors and time base errors during recording (A), and during playback. (B): High inertia drive; (C) Medium inertia drive; (D) Low inertia drive.

All can add up to substantial errors in time and frequency domains. Figure 27-3 illustrates what takes place during recording (part A) with heavy speed variations and subsequent playback on three different reproducers.

Recording on a High Inertia Drive

The speed variations are shown graphically: at time t_a the tape is slowed down by, for instance, a gyration of the recorder and at t_b the tape slips through the capstan drive with high acceleration due to a jerk or bump of the recorder. This in turn causes a loop to be formed at the supply reel, which slows down and when the loop tightens at t_c an instantaneous slowdown is caused before the tape motion stabilizes.

The data recorded during this interval is shown below as a three frequency shift keyed signal but the recorded wavelengths will be modified by the speed variations since $\lambda = v/f$. During the period t_1, for instance, there is an increase in tape speed and the recorded wavelength will consequently also increase since the input frequency is constant = f_1 during this period.

Playback on a Conventional High-Inertia Transport

In this case, we assume that reproduction is with a conventional recorder/reproducer with capstan flywheels and the synchronous motor drive in laboratory environment and the tape speed is therefore constant with exception of the inherent transport flutter. But, the average speed may be off by as much as .5-1 percent due to tolerance build up between different transports. This is indicated by ΔV and will result in an overall increase of all frequencies and faster display in the events, as shown by the deviation in the time base diagram.

With the otherwise constant speed we will now get the signal frequencies reproduced in correspondence with the recorded wavelengths ($f = v/\lambda$). This is illustrated in the middle of diagram of Fig. 27-3 under (B) where the following errors can be observed:

- The frequencies are no longer a true replica of the original frequencies, f_1, f_2, and f_3.
- The pulse lengths are no longer true replicas either, $t_1' \neq t_1$.
- All the pulses are on the average shorter, and they all follow each other faster due to the overspeed ΔV of the reproducer. This time base error Δt is plotted in the lower diagram and illustrates the cumulative error in timing if the events recorded are observed from t = 0.

How much error of one or the other kind can be tolerated depends entirely on the type of data being recorded, in what form and upon the desired accuracy. Let us, as an example, assume that a frequency analysis of the recording is required.

The frequency deviations resulting from the initial speed variations will pose a limit on how selective the analysis filters can be made for a given accuracy. The average deviation ΔV results in a corresponding frequency shift Δf that can be used as a correction to the frequency scale. This is most easily obtained by recording a fixed frequency control track on the tape. By comparing the frequency deviations during playback with a frequency equal to the one recorded one can extract a speed error signal.

The errors introduced by the transport flutter can be quite extensive. Very large variations (1 to 10 percent), as exemplified at t_a, t_b, and t_c are virtually destructive and will in most cases be categorized as loss of data. The only means to recover the recordings is by using a *servo controlled tape drive* during reproduction.

This was earliest done by recording a 17 kHz control frequency, amplitude modulated at a rate of 60 Hz. During playback this signal is demodulated and fed into a phase comparison circuit to which is also fed a precision 60 Hz voltage-controlled oscillator whose output after amplification drives the synchronous capstan motor. This technique is referred to as speed control, since the servo acts upon a high inertia drive, and the maximum error rates that can be followed in the magnitude of 1 Hz. It will therefore only handle the average speed deviation ΔV, not the rapid speed changes.

Playback on a Low Inertia Playback Transport

A great amount of research and development in the industry has recently resulted in the introduction of medium to low mass tape drive systems, employing elaborate servo circuitry

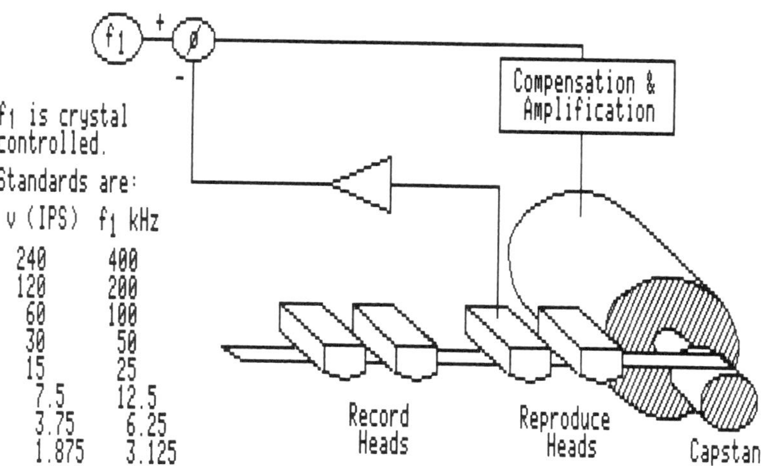

Fig. 27-4. Low inertia tape drive with servo speed control.

and DC drive motors. This is illustrated in Fig. 27-4, which also indicates the higher control frequencies used.

The increased accuracy during playback is shown in Fig. 27-3, where a distinction is made between a unit with medium inertia (C) and one with low inertia (D), assuming both are driven with identical forces. The inertia restricts the time for making speed corrections and a useful figure of merit is the *saturation acceleration* (equal to the peak torque of the drive motor divided by the inertia) in connection with the cut-off frequency f_s for the servo controlled system. This is exemplified by the acceleration dv/dt required at time t_b. The low inertia drive is capable of handling this acceleration while the medium inertia drive cannot. It loses synchronization with the control frequency on the tape and the "real" time is lost in addition to severe frequency disturbances (as illustrated at the time corresponding to t_b).

A typical example for the required acceleration is found for a tape recorded at 60 IPS with 1 percent peak to peak flutter at a 15 Hz rate:

$$\begin{aligned} v &= (60 + .3 \sin\omega t) \text{ IPS} \\ dv/dt &= 0.3\omega\cos\omega t \text{ in/sec}^2 \\ (dv/dt)_{max} &= 0.3 * 2\pi * 15 \\ &= 28 \text{ inches/sec}^2 \end{aligned}$$

This is far in excess of what a conventional recorder with speed control could handle—it corresponds to a start time of 1 second for 30 IPS operation with no over shoot. The flutter is never purely sinusoidal and the maximum acceleration for the example may therefore be in the order of several hundred inches/sec^2 which the reproducer must follow not to lose synchronization.

The specification for speed accuracy is named TDE = *Time Displacement Error* or TBE = *Time Base Error* (also time displacement error), which gives the residual dynamic time difference from "real" time. In a low TBE recorder-reproducer the timing error is reduced

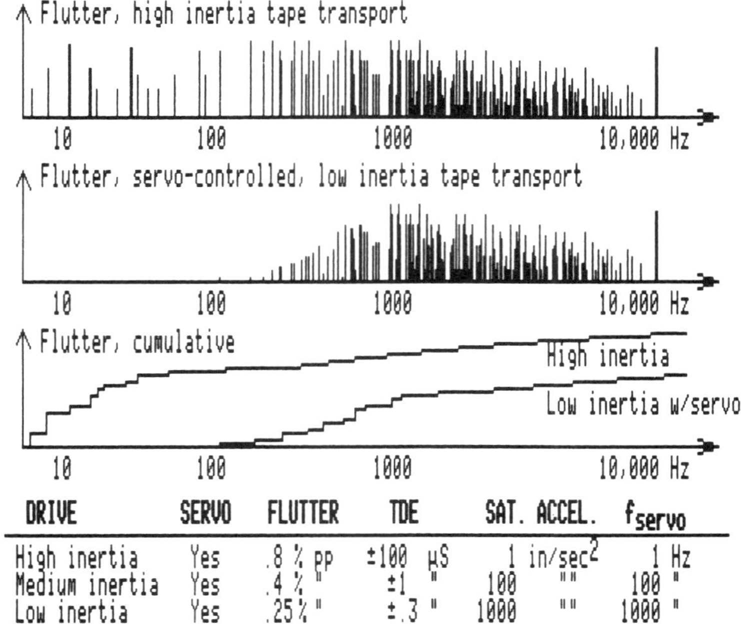

DRIVE	SERVO	FLUTTER	TDE	SAT. ACCEL.	f_{servo}
High inertia	Yes	.8 % pp	±100 µS	1 in/sec^2	1 Hz
Medium inertia	Yes	.4 % "	±1 "	100 " "	100 "
Low inertia	Yes	.25 % "	±.3 "	1000 " "	1000 "

Fig. 27-5. Tape transport flutter spectra and typical specifications.

by an order of two magnitudes, but the flutter is not reduced by the same amount. This is illustrated in Fig. 27-5, where a typical flutter spectrum is shown for a conventional transport, consisting of a number of discrete components and random distributed scrape flutter. When a servo-mechanism is employed it removes the components below f_s, which is the upper frequency response of the electro-mechanical servo. When the cumulative flutter is measured (DC-10 kHz) there is a reduction of flutter, but it does not correspond to the reduction in TBE, except below f_s.

In spectrum analysis work, it has been shown that the criterion for full recovery of the power in the signal is:

$$BW => 3\ (af_1) \tag{27.4}$$

where:

BW = Filter bandwidth.
a = peak flutter as a fraction of the average speed.
f_1 = frequency of flutter component.

Below f_s the flutter components are removed, and the bandwidth BW should not be narrower than that stated above, where f_1 corresponds to the analysis frequency and can be taken as the cumulative peak flutter up to that frequency. The design of a low inertia low TBE recorder-reproducer should therefore not trade off flutter for low TBE, which easily can occur if the servo system has resonances in the scrape flutter range. Figure 27-5 also lists representative tape transport specifications to serve as a guideline in establishing accurate requirements for recorder/reproducers.

Out of numerous applications two examples are shown in Fig. 27-6 where the top portion shows the complete loss of frequency shift modulation data if reproduced on a conventional drive. This will also be the case in, for instance, SSB recordings. The lower portion shows

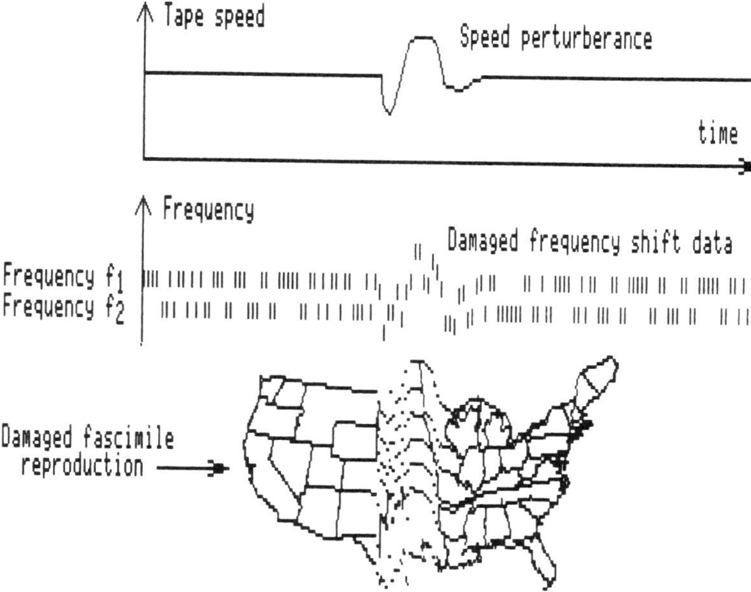

Fig. 27-6. Signal distortion in picture transmission via tape recorder, due to time base errors.

the drift problem that can occur during the reproduction of facsimile data, such as weather maps, during speed perturbations that are not corrected by means of a servo.

TIMING ERRORS BETWEEN TRACKS (SKEW)

A multichannel recorder-reproducer exhibits some predictable time differences between its various data channels. These may or may not cause problems in correlating time events between the tracks. There are several causes for the interchannel timing errors, generally called *skew*:

- Gap-scatter)
- Head-stack spacing) from mechanical tolerances
- Head tilt) in the head assembly

Dimensional changes with temperature and from excessive head wear
Tape motion irregularities and tape slitting tolerances

A worst case analysis provides the following numbers to assist in an assessment of skew error relating to a 1.5 inch *staggered head* set:

• Gap Scatter	5 μm
• Stack spacing	50 μm
• Head tilt	14 μm
• Tape tension change 1 ounce	6 μm
50 F temperature change 1 ounce	19 μm
100 percent humidity change	41 μm
Dynamic tape skew	6 μm
Total:	141 μm

which is equivalent with a TDE of 47 μsec at 120 IPS.

The errors from the 1.5 inch tape span in a staggered head assembly can be eliminated by using only *one head stack*:

• Gap scatter	5 μm
• Head tilt	14 μm
Dynamic tape skew	6 μm
Total	25 μm

or TDE = 8.5 μsec at 120 IPS.

The items marked with • can, upon playback, be corrected by means of fixed delay lines. The dynamic skew from the tape motion irregularities can be reduced by means of electrically variable delay lines. This is shown in Fig. 27-7. A constant frequency is recorded on tracks 1 and 7. The phases of the two signals are, upon playback, compared and will generate a correction signal, proportional to any phase difference. This signal is amplified and controls two current variables delay lines driven in push-pull. This will align the two signals in phase.

This technique has been used in playback predetection recordings, where the signal was recorded on two adjacent tracks in the same head stack. The carriers were filtered out and served to produce the correction signal from the phase detector. The dynamic skew was typically reduced 40 dB (100 times) and the two signals could then be added to provide protection against single track drop-outs.

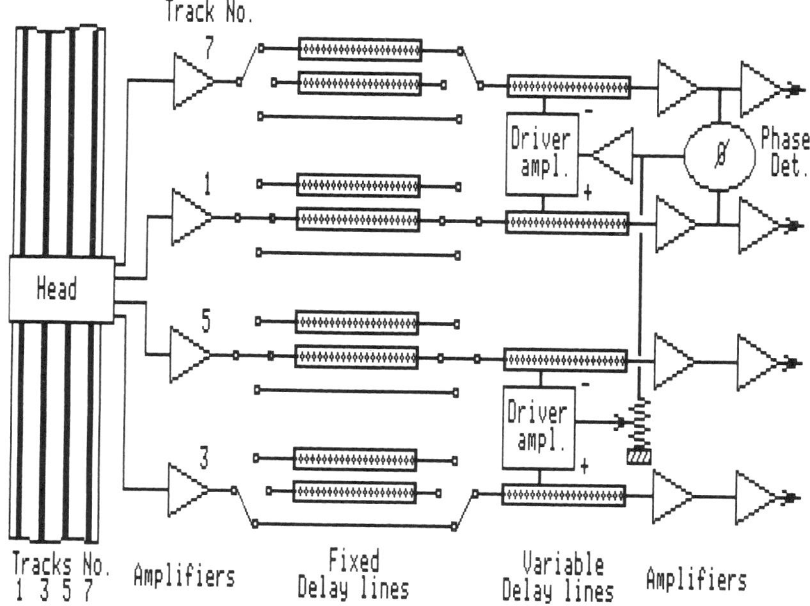

Fig. 27-7. Dynamic skew correction of four channels in a tape transport. The skew errors between tracks 1 and 7 are detected, and controls the delays lines for all tracks.

If the signals have no carriers or are different in nature, additional delay lines are required. (The two signals could both be digital, or one analog and the other digital.) These signals are recorded on tracks 3 and 5 while a control frequency, e.g., 240 kHz is recorded on tracks 1 and 7. The skew information is derived from the latter tracks during playback.

A digital system for the reduction of flutter and skew has been developed by Eastman Kodak's Datatape Division. This technique utilizes a known signal recorded with the data to extract time base error information. The known signal is referred to as a pilot and is recorded at very low amplitude relative to signal amplitude and at a frequency lying outside the range of the signal data (Fig. 27-8A).

During playback, the frequency and phase variations of the pilot are tracked by an agile phase locked loop which generates a clock for the A/D converter and memory input (Fig. 27-8B). The degree to which this loop can track the true variation of pilot frequency and phase and reject the degrading effects of tape noise in the vicinity of the pilot frequency determines the quality of correction obtainable.

In this way, flutter is reduced by a factor of three, dynamic skew by a factor of five, and time displacement errors by a factor of three to five.

TAPE DUPLICATION

A certain amount of signal degradation occurs during duplication of tapes. Frequency response and signal-to-noise ratio are affected along with an increase in envelope distortion which will result in distortion of complex waveforms.

Envelope distortion is a measure for the delay times for the various frequencies in a transmission system. A typical figure is 500 nanoseconds peak-to-peak for a bandwidth of 100 kHz to 1.2 MHz in an instrumentation recorder. This distortion occurs partly in the write/read process, partly in the magnetization pattern in the coating. This distortion can become quite

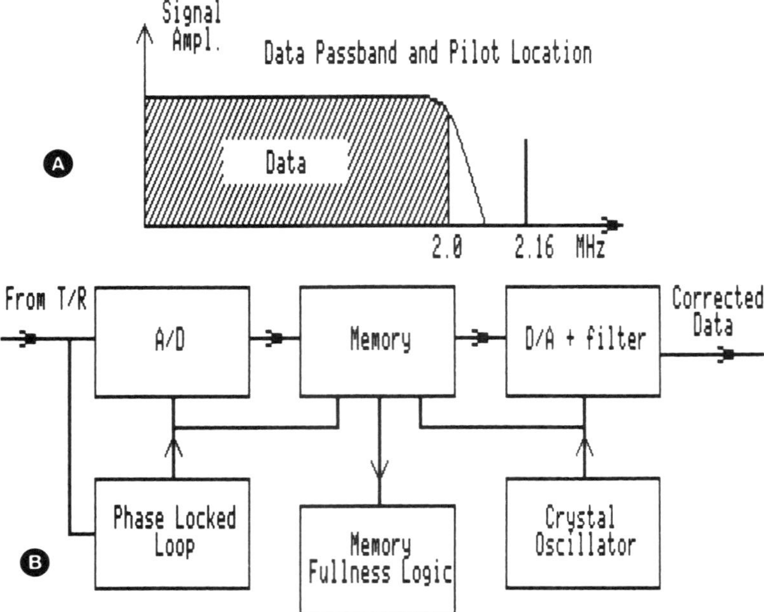

Fig. 27-8. Dynamic skew and flutter correction by digital means. (A) Data passband and pilot frequency locations; (B) reproduce electronics (courtesy Datatape).

severe when several generations of duplicates are made. The last duplicate may have transients entirely different from the original signal.

The cumulative phase errors in magnetic recording is one of the reasons why the ''direct-to-disc'' recordings have been successful in the audio industry.

The phase error can be reversed by *duplicating backwards*. First, for critical recordings, it is always advisable to begin each reel of tape with a recording of not only selected frequen-

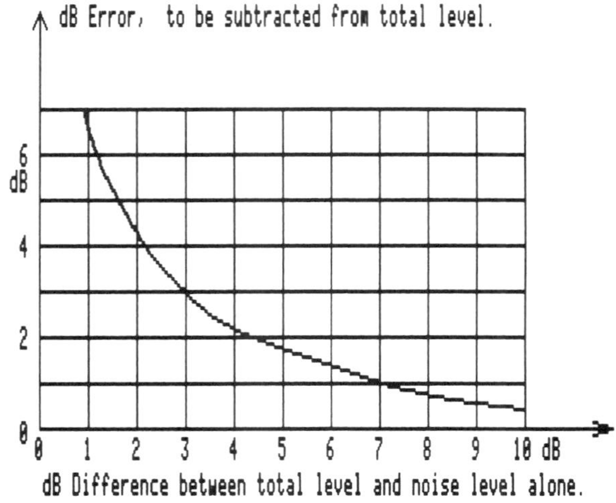

Fig. 29-9. Basic tape format and head wheel for VHS NTSC and PAL/SECAM recorders (after Sugaya).

cies for response setting, but also square waves. This will allow for proper set-up of the equalizers in the system where the tapes are copied (or reproduced).

Copies will have better fidelity if duplicated backwards. The procedure is: leave the recorded tape on the take-up reel and transfer this reel to the supply reel hub. Now duplicate by running this tape through the recorder (and the duplicate will be correct with the beginning of the recording at tape start). For additional copies, rewind the master tape each time and duplicate backwards again.

We close this section with a useful graph that should be applied to correct meter readings when the signal-to-noise ratio of a signal is less than 10 dB (Fig. 27-9).

FM AND PCM RECORD TECHNIQUES

Signals that contain DC-components cannot be played back by ordinary read heads. Some form of modulation is required. The most straightforward method is the use of flux sensitive heads to sense the flux from the tape, but it is seldom used due to several limitations (lack of sensitivity, response and noise). Note that it is a simple matter to record DC on a tape using a record amplifier without coupling capacitors or transformers.

Two basic modulation methods for recording and playback that extend to DC are available: frequency modulation and pulse modulation. This chapter will explain the principles of both and how they are employed in recording equipment. The prime area for FM is in instrumentation and television, while PCM is a fast emerging technique in instrumentation (HDDR) and in digitized audio and television.

The material in this section is limited to the fundamental aspects of the technology without details or discussions of the special disciplines of television recording and digitized recording.

Principle of FM Recording

FM recording provides response to DC, and it is immune to the signal amplitude variations in direct recording. Amplitude noise is reduced in proportion to the modulation index β, which we will soon introduce.

Disturbances in the tape motion (flutter) will act as a modulation of the FM-signal and they will consequently be detected as noise. A non-linear phase-versus-frequency response will likewise introduce distortion.

The bandwidth of the recorder channel depends upon the selected value of β. It will in general be several times wider than the bandwidth of the data. (The signal-to-noise ratio is proportional to β. The channel bandwidth is also proportional to β.)

FM-recording consists of recording a signal that varies in frequency in accordance with the data input. An oscillator in the record amplifier provides an output signal of a frequency that depends upon the signal input voltage. This is called a *voltage controlled oscillator*, or FM-modulator (Fig. 27-10). For zero input, the frequency of the signal is constant and we are recording what we call the *carrier*. A plus or minus voltage to the FM-modulator will change the frequency up or down. When these frequencies are played back, we can detect and *demodulate* them, which results in an output voltage that is identical to the input signal transmitted to the FM-modulator.

The expression for a periodic input signal may be:

$$f(t) = A * \cos 2\pi f_c t$$

where A is the signal amplitude and f_c the signal frequency, also called the carrier frequency. When f_c is varied by a data or video signal we obtain frequency modulation. In general, this can be expressed as:

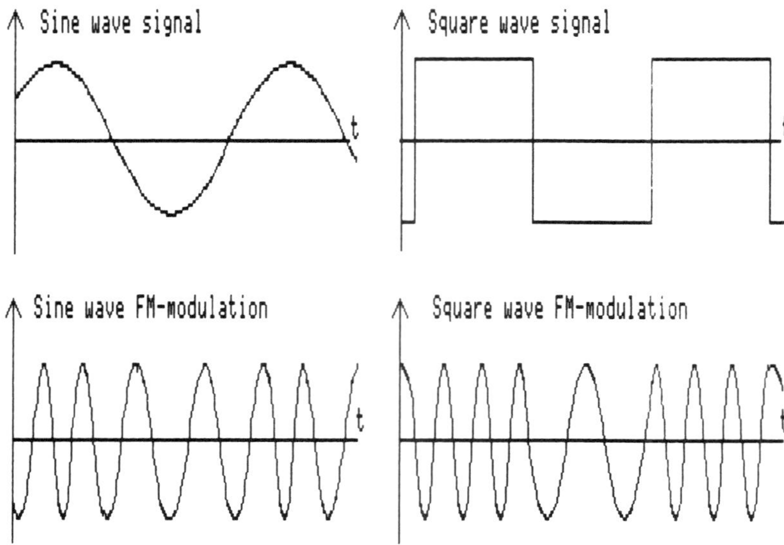

Fig. 27-10. Principle of frequency modulation.

$$f(t) = A \cos (2\pi_{fc}t + \beta * \sin 2\pi f_s t) \quad \textbf{(27.5)}$$

where:

f_c = carrier frequency
f_s = signal frequency
β = modulation index

The modulation index β is a factor that results in a maximum deviation of $2\pi\ \Delta f_c t$ (when $\sin 2\pi f_s t = 1$), where Δf_c is the frequency deviation from f_c.

We can normalize f_s to Δf_c by:

$$\beta = \Delta f_c / f_s$$

or

$$\Delta f_c = \beta * f_s.$$

We can similarly normalize f_c to Δf_c by introducing the *deviation ratio D:*

$$D = \Delta f_c / f_c.$$

β expresses how "hard" we modulate the carrier frequency f_c, and is proportional to the amplitude of the data signal; just how fast f_c is varied is of course determined by the term $\sin 2\pi f_s t$. We could also write $\beta = B * \beta'$, where B is the amplitude of the data signal and β' is the sensitivity of FM-modulator; β' is then equal to:

$$\Delta f_c / f_c.$$

The FM Signal Spectrum

Our formula for frequency modulation can be expressed as a carrier signal of frequency f_c, surrounded by an often large number of sidebands of frequencies that are equal to f_c plus and minus f_s, $2f_s$, $3f_s$, etc. Their amplitudes vary in a very complex fashion according to a series of *Bessel functions*, and (27.5) becomes:

$$\begin{aligned} f(t) = A\,(&J_o(\beta)\cos\omega_c t \\ &+ J_1(\beta)\cos(\omega_c + \omega_s)t - J_1(\beta)\cos(\omega_c - \omega_s)t \\ &+ J_2(\beta)\cos(\omega_c + 2\omega_s)t - J_2(\beta)\cos(\omega_c - 2\omega_s)t \\ &+ J_3(\beta)\cos(\omega_c + 3\omega_s)t - J_3(\beta)\cos(\omega_c - 3\omega_s)t \\ &+ J_4(\beta)\cos(\omega_c + 4\omega_s)t - J_4(\beta)\cos(\omega_c - 4\omega_s)t \ldots) \end{aligned} \qquad \textbf{(27.6)}$$

where $J_n(\beta)$ is the Bessel function of the first kind with argument β and order n; n being an integer. The values of $J_n(\beta)$ are tabulated in all books on FM and in mathematics handbook. Figure 27-11 shows the variation of $J_o(\beta)$ = the carrier signal, $J_1(\beta)$ = first set side bands, $J_2(\beta)$ = second set side bands, $J_3(\beta)$ = third set side bands, etc.

The distance between the individual sidebands and the carrier is equal to ω_s, the signal frequency. The number of sidebands is proportional to the modulation index. Typical sideband spectra are shown in Fig. 27-11.

It is apparent that there is an infinite number of sidebands around the carrier. We do not need all of them for demodulation and we can limit the transmission bandwidth (of the recorder) to include only those sidebands that are greater than one percent of the unmodulated carrier amplitude.

We can then determine the bandwidth (BW) of a sinusoidally modulated FM carrier for any modulation index β by determining the number of Bessel functions $J_n(\beta)$ which exceed 0.01 and then multiplying by $2\Delta f_s$. This bandwidth is plotted in a universal form in Fig. 27-12.

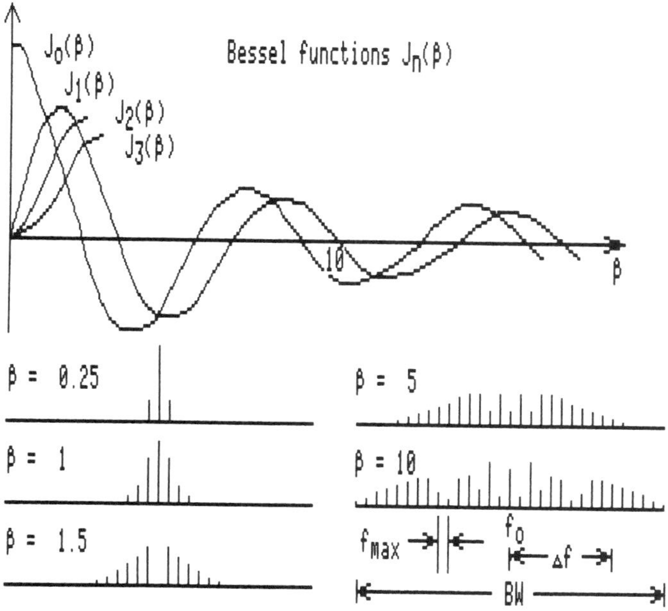

Fig. 27-11. Bessel functions, and spectra of a frequency modulated signal carrier.

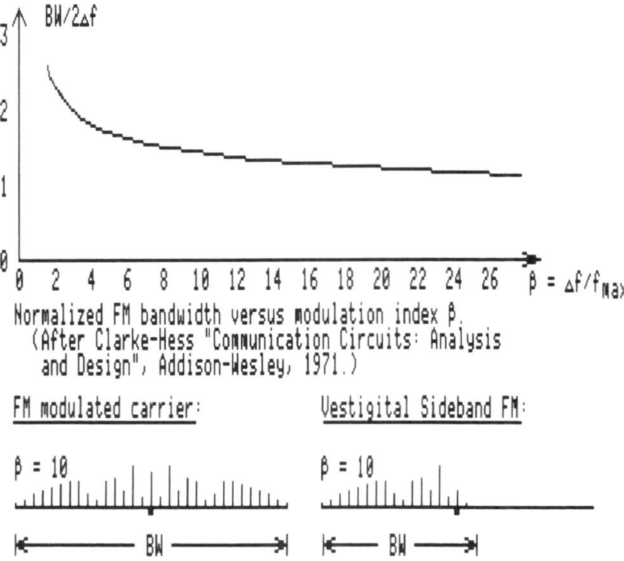

Fig. 27-12. Bandwidth requirements in FM systems (after Clarke-Hess).

A familiar example is FM-broadcast and we are all familiar with its excellent quality. The maximum modulation frequency is 15 kHz and the broadcast stations may use a maximum deviation of 75 kHz for this signal. The β is therefore = 5 : BW/2Δf is found from the curve to be 1.6. Consequently a bandwidth of 2Δf * 1.6 = 150 * 1.6 = 240 kHz is required for transmission of the 15 kHz signal.

In order to record this signal a recorder bandwidth of 250 kHz minimum would be required. This alone makes FM a poor candidate for audio recordings. Add the problem of flutter, which would not only be heard as a wobble of pure tones but would also produce signals at any distinct flutter frequencies.

Instrumentation recorders are often used to record FM/FM telemetry and they use a deviation ratio of about five. In wideband, FM recording deviation ratios in the range of one to two are common.

DIGITAL (PCM) RECORDING PRINCIPLE

Recordings with high accuracy, high signal-to-noise ratio, and reasonable bandwidth are accomplished through a combination of digital coding and the recording technology from instrumentation. Rather than recording and reproducing a signal that continuously varies in amplitude and duration (analog), the signal is changed in an *analog-to-digital converter* so the record signal has discrete levels only.

The output signal from an A/D converter is a digital *PCM signal* (Pulse Code Modulated) in the NRZ-L format (No-Return-to-Zero-Low). (This implies that all ''1''s are at a high voltage level and all ''0''s are a zero voltage.) This NRZ-L signal contains a strong DC-component, that should be completely removed by recoding the signal into one of the codes from Chapter 24.

Characteristics of PCM Recording

We can list several advantages that PCM-recording, or HDDR, has over analog recording-playback:

- A high degree of linearity.
- Theoretical limitless signal-to-noise ratio.
- Immunity against data degradation due to changes in the overall tape flux level which occur after repeated playback passes or duplication.
- Errors caused by single drop-outs can often be corrected.
- Immunity against crosstalk.
- Excellent phase and transient response.
- Complete removal of flutter by clocking the PCM-data out of the buffers.
- Computer compatible format.
- Digital signal processing (prior to D/A conversion): Synchronization, filtering, data enhancement, etc.
- Few operational adjustments.

This is truly an impressive improvement—but everything has its price: HDDR recorders consume a large amount of tape in order to accommodate the wide bandwidth of PCM-data. Digitized audio, for example, requires 6-800 kHz bandwidth for 20-20,000 Hz signal with a 90 dB SNR (signal-to-noise ratio).

This PCM signal could be handled by a home VTR (video tape recorder). Digitized data of very high bit rates (several hundred MBPS) is divided into many parallel channels of corresponding lesser bandwidths. These channels are then recorded on a multitrack machine (HDDR) and, upon playback, clocked into step with each other (through buffer storage) and then recombined into a serial stream and converted back to analog.

Analog-to-Digital Conversion

Analog-to-Digital converters (ADC) translate from analog values into digital words. Each word contains n bits (''0''s or ''1''s) and each represents a region of the analog signal range.

There are, using words with n bits, a total of 2^n uniquely different words. If n is one, then the single bit word can be a ''1'' or a ''0.'' This is like a coin showing heads (H) or tails (T). If n = 3 then eight combinations of ''1''s and ''0''s are possible, just like the tossing of three coins have eight possible outcomes: HHH, HHT, HTH, HTT, THH, THT, TTH, and TTT.

A 4-bit word has 16 possible word values, and each is assigned to a region of the analog signal range. This mapping of a voltage range into a digital word is called *quantization*. The resolution of the mapping increases in direct proportion to n.

Polarity of a voltage requires an additional bit, the ''sign bit.'' This bit is always the first in the word, and therefore called the ''most significant bit,'' abbreviated MSB. The next bit informs whether the voltage is in the upper or lower half of the voltage range. The following bit again divides this assigned range into two, and so forth. The last bit, or ''least significant bit''(LSB), is the final location of the quantization interval, which then has a size corresponding to the voltage interval = voltage range divided by 2^{n-1}. This subdivision of ranges is shown in Fig. 27-13 for a 6-bit word.

The MSB, the sign bit, is not used for the value of a voltage, only the remainder $n-1$ is used. The ratio of the LSB to the full scale voltage is therefore 2^{n-1} (2^{-5}, not 2^{-6}, in the example). The minimum voltage is therefore $1 * Q$, where Q is the final quantization interval. The maximum is $2^{n-1} * Q$, and the corresponding rms value:

$$V_{signal}\ (\text{rms}) = Q * 2^{n-1}/\sqrt{2}$$

The error in quantizing a voltage into a range $\pm$ Q/2 can be calculated to :

$$V_{noise}\ (\text{rms}) = Q/\sqrt{12.}$$

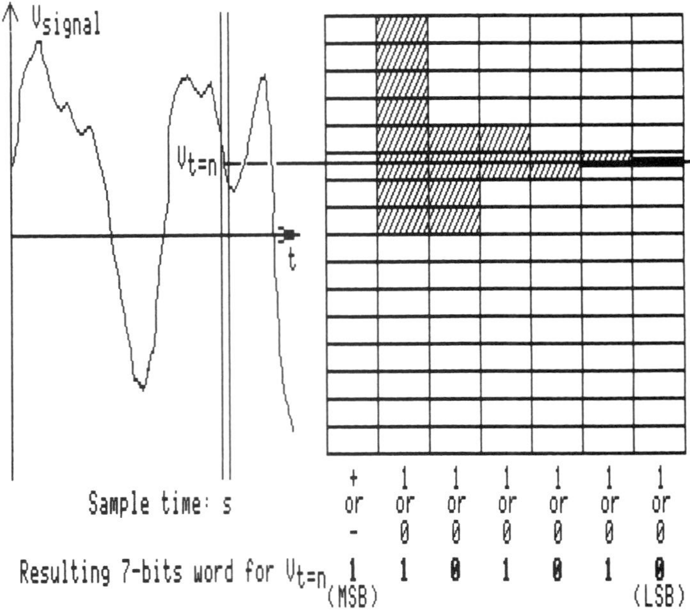

Fig. 27-13. Quantization (analog-to-digital conversion) of voltage level Vt.

We then find for the *signal-to-noise ratio*:

$$\text{SNR} = \sqrt{1.5} * 2^{n}$$

or

$$\boxed{\text{SNR} = 6 * n + 1.76\ dB} \qquad \textbf{(27.7)}$$

Each bit will therefore contribute 6 dB to the system's performance. For the 6-bit word the SNR is 36 + 1.76 dB.

It appears possible to achieve any desired SNR by selecting a word long enough. The A/D converters contain resistance dividers which must be increasingly more accurate for higher values of n. Practical tolerances limit n to the order of 15 to 17.

The word generation can be serial or parallel. Serial, or successive approximation, starts with the assignment of the MSB and it continues, in order of descending bit weights, until the LSB has been assigned. The conversion time is fixed and it is in the range of microseconds, or smaller (8 bits in 1 microsecond or 12 bits in 3.5 microseconds are typical today for standard converters; faster converters are available, at a cost).

The input voltage is, in parallel conversion, applied to $2^{n} - 1$ comparators in parallel and the conversion is as fast as the comparators can switch. Unfortunately, the number of elements in these converters increases geometrically with resolution, limiting their application.

Digital-to-Analog converters (DAC) produce output voltages that correspond to the assigned range of digital words. The resolution and speed of DAC's are similar to ADC's. As a matter of fact, ADC's make use of a DAC in their operation.

The conversion time limits the upper frequency of the analog signal that can be converted. It should be sampled a minimum of one per half cycle (*Nyquist rate*), preferably at a slightly higher rate, say, 2.5 * f, where f is the highest analog signal frequency.

The converted signal consists of a n-bit word (plus an enable signal). The sample time is therefore $1/(2.5 * f)$, which gives a maximum time of $1/(2.5 * f * n)$ per bit. This equals a transfer rate of:

$$2.5 * f * n$$

and with a minimum of 2 bits per cycle the *PCM bandwidth* is:

$$BW = 1.25 * f * n \text{ Hz} \qquad \textbf{(27.8)}$$

(The theoretical minimum is the Nyquist BW = f * n.)

The price we pay for the advantages of PCM is an increase in required BW, or transfer rate, proportional to n (or the SNR achieved, 6n dB).

HDDR—HIGH DENSITY DIGITAL RECORDING

The continuous flow of analog data is, after A/D conversion, formed into a stream of two-level bits, "0"s or "1"s. The data rate has simultaneously increased from f cycles per second to a new bandwidth of 1.25 Hertz. Many experimental data and their associated resolution (n) will produce bit streams upwards from 20 MHz to 600 MHz.

Such high data rates clearly exceed the bandwidth of existing instrumentation recorders and the problem of recording them was solved in a straightforward manner by converting the serial stream into parallel streams that all would be within the recorders bandwidth (Fig. 27-14).

The number of tracks required is determined in a straightforward fashion once the bit rated per channel has been determined. The data stream is normally in the NRZ-L digital logic format, which is not well suited for recording onto an analog recorder channel. The most severe limitation is the absence of response to DC, and it becomes necessary to encode the data stream into a format that is free from DC, while keeping the new bit rate as low as possible (see Chapter 24 on Modulation Codes).

The maximum bit rate per recording channel is limited by its upper frequency limitation and its signal-to-noise ratio. The Bit-Error-Rate (BER) is a function of the latter and if 20 dB SNR is chosen as a minimum, then a BER of better than 10^{-7} can be achieved, assuming additive white gaussian noise and not counting amplitude variations, including drop-outs.

The BER can be much improved by applying error correcting codes to the data signal (see Chapter 25). Tape and disk systems often lose data in chunks of many bits, called burst errors. A long and complicated error code is normally required to correct for these bursts, but the multi-channel HDDR recorders have a built-in cure. When the data stream is changed from a serial to many parallel streams then the data will be recorded *interleaved*, which is a very powerful assist to any error correcting code used (see Chapter 25).

The current packing densities are 66 kilobits-per-inch (kbpi), which equals 33 kHz/inch (4 mHz at 120 IPS). Higher densities (up to 140 kbpi) are at the laboratory stage, using refined heads (very narrow gaps) and modern tapes.

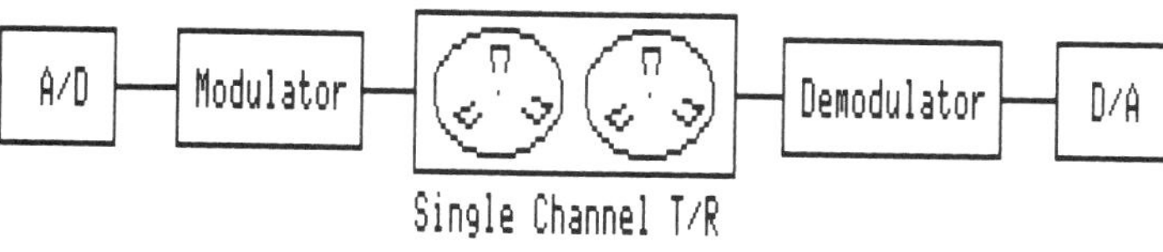

Fig. 27-14. High Density Digital Recording (HDDR) methods.

See Figure 25-10 for a multichannel HDDR system.

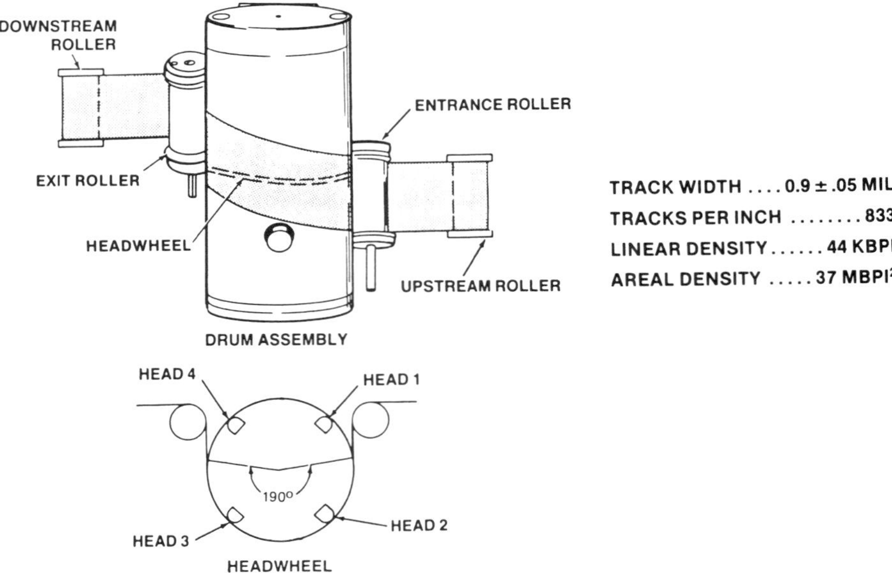

Fig. 27-15. Rotary digital recorder, scanner configuration (courtesy Datatape, Inc.).

Two methods of recording are used, direct pulse recording (saturation) and AC bias recording. At the band edge there is very little difference between the two methods and the choice is often for no-bias direct recording because of simplicity (and lower cost). The recording level of the pulse signals is then much higher (approx. 20 dB) than with AC bias and it becomes a source of channel-to-channel crosstalk during recording.

When crosstalk must be minimal, then the AC-bias is used. This is for instance the case when HDDR tracks are interleaved with other analog tracks (direct or FM). Proponents of AC-bias also claim to have better control over the phase response, which in some PCM systems is a more important parameter than the amplitude response.

Data of moderate bit rates (8 - 30 MBPS) are now often recorded on a rotary digital recorder. These recorders are derivatives of video recorders, and a typical path for the one inch wide

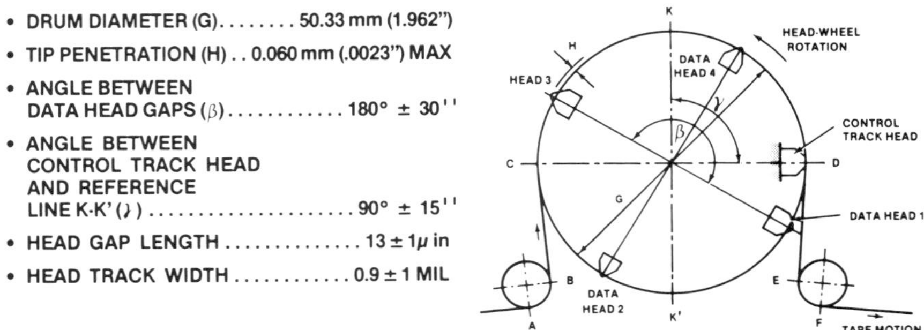

Fig. 27-16. Head wheel in rotary digital recorder (courtesy Datatape, Inc.).

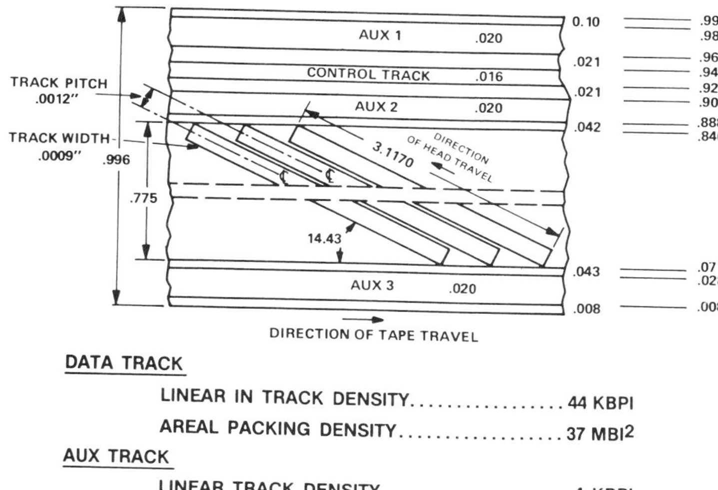

Fig. 27-17. Track format in digital rotary recorder (courtesy Datatape, Inc.).

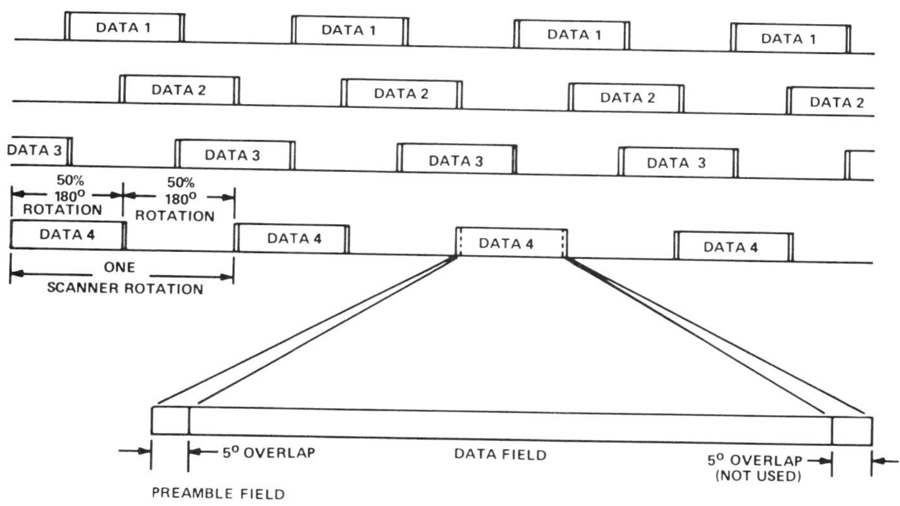

Fig. 27-18. Timing for recording of two data channels (1 + 3) and (2 + 4) (courtesy Datatape, Inc.).

tape is shown in Fig. 27-15. The head wheel may have four heads, working together in pairs (Fig. 27-16), laying down interleaved, slanted tracks (Fig. 27-17).

The result is a two channel data recorder where the digitized data stream is sliced into sequences data1 + data3 and data2 + data4. See Fig. 27-18. Upon playback data buffers reclock and combine the data to two unbroken streams of signal, with freedom from any timing error, and very low BER (pending error correcting scheme).

STANDARDS

Standards on *instrumentation recording* covers tapes, reels, track configurations, frequency response and test methods. They are issued by ANSI, ISO (International Standards Organization) and IRIG. Write to ANSI for theirs and ISO standard. IRIG's standard 106-86 (telemetry) and 118-79 (test methods) can be obtained from:

National Technical Information Service
U.S. Department of Commerce
5285 Port Royal Road
Springfield, Virginia 22161

BIBLIOGRAPHY TO CHAPTER 27

Introduction

Clarke, K.K, and Hess, D.T., *Communication Circuits: Analysis and Design*, Addison-Wesley, 1971, 658 pages.

Moulin, Pierre, *L'Enregistrement Magnetique d'Instrumentation*, Editions Radio, January 1975.

Modern Instrumentation Tape Recording, EMI Technology, Inc., January 1978, 140 pages.

Parallel Mode High Density Digital Recording, Technical Fundamentals, Datatape Division, Eastman Kodak Co., 1981, 80 pages.

Magnetic Tape Recording, Technical Fundamentals Gi, Datatape Division, Eastman Kodak Co., 1982, 111 pages.

Kalil, F., and Buschman, A., (editors), *High-Density Digital Recording*, NASA Ref. Publ. No. 1111, 1986, 313 pages. (Order from Superintendent of Documents, U.S. Government Printing Office, Washington D.C. 20402).

Lemke, J.U., "Instrumentation Recording", Chapter 4, *Magnetic Recording*, Vol. 3, McGraw-Hill, 1988.

Amplitude Variations

Voelz, H., "Spurhoehe, Spurzahl and Kanalkapacitaet bei der magnetischen Speicherung," *Hochfreq. and Elektroakustik*, 1967, Vol. 76, pp. 172-175.

Tape Speed Variations; TDE

Mullin, J.T., "Flutter Compensation for FM/FM Telemetering Recorder," *Conv. Rec. IRE*, March 1953, Pt. 1-Radar Telemetry, pp. 57-65.

Ratz, A.G., "The Effect of Tape Transport Flutter on Spectrum and Correlation Analysis," *IEEE Trans. Space Electr. and Telemetry*, December 1964, pp. 129-134.

Chao, S.C., "Flutter and Time Errors in Instrumentation Magnetic Recorders," *IEEE Trans. Aerospace and Electr. Syst.*, March 1966, Vol. AES-2, No. 2, pp. 214-223.

Moore, L., "The Effects, Measurement, and Analysis of Flutter in Instrumentation Recorders," *ITC Conf. Proc.*, January 1968, pp. 392-398.

Kashin, F.A., Nemeni, T.M., and Slepov, N.N., "Errors in Magnetic Tape Recording Caused by Tape Speed Instability," *Telecomm. Radio. Engr. Part II*, October 1968, Vol. 23, No. 10, pp. 98-101.

Frequency Modulation

Shaper, H.B., "Frequency-Modulated Magnetic Tape Transient Recorder," *Proc. I.R.E.*, November 1945, Vol. 22, No. 11, pp. 118-121.

Oliver, B.M., Pierce, J.R. and Shannon, C.E., "The Philosophy of PCM," *Proc. IRE*, November 1948, Vol. 36, pp. 1324-1332.

Gerlach, A.A., "FM Recording in Guided Missiles," *Electronics*, January 1953, Vol. 26, pp. 108-111.

Selsted, W.T., "A Low Noise FM Recording System," *JAES*, April 1953, Vol. 1, pp. 213-215.

Stewart, J., "The Power Spectrum of a Carrier Frequency-Modulated by Gaussian Noise," *Proc. IRE*, October 1954, Vol. 42, pp. 1539-1542.

Broch, J.T., "FM Magnetic Tape Recording," *Bruel and Kjaer Tech. Rev., No. 1*, 1966, 15 pages.

Wachsman, R.H., and Baghdady, E.J., "Effects of Random Fluctuation Noise on FM and FDM/FM Reception," *ITC Conf. Proc.*, January 1967, pp. 1-25.

Jorgensen, F., "Distortion in Wideband FM Magnetic Tape Recording Systems," *Telemetry Jour.*, May 1968, Vol. 3, No. 3 pp. 53-56.

Robinson, J.F. *Videotape Recording*, Focal Press, 1981, 362 pages. See Chapter 6, FM Theory.

Pulse Code Modulation (Digital Recording)

Jayant, N.S. (editor), *Waveform Quantization and Coding*, IEEE Press, 1976, 611 pages.

Bylanski, P., and Ingram, D.G.W., *Digital Transmission Systems*, Peter Perigrinus Ltd, on behalf of IEE, Cambridge, England, 1980, 431 pages.

Loriferne, B., *Analog-Digital and Digital-Analog Conversion*, Heyden, 1982, 196 pages.

Sheingold, D.H. (editor), *Analog-Digital Conversion Handbook*, Prentice-Hall, 1986, 669 pages. (Written by the staff of Analog Devices, Inc.).

Michaels, S.R., "Getting the Best from A-D Converters," *Electronic Design*, February 1982, pp. 191-199.

Fujiwara, T., and Yamamori, K., "Perpendicular Magnetic Recording of Analog Signals by Means of Pulse Width Modulation," *IEEE Trans. Magn.*, November 1982, Vol. MAG-18, No. 6, pp. 1244-1247.

Law, E.L., "Predetection Recording of PCM Telemetry Signals," *THIC Meeting*, June 1984, 20 pages.

HDDR Techniques

Spitzer, C.F., "Digital Magnetic Recording of Wideband Analog Signals," *Comput. Design*, October 1973, Vol. 12, pp. 83-90.

Miller, J., "The TBM Memory System and Future Projections in Transverse Digital Recording," *IEEE Trans. Magn.*, September 1974, Vol. MAG-10, No. 3, pp. 498-501.

Breifss, I.P., "High Density Data Recording," *IEEE Spectrum*, May 1975, pp. 58-62.

Mallinson, J.C., "A Unified View of High Density Digital Recording Theory," *IEEE Trans. Magn.*, Sept. 1975, Vol. MAG-11, pp. 1166-1169.

Mallinson, J.C., "Design Philosophy and Feasibility of a 750 Mega-Bit per Second Magnetic Recorder," *IEEE Trans. Magn.*, Sept. 1978, Vol. MAG-14, No. 5, pp. 638-643.

Mallinson, J.C., "Towards 1000 Mega-Bit Per Second Magnetic Recording," *Intl. Conf. Video and Data 79*, July 1979, IERE Publ. No. 43, pp. 267-275.

Drury, G.M., "A Tape Interface System for High Density Digital Recording," *Intl. Conf. Video and Data 79*, July 1979, IERE Publ. No. 43, pp. 241-261.

Lemke, J.U., "Ultra-High Density Recording with New Heads and Tapes," *IEEE Trans. Magn.*, November 1979, Vol. MAG-15, No. 6, pp. 1561-1564.

Wood, R.W., and Donaldson, R.W., "Signal Processing and Error Control on an Experimental High-Density Digital Magnetic Tape Recording System," *IEEE Trans. Magn.*, Sept. 1980, Vol. MAG-16, No. 5, pp. 1255-1265.

Zoeller, M., "An Ultra-High Data Rate Mass Storage System," *IEEE Trans. Magn.*, July 1981, Vol. MAG-17, No. 4, pp. 1426-1431.

Rumball, D., Maclean, D., and Clark, D.R., "Gigabyte Data Densities from an Unmodified Video-Cassette Recorder," *Intl. Conf. Video and Data 82*, April 1982, IERE Publ. No. 54, pp. 315-321.

Newby, P.S., and Yen, J.L., "High Density Digital Recording Using Videocassette Recorders," *IEEE Trans. Magn.*, September 1983, Vol. MAG-19, No. 5, pp. 2245-2252.

Coleman, C., Lindholm, D., Peterson, D., and Wood, R., "High Data Rate Magnetic Recording in a Single Channel," *Intl. Conf. Video and Data 84*, April 1984, IERE Publ. No. 59, pp. 151-157.

Morgan, D.P., "Recording PSK on a Rotary-Head Recorder," *Datatape Inc.*, A Kodak Company (THIC Meeting), October 1984, 40 pages.

Standards

Ruigrok, J.J.M., "Cross Measurements in Magnetic Recording," *IEEE Trans. Magn.*, Sept. 1984, Vol. MAG-20, No. 5, pp. 875-877.

Bender, J.C., "A Proposed Standard for Tape Recorder IEEE-488 Interfaces," *THIC Meeting*, October 1984, 16 pages.

Staff, "Remote Interface Recorder Standards of the Future," *THIC Meeting*, October 1984, 10 pages.

Telemetry Standards, IRIG Standard 106-86, issued by Telemetry Group, Range Commanders Council, 1986, 138 pages. (Write to Secretariat, Range Commanders Council, U.S. Army White Sands Missile Range, New Mexico 88002.)

Chapter 28
Audio Recording

Audio recording is the grand daddy in magnetic recording, and has only recently been taken over by the home video recorders in terms of business volume. The techniques applied in modern recording equipment are on par with what a few years ago was reserved for the much more expensive instrumentation recorder.

This chapter will deal with some aspects peculiar to the recording and playback of sound—music and speech, in particular. Sound phenomena in acoustical research are recorded on special instrumentation recorders.

THE AUDIO SIGNAL, EQUALIZATION

The allowable pre-equalization in audio recordings depends on the type of music to be recorded since different instrument groupings and musical compositions have different energy spectra (ref. Sivian et al.). In order to have interchangeable tape recordings the need for standards is obvious. There is a general agreement that any standard must pertain to the remanent flux on the recorded tape and it is common among the standards that there is adherence to a flux level versus frequency response that essentially follows the coating thickness loss. This is not frequency but wavelength dependent, which in turn means that equalization networks (or their component values) must change with different tape speeds. The variance in standards is found in the selected cross-over frequency values for tape flux, or the equivalent time constant (70, 120 μS etc.).

The set of curves in Figs. 28-1 and 28-2 illustrates three widely used standards at the tape speed of 19 CM/S (7.5 IPS) with a spread in the cross-over frequency of 2:1, which gives a 6 dB level difference at high frequencies. A few recorders offer a selector switch between, for example, CCIR (Europe) and NAB (U.S.A.) standards, although this may be rather academic since most recordings are tailored to the individual recording engineer's tastes, which

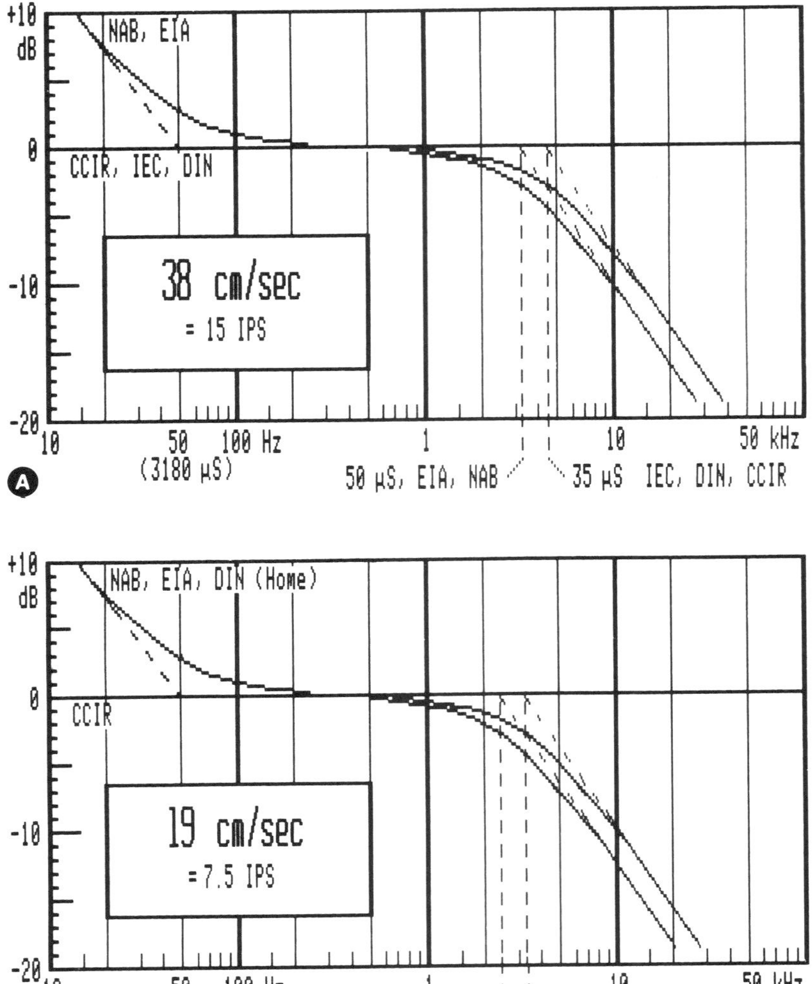

Fig. 28-1. Playback equalization, audio, (A) at 38 and (B) 19 cm/sec.

includes his control of bass and treble. The choice of standardizing organizations is also evident in the fact that NAB in the U.S.A. recommends a low-frequency boost during recording with the argument that less post-equalization is required with less amplifier hum and low-frequency noise, therefore, a lower general noise level (ref. McKnight, three).

On the other hand, CCIR in Europe argues that this very easily results in over-recording and intermodulation in program materials rich in low-frequency energy, such as organ music. European broadcast stations have not to this date recognized the NAB standard (ref. Borwick).

Improved tapes may result in different flux curves, but the different standards are nevertheless inconsistent: different concepts are used and the same terms have conflicting uses, resulting in confusion when standards are compared. And the standards should at times be updated

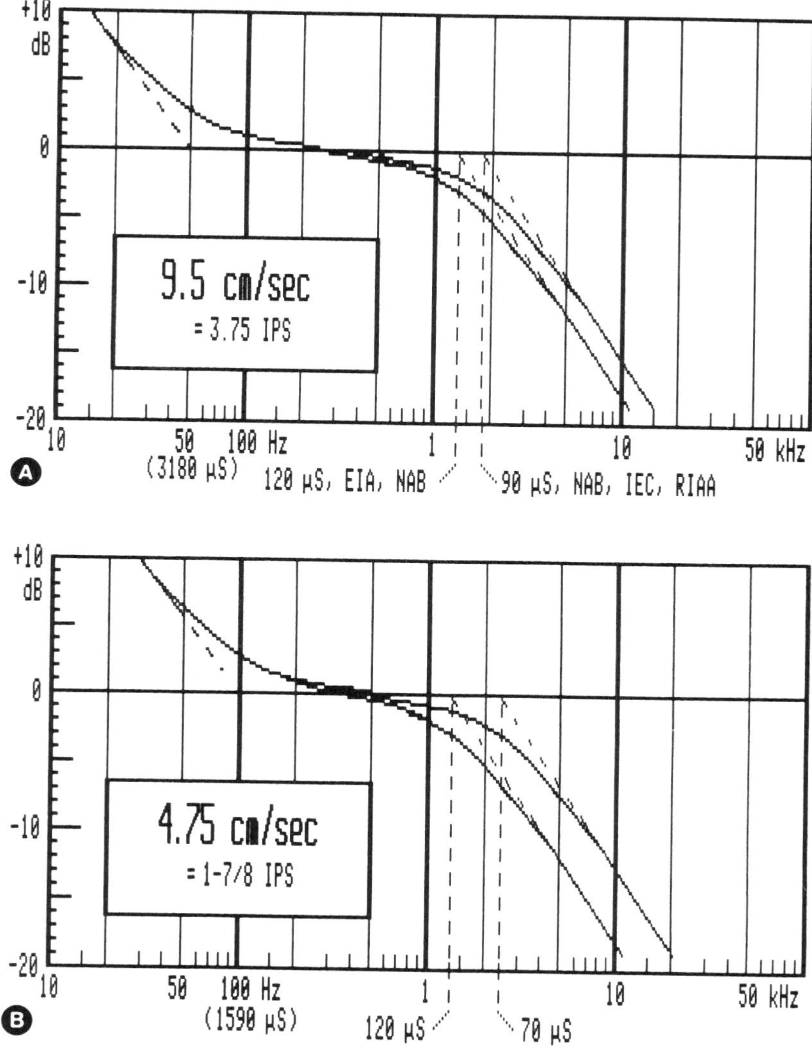

Fig. 28-2. Playback equalization, audio, (A) at 9.5 and (B) 4.75 cm/sec.

as tapes and tape recorders are improved through the ongoing evolution of about +1 dB/year. Certainly, current system's SNR can be improved by as much as 10 dB with proper use of external pre- and postequalizers (ref. Fielder).

Post-equalization in a tape recorder is most expediently adjusted by the use of a *standard alignment tape*, which contains first a high frequency tone for playback head azimuth adjustment, then a normal record level tone (0 dB), and finally a series of different frequency tones for equalizer adjustment. The next step is setting the bias level for the selected tape and finally the trimming of the pre-equalizer network to obtain an overall flat level versus frequency response. These steps were covered in detail in Chapter 23.

The design of equalizer networks must be done so overall phase distortion in the record-tape-playback channel is minimized (ref. Preis). Variations in tape characteristics require the use of one selected type only, unless the user is prepared to readjust bias and possibly record equalizer for each tape type; this operation may also be a built-in feature of the recorder (ref. Sakamoto et al.).

NOISE, PRINT-THROUGH

Several noise sources exist within the magnetic record-playback process, originating in the heads, the electronics, as well as from the tape itself (ref. Skov). Ideally, one would like the recorder to be absolutely quiet with the tape as the only noise source.

Electronics noise is quite trivial, originating from power supply hum, low-frequency noise in the input stage, and resistor noise (hiss). These noise generators can be fairly well controlled by careful design and component selection, such as low-noise resistors and high-grade transistors. If, for example, an insertion-type equalizer is used after the preamplifier, great care must be taken in optimizing the signal level and equalizer impedance in order to avoid added noise from the resistive components.

High frequency noise in a wideband instrumentation recorder (1 to 2 MHz) is often generated in the reproduce head itself. It is caused by eddy-current losses and the equivalent loss resistance of a playback head may be higher than the equivalent noise resistance of the preamplifier (50-100 ohms). That situation should be corrected by selecting a thinner lamination for the head core or by using a ferrite core, possibly with hard tips.

Electronic noise sources are aggravated by the required post-equalization boost of low and high frequencies. And here a word of caution when evaluating a recorder's signal-to-noise ratio: it is normally measured without filters or a weighting network. The measured noise voltage is representative for the low- and high-frequency noise and does not show that the mid-band noise is much lower, in the order of 20 to 40 dB for an audio recorder, and as much as 50 dB for a constant-flux instrumentation recorder. It would be entirely proper to use a *weighting network* when evaluating an audio recorder, since the shape of the noise spectrum happens to be close to Fletcher-Munson's curves for the ear's sensitivity (ref. McKnight). The subjective signal-to-noise ratio, therefore, is closer to 70 to 80 dB for a recorder that measured 50 dB on a VTVM, without filtering. It is likewise more useful to evaluate an instrumentation recorder from a spectrum of its noise, rather than by a single broadband measurement.

Another major noise source in a tape recorder is the tape itself and it shows up in various ways that are characteristic for magnetic recording. If a *virgin tape* is played in an audio recorder, one will notice a hissing sound. This noise is caused by the limited and varying number of particles that the playback head "sees." Each particle is a minute permanent magnet and if a sufficiently large and constant number were "seen" by the playback head, their fields would cancel out and no noise voltage would be generated.

If the same tape is again played on the recorder, this time with the record button pressed, the noise level will be several dB higher. This *bias induced* noise remains a limitation of the signal-to-noise ratio in modern recorders, and can at best be reduced to a level of 3 dB higher than virgin tape (or bulk degaussed tape). Investigations of this noise indicate two categories of bias noise.

When the bias frequency is higher than that resulting in a wavelength of 2 μm (80 μin), noise appears to arise from a combination of the bias field and the interaction field between particles. A lower bias frequency does not only increase the noise level, but it is now generated as *AM-modulation noise*. This is illustrated in Fig. 28-3A.

AM-modulation noise (or noise behind the signal) is best understood if for a moment we consider the DC-magnetization noise of a magnetic tape. A DC-magnetized tape can have a noise level 20 to 30 dB higher than a virgin tape. This is due to the non-uniformity of the coating dispersion, the backing surface, and the coating surface. An AC signal likewise generates noise and can be considered a slowly varying DC signal. The noise is formed as sidebands around

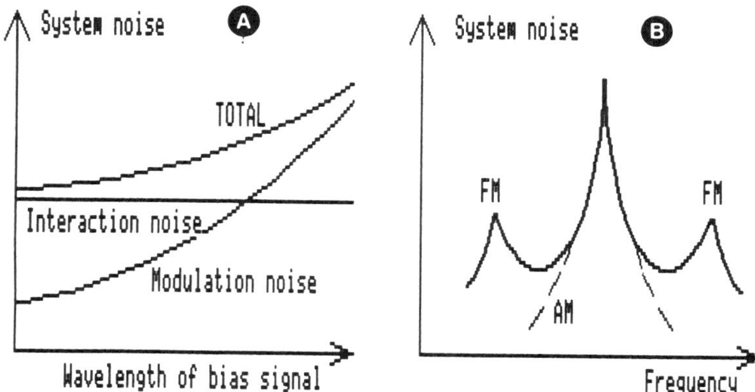

Fig. 28-3. (A) Bias noise shown as function of bias wavelength. (B) AM versus FM modulation noise.

the signal frequency (Fig. 28-3B). This noise is not very noticeable, since it is masked by the signal. Only when pure tones are recorded and played back can the ear detect AM noise.

Returning to the bias induced noise, it is likely that a low frequency bias generates AM noise with sidebands reaching into the pass band of the recorder. These findings were made with a distortion free bias oscillator (ref. Ragle et al.). Additional noise will be generated if the bias signal contains even harmonic components, which in essence constitutes a DC signal for which we have already explained the noise mechanism. The intermittent, gurgling sound of this noise is quite different from hiss. It is best avoided by using a push-pull type oscillator, possibly with filtering. Some recorders use a bugging DC-current through the record head winding, but that is only a cure, not a solution. AM-modulation is often mistaken for FM-modulation noise, which is caused by scrape flutter in the tape transport mechanism. This generates sideband, as shown in Fig. 28-3B, and can be very disturbing in changing the sound of a pure tone into a harsh sounding note.

There are several sources for noise in recording equipment. Some have already been discussed, such as the ones above, plus the section on print-through in Chapter 14. Another noise-like effect originates in surface imperfections, called *drop-outs*.

Noise from Drop-Outs

The noise from drop-outs may be quite annoying. The nature of this noise is really a momentary reduction (attenuation) of the playback signal and its value can be rated in terms of the duration and the amount of the attenuation (ref. Bauer et al.).

The drop-outs vary in duration. The average value is typically represented by an equivalent tape length of 10 mils, which in time corresponds to 5 milliseconds at a speed of 4.75 cm/s (1 ⅞ IPS).

The momentary loss of signal is an amplitude modulation, with associated sidebands in the frequency domain. When a sinusoidal tone drops out and recovers quickly enough clicks can be heard, which are loudest during the steepest transitions.

Measurements have established that drop-outs shorter than a few mS are in practice inaudible (ref. Trendell, Admiraal). Admiraal established an *annoyance scale* for use in listening tests:

h = 0 - no annoyance
= 1 - audible, but not annoying
= 2 - annoying
= 3 - very annoying

Drop-outs longer than 20-50 mS were rated h = 0 for a 3 dB attenuation, h = 1 for a 5 dB attenuation, and h = 2 for a 8-10 dB attenuation. They were very annoying only for attenuations near 20 dB. Repeat drop-outs were more annoying than single drop-outs.

A 25 mS dropout at 4.75 cm/s corresponds to a tape defect that is about 50 mils long. This can easily be seen without a magnifying glass and such tape defects should not exist in even the cheapest tapes manufactured today. They are much more likely to occur because of leaving tapes laying around without a protective case.

It appears that the earlier described *modulation noise* (see Chapter 12) is more annoying in recordings, lending support to the success of ''Direct-to-Disc'' recordings, and recently to the optical disks. Admiraal noted that flute music is the most sensitive of all to modulation noise. For flute music the modulation noise is perceptible as soon as the signal-to-modulation-noise ratio goes below about 40 dB. With orchestral music, however, the modulation noise does not become audible until the signal-to-modulation-noise ratio is less than about 20 dB.

Crosstalk

The leakage of unwanted signals from recordings on adjacent tracks should be non-existent. Some crosstalk can be tolerated in stereo recordings where the difference between channels is small.

For details on cross-talk see Fig. 10-28.

Head Noise

The noise voltage from the losses in audio heads may be limiting for the SNR (see Chapter 10).

Amplifier Noise

A number of factors contribute to the noise in amplifiers modified by equalization.

System Noise

The sum total of the noise voltage listed above will limit the signal-to-noise ratio of any high fidelity sound system. The effects depend heavily upon the specified record and playback equalizations and the allowed maximum distortion level.

The designer and user of analog recorders must bear in mind that the concepts of system grounding, shielding, cable type (often two-conductor, overall shield) and potential ground loops are very applicable.

NOISE REDUCTION

Noise reduction methods are today equated with a Dolby system. There are several versions of this method, which essentially raises faint record levels during recording to bring the magnetization well above the noise levels. Upon playback a complementary reduction in playback gain is made. This brings about a higher SNR. Something must tell the playback amplifier to ride gain, and this is what the Dolby system does (ref. Dolby). Since the details are beyond the scope of this handbook, the reader is referred to the special literature.

MAKING RECORDINGS, EDITING AND COPYING

The operation of any particular recorder is described in its instruction booklet or operating manual, which should be studied carefully. Familiarity with the manufacturer's instructions will

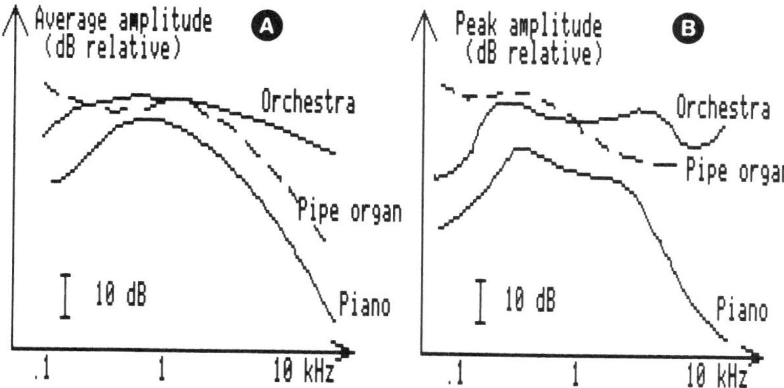

Fig. 28-4. (A) Average, and (B) peak amplitudes for music programs (after Sivian, Dunn and White).

prevent improper operation and you will be aware of the recorder's capabilities and limitations. However, there are certain fundamental rules that apply to most recorders, and one is on the proper recording level. This applies in particular to music recording.

It is essential to a good recording that it be free of distortion and that it have a maximum signal-to-noise ratio. All recorders are limited by amplifier and tape noise, and the optimum signal-to-noise ratio is, therefore, attained by recording with as strong a signal level on the tape as possible. Recording at too high a level, however, will cause both harmonic and intermodulation distortion. In music and speech the natural sound is distorted and in instrumentation recording the playback data will be erroneous.

The proper recording level can be attained only with a certain knowledge of the program material and some form of monitoring the recording level, for instance with a VU meter. A magnetic tape's overload sensitivity is essentially the same at all frequencies. In almost all cases, by contrast, the program material has an uneven amplitude versus frequency distribution and the highest level that can be tolerated is represented by the peak level in the amplitude distribution. For example, a male voice is rich in low frequencies, while a female voice is rich in high frequencies. Figure 28-4 illustrates both the average amplitudes and the peak amplitudes for an orchestra, a pipe organ, and a piano. It is clearly seen that overload during recording of a pipe organ will be caused by the low frequencies, while for a piano it will be the mid-frequencies. These examples also show that the sound level rolls off rather fast toward high frequencies and audio recorders incorporate pre-equalization which boosts the record current at high frequencies to more fully utilize the capabilities of the tape.

The curves in Fig. 28-4 are by no means universal, since they will change with orchestral composition, the type of music being played, and with the conductor. They will be further enhanced by the studio acoustics and the applied microphone technique. In order to provide interchangeable tapes, standards have been established for the amount of pre-equalization that can be applied at the various speeds.

The Correct Recording Level

Referring again to Fig. 28-4, the reader may question why two sets of graphs are shown for music amplitudes. The left set of curves shows the average amplitude as it would be measured with a VU meter, while the right set of curves shows the peak amplitude which could be monitored with a peak-reading instrument. The recording level is monitored by various devices, such as an inexpensive neon lamp, a VU meter, a peak-reading meter, or the earlier type "magic eye." In some instrumentation recorders a series of small cathode ray oscilloscopes are installed for monitoring the input signal to each channel.

The most annoying type of distortion in tape recordings, both music and instrumentation, is intermodulation distortion which occurs at the instant the peak level exceeds a certain amplitude. The peak-reading devices are clearly advantageous over the average type indicators by showing the operator when a recording level is too high. The VU meter is useful only when the operator or recording engineer has previous experience with the program material to be recorded. This experience can be obtained by rehearsing the record level setting prior to the actual recording. This is relatively easy to do on a recorder that has separate record and reproduce electronics by alternately listening to the input signal and the playback signal and comparing the sound quality. The gain control for the recording input is adjusted upwards to a level that does not result in intolerable distortion. If the recorder does not have separate record and reproduce electronics, it is necessary to record a small portion, rewind the tape and listen to it, and by memory compare the quality of the input signal with the playback signal. The next step is to repeat this process and so by trial and error obtain the proper record level.

Orchestral music often has a dynamic range that exceeds that of the recorder's and, in this case it is common practice to "ride gain." The recording engineer will be familiar with the program and can anticipate loud passages and very slowly reduce the gain prior to such a passage. This requires great skill and experience and should not be attempted by a novice. He would be better off with a recorder with an *automatic gain control*. Such a unit can be purchased as an accessory for the recorder. But the true value of an automatic gain control is attained only if, upon playback, the AGC action is reversed, which in turn requires an additional control track on the tape. Recorded on the control track is a signal that represents the AGC action and upon playback is fed to the control of another AGC unit. This method is employed in the Dolby systems.

The rapid rise of distortion versus input level is illustrated in Fig. 28-5. There is only a 10 dB margin between the one percent harmonic distortion level and full saturation of the tape. Intermodulation distortion, as a rule of thumb, is three times higher than the harmonic distortion level.

As a gauge audio recording engineers use a level referred to as MOL, which is the *maximum operating level*. It is normally defined as being the magnetization in nWB/m that corresponds to three percent harmonic distortion of a one kHz fundamental tone. Figure 28-6 shows how the MOL of a typical modern tape (EMI 816) varies with bias current (ref. Borwick).

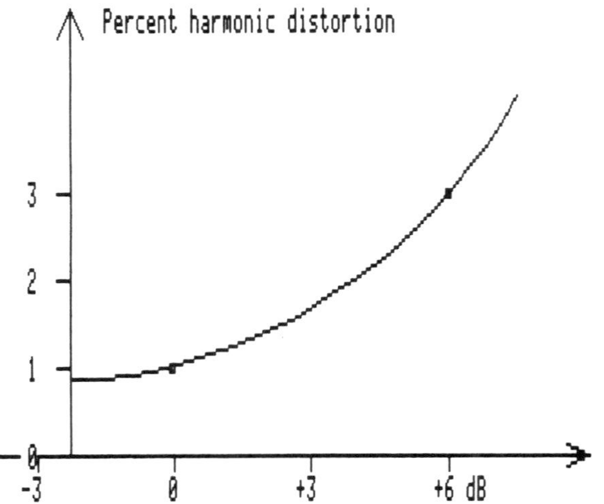

Fig. 28-5. Harmonic distortion versus record level (normal record level is at 1 percent harmonic distortion).

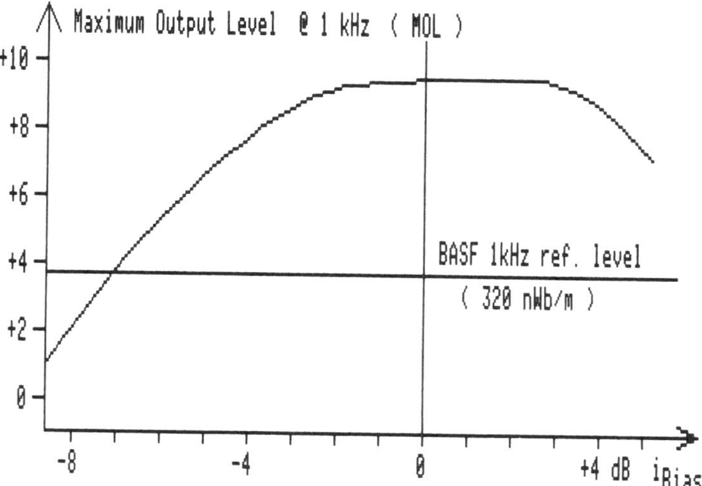

Fig. 28-6. MOL as function of bias level in a modern audio tape (after Borwick).

Microphone Techniques

The quality of a recording depends heavily on room acoustics, since the microphone will pick up the DIRECT SOUND from the sound source as well as the INDIRECT SOUND coming from reflections off the walls in the room (reverberation). The contribution from the indirect sound plays a major role in the recorded sound quality since it "colors" the sound. A recording made in a well-damped room (carpets, upholstered furniture, drapes, etc.), where the reverberation is small, will sound dry and unnatural. A recording made in an empty room with hard walls, on the other hand, will contain a large amount of reverberation, which in speech will mask the intelligence and in music will contribute heavy echo effects.

The ratio between the direct sound and the indirect sound can be varied by changing the distance between the sound source (for instance, a speaker) and the microphone. It will be necessary to experiment to find the best microphone position, where the ratio between the direct and indirect sound is the best possible and the recording sounds most natural. The reader is encouraged to undertake a simple experiment to observe this phenomenon.

Make four consecutive recordings, each time repeating the same sentence. Position the microphone eight inches from your mouth during the first recording, one yard during the second recording, two yards during the third, and four yards during the fourth recording. In each case the record gain control on the recorder must be adjusted for a normal record level (increased each time). Now rewind the tape and play back the four recordings. You will notice a striking difference in the sound. The first recording will sound dry and unnatural; the next recording will be much more alive, and at the same time the listener senses the presence of the room the recording was made in. During the third and the fourth recordings, you will not only get the impression that the distance to the microphone has been increased but also that the distance to the microphone has been increased, while the room also seems to have grown larger. This simple experiment clearly shows the importance of a proper microphone position to preserve the natural sound and to emphasize or de-emphasize the acoustical quality of the recording room.

An additional effect is obtained by using the microphone's directional characteristics. If an omni-directional microphone is used in an acoustically poor room (large degree of reverberation), it becomes necessary to position the microphone near the speaker. A cardioid type microphone is preferable in this case. If the recording is planned for a small orchestra group, the

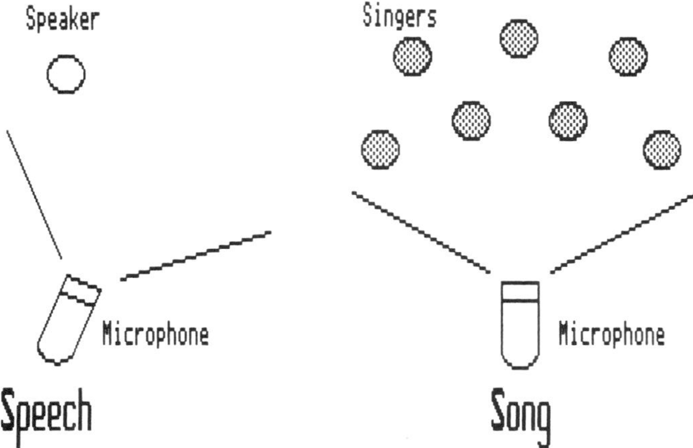

Fig. 28-7. Microphone placement for recording of speech, and song.

microphone position becomes more difficult, since different instruments produce different levels. It may in this case be necessary to try not only several microphone positions but also to place the musicians in various positions. A better solution may be to employ two or more microphones and use a mixer to obtain proper balance between the microphone inputs.

The following rules applied with reasonably flexibility will result in good recordings:

Speech: Place the microphone at a distance of 12 to 24 inches from the speaker. If any "s" sounds are annoyingly loud, turn the microphone slightly, as shown in Fig. 28-7. When recording a chorus where individual voices may have varying levels, the singers should be spaced from the microphone as shown in Fig. 28-7, right. If an intimate effect is required, the speaker should be close to the microphone and talk in a soft voice. The impression of a large room is obtained by placing the microphone away from the speaker.

Piano music is best recorded with the microphone positions shown in Fig. 28-8, left. An alternate technique is to place the microphone very close to the sound board in a grand

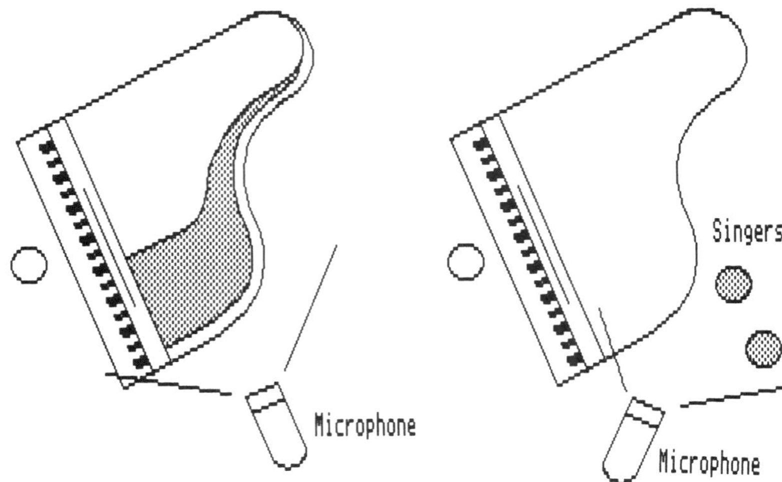

Fig. 28-8. Microphone placement for recording of piano, and piano and vocal.

piano. Piano music accompanied by song is more difficult to record. The piano will normally have a louder sound than the singer and the choice is either to use two or more microphones with mixers or to position the singer between the microphone and the piano, as shown in Fig. 28-8, right. If the piano player is singing along to the music, the microphone should be placed next to the player, never on the piano, since the microphone may pick up vibrations from the piano.

Pipe organ recordings place a great demand on both the microphones and the tape recorder. It is advisable to place the microphone at a rather large distance from the organ so as to obtain a good balance between the direct and the indirect sound. Caution should be exercised in selecting a moderate record level, particularly with recorders having NAB pre-equalization which boosts the low frequencies and easily causes intermodulation distortion.

Orchestral music is in general recorded by two general techniques. Symphonic music may be recorded with one microphone positioned over and behind the conductor. Dance or popular music is recorded with several microphones which are mixed in a suitable ratio to give the desired effect.

Stereo recordings are made with two microphones; the output of each is recorded on separate tracks. By playing back these two tracks through separate speakers, a special effect is obtained which gives the listener a sound picture, as if he were placed in front of the orchestra. This feeling is obtained by the phase difference between the sound from the two speakers and also by the intensity of the sound. The amplification, gain settings, and frequency response of each channel must be identical (plus or minus one dB), and this demands that the selected microphones be of the same type. The speakers used for playback should also be a matched pair. The listener ideally should be exactly at the midpoint in front of the two speakers in order to realize the optimum stereo effect. A more realistic spacial sound picture is obtained by *binaural* recording, a technique whereby the recordings are made with an artificial head with the two microphones placed at positions corresponding to the ears. Listening to the playback of such recordings via a pair of head phones gives strikingly realistic sound results.

RECORDING FROM RADIO AND TELEVISION SETS

The simplest (but poorest) method for recording from a radio or TV set is merely to place the tape recorder's microphone in front of the loudspeaker. The microphone picks up not only the sound from the program material, but also the background noise and any indirect sound from the room. This necessitates that the microphone be placed immediately in front of the loudspeaker, but even then a poor recording results.

A better method is to solder or clip two wires to the speaker terminals and connect them to the tape recorder input (not the microphone input, which has far too much sensitivity for the signal coming from the loudspeaker terminals). This eliminates any background noise and indirect sound from the room, but it most likely will result in a distorted amplitude versus frequency response due to tone-frequency or base control on the radio set is turned down somewhat.

Both methods are accompanied by a disadvantage: any change in listening level will require resetting the recording level. Therefore, in recent years it has become a practice to equip the tuners or amplifiers with a special output connection for tape recording, where the signal is tapped off immediately after detection or the early amplification stage and thus before any gain controls or frequency emphasizing networks can affect it.

Tape duplications are made by connecting the output terminals from one recorder to the input of another recorder. The best sound quality is obtained by playing the tape to be duplicated backwards. Simply play the tape through, then switch takeup and feed reels. Now, the tape will be playing backward as it is duplicated on the other machine. This technique will not work with dual- or multi-track recorders though. There are minute phase shifts inherent in all tape recordings and these can be reduced by the backward duplication process. The backward process should be of little concern to the user, since it is a good rule to leave the tape on the take-up reel after playing it and then to store it in that fashion. This prolongs the tape's life and minimizes dust pickup and edge damage.

SOUND EFFECTS

The advent of two-channel recorders for stereo recording and playback allowed a couple of extra features to be incorporated into the art at practically no extra cost. One of these is the echo effect which is achieved by connecting the playback head output to the record head input thereby re-recording a recorded signal to obtain an echo effect. (This should not be confused with reverberation, since the echoes are distinct and often spaced several hundred milliseconds apart.)

A more interesting and useful feature in many of these machines is the *sound-on-sound* technique. This is useful, for instance, in the recording of piano music accompanied by a vocal. A recording of the piano is first made on track one and possibly repeated until the quality and performance is satisfactory. The sound-on-sound selector switch is now turned on and the vocal is recorded. When the tape is played back, the first recorded piano music can be heard using a set of head phones, but now accompanied by the vocal. The singer is recorded on track number two and it can be repeated until the vocalist is satisfied with the quality. It is possible to add, for example, a guitar by flicking the sound-on-sound switch back to its original position. The recording will be made on track number one of the mixed input from the guitar microphone and the piano, plus the song recording on track number two. This feature has many possibilities and may be used in multiple guitar recordings (as employed by Les Paul in the 1950's).

Sound effects can be dubbed in too. They can either be purchased prerecorded on special records or tapes, or can be "homemade," as listed below. In most cases, the microphone should be placed close to the sound image.

BREAKING THROUGH A DOOR. Break a thin piece of wood with a hammer stroke.

CAR BRAKES. Move a metallic item (for example, a fork) over a glass sheet.

CAR CRASH. Let a few metal sheets fall onto the floor.

CAR DOOR. Close a thick book.

ELEVATOR. Start (or stop) a vacuum cleaner.

FIRE. Slowly squeeze cellophane wrapping in front of the microphone.

HORSE STEPS. Cut a coconut in half and clap the two halves against each other. (Cover one of the halves with, say, a sock if the sound should be of horses riding on a wooden bridge.)

JET PLANE. Hold a hairdryer in front of the microphone and let the airstream howl over a sheet of cardboard.

LOCOMOTIVE. Glue two sheets of abrasive paper onto two pieces of wood; rub them against each other.

MACHINE GUN. Squeeze a piece of aluminum wrapping foil.

OCEAN LINER. Blow downwards into an empty bottle.

POWER BOAT. Hold the microphone close to a household mixer.

RAIN. Shake 20 to 30 dry peas in a net sieve over a microphone, or let a handful of sand slowly drizzle onto a sheet of paper or cardboard.

RIFLE. Slap a flat ruler against a table surface.

ROWBOAT. Move a couple of pieces of wood in and out of a bucket filled with water.

STEPS. Squeeze a sheet of paper in a rhythmic fashion.

STEPS ON SNOW. Squeeze a handful of starch in a rhythmic fashion.

SURF. Brush a metal (or cardboard) sheet in a rhythmic fashion with two brushes.

THUNDER. Hold a metal sheet by one corner and shake it violently.

VOICE OVER A TELEPHONE. Speak into a paper cup.

WAVES. Fill a tray with water and move one hand around in the water. Hold the microphone close by.

WIND. Move a piece of silk over a table edge or imitate the sound by whistling through your teeth.

The above suggestions should provoke your imagination to devise many more.

SOUND FILM

Sound film recording is often made on a narrow magnetic strip on the photographic film. This old technique has in recent years become quite popular in the Super-8 film market. It should remain so for the serious film maker due to the ease of splicing film, as compared with difficulties in editing the new 8mm home video recordings.

Examples of striped films are shown in Fig. 28-9. The recording and playback principles are the same as described in any AC-bias audio recording system. The operating conditions are special since the recording is done on a somewhat thicker and therefore stiffer film base. The sprocket holes in the film affect its continuous motion over the heads, with a potential disturbing flutter frequency equal to the number of sprocket holes passing per second (18 or 24 Hz for Super-8 and 16mm, 96 Hz for 35mm movie films).

The film is advanced in steps past the optical system and the film must next change this motion pattern to a flawless uniform speed past the sound heads. This places a great demand upon the mechanical filtering system in the camera as well as in the projector. The distance

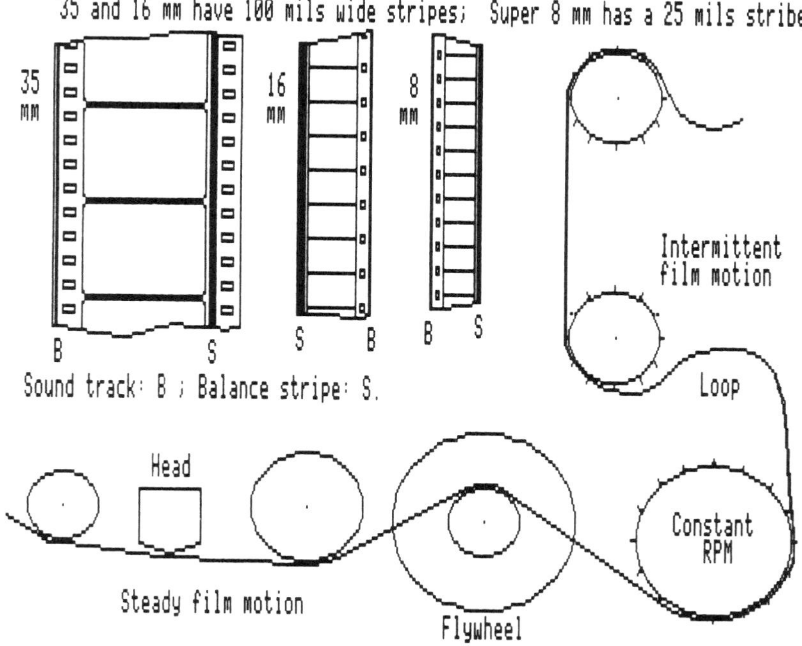

Fig. 28-9. Top: Striping of magnetic coating on sound films. Bottom: Sprocketed film transport with flywheel filtering to obtain a stable film motion past the magnetic head assembly.

between a sound and the corresponding picture is only 18 frames in Super-8, equal to a film length of 3 inches. The isolation is better in 16 and 35mm films, where the distance is up to 28 picture frames.

Recording sound directly onto the magnetic stripe on the film is only done in the Super-8 field (ref. Lovick et al., Carr). This method naturally offers great simplicity while filming, but it has disadvantages when editing. The sound is 18 frames behind the picture and all spliced scenes must therefore have about a second's silence at the end (or ambient sound: no speech).

Many projectors offer the record/play capability for the balance stripe in addition to the main sound track. This greatly facilitates the addition of music and sound effects without interfering with the original sound.

The sound quality of Super-8 film may not satisfy the critical Hi-Fi enthusiast—in particular when a series of film splices upsets the steady film motion past the playback head. In such cases a *double sound system* is used where the camera equally spaces pulses on the sound stripe and simultaneously feeds these pulses via a cable (or wireless) to a quality sound recorder. These synchronization pulses are recorded on one track while the sound is recorded on another track (or 2 tracks in the case of stereo).

The sound and sync pulses may be later transferred to a full-coated, sprocketed film. This can be banged together with the photographic film on a special editing table and they will be in perfect synchronism. Equal lengths of photographic film and sound film are added or deleted in the editing process and the sound may, after the editing is completed, be recorded onto the magnetic stripe on the film. Or it may be transferred (with sync pulses) back to a regular tape (¼" reel or cassette). In the latter case a synchronizer between the playback unit and the projector is required.

Sound recording on film has grown with the film into the 70mm format (ref. Mosely). Eastman Kodak Co. has recently made a most interesting application of magnetic recording in the fabrication of 35mm film: a magnetic control surface is coated onto the film on the side opposite the emulsion (ref. Compton and Dimitri). This layer is completely opaque, but does contain a minute amount of small magnetic particles. This makes possible the recording of about 100 bits along a track, per frame, more than enough for the standard 8-digit SMPTE time code.

DIGITAL TECHNIQUES

Digital recording of sound was first made available to the consumer in the form of the Philips optically recorded disk, while in the recording studios it was implemented in the form of multitrack tape recorders, much like the HDDR equipment described under digital applications of instrumentation recorders. These audio digital recorders use the same design philosophy in regard to selection of interleaving, error correcting codes and modulation codes, as do HDDR instrumentation recorders.

There are some special considerations; selection of the sampling frequency has been of concern to assure compatibility with the many interfacing systems (radio, TV, film, other studios etc. (ref. Mruaoka et al.)). The particular application of digital techniques to audio recording is covered in two books (Thomsen, and Nakajima et al.), and three papers (Willcocks; Blesser; and Bellis).

Editing of digital recordings is not as straightforward as the splicing method for ordinary audio tapes. The methods are electronic, use delay lines and sophisticated controls (ref. Youngquist).

The aspect of distortion and quantization noise has been treated and shown to become minimized when the signal to become quantized contains a wide-band noise dither with an amplitude of approximately the step size (ref. Vanderkooy et al.).

High quality audio can now be enjoyed not only by the means of the optical digital disk, but in the form of VCR tapes on newer machines with facilities for the recording and playback of FM audio. The next in digital sound appears to be the use of a small cassette with 3.81mm wide tape (ref. Sakamoto), or the new 8mm VCR format (ref. Watani et al.).

The audio signal is in the 8mm recorders first converted into a digital signal, and then time compressed by a factor of 6. Thus 6 channels can be recorded by dividing the 221 degrees tape wrap around the capstan into 6 sectors of each 36 degrees, as shown in Fig. 28-10 (ref. Itoh et al.).

The new recorders using the 3.81mm tape format are named *R-DAT* (Rotating (Head)—Digital Audio Tape). The proposed cartridge measures only 73 × 54 × 10 mm, and the tape is wrapped 90 degrees around the capstan drum, see Fig. 28-11. The relative velocity between the head gap and the track is approx. 5 m/sec.

The audio signal is sampled at a rate of 44.1 kHz, and must be time compressed due to the use of only two heads and the wrap angle of 90 degrees. The signal goes into a buffer memory, and is released to the head in blocks when either of the two heads are in contact with the tape. During playback a similar buffer memory stretches the data back into a continuous digital bit stream.

Synchronization between the rotating head and the tape is ordinarily done via a longitudinally recorded control track. The R-DAT recorders use a method developed by Philips, named ATF, for Automatic Track Finding. A short section of each track is recorded with a relatively low-frequency signal, that has a frequency that alternates between four values for each half rotation of the head assembly. During playback, the heads pick up more than the one track at a time (side reading at low frequencies), and proper alignment is obtained when the levels of the pilot tones from each track adjacent to the one being scanned are equal. This system makes a tracking control track unnecessary (ref. Feldman, Ranada).

These new recorder developments have caused great concern in the audio industry that fears widespread uses of the R-DAT in unlawful duplications of optical audio disks, with subse-

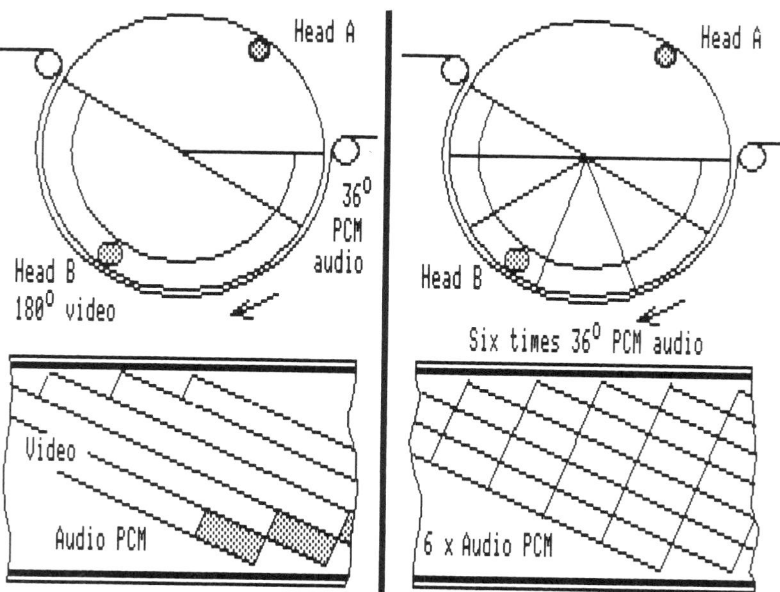

Fig. 28-10. Left: An additional 36 degrees wrap in the 8mm camcorders afford the recording of PCM sound. Right: Six sectors allow for 6 channel recording of PCM sound on an 8mm camcorder.

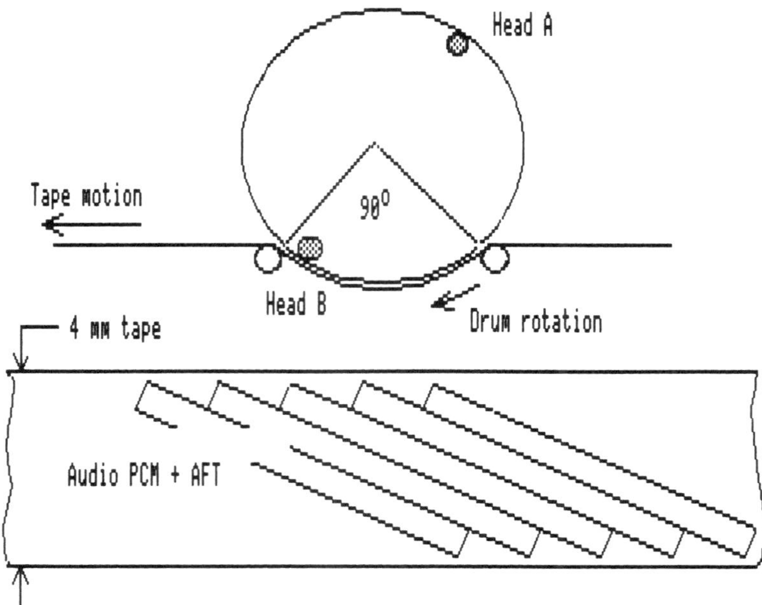

Fig. 28-11. R-DAT recorders use a 90 degrees tape wrap around the head cylinder. Tracks are recorded with AFT for track following control.

quent reduced revenues for the recording industry. In 1987 this led to an import restriction of the R-DAT into the United States.

STANDARDS

Several standards for the recorded flux level on magnetic tapes have been established in Europe and in the United States, standards that impose design restrictions on both recording and playback amplifiers. In transforming a record current to a remanent flux on the tape, a certain amount of pre-emphasis is necessary to overcome record head losses. Likewise, in transforming the flux from the tape to a voltage at the playback amplifier input, reproduce head core losses and gap losses affect the overall frequency response. It is therefore important that the designer know not only the head impedances but also has the data that tells him how the head losses vary with frequency.

The audio standards were shown in Figs. 28-1 and 28-2. Details pertaining to these, and other standards (reel, cassette and cartridge sizes; track configurations and dimensions) can be obtained from:

ITDA-International Tape/Disk Association
10 Columbus Circle, Suite 2270
New York, N.Y. 10019

ANSI-American National Standards Institute
1430 Broadway
New York, N.Y. 10018

NAB-National Association of Broadcasters
1771 N. Street N.W.
Washington, D.C. 20036

EIA-Electronic Industries Association
Engineering Department
2001 Eye Street N. W.
Washington, D.C. 20066

CCIR-International Radio Consultative Committee
International Telecommunication Union
Place des Nations
Geneva, Switzerland

Other standard organizations are:

- SMPTE—Society of Motion Picture and Television Engineers
- DIN—Deutscher Normenausschlus (DNA)
- BSI—British Standards Institute
- JIS—Japanese Standards Association
- IEC—International Electrotechnical Commission
- (Write to ANSI for information on these standards.)

REFERENCES TO CHAPTER 28

General

Borwick, J. (Editor), *Sound Recording Practice, A Handbook compiled by the Association of Professional Recording Studios*, Oxford University Press, London, 1980, 503 pages.

The Audio Signal, Equalization

McKnight, J.G. "Signal-to-Noise Problems and a New Equalization for Magnetic Recording of Music," *Jour. AES.*, Jan 1959, Vol. 7, No. 1, pp. 5-12.

Sivian, L.J., Dunn, H.K., and Write, S.D., "Absolute Amplitudes and Spectra of Certain Musical Instruments and Orchestras," *IRE Trans. Audio*, July 1959, Vol. AU-7, No. 3, pp. 1-29.

McKnight, J.G., "The Frequency Response of Magnetic Recorders for Audio," *Jour. AES*, July 1960, Vol. 8, pp. 146-153.

McKnight, J.G. "The Case Against Low-Frequency Pre-Emphasis in Magnetic Recording," *Jour. AES*, April 1962, Vol. 10, No. 4, pp. 106,107.

Preis, D. "Phase Distortion and Phase Equalization in Audio Signal Processing—A Tutorial Review," *Jour. AES*, Nov. 1982, Vol. 30, No. 11, pp. 774-794.

Sakamoto, N., Kogure, T., Ogino, M., and Kitagawa, H., "A New Magnetic Tape Recorder With Automatic Adjusting Functions for Bias and Recording Conditions," *Jour. AES*, Sept. 1982, Vol. 30, No. 9, pp. 596-606.

Fielder, L.D. "Pre- and Postemphasis Techniques as Applied to Audio Recording Systems," *Jour. AES*, Nov. 1985, Vol. 33, No. 9, pp. 649-658.

Noise, Drop-Outs

Bauer, B.B., Comerci, F.A., Foster, E.J., and Rosenheck, A.J., "Audibility of Tape Dropouts," *Jour. AES*, Feb. 1967, Vol. 15, No. 2, pp. 147.

Admiraal, D.J.D., Cardozo, B.L., Domburg, G., and Neelen, J.J.M., "Annoyance due to Modulation Noise and Drop-Outs in Magnetic Sound Recording," *Philips Tech. Review*, Jan. 1977, Vol. 37, No. 2/3, pp. 29-37.

Noise

McKnight, J.G, "A Comparison of Several Methods of Measuring Noise in Magnetic Recorders for Audio Applications," *IRE Trans. Audio*, Mar. 1960, Vol. AU-8, pp. 39-42.

Skov, E.P., "Noise Limitations in Tape Recorders," *Jour. AES*, Apr. 1964, Vol. 12, No. 4, p. 280.

Ragle, H.U., and Smaller, P., "An investigation of high-frequency bias-induced tape noise," *IEEE Trans. Magn.*, June 1965, Vol. MAG-1, No. 2, pp. 105-110.

Dolby, R.M., "An Audio Noise Reduction System," *Jour. AES*, Oct. 1967, Vol. 15, No. 4, pp. 383-388.

Trendell, E.G., "The Measurement and Subjective Assessment of Modulation Noise in Magnetic Recording," *Jour. AES*, June 1969, Vol. 17, No. 6, pp. 644-653.

Sound Film

Lovick, R.C., Bartow, R.E., and Scheg, R.F., "Recording and Calibration of Super-8 Magnetic Reproducer Test Films," *Jour. SMPTE*, July 1969, Vol. 78, No. 6, pp. 473-487.

Carr, D.L., "Audio Performance of Magnetic Prestriped Super-8-Motion Picture Films," *Jour. SMPTE*, Apr. 1975, Vol. 84, No. 4, pp. 249-257.

Mosely, J., "A New Look at 70mm Sound: Proposed Revisions of Magnetic Sound," *Jour. SMPTE*, May 1984, Vol. 93, No. 5, pp. 502-510.

Compton, D.M.J., and Dimitri, D.S., "Implementation of Time Code Using Datakode Magnetic Control Surface Film," *Jour. SMPTE*, July 1986, Vol. 95, No. 7, pp. 727-732.

Digital Techniques

Thomsen, D., *Digitale Audiotechnik*, Franzis' Verlag, Muenchen, 1983, 192 pages.

Nakajima, H., Ooi, T., Fukuoa, J., and Iga, R., *Digital Audio Technology*, TAB No. 1451, 1983, 312 pages.

Willcocks, M., "A Review of Digital Audio Techniques," *Jour. AES*, Feb. 1978, Vol. 26, pp. 56-64.

Mruaoka, T., Yamada, Y., and Yamazaki, M., "Sampling-Frequency Considerations in Digital Audio," *Jour. AES*, Apr. 1978, Vol. 26, pp. 252-256.

Blesser, B.A., "Digitization of Audio: A Comprehensive Examination of Theory, Implementation, and Current Practice," *Jour. AES*, Oct. 1978, Vol. 26, pp. 739-771.

Youngquist, R.J., "Editing Digital Audio Signals in a Digital Audio/Video System," *Jour. SMPTE*, Dec. 1982, Vol. 91, No. 12, pp. 1158-1160.

Bellis, F.A., "Introduction to Digital Audio Recording," *The Radio and El. Engr.*, Oct. 1983, Vol. 53, No. 10, pp. 361-368.

Vanderkooy, J., and Lipshitz, S.P., "Resolution Below the Least Significant Bit in Digital Systems with Dither," *Jour. AES*, Mar. 1984, Vol. 32, No. 3, pp. 106-112.

Sakamoto, N., Kogure, T., Kitagawa, H., and Shimada, T., "On High-Density Recording of the Compact-Cassette Digital Recorder," *Jour. AES*, Sept. 1984, Vol. 32, No. 9, pp. 640-645.

Watatani, Y., Itoh, S., Shibata, A., and Mohri, K., "The FM Audio Signal Recording System for 8mm Video," *IEEE Trans. Cons. Elect.*, Nov. 1984, Vol. 30, No. 4.

Feldman, L., "Digital Audio Tape Decks," *Modern Electronics*, July 1986, pp. 18-22.

Itoh, S., Watatani, Y., Azuma, N., Kaniwa, K., Masui, H., Nakama, T., Shibata, A., Watanabe, K., and Mogi, Y., "Multi-Track PCM Audio utilizing 8mm Video System," *IEEE Trans. Cons. Elec.*, Aug. 1985, Vol. 31, No. 3, pp. 438-446.

Ranada, D., "Why 8mm?", *High Fidelity*, Feb. 1987, pp. 47-52.

BIBLIOGRAPHY TO CHAPTER 28

Olson H.F., *Music, Physics and Engineering*, Dover, New York, 1967, 460 pages.

Masscho, R., *Technique du Magnetophone*, Editions Radio, Paris, 1977, 328 pages.

Monforte, J., "The Digital Reproduction of Sound," *Scien. Amer*, Dec. 1984, Vol. 251, No. 6, pp. 78-84.

Miyaoka, S., "Digital Audio is Compact and Rugged," *IEEE Spectrum*, March 1984, Vol. 21, No. 3, pp. 35-39. (Topic: optical disk).

Rundstein, R.E., and Miles Huber, D., *Modern Recording Techniques*, Howard W. Sams and Co., 1986, 384 pages.

Digital Audio, Collected Papers, AES Conv. June 1982, 262 pages.

Present and Future of Digital Audio, Proceedings of the AES 3rd Intl. Conf., June 1985, 221 pages.

"AES Recommended Practice for Digital Audio Engineering—Serial Transmission Format for Linearly Represented Digital Audio Data," *Jour. AES*, Dec. 1985, Vol. 33, No. 12, pp. 975-984.

Daniel, E. and Watkinson, J.R., "Audio Recording", Chapter 3, *Magnetic Recording*, Vol. 3, McGraw-Hill, 1988.

Watkinson, J.R., *The Art of Digital Audio*, Focal Press, 1988, 849 pages.

Camras, M., *Magnetic Recording Handbook*, VNR, 1988, 718 pages.

Chapter 29
Video Recording

Magnetic recording of video, or television pictures, is a technology only thirty years old. Recording and playback of television pictures place requirements different from that of audio and digital signals on the tape transports and the electronics. The signal channel must cover from very low frequencies (essentially DC) to several MHz, the SNR must be 40 dB (or better), and the time base errors must be within fractions of a micro-second.

The field of video recording is vast, with numerous tape and machine formats in operation. This chapter is written to serve as an introduction to the topic, and the reader is referred to the book edited by Benson to learn more about television signals, systems and recorders. The SMPTE Journal is a must for keeping abreast of new developments (see listed index and bibliography listing).

THE TELEVISION SIGNAL

In the United States the basic television picture is made up of 525 lines, as shown in Fig. 29-1. The front of the picture tube is scanned 60 times per second by the electron beam, and the picture is formed by varying the intensity of the beam. A picture is actually displayed only every 1/30th of a second since each picture is created by two scans and is updated thirty times a second. These two scans are in turn made by a 262½ lines picture, followed by another 262½ lines picture interlaced with the first.

We need to estimate the characteristics of the signal required to produce a satisfactory television picture. About 40 lines are never displayed, as indicated in Fig. 29-1, and are called overscan. Of the balance only 70 percent is normally judged useable, corresponding to 340 lines.

The aspect ratio (picture width W to height H) is 4:3, from which we can arrive at a corresponding number for vertical lines of $4*340/3 = 450$ lines. If the electron beam was to trace out these vertical lines there could at best be 225 white dots alternating with 225 black dots,

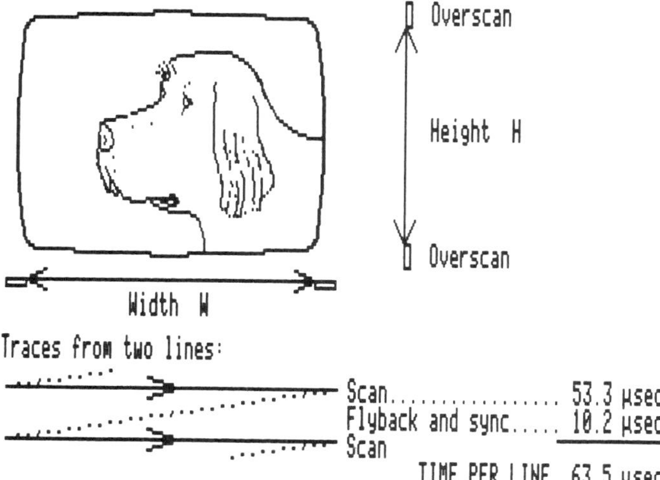

Fig. 29-1. A television picture is refreshed 30 times per second by two interleaved scans, each with 262½ lines.

making a total of 450*340 = 153,000 dots on the screen. These dots are often referred to as pixels, for picture elements. This will be the maximum resolution of a displayed television picture.

This display occurs 30 times per second, and the signal channel must therefore be capable of processing 30*153,000 = 4,590,000 dots per second; we must further consider that we have available only 53.3 μS of each 63.5 μS allocated per single line (see Fig. 29-1); the balance of 10.2 μS is used for flyback (to start a new line) plus synchronization signals. Further, the display is only a useable window for 70 percent of the total number of lines.

The signal requirements are therefore equivalent to 4,590,000 * (63.5/53.3) * (100/70) = 7,812,000 dots per second. Since one recorded wavelength corresponds to two bits we arrive at a signal processing speed of 3,906,000 wavelengths per second, or 3.9 MHz for a useable picture with a 153,000 dots resolution.

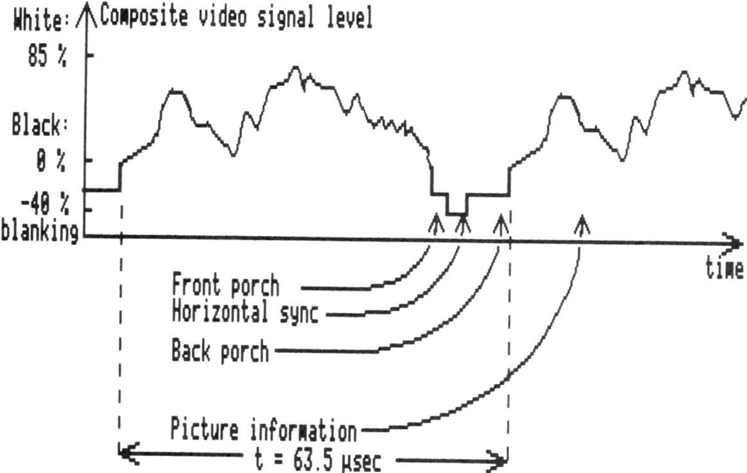

Fig. 29-2. Waveforms and levels of 1½ lines.

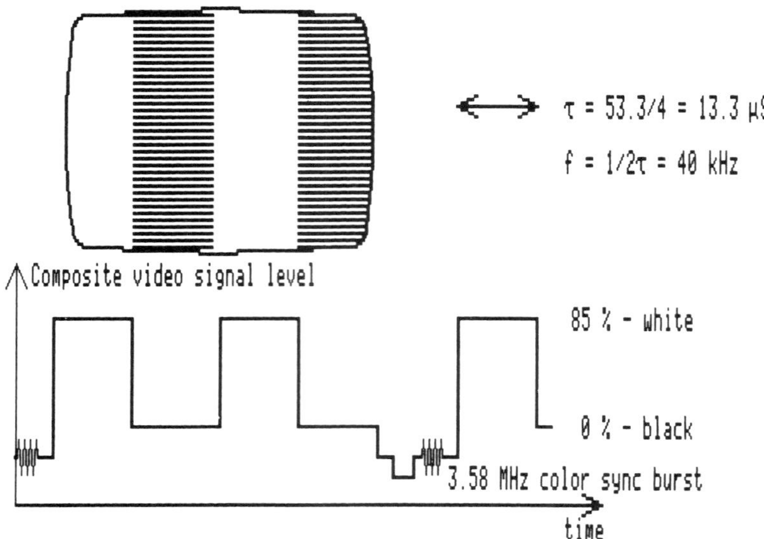

Fig. 29-3. Four bars (2 white and 2 black) are displayed by the video line voltage waveform shown.

This resolution corresponds to standard 16mm motion-picture film where each frame has about 125,000 picture elements. It is not the quality of 35mm film where each frame has 500,000 picture elements. This is why some countries have television standards with 625 lines (182,000 dots) or even 800 lines (233,000 dots).

Commercial TV calls for a black-and-white video information signal bandwidth of 3 MHz. Figure 29-2 shows the signal levels for 1.5 lines. The video signal level varies between 0 percent (black) and 85 percent (white), with the continuous full range of gray scale levels in between. A horizontal synchronization signal follows each line so they all start in synchronism.

Figure 29-3 shows the video signal required to produce four bars (white-black-white-black) on the screen. This illustration should help us appreciate that the average picture level produces a DC-signal (here at the 42.5 percent level) without which the picture would be shown with whiter-than-white or blacker-than-black levels. Hence the signal channel must provide for response to DC. The high frequency components are necessary to produce the sharp transitions between bars.

For the sake of completeness we should add that synchronization pulses are generated at the end of each set of 262.5 lines in order to synchronize the picture sequence by signalling when to start a new scan. This is a more complex topic and is deferred because it is of little consequence to the material in this chapter.

The signals and frequencies used for present color television were created with the restriction that existing black and white receivers could still use the signal and that color receivers could also use a black and white signal. Thus, color pictures, which are created by adding red, green, and blue video signals, must use the same bandwidth of only one video signal. This is done with a clever combination of the three signals into a luminance signal and two chrominance signals, which fit in a single video signal bandwidth.

The luminance signal by itself gives a black and white picture. The chrominance signals, both arriving on a 3.58 MHz subcarrier, and the luminance signal are used to recreate the red, green, and blue video signals for color televisions. In order to correctly extract the chrominance signals, a phase reference is provided by placing 8 to 11 cycles of the 3.58 MHz subcarrier on the back porch of the color video signal. This is known as the color burst.

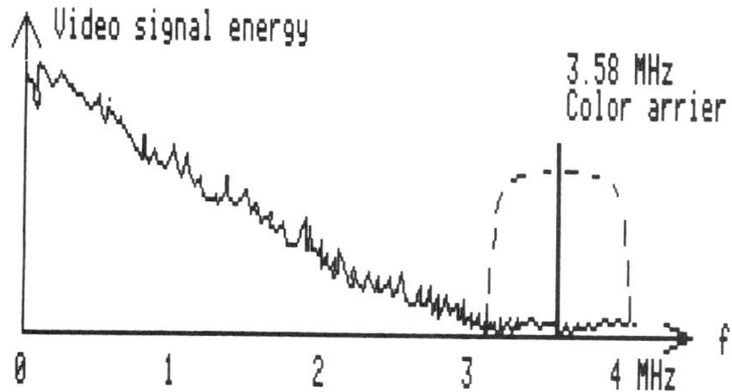

Fig. 29-4. Information in a video signal extends from DC to several MHz. The color information modulates a carefully selected carrier, which requires a color burst on the back porch to be demodulated.

The bandwidth of the chrominance or color signal, is generally only about 0.5 MHz, which corresponds to a resolution of 8 dot wide patterns. This appears satisfactory for most color pictures transmitted to the home user, and is therefore also the color resolution used in home VCRs (Video Cassette Recorders). There are VCRs with wider color bandwidth in the broadcast and studio fields.

The total signal consisting of video (luminance signal), synchronization pulses, color subcarrier and color burst is called the composite video signal; the frequency spectrum of this signal is shown in Fig. 29-4. When it is received (or played back) the video and the color signals are separated. The color burst is used to synchronize a local 3.58 MHz oscillator that serves as reference for extracting the color carrier's information. The synchronization is refreshed each line, or each 63.5 μS. A noticeable change in hue (color) occurs for just a few degrees deviation from the 3.58 MHz sinusoid, and a reasonably good reproduction requires a phase accuracy of plus or minus 4 degrees, corresponding to $(1/3.58)*(4/360) = 0.0039\ \mu S = 3.9$ nS time differential during each 63500 nS interval. There are no problems in building a reference oscillator with that stability—the problem will lie in providing a time stable tape transport for recording and reproducing the signal.

Let us for example assume that we have recorded an alternating white-black dot pattern, similar to the one we used to analyze the picture resolution. Assuming that two consecutive scans are synchronized at their start, what is the maximum time difference we can tolerate? This is a subjective question and let us say that a one bit shift at the end of the lines is unacceptable (the vertical line pattern would become a checker board on the right hand side of the screen). Such a shift corresponds to the duration of one bit, or $53.3/450 = 0.118\ \mu S$.

From this we can predict a reasonable limit for the time base error allowed for the tape transport:

$$\text{Maximum short term TBE} = \bullet\ 50 \text{ nS.}$$

This corresponds well with the maximum error output of TV recorder/reproducers with time base error correction. The long term error in the transport speed can be quite large, provided the TV monitor will stay in synchronization: Maximum long term TBE = + or − 2000 nS.

SIGNAL MODULATION AND EQUALIZATION IN A HOME VCR

The required stability of the reproduced color signal presented an even more serious problem. The time base error of a transport should be less than 4 nS in order to assure that the phase of color subcarrier is changed less than five percent, which is the maximum allowable.

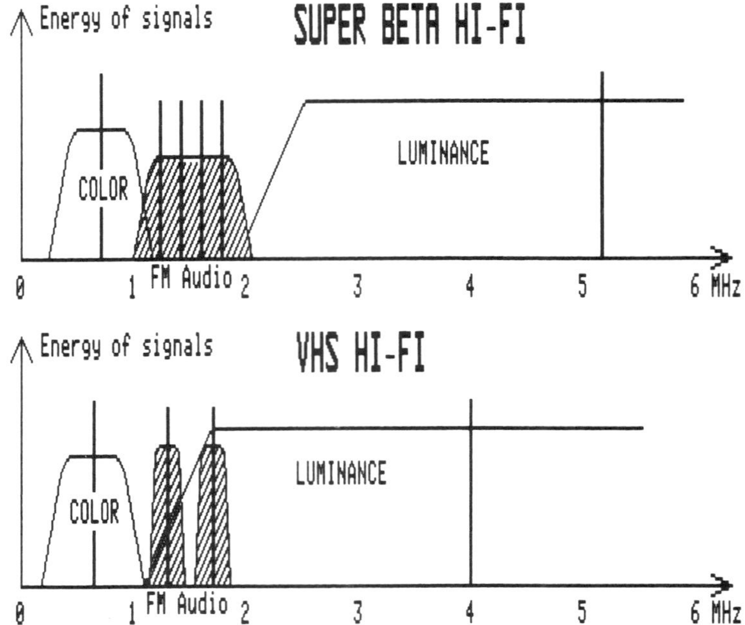

Fig. 29-5. The frequency spectra of home VCR systems.

The 4 nS applies when the color carrier is at 3.58 MHz; it can be higher if the color carrier is shifted down in frequency, which is done in the Beta and VHS machines.

This shifting down in frequency, called the "color-under" principle, extracts the 3.58 MHz color information from the composite video signal and places it at 562.5 kHz by a modulation process; the color carrier plus sidebands occupy about 1 MHz.

The video, also called luminance signal, FM modulates a carrier located at 4.05 (VHS) or 5.17 (Beta) MHz, with maximum excursions of ∓ 3¼ MHz. The resulting spectra are shown in Fig. 29-5. It is important to place the FM band as high as possible, and further adjust the carrier level so it serves as a bias for the color signal since this improves the linearity of the color levels. The two signal bands may overlap slightly to allow for a luminance signal bandwidth of up to 2.7 MHz, corresponding to a 235 line resolution in the reproduced picture.

The signal-to-noise ratio (SNR) depends upon a great number of factors such as trackwidth, system resolution, and tape quality. And it is always the case that the VTRs are pushed to provide as long a playing time as possible, which means narrow tracks and slow head-to-tape speeds.

The SNR for video transmission is normally expressed as:

$$\text{SNR} = (\text{peak signal})/(\text{rms noise})$$

Noise in a television picture displays as excessive graininess or a pattern like snow. TASO (Television Allocation Study Organization) has provided a picture rating table for pictures of a 4 MHz bandwidth:

1. Excellent (no perceptible snow in picture) .... 45 dB
2. Fine (snow just perceptible) .... 35 dB
3. Passable (snow definitely perceptible) .... 29 dB
4. Marginal (snow somewhat objectionable) .... 25 dB

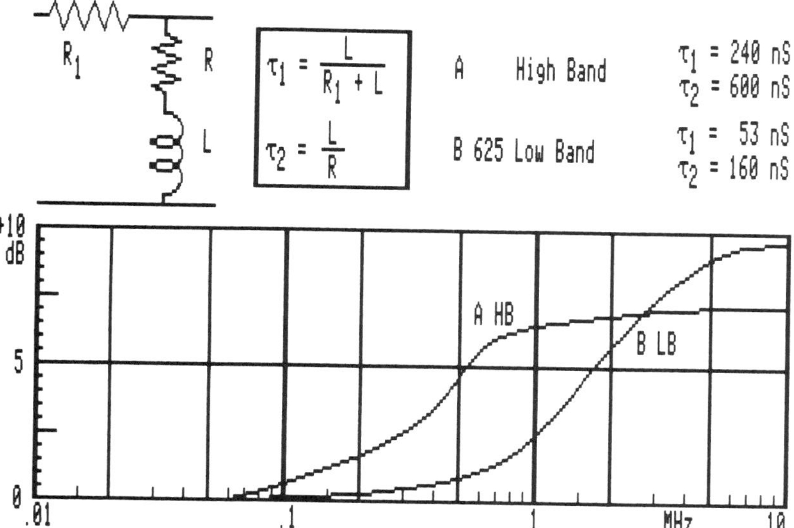

Fig. 29-6. Pre-emphasis curves for broadcast video recorders (after Robinson).

Amplitude uniformity must also be considered, and the low frequencies (1-300 kHz) should be within a 0.2 dB limit, while a 0.8 dB envelope applies for higher frequencies.

The latter is achieved by using FM-modulation for the video signal. The signal-to-noise ratio requires special equalization, much like in audio recordings. When we look at Fig. 29-4 we observe that the average video picture has very little signal energy at high frequencies, and a direct recording thereof would fail to utilize the tape magnetization at the corresponding short wavelengths.

It is therefore common practice to boost the high frequencies during recording and de-emphasize them during playback. The pre-equalizers used in the broadcast field are shown in Fig. 29-6, and a complementary de-emphasis circuit is used during playback.

Additional equalization is needed during playback due to the read losses (see Chapter 6). They are: coating thickness, spacing, and gap losses. The reader will recall that none of these losses have associated phase shifts. They must therefore be compensated with the cosine equalizer described in Chapter 22. This equalizer was first used in sound film and in television, and was called an aperture equalizer (from sound film tracks scanned by a slit, or an aperture). The reader is referred to Robinson's book for a detailed analysis of the circuit function. See also Chapter 22.

Video recording technology advances at a rapid pace, and papers by Sadashige (1973, 1975 and 1984), Shiraishi (1985), and Sugaya (1986) are recommended for review.

The push for higher packing densities often forces the system to operate at shorter wavelengths, with poor carrier-to-noise ratio and increased drop-out activity. An analog component recording principle where the luminance and color are recorded as two independent FM spectra offers some performance improvements; see the 1985 paper by Sadashige.

VIDEO RECORDERS

The first attempts to record television pictures were made in the mid-fifties by companies like RCA (ref. Olson) and Bing Crosby Enterprises (ref. Mullin), who both used beefed-up audio transports running anywhere from 120 to 360 IPS. This provided the required high frequency response, and the DC-response was either carried through on a separate FM channel,

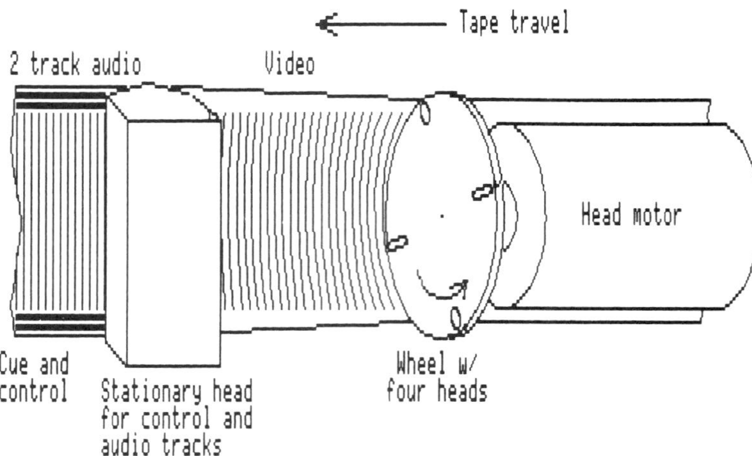

Fig. 29-7. Transverse scanning rotating head assembly in an early broadcast VTR for 2″ wide tape.

or restored by conventional DC-restorer circuits. But both attempts gave pictures that suffered badly from the imperfect tape transports. Flutter was fairly high in those days, and flutter translates into TBE.

Ampex succeeded in building a TV recorder/reproducer by using a transport where the tape moved at a reasonable speed of 15 IPS; it was two inches wide, though, and the information was recorded (and later played back) by a rotating head assembly that laid down tracks across the tape, from edge to edge (ref. Ginsburg).

The signal was FM-modulated by a clever method, that defied all conventions for FM-modulation (such as described in Chapter 27). The resounding "it will not work" from theoreticians was met with undeniable demonstrations of the system. The interested reader is referred to the papers by Anderson or Felix that analyze the video recorder FM system.

These machines were to dominate the broadcast field for the following twenty years, with about 11,000 units placed in services by American, European and Japanese manufacturers. Figure 29-7 shows the arrangements of the tape and heads: A rotating drum carries four heads, placed *exactly* 90 degrees apart. This means that the composite video signal is chopped up, and that the wheel must run in synchronism with the video signal. This presented problems upon playback (some readers may remember the venetian blind effect in TV pictures).

The audio information was recorded on an edge track on the tape. Also synchronization signals for the capstan speed control were recorded on another edge track so the head wheel would engage with the tape at the proper locations.

During the early sixties two British engineers, Rutherford and Turner, entered the stage with the announcement of a home video unit, costing around 200 pounds. The details were "proprietary," but enough was said to trigger a feverish effort within several companies in order to revive the earlier attempts to longitudinally record and play back at a speed of 120 IPS or so. The picture bandwidth was achieved with improved heads and tapes (first CrO_2), and color was tried. The sound was in one instance carried by an amplitude modulated dublet on each back porch (by Wayne Johnson), i.e., time multiplex, similar to current approaches in some home video recorders (in particular the 8mm format).

All attempts to develop such equipment failed, again because of excessive TDE in the longitudinal tape transports; the reproduced pictures were time distorted with an effect that is similar to that one gets looking at a picture through a thin sheet water fall. It is nevertheless the author's understanding that attempts at recording video longitudinally persist.

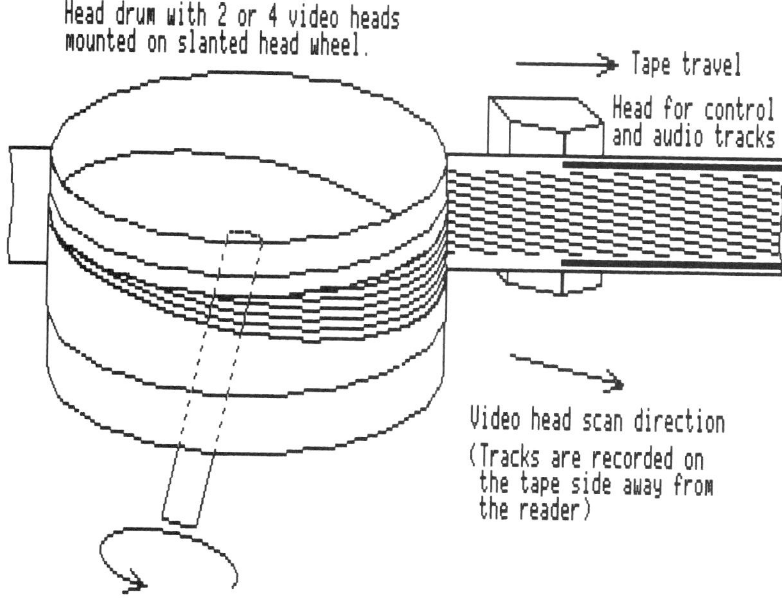

Fig. 29-8. Helical scan rotating head arrangement.

A different recording format surfaced during the early sixties, a helical scan principle where the tape is wrapped around a drum that is provided with a circumferential slit through which a head assembly contacts the tape. As the tape moves past the drum the head rotates at high speed and lays down long tracks of recorded information on the tape, see Fig. 29-8.

The helical principle is used today in all video recorders. The broadcast machines operate with a one inch wide tape, while the Beta and VHS formats use ½ inch wide tape. The most recent unit uses an 8mm wide tape.

The number of heads on the head wheel in a helical machine ranges from one to four, and the corresponding wrap angle of the tape from almost 360 degrees to 90 degrees. One scan of a head along the tape will generally correspond to one frame, or one picture of video information. This makes it possible to still frame, i.e., show just one 525 line picture on the screen at a time.

Most machines are two-headed, such as the Beta, VHS, and the new 8mm formats. The count of heads in advertisements includes stationary erase, sound and synchronization tracks. There are new schemes underway where four heads are mounted on the head wheel: One set with ∓ 15 degrees azimuth offset for recording/playback of modulated sound (FM or digital), and the other set with ∓ 6 or 7 degrees azimuth offset for recording video. The sound system uses a record level that penetrates the coating while the video information is recorded in the surface; then the two systems do not interfere with each other.

RECORDING FORMAT DESIGN AND STANDARDS

In designing a helical scan VTR (video tape recorder), first of all, the minimum recording wavelength is determined by the latest magnetic recording technology. The head wheel diameter is determined by the maximum video frequency which corresponds to the picture quality. After the tape width is determined, the video track-width and the tape speed are determined in order to obtain continuity of video signals from one track to another. It is usually necessary

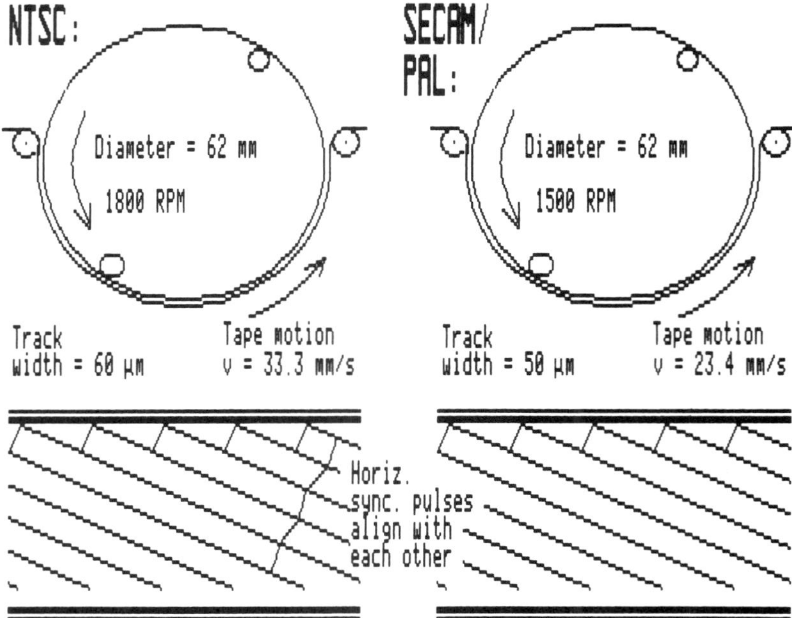

Fig. 29-9. Basic tape format and head wheel for VHS NTSC and PAL/SECAM recorders (after Sugaya).

to align the horizontal sync pulses for the PAL and SECAM systems (European television standards, ref. Sugaya). See Figs. 29-9 and 29-10.

The VHS format, using "no-guard-band recording" or "azimuth recording" (see Fig. 6-15) can record and play back the three basic television systems by switching only the electronic circuitry. In this way one basic mechanical transport assembly can be used all over the world. Caution: Recorded tapes are NOT interchangeable.

	NTSC		PAL/SECAM
PLAYING TIME (hrs.)	4 (6)	2 (3)	3 (4)
TAPE LENGTH (m)	250 (370)	250 (370)	250 (345)
TOTAL TAPE THICKNESS (μm)	20 (13)	20 (13)	20 (15)
TAPE SPEED (mm/sec)	16.7	33.3	23.4
TAPE CONSUMPTION (m^2/hour)	0.75	1.5	1.0
VIDEO TRACK WIDTH (μm)	30	60	50
VIDEO WRITING SPEED (m/sec)	5.8	5.8	4.9
VIDEO HEAD ANGLE (degrees)	± 6	± 6	± 6

Fig. 29-10. VHS global specifications (after Sugaya).

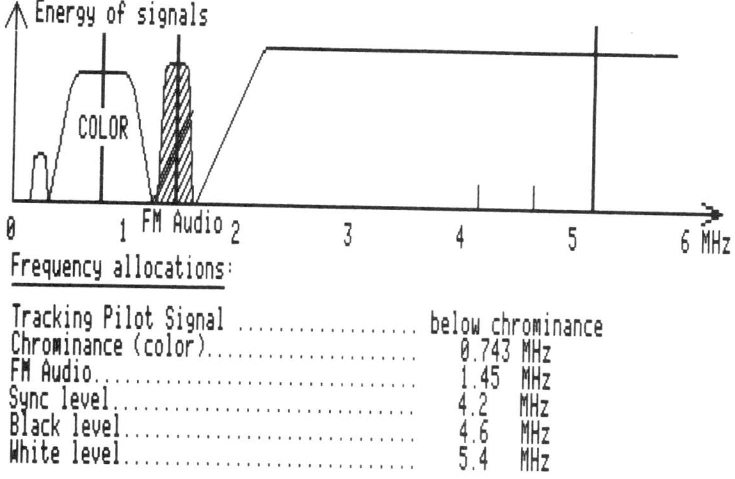

Fig. 29-11. Chrominance, audio and luminance in the 8mm camcorders.

The new 8mm format signal spectrum is shown in Fig. 29-11 (ref. Schubin). The tape is contained in a cassette measuring 95 by 62.5 by 15 mm. The drum (head wheel) diameter is only 40 mm, so each track is 40 $\pi/2$ = 62.8 mm long. From our earlier considerations we find that the shortest wavelength is in the order of 62.8/(4,000,000/30) = .47 μm = 18.8 μin. Adequate performance can only be achieved by using a short gap length head in combination with a high coercivity tape, such as a metal or Ba-ferrite.

The sound quality of home VCRs has always been pretty bad due to the fact that the tape speed is one half to one quarter of the normal audio cassette speed, and also due to the use

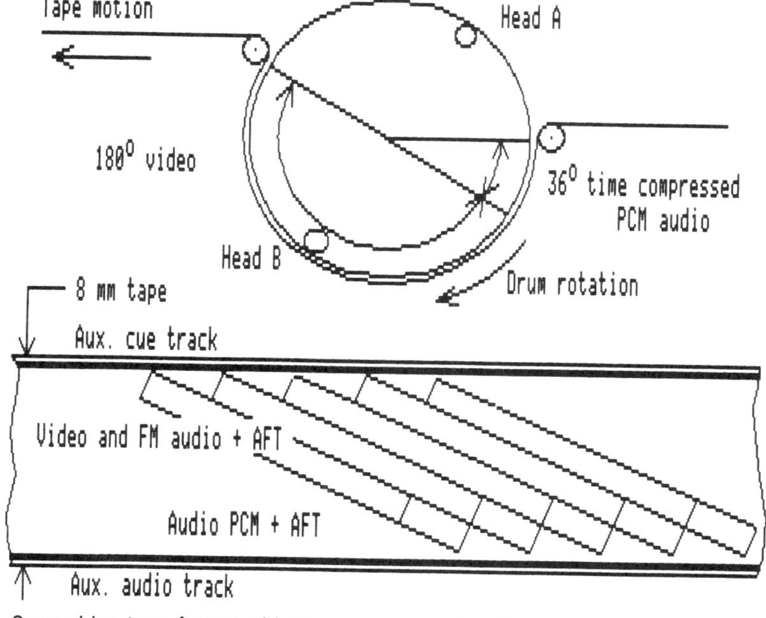

Fig. 29-12. 8mm video tape format with time compressed audio.

of a narrow track. The slow speed also falls in the danger zone of erratic tape motion due to stick-slip friction against the heads.

Recent developments have greatly improved the sound signal by not recording it in baseband or direct analog form along an edge track, but by FM modulating a signal carrier and multiplexing this signal in with the video as indicated in Fig. 29-11. And the latest is to digitize the audio signal and time multiplex it in with the video; an example is the 8mm system, where the tape wrap has been increased by 30 degrees to allow for the recording of digitized sound, see Fig. 29-12 (ref. Mitchell).

Several papers present excellent reviews of the exciting developments of magnetic recording equipment for television (Schubin, Shiraishi and Sugaya).

DUPLICATION

Duplication of prerecorded tapes is a time and equipment consuming process for video tapes. Audio tapes can be duplicated at high speeds, which is impossible with video tapes due to the rotating head assemblies and the much higher signal frequencies. The method is still one of loading a master tape onto a duplicator system with a large number of slave recorders. The duplication center is faced with high labor and maintenance cost.

The difference in Curie temperature of tapes has long been recognized as a potential solution for fast in-contact duplication of tapes (ref. Morrison). The process is fundamentally based upon using a master tape made from high Curie temperature material (any iron oxide material) that is brought into contact with a low Curie temperature slave tape, such as one made from CrO_2 particles. They are heated together to a temperature between the two Curie temperatures, at which point the CrO_2 becomes nonmagnetic.

When cooled the CrO_2 becomes magnetic, and each particle now becomes magnetic with a polarity that opposes any external field present. This will result in a mirror-image magnetization of the higher temperature tapes recording. If that recording is made in the proper format the result will be a duplicated tape; the format must correspond to the convention used in pressing records, where the pressing matrix has ridges instead of grooves.

The problems have in the past been mechanical misalignments, and the difficulties were in heating a base film (PET) to a temperature above the CrO_2 Curie temperature of about 110° C without permanent damage to the PET base film. Expansion and contractions are unavoidable which in turn make short wavelength duplication impossible.

Hope lingers on in finding a method of heating the Cro_2 particles only (micro-wave? ultraviolet?), and to apply the heat for only a very short time period so slip problems between master and slave are minimized (ref. Cole, Odagri).

EDITING OF TAPES

Editing of video tapes presents another formidable problem. Film editing is so simple: Cut and splice at the frame line. This method was applied in the transverse scan machines, and required great skill and care (ref. Kershaw, Simon).

In helical machines a cut and splice across the tape width will lead to a transition where the next sequence of pictures rolls up or down in a wiping motion. Only electronic editing will give a satisfactory result. This may be accomplished on-the-fly by operating two machines, or better yet by using some sort of memory device for temporary storage of frame sequences to be spliced. This all points to costly equipment that is not for the amateur recording person; he/she will do best with the two machine on-the-fly approach, or rent time at a local editing facility.

DIGITAL VIDEO RECORDING

The video tape recorders described earlier are about to be challenged by the rapidly developing digital signal technology. The advantages of digital formats are consistently good quality pictures and ease of copying without degradation, while the disadvantages mainly rest with the much higher packing density required (on the order of 100 MBS (megabits per second) at reasonably low bit error rates.

The present tape and transport formats are not optimum for digital video (ref. Dolby), coding schemes must be optimized (ref. Heitmann), and future developments must tie in with the final standards for digital television itself (ref. Remley, Davies).

VIDEO DISK RECORDING; STILL CAMERAS

The first video disks were large metal plated aluminum disks used for still frame as well as slow scan projections (ref. Bock). These disk record/players are still in wide use while being challenged by optical disks (ref. Toyama). See Fig. 29-13.

Another disk application may enter the consumer market in the near future, the electronic still video camera (ref. Kihara, Rivlin). A prototype was introduced by Sony in 1982, and Eastman Kodak, Fuji and Canon have since made overtures to this potentially large market for micro floppies. The camera may conceivably be configured as shown in Fig. 29-14, using a 1.85 square inch, high-resolution diskette format. A charge-coupled device (CCD) array is used to gather the picture information; the number of sensing elements will determine the picture resolution.

From the early part of this chapter we recall that hundreds of thousands of elements are required to approach the quality of 35mm film and prints therefrom. Certainly that quality must

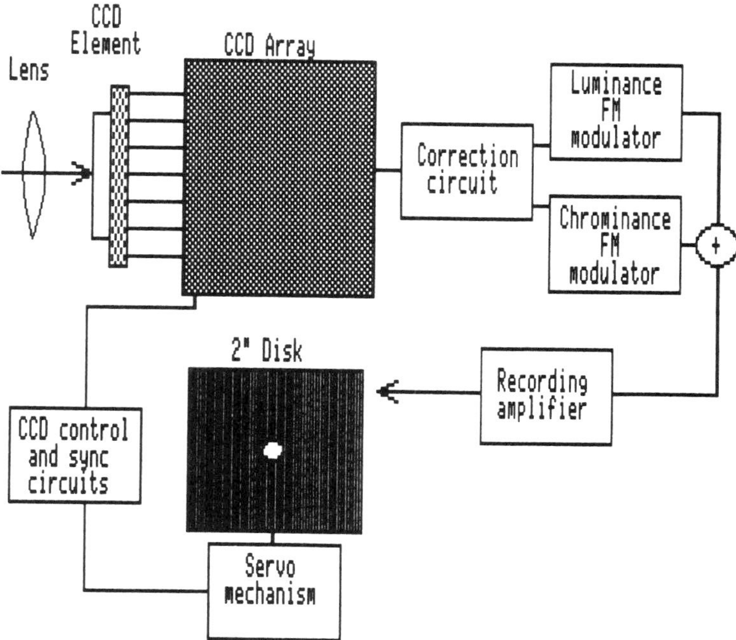

Fig. 29-13. Configuration of video disk recording.

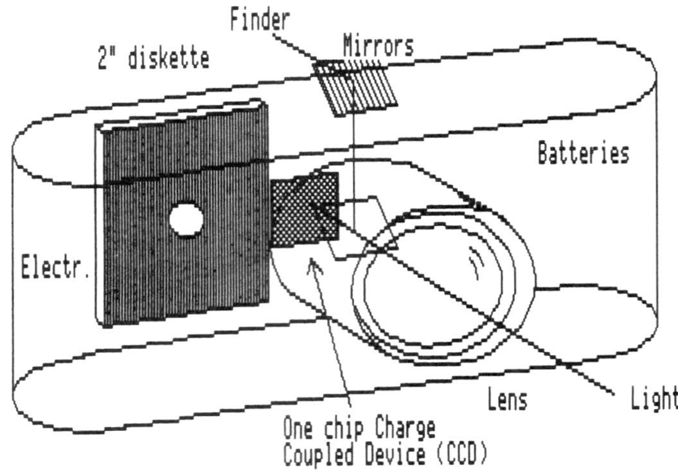

Fig. 29-14. Electronic still camera.

be met, at a reasonable price, before the consumer will embark upon a new device for his hobby. In terms of recording density a single track storage of a 500,000 elements picture will require a packing density of well over 100 kbpi, possibly reduced by picture processing techniques.

STANDARDS

Standards have been set and new ones are continuously being developed. A reference list would be truly voluminous so the reader is referred to the literature from SMPTE for particular needs.

REFERENCES TO CHAPTER 29

Robinson, J.F., *Videotape Recording,* Focal Press, London, June 1981, 362 pages.

Benson, K.B., (Editor), *Television Engineering Handbook,* McGraw-Hill, 1986, 1478 pages.

Mullin, J.T., "Video Magnetic Tape Recorder," *Tele-Tech,*, May 1954, Vol. 13, pp. 127-129.

Olson, H.F., Hougton, W.D., Morgan, A.R., Zenel, J., Artzt, M., Woodward, J.G., and Fischer, J.T., "A System for Recording and Reproducing Television Signals," *RCA Rev.,* Mar. 1954, Vol. 15, pp.3-17.

Ginsburg, G.P., "Video Tape Recorder Design: Comprehensive Description of the Ampex Video Tape Recorder," *Jour. SMPTE,* Apr. 1957, Vol. 66, pp. 177-182.

Sugaya, H., "Recent Advances in Video Tape Recording," *IEEE Trans. Magn.,* Aug. 1978, Vol. MAG-14, No. 5, pp. 632-637.

Sugaya, H., "Home Video Tape Recording and its Future Prospects," *INTL. Conf. Video and Data 82,* Apr. 1982, IERE Publ. No. 54., pp. 75-83.

Odagri, Y., and Sato, T., "High-Speed Video Tape Duplication Using Contact Printing," *IEEE Trans. Consum. El.,* Aug. 1984, Vol. 30, No. 3, pp. 397-401.

Cole, G.R., Bancroft, L.C., Chouinard, M.P., and McCloud, J.W., "Thermomagnetic Duplication of Chromium Dioxide Video Tape," *IEEE Trans. Magn.,* Jan. 1984, Vol. MAG-20, No. 1, pp. 19-23.

Sadashige, Koichi, "An Introduction to Analog Component Recording," *Jour. SMPTE,* May 1985, Vol. 93, No. 5, Pt. 1, pp. 477-485.

Schubin, M., "Camcorders: Home Movies Made Simple," *High Technology,* Oct. 1985, Vol. 5, No. 10, pp. 16-19.

BIBLIOGRAPHY TO CHAPTER 29

Introduction

Grob, B., *Basic Television, Principles and Servicing,* McGraw-Hill, 1975, 732 pages.

Kybett, H., *Video Tape Recorders,* Howard W. Sams Co., Jan. 1978, 400 pages.

Stanton, J.A., and Stanton, M.J., "Video Recording: A History," *Jour. SMPTE,* March 1987, Vol. 96, No. 3, pp. 253-263.

Five Year Index 1981-1985, Jour. SMPTE, Jan. 1986, Vol. 95, No. 1, Pt. II, 23 pages.

"Bibliography: New Technology in Video and Related Fields," *Jour. SMPTE,* Feb. 1986, Vol. 95, No. 2, pp. 239-246.

Sugaya, H., "Video Recording," Chapter 2, *Magnetic Recording,* Vol. 3, McGraw-Hill, 1988.

The Video Signal; Modulation; Equalization

Anderson C.E., "The Modulation System of the Ampex Video Tape Recorder," *Jour. of SMPTE,* Apr. 1957, Vol. 66, No. 4, pp. 182-184.

Felix, M.O., and Walsh, H., "F. M. Systems of Exceptional Bandwidth," *Proc. IEEE,* Sept. 1965, Vol. 112, No. 9, pp. 1659-1668.

Mitchell, P.W., "FM Recording Comes Home," *High Technology,* Oct. 1983, Vol. 3, No. 10, pp. 34-37.

Video Recorders (Studio, Industry, Home)

Sadashige, K., "Selected Topics on Modern Magnetic Video Recording Technology," *IERE Proc. on Video and Data Recording,* Jan. 1973, pp. 1-26.

Sadashige, K., "Improving the Performance of Quadruplex Videotape Recorders Operating on the U.S. Domestic Standard," *Jour. SMPTE,* Aug. 1975, Vol. 84, pp. 714-719.

Martens, G.J.A., and Schuddemat, J.P., "A Versatile PCM IC Set for 8mm Video," *IEEE Trans. Consum El.,* Sept. 1984, Vol. CE-30, No. 4, pp. 575-580.

Sadashige, K., "Developmental Trends for Future Consumer VCR's," *SMPTE Journ.,* Dec. 1984, Vol. 93, No. 12, pp. 1138-1146.

Shiraishi, Y., "History of Home Videotape Recorder Development," *Jour. SMPTE,* Dec. 1985, Vol. 94, No. 12, pp. 1257-1263.

Sugaya, H., "The Videotape Recorder: Its Evolution and the Present State of the Art of VTR Technology," *Jour. SMPTE,* Mar. 1986, Vol. 95, No. 3, pp. 301-309.

Tape Duplication

Morrison, J.R., and Speliotis, D.E., "The Magnetic Transfer Process," *IEEE Trans. Magn.,* Sep. 1968, Vol. MAG-4, No. 3, pp. 290-295.

Sugaya, H., Kobayashi, F., and Ono, M., "Magnetic Tape Duplication by Contact Printing at Short Wavelengths," *IEEE Trans. Magn.,* Sept. 1969, Vol. MAG-5, No. 3, pp. 437-442.

Mallinson, J.C., Bertram, H.N., and Steele, C.W., "A Theory of Contact Printing," *IEEE Trans. Magn.,* Sept. 1971, Vol. MAG-7, No. 3, pp. 524-528.

Tjaden, D.L A., and Rijckaert, A.M.A., "Theory of Anhysteretic Contact Duplication," *IEEE Trans. Magn.,* Sept. 1971, Vol. MAG-7, No. 3, pp. 532-536.

Hokkyo, J., and Satake, S., "A Theory of Anhysteretic Contact Printing," *IEEE Trans. Magn.,* Sept. 1979, Vol. MAG-8, No. 3, pp. 397-399.

Ono, M., Kobayashi, F., and Sugaya, H., "A Study of the Thermal Transfer Process," *IEEE Trans. Magn.,* Sept. 1979, Vol. MAG-8, No. 3, pp. 399-402.

Editing

Kershaw, D.L., "Video Tape Editing, Why the Mystique?," *Intl. Conf. Video and Data 79,* July 1979, No. 43, pp. 305-313.

Simon, G., and Roizen, J., "Novel Human Interfaces for Videotape Editing," *SMPTE,* Dec. 1984, Vol. 93, No. 12.

Digital Recording of Video

Dolby, D., Lemoine, M., and Felix, M., "Formats for Digital Video Tape Recorders," *The Radio and Electronic Engineer,* Jan. 1982, Vol. 52, No. 1, pp. 31-36.

Heitmann, J.K.R., "Digital Video Recording: New Results in Channel Coding and Error Protection," *Jour. SMPTE,* Dec. 1983, Vol. 93, No. 2, pp. 140-144.

Remley, F.M., "Digital Television Tape Recording: A Report of Progress Toward a Standard," *SMPTE Jour.*, Aug. 1985, Vol. 94, No. 9, pp. 914-920.

Davies, K. P., "The Digital Television Tape Recorder—Audio and Data Recording Aspects," *Jour. SMPTE,* Jan. 1986, Vol. 95, No. 1, pp. 4-12.

Baldwin, J.L.E., "Digital Television Recording—History and Background," *Jour. SMPTE,* Dec. 1986, Vol. 95, No. 12, pp. 1206-1214.

Video Disks

Bock, W., "Videodisk Standards: Software View of the Technology," *Jour. SMPTE,* May 1983, Vol. 92, No. 5, pp. 571-576.

Kihara, N., "An Electronic Still Camera and Recording System," *Intl. Conf. Video and Data 84,* Apr. 1984, *IERE Publ.* No. 59, pp. 133-136.

Rivlin, R., "The Birth of Electronic Photography," *High Technology,* Dec. 1985, Vol. 5, No. 12, pp. 60-61.

Toyama, T., Morita, Y., Hioko, T., Ohta, O., Ishii, Y., Ninomiya, Y., Ohtsuka, Y., Izimu, Y., and Goushi, S., "Optical Videodisc for High-Definition Television by the MUSE," *Jour. SMPTE,* Jan. 1986, Vol. 95, No. 1, pp. 25-30.

Chapter 30
Care and Maintenance

A tape recorder or disk drive, like other instruments (for example, a motion picture camera), is subject to malfunctions that are best prevented by proper maintenance. In addition, a reel of magnetic tape (like a roll of photographic film) or a disk is sensitive to handling, storage, and shipment hazards. Routine maintenance of the equipment and the observance of a few fundamental rules about tape and disk handling will assure the user of better performance and longer life for both equipment and media.

There are several ways in which media can be damaged, either mechanical (by handling) or chemical (environment). These failure mechanisms are analyzed, and will dictate a few guidelines for handling and storage of media.

Equipment can (and will) malfunction, causing not only signal degradation but also possibly permanent media damage. Certain maintenance procedures are common to all tape and disk drives, while others are dependent upon the particular drive. An outline of general maintenance procedures is given, while the manufacturer's procedures are found in their respective manuals.

The chapter concludes with a listing of potential failures, causes, and remedies.

CAUSES OF FAILURE IN MEDIA

Improper Winding of Magnetic Tape

The most common way in which tapes are damaged occurs during machine operation: When winding or rewinding a tape, the recorder may produce an uneven "pack" or "wind" with protruding layers which are subject to damage in handling. Holding the reel by the flanges will squeeze them against the tape, and protruding layers will be nicked and permanently deformed.

A transport winding mechanism may also wind the tape with insufficient tension. Later handling will then cause the pack to shift from side to side against the flanges, leading to later

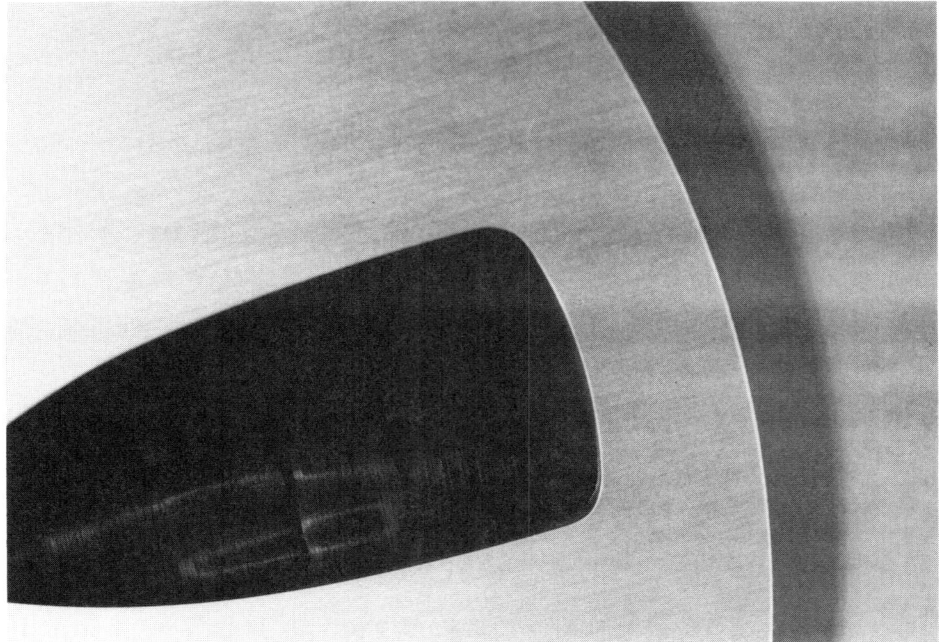

Fig. 30-1. Cinching in a wound reel of tape.

edge damage. A loose pack is also subject to tangential slippage between layers, called cinching. *Cinching* is likely to occur in a reel of tape with one or more regions of too low tension, especially if subjected to a rapid angular acceleration or deceleration which occurs during the starting and stopping of a tape handler.

Cinching is shown in Fig. 30-1. During such slippage the tape may actually fold over on itself so that permanent creases form immediately or perhaps later when tension is applied and the tape attempts to return to its original position. Creases cause dropouts by introducing a separation between tape and heads.

Prolonged storage may cause other mechanical changes in a tape or disk coating surface, such as its *wearability* (durability) and depletion of its lubricant content, which is not firmly locked in with the molecular structure of the coat.

Careless Clamping of Diskettes

The center hole in a diskette (5¼ or 8 inches) is vulnerable to damage by improper clamping. This occurs when the disk is off center in the protective jacket, inserted in the disk drive and quickly clamped. The hole may be severely stressed in one side, causing a permanent wrinkle that makes subsequent centering difficult.

Hydrolysis in Tape/Disk Coatings

High temperatures have long been suspected of causing tape degradation, a process which is accelerated in high humidity (ref. Cuddihy's papers). Almost all commercial magnetic tapes employ an oxide binder made from *polyester urethanes*, which are employed either by themselves or in combination with other polymeric materials intended to achieve special properties.

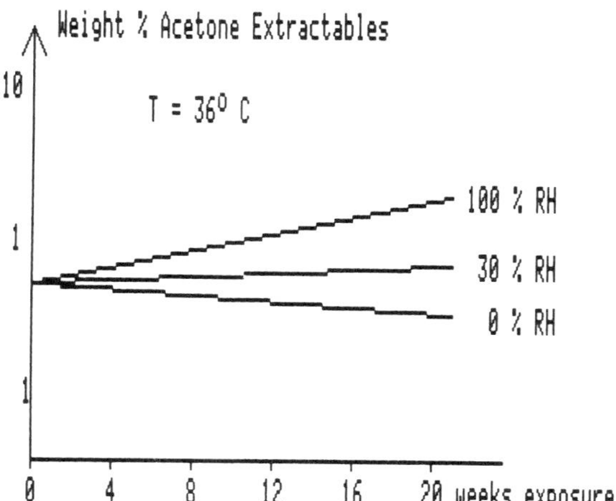

Fig. 30-2. Tape aging in humidity (after Cuddihy).

Such co-polymers vary in mechanical and physical properties as well as in processing requirements.

However, all polyester urethanes and all polyester polymers are susceptible to degradation by chemical reaction with water, a chemical reaction called *hydrolysis* (ref. Cuddihy). The process is complex but can be simplified expressed as (ref. Bertram):

Ester + Water <—> Carboxylic acid + Alcohol

The degradation products (debris = carboxylic acid) can be extracted from the tape by acetone, and the nature and rates of the chemical reaction can then be monitored. Figure 30-2 shows one set of results covering a three month span. The debris formation occurs fastest at 100 percent RH, is about zero at 20-24 percent RH—and is reversed at zero relative humidity. The latter observation verifies the reversible process as indicated by the double arrow in the expression; it also points to the possible recovery of tapes damaged from hydrolysis during long storage.

PET films used for tape and disk base are essentially stable against environmental degradation at ordinary use conditions (ref. Cuddihy).

HANDLING OF MEDIA

Cinching and nicked tape edges can be avoided if the user follows a simple advice:

Do not rewind a tape after recording or playback, but store it immediately in its container, standing on end.

Numerous tape failures and dropouts are the result of not following this practice.

A tape that is left on a recorder or placed on a shelf outside of its container, as is often the case, will collect dust within a very short time. When the tape is later played back, the dust particles will move onto the coating surface by the airflow that is caused by lifting the tape from the tape pack. This causes dropouts and may permanently damage the tape, and potentially scratch the magnetic heads. Dust particles can combine with debris from the tape and deposit it on the guides and heads. These protruding particles will scratch the tape surface and further aggravate the dropout situation. There also may be a continuous scratch in the tape surface or backing throughout an entire reel. The slight ridge thus produced can multiply through the layers of tape wound on the reel to produce a much larger ridge in the outer layers of the tape—often large enough to cause permanent deformation in the outer layers.

A word of caution applies in particular to diskettes. Hands and fingers have body oils and salts which attract foreign particles. The oils alone can leave a fingerprint on the exposed surface of a diskette that renders a readout impossible, aggravated of course by any foreign particles attracted to the oil.

Never touch a diskette in the window area!

The same advice applies to the handling of tapes, in particular reel-to-reel video tapes. Each time a tape is threaded on a video transport, the operator may transfer finger oils and salts to the tape or to the tape guides. This, naturally, is a problem whenever a tape is spliced. During a normal splicing or editing session, the tape must necessarily be handled a great deal, and it is a good idea to **wear white cotton gloves.**

In order to make a perfect video splice during an editing session, a solution such as *"Mag-View"* or *"EdiView"* is used. When this is sprayed onto the tape a gray powder remains, revealing the tape tracks and edit pulses, which greatly assists the editor. After editing, the material must be removed completely from the tape with a tissue, or dropouts are inevitable.

STORAGE OF MEDIA, GENERAL

For storage of tape, the following rules apply:

- A tape should always be stored in its container with the reel on edge rather than in a flat position. This will tend to eliminate the sideways shifting of the pack against the flanges.
- A tape should be stored under controlled environmental conditions, see Fig. 30-3. Large or sudden changes in environment should be avoided.
- A tape which has been stored under less than ideal environmental conditions should be conditioned by allowing it to remain in a suitable environment for at least 24 hours prior to use.
- When large changes in temperature cannot be avoided, the probability of damage to the tape can be minimized if the reel hub has a thermal coefficient of expansion similar to that of the base film. Most plastic reels have thermal coefficient about twice that of the polyester base film, while the thermal coefficient of aluminum is nearly equal to that of polyester.

During **shipments** of a tape, mechanical agitation will tend to shift the tape pack, especially if wound under improper tension. Any abrupt temperature change during shipment should be avoided and this is best done by placing the reel in special fiber-board shipping containers. This will also protect the reel flanges from being bent to a point where the edges of the tape run on the flange.

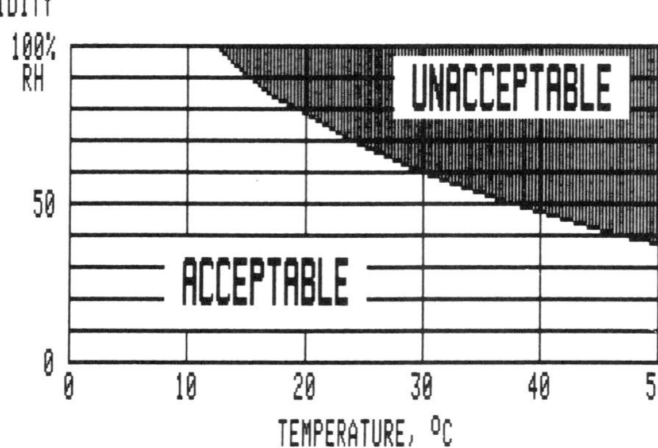

Fig. 30-3. Tapes should be stored at conditions corresponding to "acceptable," as indicated in the illustration (after Cuddihy).

Stray magnetic fields may cause some degree of erasure of the information recorded on the tape. There are a few cases on record where tape has been completely erased during shipment and if such fields are known to exist, special shielding containers are available.

Neither tapes nor disks are affected by the security check equipment in airports. Example 3.2 from Chapter 3 shows the large field required to affect recorded magnetization. Reports about data disks that supposedly have been damaged during travel do nevertheless persist. Here again: Has care been taken in sealing the disks from dust, lint etc.? Were they under any mechanical stress, or exposed to excessive temperature?

A word of caution: Get rid of all magnetic gadgets on your desk. Never place one of those magnetic holders for notes etc. on a disk, your data will be erased!

STORAGE OF MEDIA, ARCHIVAL

Precautions in addition to the ones just listed are needed when tapes or disk are to be stored for extended periods (i.e., many years). A recent study (ref. Bertram) has produced recommendations for the storage conditions for γFe_2O_3 particles in a polyester-urethane binder system:

$$T = 65° \text{ F, RH} = 40 \text{ percent.}$$
$$(\text{both } \pm 5 \text{ percent})$$

These binder systems degrade due to hydrolysis. The normal degradation for a new, conditioned tape is typically at a 7 percent hydrolysis state. Taking that value as a ''zero'' reference the chart in Fig. 30-3 was prepared. Any temperature/humidity condition that lies on the ''zero'' hydrolysis contour will preserve the binder at the 7 percent hydrolysis state; this appears to be a satisfactory condition for most applications.

An increase due to high temperatures and/or high humidity accelerates the hydrolysis, and when a value of 14 percent has been reached a tape is damaged, i.e. it produces excessive debris and sticky gunk that deposits onto heads and guides with resulting signal spacing loss and potential tape/disk damage.

The quality of the stored tape also depends upon the internal stresses that build up during winding. Excessive tension during winding may result in deformation of the tape, while a loose wind entraps air and causes cinching. Internal stresses will eventually decrease (ref. Cuddihy, van Keuren, Bertram), provided the temperature is constant. This is rarely the case, and the study makes one further recommendation:

Rewind stored tapes at regular intervals.

Recommended intervals between rewindings are shown in Fig. 30-4 (solid line). The rewinding is best when done on a transport having constant reel torque, not too fast speed, and a pack-follower wheel to prevent air entrapment. Such transports are called rewinders. It may be equipped with a cleaning device (razor blade) that scrapes excess products from the hydrolysis, whereafter the tape passes over/through a lint-free tissue wiper to remove all dirt and dust. It was found that this process even acted as a reconditioner for a damaged tape and improved its BER (bit-error-rate).

The effect of elevated temperature is strongly evident in the study's crude rule of thumb estimate for the combined effects of hydrolysis and induced thermal stresses:

1 day at 85° F is equivalent to 15 days at 65° F.
1 day at 105° F is equivalent to 250 days at 65° F.

All above recommendations and discussions apply only to tapes with γFe_2O_3. For other tapes detailed information must be obtained from the product's manufacturer. Tapes with Fe

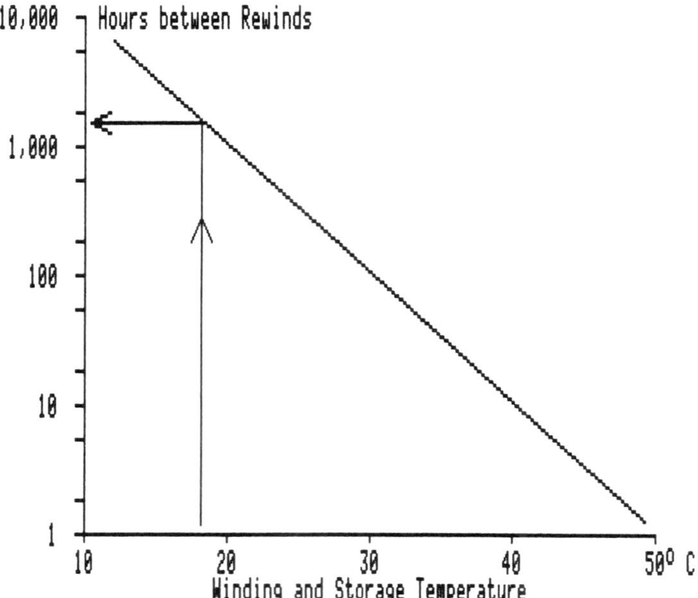

Fig. 30-4. Recommended periods between rewinds as a function of storage temperature (after Bertram).

particles are known to be sensitive to humidity and subsequent corrosion (ref. Olsen), and little is known about long time storage of the other particle tapes.

MAINTENANCE OF RECORDING EQUIPMENT

Cleanliness is fundamental to the proper operation of any magnetic recorder or disk drive. Dust particles will not only cause dropouts, but also shorten the life of the recording equipment, especially the heads and bearings. Good cleaning rules apply to home sound recording equipment, but the demands are even greater for instrumentation, video, and computer recording equipment. These machines should ideally be operated in a clean-room environment to avoid the accelerated generation of dropouts. Humidity and temperature conditions should be closely controlled, and the floor area should be cleaned daily using a damp mop. Many professional facilities also have restrictions on the use of food and tobacco, since tobacco ashes can easily accumulate on tapes and disks.

Any part touching the magnetic tape on its pass through the transport should be cleaned at regular intervals—tape guides, heads, capstans, and rubber pucks. If this practice is not followed, dirt can accumulate on the tape guides and heads, and act as an abrasive agent and scrape the oxide coating on the tape. Soon the oxide buildup from the scratching may break away and be redeposited elsewhere on the tape. As the tape is wound tightly onto the take-up reel any loose oxide may be firmly embedded into the tape surface and cause dropouts the next time the tape is used.

Care should be exercised in the selection of a *cleaning solution,* since some agents do more harm than good. About the only cleaning solution recommended for video-tape equipment is Freon TF. The reason is that Freon TF will flush off oxide particles and debris without softening the oxide or the backing. Also Freon TF will not damage the rubber capstan idler. Table 30-1 states the effects of various cleaning solutions. Cleaning agents and tools are available from companies such as TEXWIPE, Hillsdale, N.J. 07642.

Table 30-1. Table of Cleaning Agents.

CLEANING SOLVENT	HEALTH HAZARD	EFFECT ON VIDEO MAGNETIC TAPE	FLAMMABLE	EFFECT ON RUBBER
FREON TF	Very slight	None or negligible	No	None or negligible
ACETONE	Very slight	Soluble	Yes	None or slight
CARBON TETRACHLORIDE	Great	Negligible	No	Slight
ETHYL ALCOHOL	Very slight	Negligible	Yes	None or slight
HEPTANE	Slight	Softens	Yes	Swells
METHYL ALCOHOL	Some	Negligible	Yes	None or slight
NAPHTHA	Slight	Softens	Yes	Swells
MEK	Some	Soluble	Yes	None or slight
TRICHLOROETHYLENE	Some	Soluble	No	Slight
XYLENE	Some	Soluble	Yes	Swells

Modern polyurethane coatings are known to deposit debris products, probably formed by hydrolysis (see earlier this chapter). They will collectively form a clear film, invisible to the eye but noticeable in producing dropout and aggravating stick-slip phenomena. This debris film is often referred to as *brown stain* or *varnish,* and cannot be removed with any permissible head cleaner, short of a mildly abrasive tape.

This film also forms on diskette drive heads and will in some cases cause severe stick-slip actuated vibrations of the head/mounting arm/flexible disk system. The disk drive will ''sing''; when this occurs beware—either the head or the disk, or both, have a short life ahead. Stop operations and perform a head cleaning with a cleaner disk that likely has enough abrasive powder imbedded to remove the debris on the head.

Another trouble source, often neglected, is **head magnetization.** A magnetized record head will in instrumentation and audio recordings increase second-order harmonic distortion and the overall noise level. In computer tape and disk drives asymmetrical waveforms will be written. Therefore, a small head degausser should be used at regular intervals (for example, every eight running hours).

Even then the danger of magnetized heads persists: Any nearby DC-fields from permanent magnet motors (rotating in tape drives, linear motion in disk drive head actuators) may link through the head core and act as a ''data'' signal being recorded onto the head core while the alternating degaussing field acts as a decaying AC-bias field! Even the earth field may prevent complete erasure of a head core, and the outcome is high even-harmonic distortion in audio and instrumentation recorders, and asymmetry in digital equipment. The only remedy is shielding, or when the earth field enters, repeated degaussing with a different orientation of the equipment.

With regard to **lubrication** and **head alignment,** refer to your recorder manual. Many recorders have bearings that are lubricated for life. But, if lubrication is required in certain parts of the recorder, care should be exercised not to spill any oil on the capstans, rubber pucks, or other sensitive areas.

Preventive maintenance of the **tape transport** should be made on a regular basis. This should not just be cleaning of guides, rollers, etc., but also a check of the tape tension and tracking plus flutter and TDE.

Magnetic heads may change sensitivities as wear progresses; this mandates a periodic check of the **electronics and associated controls.** The reproduce level and response should be checked with a standard test tape, and the SNR verified. Bias level in the record chain may need readjustment (see equipment manual for details), and the standard record level may need recalibration, in addition to re-setting of the pre-equalization.

Additional details on care and handling are found in a paper by Ford and in two NASA publications, listed in the references under Waites, Heard, Davis, Townsend, and by THIC's Head Advisory Committee.

TROUBLESHOOTING TAPE RECORDING EQUIPMENT

The following troubleshooting guide is applicable to most recorders. However, it is necessary that the operating manual and/or the service manual be consulted for a particular recorder prior to repair. For troubleshooting the recorder electronics, instruments similar to those used for amplifiers are normally required (tone generator, voltmeters, oscilloscope). It is also useful to have a flutter meter and test or alignment tapes which have signals recorded for alignment of the reproduce head, setting of the record level indicator, frequency-response test, and a tone for wow and flutter tests. To clean various parts in the recorder, it is useful to have on hand a proper cleaning fluid and Q-tips (see Table 30-1). Listed below in tabular form are typical troubles, their possible causes, and steps for their correction.

PROBLEM—(POSSIBLE CAUSE)—CORRECTION

Noise

(DC magnetized heads)—Degauss the magnetic heads with a suitable head degausser, and do it very slowly.

(Input amplifier)—A faulty resistor will cause noise in the input stage and should be replaced. Faulty electrolytic capacitors may also cause noise and should be replaced. Also check for noisy transistors or chips.

(Distorted AC-bias)—Check the oscillator and bias amplifiers for proper operating voltages and freedom from even-order distorted waveforms.

Hum

(Faulty shields or ground connections)—Ground and shield connections may become corroded and should be scraped clean.

(Open-circuited reproduce head)—Replace the reproduce head.

(Poor power supply decoupling)—Replace faulty electrolytic capacitors.

Distortion

(Tape overload)—Carefully monitor the record level. If the distortion persists, check the setting of the level indicator on the recorder, which is easiest done with a test tape. For details, see Chapter 23.

(No or too little AC-bias)—Check the oscillator and amplifier. Also check that the proper tape, for which the bias level has been set, is used on the recorder.

No or poor erasure

(Faulty bias oscillator)—Check the oscillator for proper operation. If the oscillator has been malfunctioning, this would also have resulted in a high distortion level (as above).

(Debris on erase head)—Debris will lift the magnetic tape away from the erase gap. The erase head should be cleaned with a proper solvent and a Q-tip.

Poor frequency response

(Debris on heads)—Clean all magnetic heads with a suitable solvent and Q-tips.

(Misaligned heads)—First check the alignment of the reproduce head using a test tape. Next, check the alignment of the record head by recording a high-frequency note and then adjust the record-head azimuth screw for maximum output level. The azimuth adjustment should be made only by a person familiar with the recorder.

(Faulty equalizer)—Poor contacts in the equalizer switching circuit may cause the equalizer to function improperly. Faulty components may likewise cause this problem.

(Skew)—This condition will normally manifest itself as a variation in the high-frequency output level and is most likely caused by a worn or misaligned capstan rubber idler. A new idler should be installed and/or realigned.

(Smeared head gaps)—Foreign particles (for example, dust) may cause excessive scratches on the head surfaces which will cause the material to cold flow across the gap and thereby destroy otherwise parallel gap edges. This phenomenon also may appear if the mu-metal cores wear away faster than the head shell; a thicker tape will then no longer conform to the head contour and make proper contact with the gap.

(Wrong tape)—Tape from different manufacturers requires a slightly different bias setting for optimum performance. Once selected, the same type tape should be used in future recordings.

No recording

(Faulty amplifier)—Follow normal amplifier troubleshooting procedures. The record head current is normally referred to in the service manual and can usually be measured across a 10-ohm resistor in the ground leg of the record head.

(Faulty record head)—An open record head will result in no voltage across the 10-ohm resistor. (A shortcircuited record head or cable connection will allow the current to flow through the 10-ohm resistor and should be checked very carefully.)

No playback

(Faulty amplifier)—Troubleshoot the reproduce amplifier.

(Faulty reproduce head)—This is easy to verify; run a pre-recorded test tape through the recorder and check for proper output levels. A shorted reproduce head will result in no output voltage, while an open reproduce head will introduce excessive hum.

TRANSPORT PROBLEMS

PROBLEM—(POSSIBLE CAUSE)—CORRECTION

Flutter

(Debris on capstan)—Clean the capstan and rubber puck with a suitable solvent.

(Damaged rubber puck)—If a power failure occurs while the recorder is in the record or play mode, the rubber puck will remain engaged against the capstan and indent the rubber puck. Such an indentation may be removed by letting the recorder run in the play mode several hours without tape. Otherwise, the rubber puck must be replaced.

(Worn belts, pucks)—Replace worn-out parts.

(Tape scraping on reel flanges)—Rewind the tape onto a new reel that has no bent or damaged flanges. Also make sure that the inside edges of the reel flanges are free from any nicks or scratches. Reels must also be properly centered and perfectly running.

(Heavy oil and dirt in bearings)—Clean all bearings as outlined in the service manual and relubricate.

Too slow speed (drift)

(Reel tension too high)—Adjust reel tension in accordance with the service manual instructions.

(Debris on rubber puck)—Debris may cause a rubber puck to become excessively smooth, in which case it should be cleaned with a suitable solvent, possibly plain water (solvent may damage rubber).

(High bearing friction)—Clean and lubricate all bearings as described in the service manual.

(Speed control error)—Many portable recorders and most instrumentation recorders maintain their speed accuracy by an electronic servo system. There may be several causes for malfunction in a servo system and the reader is referred to the service manual for the particular recorder.

Squeal

(Debris on heads and guides)—Any disturbance in the tape path through the recorder may cause excessive scrape flutter. Heads and guides should be cleaned with a suitable solvent.

(Excessive tape tension)—Adjust the tape tension devices in accordance with the service manual.

(Worn felt pads; excessive felt pad pressure)—Replace pad(s). Excessive felt pad pressure, either against a guide or against a head, will cause excessive scrape flutter. Adjust the felt pad pressure in accordance with the service manual or by successive experiments.

No tape motion

(Broken belts or damaged mechanical parts)—Replace damaged parts.

(Blown fuse)—Check all fuses and replace if burned out. If a fuse repeatedly burns out, the recorder should be overhauled.

Tape breakage

(Worn brake pads)—Replace worn brake parts and adjust in accordance with service manual.

(Maladjusted brakes)—Readjust brakes in accordance with service manual.

Tape throws a loop

(Maladjusted brakes)—Adjust the brake system in accordance with the manual instructions.

REFERENCES TO CHAPTER 30

Cuddihy, E. F., and Keuren, W., "Mathematical Description of Heat Transfer in Packs of Magnetic Recording Tapes," *IFT Journal*, March 1974, pp. 5-7.

Olsen, K. H., "Microstructure Analysis of Corrosion in Iron Based Recording Tapes," *IEEE Trans. Magn.*, Sept. 1974, Vol. MAG-10, No. 3, pp. 660-662.

Cuddihy, E.F., "Hygroscopic Properties of Magnetic Recording Tape," *IEEE Trans. Magn.*, Mar. 1976, Vol. MAG-12, No. 2, pp. 126-135.

Bertram, N.H., and Eshel, A., *Recording Media Archival Attributes*, Rome Air Devel. Center, Final Tech. Report RADC-TR-80-123, April 1980, 112 pages.

Cuddihy, E. F., "Aging of Magnetic Recording Tape," *IEEE Trans. Magn.*, July 1980, Vol. MAG-16, No. 4, pp. 558-568.

Waites, J. B., "Care, Handling, and Management of Magnetic Tape," *NASA Ref. Publ. 1075*, April 1982, pp. 45-59.

Heard, F., "A Care and Handling Manual for Magnetic Tape Recording," *NASA Ref. Publ. 1075*, April 1982, pp. 127-147.

Bertram, N. H., and Cuddihy, E. F., "Kinetics of the Humid Aging of Magnetic Recording Tape," *IEEE Trans. Magn.*, Sept. 1982, Vol. 18, No. 5, pp. 993-999.

Davis, R., "Cleaning, Packing and Winding of Magnetic Tape," *NASA Ref. Publ. 1075*, April 1982, pp. 61-75.

Townsend, K., "Tape Reels, Bands and Packaging," *NASA Ref. Publ. 1075*, April 1982, pp. 77-84.

Ford, H., "Handling and Storage of Tape," *Studio Sound and Bcast. Engr.*, Dec. 1984, Vol. 26, No. 12, pp. 64-72.

Head Wear Advisory Committee, THIC, "Care and Handling of Magnetic Tape Heads," *NASA Ref. Publ. 1111*, Sept. 1985, pp. 293-296.

Appendix A
Tutorial Programs

Four small programs are listed in order to illustrate how some of the processes in magnetic recording can be modelled. The programs are:

1. DIFF_EQ—plots the graphical solutions (trajectories) to first order differential equations. An equation set is typed into the program as indicated; the set shown will draw circles (U = Y, V = −X), and represents the field lines from a current carrying conductor located at (x,y) = (0,0). (See further in Chapter 8, "Fields from Magnets and Currents".)

2. VECTRMAG—computes and displays the magnetization vectors inside a recorded medium. See further Chapter 20 ("Remanent Magnetization," formulas (20.7) and (20.8). This program is also useful to establish the recorded vectors for computation of read voltages. (See further Chapter 21, "Magnetization Patterns".)

3. DEL_DRAW—computes and displays the transition patterns between bits in a recorded medium. This program was used to produce illustrations (Figs. 5-17, 5-19, and 5-22). A modified version was used for Fig. 5-23.

4. SFD_LOSS—computes the loss in signal level due to the finite length of the recording zone, and performs a convolution to determine the magnetization pattern for a sine or square wave write current. This program was used for the computations in Chapter 19.

The programs are written in TrueBASIC, and will therefore run on an IBM PC, MacIntosh or Amiga computer, provided the reader has the TrueBASIC editor and compiler. The reader may otherwise wish to convert it to any of the other available BASICs.

The programs are self-explanatory through module definitions and REM statements (the acclamation sign ! tells the compiler that the following line is a remark). The model-

```
!       Program number 1.
!       *****************************************************************
!       *                                                               *
!       *   PROGRAM NAME:  DIFF_EQ                                      *
!       *   WRITTEN BY  :  Finn Jorgensen                               *
!       *                                                               *
!       *   DATE:          MAR 15, 1987                                 *
!       *                                                               *
!       *   This program plots the trajectories for first order         *
!       *   differential equations.                                     *
!       *                                                               *
!       *                    GLOBAL DATA                                *
!       *                                                               *
!       *    INPUT VARIABLES:                    OUTPUT VARIABLES:      *
!       *    Differential Equations              Graphics               *
!       *      (Place in module diff.eq)                                *
!       *    Borders for plot area                                      *
!       *    Number of traces                                           *
!       *    Number of steps per trace                                  *
!       *    Step size                                                  *
!       *                                                               *
!       *    The program has   internal subroutines:                    *
!       *                 INPUT                                         *
!       *                 START_POINT                                   *
!       *                 DIFF_EQ                                       *
!       *                 STEP                                          *
!       *                                                               *
!       *****************************************************************

!       *****************************************************************
!       *                                                               *
!       *    Mainprogram starts here                                    *
!       *                                                               *
!       *****************************************************************

OPTION ANGLE RADIANS
CALL INPUT
OPEN #1: SCREEN 0,1,0,1
SET WINDOW A, B, C, D
LET N = 0

DO WHILE N < N0
   LET N = N + 1
   CALL START_POINT
   LET E = 1                           ! REM Initial sign of slope
   SET COLOR (MOD(N,3) + 1)
   LET X_KEEP = X0                         ! REM Preserve start_point
   LET Y_KEEP = Y0                         ! REM Preserve start_point
   LET R = 1                           ! REM Step counter

   DO WHILE R < R0
      SET CURSOR 1,1
      PRINT USING "#####": R
      LET X = X0
      LET Y = Y0
      CALL DIFF_EQ                     ! REM Returns U, V and W for slope at X0, Y0
      LET U1 = U                       ! REM Preserve U for mean step (below)
      LET V1 = V                       ! REM Preserve V for mean step (below)
      CALL STEP                        ! REM Returns next X, Y
      CALL DIFF_EQ                     ! REM Returns U and V for slope at X, Y
      LET U = (U1 + U)/2               ! REM Mean step in dx direction
      LET V = (V1 + V)/2               ! REM Mean step in dy direction
      LET W = SQR(U^2 + V^2)           ! REM Step length
      CALL STEP                        ! REM Returns next start point X, Y
      PLOT LINES : X0, Y0; X, Y
      LET X0 = X
      LET Y0 = Y
      LET R = R + 1
```

```
      IF (X<A) OR (X>B) THEN EXIT DO
      IF (Y<C) OR (Y>D) THEN EXIT DO
   LOOP
LOOP

!       ************************************************************
!       *     MODULE: INPUT                                        *
!       *                                                          *
!       *     This internal subroutine receives dimensions for     *
!       *     the plot area, and the nominal step size.            *
!       *                                                          *
!       *     INPUT PARAMETERS:    OUTPUT PARAMETERS:  LOCAL DATA  *
!       *                         Minimum X (left)    A            *
!       *                         Maximum X (right)   B            *
!       *                         Minimum Y (bottom)  C            *
!       *                         Maximum Y (top)     D            *
!       *                         Number of traces    N0           *
!       *                         Number of steps     R0           *
!       *                         Step size           H            *
!       *                                                          *
!       ************************************************************

SUB INPUT

    CLEAR
    SET CURSOR 5,11
    PRINT "Input Data for plot of Diff. Eq. Trajectories"
    SET CURSOR 8,14
    INPUT PROMPT "Have you typed-in the equations in Module DIFF_EQ (Y/N): ": A$
    IF (UCASE$(A$)<>"Y") THEN PRINT "Press Ctrl-Scroll-Lock; Enter equations in
program"
    SET CURSOR 10,20
    INPUT PROMPT "Left border of plot area.................: ": A
    SET CURSOR 12,20
    INPUT PROMPT "Right border of plot area................: ": B
    SET CURSOR 14,20
    INPUT PROMPT "Bottom border of plot area...............: ": C
    SET CURSOR 16,20
    INPUT PROMPT "Top border of plot area..................: ": D
    SET CURSOR 18,20
    INPUT PROMPT "Number of traces.........................: ": N0
    SET CURSOR 20,20
    INPUT PROMPT "Number of steps per trace............max.: ": R0
    SET CURSOR 22,20
    INPUT PROMPT "Nominal step size for plots..............: ": H

END SUB


!       ************************************************************
!       *     MODULE: START_POINT                                  *
!       *                                                          *
!       *     This internal subroutine computes a randomly         *
!       *     selected start point for each trajectory.            *
!       *                                                          *
!       *     INPUT PARAMETERS:    OUTPUT PARAMETERS:  LOCAL DATA  *
!       *     A, B, C, D           X0, Y0                          *
!       *                                                          *
!       ************************************************************

SUB START_POINT

    RANDOMIZE
    LET X0 = A + (B - A)*RND
    LET Y0 = C + (D - C)*RND

END SUB
```

```
!       ************************************************************
!       *     MODULE: DIFF_EQ                                      *
!       *                                                          *
!       *     This internal subroutine computes the values of      *
!       *     dx/dt and dy/dt at point X, Y.                       *
!       *                                                          *
!       *     INPUT PARAMETERS:    OUTPUT PARAMETERS:  LOCAL DATA  *
!       *            X             U = dx/dt   ( = dx )            *
!       *            Y             V = dy/dt   ( = dy )            *
!       *                                                          *
!       ************************************************************

SUB DIFF_EQ

    LET U = Y
    LET V = -X
    LET W = SQR(U^2 + V^2)          ! REM  Total Step Size

END SUB

!       ************************************************************
!       *     MODULE: STEP                                         *
!       *                                                          *
!       *     This internal subroutine computes the next point on  *
!       *     a trajectory, using the modified Euler method.       *
!       *                                                          *
!       *     INPUT PARAMETERS:    OUTPUT PARAMETERS:  LOCAL DATA  *
!       *     E, H, U, V, W,       X1, Y1                          *
!       *     X, Y                                                 *
!       *                                                          *
!       ************************************************************

SUB STEP

    LET K = E * H * U / W
    LET L = E * H * V / W
    LET X = X0 + K
    LET Y = Y0 + L

END SUB

END
```

```
!         Program number 2
!       ************************************************************
!       *                                                          *
!       *   PROGRAM NAME: VECTRMAG                                 *
!       *   WRITTEN BY  : Finn Jorgensen                           *
!       *                                                          *
!       *   DATE: Feb 10, 1987                                     *
!       *                                                          *
!       *   This program computes and displays the magnetization   *
!       *   vector patterns in recorded media, as it will be between *
!       *   transitions.                                           *
!       *                                                          *
!       *   The program does not include effects of interaction or *
!       *   demagnetization.                                       *
!       *                                                          *
!       *   The write head field is modelled by the field from a   *
!       *   single current carrying conductor placed in the gap    *
!       *   center.                                                *
!       *                                                          *
!       *                      GLOBAL DATA                         *
!       *                                                          *
!       *    INPUT VARIABLES:               |   OUTPUT VARIABLES:  *
!       *                                   |                      *
```

```
!          *    Media_Thickness  in microns    |  Graphic display      *
!          *    Media_Spacing, in microns      |  Table of JX and JY   *
!          *    Square_Long (longitudinal squareness. fraction)        *
!          *    Square_Perp (perpendicular squareness, fraction)       *
!          *    Write_Level, in %  ( See 1) Note )                     *
!          *    No_Layers, the number of lamina the recorded portion   *
!          *               of the media is divided into.               *
!          *                                                           *
!          *  The program has with 3 internal subroutines:             *
!          *                      Data_Inputs                          *
!          *                      Mag_Vectors                          *
!          *                      Draw_Vectors                         *
!          *                                                           *
!          *  There are no external subroutines.                       *
!          *                                                           *
!          *  1) Note: A 100 % level corresponds to the situation      *
!          *           where a field strength of Hc exists at the      *
!          *           back side of the coating. This level is         *
!          *           located halfway between the head surface and    *
!          *           the backside of the coating at the 50 % level.  *
!          *                                                           *
!          *           Levels at higher than 100 %, 150 % for instance *
!          *           corresponds to 50 % increase in level relative  *
!          *           to 100 %, and will therefore be located half    *
!          *           of (coating thickness + spacing) behind the     *
!          *           coating back.                                   *
!          *                                                           *
!          *           This definition of level is therefore identical *
!          *           to write current levels, not the depth of       *
!          *           magnetization (except when the Media_Spacing=0) *
!          *                                                           *
!          *************************************************************


!          *************************************************************
!          *                                                           *
!          * Programs' mainline starts here.                           *
!          *                                                           *
!          *************************************************************

DIM MAG_LONG(20)
DIM MAG_PERP(20)
DIM MAG_TOTAL(20)
DIM MAG_ANGLE(20)
DIM Y(20)
CALL Data_Inputs
CALL Mag_Vectors
CALL Draw_Vectors


!          *************************************************************
!          *  MODULE:  Data_Inputs                                     *
!          *                                                           *
!          *  This internal subroutine accepts input variables.        *
!          *                                                           *
!          *************************************************************

SUB Data_Inputs
    CLEAR
    SET CURSOR 5,11
    PRINT"Input data for plot and computation of magnetization vectors"
    SET CURSOR 8,20
    INPUT PROMPT "Media Thickness.........(in microns): ": Media_Thickness
    SET CURSOR 10,20
    INPUT PROMPT "Media Spacing...........(in microns): ": Media_Spacing
    SET CURSOR 12,20
    INPUT PROMPT "Squareness, longitudinal..(fraction): ": Square_Long
    SET CURSOR 14,20
    INPUT PROMPT "Squareness, perpendicular.(fraction): ": Square_Perp
    SET CURSOR 17,23
    PRINT USING "Level must exceed ###.# percent":(100*Media_Spacing/(Media_Thic
kness+Media_Spacing))
```

```
    SET CURSOR 16,20
    INPUT PROMPT "Write Level...................... % : ": Write_Level
    SET CURSOR 19,20
    INPUT PROMPT "Number of Layers............(max 18): ": No_Layers
END SUB

!        ***********************************************************
!        *  MODULE:  Mag_Vectors (magnetization vectors)            *
!        *                                                          *
!        *  This internal subroutine computes the longitudinal and  *
!        *  the perpendicular magnetization vectors recorded in a   *
!        *  media coating.                                          *
!        *                                                          *
!        *  INPUT PARAMETERS:      OUTPUT PARAMETERS:   LOCAL DATA  *
!        *                                                          *
!        *  Media_Thickness        MAG_LONG()           n/a         *
!        *  Media_Spacing          MAG_PERP()                       *
!        *  Square_Long            MAG_TOTAL()                      *
!        *  Square_Perp            MAG_ANGLE()                      *
!        *  Write_Level            Y() (position of layers)         *
!        *  No_Layers                                               *
!        *                                                          *
!        ***********************************************************

SUB Mag_Vectors

    LET R = (Media_Thickness + Media_Spacing) * Write_Level/100
    LET Y_BACK = R
    IF Y_BACK > (Media_Thickness + Media_Spacing) THEN LET Y_BACK = Media_Thickn
ess + Media_Spacing

    ! REM    If Y_BACK < Media_Spacing then program will stop
    ! with error message : "division with zero".  Check that
    ! write level reaches into coating.

    LET DY = (Y_BACK - Media_Spacing)/No_Layers

    OPTION ANGLE RADIANS
    LET N = 0
    LET MAG_AVER = 0

    FOR M = No_Layers TO 1 STEP -1
        LET N = N + 1
        LET Y(N) = Y_BACK - (N - .5)*DY
        LET MAG_LONG(N) = Square_Long * Y(N)/R
        LET MAG_PERP(N) = Square_Perp * SQR(R^2 - Y(N)^2)/R
        LET MAG_TOTAL(N) = SQR(MAG_LONG(N)^2 + MAG_PERP(N)^2)
        LET MAG_ANGLE(N) = (180/PI)*ATN(MAG_PERP(N)/(MAG_LONG(N)+.00001))
        LET MAG_AVER = MAG_AVER + MAG_TOTAL(N)
    NEXT M

    LET MAG_AVER = MAG_AVER/N

    ! REM  All calculated magnetization levels are relative to a
    ! maximum value of one.  For absolute numbers multiply with the
    ! magnetization level that corresponds to the volume of coating
    ! under consideration, i.e. DY by DX (possibly equal to DY) by
    ! the track width, multiplied by the coatings saturation
    ! magnetization

END SUB


!        ***********************************************************
!        *  MODULE:  Draw_Vectors                                   *
!        *                                                          *
!        *  This internal subroutine draws a picture of the         *
!        *  longitudinal and the perpendicular magnetization vectors *
!        *  and their resultants.  Computed value are tabulated.    *
!        *                                                          *
```

```
!        *  INPUT PARAMETERS:     OUTPUT PARAMETERS:    LOCAL DATA   *
!        *                                                           *
!        *  MAG_LONG()              Graphics, data        n/a        *
!        *  MAG_PERP()                                               *
!        *  MAG_TOTAL()                                              *
!        *  MAG_ANGLE()                                              *
!        *  Y()                                                      *
!        *  No_Layers                                                *
!        *                                                           *
!        *************************************************************



!        *************************************************************
!        *                                                           *
!        *  DISPLAY COORDINATES AND GEOMETRY:                        *
!        *                                                           *
!        *  Backside of recorded thickness (= media backside for     *
!        *                  a recorded level of 100 %, or more)      *
!        *                                                           *
!        *                  ------------------- -Media Back          *
!        * Y = N*DY/2     |   Screen Plot    |                       *
!        *                |       of         |                       *
!        *                |  Magnetization   |                       *
!        *                |   Vectors and    |                       *
!        * Y = DY/2       |     Values       |                       *
!        *                  ------------------- -Media Front         *
!        *                                                           *
!        *************************************************************


SUB Draw_Vectors

    LET N = No_Layers
    LET SCALE = DY/SQR(Square_Long^2 + Square_Perp^2)
    ! REM DY for scaling to graphics and /"Squareness magnitude"
    ! to limit drawing size
    OPEN #1: SCREEN .1, .3, 0, 1
    SET WINDOW 0, (N+1.5)*(1/5)*(4/3)*DY, 0, (N+1.5)*DY
    ! REM 1/5 for screen size ratio, 4/3 for display ratio
    FOR M=1 TO N STEP 1
        SET COLOR "RED"
        ! REM Longitudinal and Perpendicular components
        PLOT 0.1*DY, (N-M+2)*DY; (0.1*DY + MAG_LONG(M)*SCALE), (N-M+2)*DY
        PLOT 0.1*DY, (N-M+2)*DY; 0.1*DY,((N-M+2)*DY - MAG_PERP(M)*SCALE)
        SET COLOR "BLUE"
        ! REM Vectorsum magnetization
        PLOT 0.1*DY, (N-M+2)*DY; (0.1*DY + MAG_LONG(M)*SCALE),((N-M+2)*DY - MAG_
PERP(M)*SCALE)
    NEXT M

    OPEN #2: SCREEN .35, 1, 0, 1
    SET CURSOR 1,3
    PRINT "MAGNETIZATION LEVELS"
    PRINT "relative to a value of 1"
    PRINT
    PRINT "   Long   Perp   TOTAL"
    PRINT
    FOR M = 1 TO N STEP 1
        PRINT USING "###.###": MAG_LONG(M), MAG_PERP(M), MAG_TOTAL(M)
    NEXT M
    PRINT USING "Average level = #.###": MAG_AVER

END SUB

END
```

```
Program number 3
*****************************************************************
*                                                               *
*  PROGRAM NAME: DEL_DRAW                                       *
*  WRITTEN BY  : Finn Jorgensen                                 *
*                                                               *
*  DATE: Jan 17, 1987                                           *
*                                                               *
*  This program computes and displays the magnetization         *
*  patterns in recorded transitions, also called delta-x,       *
*  between bits written onto a PARTICULATE media.               *
*                                                               *
*  The program does not include effects of interaction or       *
*  demagnetization.  The distribution of the particles'         *
*  coercivities (the switching field distribution) is           *
*  assumed Gaussian normal.                                     *
*                                                               *
*  The write head field is modelled by the field from a         *
*  single current carrying conductor placed in the gap          *
*  center.                                                      *
*                                                               *
*                       GLOBAL DATA                             *
*                                                               *
*    INPUT VARIABLES:                     OUTPUT VARIABLES      *
*                                                               *
*    Media_Thickness  in microns          Graphic display       *
*    Media_Spacing    in microns                                *
*    Bit_Length       in microns                                *
*    Switch_Field_Distr     in %                                *
*    Write_Level            in %   1) Note                      *
*                                                               *
*  The program has six internal subroutines:                    *
*                                                               *
*                     Data_Inputs                               *
*                     Bit_Centers                               *
*                     Draw_Media                                *
*                     Particle_Coercivity                       *
*                     Write_Field                               *
*                     Checker_Pattern                           *
*                                                               *
*  There are no external subroutines.                           *
*                                                               *
*  1) Note: A 100 % level corresponds to the situation          *
*           where a field strength of Hc exists at the          *
*           back side of the coating. This level is             *
*           located halfway between the head surface and        *
*           the backside of the coating at the 50 % level.      *
*                                                               *
*           Levels at higher than 100 %, 150 % for instance     *
*           corresponds to 50 % increase in level relative      *
*           to 100 %, and will therefore be located half        *
*           of (coating thickness + spacing) behind the         *
*           coating back.                                       *
*                                                               *
*           This definition of level is therefore identical     *
*           to write current levels, not the depth of           *
*           magnetization (except when the Media_Spacing=0      *
*                                                               *
*****************************************************************

*****************************************************************
*                                                               *
* Programs' mainline starts here.                               *
*                                                               *
*****************************************************************

CALL Data_Inputs
CALL Bit_Centers
CALL Draw_Media
```

```
FOR X = 313 TO 50 STEP -1
    FOR Y = 158 TO ( Media_Front + 1 ) STEP -1
        LET Bit_No = INT(( X - 50 )/Bit_Length + 1)
        LET XN = 0
             CALL Particle_Coercivity          ! Has this particle
             CALL Write_Field                  ! become magnetized?
                IF Particle_hci > Writ_Fld THEN
                     CALL Checker_Pattern
                ELSE                           ! -to see if magnetized
                     CALL Left_Bits            ! by the bit to the left
                END IF
        PLOT X,Y
    NEXT Y
NEXT X

        *************************************************************
        *                                                           *
        *  DISPLAY COORDINATES AND GEOMETRY:                        *
        *                                                           *
        * Y=199                          Percent write level: 129 *
        *  Media back                                         114 *
        * Y=159--xooox--------------- xooox---------------xoox- 100 *
        *           xoox                  xoox              xoo  86 *
        *              xox                   xox                 72 *
        *                xox                   xox               57 *
        *  Media front    xox <- Bit Length -> xox               43 *
        *  ---------------xox-------------------xox------------  29 *
        *                                                        14 *
        * Y=39-   gap   -------------------------------------     0 *
        *      | center  |                                        *
        * X=0 | at X=50 |                              X=319      *
        *                                                           *
        *  Sketch of head/media interface.                          *
        *                                                           *
        *************************************************************

        *************************************************************
        *  MODULE:  Data_Inputs                                     *
        *                                                           *
        *  This internal subroutine accepts input variables.        *
        *                                                           *
        *************************************************************

SUB Data_Inputs
    CLEAR
    SET CURSOR 5,16
    PRINT"Input data for plotting of Transition Zones"
    SET CURSOR 8,20
    INPUT PROMPT "Media Thickness......(in microns): ": Media_Thickness
    SET CURSOR 10,20
    INPUT PROMPT "Media Spacing........(in microns): ": Media_Spacing
    SET CURSOR 12,20
    INPUT PROMPT "Bit Length...........(in microns): ": Bit_Length
    SET CURSOR 14,20
    INPUT PROMPT "Switching Field Distribution.. % : ": SFD
    SET CURSOR 16,20
    INPUT PROMPT "Write Level................... % : ": Write_Level
END SUB

!       *************************************************************
!       *  MODULE:  Bit_Centers                                     *
!       *                                                           *
!       *  This internal subroutine converts Bit_Length to number   *
!       *  of pixels, and computes the centers for the circular     *
!       *  transition zone patterns.                                *
!       *                                                           *
!       *  INPUT PARAMETERS:     OUTPUT PARAMETERS:    LOCAL DATA   *
!       *                                                           *
```

```
!         *  Media_Thickness         No_Bitlengths          n/a           *
!         *  Media_Spacing           Bit_Center(N)                        *
!         *  Bit_Length                                                   *
!         *  Write_Level                                                  *
!         *                                                               *
!         *****************************************************************

SUB Bit_Centers
    LET Bit_Length = Bit_Length * 120 / ( Media_Thickness + Media_Spacing )
    LET No_Bitlengths = INT ( ( 319 - 50 ) / Bit_Length  + 1 )
    DIM Bit_Center( 20 )
    FOR N = 1 to No_Bitlengths
        LET Bit_Center( N ) = 50 + ( N - 1 ) * Bit_Length
    NEXT N
END SUB

!         *****************************************************************
!         *  MODULE:  Draw_Media                                          *
!         *                                                               *
!         *  This internal subroutine draws a picture of the head/        *
!         *  media interface.  The distance from the head surface         *
!         *  to the media bask side is constant, equal to 120 pixels.     *
!         *                                                               *
!         *  INPUT PARAMETERS:      OUTPUT PARAMETERS:     LOCAL DATA     *
!         *                                                               *
!         *  Media_Thickness        Graphics               n/a            *
!         *  Media_Spacing                                                *
!         *                                                               *
!         *****************************************************************

SUB Draw_Media
    LET Media_Front = 159 - 120 * Media_Thickness / ( Media_Thickness + Media_Sp
acing)
    LET Media_Front = INT(Media_Front)
    SET WINDOW 0, 319, 0, 199
    SET BACKGROUND COLOR 8
    SET COLOR "WHITE"
    BOX  AREA   0, 35, 0, 39            ! REM  Left pole piece
    BOX  AREA  65,319, 0, 39            ! REM  Right pole piece
    PLOT TEXT, AT 120, 0: " Gap Length not to scale        "
    BOX LINES 5, 314, Media_Front, 159   ! REM Media
END SUB

!         *****************************************************************
!         *  MODULE:  Particle_Coercivity                                 *
!         *                                                               *
!         *  This internal subroutine computes the hci of a randomly      *
!         *  selected particle.                                           *
!         *                                                               *
!         *   Note: First line : Standard Normal Density                  *
!         *   Note: Second line: Normal Density, SFD wide at 25 % and     *
!         *          and 75 % points on the BH-loop, converted to         *
!         *          standard.                                            *
!         *   Note: Third line : Average coercivity normalized to         *
!         *          one unit of field strength.                          *
!         *                                                               *
!         *  INPUT PARAMETERS:      OUTPUT PARAMETERS:     LOCAL DATA     *
!         *                                                               *
!         *  SFD                    Particle_hci                          *
!         *                                                               *
!         *****************************************************************

SUB Particle_Coercivity

    LET A = ( RND + RND + RND - 1.5 ) * 1.925
    LET A = A * ( SFD / 100 ) * .742
    LET Particle_hci = 1 + A

END SUB
```

```
!         *****************************************************
!         *  MODULE:  Write_Field                             *
!         *                                                   *
!         *  This internal subroutine computes the field strength  *
!         *  from bit centers along the X-axis.  A strength of one *
!         *  corresponds to the average value Hc of hci.  A 100 %  *
!         *  level corresponds to a strength of one at the back side *
!         *  of the media (The transition pattern tangents the back). *
!         *                                                   *
!         *  INPUT PARAMETERS:     OUTPUT PARAMETERS:    LOCAL DATA  *
!         *                                                   *
!         *  Write_Level           Writ_Fld                   *
!         *  Bit_Center(Bit_No)
!         *  XN                                               *
!         *  Y                                                *
!         *                                                   *
!         *****************************************************

SUB Write_Field

     LET Writ_Fld = 120*( Write_Level / 100 )/SQR(XN^2 + (Y-39)^2)

END SUB


!         *****************************************************
!         *  MODULE:  Left_Bits                               *
!         *                                                   *
!         *  This internal subroutine computes the polarity of     *
!         *  particle magnetization in a pattern of alternating bits. *
!         *                                                   *
!         *  INPUT PARAMETERS:     OUTPUT PARAMETERS:    LOCAL DATA  *
!         *                                                   *
!         *  Bit_No                Plot color COLNO           *
!         *  Bit_Length                                       *
!         *  XN                                               *
!         *                                                   *
!         *****************************************************

SUB Left_Bits
    FOR ToTheLeft_Bit = Bit_No TO 1 STEP -1
        LET XN = X - Bit_Center(ToTheLeft_Bit)
        LET COLNO = 4 + MOD( ToTheLeft_Bit, 2 )
        CALL Write_Field
        IF Particle_hci > Writ_Fld THEN EXIT FOR
        IF ToTheLeft_Bit = 1 THEN LET COLNO = 4   ! PATTERN AT GAP
    NEXT ToTheLeft_Bit
    SET COLOR COLNO
END SUB


!         *****************************************************
!         *  MODULE: Checker_Pattern                          *
!         *                                                   *
!         *  This internal subroutine computes the alternating     *
!         *  magenta/cyan pattern of erased coating (no net mag-   *
!         *  netization)                                      *
!         *                                                   *
!         *  INPUT PARAMETERS:     OUTPUT PARAMETERS:    LOCAL DATA  *
!         *                                                   *
!         *  Pixel position X,Y    Color setting              *
!         *                                                   *
!         *****************************************************

SUB Checker_Pattern

    IF MOD ((X+Y),2) > .5 THEN
       SET COLOR "CYAN"
    ELSE
```

```
        SET COLOR "MAGENTA"
      END IF

  END SUB

  END
```

```
Program number 4
**************************************************************
*                                                            *
*  PROGRAM NAME: SFD_LOSS                                    *
*  WRITTEN BY  : Finn Jorgensen, DANVIK                      *
*                                805-682-2102                *
*  DATE: Nov 17, 1987                                        *
*                                                            *
*  This program computes and displays the losses that        *
*  result from writing with a finite field gradient on       *
*  a particulate medium with a certain Switching Field       *
*  Distribution (SFD).                                       *
*                                                            *
*  The SFD is here defined as delta-h-r, which equals the    *
*  field strength range that encompasses fifty percent of    *
*  the particles coercivities, divided by the mean coer-     *
*  civity.                                                   *
*                                                            *
*  The program does not include effects of interaction or    *
*  demagnetization.  The distribution of the particles'      *
*  coercivities (the switching field distribution) is        *
*  assumed Gaussian normal.                                  *
*                                                            *
*  The write head field is modelled by the field from a      *
*  single current carrying conductor placed in the gap       *
*  center.                                                   *
*                                                            *
*                        GLOBAL DATA                         *
*                                                            *
*    INPUT VARIABLES:                   OUTPUT VARIABLES     *
*                                                            *
*    Media_Thickness  in microns        Screen dump of       *
*    Media_Spacing    in microns          Input Data         *
*    Bit_Length       in microns        Printed lists of     *
*    Switch. Fld. Distr.    in %        Loss vrs. wavelength *
*    Write Level            in % 1)     Signal waveforms     *
*    No. points per bit                                      *
*                                                            *
*  The program has six internal subroutines:                 *
*                                                            *
*                    Data_Inputs                             *
*                    SFD_Function                            *
*                    Probability                             *
*                    Convolution                             *
*                    RMS_Peak                                *
*                    Print_Data                              *
*                                                            *
*  There are no external subroutines.                        *
*                                                            *
*  1) Note: A 100 % level corresponds to the situation       *
*           where a field strength of Hc exists at the       *
*           back side of the coating. This level is          *
*           located halfway between the head surface and     *
*           the backside of the coating at the 50 % level.   *
*                                                            *
*           Levels at higher than 100 %, 150 % for instance  *
*           corresponds to 50 % increase in level relative   *
*           to 100 %, and will therefore be located half     *
*           of (coating thickness + spacing) behind the      *
*           coating back.                                    *
```

```
!         *                                                              *
!         *          This definition of level is therefore identical    *
!         *          to write current levels, not the depth of          *
!         *          magnetization (except when the Media_Spacing=0     *
!         *                                                              *
!         ****************************************************************
!

!         ****************************************************************
!         *                                                              *
!         * Programs' mainline starts here.                              *
!         *                                                              *
!         ****************************************************************
! REM   OPTION TYPO requires TrueBASIC Vers. 2.0.  It will at compile
! REM   time catch typos in the variables listed below under LOCAL.
OPTION TYPO
LOCAL Bit_Length, Prob_Sum, Delta_X,Eff_Thickness, Field_Radius
LOCAL Mag_Thickness, Lambda, Media_Thickness, Media_Spacing, No_PtsMAX
LOCAL No_Pts, Nmbr_Pts, Phase, Prob, RMS, Average
LOCAL SFD, Signl, Write_Level, SFD_pts, Conv_Pts, N, P, Q, X, X_Max, Y, Z

OPTION ANGLE RADIANS
DIM Prob_Hci(4000)              ! REM  Program may stop with the message
DIM Prob_Flip(4000)             ! REM  "Subscript out of bound": Re-run
DIM Mag_Field(4000)             ! REM  with fewer points, or increase the
                                ! REM  number 4000 in the first three DIMs.
DIM Level(11)                   ! REM  This situation may occur for small
DIM Lambd(11)                   ! REM  values of SFD and/or short bit lengths;
DIM Magnetization(256,11)       ! REM  Delta_X is made small in both cases.
                                ! REM  - See also below at REM statement
                                ! REM  under Data_Inputs.

CALL Data_Inputs
LET Lambda = Lambda*SQR(2)
FOR N = 1 TO 11                      ! REM  Five octaves
    LET Lambd(N) = Lambda/SQR(2)^N
    LET No_Pts = Nmbr_Pts            ! REM  Keep Nmbr_Pts as No_Pts will change
    LET X_Max = 4*Field_Radius
    IF X_Max => Lambd(N) THEN
       LET X_Max = Lambd(N) * INT(X_Max/Lambd(N) + 1 )
    ELSE
       LET X_Max = Lambd(N)
    END IF

    LET No_PtsMAX = 0
    LET Delta_X = Lambd(N)/No_Pts  !  REM  RESET (Possibly)
    LET Eff_Thickness = 0.22 * Lambd(N)
    IF Eff_Thickness > Mag_Thickness THEN LET Eff_Thickness = Mag_Thickness

    LET Y = Media_Spacing + 0.5 * Eff_Thickness
    LET Delta_X = 2*Delta_X
    LET No_Pts = No_Pts/2
    LET Conv_Pts = 0
    DO
       LET Delta_X = Delta_X / 2  ! REM Otherwise finer division
       LET No_Pts = No_Pts * 2    ! REM and more points.
       CALL SFD_Function          ! REM for convolution.
    LOOP UNTIL Conv_Pts > 8       ! REM Min. 8 points for SFD function
    IF No_Pts > No_PtsMAX THEN LET No_PtsMAX = No_Pts

    OPEN #1: PRINTER
    PRINT#1: "Date and time:  :";DATE$;"    ";TIME$
    PRINT#1: "Wavelength = ";Lambd(N);" microns."
    PRINT#1: "Distance from head surface = ";Y;" microns."
    PRINT#1: "Delta_X increment is = ";Delta_X;" microns."
    PRINT#1
    PRINT#1: "The SFD curve for convolution covers ";Conv_Pts*Delta_X;" microns,
 over ";Conv_Pts;" points, each of length :";Delta_X
    PRINT#1: "Prob_Sum = ";Prob_Sum
    PRINT#1
    PRINT#1
    CLOSE #1
```

```
    CALL Convolution
    CALL RMS_Signal
NEXT N

CALL Print_Data                      ! REM This end main program

!        ***********************************************************
!        *  MODULE:  Data_Inputs                                   *
!        *                                                         *
!        *  This internal subroutine accepts input variables.      *
!        *                                                         *
!        *  Wavelength Lambda is calculated, as is the number of   *
!        *  points Nmbr_Pts desired for signal waveform plot and   *
!        *  possible analysis.                                     *
!        *  Also the distance Field_Radius, where the field strength *
!        *  is normalized to a value of 1, is calculated, as is the *
!        *  magnetized thickness (which may be less than the physical*
!        *  thickness of the coating.                              *
!        *                                                         *
!        ***********************************************************

SUB Data_Inputs
    CLEAR
    SET CURSOR 5,16
    PRINT"Input data for Computation of Aperture Loss"
    SET CURSOR 6,16
    PRINT"   during recording of sine wave signals"
    SET CURSOR 8,20
    INPUT PROMPT "Media Thickness......(in microns): ": Media_Thickness
    SET CURSOR 10,20
    INPUT PROMPT "Media Spacing........(in microns): ": Media_Spacing
    SET CURSOR 12,20
    INPUT PROMPT "Bit Length...........(in microns): ": Bit_Length
    SET CURSOR 14,20
    INPUT PROMPT "Switching Field Distribution.. % : ": SFD
    SET CURSOR 16,20
    INPUT PROMPT "Write Level................... % : ": Write_Level
    SET CURSOR 21,10
    PRINT "BEFORE HITTING LAST <RET>, PRESS Shift-PrtSc FOR SCREEN COPY"
    SET CURSOR 18,20
    INPUT PROMPT "Number of points per bit (max.128) ": Nmbr_Pts

    ! REM  If program stops with message "Subscript out of bound" then
    ! REM  try to re-run with a smaller Number of Points per bit (equal
    ! REM  to half a wavelength).  ELSE look at the REM statements above,
    ! REM  at DIM

    LET SFD = SFD/100
    LET Lambda = 2 * Bit_Length
    LET Nmbr_Pts = 2 * Nmbr_Pts    ! REM Now per wavelength
    LET Field_Radius = (Media_Thickness + Media_Spacing) * Write_Level/100
    LET Mag_Thickness = Field_Radius - Media_Spacing
    IF Mag_Thickness > Media_Thickness then LET Mag_Thickness = Media_Thickness
    IF Mag_Thickness <= 0 THEN
       PRINT "Re-run with Higher Write Level"
       STOP
    END IF

END SUB

!        ***********************************************************
!        *  MODULE: SFD_Function                                   *
!        *                                                         *
!        *  This internal subroutine computes Prob(H = hci) = f(x), *
!        *  see Handbook of Magnetic Recording, Fig. 19-8.         *
!        *  The results is a number of points Prob_Flip() that has *
!        *  an area of nominally one.                              *
!        *                                                         *
!        ***********************************************************

SUB SFD_Function
```

```
    LET X = - Delta_X
    LET Q = 0
    LET SFD_pts = 0
    FOR X = -Delta_X TO X_Max STEP Delta_X
        LET SFD_pts = SFD_pts + 1
        LET Q = Q +1
        LET Mag_Field(Q) = Field_Radius/SQR(X^2 + Y^2)
    NEXT X

    LET Conv_Pts = 0
    LET Prob_Sum = 0

    FOR Q = 1 TO SFD_pts
        CALL Probability
        IF (Prob > .026) THEN
           LET Conv_Pts = Conv_Pts + 1
           LET Prob_Hci(Conv_Pts) = Prob
           LET Prob_Sum = Prob_Sum + Prob
        END IF
    NEXT Q

    FOR Q = Conv_Pts TO 1 STEP -1
        LET Prob_Flip(Conv_Pts-Q+1) = Prob_Hci(Q)/Prob_Sum
    NEXT Q

END SUB

!         ************************************************************
!         *  MODULE:  Probability                                     *
!         *                                                          *
!         *  This internal subroutine computes the probability that  *
!         *  a particle at the coordinates (X,Y) has a coercivity    *
!         *  (switching field) equal to the head field strength      *
!         *  Z = Mag_Field at that point.                            *
!         *                                                          *
!         ************************************************************

SUB Probability

    LET Mag_Field(Q) = Mag_Field(Q) * 1.5    ! REM  Normalize
    LET Mag_Field(Q) = Mag_Field(Q) - 1.5    ! REM  Offset to change
    LET Mag_Field(Q) = Mag_Field(Q)*.47/SFD  ! REM  Adjust to SFD
    LET Mag_Field(Q) = Mag_Field(Q) + 1.5    ! REM  Reset after change
    LET Z = Mag_Field(Q)

    IF (Z<0) THEN LET Prob = 0
    IF (Z>0) AND (Z<1) THEN LET Prob = (Z^2)/2
    IF (Z>=1) AND (Z<=2) THEN LET Prob = - Z^2 + 3*Z - 1.5
    IF (Z>=2) AND (Z<=3) THEN LET Prob = .5*(Z^2 - 6*Z + 9)
    IF (Z>=3) THEN LET Prob = 0

END SUB

!         ************************************************************
!         *  MODULE: Convolution                                      *
!         *                                                          *
!         *  This internal subroutine computes the signal waveform   *
!         *  that results after convolution of the Prop_Flip curve   *
!         *  and a SINE WAVE signal.                                 *
!         *  For a SQUARE WAVE, add SGN(SIN etc.  ), and change the  *
!         *  the multiplier 1.57 in RMS = Average (module RMS of     *
!         *  signal)                                                 *
!         *                                                          *
!         ************************************************************

SUB Convolution

    FOR Q = 1 TO No_Pts
```

```
        LET Phase = Q * 2 * 3.1416/No_Pts
        LET X = - Delta_X
        LET Signl = 0

        FOR P = 1 TO Conv_Pts
            LET X = X + Delta_X
            LET Signl = Signl + Prob_Flip(P) * SIN (( 2 * 3.1416 * X / Lambd(N))
 + Phase)
        NEXT P

        LET MAGNETIZATION(Q,N) = Signl

    NEXT Q

END SUB


!        ************************************************************
!        *  MODULE:  RMS of signal                                  *
!        *                                                          *
!        *  This internal subroutine computes the signal levels    *
!        *  MAGNETIZATION(,), and finds the relative level to a    *
!        *  fully magnetized coating, without write (SFD) losses,  *
!        *  expressed in decibels.                                  *
!        *                                                          *
!        ************************************************************

SUB RMS_Signal

    LET Average = 0
    FOR Q = 1 TO No_Pts
        LET Average = Average + ABS(MAGNETIZATION(Q,N))
    NEXT Q

    LET RMS = Average*1.57/No_Pts
    LET Level(N) = 20*LOG10(RMS)

END SUB


!        ************************************************************
!        *  MODULE:  PRINT DATA TABLE                               *
!        *                                                          *
!        *  This internal subroutine prints a table of signal levels *
!        *  in dB relative, at the selected wavelengths.            *
!        *  It also prints a table of signal waveforms for export   *
!        *  to plot and/or analysis programs.                       *
!        *                                                          *
!        ************************************************************

SUB Print_Data

    OPEN #1: PRINTER
    PRINT#1: "Magnetized Thickness ........... ";Mag_Thickness;" microns"
    PRINT#1:
    PRINT#1: "RECORDED SIGNAL WAVEFORM:"
    PRINT#1:
    FOR N = 1 TO 11
        PRINT#1, USING "At Lambda = ###.## the signal is --##.## dB rel. ": Lamb
d(N), Level(N)
    NEXT N
    PRINT#1:
    PRINT#1: "Signals are:"
    PRINT #1
    FOR N = 1 TO 10
        PRINT #1, USING "###.###":LAMBD(N);
    NEXT N
        PRINT#1, USING "###.###":LAMBD(11)
```

```
        PRINT#1
    FOR Q = 1 TO No_PtsMAX
        FOR N=1 TO 10
            PRINT #1, USING "###.###": MAGNETIZATION(Q,N);
        NEXT N
        PRINT#1, USING "###.###": MAGNETIZATION(Q,11)
    NEXT Q
    CLOSE #1

END SUB

END                                ! REM Program end; All routines internal.
```

---

ling is briefly explained, and the accuracy of some modelling steps can always be discussed. The importance of the programs lie in their tutorial function, and trend data that will be observed by varying input parameters will be correct, and can be correlated with measurements.

The programs are written in small module for clarity, and no external library functions are required.

Only printing of the graphics remains to be resolved by the reader for his particular system (see f. ex. *Structured Programming with True Basic* pages 517-522.) The data input screen is easily preserved to the printer by performing a Ctrl-Prt-Screen before hitting <ENTER> after the last entry. If the command GRAPHICS is performed prior to loading True BASIC, then Ctrl-Prt-Screen will also print the graphics.

These programs (and additional ones, such as READ_OUT (for Chapter 21) are available on disk, which also contain compiled versions, ready to run. The compiled versions have the graphics capabilities plus some additional features. An instruction manual with detailed explanations and examples is included.

For further information write to the author:

Finn Jorgensen
DANVIK
P.O. Box 30791
Santa Barbara, CA 93130.

POST SCRIPT

Computations of the Gaussian (standard normal) Density and Distribution Functions

Magnetic recording performance is controlled by two magnetic properties of the powder used in the coating; the individual particle's coercivity hci, and the interacting fields Δh from neighboring particles. Simple models of the magnetization process are based on the fact that an external field (the write field) must exceed h_{ci} in order to switch magnetization in a particle (see Chapter 5). A realistic model sets the external field equal to the write field plus the interacting fields, plus any demagnetizing fields.

The particle coercivities vary around a mean value H_c. We will assume a Gaussian density. In real life it may have a tendency to be a log-normal characteristic for some powders; our assumption of a Gaussian distribution will not affect the accuracy of predictions using our write process model.

The computational problem lies in obtaining random numbers that have a normal distribution. Microcomputers provide random numbers that are uniformly distributed between zero and one. A very simple solution exists in calling the RND(X) function three times and adding the numbers:

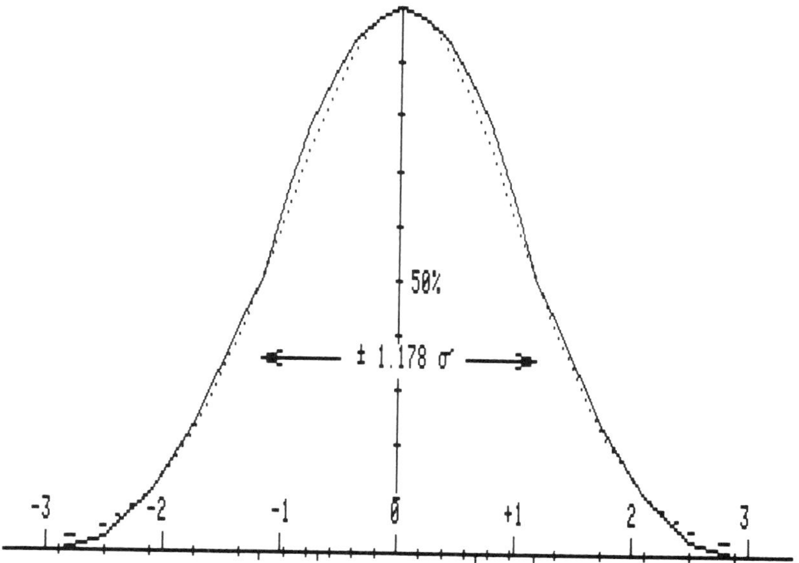

Fig. A-1. Density function ((RND(1)+RND(1)+RND(1))−1.5)*1.925; (*1.925 to make 50% width points coincide with the standard normal density, shown dashed).

$$R = RND(X) + RND(X) + RND(X) \quad \textbf{(A.1)}$$

$$R = (RND(X) + RND(X) + RND(X) - 1.5) * 1.925 \quad \textbf{(A.2)}$$

The computation of the written transition in a tape or disk coating requires an input of the switching field distribution SFD, which will further modify (A.1). There are several definitions for the measurement of the SFD, as described in Chapters 5 and 14. See Figs. A-4 and A-5.

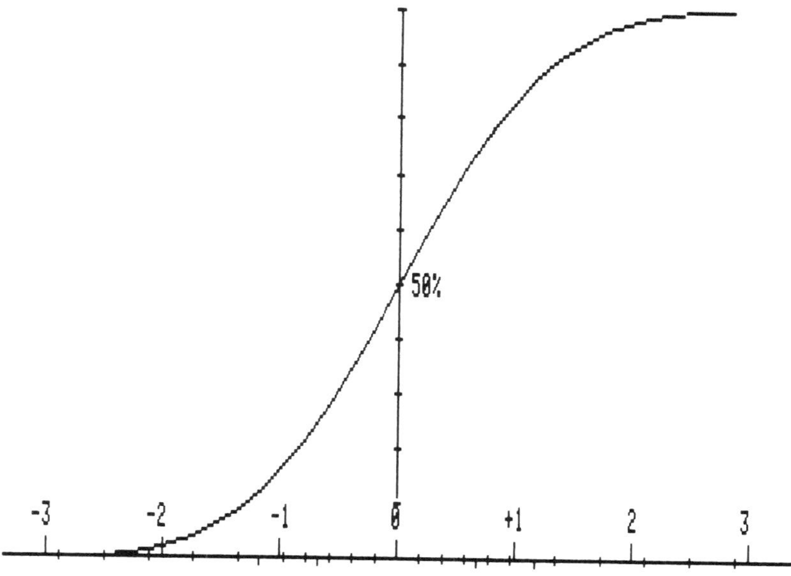

Fig. A-2. Distribution function for ((RND(1)+RND(1)+RND(1))−1.5)*1.925.

$$f_z(z) = \begin{cases} \dfrac{z^2}{2} & \text{for } 0 <= z <= 1, \\ -z^2 + 3z - 3/2 & \text{for } 1 <= z <= 2, \\ \dfrac{1}{2}(z^2 - 6z + 9) & \text{for } 2 <= z <= 3, \\ 0 & \text{for } z < 0 \text{ or } z > 3 \end{cases}$$

$$F_z(z) = \begin{cases} 0 & \text{for } z < 0, \\ \dfrac{z^3}{6} & \text{for } 0 <= z <= 1, \\ z - 1 + \dfrac{(2-z)^3}{6} - \dfrac{(z-1)^3}{6} & \text{for } 1 <= z <= 2, \\ 1 - \dfrac{(3-z)^3}{6} & \text{for } 2 <= z <= 3, \\ 1 & \text{for } z > 3. \end{cases}$$

Fig. A-3. Probability density fz(z) and distribution function Fz(z) for the function (RND + RND + RND), where RND is uniformly distributed between 0 and 1. After A.A Sveshnikov *Problems in Probability Theory*, Dover Publ., 1978, p. 130.

Figure A-4 illustrates the two most commonly used and how they relate to the standard deviation σ. The method shown left measures the difference in field corresponding to the 25 and 75 percent levels of the JH-loop; this value is very close to another measurement using the tangent to the BH-loop, resulting in a number $(1 - S*) = \text{SFD}$.

The method shown to the right uses the unintegrated voltage from the sampling coil in the BH-meter and relates SFD to its width at the 50 percent points.

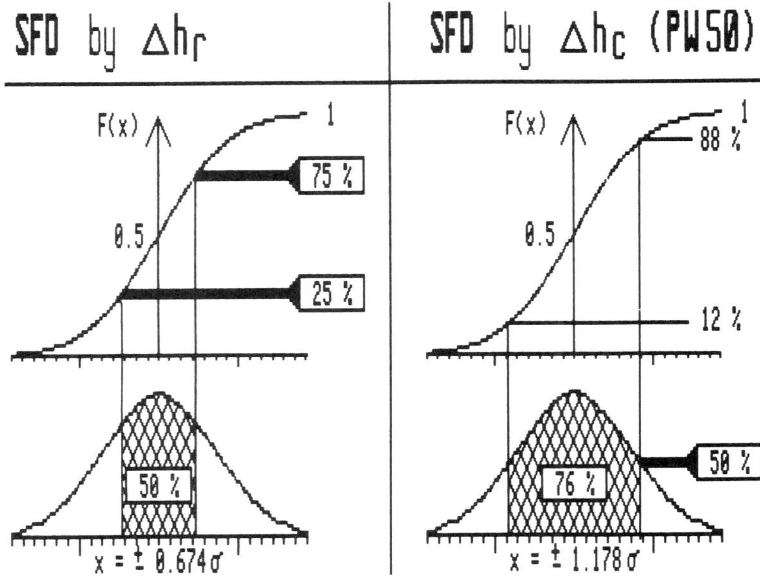

Fig. A-4. Two methods for determination of the SFD, assuming normal densities.

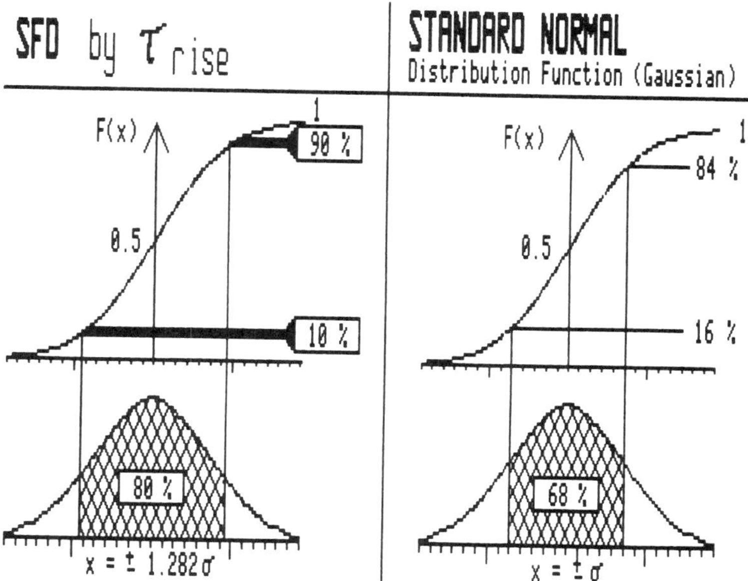

Fig. A-5. SFD by pulse rise time τ; Standard normal distribution functions.

Neither number for the SFD corresponds to the standard deviation, as the multipliers indicate. The total number of particles switched for a change in field proportional to SFD are different from each other, and from the Gaussian, shown to the right in Fig. A-5. Also shown is the SFD as a communication engineer would have defined it, corresponding to the rise time for a step voltage to rise from 10 to 90 percent of its final value.

The program in this appendix assumes that the values for the SFD are provided for by magnetic measurements, either $(1-S*)$, or a Δh_r value; $(1-S*)$ is the recommended standard (see Koester's paper, Chapter 5 references).

The computed density function (A.2), multiplied by .742, will correspond to the SFD $= (1-S*) = 1$; multiplication with a SFD value (in fraction), and a shift back to the normalized coercivity (= 1) by adding one will produce the sequence of numbers with the desired probability.

REFERENCES

Davis, W.S., *True BASIC Primer*, Addison-Wesley, 1986, 450 pages.

Elliott, B., *True BASIC Reference Manual*, Addison-Wesley, 1985, 330 pages, plus addendum for Vers. 2.0 from True BASIC, Inc., Hanover, New Hampshire 03755.

Goldstein, L.J., Moore, C.E., and Welcher, P.J., *Structured Programming with True BASIC*, Prentice-Hall, 1986, 534 pages.

Shammas, N.C., *The New BASICS*, M and T Books, 1987, 417 pages.

Artigue, M., and Gautheron, V., *Systemes Differentiels, Etude Graphique*,.Cedic, Paris, 1983, 180 pages.

Sveshnikov, A.A., *Problems in Probability Theory, Mathematical Statistics and Theory of Random Functions*, Dover, 1978, 481 pages.

Appendix B

List of Symbols

This listing covers all symbols that are commonly used throughout the book. Symbols that are used in one chapter only (in examples, proofs, derivation, etc.) are not necessarily found in this listing.

The numbers in parentheses are units and/or formula numbers where the symbol first appears, or where it was defined.

Item:	**Definition:**
a	= length (m)
A	= area (m^2)
A_C	= area of core cross section (m^2)
A_{fg}	= area of front gap in magnetic head (m^2)
A_g	= area of gap in magnetic circuit (m^2)
A_t	= irregularity loss (dB) (9.7), (9.8)
B	= signal bandwidth (Hz)
B	= flux lines per unit area = flux density (Wb/m^2) (3.2), (3.21)
BL	= bit length (m)
B_g	= flux density inside write head gap (Wb/m^2)
B_r	= remanence (Wb/m^2)
B_r'	= remanence after demagnetization (Wb/m^2) (4.10)
B_r''	= remanence at time of read-out (after demagnetization, and recoil) (Wb/m^2) (Fig. 4-14)
B_{rs}	= remanence after saturation (Wb/m^2) (11.1)

B_{rs}' = remanence after saturation, and after demagnetization (Wb/m^2)
B_{rsat} = remanence after saturation (Wb/m^2)
B_s = saturation remanence (Wb/m^2)
B_{sat} = saturation remanence (Wb/m^2)
BW = signal bandwidth (Hz)

c = coating thickness (m)
c′ = written (recorded) thickness ($c' < c$) (m) (Fig. 4-12)
c_{eff} = $0.44 * BL = 0.22 * \lambda$ = part of coating thickness that contributes 75 percent of signal output level. ($c_{eff} <= c' <= c$) (m) (6.8)
C = channel capacity (27.1)
C_e = self-capacitance of wound coil, or inductor (Farad)

d = effective spacing between surfaces of media and head (m)
d = code constraint, i.e., minimum number of zeros between transitions
D_g = depth of front gap in magnetic head assembly (m)
D_{gap} = D_g

E = voltage (volts)
E = Young's modulus of elasticity (Nt/m^2)
E_d = Young's modulus of elasticity, dynamic value (17.3)
e = induced voltage in read head (volts) (6.1)

f = frequency in Hertz (sec^{-1})
f = force (Nt) (current in B-field) (3.6)
F = force (N)
f_k = frequency in kHz
f_s = spin resonance frequency in magnetic material (MHz) (7.8)
f_u = upper signal frequency (at upper band-edge) (Hz)
f_w = wall resonance frequency in magnetic material (MHz) (7.6)

h_{ci} = coercivity of a single particle (A/m)
H = magnetic field strength (A/m) (pole: (3.5); current: (3.8); coil: (3.14))
H_a = applied magnetic field (A/m)
H_c = coercivity, or coercive force, BH-loop (A/m) (Fig. 3-32)
H_{core} = net field inside a magnetic core material (A/m) (4.3)
H_d = demagnetizing field (A/m) (4.3)
H_g = field inside gap in magnetic circuit (A/m) (4.27)
H_r = remanent coercivity, JH- (or MH-) loop (A/m) (Fig. 3-32)
H_s = sensing function during read-out (m^{-1}) (with components H_{sx} and H_{sy}) (6.5)
H_x, H_y, H_z = vector components of field H (A/m)
Δh_r = switching field distribution (Oe, or percent, or fraction) (ref. to H_r; see (5.1))
Δh_c = switching field distribution (ref. to H_c see (5.1)

i = current (A)
I = current (A)

I	=	moment of inertia
i_{write}	=	write (record) current through write head winding (4.29), (10.7)
j	=	$\sqrt{-1}$
j	=	iAn = magnetic dipole moment (Am^2) (3.16)
j′	=	magnetic dipole moment (A/m) (μ_om − from (6.5))
J	=	magnetization (J = μ_oM) (Wb/m^2) (3.23)
J_r'	=	magnetization after demagnetization (Wb/m^2) (Fig. 4-6)
J_r''	=	magnetization after demagnetization and recoil (Wb/m^2) (Fig. 4-14)
J_{sat}	=	saturation magnetization ($J_s = \mu_o M_s$) (Wb/m^2)
J_x, J_y	=	vector components of magnetization J (Wb/m^2)
k	=	Boltzmann's constant = 1.38 * 10^{-23} (J/°K)
k	=	coupling coefficient between two inductors
k	=	code constraint, i.e., maximum number of zeros between transitions
k	=	number of bits in datawords fed into an EDAC encoder
K	=	crystalline anisotropy constant (dep. on direction)
K	=	$2\pi d_{cm} \sqrt{\mu_{rdc} * f_k/\sigma}$ (7.3)
l	=	length (m)
L	=	length (m)
L	=	inductance of coil (Hy) (4.30)
L′	=	true inductance (Hy) (10.9)
L_C	=	length of element (core) in magnetic circuit (m)
l_g	=	gap length (m)
L_g	=	gap length (m) (capital L more readable in some cases)
L_{fg}	=	gap length, front (m)
m	=	number of bits in datawords fed into a modulation encoder
m	=	Bohr's magneton = 9.3 * 10^{-24} Am^2 (3.18)
m	=	pl = magnetic moment of a permanent magnet dipole (Am^2) (3.4)
m	=	amplitude modulation of recorded signal (27.3)
M	=	magnetization in A/m (M = J/μ_o) (3.24)
M	=	k $\sqrt{L_1 L_2}$ = mutual coupling (Hy) (6.2)
M_r	=	remanent magnetization (A/m)
M_r'	=	remanent magnetization after demagnetization (A/m)
M_r''	=	remanent magnetization after demagnetization, and after recoil (A/m)
M_s	=	saturation magnetization ($M_s = J/\mu_o$) (A/m)
n	=	number of turns in a winding, or coil
n	=	number of bits in codewords
N	=	magnetic North pole (+) (Am)
N	=	demagnetization factor (max = 1 in MKS, 4π in cgs) (4.10)
N_x, N_y, N_z	=	magnitudes of N in the three coordinate directions (4.14)
p	=	magnetic pole strength (Am) (3.1)

p' = magnetic pole strength per unit length in line pole (A) (3.5)
p = packing fraction of magnetic particles in a coating (11.1)
P = permeance = $1/R_m$ (Hy)
P_{eddy} = eddy current power loss in magnetic material (W) (10.13)
P_{hyst} = hysteresis power loss in magnetic material (W) (10.12)

Q = $R_p/\omega L = \omega L/R_s$ = coil quality
q = electron charge = $1.6022 * 10^{-19}$ Coulomb

r = distance (m)
R = electrical resistance in (Ω)
R = code rate = k/n (EDAC) or
= m/n (modulation coding)
R' = true electrical resistance (Ω) (10.19)
R_{bg} = magnetic resistance of back gap in a magnetic head (Hy^{-1})
R_c = magnetic resistance of core portion(s) in a magnetic head (Hy^{-1})
R_{fg} = magnetic resistance of front gap in a magnetic head (Hy^{-1}) (4.26)
R_{fgs} = R_{fg}, including stray fields (Hy^{-1})
R_m = magnetic resistance (= $L/\mu A$) (Hy^{-1})
ΣR_m = sum off all magnetic resistance in a magnetic circuit, going "once around" in the circuit (Hy^{-1}) (4.26)
ΣR_{ms} = ΣR_m, including stray fields (Hy^{-1})
R_P = loss resistance of coil, parallel equivalent (Ω)
R_s = loss resistance of coil, series equivalent (Ω)
R_s = resistance of winding of a coil (Ω)
R_{wdg} = R_{stray} magnetic resistance of stray flux (Hy^{-1}) (4.28)

S = magnetic South pole (−) (Am)
S = squareness of hysteresis loop for magnetic coating (11.1)
S^* = measure of coating switching field distribution (fraction) (see Fig. 12-6)

t = time (sec)
T = tension (dynes) (17.5)
T = torque (Nm) (on dipole (3.3) and (3.15)
T_c = Curie temperature (°C)

v = velocity (m/sec)
v_c = group velocity (m/sec)
V_{noise} = noise voltage from resistance (10.10)

w = track width (m)

x = distance (m)
Δx = length of transition zone (m)

z = distance (m)
Z = electrical impedance (Ω)
Z_m = magnetic impedance (Hy^{-1})

y = distance (m)

Z_{mreal} = magnetic resistance, real component (Hy^{-1}) (10.8)
Z_{mimag} = magnetic resistance, imaginary component (Hy^{-1}) (10.8)

GREEK SYMBOLS

α_B = slope of demagnetization line in BH-loop diagram (4.11) (4.19)
α_J = slope of demagnetization line in JH-loop diagram (4.12) (4.20)
α_M = slope of demagnetization line in MH-loop diagram (4.13) (4.21)
α_x = slope of demagnetization line, longitudinal magnetization
α_y = slope of demagnetization line, perpendicular magnetization
α = angle

β = angle between sensing function direction and the recorded magnetization direction
β = modulation index in FM modulation (27.5)
β = skin depth (7.1) (cm)

ϵ = mass in grams per cm (17.5) or per cm^2 (17.6)

η_r = read efficiency (fraction) (4.26)
η_w = write efficiency (4.28)
η_{coupl} coupling efficiency (flux linkage) (8.6) (4.29)

λ = wave length (m)
λ_{xyz} = magnetostriction constant in crystal direction xyz

o = conductivity (Siemens)
o = density in kg/cm^3 (17.3)

μ = coefficient of friction (17.1)
μ = $\mu_o\mu_r$ = absolute permeability (Hy/m)
μ_o = $4\mu * 10^{-7}$ Hy/m = permeability of vacuum, or air (3.2), (3.21)
μ_{rdc} = relative permeability at DC
μ_{init} = initial permeability, measured at very low magnetic field strength (Fig. 3-16)
μ_{max} = maximum permeability (Hy/m) (Fig. 3-16)
μ_r = relative permeability (3.20)
μ_t = relative permeability of tape or disk coating = $\sqrt{\mu_x\mu_y}$
μ_{rt} = μ_t
μ_x = relative permeability of media coating, longitudinal direction
μ_y = relative permeability of media coating, perpendicular direction
μ_d = dynamic coefficient of friction
μ_s = static coefficient of friction
μ' = real component of the permeability (Hy/m) (7.2)
μ'' = imaginary component of the permeability (Hy/m) (7.2)

ν = field strength relative to gap field strength, in front of write head
ν = Poisson's ratio (17.3)

σ	=	resistivity ($\mu\Omega$cm)
σ	=	resistivity of copper
σ^2	=	variance
σ	=	standard deviation
σ_s	=	specific magnetization (Wb/m^2, or EMU/cc) (11.1)
Θ	=	angle in rads (17.2)
ψ	=	flux, field lines (Wb = Weber = volt-seconds)
ψ_g	=	flux in write gap (Wb)
ψ_m	=	maximum flux level (Wb) (6.1)
ψ_{rsat}	=	maximum flux level after saturation (Wb) (11.1)
χ	=	susceptibility (to become magnetized)
ω	=	$2\pi f$ (s^{-1})
ϕ	=	scalar magnetic potential

GLOSSARY

1f—The lower operating frequency of a rotating memory device. In a single density system it corresponds to all clock and no data bits.

2f—The higher operating frequency of a rotating memory device. In a double density system it corresponds to clock and data bits in each bit cell.

3PM—Three-position modulation.

4/5 Code—A group-coded recording code developed by IBM in which the 16 possible combinations of 4 bits are encoded in sixteen 5-bit groups selected such that each has at least two "1" bits and no more than two "0" bits in sequence.

A/D/A—Analog to digital to analog.

A/D—Analog to digital.

abrasivity—The ability of the tape to wear the head.

absorption—Occurs when molecules successfully penetrate the entire volume of a body.

AC bias—The alternating current, usually of frequency several times higher than the highest signal frequency, that is fed to a record head in addition to the signal current. AC bias serves to linearize the recording process and is universally used in direct analog recording.

access time—The time required to move from one track to another in a disk drive. Seek settle time and the rotational latency should be included.

acicular—Needle-shaped, used to describe the shape of oxide particles.

actuator—A mechanism that positions read/write heads in a disk drive.

ADC, (A/D converter)—Analog-to-digital converter.

additive—Any material in the coating of magnetic tape other than the oxide and the binder resins; for example, plasticizers (to soften an otherwise hard or brittle binder), lubricants (to lower the coefficients of friction of an otherwise high-friction binder), fungicides (to prevent fungus growth), dispersants (to uniformly distribute the oxide particles), and dyes.

address—The location of any physical record on a disk, specified by the cylinder number, head number, and record number. The numerical location is usually given as a binary number from 0000 to FFFF (65,535) decimal.

address mark—A special byte that can be readily identified by a controller because it does not have the proper amount of clock bits.

adsorption—Occurs when molecules cannot penetrate within a body, but successfully stick to its surface.

AES—Audio Engineering Society.

algorithm—A sequence of steps which may be performed by a program or other process, which will produce a given result.

aliasing—The misinterpretation of a high-frequency tone as a lower frequency tone due to sampling at less than twice the highest frequency of interest.

amplitude/frequency response—See Frequency response.

analog—Pertaining to representation by means of continuously variable physical quantities, as contrasted to digital, or discrete, representation.

analog recording—In the broadest sense, analog recording is a method of recording in which some characteristic of the record current, such as amplitude or frequency, is continuously varied in a manner analogous to the variations of the original signal.

analog-to-digital (A/D)—The process of converting from analog format to digital format.

analog-to-digital-to-analog (A/D/A)—The process of converting from analog format to digital format and back to analog format; usually required when digital tapes are used to provide data for analog systems.

anchorage—The degree to which the magnetic tape oxide coating adhere to the base film.

ancillary frame—In image processing format, a tape format element carrying all geometric correction coefficients.

AND gate—A multiple-input, single-output gate that has a "1" output only when all inputs are "1's."

anhysteresis—The process whereby a material is magnetized by subjecting it to a unidirectional magnetic field that is superimposed on an alternating field of gradually decreasing amplitude. One form of this process is analogous to the recording process using AC bias.

anisotropy (magnetic)—Directional dependence of magnetic properties leading to the existence of easy, or preferred, directions of magnetization. Shape anisotropy is the dominant form in acicular particles.

annotation frame—In image processing format, a tape format element carrying mapping and other information.

ANSI—American National Standards Institute.

ASCII—American Standard Code (for) Information Interchange. An eight bit code that uses seven bits to represent data, and bit number eight for parity check schemes.

asperities—Small projecting imperfections on the surface of the tape coating that limit and cause variations in head-to-tape contact.

ASA—Acoustical Society of America.

average access time—Traditionally the time to access one third of the tracks, or cylinders. (Minimum step time) × (number of tracks/3) + (one seek settle time) + (one Latency) would be the average access time.

audio no. 1 track—Synonym for program audio track on video tape.

audio no. 2 track—Synonym for cue track on video tape.

azimuth alignment—Alignment of the recording and reproducing gaps so that their center lines lie parallel with one another. Misalignment of the gaps causes a loss in output at short wavelengths.

backing—See base film.

back bar—The ferrite bar that is bonded to the back of a magnetic head core to complete its magnetic path by shorting the back gap flux path.

banding—A visible difference in the reproduced characteristics in that portion of a picture associated with one head channel, when compared with adjacent areas associated with other head channels. In quadruplex video recorders, these differences occur in horizontal bands of 16 or 17 scanning lines when reproducing a 525/60 signal. Hue: Banding in which the visible difference is in the hue. Noise: Banding in which the visible difference is in noise level. Saturation: Banding in which the visible difference is in saturation.

bandwidth sensitivity—The tendency of a code to increase bit error rate with increasing bandwidth.

bandwidth—The range of frequency within which the performance of a recorder with respect to some characteristic (usually frequency response) falls within specified limits, or within which some performance characteristic (such as noise) is measured.

base film—The plastic substrate material used in magnetic tape that supports the coating.

baseband—The frequency band occupied by data before modulation a carrier.

baseline shift—A shift of the DC average of a data sequence relative to the peak value caused by the lack of DC or low-frequency AC response of the recorder.

baud—The baud, named for Emile Baudot, is the signaling rate in code elements per second; usually 1 baud equals 1 bit per second.

BCC—Block check character.

BEP—Bit error probability.

BER—Bit error rate.

BH-loop—Curve showing the relation between the flux density B and the magnetizing force H in materials.

bias (sometimes called AC bias)—A high-frequency signal, usually three to five times the highest data frequency, that is linearly mixed with the data signal and fed to the record heads to compensate for the hysteresis effect of the tape. The use of bias improves the linearity and sensitivity of the system and provides maximum distortion free output levels. The amount of bias current used represents the best compromise of low distortion, extended high-frequency response, and high output.

binary—A number system based upon the powers of 2. Only the digits 0 and 1 are used.

binder—A compound consisting of organic resins used to bond the magnetic particles to the base material. The actual composition of the binder is considered proprietary information by each magnetic tape manufacturer. The binder is required to be flexible and still maintain the ability to resist flaking or shedding during extended wear passes.

biphase encoding—Double frequency encoding in which an extra transition occurs either at the beginning or in the middle of every bit cell.

biphase, level—Also known as split-phase encoding. Level change occurs at the center of every bit period. A ''1'' is represented by a ''1'' level with a transition to the ''0'' level; ''0'' is represented by a ''0'' level with a transition to the ''1'' level.

biphase, mark—Level change occurs at the beginning of every bit period. A ''1'' is represented by a midbit level change; ''0'' is represented by no midbit level change.

biphase, space—Level change occurs at the beginning of every bit period. A ''1'' is represented by no midbit level change; ''0'' is represented by a midbit level change.

bit—As applied in magnetic recording, it represents one recorded information cell (''bit'' is a contraction of ''binary'' and ''digit'' to define a unit of information.)

bit cell time—The average bit duration during the process of recording at continuous maximum flux reversal rates.

bit crowding—A condition in magnetic recording wherein a transition placed close to another appears to migrate into the larger space available between it and the transition on its other side.

bit density—Bits per unit length, area, or volume of the recording medium; for example, the number of bits per square centimeter of magnetic tape.

bit error rate (BER)—The fraction of bits that are in error. BER = e/N where e is the number of errors and N is the total number of bits (correct bits plus erroneous bits).

bit mapping—A technique for evaluating error rate performance by dividing the recovery window into intervals and accumulating a histogram of data bit occurrences. Examination of the histogram provides a characterization of system performance as well as a margin estimate.

bit parallel—Information transfer in which all bits constituting one word are transmitted simultaneously in different parallel channels.

bit rate—The speed at which encoded information is transmitted. In digital television, where an 8-bit PCM encoding of each sample is commonly required for acceptable quality when a sampling frequency of 10.7 MHz is used, the bit rate is approximately 85 to 86 million bits per second (usually expressed as Mbit/s).

bit serial—Information transfer in which all bits are transmitted one after the other in a single channel.

bit shift—Also called peak shift. A shift in the detected signal peak caused by the influence of a neighboring flux transition.

bit slip—Reconstructed bit stream out of synchronization with the actual bit stream because the reconstructed clock has gained or lost one or more cycles with respect to the correct clock. Continuous errors result until recovery is made, because the "interpreter," such as a D/A converter, needs to know exactly which bit is the first bit of a word.

bit stream—The flow of encoded information.

bit stuffing—Insertion of overhead bits on the data tracks.

BL—Bit length.

block—A set of adjacent logical records recorded as a unit.

block check character (BCC)—Contained in the trailer field of a transmission block, such a character is generated by a checking algorithm applied to the block data.

block data—Groups of words, frames, or characters handled as single units separated by interblock characters.

blocking—The tendency of the adjacent layers of tape in a roll to stick together.

BNC—Baby "N" connector.

bonding—Joining of the pole pieces of a magnetic core with a low permeability material that fills the gap; glass or epoxy can be used. Also used in bonding the core to the magnetic recording head housing.

bookkeeping (housekeeping)—Computer coding that reserves, restores, and clears memory areas or sets up constants and variable to be used by the program.

Boolean logic—The logic resulting from the use of the AND, OR, and NOT functions (gates).

BPI—Bits per inch.

BPS—Bits per second.

break elongation—The relative elongation of a specimen of magnetic tape or base film at the instant of breaking when it has been stretched at a given rate.

brown stain—A discoloration of the head top surface, usually a chemical reaction between the head surface materials and either the tape binder, tape lubricant, or head bonding materials. The stain is usually very thin. Its origin is not well understood but is known to occur in the presence of low humidity. Brown stain is very difficult to remove with ordinary head cleaner solutions; a green abrasive tape is normally required.

buckling—Deformation of the circular form of a tape pack that may be caused by a combination of improper winding tension, adverse storage condition, and/or poor reel hub configuration.
buffered seek—Rather than tie up the controller to issue timed step commands a drive capable of buffered seek accepts a burst of very short duration step pulses. The controller can then address other drives while the drive issues it's own timed step commands.
buffer register—A temporary storage register capable of receiving, storing and transmitting data at different I/O rates.
buildup—A ''snowballing'' effect started by debris and the tape magnetic particles embedded in the contamination; the thickness of this buildup can cause increase in head-to-tape separation as well as increase in coefficient of friction. Solvent cleaning of the head top surface will usually remove the buildup, in contrast to brown stain formation.
bulk eraser (degausser)—Equipment for erasing previously recorded signals on tape a full reel at a time.
burried servo—Servo data recorded permanently on a disk; not overwritten by data.
button head—A curved or flat magnetic recording head having no freedom of rotation; used in floppy disk drives.
byte—A sequence of adjacent binary digits which is operated upon as a unit and usually shorter than a word. A byte usually is made up of 8 bits.

capability factor—As obtained from an eye pattern, the ratio of trace thickness (measured normal to the trace) to the total width (opening) of the eye at the center of the waveform.
capacity—The amount of data (typically in bytes) that can be stored in a given element of a rotating memory device. Typical elements used are track, cylinder, surface, disk , and spindle.
capstan—The driven spindle in a tape machine, sometimes the motor shaft itself, which rotates in contact with the tape and meters the tape across the tape transport.
cartridge—A rigid plastic shell that encloses and protects a rigid disk, microfloppy diskette, or tape.
catalog—A directory of files on a disk.
central processing unit—See CPU.
certification—The process during manufacture of detecting and removing from the diskette defects which may cause data errors.
certified tape—Tape that is electrically tested on a specified number of tracks with a specific drive and electronics using a defined code at a specified tape speed and data rate,and certified by the supplier to have less than a certain total number of permanent errors of a specified duration.
certifier—Equipment that evaluates the ability of magnetic tape to record and reproduce. The equipment normally counts and charts each error on the tape, including the level and duration of dropouts. In the ''certify'' mode, the equipment stops the tape at an error to allow for visual inspection of the tape to see if the cause of the error can be corrected or if it is a permanent flaw.
CGS—centimeter-gram-second.
channel—(A) A single path for transmitting electric signals, usually in distinction from other parallel paths; (B) a band of frequencies.
character—A single letter or number, usually made up of 8 bits = one byte.
check bit—A bit generated periodically to assist in error detection and correction.
chip—A piece of semiconductor material that contains a circuit element. The circuit element may be as simple as a discrete device or as complicated as a microprocessor.
chips—Cavities or voids at an edge of head track.

chrominance—Indicates both hue and saturation of a color. In color television, the 3.58 MHz color signal more specifically is the chrominance signal. Chrominance, also called *chroma*, includes all the color information, without the brightness.

cinching—Tape folds resulting from longitudinal slippage between the layers of tape in a tape pack caused by uneven tension when the roll is accelerated or decelerated.

circuit—A combination of electronic components that together perform a specific function.

clean room—An enclosed room where the air is continuously circulated and filtered to remove contaminants. Clean rooms are rated by the size, and number, or contaminant particles per cubic foot of air. Depending upon the level of cleanliness occupants may be required to wear special garments, shoe covers, masks, etc.

cleaner—See Winder/Cleaner.

clock—A timing reference required to decode digital signals; i.e., a timing reference for digital data to keep track of bits during periods of no transitions and to indicate instants when to sample data.

clock bit—A transition that is recorded to maintain synchronism between the data stream and the data separator.

clock efficiency—The ratio of data bits to total bits (data bits plus overhead bits) for a given bit error rate.

cm/s—centimeter per second.

coating—The magnetic layer of a magnetic tape consisting of oxide particles held in a binder that is applied to the base film.

coating resistance—The electrical resistance of the coating measured between two parallel electrodes spaced a known distance apart along the length of tape. On the specification sheets this is called resistivity.

coating thickness—The thickness of the magnetic coating applied to the base film.

coating-to-backing adhesion—See anchorage.

cocking—A design modification in which the magnetic head is rotated by approximately ½ degree relative to the radius of the diskette, parallel to the plane of the diskette. The purpose of the rotation is to ensure even erasure of inner tracks.

CODEC—A contraction of "coder and decoder," used to imply the physical combination of the coding and decoding circuits.

coefficient of friction—The tangential force required to maintain (dynamic coefficient) or initiate (static coefficient) motion between two surfaces divided by the normal force pressing the two surfaces together.

coefficient of hygroscopic expansion—The relative increase in the linear dimension of a tape or base material per percent increase in relative humidity measured in a given humidity range.

coefficient of thermal expansion—The relative increase in the linear dimension of a tape or base material per degree rise in temperature (usually Fahrenheit) measured in a given temperature range.

coercivity or coercive force—The field strength required to bring the flux density to zero in a magnetic material.

comb filter—A wave filter whose frequency spectrum consists of a number of equi-spaced elements. It has repetitive pass and stop bands (resembling the teeth of a comb) and is usually implemented with a transversal filter.

companding—A contraction of "compressing and expanding." Compression is used at one point in the communication path to reduce the amplitude range of the signals, followed by an expander to produce a complementary increase in the amplitude range.

compatibility—The ability of two types of recording equipment to perform similar functions and operate interchangeably.

composite head—A miniature magnetic head inserted into a non-magnetic slider body.

conductive coatings—Coatings that are specially treated to reduce the coating resistance and thus prevent the accumulation of static electrical charge.

contamination—A thin, tacky (viscous) deposit on the head top surface. This deposit causes a large increase in the effective head-to-tape coefficient of friction and may not be removable by solvent cleaning.

contour pulse (spurious response—little Joe, big Joe**)**—A secondary pulse caused by the contour of the read head. A secondary or virtual head gap is formed by the leading and trailing edges of the core material of the read head. Thus a pulse (called contour pulse) is generated when a recorded tape is passed over this secondary gap.

contouring head—This is a method of establishing the final head contour and obtaining the proper surface finish. An abrasive tape is passed over the head. The grit of the abrasive, the lubricant, the tape stiffness, the tape tension, the tape speed, and the amount of tape passed over the head are parameters that will control the final head contour and top surface finish.

control track—The area on the tape containing a recording used by a servo mechanism primarily to control the longitudinal motion of the tape during playback.

control track signal—The signal recorded on the control track.

controller—A device that interfaces one, or a number of, rotating memory devices to a computer system. The controller provides features such as data separation, CRC generation and checking, write clock, drive select, data transfer buffers, etc.

cookie—The coated circular diskette that has been punched to its final shape, but has not been sealed in it's jacket.

core—That part of a read/write head which is composed of soft magnetic material and converts write current to magnetic flux, or couples readback flux into the coil windings. The core is usually made of nickel-zinc or manganese-zinc ferrite. A non-magnetic gap in the core allows flux to fringe into the media during the write cycle, and couple into the core during the read cycle.

core material, hard—(1) ''Hard'' metal laminations bonded together to form the core; typical thickness is 0.0005 to 0.004 in.; ''hard'' metal wears much more slowly than ''soft'' lamination. (2) ''Hard'' solid metal are alfenol, sendust and metglass.

core material, soft—''Soft'' metal laminations bonded together to form the head core; typical lamination thickness is 0.0005 to 0.004 in; usually a nickel/iron alloy such as Hy Mu 800. These materials have a relatively poor wear rate.

coring—A system for reducing the noise content of circuits by removing low-amplitude noise riding on the baseline of the signal.

CPU—Central processing unit. The part of a computer system that examines and executes the instructions contained in a program.

CRC—Cyclic redundancy check.

creep—Time-dependent strain at constant stress (tape deformation.)

cross play—The ability to interchange recordings between recorders while maintaining a given level of performance.

cross talk—Magnetic coupling from one track to another track in a magnetic tape or disk system.

cue track—The area reserved on the tape for audio information relating to production requirements, electronic editing information, of a second program signal.

cupping—Curvature of a magnetic tape pack in the lateral direction. Cupping may occur because of differences between the coefficients of thermal or hygroscopic expansion of coating and base film.

cyclic redundancy check (CRC)—An error detection method in which check bits are generated by taking the remainder after dividing the data bits by a cyclic code polynomial.

cylinder—When a rotating memory device has more than one storage surface each positioner location is called a cylinder rather than a track.

DAC (D/A converter)—Digital-to-analog converter.

data—Information.

data bit—A transition recorded to represent one bit of information.

data compression—A technique for saving storage space or transmission bandwidth by eliminating gaps, empty fields, redundancies or unnecessary data to shorten the length of records or blocks. Also: Short code words are substituted for data strings whenever possible.

data density (BPI)—The number of data characters stored per unit length of tape or disk track.

data rate—The rate at which data are transferred from one part of the system to another.

data separator—A circuit that examines the bit stream, locks on to the clock bits and separates out the data bits.

dB—See decibel.

DC restorer—Circuitry designed to reduce baseline shift.

decibel—A dimensionless unit for expressing the ratio of two powers, or more usually voltages or currents, on a logarithmic scale. If A and B represent two voltages or currents, the ratio A/B corresponds to 20 log A/B decibels. One decibel represents a ratio of approximately 1.1 to 1 between A and B.

decimal—A number system based upon powers of 10. Digits range from 0 to 9.

decoding—To recover the original data stream from an encoded form of the data stream.

decoder/deskew assembly—A component of a typical N-track high-density digital recorder that accepts up to N lines of bit parallel data in whatever code was recorded, converts in to NRZ-L data, time aligns all tracks, and outputs serial or parallel NRZ-L data and clock.

dedicated servo—A servo system in diskpacks (multiple platters) where one media surface is "dedicated" to servo patterns and is not used to store data.

defect—An imperfection in a disk or tape coating surface leading to a variation in output or a dropout. The most common defects take the form of surface projections consisting of oxide agglomerates, embedded foreign matter, and redeposited wear products.

degauss—To return the magnetization in a media coating or in a head to a zero state by applying a decaying and alternating magnetic field.

delay modulation code (DM)—Code characterized by having a minimum of one transition for every other bit cell. Transitions occur at the middle of all "1" bits and between all "0" bits; (also known as delay modulation, mark (KM—M)); Delay Modulation, Space (DM-S): Transitions occur at the middle of all "0" bits and between all "1" bits.

delta modulation—The simplest form of DPCM (q.v.) in which one of only two codes is transmitted for each sample, instructing the receiver to either add or subtract a fixed unit change to or from an accumulating total signal.

demultiplexing (DMX)—Separation of independent channels from a single high-speed data stream.

density identification area—A recording in the beginning of tape marker to identify the method and density of recording.

density ratio (DR)—Data density divided by translation density.

difference signal—In communication theory models, a convenient function of two known signals on which the correlation detection process is based.

differential pulse code modulation (DPCM)—A PCM variant in which the coded value transmitted for each sample represents the quantized difference between the present sample value and some combination (e.g., the integrated sum) of all previously transmitted values. For signals having strong correlation between successive samples, fewer levels may be used to quantize differences than would be required for quantizing sample values with comparable precision.

digital data—Information in a series of digits having only two possible values, "0" or "1."
digital recording—A method of recording in which the analog information is first digitized and then recorded on tape. Usually a binary code is used and two discrete values are recorded.
digital sum variations (DSVs)—Values of the running integral of a bit sequence whose levels are assumed to be + or − 1.
dipulse—A read/write head gap test that writes two transitions close together and measures the resultant peak shift.
direct access (DAS)—Peripheral storage allowing rapid access of any piece of data, regardless of its placement on the medium. Magnetic tape is generally not considered direct access, since the entire tape must be read to locate the last byte. A diskette is direct access.
direct recording—A type of analog recording that records and reproduces data in the electrical form of its source.
directory—A catalog of files stored on a disk.
discrete component—An electronic component that performs one specific function such as a diode, transistor, light emitting diode, etc.
diskette—A flexible, circular substrate made of polyester, coated on one or both sides with magnetic oxide, and designed to receive and store electrically coded information for later retrieval.
diskette assembly—The combination of the diskette, jacket, and liner. Also called a diskette cartridge assembly.
disparity—In a code word of character, the excess number of "+1" bits over "−1" bits.
dispersion—Distribution of the oxide particles within the binder of a tape.
distortion (harmonic)—A nonlinear change in signal waveform where undesired harmonics of a sinusoidal input are generated.
dither signal—A simulated noise waveform combined with the signal before quantization (q.v.) to compensate for the contouring effects caused by quantization. It effectively reduces the number of bits required to produce an acceptable picture.
DM—Delay modulation.
DM-M—Delay modulation, mark (Miller code).
DM-S—Delay modulation, space.
DMX—Demultiplexing.
domains—A volume of a magnetic material where the combined electron spin results in a net magnetic effect. The smallest increment of magnetic storage.
DOS—Disk Operating System.
double density code—A digital recording method whereby the data bits are used for system clocking instead of alternating a clock bit with a data bit as is done in single density recording. The consequent absence of clock bits allows data bits to take their place, thus doubling the bit density of the diskette.
double-frequency clock—A clock at twice the bit rate.
downstream—Pertaining to locations on the tape longitudinally displaced from a given reference point in the direction of tape motion.
DPCM—See differential pulse-code modulation.
DR—Density Ratio.
DR—Data Rate.
drag—When the tape contacts some element in the tape path (such as the head, tape guides, tape bearings, or column walls), there is a tension differential across the contact area. This tension differential is caused by friction and is called "drag."
drive—The mechanics that rotates the media, positions the read/write head, drives the write and erase elements, amplifies the read signal, etc. Drives do not usually include controllers.
drop-in—Also called extra pulse or extra bit. A readback pulse which exceeds a prescribed

threshold on a D.C. erased track. The threshhold is usually a percentage of the "track average amplitude" (TAA) for the particular track.

dropout—Variation (reduction) in signal level of reproduced, tape-recorded data, resulting in an error in data reproduction. More specifically, a loss in output from a magnetic tape of more than a certain predetermined amount (depth), expressed in terms of the percent of reduction of decibel loss for a specified time period (length); e.g., 12 dB (75 percent) for 1 microsecond.

dropout count—The number of dropouts detected in a given length of magnetic tape.

drum—The preferred synonym for the quadruplex recording system is head wheel.

DSV—Digital sum variation.

dub—To make a copy of a recording by rerecording.

durability—Usually expressed as a number of passes that can be made before a significant degradation of output occurs divided by the corresponding number that can be made using a reference tape.

dynamic coefficient of friction—See coefficient of friction.

dynamic skew—The changes in skew caused by tape motion.

E value—The difference in inches between the radii of the outside layer of tape in a roll and the outside edge of the reel flange. Inadequate E value may prohibit the use of protective reel bands.

EBU—European Broadcast Union.

E-NRZ—Enhanced nonreturn to zero.

ECC—Error correction code.

ECD—Error correction decoder.

ECL—Emitter-coupled logic.

EDAC—Error detection and correction of recorded data using simultaneously recorded correction data either added to the data stream or recorded separately on an auxiliary track.

edge gutter (groove; cut)—Removing the head surface material where the outer edges of the tape would normally touch the head. Prevents head guidance of the tape.

edge margin—The distance between the outside edge of the highest number track and the tape edge.

edge roll or lift—Caused by head wear. The rolloff or liftup of the outer edges of the head.

edit pulse—The preferred term is Frame Pulse.

EDP—Electronic data processing.

embedded servo—A servo system where the servo information is inserted into the format of all data tracks.

encoding—To express a single character or a data stream in terms of digital bit code.

ENG—Electronic News Gathering, i.e. portable TV recording equipment.

enhanced nonreturn to zero (E-NRZ)—A modification of NRZ encoding in which the bit stream is separated into 7-bit words, bits 2, 3, 6, and 7 inverted, and a parity bit is added to minimize DC components and maintain phase lock in the playback timing oscillator.

envelope—(A) A waveform composed of the instantaneous peak values of an original wave of alternating quantity, and which indicates the variation in amplitude undergone by that quantity. (B) The outer protective covering which fits around a diskette assembly to prevent damage and contamination during shipping, storage, or mailing.

envelope modulation—Amplitude modulation in the envelope of the peak-to-peak signal amplitude at the maximum flux reversal of the desired recording method due to changing head-to-media spacing, changing media velocity, or changing magnetic characteristics of the media.

environmental conditions—External factors such as room temperature distortion of a circuit by compensating for differences in attenuation or time delay throughout the transmission band.

erase cores—The magnetic cores on the sides of a recording head that erase old data from the track edges on a diskette.

erasure—A process by which a signal recorded on a tape or disk is removed and the tape or disk made ready for rerecording.

error—In digital recording, a number expressing the count of binary digits (bits) not accurately reproduced within a larger number of bits.

error burst—Multiple errors caused by a common event such as dirt and dust between head and media. Occasionally caused by defects in the magnetic disk or tape surface.

error detection and correction—Coding schemes incorporated into the information before it is transmitted (or stored) in such a way that errors which may arise in transmission can be detected and corrected before restoration or retrieval.

error log—A capability for storing the location and number of errors found on a magnetic medium.

error multiplication—A property of code process converters whereby a single incorrect digit in the input signal can cause more than one digital error in the output signal.

error rate—The ratio of the number of bits incorrectly received to the total number of bits of information transmitted.

error recovery strategy—Selection of error detection and correction processes to minimize the probability, and catastrophic probability.

euclidean division algorithm—A formal method allowing the selection of check bits, based on the remainder formed when a data polynomial is divided by a generator polynomial.

evaluator—Equipment, usually provided as an adjunct to a winder/cleaner, that evaluates physical and magnetic quality of tape. In contrast to a certifier, an evaluator does not stop when it detects an error.

even parity—A block of data with an even number of "1" bits, either vertically or longitudinally along the tape.

extra bit—A bit with greater than 15 percent amplitude after erasure.

eye patterns—The reproduce system output is applied to the vertical input of an oscilloscope with the horizontal input synchronized to the bit rate clock. The oscilloscope display gives an insight into the quality of the reproduced data.

FCI—Flux changes per inch.

FCL—Flux change length.

ferrite—A magnetic ceramic material used in the construction of magnetic recorder heads.

FET—Field effect transistor.

FFT—Fast Fourier Transform.

field—(A) A group of contiguous (adjacent) bytes forming a single piece of data. In disk formatting, a group of bytes surrounded by gaps. (B) In television: one of the two equal parts into which a frame is divided in interlaced scanning.

file—Groups of related records stored together.

fire code—An error correction code for which the generator polynomial has a particular form.

flag—A term used to describe a track on a diskette that is unusable: a "flagged" track is one that the computer will not access.

flippy—A diskette which has been certified on both sides to be 100 percent error-free and thus able to be "flipped" over for use on either side.

floppy—A term used to describe a diskette's characteristic flexibility when free of its protective jacket.

flutter—Undesired changes in the frequency of signals during the reproduction of a magnetic tape or disk produced by speed variations of the medium during recording or reproducing.

flux—Lines of magnetic force.

flux transition density—Number of flux transitions per length unit along a written track.

FM code—A pulse code for which a flux reversal at the beginning of a cell time represents a clock bit, a flux reversal at the center of the cell time represents a "1" bit, and an absence of a flux reversal represents a "0" bit.

format—The structuring of the magnetic recording area of a diskette, either physically or electrically, into uniform addressed segments where data may be recorded, stored, and retrieved.

formatted capacity—The maximum amount of data bytes that can be stored on the media. Formatted capacity = unformatted capacity − overhead.

forward/backward ratio—Forward read signal divided by backward read signal.

fourier transform—A transformation in which the orthogonal generating functions are sets of sinusoids.

FRI—flux reversals per inch.

FRPI—flux reversals per inch.

frame—In high-density digital recording, an entity containing data bits and overhead bits. It is preceded and followed by synchronization words.

frame synchronization—The process whereby a given receiving channel is aligned in time with the corresponding transmitting channel.

frame pulse—A pulse superimposed on the control track signal to identify the longitudinal position of a video track containing a vertical synchronizing pulse. Used as an aid in editing and in the synchronization of some recorders.

frequency response—The variation of signal amplitude with signal frequency. Usually the frequency response of a tape is given in decibels relative to that of a referenced frequency output level.

frequency-division multiplexing—The process of dividing the available frequency passband into a number of narrower frequency bands, each available for a separate signal.

fringing flux—The magnetic flux that extends beyond the lateral boundaries of the magnetic core that is causing the flux. It is particularly significant in relation to erase cores.

gamma ferric oxide—The magnetic constituent of most magnetic tapes and disks in the form of a dispersion of fine acicular particles within the coating.

gap—The separation between the poles of a read/write or erase core where reading, writing and erasure of data occur.

gap azimuth—The angle between any gap and the gap azimuth line.

gap azimuth line—A line drawn through the gap centers of the outside tracks or a least squares fit line through the gap centers in tracks 1 through N (for N tracks). This can be done optically or electronically.

gap depth—The dimension of the gap of a magnetic head measured in the direction perpendicular to its surface.

gap erosion—The read or write gap will effectively increase in length and will retreat below the head surface, usually due to deterioration of core material at the edges of the gap.

gap front—Region of nonmagnetic material between the two sections of the magnetic core that contact the tape.

gap height (throat height; pole tip height)—The height of the gap.

gap height cut (pole tip cut)—A cut into the magnetic core that establishes the gap height.

gap length—The dimension of the gap of a magnetic head measured from one pole face to the other. In longitudinal recording the gap length can be defined as the dimension of the gap in the direction of tape travel.

gap loss—The loss in output attributable to the finite gap length of the reproduce head. The loss increases as the wavelength decreases.

gap material (shim material)—The material in the gap front.

gap scatter—The variation of the location of the gap position of any track from the azimuth line. This can be done optically or electronically.

gap width—The dimension of the gap of a magnetic head measured in the direction perpendicular to the direction of the tape path.

gap-to-gap distance—Distance from center of read gap to center of write gap on the mean track center line for any track.

gate—A device having one output channel and two or more input channels such that the output channel state is completely determined by contemporaneous input channel states.

gauss—The cgs unit of magnetic flux density equal to 1 Mx/cm × cm. Also equal to 0.0001 Wb/m^2.

GCR—See Group-coded recording.

gimballed head—A magnetic head having two axes of rotation, allowing it flexibility in its contact with the diskette.

glue lines or glue areas—Glue (or bonding materials) exposed on the top surface of a magnetic head.

green tape—An abrasive tape used to clean and lap heads that are unevenly worn, stained, scratched, etc. It should be used with caution and should not be used on ferrite heads. These comments also apply to gray tape.

group-coded recording (GCR) code—A code that groups characters and encodes them, using a lookup table, allowing a maximum of two "0" values in sequence.

guard band—The unused space between data tracks on a disk or diskette that safeguards against crosstalk and accidental erasures of adjacent tracks while writing.

gumball head—A unique two sided floppy diskette head system that has a small radius of curvature, hence the nickname.

hadamard transform—A transformation algorithm which may be used to encode picture signals. It lends itself to implementation in such a way as to reduce the bit rate to a level lower than that required by PCM encoding.

hamming code—A data transmission code that is correctable.

hard-decision circuit—A circuit that uses the sign of each received pulse-amplitude to decide whether a "0" or a "1" was sent.

hard disk—Also rigid disk. Rotating memory device media that has a hard, or rigid, substrate.

hard error—An error that is repeatable in the same track and sector location.

hard sector format—A configuration in which the recording surface of the diskette is physically divided into 32 sectors by an evenly-spaced series of punched holes on either the inner or outer diameter.

hardware—Equipment, as contrasted to computer programs (software), used for data processing.

harmonic distortion—Signal nonlinearity characterized by the appearance in the output of harmonics of the fundamental when the input signal is sinusoidal.

HDDR—See high density digital recording.

HDTR—High density tape recorder.

head—A transducer to convert electrical signals into magnetic signals for recording on magnetic tape and vice versa.

head aperture—A slotted area of the jacket which allows the recording head on the drive to have physical access to the diskette media. All recording and data retrieve occurs in this location.

head block—A specialized head assembly containing recording head stacks, position sensors, temperature sensors, a ballast resistor, and interface electronics.

head body—The structure that supports the magnetic cores and may form a support for the tape.

head channel—The signal path unique to each magnetic head. In a quadruplex system, the outputs of four channels combined to provide a continuous RF signal.

head clogging—The accumulation of debris on the head, the usual result of which is a loss of signal during playback, degradation, or failure to record during the record mode.

head coating—conductive—A coating on the top surface of the head to reduce "feed through."

head coating—hard (low-wear coating)—A coating on the top surface of the head to increase head life and reliability.

head coating—exposed core window—The part of the magnetic core that is not coated. It is exposed to the tape.

head contour—The complex shape of the contacting surface of a head either a result of manufacturer, head lapping, or wear. The contour of a head is always changing throughout the life of the head and in many cases is responsible for retiring the head.

head load—Some floppy diskette drives incorporate an electromechanical relay to unload the read/write head when the drive is not in use. This is called a head load system.

head settle time—The wait time that should be allowed after a head load operation to allow all read/write head motion to cease.

head stick; sticktion; stick-slip—Common words for a large increase in head-to-tape friction caused by (1) a sticky byproduct exuded by the tape under certain conditions of tape age, temperature/humidity, and head-to-tape pressure; (2) very smooth tapes coupled with large area heads.

head tilt—The angle between the plane tangent to the front surface of the head at the center line of the head segment gaps and a line perpendicular to the head reference plane.

head wheel—A wheel with magnetic heads mounted on its rim. Used in video and R-DAT recorders.

head-to-tape contact—The degree to which the surface of the magnetic coating of a tape approaches the surface of the record or reproduce heads during normal operation of a recorder.

head-to-tape-speed—The relative speed between tape and head during normal recording or replay.

helical scan—A method of video magnetic tape recording which uses one or more rotating video magnetic heads which engage a tape wrapped at least partially around a cylindrical column. The tape path describes a portion of a helix. The recorded video tracks lie at an angle substantially less than 45 degrees with respect to the length of the tape. Usually, one complete television field is recorded as one head passes across the width of the tape.

hexadecimal/HEX—A numeric system based on powers of 16. Valid hex digits range from 0 to 9 and A to F, where A is 10, B is 11,..., F is 15.

high band—1. Pertaining to those frequencies specified in SMPTE Recommended Practice RP, Practice HB. 2. Pertaining to those recordings made in accordance with Practice HB or RP6, or equipment capable of making those recordings. 3. The carrier frequencies which appear on a tape made in accordance with such practices.

high density digital recording—Recording of digital data on a magnetic medium resulting in a flux transition density in excess of 591 transitions per millimeter (15,000 transitions per inch) per track.

high energy tape—Magnetic tape having coercivity higher than 600 Oersteds.

homogeneous—A physical characteristic that does not vary from point to point.

horizontal recording—The common name for longitudinal recording which is the standard technique in current use with digital rotating memory devices.

hub access hole—The hole in the center of the diskette into which the drive hub of the transport fits.

hue—A color itself is its hue or tint. Green leaves have a green hue; a red apple has a red hue. Different hues result from different wavelengths of the light producing the visual sensation in the eye.

hysteresis—A condition where the current state of a mechanism or material is dependent upon its history. In positioners that do not servo the actual location can be related to the previous seek length and direction because of system friction. In magnetic materials the current state of a domain depends upon the strength and direction of the last field that influenced it.

Hz—Hertz (one Hz equals one cycles per second).

I/O—Input/output.

IBM compatible—A configuration wherein track and sector addresses are initialized on the diskette by the manufacturer and which allow interchangeability between IBM systems and systems compatible with them. Also referred to as industry compatible.

ID Field—Identification field. That part of a soft sector format that precedes the data field and identifies that track, side, sector, and record size.

IEEE—Institute of Electrical and Electronics Engineers (USA).

IERE—Institute of Electrical and Radio Engineers (England).

IFR—Inches per flux reversal.

image major frame—In image processing format, a tape format element that carries most of the data.

impact strength—A measure of the work done in breaking a test sample of tape or basefilm by subjecting it to a sudden stress.

index cylinder—Cylinder 00. This cylinder is used to store information about the contents of the floppy.

index hole—A physical hole in the diskette which, when detected, notifies the drive that the beginning of track is under the read-write gap.

initialization—The process of writing the addresses, index cylinder information, and other system information of the diskette. (Initialization is also used to assign alternate tracks.)

inter-frame coding—Coding techniques which involve separating the signal into segments which have changed significantly from the previous frame and segments which have not changed.

interblock gap—A section of tape separating blocks of information.

interface—A shared boundary between two systems, or parts of a system, through which data are transmitted, consisting in practice of mechanical or electronic elements, or both.

interlayer transfer—Loose material, such as oxide, that is generated by tape wear or a "head stick" condition which is transferred from the oxide to the back of the tape or from the back side to the oxide when tape is wound on a reel.

interleaver—A special circuit used to separate (displace into different data blocks) data words to minimize the effects of burst errors.

intermodulation distortion—Signal nonlinearity characterized by the appearance of frequencies in the output equal to the sums and differences of integral multiples of the component frequencies present in the input signal.

interpolation—The technique of filling in missing information in a sampled system.

interpolation, line—In television standards conversion, the technique for adjusting the number of lines in a 625-line system (and vice versa) without impairing the picture quality.

interpolation, movement—A technique used in standards conversion to compensate for the degrading effects of different field frequencies on pictures which contain movement. Different approximate proportions of successive input fields are used in each output field.

intersymbol interference—When a recording system has limited record resolution, a flux transition being recorded will extend beyond its cell boundaries adding or subtracting from the flux in the adjacent bit cells or "symbols." This type of interference will result in phase shift of the cell playback crossover point with respect to the data clock.

intertrack shields—Laminations of copper and nickel-iron or sections of ferrite placed between tracks in a head assembly to reduce crosstalk between tracks.
intrinsic coercive force—The magnetic field strength needed to switch the magnetization in a small particle.
intrinsic flux—In a uniformly magnetized sample of magnetic material, the product of the intrinsic flux density and the cross-sectional area.
IPS—Inches per second.
IPF—Image processing format; Image Processing Facility.
IRIG—Inter-Range Instrumentation Group; now Range Commanders Council (RCC).
iron oxide—See gamma ferric oxide.
isotropic—Pertaining to a material whose isotropic properties are independent of direction.

jacket—A protective vinyl covering, lined with a soft, non-woven synthetic liner which encloses the diskette but also allows access to it by the head. The diskette can rotate freely within the jacket and the jacket's design permits storage, use, and handling without contacting the diskette inside.
JH-loop—Curve showing the relation between the magnetization J and the magnetizing force H in a material.
jitters—Sudden, small, irregular departures from phase, amplitude, or pulse duration of a signal caused by the recording/reproducing mechanisms.

kBPI—Kilobits per inch.
kBPS—Kilobits per second.
kFR/CM—Kilo flux reversals per centimeter.
kilobyte—When used with the suffix byte the term kilo refers to "binary" numbering. One kilobyte is 1,024 bytes and not 1,000.
kTPI—Kilotransitions per inch.

lapping—The fine-polishing of magnetic recording heads to finish its surface to the proper contour. Also to assure minimum wear of the recording media surface.
latch—A single-input, single-output circuit whose output at a point of time is the bit that appeared on its input when the latching command was issued.
latency—The time that elapses before a specific sector passes beneath the read/write head. The average latency is one half the time of a complete rotation. For example; the latency of a drive that rotates it's media at 300 RPM (revolutions per minute) is 100 milliseconds.
lateral direction—Across the width of the tape.
layer-to-layer signal transfer—The magnetization of a layer of tape in a roll by the field from a nearby recorded layer, sometimes referred to as "print through."
layer-to-layer adhesion—The tendency for adjacent layers of tape in a roll to adhere to one another.
LBE—Lower band edge of the recorder/reproducer response.
leadscrew—A positioner system implementation that uses a screw thread extension of the rotor to convert the rotation of a stepper motor to linear travel of the read/write head carriage.
LED—Light emitting diode.
linear function—A function satisfying the relation: $f(x+y+ \ldots +n) = f(x)+ \ldots +f(n)$.
linear sequential circuits (LSCs)—Circuits constructed with modulo 2 addition (XOR gates), memory circuits (latches), and constant multipliers.
linearity—The extent to which the magnitude of the reproduced output is directly proportional to the magnitude of the signal applied to the input of the recorder. Also a property that is independent upon magnitude.

liner—A soft, non-woven synthetic material selected for its wiping and special anti-friction characteristics. The liner is precision-bonded to the interior of the jacket and provides minimum resistance to the rotation of the diskette.

logic levels—Nominal voltages that represent binary conditions in a logic circuit.

longitudinal recording—(1) Recording in which the tape tracks are parallel to the direction of tape motion, contrary to rotary recording. (2) Recording in which the magnetic vector of the tape magnetization parallels the track surface on the tape.

longitudinal curvature—Any deviation from straightness of a length of tape.

longitudinal direction—Along the length of the tape.

loose debris—Material very lightly bonded to the tape or head top surface, removable by tape motion.

lot code—A unique number which identifies the parent production lot from which an individual tape or diskette was made.

low band—(1) Pertaining to those frequencies specified in SMPTE Recommended Practice RP6, Practice LBM or LBC. (2) Pertaining to those recordings made in accordance with Practice LBM or LBC or RP6, or equipment capable of making those recordings. (3) The carrier frequencies which appear on a tape made in accordance with such practices.

LSB—Least significant bit in the PCM representation of a sample value.

LSC—Linear sequential circuit.

LSR—Linear shift register.

lubricant—See additive.

luminance—Luminance indicates the amount of light intensity, which is perceived by the eye as brightness.

LVDT—Linear variable differential transformer.

magnetic flux—The magnetic lines of force produced by a magnet or electric current.

magnetic field strength—The magnitude of a magnetic field vector, usually expressed in oersteds or ampere-turns per meter.

magnetic instability—The property of a magnetic material that causes variations in the residual flux density of a tape to occur with temperature, time, and/or mechanical flexing. Magnetic instability is a function of particle size, magnetizing field strength, and anisotropy.

magnetic media—A thin coating of magnetically hard material bonded to a substrate and which is capable of receiving and storing recorded information. Used for any configuration: tapes, disks, cards.

magnetizing field strength—The instantaneous strength of the magnetic field applied to a sample of magnetic material.

magnetomotive force—The magnetic analog of electromotive force, which, when due to a current in a coil, is proportional to the product of current in amperes and the number of turns.

magnetoresistance—The change in electrical resistance of a conductor or semiconductor due to a change in the applied magnetic field.

magnetostriction—The change in dimensions of a ferromagnetic body when placed in a magnetic field.

manchester codes—Biphase, level, and biphase, mark, codes, which are self-clocking and avoid the necessity of DC response, although at the expense of required additional bandwidth.

manganese-zinc (Mn-Zn)—A ferromagnetic ceramic having superior properties for magnetic read/write and erasure.

margin—Also window margin. A figure of merit that predicts the ability of a given drive to meet its published error rate by examining the data bit's timing distribution in the data separation window.

maximum flux—See maximum intrinsic flux.

maximum flux density—See maximum intrinsic flux density.

maximum induction—See maximum intrinsic flux density.

maximum intrinsic flux—In a uniformly magnetized sample of magnetic material, the product of the maximum intrinsic flux density and the cross-sectional area.

maximum intrinsic flux density—The maximum value, positive or negative, of the intrinsic flux density in a sample of magnetic material that is in a symmetrically, cyclically magnetized condition.

maximum-likelihood receiver—A receiver that can determine the coded signal or word that was most probably transmitted.

maxwell—A unit of magnetic flux.

MBPS—Megabits per second.

MCT—Manufacturer's center line tape; a tape selected by the manufacturer from his production, where the electrical and physical characteristics are employed as a reference standard for all production tapes.

mean track centerline—The average centerline of the read and write track.

media—The storage element of a tape recorder or a rotating magnetic memory device.

megabyte—When used with the term byte the prefix mega refers to "binary" counting. One megabyte is 1,024,000 bytes.

memory—Also called storage. A device that accurately stores and later reproduces information.

merging—In three-position modulation (3PM) code, combination of two transitions to form a single transition when there is only one transition-free position between them.

MFM (modified frequency modulation or delay modulation)—A code that has a "1" and a "0" correspond to the presence or absence, respectively, of a transition in the center of the corresponding bit cell. However, additional transitions at the cell boundaries occur only between bit cells that contain consecutive "0" values.

MHz—Megahertz (one million hertz).

missing pulse or bit—See dropout.

MKS—Meter-kilogram-second.

microcode—Computer programming using elementary machine instructions.

microdiagnostics—Data testing required to implement error correction circuits of data storage devices.

microhardness—Hardness of core material measured in knoop or vickers.

MIG—Metal in gap, for high density record and write heads. An Al-Fe alloy is deposited on the interfacing pole pieces in short gap length record heads. Al-Fe has a high saturation flux density and allows for higher flux density in the gap.

miller code—A biphase encoding scheme in which a minimum of one transition occurs in every two bit cells, minimizing signal DC components and providing run length limitation. See also delay modulation (DM) code.

minimum transition density—The least transition density that allows maintaining synchronization without extending the low-frequency response requirement below the capability of wideband direct record/reproduce systems.

misregistration—The improper interpretation of a word, while passing, for example, between a head and a track during read or write operations.

missing bit—A bit with less than 50 percent amplitude.

modem—A data station device capable of modulating the transmitted signal and demodulating the received signal, and also capable of related functions such as multiplexing.

modulation—Readback signal envelope distortion is called modulation.

modulated carrier recording—Signal information recorded in the form of a modulated carrier.

modulo 2 adder—A logic element whose output at a point of time is the modulo 2 sum of its inputs at that time.

modulus of elasticity—Average slope of the stress/strain curve of the tape when the strain is less than 3 percent and strain is applied at less than 1 percent/s.
moment of inertia—A measure of the rotational force required to accelerate or decelerate a reel of tape.
monolithic head—Winchester head constructed with a slider body made out of magnetic ferrite material.
mounting plate—A means of holding the head body and mounting it into the tape transport. It is the reference surface for most mechanical measurements.
MSB—Most significant bit in the PCM representation of a sample value.
MTBF—Mean time between failures.
MTRR—Magnetic tape recorder/reproducer.
MTTR—Mean time to repair.
MTU—Magnetic tape unit.
multiplexing—Interleaving of two or more independent channels (of various data rates) into a single high speed data stream; or, in tape recording, time or frequency sharing (i.e., time division or frequency division) of a single track on tape by two or more signals, each occupying a different time or frequency domain.
MUX—Multiplexer.
MX—Multiplexing.

NB—Narrow band, as in certain measurement sets using one, ½ or ⅓ octave bandpass filters.
nickel-zinc (Ni-Zn)—A ferromagnetic ceramic that is used for core material in magnetic heads.
noise—Any unwanted electrical disturbances other than crosstalk or distortion components that occur at the output of the read or reproduce amplifier. System noise is the total noise produced by the whole recording system including the medium. Equipment noise is the noise produced by all the components of the system, with the exception of the media, disk or tape. Media noise is the noise that can be specifically ascribed to the coating properties.
noise pulse—A spurious signal of short duration that occurs during reproduction of a tape and is of a magnitude considerably in excess of the average peak value of the ordinary system noise.
non-recoverable error—A head error whose length exceeds the correction capability of the drive system's error recovery circuit.
nonreturn to zero code (NRZ)—A binary digital code in which transition between signal levels, rather than the levels themselves, represent bits.
nonreturn to zero, inverted (NRZ-I)—Same as nonreturn to zero, mark.
nonreturn to zero, level (NRZ-L)—A binary digital code in which a "1" is represented by one level and a "0" by the other level.
nonreturn to zero, mark (NRZ-M)—A binary digital code in which a "1" is represented by a transition at the beginning of the bit cell, and a "0" is represented by no transition.
nonreturn to zero, space (NRZ-S)—A binary digital code in which a "1" is represented by no transition, and a "0" is represented by a transition at the beginning of the bit cell.
normal bit time (cell time)—The average bit time of recording at continuous maximum flux reversals.
NOT function (gate)—An inverting function (inverter).
Nyquist rate (limit)—Maximum rate of transmitting pulse signals through a channel of given bandwidth. If B is the effective bandwidth in Hertz, then 2B is the maximum number of code elements per second that can be received with certainty. The definition is often inverted, in effect, to read "the theoretical minimum rate at which an analog signal can be sampled for transmitting digitally."
Nyquist sampling theorem—A theorem which holds that the minimum sampling frequency

which can be used without introducing unwanted components into the decoded analog signal is equal to twice the highest frequency of the original analog signal.

odd parity—A property of a vector, word, or data stream with an odd number of "1" bits.
OEM—Original equipment manufacturer.
Oersted—A unit of magnetic field strength.
offtrack—A condition where the read/write head is offset from the location of the recorded track.
oligimers—Low molecular weight components of polymer films.
operating system—The software instructions used to control the work of a computer.
OR gate—A multiple-input, single-output gate that has a "0" output only when all inputs are "0's."
OR—Orientation ratio.
orientation—See particle orientation.
orientation direction—The direction in which particle orientation takes place.
orientation ratio—In a material composed of directionally oriented particles, the orientation ratio is the ratio of the residual saturation flux density in the orientation direction to the residual saturation flux density perpendicular to the orientation direction.
overhead—Information added to the data stream, such as parity bits, and cyclic redundancy check characters and synchronization words.
overwrite—Overwrite of an FM (digital) recording with the two frequencies 1f and 2f is defined as the ratio of the original 1f (or half band edge signal) to that remaining after being overwritten with a 2f (or band edge signal). Overwrite of 30 dB is considered adequate for proper system performance.
oxide—Magnetic gamma ferric oxide particles, used in coatings for tapes and disks.
oxide buildup—The accumulation of oxide or, more generally, wear products in the form of deposits on the surface of heads and guides.
oxide loading—A measure of the density with which oxide is packed into a coating. It is usually specified in terms of the weight of oxide per unit volume of the coating.
oxide shed—The loosening of particles of oxide from the tape coating during use.

P-P—Peak to peak.
packing density—The number of bits which can be stored per unit of dimension of a recording medium.
PALE—Phase alternating line encoding. A method of encoding the PCM NTSC signal by reversing the encoding phase on alternate lines to align the codewords vertically.
PAM—Pulse amplitude modulation.
parallel-mode high-density digital recording—Process of converting a bit stream to parallel bytes of N Bits, to be recorded across N tracks on tape.
parity—The property of oddness or evenness of the number of "1" bits in a word.
parity bit—An extra bit appended to an array of bits to permit subsequent checking for errors.
parity channel—A channel added or used to contain bits generated to maintain odd/even parity across all channels for detection of bit errors.
partial response coding—The binary signal is passed through some linear device, e.g. a low-pass filter, whose rise time is greater than one bit period, giving rise to a multilevel waveform.
particle orientation—The process by which acicular particles are rotated so that their longest dimensions tend to lie parallel to one another, and in the direction of the recording field.
pattern sensitivity—Vulnerability of a given code to certain patterns of "1" and "0" resulting in a higher bit error rate.
PCM—Pulse code modulation.

PDM—Pulse duration modulation. Also known as pulse width modulation (q.v.)

PE (phase encoding)—A "1" bit is a flux reversal to the polarity of the interblock gap. A "0" data bit is a flux reversal to the polarity opposite that of the interblock gap. A flux reversal shall be written at the nominal midpoint between successive "1" bits or between successive "0" bits to establish proper polarity.

peak—The highest amplitude portion of a waveform.

peak shift (pulse crowding)—The displacement of a positive or negative peak of a readback pulse from its nominal peak position as if a continuous maximum flux reversal had been written.

pel—Picture element (see also pixel).

percent peak shift—Peak shift divided by nominal bit time and multiplied by 100.

peripheral—An accessory to a CPU that provides additional memory, input, or output capability.

permalloy—A magnetic alloy, consisting of iron, nickel, and small quantities of other metals, used to make core assemblies for magnetic heads.

permanent elongation—The percentage elongation remaining in a tape or length of base film after a given load applied for a given time has been removed.

permeability, absolute—The ratio of magnetic flux density B to magnetizing force H for a given material.

permeability, relative—Magnetic property of all materials. Its value is generally around one, except for iron, nickel, cobalt, chromium dioxide and ferrites with a range from a few hundred up toward one million.

permed—Magnetized to a level that cannot be removed with a handheld degausser.

perpendicular recording—Recording in which the magnetic vector of media magnetization is perpendicular to the track surface. Also known as vertical recording.

perpendicular direction—Perpendicular to the plane of the media.

PET—Polyethylene terephthalate.

phase-locked loop—A circuit containing an oscillator whose output phase and frequency is in synchronization with a reference.

phase margin—A technique similar to bit mapping that estimates the error rate performance of a system.

physical recording density—The number of recorded flux reversals per unit length of track (FRPI).

pixel—Smallest picture element (also known as a pel) to which are assigned discrete RGB values.

plasticizer—See additive.

PLL—Phase-locked loop.

polarity reversal—A 180 degree phase shift in signal caused by each stage of amplification. Some codes are polarity sensitive, others are not.

pole tips—Those parts of the video head which protrude radially beyond the rim of the head wheel and form the magnetic path to and from the tape.

polyester—An abbreviation for polyethylene glycol terephthalate - the material most commonly used as a base film for precision magnetic tape.

polynomial error detection—Use of a polynomial expression applied to blocks of data to generate remainders for both input and output data, the comparison of which indicates whether reproduction errors have occurred.

porosity—The ratio of voids to a solid volume of magnetic material usually expressed in percent.

positioning error—Error in location of magnetic recording head in a disk drive and, therefore, of read/write and erase cores when a data track is accessed. The error is in a radial direction (toward or away from the hub of the drive) and can result in reduced sensitivity in reading a track or accidental erasures of adjacent tracks while writing.

positioner—Also called actuator. The mechanism that positions the read/write head, and in servo systems keeps it on track.

post compensation—A method of offsetting the effects of peak shift by changing the group delay of the readback signal.

postamble—Group of special signals recorded at the end of each block on tape for the purpose of electronic synchronization.

power spectrum—Signal power graphed as a function of frequency.

PPM—Pulse position modulation.

PR—Pseudorandom.

PRBS—Pseudorandom binary sequence.

Preamble—Group of special signals recorded at the beginning of each block on tape for the purpose of electronic synchronization.

preamble major frame—In image processing format, a tape format element that does not carry data but serves as a pilot tone and filler to maintain bit stream continuity.

precompensation—A method of reducing the effects of peak shift by changing the transition locations in the write data stream.

print through—See layer-to-layer signal transfer.

PRN—Pseudorandom noise.

probe head—A head suited for writing on perpendicular media.

program—The instructions, either written by the user or purchased ready to run, that tell the computer how to perform a given task.

program audio track—The area reserved on the tape for the main audio signal, usually associated with the accompanying video recording.

pseudorandom bit sequence (PRBS)—A digital code having the appearance of random bit sequence, but of finite length. It is usually achieved by the use of a feedback shift register to modify sequences or patterns or bits in a predetermined manner or to restore each modified bit pattern to their original sequence. The repeating sequences exhibit many of the statistical properties of uniformly distributed random number sequences: hence, they are called pseudorandom.

pulse code—Codes in which groups of pulses/bits represent digits/amplitude.

pulse-code modulation (PCM)—Modulation process involving the conversion of a waveform from analog to digital form by means of sampling, quantizing and coding. The peak-to-peak amplitude range of the signal is divided into a number of standard values each having its own value code. Each sample of the signal is then transmitted as the code word corresponding to the nearest standard amplitude.

pulse code modulation (PCM) multiplexing—A combination of pulse code modulation with time division multiplexing to produce a single digital signal consisting of a given number of channels.

pulse—A transient signal that is usually of short duration. In digital applications, it is of constant amplitude and polarity.

PWM—Pulse-width modulation (also known as pulse duration modulation). A form of pulse-time modulation in which the duration of a pulse is varied by the value of each instantaneous sample of the modulating wave.

quadrature—(A) The relation between two periodic functions when the phase difference between them is one-fourth of a period. (B) The geometric relationship among the four magnetic heads mounted on the head wheel of a quadruplex recorder. Ideally, the heads are precisely 90 degrees apart.

quadruplex—An adjective describing a standardized method of video magnetic tape recording which uses four magnetic heads mounted around the rim of a head wheel. The head wheel rotates in a plane perpendicular to the direction of the tape motion.

quantization—The division of a continuous range of values into a finite number of distinct values.

quantization noise—The sequence of quantization errors.

R-NRZ-L—Randomized nonreturn to zero, level.
R-NRZ—Randomized nonreturr to zero.
RAM—Random access memory: a storage device from which information may be obtained at a speed which is independent of the location of the data, and from any required location, without searching all information sequentially.
ramp signal—A test signal that changes linearly with time.
random noise—Noise caused by a large number of superimposed elementary disturbances, characterized by random occurrences in time and amplitude.
randomization—A modification of nonreturn to zero (NRZ) encoding in which the bit stream is randomized to reduce DC components before being recorded on tape.
RB—Return to bias.
R-DAT—Digital audio recording using a rotating head recorder for cassettes containing a 3.8mm wide tape.
read—To convert the transitions recorded on the media to a stream of data pulses.
read reduction (self-erase, residual erasure)—Reduction in the average peak-to-peak signal amplitude at the maximum flux reversal of the desired recording method due to residual magnetism in the read and/or write tracks of the read/write head partially erasing the written information on the tape.
read-only memory—A device in which information is stored in such a way that it may be read but not modified.
read-write core—The magnetic core on a recording head that reads or writes data on the diskette.
read/write head—A single or two gap head (read and write gap) on one body.
readback amplitude—The average peak-to-peak signal amplitude at the maximum flux reversal of the desired recording method.
real time—When the processing of a signal takes place during the time that the related physical process is actually taking place, the signal may be said to be processed in "real time."
reconstruction—The process of forming the output data stream to be an exact replica of the input data stream in the absence of bit errors.
record—The magnetic pattern on a tape or disk corresponding to a collection of related items of data, treated as a unit.
record level—See standard record level.
record margin—The change in signal-to-noise ratio achieved by reducing the record level from optimum while maintaining the reproduce level constant to reach a specific bit error rate.
recording density—Also FCI or BPI. The maximum number of transitions recorded at the innermost track of a disk.
recoverable error—A hard error whose length is less than the correction capability of the drive system's error correction circuit.
redundancy—The fraction of the gross information content of a message that can be eliminated without loss of essential information.
reel—The flanged hub, made of metal, glass, or plastic, on which magnetic tape is wound.
reference edge—On a videotape containing quadruplex recorded information, that longitudinal tape edge nearest the tracking control record.
reference tape—A tape used as a reference against which the performances of other tapes are compared.
reinforcement ring—A vinyl ring added to the hub access hole of diskettes.
reluctance—For an element of a magnetic circuit, the magnetomotive force per unit magnetic flux. Also named magnetic resistance.

remanence—The magnetic flux density that remains in a magnetic circuit after removal of applied magnetomotive force. (Note: remanence is not necessarily equal to residual flux density.)

reread—Repetition of signal receiving process to improve syndrome errors. Correction is best performed only after a consistent syndrome has been received.

residual erase signal (residual magnetic erase)—The average peak-to-peak signal amplitude at the minimum flux reversal of the desired recording method after DC erase by the write or erase head.

residual flux—In a uniformly magnetized sample of magnetic material, the product of the residual flux density and the cross-sectional area.

residual flux density—The magnetic flux density at which the magnetizing field strength is zero when a sample of magnetic material is in a symmetrically, cyclically magnetized condition.

residual-to-maximum-flux ratio—A ratio indicating the degree of particle orientation in coated particulate media.

resolution (dynamic range)—The average peak-to-peak signal amplitude at the maximum flux reversal divided by the average peak-to-peak signal amplitude at the minimum flux reversal at the desired recording method.

retentivity—The maximum value of the residual flux density corresponding to saturation flux density.

return to zero (RZ)—A binary digital code using a signal with a positive pulse to indicate a "1" and a negative or zero pulse to indicate "0." The signal level returns to zero during each bit period.

rigid disk—See hard disk.

RH—Relative humidity.

ring head—A recording head with geometry that is particularly suited for high flux-gathering efficiency.

RLL—Run length limited.

RMS—Root mean square.

roll—A reel wound with a standard length of tape.

rolloff—A gradually increasing loss or attenuation with change of frequency beyond the substantially flat portion of the amplitude-frequency response characteristic of a system.

ROM—See read-only memory.

run-length limitation—A safeguard to prevent the data stream from containing long uninterrupted strings of "1's" or "0's."

RWC—Read/write calibration.

RZ—Return to zero.

safety margin—The tolerance width within a guard band. A positioning error that remains within the safety margin will not cause accidental erasures of adjacent tracks during the write mode.

sampling theorem—The statement that 2f equally spaced samples per second completely characterize signal that is band limited at frequency f.

sampling—The process of obtaining a series of discrete instantaneous values of a signal at regular or intermittent intervals.

saturation—A condition in which increasing field strength H provides no increases in magnetization J (or M).

saturation curve—A head/media performance plot. Typically plotting readback amplitude against write current.

saturation flux density—The maximum intrinsic flux density possible in a sample of magnetic material. The intrinsic flux density approaches the saturation flux density as the magnetizing field strength is increased, B_{sat}-$\mu_o H + J_{sat}$.

saturation moment—The maximum magnetic moment possible in a sample of magnetic material.

saturation recording—Magnetic recording where the total thickness of the magnetic coating is subjected to fields strong enough to magnetize the coating completely in one or the other direction.

scatterwind—Lateral displacements of tape wound on a reel which gives an irregular appearance to the side surfaces of a tape pack. Scatterwind can result from such things as poorly controlled tape tension, guiding, static electrical charge, and poor tape slitting. May cause permanent damage to tape and edge tracks in particular, if reel flanges press against tape pack.

sector—The addressable units into which each track on a disk or diskette is divided.

sector holes—The punched holes around the inner or outer diameter of a "hard" sectored diskette which identify the beginning, end, and address of each sector.

seek—To move from one track location to another. Also called step.

seek settle time—Waiting time that must follow the end of the final step time to make sure that the positioner system and head have stopped moving.

self-clocking code—A binary digital code that permits intrinsic timing of pulses for both "0" and "1" bits, such as a biphase code.

self-demagnetization—The process by which a magnetized sample of magnetic material tends to demagnetize itself because of the opposing fields created within it by its own magnetization.

SEM—Scanning electron microscopy.

Sendust—A magnetic alloy (Al-Fe) with a high coercivity used in the construction of pole tips for recorder heads.

sensitivity—The ability or tendency of a circuit or device to respond to a low-level applied stimulus. Also: The magnitude of the output when reproducing a tape recorded with a signal of given magnitude and frequency.

separation loss—The loss in output that occurs when the surface of the coating of a magnetic tape fails to make perfect contact with the surface of either the record or reproduce head.

sequential access storage (SAS)—A mode of data storage where each byte of data is written in the order in which it is received.

serial—A communications mode which sends data bits one at a time over a single channel. As opposed to parallel.

servo—In general, a self-correcting, closed-loop control system. In magnetic tape recording, a speed correcting, closed-loop control system that generates an error signal when tape is not moving at the speed selected. Recorder can servo from either a built-in tachometer or a signal recorded on tape. Also used in disk drives to make the head follow a recorded track.

SFD—Switching field distribution.

Shannon's theorem—A criterion for estimating the theoretical limit to the rate of transmission - and correct reception of information with a given bandwidth and signal-to-noise ratio. (See C.E. Shannon, "A Mathematical Theory of Communication," *Bell System Technical Journal*, 27: 379-423, July 1948.)

shedding—The loss of oxide or other particles from the coating or backing of a tape, usually causing contamination of the tape transport and, by redeposit, of the tape itself.

shield front—A magnetic shield close to the front (top) surface of the read/write head to reduce "feed through" (cross feed).

shift register—A set of serially connected memory cells in which the stored contents of all cells may be simultaneously shifted forward or backward by one or more cell locations. At the time of shifting, new contents may enter at one end of the register while previous contents are displaced and lost at the other.

signal bits—Data stream bits other than check bits and synchronization word bits.

signal-to-noise ratio—The ratio of the power output of a given signal to the noise power in a given bandwidth. The signal-to-noise ratio is usually measured in terms of the corresponding root mean square signal and noise voltages appearing across a constant output resistance.

simultaneous bit errors—The occurrence of two isolated bit errors in the same bit column within the same data block.

single density—A digital recording method whereby each data bit is preceded by a clocking bit which is used to synchronize the internal clock of the system.

single domain particle—All ferromagnetic materials are composed of permanently magnetized regions in which the magnetic moments of the atoms are ordered. These domains have a size determined by energy considerations. When a particle is small enough, it cannot support more than one domain and is called a single domain particle.

skew—Deviation of a line connecting the average displacement of the multiple read or write track gaps from a line perpendicular to the reference edge of the tape in the direction of tape motion.

skew tape—Continuous strings of "1" values written on a properly adjusted tape drive for the entire recoverable length of the tape; an "all 1" pattern on all tracks; the write head, write delays, and tape drive adjusted to write with minimum physical skew and gap scatter.

slant track—The preferred term is Helical scan.

sliders—Sections of a magnetic-recording head that contact the diskette surface while supporting the cores.

slip agent—A compound such as ultrafine silica that is dusted on films to facilitate winding and unwinding.

SMPTE—Society of Motion Picture and Television Engineers.

SNR—Signal-to-noise ratio.

soft-decision decoder—works with real values rather than reconstituted bits.

soft error—Errors that are non-repeatable. These types of errors are caused by loose debris on the medium surface and are induced by the system/user.

soft sector format—A configuration in which the recording surface of a disk is magnetically divided into a number of sectors by an initialized coded address pattern, unique to each sector, at the time of manufacture.

software—Computer programs and associated documentation defining the operation of a data processing system.

source codes—Codes that are inputs to a given encoding system.

specific magnetic moment—The value of the saturation moment per unit weight of a magnetic material expressed in electromagnetic units per gram, or per cc. The specific magnetic moment is the most convenient quantity in which to express the saturation magnetization of fine-particle materials.

splice (mechanical)—A butt-joint between two pieces of tape held together by means of a strip of self-adhesive.

splice patch—See splice (mechanical).

splicing tape—The self-adhesive foil used to secure the butt-joint in a mechanical splice.

split-phase encoding—Double-frequency encoding similar to biphase, level, in which a transition always occurs at the center of every bit period.

spoking—A form of buckling in which the tape pack is deformed into a shape that approximates a polygon.

spool—See reel.

sputtering—A process allowing fabrication of a metallic and magnetic coating on a substrate.

squeal—See stick-slip.

staircase ramp—A test pattern in which a data word is repeated a fixed number of times per ramp step, then incremented by one for the next step, and so on, to the end of a major frame.

standard amplitude reference tape—A tape that has been selected for given properties as a standard for signal amplitude.

standard record level (IRIG)—The recorded level that produces 1 percent third harmonic distortion of the record level set frequency.

standard record level (broadcast use)—The recorded level that equals a certain predefined level measured in nanoWebers/meter.

standard reference amplitude—The average peak-to-peak output signal amplitude derived from the amplitude reference tape recorded at the standard measurement control.

standard reference current—The minimum current applied to the amplitude reference tape that causes an output signal amplitude equal to 95 percent of the maximum output signal.

standard reference tape—A tape intended for daily calibration, the performance of which has been calibrated to the amplitude reference tape.

stick-slip—Generally a low-speed phenomenon, a relationship between tension, temperature, humidity, wrap angle, head material, binder material of tape, and the elastic properties of the tape. When detected audibly, it is described as squeal.

sticktion—A term loosely used to describe the phenomenon of tape adhering to transport components such as heads or guides.

stiffness—Resistance to bending of the tape; a function of tape thickness and modulus of elasticity.

stepper motor—A converter of electrical energy to mechanical motion that moves in discrete increments in response to digital signals.

storage—See memory.

straddle erase—A design in which there is no lag between the action of the read/write core and the trimming action of the erase cores. The erase cores are adjacent to the read/write core with gaps aligned.

strain relief notches—Semi-circular notches cut into the diskette jacket for the purpose of relieving strain when the jacket is flexed. The notches direct any tendency toward creasing away from the head aperture.

sub-Nyquist sampling—A scheme for sampling at a frequency lower than that prescribed by the Nyquist sampling theorem (q.v.).

substrate—Polyester film or rigid disk, made from aluminum or glass upon which the magnetic recording coating is placed.

supply reel—A reel from which the tape is unwound during the record, reproduce, or fast-forward modes.

surface treatment—Any process by which the surface smoothness of the tape coating is improved after it has been applied to the base film.

SWDT—Synchronization word detect time.

switching field—A magnetizing field of sufficient strength to switch the magnetization of a single particle in a disk or tape coating.

switching field distribution (SFD)—The coercivity range of individual particles that determines the length of a transition zone when switching fields are applied by the head.

symptom—An element of the syndrome vector, the value of which is "0" unless a parity error has been detected.

sync. bytes—A series of bytes located before ID or data bytes to allow the data separator to synchronize with the data stream. Each time a sector is updated a write splice is created at the beginning and end of the update. The write splice can throw the data separator out of synchronism.

synchronization words—Channel words allowing deskew of channel data and reconstruction of the data to serial form. Also, words inserted at the beginning of a sequence/frame/block of bits to inform the "interpreter," such as a D/A converter, which bit is the first bit in the sequence.

synchronizer—A device used at a receiving site to achieve coherence with a transmitted signal.

synchronizing pulse—A pulse used to control the frequency of an oscillator or a system to maintain lock with a reference.

synchronizing signal—A signal used to synchronize another signal, usually in frequency.

synchronous clock—The timing source in a synchronous computer.

synchronous computer—A computer whose operations are timed by single-frequency clock signals.

synchronous inputs—In a computer flip-flop circuit, inputs that accept pulses only at the command of the clock.

synchronous transmission—Data transmission in which synchronization of characters is controlled by timing signals generated at the sending and receiving stations.

syndrome—A vector generated by taking the XOR sum of a set of parity checks generated on receive with a set of parity checks generated on transmit; a symptom of an error.

TAA—Track average amplitude. The average value of all of the individual waveform peaks on a specific track.

tape, magnetic—A tape consisting of a flexible base material usually coated on one side with a thin magnetizable layer.

tape curvature—The maximum deviation of either tape edge from a 48-inch straightedge when a sufficient length of tape is allowed to lie in a horizontal plane parallel to the straightedge under zero tension. (Refer to ANSI C98.1-1963.)

tape guides—Rollers or posts used to position the tape correctly along its path on the tape transport.

tape leader (magnetic)—The section of a tape, usually recorded ahead of the program material, which contains engineering alignment signals and production information.

tape mark—A special control block recorded on magnetic tape to serve as a separator between files and file labels.

tape pack—The form taken by the tape wound on a reel.

tape skew—The deviation of a tape from following a linear path when transported across the heads. The terms "static" and "dynamic" are used to distinguish the physically fixed and fluctuating components of total tape skew.

tape speed—The speed at which tape is transported across the read/write head during normal recording or reproduction.

tape speed—long term—The tape speed averaged over a minimum of 15 in. of tape (in cm/sec or inches per second).

tape speed—short term—The instantaneous (dynamic) tape speed (in cm/sec or inches per second).

tape tension variation—If the tape tension is reduced, the head-to-tape separation can increase. If the tape tension variation is cyclic, then the separation variation will be cyclic. A typical low-tension condition occurs during "backward start" in a high-acceleration capstan tape drive.

tape transport—The mechanism that extracts magnetic tape from a storage device, moves it across magnetic heads at a controlled speed and then feeds it into another storage device. Typical storage devices are tape loops, bins, reels, and magazines (cassettes and cartridges). The tape transport is the part of a magnetic tape recorder/reproducer system that normally consists of magnetic heads, magnetic tape, tape transport, record electronics, and reproduce electronics.

tape-to-head separation—The separation between a magnetic head and the magnetic tape caused by (1) the foil bearing effect; (2) improper head contour, which generates standing waves in the tape; and (3) surface roughness of the tape surface. These conditions are interrelated and are greatly influenced by tape tension and tape compliancy. In a properly designed system, tape roughness is the limit of head-to-tape separation, usually $< 0.1\ \mu m$.

tape-to-head separation—changes—(1) Head contamination : Debris attached to the head, which causes the tape to lift away from the head forming a tentlike deformation of the tape. This tent does not move or change shape until the contamination is removed. (2) Tape contamination: Particles attached to the tape result in a "tent" formed by particles that move across the head with the tape.

tape-to-head separation—effective—The actual distance from the magnetic storage material on the tape to the top of the active magnetic core material at the read or write gap. The effective head-to-tape separation is usually somewhat larger than the mechanical head-to-tape separation.

TBC—Time base correction.

TBE—Time base error.

tear strength—The force required to initiate and/or propagate a tear in a specially shaped specimen of tape or base film.

thermal noise—Electronic noise due to thermal excitation of electrons in conductors and semiconductors.

thermal seal—The areas where the jacket folds in a diskette holder have been heat-sealed.

THIC—Tape Head Interface Committee.

thin-film heads—Heads that are constructed using standard thin-film deposition and lithography techniques; for example, sputtering and ion etching.

three position modulation (3PM)—A coding scheme in which data are grouped into three-bit bytes. Each byte is divided into six periods. A lookup table is used to determine the location of transition, not bits, within the six periods to represent each byte.

throat height—The critical dimension of the region of a read/write head core with reduced cross sectional area through which all of the flux must flow.

time base expansion—Playback at reduced tape speed.

time base errors (TBE's)—Errors introduced because of short-term speed inaccuracy.

time base correction (TBC)—Methods employed to reduce time base errors in analog recordings.

time division multiplexing—The process of interleaving in time two or more digital signals for transmission over a common channel.

tip engagement—The preferred term is tip penetration.

tip height—Synonym for tip projection.

tip penetration—The momentary radial deflection of the tape in the vacuum guide caused by the passage of a video head pole tip.

tip projection—The measured radial difference between the pole tip and the head wheel rim.

tip protrusion—The preferred term is tip projection.

total thickness—Normally the sum of the thicknesses of the base film and the magnetic coating (symbol c) as well as backcoating when applied. The total thickness governs the length of tape that can be wound on a given reel.

TPI—Tracks per inch.

track—That portion of a recording surface available to one read/write head at each access position. Also: An area of tape surface that coincides with the location of the recorded magnetization produced by one record gap.

track curvature—The deviation from straightness of a single video track record.

track density—Number of bits stored in a track per unit length.

track spacing—The space between adjacent data tracks on a disk or tape. Track spacing can be measured between any corresponding points on adjacent tracks, such as between center lines or between their inner or outer edges.
track width—The width w of the track corresponding to a given record gap.
trailing edge, video track—The upstream edge of the video track.
transducer—Any device that converts energy from one form to another. In high-density digital recording, a magnetic recording head assembly that converts magnetic pulses into corresponding voltage or current pulses, or vice versa.
transfer function—A mathematical function that relates the input to the output of a system.
transform coding—A method of encoding a picture by dividing each picture into sub-pictures, performing a linear transformation on each sub-picture and then quantizing and coding the resulting coefficients.
transitions—(A) Short regions along a recorded track where the magnetization changes polarity. (B) Changes from one circuit condition to the other; e.g., change from mark to space ("1's" to "0's") or from north to south, and vice versa.
transition density—On a magnetic tape, the number of available magnetic transitions per unit length of tape.
transverse—Pertaining to dimensions or motions perpendicular to the tape travel.
transverse recording—The preferred term is quadruplex recording.
trimming—The action of the erase cores in erasing the edges of data tracks to avoid crosstalk between adjacent tracks.
TRIUMF—European tape head committee, similar to THIC.
TTL—Transistor-transistor logic. One of the families of integrated-circuit logic gates. Others are: emitter-coupled logic (ECL), diode-transistor logic (DTL), and resistor-transistor logic (RTL).
tunnel erase—A design in which the two flanking erase cores are placed so that they follow the read/write core relative to points on the rotating diskette. The lag distance between the read/write and erase gaps slightly reduces available data storage capacity, but this design is the industry standard.

UBE—Upper band edge of the recorder/reproduce response (usually at the − 3 dB point).
UHF—Ultra high frequencies, 300 MHz to 3 GHz.
ultimate tensile strength—The force per unit cross-sectional area required to break a tape or length of base film usually given in pounds per square inch.
uniformity—The extent to which the output remain free from variation in amplitude. Uniformity is usually specified in terms of the positive and negative deviations from the average output within a roll of tape.
unformatted capacity—The total capacity of a rotating memory device or media element. Format overhead losses are not considered. (The number of bits per track) × (the number of tracks).
update—Magnetic rotating memory devices do not erase on track information. They write on top of, or update, the existing data.
upstream—Pertaining to locations on the tape longitudinally displaced from a given reference point, in a direction opposite to tape travel.

vacuum guide—The part of the video head assembly used to maintain the tape in the correct position relative to the head wheel by means of a suction system.
VCO—Voltage-controlled oscillator.
VCR—Video cartridge (cassette) recorder.

vertical recording—Common term for perpendicular recording.
video tape—Magnetic recording tape for recording and playback of television signals.
VHF—Very high frequencies, 30 to 300 MHz.
void—An area where material is missing on the surface of a head track; also applies to tape surface.

washout or undercut—This occurs in a hard-coated head when the magnetic core material has a higher wear rate than the coating. The radius of curvature of the core material will be larger than the surrounding coating of the softer material, which could even be undercut. This could cause an increase in the head-to-tape separation.
wavelength—The distance along the length of a sinusoidally recorded tape corresponding to one cycle.
wavelength response—The record and reproduce characteristics of a magnetic medium which depends on geometry and is a function of the recorded wavelength rather than the actual frequency response.
wavyness—Irregular head top surface perpendicular to tape motion due to different wear rates in top surface materials. The "harder" material will be up. This can occur during breaking and field use. The head core is usually the harder material; therefore, there will be increased head-to-tape contact pressure at the cost of tape life.
wear ability—See durability.
wear product—Any material that is detached from the tape during use.
wear test—See durability.
wideband channel—In magnetic recording, a channel that will satisfactorily record and reproduce a signal with an on-tape wavelength of 80 micro inches.
wind—The way in which tape is wound onto a reel. An A-wind is one in which the tape is wound so that the coated surface faces toward the hub.
winder/cleaner—A device designed to wind and clean magnetic tape to restore it to a quality that approaches the condition of a new tape providing the tape has not been physically damaged.
window—The time interval during which a data separator is expecting a data pulse.
window margin—A technique for estimating the error rate performance of a system, without transferring the specified number of bits. The window is reduced until a predetermined error criteria is encountered. This figure of merit is indicative of the performance of the device under test.
windowing—The preferred term is cinching.
word—A group of bits, usually of a fixed length, in a given system.
wow and flutter—Terms used to describe changes in signal output frequency caused by tape speed variations occurring at relatively low and relatively high rates, respectively; however, the term wow is no longer used but is incorporated into the flutter measurement.
write—To convert a data stream from the controller to a pattern of transitions in the media coating.
write enable tab—A metallized, adhesive label which, when applied so as to cover the write protect notch, allows recording to occur. This applies to the older 8″ diskettes, but works *opposite* for the 5¼″ floppies: Data can NOT be written to the floppies when the tab is in place.
write feedthrough (crossfeed) (feedthrough) (crosstalk)—Magnetic coupling from the write tracks to a read track in the read/write head.
write protect notch—An optional notch in the jacket which prevents inadvertent recording of the diskette. This applies to the older 8″ diskettes, but works *opposite* for the 5¼″ floppies: Data can be written to the floppies when a tab does not cover the notch.
write splice—The discontinuity that results when a write operation is started or stopped,

especially over a formatted track. Each update write will result in a write splice in the format gaps.

XOR gate—Exclusive OR gate requires both a "0" and a "1" input for a "1" output. Two "0" bits or two "1" bits give a "0" output. A combinatorial logic circuit often used to show whether two inputs are alike or different on a bit-by-bit basis.

yield point—Stress to produce 3 percent strain.
yield strength—The minimum force per unit cross-sectional area at which the tape or base film deforms without further increase in the load. Units are pounds per square inch or pounds per tape sample of given width and base film thickness.

Listing compiled from:

Manuscript for this book.
"Glossary of Terms used in Magnetic Recording," E.D.Daniel, *Memorex Monograph No. 1*, 1963.
Journal of the SMPTE (several).
IRIG Telemetry Standards 106-86, May 1986.
NASA Reference Publication 1075, *Magnetic Tape Recording for the Eighties*, NASA, April 1982.
NASA Reference Publication 1111, *High-Density Digital Recording*, NASA, Sept. 1985.
IEEE Standard Dictionary of Electrical and Electronic Terms, 1984.

Illustrations Index

Subject Index

B

D

N

Q

R

Names Index

Edited by Eric S. Phelps